Handbook of
Neurochemistry

SECOND EDITION

Volume 6
RECEPTORS IN
THE NERVOUS SYSTEM

Handbook of
Neurochemistry
SECOND EDITION

Edited by Abel Lajtha
Center for Neurochemistry, Wards Island, New York

Handbook of
Neurochemistry

SECOND EDITION

Volume 6
RECEPTORS IN
THE NERVOUS SYSTEM

Edited by
Abel Lajtha

Center for Neurochemistry
Wards Island, New York

SPRINGER SCIENCE+BUSINESS MEDIA, LLC

Library of Congress Cataloging in Publication Data

Main entry under title:

Handbook of neurochemistry.

　　Includes bibliographical references and index.
　　Contents: v. 1. Chemical and cellular architecture—v. 2. Experimental neuro-
chemistry—[etc.]—v. 6. Receptors in the nervous system.
　　1. Neurochemistry—Handbooks, manuals, etc. 2. Neurochemistry. I. Lajtha,
Abel. [DNLM: 1. Neurochemistry. WL 104 H235 1982]
QP356.3.H36　1982　　　　　　　　　612′.814　　　　　　　　　82-493
ISBN 978-1-4684-4570-1　　　ISBN 978-1-4684-4568-8 (eBook)
DOI 10.1007/978-1-4684-4568-8

Contributors

R. D. Allan, Department of Pharmacology, University of Sydney, Sydney NSW 2006 Australia

Kenneth A. Bonnet, Department of Psychiatry, Millhauser Laboratories, New York University School of Medicine, New York, New York 10016

Anna Borsodi, Institute of Biochemistry, Biological Research Center, Hungarian Academy of Sciences, H-6701 Szeged, Hungary

S. Bourgoin, Groupe NB, INSERM U. 114, Collège de France, 75231 Paris, Cedex 05, France

David R. Burt, Department of Pharmacology and Experimental Therapeutics, University of Maryland School of Medicine, Baltimore, Maryland 21201

De-Maw Chuang, Laboratory of Preclinical Pharmacology, National Institute of Mental Health, Saint Elizabeths Hospital, Washington, DC 20032

M. Blair Clark, Department of Pharmacology, Medical College of Pennsylvania, Philadelphia, Pennsylvania 19129

E. Costa, Laboratory of Preclinical Pharmacology, National Institute of Mental Health, Saint Elizabeths Hospital, Washington, DC 20032

Frederick J. Elhert, Departments of Pharmacology, Internal Medicine, Biochemistry, Psychiatry, and the Arizona Research Laboratories, University of Arizona Health Sciences Center, Tucson, Arizona 85724

Yigal H. Ehrlich, Neuroscience Research Unit, Departments of Psychiatry and Biochemistry, University of Vermont College of Medicine, Burlington, Vermont 05405

S. J. Enna, Departments of Pharmacology, Neurobiology and Anatomy, University of Texas Medical School at Houston, Houston, Texas 77025

Marianne Fillenz, University Laboratory of Physiology, Oxford, Oxford OX1 3PT England

Arnold J. Friedhoff, Department of Psychiatry, Millhauser Laboratories, New York University School of Medicine, New York, New York 10016

Kelvin W. Gee, Departments of Pharmacology, Internal Medicine, Biochemistry, Psychiatry, and the Arizona Research Laboratories, University of Arizona Health Sciences Center, Tucson, Arizona 85724

C. Goetz, Groupe NB, INSERM U. 114, Collège de France, 75231 Paris,Cedex 05, France

Jack Peter Green, Department of Pharmacology, The Mount Sinai School of Medicine of the City University of New York, New York, New York 10029

Louise H. Greenberg, Department of Pharmacology, Medical College of Pennsylvania, Philadelphia, Pennsylvania 19129

M. Hamon, Groupe NB, INSERM U. 114, Collège de France, 75231 Paris, Cedex 05, France

Edith D. Hendley, Department of Physiology and Biophysics, University of Vermont College of Medicine, Burlington, Vermont 05405

Fusao Hirata, Laboratory of Cell Biology, National Institute of Mental Health, Bethesda, Maryland 20205

Lindsay B. Hough, Department of Pharmacology, The Mount Sinai School of Medicine of the City University of New York, New York, New York 10029

G. A. R. Johnston, Department of Pharmacology, University of Sydney, Sydney NSW, 2006 Australia

Giulio Levi, Istituto di Biologia Cellulare, Consiglio Nazionale delle Ricerche, 00196 Rome, Italy

Bertha K. Madras, Psychopharmacology Section, Clarke Institute of Psychiatry, Toronto, Ontario M5T 1R8 Canada

Mario Marchi, Istituto di Farmacologia e Farmacognosia, Università di Genova, 16146 Genoa, Italy

Guido Maura, Istituto di Farmacologia e Farmacognosia, Università di Genova, 16146 Genoa, Italy

Edith McGeer, Kinsmen Laboratory of Neurological Research, Department of Psychiatry, The University of British Columbia, Vancouver, British Columbia V6T 2A1 Canada

Patrick L. McGeer, Kinsmen Laboratory of Neurological Research, Department of Psychiatry, The University of British Columbia, Vancouver, British Columbia V6T 2A1 Canada

S. El Mestikawy, Groupe NB, INSERM U. 114, Collège de France, 75231 Paris, Cedex 05, France

Neville N. Osborne, Nuffield Laboratory of Opthalmology, University of Oxford, Oxford OX2 6AW England

Maurizio Raiteri, Istituto di Farmacologia e Farmacognosia, Universita di Genova, 16146 Genova, Italy

William R. Roeske, Departments of Pharmacology, Internal Medicine, Biochemistry, Psychiatry, and the Arizona Research Laboratories, University of Arizona Health Sciences Center, Tucson, Arizona 85724

Jack W. Schweitzer, Department of Psychiatry, Millhauser Laboratories, New York University School of Medicine, New York, New York 10016

Najam R. Sharif, Department of Pharmacology and Experimental Therapeutics, University of Maryland School of Medicine, Baltimore, Maryland 21201. *Present Address*: Department of Biochemistry, University of Nottingham, Queen's Medical Centre, Nottingham, NG7 2UH, United Kingdom

Eric J. Simon, Departments of Psychiatry and Pharmacology, New York University School of Medicine, New York, New York 10016

J. H. Skerritt, Department of Pharmacology, University of Sydney, Sydney NSW 2006 Australia

Richard F. Squires, Nathan S. Kline Institute for Psychiatric Research, Orangeburg, New York 10962

Benjamin Weiss, Department of Pharmacology, Medical College of Pennsylvania, Philadelphia, Pennsylvania 19129

Michael Williams, Nova Pharmaceutical Corporation, Baltimore, Maryland 21228

Maria Wollemann, Institute of Biochemistry, Biological Research Center, Hungarian Academy of Sciences, H-6701 Szeged, Hungary

Henry I. Yamamura, Departments of Pharmacology, Internal Medicine, Biochemistry, Psychiatry, and the Arizona Research Laboratories, University of Arizona Health Sciences Center, Tucson, Arizona 85724

Foreword

A major advance in the biological sciences in the past decade has been the biochemical identification of cell membrane receptors. The existence of receptor substances on the surface of cells that recognize and bind to extracellular molecules was proposed at the beginning of the century by the pharmacologist and immunologist Paul Ehrlich and the physiologist J. N. Langley. Since then, receptors have been found to play an important role in numerous physiological and pathological processes. Over the years many attempts have been made to physically isolate and chemically characterize receptors, but because of the receptors' extremely low concentration and membrane localization, these efforts have met with limited success. Yet, despite the failure to characterize receptor substances, the concept of the presence of such molecules has had considerable heuristic value. Using pharmacological, physiological, and immunologic approaches, researchers have identified several specific receptors, e.g., α- and β-adrenergic, nicotinic and muscarinic cholinergic, and histaminergic. With the characterization of various types of receptors on cell membranes, many drugs were developed that proved to be experimentally and therapeutically useful.

It was only in the early 1970s that methods for the specific measurement, chemical characterization, and physical isolation of cell membrane receptors were developed. These advances were made possible by the availability of ligands with high specific radioactivity that retained their biological activity and of experimental procedures that differentiated between specific and nonspecific binding of ligands. Another reason for the rapid advances in receptor research was the recognition that experimental data had to fulfill certain criteria, including the following: (1) structural and steric binding properties of agonists and antagonists should be consistent with their biological activity; (2) ligand binding to the receptor should show high affinity and saturability at concentrations that elicit a biological response; and (3) the specific binding should be greatest in tissues responsive to selective agonists.

Once suitable radioactive ligands were available, and rigorous criteria for the characterization of interaction of ligands with their receptors were established, many previously unrecognized receptors were found. Among the more significant were opiate, neuropeptide, and benzodiazepine receptors. Many of these discoveries were especially important to neurochemistry and opened a new dimension in this field.

This volume presents examples of the richness and variety of receptor research that is especially relevant to neurochemistry. In spite of the many advances concerning receptors, what we know is just the tip of the iceberg. We are just beginning to understand how receptors transduce their specific information through membranes. Receptors are in a dynamic state, continuously changing the cell's responsiveness to its environment. There is much to be learned about how these changes come about. With the development of powerful new technologies and instrumentation, the next decade in receptor research promises to be even more exciting than the past one.

Julius Axelrod

Contents

Chapter 3

β-*Adrenergic Receptors*

Maria Wollemann and Anna Borsodi

Chapter 4

Norepinephrine

Marianne Fillenz

Chapter 5

Dopamine

Bertha Kalifon Madras

Chapter 8

GABA Receptors

  G. A. R. Johnston, R. D. Allan, and J. H. Skerritt

Chapter 9

Excitatory Amino Acid Receptors

Najam A. Sharif

Chapter 12

Opiate Receptors

Eric J. Simon

Chapter 13

Peptide Receptors

David R. Burt and Najam A. Sharif

Chapter 14

Cholinergic Systems and Cholinergic Pathology

Patrick L. McGeer and Edith G. McGeer

Chapter 15

Neurotransmitter Uptake

Edith D. Hendley

Chapter 16

Release of Catecholamines, Serotonin, and Acetylcholine from Isolated Brain Tissue

Maurizio Raiteri, Mario Marchi, and Guido Maura

Chapter 17

Release of Putative Transmitter Amino Acids

Giulio Levi

Chapter 18

Transmitter Specificity in Neurons

Neville N. Osborne

Chapter 19

Phospholipid Methylation

   Fusao Hirata

Chapter 20

*Protein Phosphorylation: Role in the Function, Regulation, and
Adaptation of Neural Receptors*

Yigal H. Ehrlich

Chapter 21

Heterogeneity of Benzodiazepine Receptors

*Kelvin W. Gee, Frederick J. Ehlert, William R. Roeske,
and Henry I. Yamamura*

Chapter 22

Modulation of Catecholaminergic Receptors During Development and Aging

Benjamin Weiss, M. Blair Clark, and Louise H. Greenberg

Chapter 23

Receptor Regulation

    S. J. Enna

Chapter 24

Receptor Adaptation to Psychotropic Drugs

    Jack W. Schweitzer, Kenneth A. Bonnet, and Arnold J. Friedhoff

Mammalian Central Adenosine Receptors

Michael Williams

1. INTRODUCTION

Although adenosine was reported to cause vasodilation and to have a negative inotropic effect in the heart as long ago as 1929,[1] and ATP was both shown to cause sedation in cats[2] and proposed as a neurotransmitter candidate at primary afferent fibers[3] in the 1950s, it was not until the last decade or so, following from the seminal study of Sattin and Rall[4] on adenosine-evoked increases in cyclic AMP production in mammalian brain slices, that the role of adenosine and adenine nucleotides in CNS function has been studied to any great extent.

The major cause for this delay was the very uniquity of these compounds, adenosine and its nucleotides being intimately involved in all aspects of cellular metabolism,[5] and the dynamic nature of this involvement tending to confuse the interpretation of experimental data in regard to a role unique to nervous tissue. Also, unlike other putative neurohumoral agents such as norepinephrine or substance P, it has not been possible to measure synthesis and turnover, detect regional variations in distribution, map discrete pathways, or deplete endogenous stores of purines. These limitations have contributed to a certain degree of skepticism in regard to the concept of "purinergic" or adenosine/ATP-mediated neurohumoral events both centrally and peripherally.

The extensive studies of Burnstock and his colleagues[6–8] on "noncholinergic, nonadrenergic" neurotransmission and data from the laboratories of McIlwain,[9] Phillis,[10,11] Daly,[12] and, more recently, Snyder[13] on the central actions of these compounds have supported the concept that adenosine can modulate cellular function in a physiologically relevant manner via extracellular receptors sensitive to blockade by the alkylxanthines. The status of ATP as a neuromodulator in the CNS, as discussed below, is unclear.[10]

From a behavioral standpoint, it appears likely that adenosine functions as a potent central depressant, the central stimulatory properties of the meth-

Michael Williams • Nova Pharmaceutical Corporation, Baltimore, Maryland 21228.

ylxanthines, caffeine and theophylline, as originally suggested by Rall,[4] resulting from their antagonism of the effects of endogenous adenosine. It is also relevant to note that the N^6-substituted cyclohexyl (CHA) and *l*-phenylisopropyl (*l*-PIA) analogues of adenosine, which are metabolically stable, are psychoactive agents as potent in activity as LSD.*[14] Adenosine has also been implicated in the molecular mechanisms by which anxiolytics[15] and antidepressants[16] produce their clinical effects.

2. ADENOSINE EFFECTS IN THE MAMMALIAN CENTRAL NERVOUS SYSTEM

2.1. Cyclic Nucleotide Metabolism

Adenosine can stimulate the formation of both cyclic AMP[4,12,17] and cyclic GMP[18,19] in brain tissue, the purine being some 5–50 times less active in stimulating guanylate than adenylate cyclase. Adenosine-elicited increases in rat and guinea pig brain slice cyclic AMP levels can be antagonized by the alkylxanthines caffeine, theophylline, and IBMX†[4,12] but are potentiated by the phosphodiesterase inhibitor Ro 20-1724.[12] Purine effects on cyclic AMP production are GTP dependent.[17]

Adenosine can serve as a precursor for cyclic AMP by uptake and subsequent phosphorylation to ATP, thus increasing the available substrate pool for adenylate cyclase; however, this contribution is only 20–40% of the total cyclic nucleotide response to the purine.[20,21] This observation and the finding[22] that the effects of low concentrations of adenosine can be potentiated by inhibitors of high-affinity adenosine uptake such as dipyridamole provided good evidence that the purine effects on cyclic AMP production were mediated by an interaction with specific extracellular recognition sites.

In addition to its direct effects on cyclic AMP production, adenosine in combination with other putative neurotransmitters such as histamine and norepinephrine can elicit synergistic increases in cyclic nucleotide levels, the purine interaction with catecholamines involving an α-adrenergic component.[23,24] Adenosine-elicited increases in cyclic AMP have also been observed in the retina,[25] spinal cord,[26] and homogenates of rat brain.[17,27] In the latter preparation, loss of cellular integrity and the concomitant availability of large quantities of endogenous adenosine necessitated the use of the catabolic enzyme, adenosine deaminase (ADA) to demonstrate cyclase stimulation. This in turn has required the use of a stable analogue of adenosine, in this case 2-CADO,‡ as an agonist for the enzyme.

In striatal homogenates, adenosine-sensitive cyclase is localized in nuclear and mitochondrial subfractions.[27,28] In the latter, the synaptic membrane fraction was most enriched in enzyme activity. A limited regional distribution was

---

* LSD, lysergic acid diethylamide.
† IBMX, isobutylmethylxanthine.
‡ 2-CADO, 2-chloroadenosine.

observed in homogenates of rat brain,[28] with the distribution of adenosine-stimulated cyclase in striatum paralleling that of the dopamine-sensitive enzyme.[28] No activity was detected in the cerebral cortex, anterior and central cerebellar cortex, or hippocampus. These findings, which contrast with those from brain slice preparations,[12] may indicate a functional uncoupling of the enzyme in homogenate preparations. Adenosine-sensitive cyclase in homogenates disappears after destruction of neuronal elements of the striatum with kainic acid,[29] which suggests that the enzyme is localized on neurons rather than glial cell elements.

Although limited structure–activity studies have been carried out in nervous tissue, Bruns[29,30] has examined over 100 adenosine agonists and antagonists for their effects on the adenosine-stimulated adenylate cyclase in the VA 13 human fibroblast cell line and concluded that the N^6-amino group of purines is probably involved in the change in receptor conformation resulting in agonist activation of the enzyme and that a spatial "anti" conformation of the molecule was necessary to bind to the receptor.

In addition to the well-documented effects of the purine in increasing adenylate cyclase activity, studies in fat cells[31,32] and other peripheral tissues[33-35] have revealed the presence of purine recognition sites that cause inhibition of the enzyme. Such inhibition can occur either via an interaction with the catalytic subunit of the enzyme, an intracellular event not sensitive to methylxanthine blockade, or by interaction with an extracellular recognition site that can be blocked by theophylline.[32-36] Studies in mouse fetal brain cell cultures[36,37] and rat brain striatal homogenates[38] have confirmed the existence of both stimulatory and inhibitory extracellular receptors for adenosine in nervous tissue. In the adipocyte membrane, the methylxanthine-antagonized inhibitory effects of PIA on cyclic AMP production have been related to stimulation of a high-affinity GTPase,[39] formation of GDP from GTP resulting in an inactivation of the cyclase.

The adenosine receptors mediating inhibition and stimulation of brain adenylate cyclase have been termed A_1 and A_2, respectively.[36,37] A similar subclassification of receptors in terms of their interaction with adenylate cyclase has been described for dopamine,[40] the amine inhibiting the enzyme in the pituitary via a D-2 receptor and stimulating it in the striatum via a D-1 type receptor.

2.2. Electrophysiological Studies

Iontophoretic application of adenosine and adenine nucleotides onto brain cells *in situ* causes a marked depression of cell firing,[10,41-44] which, like the effects of the purine on brain slice cyclic AMP production, can be antagonized by methylxanthines.[10] The effects of the purines on cell firing can be enhanced by the adenosine uptake inhibitor dipyramidole[10] or by inhibitors of ADA such as deoxycoformycin. Adenosine deaminase itself or alkylxanthines can increase spontaneous cell activity,[44a] presumably via antagonism of endogenous purines, an observation that would indicate that central neurons are under a purinergic inhibitory tone.[45,46]

In both *in situ* and *in vitro* brain slice preparations,[10,48-50] decreases in spontaneous and evoked synaptic potentials are accompanied by hyperpolarization with no significant change in resistance. In guinea pig hippocampal slices,[50] neither the membrane conductance nor the spike-generating mechanisms are affected by adenosine. The general consensus from these findings is[10,45-50] that the effects of the purine on cell firing occur presynaptically via a mechanism that differs from that observed with more convential inhibitory neurotransmitters: from data obtained in the rat phrenic nerve and diaphragm[51] and spinal cord[52] and on the effects of adenosine on neurotransmitter release (see below) to inhibition of excitatory transmitter release. More recent studies using intracellular recording techniques[52a] have also indicated a postsynaptic action of the purine in relation to its effects on K^+ conduction.

Early studies in the mammalian superior cervical ganglion[53] and cerebellar Purkinje cell system[54] indicated that cyclic AMP might function as a second messenger for the effects of neurohumoral agents on the electrophysiological parameters of these systems. Although it was initially thought[47] that the effects of adenosine might be mediated via the cyclic nucleotide, the magnitude and dose–response relationships of the effects of adenosine and 5'-AMP in hippocampal slices in terms of cyclic AMP accumulation and depression of cell firing did not correlate.[55] Furthermore, the dibutyryl, monobutyryl,[48] and 8-bromo[48,56] analogs of cyclic AMP enhanced rather than decreased postsynaptic potentials in olfactory cortex and hippocampal slice preparations, a finding opposite to that which would be anticipated for an involvement of the cyclic nucleotide in the depression of cell firing. Also, in slices of superior colliculus, although adenine nucleotides elevate cyclic AMP, they have no inhibitory action on postsynaptic potential generation.[57]

Discrepancies in the molar effectiveness of both *l* and *d*-PIA and 2-CADO in evoking electrophysiological responses and their effect on neurotransmitter release (see below) and cyclic AMP production have led Phillis[11] to suggest that there might be a third type of extracellular adenosine receptor, distinct from those linked to adenylate cyclase, which mediates the electrophysiological actions of adenine nucleosides and nucleotides. This hypothesis, which has yet to be fully evaluated, must be contrasted with the findings of Reddington *et al.*,[57a] who have shown an excellent correlation between the effects of several adenosine analogs in displacing [^{3}H]CHA from A_1 type receptors and their ability to depress evoked potentials in rat hippocampal slices. Irrespective of the adenosine receptor subtype involved in the electrophysiological actions of the purine, it is apparent that the evidence relating purine-evoked changes in cyclic AMP to its depressant effects on cell firing is poor. An involvement of cyclic nucleotides in the depolarizing responses to purine and pyrimidine nucleotides in bullfrog sympathetic ganglia[58] has not been discounted, however.

Iontophoretic application of ATP to cortical neurons causes a transient excitatory response that is inversely related to their spontaneous activity[10] followed by a longer-lasting depression of cell firing. This biphasic response has been attributed to an initial effect of the nucleotide on cell excitability mediated either by its ability to chelate Ca^{2+} [10,11,59] or by an excitatory action of the polyphosphate side chain[11] prior to formation of the nucleoside by ectoenzyme systems in the synaptic cleft, the adenosine thus formed depressing cell firing. Studies with diadenosine polyphosphates[60] have also discounted the possibility of there being neuronal receptors in the CNS for ATP.

The adenosine cofactor S-adenosylmethionine, which lacks depressant activity in olfactory cortex slices,[61] has been reported[62] to have transient excitant effects in the rat sensoriomotor neocortex, an action attributed to possible methylation of the cell membrane. Coenzyme A and NAD, which also contain an adenosine moeity, have depressant effects in slices[61] and *in situ*.[10] This latter effect of NAD contrasts with its lack of effect on tissue cyclic AMP levels[61] and provides additional evidence for the lack of a direct involvement of cyclic AMP in purine-evoked electrophysiological responses.

2.3. Effect of Adenosine on Neurotransmitter Release

Adenosine can depress spontaneous and evoked acetylcholine (ACh) release from the rat phrenic nerve–diaphragm preparation.[51] In the neuromuscular junction, phosphodiesterase inhibitors and lipid-soluble analogues of cyclic AMP increase transmitter release,[63] from which it has been suggested that the cyclic nucleotide, via a protein phosphorylation mechanism, may regulate Ca^{2+} fluxes and thus stimulus–secretion coupling. Adenosine can inhibit protein kinase[64] and would therefore block release by preventing phosphorylation of proteins involved in the Ca^{2+} system. Alterations in the phosphorylation of specific synaptosomal proteins have been implicated in both calcium movement and stimulus–secretion coupling,[65,66] and in slices of rat striatum the purine can decrease phosphorylation.[67] It is also of interest in regard to this hypothesis that nearly all of the cyclic-AMP-stimulated protein kinase activity in rat brain is located presynaptically.[68] Such findings must, however, be contrasted with the observation that adenosine is a potent stimulator of presynaptic cyclic AMP formation.[69] Also, in adrenal medulla and several central and peripheral nerve tracts, cyclic AMP and phosphodiesterase inhibitors increase spontaneous release of catecholamines.[69-74] The involvement of the cyclic nucleotide in the effects of adenosine on transmitter release would appear to be complex.

The purine can inhibit the release of norepinephrine, ACh, dopamine, serotonin, and GABA* from prelabeled brain slice preparations.[75-78] Adenosine can also inhibit the spontaneous release of ACh from guinea pig neocortex *in situ* via a theophylline-sensitive process,[79] whereas the methylxanthine, in contrast, can enhance resting ACh release from neocortical slices[80] and the rat cortex *in situ*,[81] supporting the concept of a purinergic inhibitory tone in the CNS.[11,45] Dopamine release from striatal synaptosomes is potently inhibited by 2-CADO with an IC_{50} of 10 nM,[77] and *in vivo* this stable adenosine analog can decrease ACh turnover in rat hippocampus and parietal cortex.[82] In a separate study,[83] the adenosine effects on ACh turnover were shown to be stereoselective by the use of the diastereomers of PIA and were blocked by theophylline. Neither of the purine analogs altered ACh content.

---

* GABA, γ-aminobutryric acid.

Adenosine may inhibit transmitter release by affecting Ca^{2+} transport. In K^+-depolarized synaptosomal preparations, adenosine can reduce the uptake of $^{45}Ca^{2+}$,[84] and in olfactory cortex slices,[47] the purine inhibition of postsynaptic potential generation can be antagonized by elevated (5.6 mM) concentrations of Ca^{2+}. Further examination of the K^+-depolarized synaptosomal Ca^{2+} uptake system[84a] using a number of adenosine analogs has shown the process to be theophylline sensitive and, on the basis of the low separation in activity of the diasteromers of PIA, to probably be mediated via an extracellular A_2 type receptor. In the periphery, adenosine can depress the calcium-mediated slow action potential in heart muscle cells[85] and antagonize Ca^{2+}-dependent potentials in sympathetic ganglion cells.[86] More recently, adenosine has been shown to interact with calcium channels as assessed by the effect of the dihydropyridine calcium channel blockers[87] with the binding of [^{3}H]CHA in rat brain membranes.[88] In the *Aplysia*, cyclic AMP is thought to modulate presynaptic calcium currents, increasing transmitter release by enhancing the voltage-dependent Ca^{2+} current during presynaptic action potential generation.[89]

Adenosine can also modulate the activity of the catecholamine-synthesizing enzyme, tyrosine hydroxylase.[69,90–91] In pheochromocytoma cells,[91] ADA decreases basal tyrosine hydroxylase activity, and in ADA-pretreated cells, 2-CADO can cause a two- to fivefold increase in enzyme activity which is accompanied by an increase in cyclic AMP. Tyrosine hydroxylase can be activated by cyclic AMP[92] by a mechanism that is protein kinase dependent.[93] On the basis of these findings, a positive feedback system has been proposed[69]: ATP is released together with the catcholamines from the nerve terminal; adenosine is formed by ectoenzyme activity and increases cyclic AMP presynaptically, which then activates tyrosine hydroxylase to replace the catecholamine released. The discrepancy in the effects of cyclic AMP at catecholaminergic and cholinergic nerve terminals (see above) may be related to this sensitivity of the anabolic enzyme to the cyclic nucleotide.

Studies on the presynaptic effects of adenosine are, however, complicated by the presence of high-affinity uptake systems for the purine, which tend to cause an underestimate of the efficacy of the nucleoside.[94] The use of analogs such as 2-CADO, which are not substrates for uptake, or uptake inhibitors such as dipyridamole or NBTG*[95] may allow for a more systematic evaluation of the presynaptic effects of adenosine.

The potential physiological effect of the "purinergic inhibitory tone" on neurotransmitter release remains to be fully evaluated, especially in regard to the autoreceptors for such transmitters on cell bodies and presynaptic nerve terminals, which are also thought to modulate transmitter release.[95a] Presumably, the mechanisms operating in physiological situations would closely regulate the amount of purine available in the synaptic cleft region such that the autoinhibitory actions of a given transmitter are not negated by an already active, nonspecific, purinergic inhibition.

---

* NBTG, 6-(4-nitrobenzyl)-thioguanosine.

2.4. Behavioral Effects of Adenine Nucleosides and Nucleotides

Adenosine, several of its stable analogs, and ATP have sedative and anticonvulsant properties when administered either centrally or peripherally.[2,14,96-101] In chicks, behavioral and electrocortical sleep have been induced by administration of adenosine into the third ventricle or hypothalamus,[96] and in dog, the purine has hypnogenic and sedative effects when injected into the lateral ventricle.[97] Following peripheral administration, the *d* and *l* diastereomers of PIA have stereospecific effects in depressing mouse locomotor activity.[98] CHA and 2-CADO at 500 μg/kg i.p. significantly decrease movement, increase pause duration, and decrease mean exploratory distance per move in mice.[101] Defects in rotorod motor coordination and hotplate reaction time were also observed. CHA, *l*-PIA, and 2-CADO also increase seizure latency when given prior to convulsants such as pentylenetetrazol, strychnine, and kainic acid[102] and can induce hypothermia. Although the sedative, hypothermic, and anticonvulsant effects of CHA and 2-CADO are antagonized by theophylline, the anticonvulsant actions of *l*-PIA are not. Furthermore, its anticonvulsant profile differs from that of either CHA or 2-CADO. This has led to the suggestion[102] that the phenylisopropyl analog may interact with "nonpurinergic receptors." This analog also has some antinociceptive activity,[14] but, like its anticonvulsant effects, this action occurs at doses some ten times higher than that (50 nmol/kg i.p.) evoking behavioral depression in mice. Adenosine has also been reported to facilitate learning.[103]

The alkylxanthine antagonism of the behavioral effects of the purine and its various analogs is in line with the concept that these compounds are adenosine antagonists[4,11,46,104-105] rather than inhibitors of phosphodiesterase[106] or calcium mobilizers.[107]

Increased locomotor activity[108] and increased sensitivity to nocioceptive stimuli[109] have been observed following theophylline administration. A more in-depth study[14] using ten alkylxanthines showed a good correlation between their affinity for adenosine binding sites (see below) and their effects as central stimulants. Brain levels following administration of behaviorally effective doses of the antagonists were also found to correlate well with their receptor affinities.[14]

One problem encountered in considering the behavioral effects of purines when they are administered peripherally relates to their well-characterized hypotensive actions,[110] which may contribute significantly to their observed effects on behavior.[10] The lack of toxicity of high doses of stable adenosine analogs[14] has been taken as an indication of the lack of involvement of peripheral blood flow in behavioral responses. However, in view of the proposed involvement of adenosine in the etiology of migraine,[111] a disease state involving cerebral blood flow, the effects of either endogenous or applied adenosine on central and peripheral blood flow should not be discounted in the interpretation of whole-animal responses to purines. Indeed, Phillis[111a] has demonstrated a close correlation between the effects of a number of adenosine analogs on rat cortical cell firing when applied iontophoretically and their hypotensive actions when given intravenously. Furthermore, induction of hy-

potension by the removal of blood from the rat femoral vein[111b] caused a reduction in neuronal firing rate when the fall in blood pressure exceeded 40–50
mm Hg, indicating that a depression of CNS activity can occur secondarily to
a fall in blood pressure.

2.5. Endogenous Adenosine in the Mammalian Central Nervous System

The data previously discussed, although indicating that exogenous adenosine has biochemical, electrophysiological, and behavioral effects in the CNS,
do not directly address the possible physiological relevance of the purine in
nervous tissue, i.e., actions mediated by endogenously occurring adenine nucleosides. Such a role can, however, be inferred from the excitant effects of
ADA and the methylxanthines,[11,56] removal or antagonism of the endogenous
purine increasing cell activity. It would also appear from the necessity to use
ADA to demonstrate adenosine-stimulated adenylate cyclase activity in homogenates (see above) that the receptors have a high potential for occupancy
in vivo.

Brain concentrations of adenosine are up to 2 nmol/g wet weight[112]; however, regional variations in adenosine and ATP are not readily observable,[11,113]
although differences have been found with use of a luciferin–luciferase assay.[114]

More direct evidence for a physiological, as opposed to pharmacological,
role for adenosine has come from studies on its release from nerve terminals
and subsequent inactivation, either by uptake or degradation. Such studies on
purine release are extensive; they are, however, also confusing in terms of the
mechanisms involved and of the identity and origin of the purines released, a
subject of some controversy[115,116] since release is often expressed only in terms
of radioactivity released from prelabeled tissue rather than identified compounds.

2.5.1. Release

Noncholinergic, nonadrenergic inhibitory responses in peripheral tissues
are mimicked by ATP, and the nucleotide is known to undergo Ca^{2+}-dependent
release following electrical stimulation.[117,118] The effects of adenine nucleosides and nucleotides in the periphery led Burnstock[6–8] to propose the concept
of purinergic neurotransmission. An impressive amount of evidence has accumulated to support both this hypothesis and a role for purines as cotransmitters. In the periphery, ATP is released together with norepinephrine[119] and
ACh[120] and is thought to be involved in the phenomenon of retrograde feedback
transmission.[121] In the *Torpedo* electric organ, however, ATP may have a post-
rather than presynaptic origin.[122]

In passing, it may be noted that although the concept of cotransmitters
was a somewhat controversial topic in the past because of its violation of what
is termed "Dale's Principle," i.e., the release of a single neurohumoral agent
from one nerve ending, an ever-growing body of evidence, especially in the

area of brain peptides, has indicated that neurotransmitter release can be dualistic (and perhaps bifunctional) in nature.[123,123a]

In the CNS, as previously mentioned, it appears unlikely from available evidence that ATP plays any role in purine-modulated neurohumoral events. This does not, however, preclude the possibility that the purine actively released from presynaptic nerve terminals is in the form of the nucleotide, which is subsequently degraded to adenosine.

In brain slice and synaptosomal preparations prelabeled with either [^{14}C]- or [^{3}H]adenine or adenosine, release of radioactive purine derivatives has been evoked by electrical stimulation, elevated K^+, ouabain, and veratridine.[124–127] Only a small fraction of labeled material, identified as approximately 50% adenosine and the remainder its catabolic products, inosine and hypoxanthine, is released in response to $K.^+$ In guinea pig cortical synaptosomes, K^+-evoked purine release was calcium dependent,[127] whereas in hypothalamic synaptosomes, it was largely Ca^{2+} independent.[128] As noted by Stone,[115] the amount of purine released in response to K^+ is low in comparison to that evoked by veratridine and ouabain and in comparison to the release of other, more conventional neurotransmitters. Furthermore, the apparent direct depolarizing effects of K^+ on purine release are temporally different from the effects of the cation on the release of other more conventional neurotransmitters.[129] In rat cortical slices, K^+-evoked release of [^{3}H]purine lags behind that of [^{3}H]GABA,[130] suggesting that purine release is secondary to elevation of cation concentrations.

On the basis of studies on ouabain-evoked release, which, from the effects of the metabolic inhibitor *p*-hydroxymercuribenzoate and the lack of effect of phenytoin and ethacrynic acid, would not appear to involve ATPase inhibition, it has been suggested that cyclic AMP may be the major source of the adenosine released.[115] This would be consistent with studies showing that increases in adenosine levels following brain slice stimulation occur after cyclic AMP[112] and also with the observations[131] that a) adenosine and not ATP is released from brain slices, and b) that purine nucelotide is present in high concentrations intracellularly.[115] Adenosine has also been identified as the major constituent (86–99%) of the labeled purine released from [^{3}H]adenosine-prelabeled synaptosomes[132] even though nucleotides represent 60–80% of the labeled material actually in the synaptosomes. Thus, the majority of evidence obtained using *in vitro* brain tissue preparations would indicate that adenosine is the major purine undergoing depolarization-induced release.

Labeling of the entorhinal cortex *in vivo* and study of the release of labeled purines from subsequently prepared hippocampal slices *in vitro* have shown that purine release is a specific event, with stimulation of the perforant path but not of the fimbria resulting in labeled purine release.[133]

Potassium- and veratridine-induced ATP release has also been reported in synaptosomes,[134] the former but not the latter being Ca^{2+} dependent. Regional variations in K^+-evoked ATP release have been observed[135] using the luciferin–luciferase technique. Release was greatest in the caudate and cerebral cortex and lowest in the cerebellum. In another study,[128] however, in which labeled purines were identified chromatographically, no ATP release could be detected in the superfusate from the labeled synaptosomes even though the nucleotide could be detected following hypoosmotic shock.

The involvement of Ca^{2+} in purine release in the CNS is therefore unclear. Synaptosomal K^+-evoked release was largely Ca^{2+} independent,[127] although the Ca^{2+} chelator EGTA can enhance basal and veratridine-evoked release from prelabeled rat synaptosomal preparations.[132] Thus, calcium-independent mechanisms may contribute significantly to synaptosomal adenosine release, possibly from glial components.[134] The release of adenosine from two synaptosomal pools[132] further complicates data interpretation, and one may question whether the calcium-independent release of purine reflects nonvesicular release. The use of high-performance liquid chromatographic (HPLC) techniques to measure release of endogenous rather than exogenously prelabeled purines may help to resolve the confusion surrounding the present *in vitro* release studies.

Up to 70% of the purine released *in situ* from exposed cat and rat cerebral cortex[81,136] is in the form of nucleotides. Release is calcium dependent, and use of the ADA inhibitor erythro-9-(hydroxy-3-nonyl)adenine (ENHA) results in an increase in the amount of released adenosine[137] with a corresponding decline in the amounts of inosine and hypoxanthine. Release *in situ* can be stimulated by the sodium channel blocker tetrodotoxin,[137] a phenomenon that may indicate the actual presence of a tonic system inhibiting purine release.

It is of interest in this regard and to the previously discussed effects of adenosine on neurotransmitter release that evidence has been obtained in the pulmonary artery for the existence of adenosine autoreceptors.[138] The differences in the release characteristics *in vitro* and *in situ* may therefore reflect the metabolic states of the preparations. Cerebral ischemia causes marked increases (as high as 1 M) in the amount of adenosine released into dog cerebrospinal fluid,[110] and metabolic inhibitors[115,129] and glucose deprivation[128] also increase purine release.

2.5.2. Inactivation

Termination of the actions of adenosine and adenine nucleotides can occur either by degradation via ectoenzyme systems or by reuptake, the latter system being specific for the nucleoside.

In cultured rat glioma and neuroblastoma cells, ecto nucleoside triphosphates[139,140] and 5'-nucleotidases[141,142] have been reported. The topographical distribution of the 5'-nucleotidases in hippocampus has been correlated with the transneuronal transfer of adenosine[143] and may reflect the necessity to convert ATP to its nucleoside to ensure postsynaptic uptake, since nucleotides do not cross the plasma membrane.[11] The physiological significance of purine transport, both inter- and transneuronal,[143–147] has not yet been established, but it may be of teleological significance in regard to the "rheoseme" concept of McIlwain.[148]

Adenosine deaminase does not appear to be involved in the termination of the actions of adenosine; it is primarily a cytoplasmic and not an ectoen-

Table I

Classification of Receptors Sensitive to Adenosine and ATP[a]

Re-ceptor	Location	Agonist		Antagonist	Response
		Pharmacology	Affinity (M)		
P_1	Extracellular	Ado[b] > AMP > ADP > ATP	10^{-6}	MeX[b]	↑ Cyclic AMP
P_2	Extracellular	ATP > ADP > AMP > Ado	—	None known	↑ PG formation
R_a	Extracellular	NECA > Ado ≥ PIA	10^{-6}	MeX	↑ Cyclic AMP
R_i	Extracellular	CHA > *l*-PIA > 2-CADO, Ado > N^6 methyl ado > *d*-PIA NECA	10^{-9}	MeX	↓ Cyclic AMP
P	Intracellular	2'5'-dideoxyAdo ≫ Ado	10^{-5}	—	↓ Cyclic AMP formation
A_1	Extracellular	CHA = *l*-PIA > 2-CADO > NECA > *d*-PIA > Ado	10^{-9}	MeX	Inhibits cyclic AMP formation
A_2	Extracellular	NECA > 2-CADO > Ado > PIA	10^{-6}	MeX	↑ Cyclic AMP

[a] For further details see references 7, 34, 36, 37, 104, and 116.
[b] MeX, methylxanthine; Ado, adenosine; PG, prostaglandin. See text for remainder of abbreviations.

zyme.[149] Thus, the appearance of extracellular inosine results either from its formation from 5'-AMP via adenylate deaminase[149] or from direct release.

High-affinity purine uptake occurs in guinea pig[150] and mouse[151] cortical slices, cultured astrocytes,[152,153] and guinea pig[127] and rat cortical synaptosomes.[132,154] In the latter system, Phillis and co-workers have demonstrated the presence of two uptake systems, one rapid, with saturation occuring within 1 min, and a second slower system, requiring 30 min to reach equilibrium. Uptake is apparently a carrier-facilitated diffusion process; the slow uptake system comprises two high-affinity (K_m = 1 and 5 μM) mechanisms, which may be glial and neuronal, respectively. The rapid uptake system had a K_m of 0.9 μM and could be completely inhibited by several purines and pyrimidines, implying the presence of a common nucleoside carrier system.

3. ADENOSINE RECEPTORS

Over the past 6 years, various systems of nomenclature for the receptors sensitive to adenine nucleosides and nucleotides have arisen. The initial classification, that by Burnstock,[7] termed receptors sensitive to adenosine P_1 and those sensitive to ATP, P_2. Independently, Londos and Wolff,[34] classifying adenosine-sensitive receptors in terms of their interaction with adenylate cyclase systems in adipocytes, hepatocytes, and Leydig cell lines, proposed the R/P classification (Table I). Those termed R require an intact ribose ring for activity and are methylxanthine sensitive and extracellular. Those termed P are intracellular, are not blocked by the methylxanthines, and require an intact purine ring for activity. The R receptor was further subdivided into R_a and R_i

on the basis of whether interaction of adenosine agonists caused stimulation or inhibition of adenylate cyclase activity. Interaction of agonists with the P site at micromolar concentrations can also cause cyclase inhibition. However, nanomolar amounts are effective at the R_i site.

Van Calker, Hamprecht, and colleagues,[36,37] using mouse fetal brain cells, also defined two extracellular receptors, termed A_1 and A_2, which inhibit and activate adenylate cyclase, respectively.

Although there is no concensus at the present time as to a uniform nomenclature, literature related to the CNS uses the A_1/A_2 classification, whereas the R_a, R_i, P_1, and P_2 classifications are used in the periphery. It is probable that the R_a, A_2, and P_1 receptors are identical, as they are all sensitive to methylxanthine inhibition and require micromolar concentrations of agonist to elicit increases in cyclic AMP. Likewise, the R_i and A_1 classifications may refer to the same receptor. Both are antagonized by methylxanthines and, when activated by nanomolar concentrations of agonist, cause cyclase inhibition. The P receptor classification, which should not be confused with either the P_1 or P_2 receptor, is in common usage.

Categorical identification of each of these individually defined receptor systems as being identical is premature at the present time. Studies using radioligand binding techniques (see below) and behavioral analysis of the effects of agonists at the A_1 receptor[102] have indicated that there are probably species differences in the A_1 receptor, and there may also be A_1 receptor subtypes.

At the present time, there is also some controversy regarding the delineation of A_1 and A_2 receptor subtypes on the basis of their responses to the diastereomers of PIA. Although it was originally noted[167] that there was a 50- to 100-fold difference in the potency of *d*- and *l*-PIA on evoked potential generation in rat hippocampus (an A_1 response) and only a fourfold difference in the effects of the isomers on cyclic AMP formation in guinea pig hippocampal slices (an A_2 response), more recent work[154a] has indicated that this is probably a species- rather than a receptor-mediated phenomenon. Thus, this confuses the distinction between A_1 and A_2 receptors in terms of radioligand binding, and the categorical assessment (see below) of all the currently known binding assays as being of the A_1 type is somewhat premature.

4. RADIOLIGAND BINDING TO CENTRAL ADENOSINE RECEPTORS

Since the advent of radioligand binding techniques, the demonstration of high-affinity, specific, reversible binding of a suitable radioligand to brain membranes has become an additional criterion for the identification of an endogenous substance as a putative neuromodulator. Adenosine is involved in nearly all aspects of cellular metabolism,[5] a fact that has made the search for a suitable radioligand difficult in view of the possibility of extensive metabolism and binding to cellular components unrelated to those involved in cellular communication.

Table II
Parameters of Radioligand Binding to Adenosine Receptors in the Mammalian
Central Nervous System

Ligand	K_d (nM)	B_{max} (pmol/mg protein)	Membrane preparation
CHA	0.3	0.34[a]	Bovine brain[161]
	1.8	0.20[a]	Bovine brain[161]
	6.0	0.37[a]	Guinea pig brain[161]
	4.5	0.20	Rat striatum[172]
	0.29	0.31[a]	Calf neocortex[165]
	5.2	0.92[a]	Human neocortex[165]
l-PIA	5.1	0.8	Rat brain P$_2$[163]
	1.8	0.2[a]	Guinea pig neocortex[165]
	0.7	0.2[a]	Rat neocortex[165]
	0.9	0.2[a]	Rabbit neocortex[165]
2-CADO	1.3	0.2	Rat brain CSM[162b]
	16.0	0.4	Rat brain CSM[162]
	24.0	0.5	Rat brain[173c]
NECA	2–4	0.1	Rat striatum[172]
	30–40	0.4	Rat striatum[172]
DPX	5.0	1.0[a]	Bovine forebrain[161]
	70.0	0.5[a]	Guinea pig cortex[161]
Adenosine	1,700	31.0	Rat brain P$_2$[158c]
	13,600	165.0	Rat brain P$_2$[158c]
	50–13,000	—	Rat brain[157c]

[a] Data for B_{max} recalculated on basis of 100 mg protein/g tissue.
[b] CSM, crude synaptic membranes.
[c] Tissue not pretreated with ADA.

Early studies using [^{3}H]adenosine[155–158] (Table II) reflect this problem. Dissociation constants (K_d) of 0.5[155] to 10[157] μM and high densities of binding sites were observed,[157,158] and although binding was theophylline sensitive, subcellular localization[157] and pharmacological studies[158] gave data inconsistent with those expected for a purine recognition site involved in synaptic transmission. Similar low-affinity binding was also observed in fat cells.[159] Such studies have led to the conclusion[159a] that [^{3}H]adenosine is not a suitable ligand to measure adenosine receptors linked to adenylate cyclase.

High-affinity (K_d = 14 nM) binding of [^{3}H]adenosine 5′-cyclopropyl carboxamide was also reported[160] but had pharmacology inconsistent with an extracellular binding site.

The use of the catabolic enzyme ADA in conjunction with metabolically stable analogs of adenosine has, however, enabled the detection of high-affinity (K_d = 10^{-9}–10^{-8} M) binding sites in mammalian brain. Three agonist ([^{3}H]CHA,[161] [^{3}H]2-CADO,[162] and [^{3}H]*l*-PIA[163]) and one antagonist radioligand ([^{3}H]DPX*[161]) have been studied in ADA-treated membrane preparations. As with many radioligand binding assays, the time for equilbrium binding to

* DPX, 3-diethyl-8-phenylxanthine.

occur bears little relationship to the physiological events in which adenosine is thought to be involved. At 37°C, [³H]*l*-PIA reaches apparent equilibrium at 12 min,[163] whereas at ambient (22–26°C) temperature, [³H]CHA and [³H]2-CADO have equilibrium binding times of 60 and 120 min, respectively.[161,162]

Subcellular distribution studies with [³H]2-CADO[162] and [³H]CHA[164] showed enhancement of specific binding in enriched synaptosomal subfractions with low amounts of binding in the myelin, mitochondrial, nuclear, and microsomal subfractions. A limited regional distribution of the binding of [³H]2-CADO[162], [³H]CHA,[164] and [³H]DPX[165] has also been observed, although only a maximal threefold variation was noted. Binding was highest in the caudate and hippocampus and lowest in the hypothalmaus and spinal cord.[162] In peripheral tissues, significant agonist binding has only been observed in the testes,[162,166] such binding being to the spermatozoa rather than to P_1 receptors in the associated vasa deferentia.

Specific binding can be destroyed by boiling or protease pretreatment[162] and by the alkylating agent, iodoacetamide,[164] indicating the involvement of a protein moiety in binding. High concentrations of sulfhydryl group reagents such as β-mercaptoethanol can also reduce binding.[164] Agonist binding is favored at high temperatures,[161,165] whereas binding of the antagonist DPX occurs most readily at 0°C.

The binding pharmacology of all three agonist ligands, with the caveat mentioned in Section 3, is consistent with labeling of the A_1 receptor (Table III). Thus, *l*-PIA is 28–41 times more potent in displacing [³H]CHA, [³H]2-CADO, and [³H]*l*-PIA than its *d* diastereomer. Unlabeled CHA and 2-CADO have approximately the same activity as *l*-PIA, whereas the 5'-substituted adenosine 5'-ethylcarboxamide (NECA) is slightly less active. Adenosine itself, which was studied only in the 2-CADO binding assay, was a weak displacer of specific binding ($K_i = 11,400$ nM), a fact attributed to the presence of residual ADA in the membrane preparations.[162] Cyclic AMP was considerably less active than *l*-PIA in displacing [³H]2-CADO, indicating that binding was probably not through an interaction with some cyclic nucleotide binding protein. In the *l*-PIA binding assay, however, cyclic AMP was equal in activity to 2-CADO. The methylxanthines IBMX, theophylline, and caffeine were weak displacers of [³H]CHA and [³H]2-CADO binding with K_is between 2500 and 26,000 nM. Theophylline and IBMX were 2–15 times more active in the *l*-PIA assay, with caffeine having a K_i of 26,800 nM. The potent adenosine antagonist[167] 8-phenyltheophylline had a K_i of 857 nM in the CHA binding assay and one of 66 nM in the 2-CADO assay. The phosphodiesterase inhibitor Ro 20-1724 was, however, inactive. Adenine and inosine were essentially inactive in all three assays, whereas the uptake inhibitor dipyridamole was fairly active in the *l*-PIA and 2-CADO assays.

The stable GTP analog GppNHp was reasonably active in the CHA and 2-CADO assays. Examination of the effects of guanine nucleotides on CHA binding showed [161] a rank order of potency similar to that observed[168] for other neurotransmitter radioligands that bind to a receptor linked to a GTPase/adenylate cyclase complex. Guanine nucleotides decrease CHA binding to guinea pig and bovine brain membranes with an IC_{50} of 1–3 μM, whereas

Table III
Pharmacology of Adenosine Radioligand Binding^a

Compound	Radioligand K_i (nM)^b			
	[^{3}H]CHA	[^{3}H]2-CADO	[^{3}H]l-PIA	[^{3}H]DPX^c
l-PIA	3.4	0.6	24.0	140,000
CHA	5.1	0.7	—	74,600
2-CADO	8.6	1.5	1,000.0	1,867
NECA	8.6	24.3	—	93,333
d-PIA	129	17.0	976	93,333
Cyclic AMP	—	424	>27,100	—
GppNHp	257	660	—	—
8-Phenyltheophylline	857	66	—	747
IBMX	8,571	2,217	867	2,800
Theophylline	12,860	4,960	2,060	18,667
Caffeine	85,700	15,340	26,800	186,667
Adenosine	—	11,400	—	—
Adenine	685,700	>56,500	>27,100	186,667
Dipyridamole	—	564	7,860	—
Ro 20-1724	—	>56,500	>27,100	—

^{a} For further details, see text and references 161–163.
^{b} K_i values were calculated according to the relationship $K_i = IC_{50} + c/K_d$, where c is concentration of radioligand, and K_d is the dissociation constant. For CHA, $c = 1$ nM; $K_d = 6$ nM. For 2-CADO, $c = 1$ nM; K_d
= 1.3 nM. For l-PIA, $c = 13.7$ nM; $K_d = 5.1$ nM. For DPX, $c = 5.0$ nM; $k_d = 70$ nM.
^{c} K_i values were derived from IC_{60} data.

bovine DPX binding is unaffected.[169] GTP can reduce agonist but not antagonist binding potency at DPX sites, an observation that may provide a means to delineate agonist and antagonist activity at the receptor level. The GTP effects on purine binding are maintained when the receptor is solubilized.[170]

The rank order of potency of the compounds studied at each of agonist binding sites differs (Table III). This is most notable with l-PIA binding, for which IBMX is some 36 times less active than unlabeled l-PIA, whereas the xanthine is 2500–3700 times less active than the agonist in the other two assays. Similar differences exist for caffeine, theophylline, and adenine. 2-Chloroadenosine is some 17 times less active in the l-PIA assay, whereas the analog itself is 7–45 times less active in displacing itself than in displacing either CHA or 2-CADO. The density of l-PIA binding sites (Table II) is two to four times greater than that observed for either CHA or 2-CADO. More prolonged incubation times (120 $vs.$ 12 min) have, however, given data showing l-PIA binding to be similar to that observed for the other agonist radioligands.[165] The phenylisopropyl analog may in fact bind to more than one adenosine recognition site depending on the assay condition. The behavioral data previously discussed[102] may support this suggestion.

The binding of the antagonist DPX also differs markedly from that of the agonists (Table III). Agonists give multiphasic displacement curves with this ligand, whereas xanthines have Hill coefficients around 1.0.[161] 8-Phenyltheophylline and 2-CADO are the most active compounds in this assay, but the order of potency of the diasteromers of PIA is reversed.

Examination of the binding of radioactive CHA, *l*-PIA, and DPX in calf, rat, rabbit, human, and guinea pig brain membranes has raised the distinct possibility of there being a heteogeneity in A_1 receptors.[165] The affinity of CHA in calf neocortex differed markedly from that observed in human (K_d = 0.3 *vs.* 5.2 nM; Table II), but there were one-third the number of binding sites in human cortex as compared to calf. The pharmacology of CHA binding also showed marked species differences. In calf and rabbit brain, *l*-PIA is eight times more active than in rat, guinea pig, or human brain. Cyclohexyladenosine is two to six times more active in calf than in the other species, whereas *d*-PIA is 17–57 times more active in calf in displacing [³H]CHA than in rabbit, rat, guinea pig, and human. The stereospecificity of *l*- and *d*-PIA interactions with CHA binding sites also shows species differences. A tenfold separation is seen in calf, a 34-fold separation in guinea pig and human cerebral cortex, a 78-fold separation in rat, and a 132-fold separation in rabbit.[165] 3-Diethyl-8-phenylxanthine shows a 260-fold separation in displacing [³H]CHA in calf versus guinea pig, whereas the pyrazolopyridine putative anxiolytic tracazolate shows a tenfold separation among the species studied. IBMX is equipotent in guinea pig and human brain and some two to three times less active in calf, rabbit, and rat brain. [³H]DPX binding can be observed in rat, calf, and rabbit but not human or guinea pig brain.

Thus, there is a growing body of evidence to suggest that there are receptor subtypes. This may be reflected in the discrepancies in adenylate cyclase responses to the purine between guinea pig and rat.[27] It is of interest in regard to the reported interaction between adenosine and the dihydropyridine calcium entry blockers[88] that CHA binding to membrane-bound but not solubilized receptors can be modulated by EDTA and divalent cations.[169,170] Adenosine/calcium interactions are well documented[171] and may provide some additional basis for the inhibitory effects of the purine on neurotransmitter release.

In addition to the high-affinity binding sites already described for 2-CADO and CHA, Scatchard analysis has also revealed the presence of lower-affinity sites, K_d = 16 nM for 2-CADO[162] and K_d = 2.4 nM for CHA.[164] From a pharmacological standpoint, these sites have not been examined in any great detail, but the possibility does arise that such sites may represent the A_2 receptor. As discussed in Section 3, the delineation of A_1 and A_2 receptors on the basis of the separation in activity of the diastereomers of PIA may no longer be valid[154a]; thus, until further criteria are proposed and undergo experimental evaluation, the receptors labeled by the various radioligands could be of either subtype. It is of interest in this regard that computer analysis of the binding of [³H]NECA in rat striatum[172] suggests that this adenosine analog binds to both A_1 and A_2 receptors.

Examination of the binding of 2-CADO in the absence of ADA pretreatment has yielded binding data markedly different from those described above. A site with a K_d of 23 nM was observed.[173] Adenosine had a K_i of 52 nM,[11] whereas 2-CADO was active at 43 nM. Both IBMX and theophylline were extremely active in the untreated tissue, with K_i values of 56 and 151 nM, respectively. The diastereomers of PIA showed a 500-fold separation in displacing [³H]2-CADO, whereas regional binding showed an eightfold variation, being highest in the cerebral cortex and lowest in the hippocampus; this latter finding is in marked contrast to the regional distribution observed for radio-labeled CHA,[164] 2-CADO,[162] and DPX[165] binding in ADA-pretreated brain membranes. The significance of the 2-CADO binding site in untreated tissue remains unclear at the present time but has led Phillis[11] to consider whether the binding seen following ADA pretreatment represents adsorption of the enzyme onto the brain membranes. Varying the amount of ADA used does not increase binding to any significant extent.[155,164]

Removal of the enzyme by washing does, however, dramatically reduce the amount of specific radioligand binding,[155,164] an effect that can be reversed by readdition of the enzyme. This phenomenon and the effects of the ADA inhibitors deoxycoformycin[161] and EHNA[162] on binding in ADA-pretreated tissue suggest that the membrane preparations used to measure binding are capable of generating considerable amounts of the purine. This may be a situation analogous to the appearance of the purine following ischemia[110] and consequently indicate a high degree of receptor occupancy.[45] The 2-CADO binding site in non-ADA-treated tissues has been suggested[173a] to be an A_1 receptor distinct from that labeled by either CHA or 2-CADO in ADA-treated tissue. The kinetic conditions occurring in the presence of copious amounts of endogenous adenosine may, however, be responsible for this interpretation. It is of additional interest in this regard to note that *in vivo* ADA has been reported to increase the amount of specific radioligand binding.[174]

5. AUTORADIOGRAPHIC LOCALIZATION OF ADENOSINE RADIOLIGAND BINDING

The elegant autoradiographic techniques of Kuhar[175] have enabled the direct visualization of adenosine receptors using [^{3}H]CHA as ligand. The highest density of sites occurs in the hippocampus,[176] a finding consistent with the regional distribution studies mentioned above. Such binding is highest in the molecular and polymorphic areas, with lower levels of binding over the pyramidal cell layer. The medial, gelatinosus, and lateral nuclei of the thalamus have high densities of binding, as do the lateral septum and medial geniculate body.

The superficial layer of the superior colliculus, the piriform cortex, the olfactory tubercule, layers I, IV, and VI of the cerebral cortex, the granule cell layer of the cerebellum, the nucleus accumbens, the caudate–putamen, the ventral thalamic nuclei, and the central amygdaloid nucleus have a moderate density of [^{3}H]CHA binding sites. Layers II, III, and V of the cerebral cortex, the trigeminal nerve, the spinal tract nucleus, the periaqueductal gray matter, the inferior collicular pontine nuclei, the corpus callosum, and the pontine and medullary reticular formation show low densities of binding sites, whereas the hypothalamus, anterior commisure, and cerebellar white matter also have low densities of specific binding sites. No detectable binding could be observed in the spinal or pyramidal tracts, the spinal tract of the trigeminal nerve, or the superior cerebellar peduncle. A more limited study[177] has confirmed certain of these observations. It has been noted[176] that the autoradiographic localization of CHA binding sites correlates with the histochemical localization of the enzyme 5'-nucleotidase,[144] which can generate adenosine, suggesting the possibility of a functional relationship between the enzyme and the recognition sites for its product.

6. ALTERATIONS IN ADENOSINE RADIOLIGAND BINDING

Understanding of the physiological role of adenosine in the CNS can be greatly enhanced by studying changes in binding as a result of animal age, genetic characteristics, or various surgical and *in vivo* pharmacological procedures. Although such studies are limited for the adenosine receptor, it has been found that the ontogeny of binding is biphasic in rat forebrain and cerebellum[178] with a rapid increase from E_{18} to birth followed by a slower increase up to 24 days, at which time adult levels are attained. The appearance of adenosine receptors has a time course similar to that seen for muscarinic cholinergic,[179] dopamine,[180] and GABA[181] receptors and is consistent with the process of neuronal differentiation. This ontogeny is in marked contrast to that of benzodiazepine (BZ) receptors, over half of which are present at birth.[182]

Chemical and surgical lesioning studies[183] have shown no change in specific [³H]CHA binding following destruction of striatal neurons, hippocampal pyramidal cells, and the glutaminergic and dopaminergic inputs to the striatum. When the cholinergic innervation to the hippocampus is chronically removed, however, a significant 57% increase in binding was observed, implying a functional supersensitivity. Overall, however, this preliminary data would indicate that the adenosine normally interacting with the [³H]CHA binding sites does not originate from any discrete "purinergic" innervation. It may be noted, however, that certain lesioning procedures can alter adenosine A_2 receptor-mediated responses.[183a] Following pentylenetetrazol-induced convulsions, a significant decrease (21%) in the number of cerebellar but not cortical [³H]CHA binding sites occurs[184] with no change in binding affinity, suggesting an involvement of endogenous adenosine in the seizure process. Chronic treatment with caffeine, on the other hand, has been reported to increase adenosine receptor density.[184a,184b]

7. POSSIBLE PHYSIOLOGICAL ROLES OF ADENOSINE IN THE CENTRAL NERVOUS SYSTEM

7.1. Adenosine and Anxiolytics

In addition to their anxiolytic actions, the benzodiazepines (BZs) are also anticonvulsants and muscle relaxants. As antianxiety agents they are the most widely used centrally acting therapeutic agents in the world. These compounds bind to specific receptor sites in the CNS,[15] of which two have been postulated,[185] the BZ_1 receptor, which is not coupled to a GABA receptor and is thought to mediate the anxiolytic actions of BZs, and the BZ_2 receptor, which is linked to GABA and is involved in the sedative and ataxic side effects of this class of drugs.

Purine agonists and antagonists can displace [^{3}H]diazepam specifically bound to rat brain membranes,[186,187] albeit at rather high concentrations (10^{-4}–10^{-3} M). Based on these findings, it has been postulated[186] that the endogenous ligand for the BZ receptor, analogous to the endorphins and enkephalins for the opiate receptors, may be a purine. The enhancement of the binding of [^{3}H]BZ by the N^6-substituted adenosine analog EMD 28422[188] and the increased potency of the purines in GABA-stimulated[189] BZ binding [190] would support the concept of a purine interaction with this class of drugs. However, since the β-carbolines,[191] several proteins,[192] and nicotinamide[193] have also been postulated as endogenous BZ-like agents, the area is somewhat controversial[193a]. Furthermore, there is no correlation between the effect of a series of BZ and non-BZ anxiolytics in displacing [^{3}H]diazepam and [^{3}H]2-CADO binding.[194] Also, although the BZ antagonist CGS 8216 does have some adenosinelike activity,[195] it is a weak displacer of specific [^{3}H]CHA binding.[196] This does not, however, preclude some other form of interaction, perhaps with an A_2-receptor-mediated mechanism.

Following from the observation that diazepam can inhibit adenosine uptake,[197] Phillis (see ref. 11 for review) has presented evidence that the BZs may produce certain of their pharmacological actions (i.e., sedation) by blocking adenosine uptake and thus increasing the availability of endogenous adenosine.

7.2. Adenosine and Antidepressants

At high concentrations ($>10^{-4}$ M), tricylic and atypical antidepressants can increase guinea pig brain cyclic AMP levels via a mechanism that is susceptible to blockade by theophylline.[198] Subsequent studies[16] suggested that these compounds might be releasing adenosine by an effect on cellular energy metabolism, the purine then acting to stimulate the cyclase. Following electroconvulsive shock therapy in experimental animals, the cyclic AMP response to adenosine is significantly enhanced,[16] a phenomenon that may involve α-adrenoceptor systems in the cortex.

The concentrations of antidepressants used and the inconsistent dose–response relationships observed[16,198] suggest the need for a more rigourous evaluation of the relationship between the purine and the drug. However, in light of the apparent supersensitivity of the adenosine response following electroconvulsive shock therapy and the known depressant[105] effects of adenosine, it is possible that treatment consistent with clinical antidepressant therapy may attenuate purine-mediated mechanisms in the CNS. Chronic treatment with the antidepressants mianserin and desmethylimpramine, a procedure that causes a decrease in β-adrenoceptor and serotonin-2 receptors in rat frontal cortex that is temporally related to the clinical actions of this class of compound, does not, however, alter adenosine radioligand binding.[198a]

7.3. Adenosine and Analgesics

Methylxanthines may act as morphine antagonists. Theophylline reduces morphine analgesia in mice,[199] and aminophylline reduces the depressant effects of morphine on rat striatal neurons.[200] Methylxanthines can also increase sensitivity to nocioceptive stimuli and can elicit a "quasi-morphine-withdrawal syndrome" in morphine-dependent rats.[201] The depressant actions of the analgesic on *in situ* release of ACh can be antagonized by methylxanthines,[202] suggesting that morphine can exert some of its pharmacological actions by increasing extracellular levels of adenosine. The effects of stable adenosine analogs on hot-plate reaction time[98,101] would support this concept, as would the observation that morphine can increase veratridine-induced purine release from rat neocortical slices,[203] a finding that has been confirmed *in situ*.[81] This has, like the interactions between the BZs and adenosine, been attributed to an effect on uptake.[11]

8. CONCLUSIONS

It is apparent from the data reviewed above that adenosine and its related analogs have potent effects on CNS function. The precise nature of such effects and their relationship to the overt functioning of the CNS and its interaction with various therapeutic agents remain unknown at the present time. Although there is no apparent regional distribution of the purine in the CNS, the organization of CHA binding sites in discrete domains[176] would point to a specific, rather than general, function of this putative neuromodulator. This observation must, however, be contrasted against the confusion surrounding the precise mode by which the purine is made available to the extracellular milieu. If release does occur as a consequence of increased intracellular cyclic AMP[115] rather than a defined stimulus–secretion coupling event, purinergic interactions in the CNS may to some extent be serendipitous, depending on receptor availability rather than the specific juxtaposition of transmitter and receptor usually associated with cellular communication. In response to specific neuronal activation, adenosine may, however, function as a cotransmitter, acting as a feedback inhibitor, this being a local as opposed to global (purinergic tone) phenomenon.

Further clarification of the mechanisms and potential for adenosine release and the characterization of the actual adenosine receptor subtypes being labeled by radioligands should provide new information on the role of adenosine in the CNS and replace much of the present speculation with fact. New, more potent, and perhaps specific A_1 and A_2 antagonists[204] should also help to provide new tools for the study of these systems. One other facet of adenosine action in the CNS that has yet to be explored is the possibility that there may be receptor subtypes that do not mediate their effects by an adenylate cyclase system.

From a therapeutic viewpoint, it seems probable that a more potent A_1 antagonist would be a potential central stimulant useful as an analeptic. The possibility that purines (or their antagonists) may represent novel analgesics, anxiolytics, anxiogenic/analeptics, or antidepressants is an exciting area for future research efforts. As Daly[116] has indicated, the 1980s will be a turning point in research on the physiological function(s) of adenosine.

ACKNOWLEDGMENTS. The author would like to thank John Daly, Solomon Snyder, John Phillis, Ken Murphy, Robert Goodman, and Richard Green for making available preprints of their work and also the many "adenophiles" at the Second International Symposium on Adenosine in Charlottesville, Virginia for sharing their thoughts and their work.

REFERENCES

1. Drury, A. N., and Szent-Gyorgyi, A., 1929, *J. Physiol. (Lond.)* **68**:213–217.
2. Feldberg, W., and Sherwood, S. L., 1954, *J. Physiol. (Lond.)* **123**:148–167.
3. Holton, F. A., and Holton, P., 1954, *J. Physiol. (Lond.)* **126**:124–140.
4. Sattin, A., and Rall, T. W., 1970, *Mol. Pharmacol.* **6**:13–23.
5. Arch, J. R. S., and Newsholme, E. A., 1978, *Essays Biochem.* **14**:82–123.
6. Burnstock, G., 1972, *Pharmacol. Rev.* **24**:509–581.
7. Burnstock, G., 1978, *Cell Membrane Receptors For Drugs and Hormones* (R. W. Straub and L. Bolis, eds.), Raven Press, New York, pp. 107–118.
8. Burnstock, G., 1979, *Physiological and Regulatory Functions of Adenosine and Adenine Nucleotides* (H. P. Baer and G. I. Drummond, eds.), Raven Press, New York, pp. 3–32.
9. McIlwain, H., 1972, *Biochem. Soc. Symp.* **36**:69–85.
10. Phillis, J. W., Edstrom, J. P., Kostopoulos, G. K., and Kirkpatrick, J. P., 1979, *Can. J. Physiol. Pharmacol.* **57**:1289–1312.
11. Phillis, J. W., and Wu, P. H., 1981, *Prog. Neurobiol.* **16**:187–239.
12. Daly, J. W., 1977, *Cyclic Nucleotides and the Central Nervous System*, Plenum Press, New York.
13. Snyder, S. H., Bruns, R. F., Daly, J. W., and Innis, R. B., 1981, *Fed. Proc.* **40**:142–146.
14. Snyder, S. H., Katims, J. J., Annau, Z., Bruns, R. F., and Daly, J. W., 1981, *Proc. Natl. Acad. Sci. U.S.A.* **78**:3260–3264.
15. Tallman, J. F., Paul, S. M., Skolnick, P., and Gallager, D. W., 1980, *Science* **207**:275–281.
16. Sattin, A., 1981, *Chemisms of the Brain* (R. Rodnight, H. S. Bachelard, and W. S. Stahl, eds.), Churchill-Livingstone, Edinburgh, pp. 265–275.
17. Premont, J., Perez, M., and Bockaert, J., 1977, *Mol. Pharmacol.* **13**:662–670.
18. Ohga, Y., and Daly, J. W., 1977, *Biochim. Biophys. Acta* **498**:46–60.
19. Saito, M., 1977, *Biochim. Biophys. Acta* **498**:316–325.
20. Schultz, J., and Daly, J. W., 1973, *J. Biol. Chem.* **248**:843–852.
21. Skolnick, P., and Daly, J. W., 1974, *Brain Res.* **73**:513–525.
22. Huang, M., and Daly, J. W., 1974, *J. Neurochem.* **23**:393–404.
23. Daly, J. W., McNeal, E., Partington, C., Neuwirth, M., and Creveling, C. R., 1980, *J. Neurochem.* **35**:326–337.
24. Daly, J. W., Padgett, W., and Seamon, K., 1982, *J. Neurochem.* **38**:532–544.
25. Paes De Carvalho, R. and De Mello, F. G., 1982, *J. Neurochem.* **38**:493–500.
26. Jones, D. G., 1981, *J. Pharmacol. Exp. Ther.* **219**:370–376.
27. Premont, J., Tassin, J. P., Blanc, G., and Bockaert, J., 1979, *Physiological and Regulatory Functions of Adenosine and Adenine Nucleotides* (H. P. Baer and G. I. Drummond, eds.), Raven Press, New York, pp. 259–269.
28. Premont, J., Perez, M., Blanc, J., Tassin, J. P., Thierry, A. M., Herve, D., and Bockaert, J., 1979, *Mol. Pharmacol.* **16**:790–804.
29. Bruns, R. F., 1980, *Can. J. Physiol. Pharmacol.* **58**:673–691.
30. Bruns, R. F., 1981, *Biochem. Pharmacol.* **30**:325–333.
31. Fain, J. N., 1973, *Pharmacol. Rev.* **25**:67–118.

32. Fain, J. N., and Malbon, C. C., 1979, *Mol. Cell. Biochem.* **23**:1–17.
33. Londos, C., and Wolff, J., 1977, *Proc. Natl. Acad. Sci. U.S.A.* **24**:5482–5486.
34. Londos, C., Wolff, J., and Cooper, D. M. F., 1979, *Physiological and Regulatory Functions of Adenosine and Adenine Nucleotides* (H. P. Baer and G. I. Drummond, eds.), Raven Press, New York, pp. 271–281.
35. Londos, C., Cooper, D. M. F., and Wolff, J., 1980, *Proc. Natl. Acad. Sci. U.S.A.* **77**:2551–2554.
36. Van Calker, D., Muller, M., and Hamprecht, B., 1978, *Nature* **273**:839–841.
37. Van Calker, D., Muller, M., and Hamprecht, B., 1979, *J. Neurochem.* **33**:999–1005.
38. Cooper, D. M. F., Londos, C., and Rodbell, M., 1980, *Mol. Pharmacol.* **18**:598–601.
39. Aktories, K., Schultz, G., and Jakobs, K. H., 1982, *Life Sci.* **30**:269–275.
40. Creese, I., 1982, *Trends Neurosci.* **5**:40–43.
41. Phillis, J. W., Kostopoulos, G. K., and Limacher, J. J., 1975, *Eur. J. Pharmacol.* **30**:125–129.
42. Phillis, J. W., and Kostspoulos, G. K., 1975, *Life Sci.* **17**:1085–1094.
43. Kostopoulos, G. K., Limacher, J. J., and Phillis, J. W., 1975, *Brain Res.* **88**:162–165.
44. Kostopoulos, G. K., and Phillis, J. W., 1977, *Exp. Neurol.* **55**:719–724.
44a. Dunwiddie, T. V., Hoffer, B. J., and Fredholm, B. B., 1981, *Naunyn Schmiedebergs Arch. Pharmacol.* **316**:326–330.
45. Harms, H. H., Wardeh, G., and Mulder, A. H., 1978, *Eur. J. Pharmacol.* **49**:305–308.
46. Fredholm, B., 1980, *Trends Pharmacol. Sci.* **1**:129–132.
47. Kuroda, Y., Saito, M., and Kobayashi, K., 1976, *Brain Res.* **109**:196–201.
48. Kuroda, Y., 1978, *J. Physiol. (Paris)* **74**:463–470.
49. Scholfield, C. N., 1978, *Br. J. Pharmacol.* **63**:239–244.
50. Okada, Y., and Ozawa, S., 1980, *Eur. J. Pharmacol.* **68**:483–493.
51. Ginsborg, B. L., and Hirst, G. D. S., 1972, *J. Physiol. (Lond.)* **224**:629–645.
52. Lekic, D., 1977, *Can. J. Physiol. Pharmacol.* **55**:1391–1393.
52a. Segal, M., 1982, *Eur. J. Pharmacol.* **79**:193–199.
53. McAfee, D., and Greengard, P., 1972, *Science* **178**:310–312.
54. Bloom, F. E., Siggins, G. R., Hoffer, B. J., Segal, M., and Oliver, A. P., 1975, *Adv. Cyclic Nucleotide Res.* **5**:603–617.
55. Reddington, M., and Schubert, P., 1979, *Neurosci. Lett.* **14**:37–42.
56. Dunwiddie, T. V., and Hoffer, B. J., 1980, *Br. J. Pharmacol.* **69**:59–68.
57. Okada, Y., and Saito, M., 1979, *Brain Res.* **160**:368–371.
57a. Reddington, M., Lee, K. S., and Schubert, P., 1982, *Neurosci. Lett.* **28**:275–279.
58. Siggins, G. R., Gruol, D. L., Padjen, A. L., and Forman, D. S., 1977, *Nature* **270**:263–265.
59. Krenjevic, K., 1974, *Physiol. Rev.* **54**:418–540.
60. Stone, T. W., and Perkins, M. N., 1981, *Brain Res.* **229**:241–245.
61. Okada, Y., and Kuroda, Y., 1980, *Eur. J. Pharmacol.* **61**:137–146.
62. Phillis, J. W., 1981, *Brain Res.* **213**:223–226.
63. Standaert, F., and Dretchen, K. L., 1979, *Fed. Proc.* **38**:2183–2192.
64. Weller, M., 1979, *Protein Phosphorylation*, Pion, London.
65. Weller, M., and Morgan, I. G., 1977, *Biochim. Biophys. Acta* **465**:527–534.
66. Williams, M., and Rodnight, R., 1977, *Prog. Neurobiol.* **8**:182–250.
67. Williams, M., 1976, *Brain Res.* **109**:190–195.
68. Weller, M., 1977, *Biochim. Biophys. Acta* **469**:350–354.
69. Kobayashi, K., Kuroda, Y., and Yoshioka, M., 1981, *J. Neurochem.* **36**:86–91.
70. Berkowitz, B., Tarver, J. H., and Spector, S., 1970, *Eur. J. Pharmacol.* **10**:64–71.
71. Peach, M. J., 1972, *Proc. Natl. Acad. Sci. U.S.A.* **69**:834–836.
72. Goldberg, A. L., and Singer, J. J., 1969, *Proc. Natl. Acad. Sci. U.S.A.* **64**:134–141.
73. Wooten, F. G., Thoa, N. B., Kopin, I. J., and Axelrod, J., 1973, *Mol. Pharmacol.* **9**:178–183.
74. Wilson, D. F., 1974, *J. Pharmacol. Exp. Ther.* **188**:447–452.
75. Juel, C., 1980, *Comp. Biochem. Physiol. [C]* **68**:21–27.
76. Harms, H. H., Wardeh, G., and Mulder, A. H., 1979, *Neuropharmacology* **18**:577–580.
77. Michaelis, M. L., Michaelis, E. K., and Myers, S. L., 1979, *Life Sci.* **24**:2083–2092.

78. Hollins, C., and Stone, T. W., 1980, *Br. J. Pharmacol.* **69:**107–112.
79. Sawynok, J., and Jhamandas, K. H., 1976, *J. Pharmacol. Exp. Ther.* **197:**179–190.
80. Vizi, E. S., and Knoll, J., 1976, *Neuroscience* **1:**391–398.
81. Phillis, J. W., Ziang, Z. G., Chelack, B. J., and Wu, P. H., 1980, *Pharmacol. Biochem. Behav.* **13:**421–427.
82. Haubrich, D. R., Williams, M., Yarbrough, G. G., and Wood, P. L., 1981, *Can. J. Physiol. Pharmacol.* **59:**1196–1198.
83. Murray, T. F., Cheney, D. L., and Costa, E., 1981, *Soc. Neurosci. Abstr.* **7:**495.
84. Riberio, J. A., Sa-Almeida, A. M., and Namorodo, J. M., 1979, *Biochem. Pharmacol.* **28:**1297–1300.
84a. Wu, P. H., Phillis, J. W., and Thierry, D. L., 1982, *J. Neurochem.* **39:**700–708.
85. Schrader, J., Rubio, R., and Berne, R. M., 1975, *J. Mol. Cell. Cardiol.* **7:**427–433.
86. Henon, B. K., Turner, D. K., and McAfee, D. A., 1980, *Soc. Neurosci. Abstr.* **6:**257.
87. Triggle, D. J., 1981, *Calcium Antagonists* (G. B. Weiss, ed.), American Physiological Society, Bethesda, pp. 1–18.
88. Murphy, K. M. M., and Snyder, S. H., 1982, *Calcium Entry Blockers, Adenosine and Neurohumors: Recent Advances* (G. F. Merill and H. R. Weiss, eds.), Urban & Schwarzenberg, Baltimore pp. 295–306.
89. Klein, M., and Kandel, E. R., 1978, *Proc. Natl. Acad. Sci. U.S.A.* **75:**3512–3516.
90. Kuroda, Y., and Kobayashi, K., 1978, *Proc. Jpn. Acad. [Physiol. Biol. Sci.]* **54:**243–247.
91. Erny, R. E., Berezo, M. W., and Perlman, R. L., 1981, *J. Biol. Chem.* **256:**1335–1339.
92. Harris, J. E., Morgenroth, V. H., Roth, R. H., and Baldessarini, R. J., 1975, *Nature* **260:**101–103.
93. Edelman, A. M., Raese, J. D., Lazar, M. A., and Barchas, J. D., 1978, *Commun. Psychopharmacol.* **2:**461–465.
94. Muller, M. J., and Paton, D. M., 1979, *Naunyn Schmidebergs Arch. Pharmacol.* **306:**23–28.
95. Daly, J. W., 1979, *Physiological and Regulatory Functions of Adenosine and Adenine Nucleotides* (H. P. Baer and G. I. Drummond, eds.), Raven Press, New York, pp. 229–241.
95a. Starke, K., 1981, *Annu. Rev. Pharmacol. Toxicol.* **21:**7–30.
96. Marley, E., and Nistico, G., 1972, *Br. J. Pharmacol.* **46:**619–636.
97. Maitre, M., Ciesielski, L., Lehmann, A., Kempf, E., and Mandel, P., 1974, *Biochem. Pharmacol.* **23:**2807–2816.
98. Vapaatalo, H., Onken, D., Neuvonen, P., and Westerman, E., 1975, *Arnzeim. Forsch.* **25:**407–410.
99. Baird-Lambert, J., Marwood, J. F., Davies, L. P. and Taylor, K. M., 1980, *Life Sci.* **26:**1069–1077.
100. Haulica, I., Ababei, L., Branisteanu, D., and Topoliceanu, F., 1973, *J. Neurochem.* **21:**1019–1020.
101. Crawley, J. N., Patel, J., and Marangos, P. J., 1981, *Life Sci.* **29:**2623–2630.
102. Dunwiddie, T. V., and Worth, T., 1982, *J. Pharmacol. Exp. Ther.* **220:**70–76.
103. Mascherpa, P., 1971, *Arnzeim. Forsch.* **21:**25–26.
104. Daly, J. W., Bruns, R. F., and Snyder, S. H., 1981, *Life Sci.* **28:**2083–2097.
105. Snyder, S. H., 1981, *Trends Neurosci.* **4:**242–244.
106. Fredholm, B. B., Fuxe, K., and Agnati, L., 1976, *Eur. J. Pharmacol.* **38:**31–38.
107. Kuba, K., and Nishi, S., 1976, *J. Neurophysiol.* **39:**547–563.
108. Thithapandha, A., Maling, H. M., and Gillette, G. R., 1972, *Proc. Soc. Exp. Biol. Med.* **139:**582–586.
109. Paazlow, G., and Paazlow, L., 1973, *Acta Pharmacol. Toxicol.* **32:**22–32.
110. Berne, R. M., Rubio, R., and Curnish, R. R., 1974, *Circ. Res.* **35:**262–271.
111. Burnstock, G., 1981, *Lancet* **i:**1397–1399.
111a. Phillis, J. W., 1982, *J. Pharm. Pharmacol.* **34:**453–454.
111b. Phillis, J. W., and Wu, P. H., 1982, *Physiology and Pharmacology of Adenosine* (J. W. Daly, Y. Kuroda, J. W. Phillis, H. Shimizu, and M. Ui, eds.), Raven Press, New York, pp. 219–236.
112. Newman, M., and McIlwain, H., 1977, *Biochem. J.* **164:**131–137.
113. Wu, P. H., Moore, K. C., and Phillis, J. W., 1979, *Experientia* **35:**881–883.

114. Kogure K., and Alonso, D. F., 1978, *Brain Res.* **154:**273–284.
115. Stone, T. W., 1981, *Neuroscience* **6:**523–555.
116. Daly, J. W., 1982, *J. Med. Chem.* **25:**197–207.
117. Burnstock, G., Campbell, G., Satchell, D., and Smythe, A., 1970, *Br. J. Pharmacol.* **40:**668–688.
118. Su, C., Bevan, J., and Burnstock, G., 1971, *Science* **173:**337–339.
119. Stjarne, L., Hedqvist, P., and Lagercrantz, H., 1970, *Biochem. Pharmacol.* **19:**1147–1148.
120. Silinsky, E. M., 1975, *J. Physiol. (Lond.)* **247:**145–162.
121. Israel, M., Lesbats, B., Manaranche, R., Muenier, F. M., and Frachon, P., 1980, *J. Neurochem.* **34:**923–932.
122. Israel, M., Lesbats, B., Muenier, F. M., and Stinnakre, J., 1976, *Proc. R. Soc. Lond. [Biol.]* **193:**461–468.
123. Potter, D. D., Furshpan, E. J., and Landis, S. C., 1981, *Neurosci. Commun.* **1:**1–9.
123a. Burnstock, G., 1983, *Dale's Principle and Communication Between Neurones.* (N. N. Osborne, ed.), Pergamon Press, New York, pp. 7–35.
124. Shimizu, H., Creveling, C. R., and Daly, J. W., 1970, *Proc. Natl. Acad. Sci. U.S.A.* **65:**1033–1040.
125. Pull, I., and McIlwain, H., 1972, *Biochem. J.* **136:**893–901.
126. Heller, I. H., and McIlwain, H., 1973, *Brain Res.* **33:**105–116.
127. Kuroda, Y., and McIlwain, H., 1973, *J. Neurochem.* **21:**889–900.
128. Fredholm, B. B., and Vernet, L., 1979, *Acta Physiol. Scand.* **106:**97–107.
129. Daval, J. L., Barberis, C., and Gayet, J., 1980, *Brain Res.* **181:**161–174.
130. Stone, T. W., Hollins, C., and Lloyd, H., 1981, *Brain Res.* **207:**421–431.
131. Pons, F., Bruns, R. F., and Daly, J. W., 1980, *J. Neurochem.* **34:**1319–1323.
132. Bender, A. S., Wu, P. H., and Phillis, J. W., 1982, *J. Neurochem.* **36:**651–660.
133. Lee, K., Schubert, P., Gribkoff, V., Sherman, B., and Lynch, G., 1982, *J. Neurochem.* **38:**80–83.
134. White, T. D., 1978, *J. Neurochem.* **30:**329–336.
135. Potter, P., and White, T. D., 1980, *Neuroscience* **5:**1351–1356.
136. Sulahke, P. V., and Phillis, J. W., 1975, *Life Sci.* **17:**551–556.
137. Jhamandas, K., and Dumbrille, A., 1980, *Can. J. Physiol. Pharmacol.* **58:**1262–1278.
138. Katsuragi, T., and Su, C., 1982, *J. Pharmacol. Exp. Ther.* **220:**152–156.
139. Trams, E. G., Lauter, C. J., and Banfield, W. G., 1976, *J. Neurochem.* **27:**1035–1042.
140. Stefanovic, V., Ciesielski-Treska, T., Ebel, A., and Mandel, P., 1974, *Brain Res.* **81:**427–441.
141. Stefanovic, V., Mandel, P., and Rosenberg, A., 1976, *J. Biol. Chem.* **251:**3900–3905.
142. Scott, T. W., 1965, *J. Histochem. Cytochem.* **13:**657–667.
143. Schubert, P., Komp, W., and Kreutzberg, G. W., 1979, *Brain Res.* **168:**419–424.
144. Hunt, S. P., and Kunzle, H., 1976, *Brain Res.* **112:**127–132.
145. Wise, S. P., and Jones, E. G., 1976, *Brain Res.* **107:**127–131.
146. Rose, G., and Schubert, P., 1977, *Brain Res.* **121:**353–357.
147. Kruger, L., and Saporta, S., 1977, *Brain Res.* **122:**132–136.
148. McIlwain, H., 1978, *Prog. Neurobiol.* **11:**189–203.
149. Pull, I., and McIlwain, H., 1974, *Biochem. J.* **144:**37–41.
150. Shimizu, H., Tanaka, S., and Kodama, T., 1972, *J. Neurochem.* **19:**687–698.
151. Kuroda, Y., and McIlwain, H., 1974, *J. Neurochem.* **22:**691–699.
152. Lewin, E., and Bleck, V., 1979, *J. Neurochem.* **33:**365–367.
153. Hertz, L., 1978, *J. Neurochem.* **31:**55–62.
154. Bender, A. S., Wu, P. H., and Phillis, J. W., 1980, *J. Neurochem.* **35:**629–640.
154a. Fredholm, B. B., Jonzon, B., Lindgren, E., and Lindstrom, K., 1982, *J. Neurochem.* **39:**165–175.
155. Williams, M., 1981, *Chemisms Of The Brain* (R. Rodnight, H. S. Bachelard, and W. S. Stahl, eds.), Churchill-Livingstone, Edinburgh, pp. 88–99.
156. Shimizu, H., 1979, *Physiological and Regulatory Functions Of Adenosine and Adenine, Nucleotides* (H. P. Baer and G. I. Drummond, eds.), Raven Press, New York, pp. 243–248.
157. Newman, M. E., Patel, J., and McIlwain, H., 1981, *Biochem. J.* **194:**611–620.

158. Schwabe, U., Kiffe, H., Puchstein, C., and Trost, T., 1979, *Naunyn Schmidebergs Arch. Pharmacol.* **310**:59–67.
159. Malbon, C. C., Hert, R. C., and Fain, J. N., 1978, *J. Biol. Chem.* **253**:3114–3122.
159a. Newman, M., and Levitzki, A., 1982, *Biochim. Biophys. Acta.* **685**:129–136.
160. Daly, J. W., Nimitkitpaisan, Y., Pons, F., Bruns, R. F., Smellie, F., and Skolnick, P., 1979, *Pharmacologist* **21**:253.
161. Bruns, R. F., Daly, J. W., and Snyder, S. H., 1980, *Proc. Natl. Acad. Sci. U.S.A.* **77**:5547–5551.
162. Williams, M., and Risley, E. A., 1980, *Proc. Natl. Acad. Sci. U.S.A.* **77**:6892–6896.
163. Schwabe, U., and Trost, T., 1980, *Naunyn Schmiedebergs Arch. Pharmacol.* **313**:179–187.
164. Patel, J., Marangos, P. J., Stivers, J., and Goodwin, F. K., 1982, *Brain Res.* **237**:203–208.
165. Murphy, K. M. M., and Snyder, S. H., 1982, *Mol. Pharmacol.* **22**:250–257.
166. Murphy, K. M. M., and Snyder, S. H., 1980, *Life Sci.* **28**:917–920.
167. Smellie, F. W., Daly, J. W., Dunwiddie, T. V., and Hoffer, B. J., 1979, *Life Sci.* **25**:1739–1748.
168. Rodbell, M., 1980, *Nature* **284**:17–22.
169. Goodman, R. R., Cooper, J. H., Gavish, M., and Snyder, S. H., 1982, *Mol. Pharmacol.* **21**:329–335.
170. Gavish, M., Goodman, R. R., and Snyder, S. H., 1982, *Science* **215**:1633–1635.
171. Partington, C. R., Edwards, M. N., and Daly, J. W., 1980, *J. Neurochem.* **34**:76–82.
172. Yeung, S. M., and Green, R. D., 1981, *Pharmacologist* **23**:184.
173. Wu, P. H., Phillis, J. W., Balls, K., and Rinaldi, B., 1980, *Can. J. Physiol. Pharmacol.* **58**:576–579.
173a. Wu, P. H., and Phillis, J. W., 1982, *Int. J. Biochem.* **14**:399–404.
174. Schwabe, U., 1981, *Trends Pharmacol. Sci.* **2**:299–303.
175. Kuhar, M. J., 1981, *Trends Neurosci.* **4**:60–64.
176. Goodman, R. R., and Snyder, S. H., 1982, *J. Neurosci.* **2**:1230–1241.
177. Lewis, M. E., Patel, J., Moon Edley, S., and Marangos, P. J., 1981, *Eur. J. Pharmacol.* **73**:109–110.
178. Marangos, P. J., Patel, J., and Stivers, J., 1982, *J. Neurochem.* **39**:267–270.
179. Coyle, J. T., and Yamamura, H. I., 1976, *Brain Res.* **118**:429–440.
180. Prado, J. V., Creese, I., Burt, D., and Snyder, S. H., 1977, *Brain Res.* **125**:376–382.
181. Coyle, J. T., and Enna, S. J., 1976, *Brain Res.* **111**:119–133.
182. Braestrup, C., and Nielsen, M., 1978, *Brain Res.* **147**:170–173.
183. Williams, M., and Wood, P. L., 1981, *Pharmacologist* **23**:184.
183a. Wojcik, W. J., and Neff, N. H., 1981, *Fed Proc.* **41**:1046.
184. Wybenga, M. P., Murphy, M. G., and Robertson, H. A., 1981, *Eur. J. Pharmacol.* **75**:79–80.
184a. Fredholm, B., 1982, *Acta Physiol. Scand. (Suppl.)* **508**:31.
184b. Boulenger, J-P., Patel, J., Post, R. M., Parma, A. M. and Marangos, P. J., 1983, *Life Sci.* **32**:1135–1142.
185. Klepner, C. A., Lippa, A. S., Benson, D. I., Sano, M. C., and Beer, B., 1979, *Pharmacol. Biochem. Behav.* **11**:457–462.
186. Skolnick, P., Marangos, P. J., Goodwin, F. K., Edwards, M., and Paul, S., 1978, *Life Sci.* **23**:1473–1480.
187. Asano, T., and Spector, S., 1979, *Proc. Natl. Acad. Sci. U.S.A.* **76**:977–981.
188. Skolnick, P., Lock, K. L., Paul, S. M., Marangos, P. J., Jonas, R., and Irmscher, K., 1980, *Eur. J. Pharmacol.* **67**:179–186.
189. Williams, M., and Risley, E. A., 1979, *Life Sci.* **24**:833–842.
190. Marangos, P. J., Paul, S. M., Parma, A. M., and Skolnick, P., 1981, *Biochem. Pharmacol.* **30**:2171–2174.
191. Braestrup, C., Nielsen, M., and Olsen, C. E., 1980, *Proc. Natl. Acad. Sci. U.S.A.* **77**:2288–2291.
192. Davis, L. G., and Cohen, R. K., 1980, *Biochem. Biophys, Res. Commun.* **92**:141–148.
193. Mohler, H., Polc, P., Cumin, R., Pieri, L., and Kettler, R., 1979, *Nature* **278**:563–565.
193a. Williams, M., 1983, *J. Med. Chem.* **26**:619–628.

194. Williams, M., Risley, E. A., and Huff, J. R., 1981, *Can. J. Physiol. Pharmacol.* **59:**897–900.
195. Czernik, A. J., Petrack, B., Kalinsky, H. J., Psychoyos, S., Cash, W. D., Tsai, C., Rinehart, R. K., Granat, F. R., Lovell, R. A., Brundish, D. E., and Wade, R., 1982, *Life Sci.* **30:**363–372.
196. Williams, M., and Risley, E. A., 1982, *Arch. Int. Pharmacodyn.* **260:**50–53.
197. Mah, H. D., and Daly, J. W., 1976, *Pharmacol. Res. Commun.* **8:**65–79.
198. Sattin, A., Stone, T. W., and Taylor, D. A., 1978, *Life Sci.* **23:**2621–2626.
198a. Williams, M., Risley, E. A., and Robinson, J. L., 1983, *Neurosci. Lett.* **35:**47–51.
199. Ho, I. K., Loh, H. H., and Way, E. L., 1973, *J. Pharmacol. Exp. Ther.* **185:**336–346.
200. Stone, T. W., and Perkins, M. N., 1979, *Nature* **281:**227–228.
201. Collier, H. O. J., Cuthbert, N. J., and Francis, D. L., 1981, *Fed. Proc.* **40:**1513–1518.
202. Jhamandas, K., Sawynok, J., and Sutak, M., 1978, *Eur. J. Pharmacol.* **49:**309–312.
203. Fredholm, B. B., and Vernet, L., 1978, *Acta Physiol. Scand.* **104:**502–504.
204. Fredholm, B. B., and Persson, C. G. A., 1982, *Eur. J. Pharmacol.* **81:**673–678.

α-Adrenergic Receptors

Anna Borsodi and Maria Wollemann

1. INTRODUCTION

Adrenergic receptors are discrete recognition sites located on the plasma membrane of cells. Their primary function is to recognize and to bind epinephrine and related molecules, which can initiate several biochemical processes resulting in a physiological response. In addition to catecholamines, several other hormones of different structure are also able to change the properties of these receptors, reflecting their dynamic nature.

The existence of two discrete types of adrenergic receptors—α and β—was proposed by Ahlquist[1] in 1948. The distinction between them was originally based on their relative sensitivity to various agonists, inducing a response. For responses controlled through α-adrenergic receptors (e.g., smooth muscle contraction), epinephrine is the most potent agonist, followed by norepinephrine and then isoproterenol. For β-adrenergic receptors, almost an opposite order of potency was observed; i.e., isopropterenol had the highest potency of all. This classification was later confirmed by use of highly specific antagonists (drugs capable of intereaction with the receptor without eliciting a response) for each class of receptors.[2–7] Alpha- as well as β-adrenergic receptors exhibit stereospecificity with the *l* or (−) stereoisomers being more potent than *d* or (+) stereoisomers.

Alpha- and β-adrenergic receptors located in the same tissues may elicit similar or different, even opposite, effects. The interconversion between the receptors was proposed by Kunos,[8,9] but it has recently been shown that α- and β-adrenergic receptor sites do not reside on the same macromolecule.[10] Nevertheless, a possible interaction between these types of receptors cannot be ruled out,[11] and the overall response may be dependent on the relative activity of the two receptors.

The most extensively studied and best understood of the catecholamine receptors is the β-adrenergic system. Since 1976, when the first successful

Anna Borsodi and Maria Wollemann • Institute of Biochemistry, Biological Research Center, Hungarian Academy of Sciences, H-6701 Szeged, Hungary.

Table I
Agonist Binding Studies of α-Adrenergic Receptors

Radiolabelled ligands	Tissue source	Species	References
Epinephrine	Brain	Calf	20,22–27
		Rat	22
	Liver	Rat	28–31
	Platelet	Human	32,33
Norepinephrine	Brain	Calf	20,23–27,33
		Rat	24,33
	Liver	Rat	29,34,35
Clonidine	Brain	Calf	24,33,36,37
		Rat	13,33,36–43
	Duodenum	Rat	33
	Kidney	Rat	33,36,41
		Guinea pig	44
	Neuroblastoma–glioma		40
	Pancreas	Rat	33
	Platelet	Human	45
	Spleen	Rat	33,36
	Submaxillary gland	Rat	33

direct radioligand binding studies of α-adrenergic receptors were carried out,[12–14] considerable information has been obtained, and this had led to great interest in this type of membrane receptor.

2. BINDING OF RADIOACTIVE LIGANDS TO α-ADRENERGIC RECEPTORS

For the direct identification of α-adrenergic receptors, several new and sensitive techniques have been developed, and a number of radioligand agonists as well as antagonists are now available. These methods permit the detection of receptors by ligand concentrations as low as 5×10^{-11} M. As a result, these direct binding studies have provided important new data including numbers of binding sites, affinities for different drugs, effects of regulatory factors, etc.

2.1. Binding of Agonists

Physiologically relevant adrenergic receptors were initially studied using tritiated catecholamines.[15–17] In those cases, the binding did not reveal all of the characteristics of the true receptors such as stereospecificity or the appropriate relationship of pharmacological activity and binding potency. A great deal of nonspecific binding was also observed.[18] Several laboratories [19–21] succeeded in overcoming these problems by changing the experimental conditions in the binding assay (e.g., by the use of antioxidants and catechol). Agonist binding studies have been carried out in a variety of tissues, as depicted in Table I.

Table II
Antagonist Binding Studies of α-*Adrenergic Receptors*

Radiolabelled ligands	Tissue source	Species	References
Dihydroergocriptine	Adipocytes	Hamster	49,50
	Brain	Calf	23,26,27,33,51–53
		Rat	22,33,38,41,51,54,55
	Heart	Rat	56–58
	Kidney	Rat	33,41
	Liver	Rat	14,28,29,31,35,59–63
	Neuroblastoma–glioma		64
	Parotid cells	Rat	65,66
	Platelet	Human	31,59,67–74
	Smooth muscle	Rabbit	75
	Urinary bladder	Rabbit	76
	Uterus	Rabbit	12,53,59,67,75,77–81
	Vas deferens	Guinea pig	82
		Rat	33
		Rat	13,33,38–42,51,83
Prazosin	Brain	Rat	42,84,85
	Heart	Guinea pig	86
	Liver	Rat	35,87
	Lung	Guinea pig	88
WB 4101	Brain	Calf	23–27,33,53
		Rat	13,33,38–42,51
	Heart	Rat	33,41
	Kidney	Rat	33,41
	Liver	Rat	10
	Spleen	Rat	33
	Submaxillary gland	Rat	33
	Uterus	Rabbit	53
	Vas deferens	Rat	33,41
Yohimbine	Liver	Rat	87
	Platelet	Human	32,89,90

Clonidine (an iminoimidazolidine derivative) was originally designed as a vasoconstricting agent. It shows a wide spectrum of pharmacodynamic and therapeutic effects, and almost all of them appear to be related to the stimulation of α-adrenergic receptors.[46–48]

2.2. Binding of Antagonists

Tritiated antagonists were also used to identify α-adrenergic receptors. The most often used ligands are listed in Table II. The specificity of these ligands is discussed in Section 3.

3. HETEROGENEITY OF α-ADRENERGIC RECEPTORS

3.1. Subtypes of α-Adrenergic Receptors

Subtypes of both α- and β-adrenergic receptors have been described. Lands *et al.*,[91] in 1967, provided evidence for the existence of two subtypes

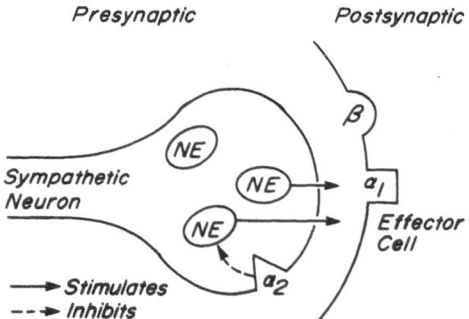

Fig. 1. Schematic representation of the synapse of the sympathetic neuron and effector cell showing presynaptic and postsynaptic receptors. (From ref. 96. Reprinted by permission of the *New England Journal of Medicine*.)

of β-adrenergic receptors designated β_1 and β_2. Subsequently, data became available suggesting that at least two subtypes of α-adrenergic receptors also exist.[92-95] The autoregulatory α receptors that inhibit norepinephrine release from nerve terminals have been called presynaptic or α_2 receptors, whereas α receptors located on effector cells have been termed postsynaptic or α_1 receptors (Fig. 1). The recently favored terminology based on α_1 and α_2 was proposed by Berthelson and Pettinger.[97] Their classification is functional rather than anatomic. This is preferred, because it was shown that α_2 receptors can be found in several tissues other than at the presynaptic sites of the neurons.[92,97-99] Among these, human platelets have recently been extensively studied (see Tables I and II).

Agonist drugs have not been successful in distinguishing α_1 and α_2 receptors because of lack of proper selectivity and variability in different systems. The most common antagonists presently used for differentiating α_1 and α_2 receptors are prazosin and yohimbine. The former is a much more potent antagonist of α_1 than α_2 responses.[41,59,67,100] Yohimbine is generally considered to be selective for α_2 receptors.[67] WB 4101 (2-[2(2',6'-dimethoxy)phenoxyethylamino]methylbenzodioxane) appears to be α_1 selective in brain.[33] The irreversible α blocker phentolamine and the ergot alkaloids exert equal affinity at α_1 and α_2 receptors.

Heterogeneity of α-adrenoreceptors in several tissues (liver, heart, brain, adipocytes) has been described.[29,35,50,57,100] Snyder and his collaborators extensively studied the heterogeneity of α receptors in rat brain membranes.[13,20,27,38,51] Originally, they suggested the existence of discrete agonist and antagonist states of the receptor. Recently, they interpreted the data as corresponding to α_1 and α_2 receptors.[33,41,84] The situation is further complicated by the fact that the α-adrenergic receptors may vary in their characteristics from tissue to tissue.[55,93,101] The proportion of the subtypes of α-adrenergic receptors in a given system can be estimated by direct ligand binding studies[92] either by using non-subtype-selective[53,67,71,79,100] or subtype-selective radioligands.[33,84,97] Furthermore, computer modeling of competition curves has provided an important tool in analyzing data obtained from these kinds of experiments.[67,102,103]

The subclassification of α-adrenergic receptors to just α_1 and α_2 appears to be an oversimplification. There are several data indicating more extensive

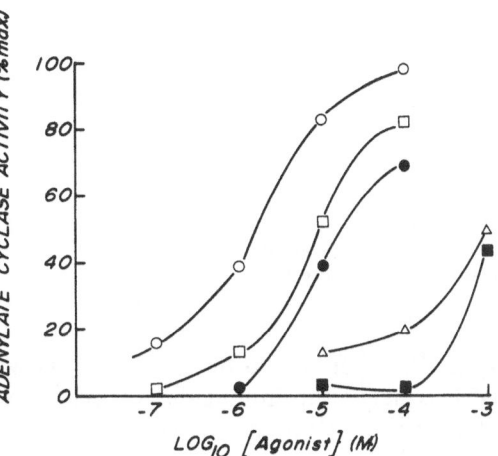

Fig. 2. Inhibition of prostaglandin E₁-stimulated adenylate cyclase activity in human platelet lysates by adrenergic agonists. ○, (−)epinephrine; ●, (+)epinephrine; □, (−)norepinephrine; ■, (−)norepinephrine; △, isoproterenol. (From ref. 68 courtesy of Drs. Newman *et al.* Reprinted by kind permission of the *Journal of Clinical Investigation*.)

complexity. Data from several laboratories suggest that heterogeneity may also exist in each subclass.

3.2. *Heterogeneity in Subtypes of* α-*Adrenergic Receptors*

Heterogeneity of α_2-adrenergic receptor sites in human platelet and rabbit uterus membrane was shown by Lefkowitz and collaborators.[31,59] They claim the existence of high- and low-affinity binding sites in the presence of agonist and give a similar interpretation to previous data of El-Refai *et al.* obtained with rat liver plasma membranes.[29] The two distinct binding states referred to as α_{2H} and α_{2L} appear to be under different regulation. Only the high-affinity state is affected by guanine nucleotides,[59,74] resulting in the formation of a unique low-affinity binding site.

Hanoune and collaborators[35] have suggested the existence of two distinct states of a single α_1-adrenergic receptor in rat liver plasma membranes. One of these might be in an inactive uncoupled state of the physiological α-adrenergic receptor. It has also been reported that rat brain membranes exhibit multiple α_2-noradrenergic binding sites[33,43] with different affinities, drug specificities, and regulatory patterns.

4. *INHIBITION OF ADENYLATE CYCLASE THROUGH* α-*ADRENERGIC RECEPTORS*

In certain cells (e.g., platelets, adipocytes), some of the α-adrenergic (α_2) responses seem to be mediated through the inhibition of the activity of the enzyme adenylate cyclase.[68,69,98,99,104–110] Like other receptors known for their inhibitory action on adenylate cyclase, α-adrenergic agonists are able to reduce enzyme activity (Fig. 2), and this is reversed by antagonists.

The inhibition of adenylate cyclase activity exhibits a requirement for GTP and sodium ions[99,104–107,111] similar to the hormonal stimulation of the en-

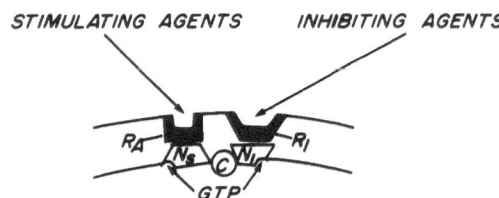

Fig. 3. Schematic representation of dual regulation of adenylate cyclase systems by stimulatory and inhibitory agents. Depicted in the model are two classes of receptor (R), one (R_a) mediating hormone effects through stimulatory nucleotide regulatory units (N_s) and another (R_i) mediating inhibitory effects through linkage with an N_i unit that binds GTP and inhibits adenylate cyclase activity. (Modified from Ref. 116 courtesy of Dr. Rodbell. Reprinted by kind permission of *Nature*.)

zyme.[112–115] It should be mentioned, however, that a higher concentration of GTP is needed to inhibit the adenylate cyclase than is required for the stimulatory effect.[111] It has been suggested that two distinct regulatory components are involved in the stimulatory and inhibitory effects on adenylate cyclase[111,116] (Fig. 3). Hoffman *et al.* recently reported the preferential uncoupling of α-adrenergic receptor-mediated inhibition of adenylate cyclase by manganase in human platelets.[89] Consequently, the presence of distinct regulatory components for stimulation and inhibition of adenylate cyclase seem likely.

5. SOLUBILIZATION AND PURIFICATION OF α-ADRENERGIC RECEPTORS

Efforts have been made in the last few years to purify α-adrenergic receptors. Successful solubilization of rat liver and human platelet α-adrenergic receptors has been carried out.[10,32] Alpha- and β-adrenergic receptors have been separated by affinity chromatography.[10]

Guellaen *et al.* reported the solubilization and partial purification of [^{3}H]phenoxybenzamine-labeled α-adrenergic receptors from rat liver plasma membrane.[117] The hydrodynamic parameters of the complex have been determined, and the estimated molecular weight was found to be approximately 96,000. The recent development of specific antibodies raised against phenoxybenzamine[118] might provide a useful tool for purification of α-adrenergic receptors prelabeled with phenoxybenzamine.

6. REGULATION OF α-ADRENERGIC RECEPTORS

6.1. Regulation by Guanine Nucleotides

Guanine nucleotides have been shown to exert an agonist-specific regulatory effect on α-adrenergic receptors,[23,43,59,64,74,87] similar to that observed for β-adrenergic receptors.[119,120] In the presence of guanine nucleotides, the affinity of the binding sites for agonists is markedly reduced, whereas no change can be seen in antagonist binding.[74,87]

An agonist-promoted increase in apparent molecular size of α-adrenergic receptors has also been described.[32] This was similar to that previously ob-

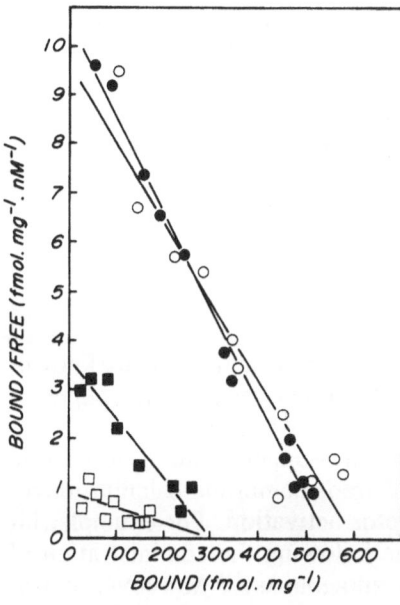

Fig. 4.   Scatchard analysis of tritiated norepineph-rine binding data obtained with untreated (■, □) and α-chymotrypsin-treated (●, ○) rat liver plasma membranes in the absence (closed symbols) and in the presence (open symbols) of guanyl-5'-ylimi-dodiphosphate [Gpp(NH)p]. Note the changes in binding properties and nucleotide regulatory effect following proteolysis. (From ref. 34 courtesy of Drs. Geynet *et al.* Reprinted by kind permission of *Biochemical and Biophysical Research Commu-nications.*)

served for β-adrenergic receptors.[121,122] There is a possibility that this effect might be caused by the action of the α_2-receptor guanine nucleotide regulatory protein. The α_2-adrenergic receptor-coupled adenylate cyclase activity regu-lation by guanine nucleotides has been discussed in Section 4.

6.2. Regulation by Cations

Divalent cations such as magnesium seem to be necessary for the agonist-specific regulatory effect of guanine nucleotides.[43,74,90] Monovalent cations such as sodium are able to regulate some α receptors and coupled adenylate cyclase in an agonist-specific fashion.[23,26,85,107] This phenomenon is analogous to that previously reported for opiate receptors.[123]

6.3. Physiological Regulation

A decrease in responsiveness of α-adrenergic receptors after prolonged exposure to agonists (desensitization) has been described.[66,70] This feature may be caused, at least in part, by a decrease in the number of binding sites.

Some data demonstrate the effect of steroid hormones on α-adrenergic receptors.[81,124] Estrogen and progesterone administration appears to induce opposite changes in number of α-adrenergic receptors in rabbit uterus. There are conflicting reports on the possible changes in α-adrenergic receptor prop-erties in altered thyroid states.[56,58,125–128]

It has also been reported that the number of α-adrenergic agonist binding sites can be enhanced by limited proteolysis[28,34,35] with concomitant changes in binding and regulatory properties (Fig. 4). The possible regulatory role of proteolytic enzymes in certain hormone actions as well as their relationship to the previously reported effect on adenylate cyclase and guanylate cyclase systems [129–132] remain to be established.

Although there are obviously quite a number of similarities between the regulation of α- and β-adrenergic receptors, considerably less is known about the α-adrenergic system.

7. MECHANISM OF α-ADRENERGIC ACTION

Although not all of the consequences of α-adrenergic activation are known, there are several lines of evidence indicating that a number of events might be effected.

The possibility of Ca^{2+} being the second messenger in the α-adrenergic system has been considered.[133,134] Increased transmembrane calcium movements have been observed following α-receptor activation. For example, increased ^{45}Ca uptake was shown in liver cells following the administration of an agonist, and this response was blocked by antagonists.[135] Moreover, extracellular calcium levels seem to affect a number of α-adrenergic receptor-mediated tissue responses.[135,136] In several systems, calcium ionophores have been reported to mimic the effects of α-adrenergic receptor activation.[135,137,138] In rat parotid cells, the increased calcium uptake was accompanied by an α-adrenergic receptor-mediated release of potassium.[65,137] Chan and Exton proposed that calcium regulates the coupling between the α-adrenergic receptor and adenylate cyclase.[139]

Besides adenylate cyclase (see Section 4), several other enzymes also appear to be effected by α-adrenergic stimulation. The regulation of carbohydrate metabolism in rat hepatocytes seems to be predominantly mediated through α-adrenergic receptors.[61,140,141] At least some of these effects might be secondary to the influx of calcium and increase in cyclic AMP or GMP level, which might also occur in certain systems following α-adrenergic stimulation.[133]

An α-adrenergic receptor-mediated increase in phosphatidylinositol turnover has also been demonstrated.[142] In a recent review, the $α_1$ specificity of this effect was claimed.[143] It has also been suggested that α-adrenergic receptors may play an important role in the regulation of lipolysis.[144]

8. CONCLUSIONS

The development of direct radioligand binding studies has provided a useful tool with which to probe the molecular mechanism of α-adrenergic receptor activation. In the next few years considerable progress might be expected in the solubilization and purification of these receptors, and it is hoped that a reconstituted system will be developed. Further investigation is also required for better understanding of the physiological role of α-adrenergic receptors.

REFERENCES

1. Ahlquist, R. P., 1948, *Am. J. Physiol.* **153**:586–688.
2. Powell, C. E., and Slater, I. H., 1958, *J. Pharmacol. Exp. Ther.* **122**:480–488.
3. Moran, N. C., and Perkins, M. E., 1958, *J. Pharmacol. Exp. Ther.* **124**:223–237.
4. Moran, N. C., 1967, *Ann. N.Y. Acad. Sci.* **139**:649–660.
5. Nickerson, M., and Hollenberg, N. K., 1967, *Physiological Pharmacology*, Volume 4, *The Nervous System*, Part D: *Autonomic Nervous System Drugs*, (W. S. Root and F. G. Hofmann, eds.), Academic Press, New York, pp. 243–305.
6. Furchgott, R. F., 1967, *Ann. N.Y. Acad. Sci.* **139**:553–570.
7. Furchgott, R. F., 1972, *Handbook of Experimental Pharmacology*, Volume 33: *Catecholamines* (H. Blaschko and E. Muscholl, eds.), Springer-Verlag, Berlin, Heidelberg, New York, pp. 283–334.
8. Kunos, G., 1978, *Annu. Rev. Pharmacol. Toxicol.* **18**:291–311.
9. Kunos, G., 1980, *Trends Pharmacol. Sci.* **1**:282–284.
10. Wood, C. L., Caron, M. G., and Lefkowitz, R. J., 1979, *Biochem. Biophys. Res. Commun.* **88**:1–8.
11. Woodcock, E. A., and Johnston, C. I., 1980, *Nature* **286**:159–160.
12. Williams, L. T., and Lefkowitz, R. J., 1976, *Science* **192**:791–793.
13. Greenberg, D. A., U'Prichard, D. C., and Snyder, S. H., 1976, *Life Sci.* **19**:69–76.
14. Guellaen, G., Yates-Aggerbeck, M., Vauquelin, G., Strossberg, D., and Hanoune, J., 1978, *J. Biol. Chem.* **253**:1114–1120.
15. Lefkowitz, R. J., Sharp, G., and Haber, E., 1973, *J. Biol. Chem.* **248**:342–349.
16. Lefkowitz, R. J., and Haber, E., 1971, *Proc. Natl. Acad. Sci. U.S.A.* **68**:1773–1777.
17. Ruffolo, R. R., McCreery, R. L., and Patil, P. N., 1976, *Eur. J. Pharmacol.* **38**:221–232.
18. Cuatrecasas, P., Tell, G. P. E., Sica, V., Parikh, I., and Chang, K. J., 1974, *Nature* **247**:92–97.
19. Lefkowitz, R. J., and Williams, L. T., 1977, *Proc. Natl. Acad. Sci. U.S.A.* **74**:515–519.
20. U'Prichard, D. C., and Snyder, S. H., 1977, *Life Sci.* **20**:527–533.
21. Pairault, J., and Laudat, M. H., 1975, *FEBS Lett.* **50**:61–65.
22. U'Prichard, D. C., and Snyder, S. H., 1977, *Nature* **270**:261–263.
23. U'Prichard, D. C., and Snyder, S. H., 1978, *J. Biol. Chem.* **253**:3444–3452.
24. U'Prichard, D. C., and Snyder, S. H., 1977, *J. Biol. Chem.* **252**:6450–6463.
25. U'Prichard, D. C., Greenberg, D. A., Sheehan, P., and Snyder, S. H., 1977, *Brain Res.* **138**:151–158.
26. Greenberg, D. A., U'Prichard, D. C., Sheehan, P., and Snyder, S. H., 1978, *Brain Res.* **140**:378–384.
27. Peroutka, S. J., Greenberg, D. A., U'Prichard, D. C., and Snyder, S. H., 1978, *Mol. Pharmocol.* **14**:403–412.
28. El-Refai, M. F., and Exton, J. H., 1980, *J. Biol. Chem.* **255**:5853–5858.
29. El-Refai, M. F., Blackmore, P. F., and Exton, J. H., 1979, *J. Biol. Chem.* **254**:4375–4386.
30. Blackmore, P. F., El-Refai, M. F., Dehaye, J. P., Strickland, W. G., Hughes, B. P., and Exton, J. H., 1981, *FEBS Lett.* **123**:245–248.
31. Hoffman, B. B., Michel, T., Mullikin-Kilpatrick, D. B., Lefkowitz, R. J., Tolbert, M. M., Gilman, H., and Fain, J. N., 1980, *Proc. Natl. Acad. Sci. U.S.A.* **77**:4569–4573.
32. Michel, T., Hoffman, B. B., Lefkowitz, R. J., and Caron, M. G., 1981, *Biochem. Biophys. Res. Commun.* **100**:1131–1135.
33. U'Prichard, D. C., and Snyder, S. H., 1979, *Life Sci.* **24**:79–88.
34. Geynet, P., Borsodi, A., Ferry, N., and Hanoune, J., 1980, *Biochem. Biophys. Res. Commun.* **97**:947–954.
35. Geynet, P., Ferry, N., Borsodi, A., and Hanoune, J., 1981, *Biochem. Pharmacol.* **30**:1665–1675.
36. Rouot, B. R., and Snyder, S. H., 1979, *Life Sci.* **25**:769–774.
37. Smith, C. B., Garcia-Sevilla, J. A., and Hollingsworth, P. J., 1981, *Brain Res.* **210**:413–418.
38. U'Prichard, D. C., Greenberg, D. A., and Snyder, S. H., 1976, *Mol. Pharmacol.* **13**:454–473.

39. U'Prichard, D. C., Bechtel, W. D., Rouot, B. M., and Snyder, S. H., 1979, *Mol. Pharmacol.* **16**:47–60.
40. Atlas, D., and Adler, M., 1981, *Proc. Natl. Acad. Sci. U.S.A.* **78**:1241–1273.
41. U'Prichard, D. C., Charness, M. E., Robertson, D., and Snyder, S. H., 1978, *Eur. J. Pharmacol.* **50**:87–89.
42. Hamburg, M., and Tallman, J. F., 1981, *Nature* **291**:493–495.
43. Rouot, B. M., U'Prichard, D. C., and Snyder, S. H., 1980, *J. Neurochem.* **34**:374–384.
44. Jarrott, B., Louis, W. J., and Summers, R. J., 1979, *Br. J. Pharmacol.* **65**:663–670.
45. Shattil, J. S., McDonough, M., Turnbull, J., and Insel, P. A., 1981, *Mol. Pharmacol.* **19**:179–183.
46. Andén, N. E., Corrodi, H., Fuxe, K., Larsson, K., Olson, L., and Ungerstedt, U., 1970, *Life Sci.* **9**:513–523.
47. Schmitt, H., 1977, *Handbook of Experimental Pharmacology*, Volume 39, (F. Gross, ed.), Springer-Verlag, Berlin, Heidelberg, New York, pp. 299–396.
48. Kobinger, W., 1978, *Rev. Physiol. Biochem. Pharmacol.* **81**:39–100.
49. Pecquery, R., Malagrida, L., and Giudicelli, Y., 1979, *FEBS Lett.* **98**:241–246.
50. Pecquery, R., and Giudicelli, Y., 1980, *FEBS Lett.* **116**:85–90.
51. Greenberg, D. A., and Snyder, S. H., 1977, *Mol. Pharmacol.* **14**:38–49.
52. Titeler, M., and Seeman, P., 1978, *Proc. Natl. Acad. Sci. U.S.A.* **75**:2249–2253.
53. Hoffman, B. B., and Lefkowitz, R. J., 1980, *Biochem. Pharmacol.* **29**:1537–1541.
54. Greenberg, D. A., and Snyder, S. H., 1977, *Life Sci.* **20**:927–932.
55. Haga, T., and Haga, K., 1980, *Life Sci.* **26**:211–218.
56. Sharma, V. K., and Banerjee, S. P., 1978, *J. Biol. Chem.* **253**:5277–5279.
57. Guicheney, P., Garay, R. P., Levy-Marchal, C., and Meyer, P., 1978, *Proc. Natl. Acad. Sci. U.S.A.* **75**:6285–6289.
58. Schümann, H. J., and Brodde, O. E., 1979, *Naunyn Schmiedeberg's Arch. Pharmacol.* **308**:191–198.
59. Hoffman, B. B., Mullikin-Kilpatrick, D., and Lefkowitz, R. J., 1980, *J. Biol. Chem.* **255**:4645–4652.
60. Clarke, W. R., Jones, L. R., and Lefkowitz, R. J., 1978, *J. Biol. Chem.* **253**:5975–5979.
61. Aggerbeck, M., Guellaen, G., and Hanoune, J., 1980, *Biochem. Pharmacol.* **29**:643–645.
62. Aggerbeck, M., Guellaen, G., and Hanoune, J., 1980, *Biochem. Pharmacol.* **29**:1653–1662.
63. Butler, D., Guillon, G., Cantau, B., and Jard, S., 1980, *Mol. Cell. Endocrinol.* **19**:275–289.
64. Haga, T., and Haga, K., 1981, *J. Neurochem.* **36**:1152–1159.
65. Strittmatter, W. J., Davis, J. N., and Lefkowitz, R. J., 1977, *J. Biol. Chem.* **252**:5472–5477.
66. Strittmatter, W. J., Davis, J. N., and Lefkowitz, R. J., 1977, *J. Biol. Chem.* **252**:5478–5782.
67. Hoffman, B. B., DeLean, A., Wood, C. L., Schocken, D. D., and Lefkowitz, R. J., 1979, *Life Sci.* **24**:1739–1745.
68. Newman, K. D., Williams, L. T., Bishopric, N. H., and Lefkowitz, R. J., 1978, *J. Clin. Invest.* **61**:395–402.
69. Alexander, R. W., Cooper, B., and Handin, R. I., 1978, *J. Clin. Invest.* **61**:1136–1144.
70. Kafka, M. S., Tallman, J. F., and Smith, C. C., 1977, *Life Sci.* **21**:1429–1438.
71. Cooper, B., Handin, R. I., Young, L. H., and Alexander, R. W., 1978, *Nature* **274**:703–706.
72. Scrutton, M. C., and Grant, J. A., 1979, *Nature* **280**:700.
73. Insel, A. P., Nirenberg, P., Turnbull, J., and Shattil, S. J., 1978, *Biochemistry* **17**:5269–5274.
74. Tsai, B. S., and Lefkowitz, R. J., 1979, *Mol. Pharmacol.* **16**:61–68.
75. Kunos, G., Hoffman, B., Kwok, Y. N., Kan, W. H., and Mucci, L., 1979, *Nature* **278**:254–256.
76. Levin, R. M., and Wein, A. J., 1979, *Mol. Pharmacol.* **16**:441–448.
77. Williams, L. T., Mullikin-Kilpatrick, D., and Lefkowitz, R. J., 1976, *J. Biol. Chem.* **254**:6915–6923.
78. Williams, L. T., and Lefkowitz, R. J., 1977, *Mol. Pharmacol.* **13**:304–313.
79. Hoffman, B. B., and Lefkowitz, R. J., 1980, *Biochem. Pharmacol.* **29**:452–454.
80. Leonard, J. P., Desager, J. P., Van Der Linden, L., and Harvengt, C., 1980, *Life Sci.* **27**:1875–1880.
81. Roberts, J. M., Insel, P. A., Goldfien, R. D., and Goldfien, A., 1977, *Nature* **270**:624–625.

82. Holck, M. I., Marks, B. H., and Wilberding, C. A., 1979, *Mol. Pharmacol.* **16**:77–90.
83. Davis, J. N., Arnett, C. D., Hoyler, E., Stalvey, L. P., Daly, J. W., and Skolnick, P., 1978, *Brain Res.* **159**:125–135.
84. Greengrass, P., and Brenner, R., 1979, *Eur. J. Pharmacol.* **55**:323–326.
85. Glossmann, H., and Hornung, R., 1980, *Naunyn Schmiedebergs Arch. Pharmacol.* **312**:105–106.
86. Karliner, J. S., Barnes, P., Hamilton C. A., and Dollery, C. T., 1979, *Biochem. Biophys. Res. Commun.* **90**:142–149.
87. Hoffman, B. B., Dukes, D. F., and Lefkowitz, R. J., 1981, *Life Sci.* **28**:265–272.
88. Barnes, P., Karliner, J., Hamilton, C., and Dollery, C., 1979, *Life Sci.* **25**:1207–1214.
89. Hoffman, B. B., Yim, S., Tsai, B. S., and Lefkowitz, R. J., 1981, *Biochem. Biophys. Res. Commun.* **100**:724–731.
90. Motulsky, H. J., Shattil, S. J., and Insel, P. A., 1980, *Biochem. Biophys. Res. Commun.* **97**:1562–1570.
91. Lands, A. M., Arnold, A., McAuliff, J. P., Luduena, F. P., and Brown, T. G., Jr., 1967, *Nature* **214**:597–598.
92. Wood, C. L., Arnett, C. D., Clarke, W. R., Tsai, B. S., and Lefkowitz, R. J., 1979, *Biochem. Pharmacol.* **28**:1277–1282.
93. Langer, S. Z., 1974, *Biochem. Pharmacol.* **23**:1793–1800.
94. Langer, S. F., 1976, *Clin. Sci. Mol. Med.* **51**:423S–426S.
95. Starke, K., 1977, *Rev. Physiol. Biochem. Pharmacol.* **77**:1–124.
96. Hoffman, B. B., and Lefkowitz, R. J., 1980, *N. Engl. J. Med.* **302**:1390–1396.
97. Berthelson, S., and Pettinger, W. A., 1977, *Life Sci.* **21**:595–606.
98. Woodcock, E. A., Johnston, C. I., and Olsson, C. A., 1980, *J. Cyclic Nucleotide Res.* **6**:261–269.
99. Sabol, S. L., and Nirenberg, M., 1979, *J. Biol. Chem.* **254**:1913–1920.
100. Miach, P. J., Dausse, J. P., and Meyer, P., 1978, *Nature* **274**:492–494.
101. Barker, K. A., Harper, B., and Hughes, I. E., 1977, *J. Pharm. Pharmacol.* **29**:129–134.
102. Hancock, A. A., DeLean, A. L., and Lefkowitz, R. J., 1979, *Mol. Pharmacol.* **16**:1–9.
103. Kent, R. S., DeLean, A., and Lefkowitz, R. J., 1980, *Mol. Pharmacol.* **17**:14–23.
104. Aktories, K., Schultz, G., and Jakobs, K., 1979, *FEBS Lett.* **107**:100–104.
105. Jakobs, K. H., Saur, W., and Schultz, G., 1976, *J. Cyclic Nucleotide Res.* **2**:381–392.
106. Jakobs, K. H., Saur, W., and Schultz, G., 1978, *FEBS Lett.* **85**:167–170.
107. Tsai, B. S., and Lefkowitz, R. J., 1978, *Mol. Pharmacol.* **14**:540–548.
108. Salzman, E. W., and Neri, L. L., 1969, *Nature* **224**:609–610.
109. Brown, E. M., and Hurwitz, S. H., 1978, *Endocrinology* **103**:893–899.
110. Jakobs, K. H., 1978, *Molecular Biology and Pharmacology of Cyclic Nucleotides* (G. Folco and R. Paoletti, eds.), Elsevier/North Holland, Amsterdam, pp. 265–278.
111. Steer, M. L., and Wood, A., 1979, *J. Biol. Chem.* **254**:10791–10797.
112. Jakobs, K. H., and Schultz, G., 1980, *Trends Pharmacol. Sci.* **1**:331–333.
113. Rodbell, M., 1978, *Molecular Biology and Pharmacology of Cyclic Nucleotides* (G. Folco and R. Paoletti, eds.), Elsevier/North Holland, Amsterdam, pp. 1–12.
114. Abramowitz, J., Iyengar, R., and Birnbaumer, L., 1979, *Mol. Cell. Endocrinol.* **16**:129–146.
115. Rodbell, M., Lin, M. C., Salomon, Y., Londos, C., Harwood, J. P., Martin, B. R., Rendell, M., and Bermann, M., 1975, *Adv. Cyclic Nucleotide Res.* **5**:3–29.
116. Rodbell M., 1980, *Nature* **284**:17–22.
117. Guellaen, G., Aggerbeck, M., and Hanoune, J., 1979, *J. Biol. Chem.* **254**:10761–10768.
118. Goodhardt, M., Guellaen, G., and Hanoune, J., 1981, *Biochem. Pharmacol.* **30**:1685–1692.
119. Lefkowitz, R. J., Mullikin-Kilpatrick, D., and Caron, M. G., 1976, *J. Biol. Chem.* **251**:4686–4692.
120. Maguire, M. E., Van Arsdale, P. M., and Gilman, A. G., 1976, *Mol. Pharmacol.* **12**:335–339.
121. Limbird, L. E., and Lefkowitz, R. J., 1978, *Proc. Natl. Acad. Sci. U.S.A.* **75**:228–232.
122. Limbird, L. E., Gill, D. M., and Lefkowitz, R. J., 1980, *Proc. Natl. Acad. Sci. U.S.A.* **77**:775–779.
123. Pert, C. B., and Snyder, S. J., 1976, *Mol. Pharmacol.* **10**:868–879.
124. Williams, L. T., and Lefkowitz, R. J., 1977, *J. Clin. Invest.* **60**:815–818.

125. Ciaraldi, T., and Marinetti, G. V., 1977, *Biochem. Biophys. Res. Commun.* **74:**984–991.
126. Ciaraldi, T. P., and Marinetti, G. V., 1978, *Biochem. Biophys. Acta* **541:**334–346.
127. Williams, R. S., and Lefkowitz, 1979, *J. Cardiovasc. Pharmacol.* **1:**181–189.
128. Kunos, G., Vermes-Kunos, I., and Nickerson, M., 1974, *Nature* **250:**779–781.
129. Hanoune, J., Stengel, D., Lacombe, M. L., Feldmann, G., and Coudrier, E., 1977, *J. Biol. Chem.* **252:**2039–2045.
130. Lacombe, M. L., and Hanoune, J., 1979, *J. Biol. Chem.* **254:**3697–3699.
131. Lacombe, M. L., Stengel, D., and Hanoune, J., 1977, *FEBS Lett.* **77:**159–163.
132. Lacombe, M. L., Haguenauer-Tsapis, R., Stengel, D., Ben Salah, A., and Hanoune, J., 1980, *FEBS Lett.* **116:**79–84.
133. Jones, L. M., and Michell, R. H., 1978, *Biochem. Soc. Trans.* **6:**673–688.
134. Exton, J. H., 1979, *Biochem. Pharmacol.* **28:**2237–2240.
135. Assimacopoulos-Jeannet, F. D., Blackmore, P. F., and Exton, J. H., 1977, *J. Biol. Chem.* **252:**2662–2669.
136. Blackmore, P. F., Brumley, T. F., Marks, J. L., and Exton, J. H., 1978, *J. Biol. Chem.* **253:**4851–4858.
137. Selinger, Z., Eimerl, S., and Schramm, M., 1974, *Proc. Natl. Acad. Sci. U.S.A.* **71:**128–131.
138. Rodan, G. A., and Feinstein, M. B., 1976, *Proc. Natl. Acad. Sci. U.S.A.* **73:**1829–1833.
139. Chan, T. M., and Exton, J. H., 1977, *J. Biol. Chem.* **252:**8645–8651.
140. Hutson, N. J., Brumley, F. T., Assimacoupoulos, F. D., Harper, S. C., and Exton, J. H., 1976, *J. Biol. Chem.* **251:**5200–5208.
141. Scmelck, P. H., and Hanoune, J., 1980, *Mol. Cell. Biochem.* **33:**35–48.
142. Michell, R. H., 1975, *Biochem. Biophys. Acta* **415:**81–147.
143. Fain, J. N., and Garcia-Sainz, J. A., 1980, *Life Sci.* **26:**1183–1194.
144. Lafontan, M., and Berlan, M., 1981, *Trends Pharmacol. Sci.* **2:**126–129.

β-Adrenergic Receptors

Maria Wollemann and Anna Borsodi

1. INTRODUCTION

The concept of α- and β-adrenergic receptors was introduced by Ahlquist[1] in 1948, but the fact that sympathetic excitatory actions of epinephrine are abolished by ergot alkaloids whereas inhibitory actions of epineprine are not was discovered by Dale[2] in 1906.

At the time Ahlquist made his classification of adrenergic receptors, appropriate β-receptor blockers were yet not known, and therefore his classification was based on the use of such different agonists as epinephrine, norepinephrine, their methyl derivatives, and isoproterenol. The original distinction he made is shown in Table I.

Curiously, Ahlquist concluded from his experiments that there is only one adrenergic transmitter and that it is epinephrine instead of the two compounds sympathin I and E (inhibitory and excitatory) as had been claimed in 1937 by Cannon and Rosenblueth.[3]

Ahlquist's new idea was that instead of two neurotransmitters two receptors are responsible for the different actions. Although norepinephrine proved to be the real adrenergic transmitter as shown a year later by von Euler,[4] Ahlquist was basically correct in his conclusion. The first β-adrenergic receptor blocker, dichlorisoproterenol (DCI), was synthesized by Moran and Perkins[5] in 1958, and the number of β blockers is still growing. Meanwhile, the classification of Ahlquist was revised, and new subgroups of α and β receptors became known: first β_1 and β_2,[6] then α_1 and α_2.[7] The most significant step in the understanding of their action on the molecular level, however, was the discovery of the relationship of the β-adrenergic receptor to adenylate cyclase.[8] The number of components of this system known to be necessary for hormonal stimulation has increased since its discovery from a simple two-subunit enzyme to a multicomponent complex that is still the most favored model of hormone–receptor interactions at the molecular level (Figs. 1 and 2).

Maria Wollemann and Anna Borsodi • Institute of Biochemistry, Biological Research Center, Hungarian Academy of Sciences, H-6701 Szeged, Hungary.

Table I
Summary of the Relative Order of Activity of the Amines[a]

Receptor	Order of activity	
	Most active	Least active
Vasoconstrictor	*l*-epi. *dl*-epi. art. methyl-art. methyl-epi. N-iso-art.	
Uterine excitatory	*l*-epi. *dl*-epi. art. methyl-art. methyl-epi. N-iso-art.	
Nictitating membrane excitatory	*l*-epi. *dl*-epi. art. methyl-art. methyl-epi. N-iso-art.	
Dilator pupillae excitatory	*l*-epi. *dl*-epi. art. methyl-art. methyl-epi. N-iso-art.	
Ureteral excitatory	*l*-epi. *dl*-epi. art. methyl-art. methyl-epi. N-iso-art.	
Intestinal inhibitory	*l*-epi. *dl*-epi. art. methyl-art. methyl-epi. N-iso-art.	
Vasodilator	N-iso-art. *l*-epi. methyl-epi. *dl*-epi. methyl-art. art.	
Uterine inhibitory	N-iso-art. *l*-epi. methyl-epi. *dl*-epi. methyl-art. art.	
Myocardial excitatory	N-iso-art. *l*-epi. methyl-epi. *dl*-epi. methyl-art. art.	

[a] Reprinted from ref. 1 by permission of the American Physiological Society and courtesy of Dr. Ahlquist.

2. β-ADRENERGIC RECEPTOR

2.1. β-Adrenergic Receptor Binding Sites

2.1.1. Binding of Agonists

Binding studies of β-adrenergic receptors started with labeled agonist (norepineprine) binding.[9] This method was heavily criticized by Cuatrecasas,[10] who first insisted on certain specific requirements for the study of β-adrenergic receptor binding and activation of adenylate cyclase, including stereospecificity and the use of specific β-receptor blockers. More recently, binding studies with the labeled agonist [3H]hydroxybenzylisoproterenol have also been successfully carried out.[13]

2.1.2. Binding of Antagonists

The labeling of specific β-adrenergic receptors was first achieved with the use of labeled antagonists for receptor binding; among these,

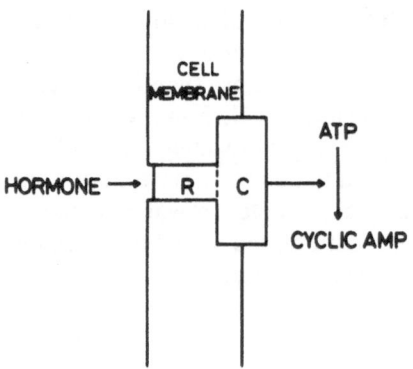

Fig. 1. Model of adenylate cyclase complex. R, hormone receptor; C, catalytic portion of enzyme. (From ref. 8a courtesy of Drs. Robison *et al.* and by kind permission of Academic Press.)

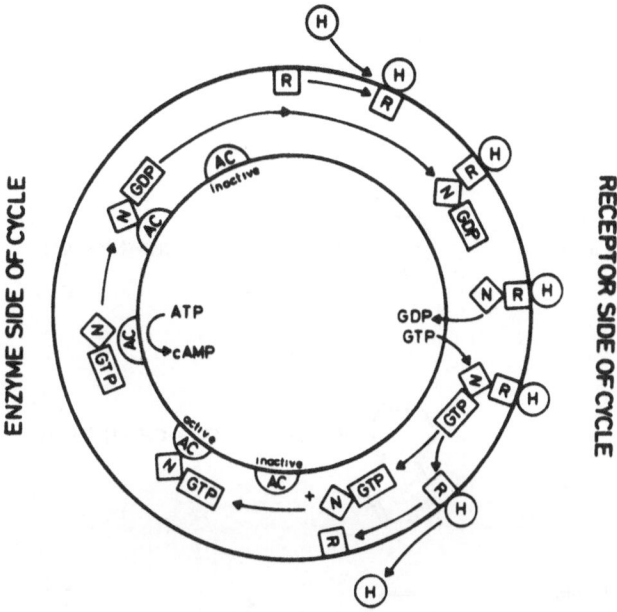

Fig. 2. Schematic representation of the role of the guanine nucleotide binding protein "shuttle" between receptor and adenylate cyclase. In effect, one has three interconnecting cycles: one for the receptor; one for the "G-binding protein"; and one for the cyclase. H, hormone; R, receptor; N, nucleotide binding protein; AC, adenylate cyclase. (From ref. 8b by kind permission of Elsevier/ North Holland Biomedical Press.)

[³H]dihydroalprenolol[11] and [¹²⁵I]hydroxybenzylpindolol[12] are used most widely.

Interest in new β-receptor-blocking drugs is still growing, since their therapeutic applications from tachycardia and hypertension to schizophrenia are promising. The most widely used β-adrenergic receptor blockers are depicted in Fig. 3. These are all competitive inhibitors of β-receptor agonists. Data on noncompetitive and irreversible inhibitors are scarce.[14,15]

Attempts have also been directed to the synthesis of selective β_1 (cardiac and adipose tissue) and β_2 (vascular and bronchial smooth muscle) receptor blockers. Examples are enumerated in Table II.

2.2. Activation and Inhibition of Adenylate Cyclase through β-Adrenergic Receptors

Sutherland and co-workers[16] first reported that stimulation of adenylate cyclase by catecholamines was mediated by β-adrenergic receptors, which means that the order of potency of catecholamines in activating the enzyme was isoproterenol > epinephrine > norepinephrine. They also demonstrated that DCI inhibited the stimulation by catecholamines of adenylate cyclase activity. Subsequently, the correlation of β_1 and β_2 agonists and antagonists[6] with stimulation and inhibition, respectively, of adenylate cyclase was also dem-

(±)-Sotalol

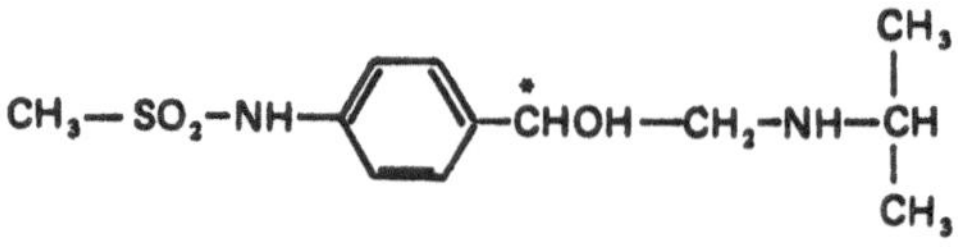

(±)-Alprenolol

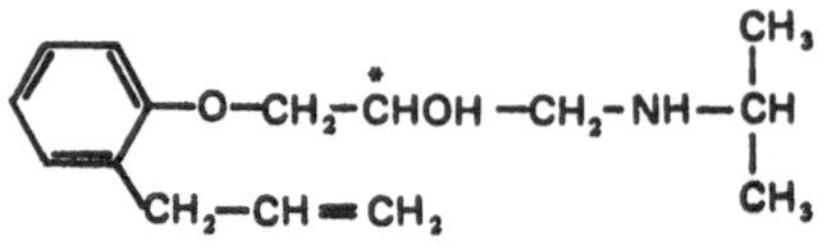

(±)-Propranolol

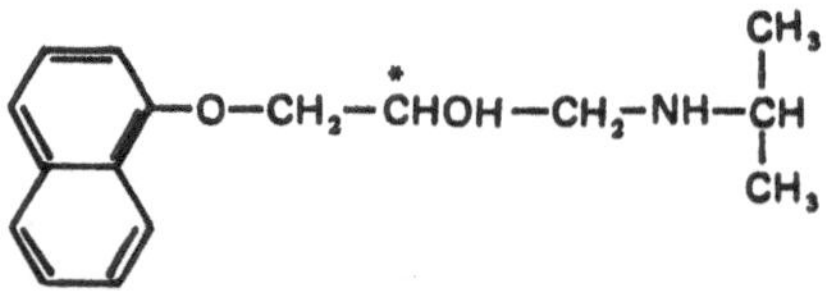

(-)-Oxprenolol

(+)-Oxprenolol

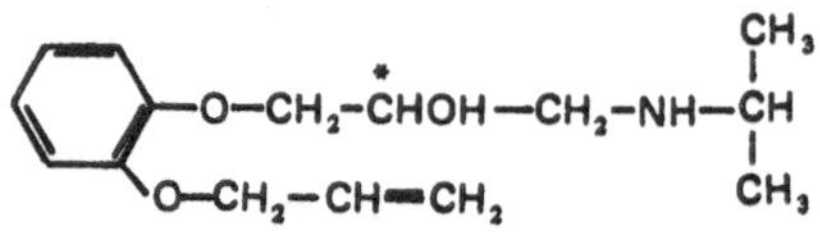

(±)-Pindolol

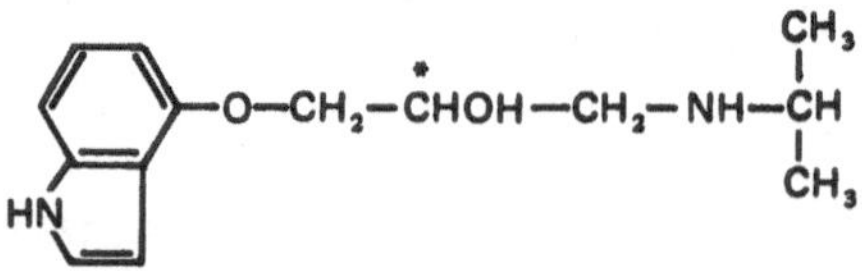

(±)-Practolol

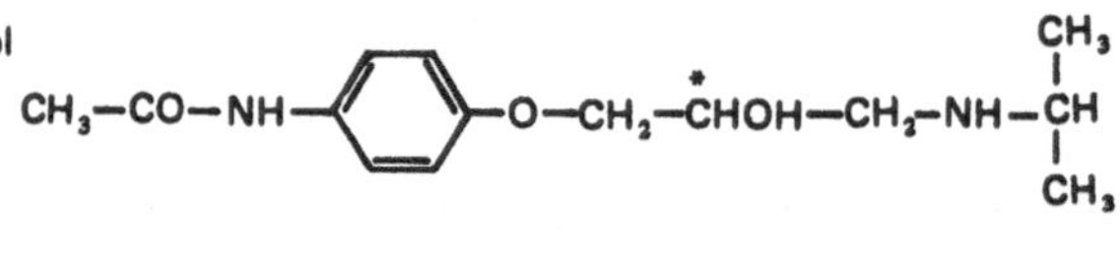

Fig. 3. Widely used β-adrenergic receptor blocking agents.

onstrated in different tissues such as heart (β_1[17]), skeletal muscle (β_2[18]), and lung (β_2[19,20]). Radioactive binding methods and nonlinear curve-fitting techniques have further revealed that β_1 and β_2 receptors coexist in the same tissue but that the proportions are different in each tissue (Table III).

The role of the coupling factor, G/F protein, which promotes activation of adenylate cyclase by β-adrenergic receptors, is emphasized in the following sections.

2.3. Solubilization and Purification of β-Adrenergic Receptor and Adenylate Cyclase

Over 50 articles dealing with solubilization and purification of β-adrenergic receptor and adenylate cyclase have appeared in the literature during the last

Table II

K_i Values Obtained from the Inhibition of [^{3}H]Dihydroalprenolol Binding on Heart and Lung Membrane Preparations[a]

		Heart K_i (nM)	Lung K_i (nM)	β_1/β_2 activity
Practolol	Para	500	5000	10
	Ortho	225	1385	6
Oxprenolol	Para	205	615	3
	Ortho	2.6	1.7	0.6
Alprenolol	Para	330	923	2.8
	Ortho	9	1.8	0.2

[a] Reprinted from ref. 15a courtesy of Drs. Leclerc *et al.* and by kind permission of Elsevier/North Holland Biomedical Press.

10 years, evidence that interest in the subject has not diminished, although solution of the problem remains elusive.

In some experiments different detergents were used to solubilize the β receptor (for example, digitonin[21]) and the adenylate cyclase (for example, Triton-X 100 and Lubrol PX[22,23]), but it later appeared that coupling factors between receptor and cyclase had been lost during solubilization.[24–26] Deter-

Table III

Tissue Distribution of β-Adrenoceptor Subtypes[a,b]

Tissue	Total β-adrenoceptor density (fmol·mg protein^{-1})	Percentage	
		β_1	β_2
Rat lung	400	20	80
Rabbit lung	350	80	20
Bovine lung	250	25	75
Rat ventricle	50	65	35
Rat spleen	250	35	65
Rat uterus (estrogen dominated)	100	20	80
Rat uterus (progesterone dominated)	100	0	100
Rat erythrocyte	100	0	100
Rat reticulocyte	600	0	100
Rat cerebral cortex	120	65	35
Rat cerebellum	50	0	100
Rat striatum	100	65	35
Rat limbic forebrain	70	55	45

[a] All studies were performed using the ligand (−)-[^{3}H]dihydroalprenolol. Proportions of β_1 and β_2 sites were estimated by computer-assisted curve fitting of the atypical displacement curves generated by highly selective β_1 or β_2 agents.

[b] Reprinted from ref. 20 courtesy of Dr. Nahorski and by kind permission of Elsevier/North Holland Biomedical Press.

gents with a high hydrophilic/lipophilic ratio exhibited a low yield of adenylate cyclase activity, but some of the hormonal enzyme activation was retained.[27,28] Higher hydrophobicity resulted in a higher yield of enzyme activity, but hormonal activation of enzyme activity was lost in spite of unimpaired hormone binding.[24–26]

Further substantial purification of β-receptors succeeded only with affinity chromatography of Sepharose 4B alprenolol spacer;[29] 15,000-fold purification was obtained from frog erythrocytes, and 2000-fold purification from turkey erythrocytes.[30] Highest enzyme purification activities were relatively low by conventional methods. These resulted only in 10- to 20-fold purification because of difficulties in removing the detergents.[23,31,32] However, Homcy *et al.*[33] and Stockton *et al.*[34] introduced hydrophobic resolution on an uncharged resin followed by affinity chromatography, which might result in several-thousand-fold purification of the enzyme activity. The molecular weight of the β-adrenergic receptor purified from frog erythrocytes was 150,000.[35] From turkey erythrocytes, after SDS-polyacrylamide and affinity labeling, two subunits of 37,000 and 41,000 daltons were reported.[36] Molecular weights for the solubilized adenylate cyclase have been reported between 160,000 and 200,000.[37–39]

2.4. Attempts at Reconstitution of the β-Adrenergic Receptor Activation of Adenylate Cyclase

2.4.1. The Role of Lipids

Attempts at reconstitution of the β-adrenergic receptor activation of adenylate cyclase first focused on the role of lipids. Addition of acidic phospholipids (phosphatidylcholine, -serine, and -inositol) partially restored the glucagon, epinephrine, and histamine sensitivity of the solubilized or phospholipase-A- and C-treated heart, brain, and liver adenylate cyclase activity[40–45] (N. J. Swislocki, unpublished observations cited in ref. 31). Recent results have demonstrated that acidic phospholipids also enhance β-adrenergic receptor binding in solubilized heart preparation.[25,28]

The important role of phosphatidylcholine was also recently demonstrated by a different approach.[46–48] In the first case, it was shown in rat reticular ghosts that the number of β-adrenergic receptors increased in parallel with the increase in phospholipid methylation. Conversely, catecholamine stimulation also caused an increase in phospholipid methylation. Increased phospholipid methylation enhances lateral mobility in membranes, whereas a decrease in phosphatidylcholine increases the rigidity of membranes; thus, the receptor–cyclase coupling is facilitated in the first case and inhibited in the second.

On the other hand, brief phospholipase C treatment enhanced basal adenylate cyclase activity but decreased isoproterenol activation. Phosphatidylcholine addition restored isoproterenol activation of rat liver plasma membranes.[48]

2.4.2. The Role of Protein Factors and Nucleotides

Activating and inhibitory regulating factors were soon established in studies on reconstitution of adenylate cyclase hormonal activation,[49,50] but the first decisive results were presented from a genetic approach in a series of papers by Orly and Schramm,[51-56] who succeeded in fusing cell variants lacking either β-adrenergic receptors or adenylate cyclase activity and restored the adenylate cyclase sensitivity to β-adrenergic antagonists. Meanwhile, it became evident that the hypothetical coupling factor of Birnbaumer[57] binds GTP and F[58,59]; it was therefore also called the G/F, G, or N protein, and the activating effects of GTP and F on adenylate cyclase activity are based on this protein. The G/F protein was first identified by Cassel and Selinger[60,61] in turkey erythrocytes as a GTPase and it has since been purified as a 126,000-dalton protein consisting of 42,000-dalton subunits.[62,63]

This regulatory protein binds GTP and is ADP-ribosylated by cholera toxin.[64] Hormones facilitate the displacement of inactive GDP by free GTP.[59,65] The slow displacement of GDP by GTP in the absence of hormones is measured as the basal activity of adenylate cyclase. Cholera toxin, by inhibiting the degradation of GTP or the exchange of GTP for GDP, permanently activates[61,63] adenylate cyclase. A similar effect is achieved by the nonhydrolyzable derivatives of GTP such as Gpp(NH)p or Gpp(CH$_2$)p.

Recently, another action of guanine nucleotides became known, i.e., a direct action on agonist binding. Thus, GTP shifts the agonist displacement curve to the right by increasing the IC$_{50}$ or ED$_{50}$ and changes the negative cooperativity from a shallow curve slope to a deep curve slope, whereas antagonist binding is not affected.[66]

These experiments demonstrated that antagonist binding, although competitive in kinetic experiments, is different from agonist binding. However, if preincubation of antagonist is carried out in the presence of GTP following washings, the subsequent binding of labeled alprenolol is enhanced[67] (Figs. 4 and 5).

3. REGULATION OF THE β-ADRENERGIC RECEPTOR

3.1. Supersensitivity of β-Adrenergic Receptor and Adenylate Cyclase Activation

Supersensitivity of β-adrenergic receptor and adenylate cyclase can be achieved by surgical denervation or chemical denervation. Chemical denervation was used in the form of reserpine, 6-hydroxydopamine, or guanethidine treatment in the pineal gland, CNS, and heart.[68-75] In cases in which the binding of β-adrenergic receptors was determined, an increase in their number was found, without any change in affinity.[74,75]

3.2. Desensitization of β-Adrenergic Receptor and Adenylate Cyclase Activation

Desensitization or hyposensitivity of β-adrenergic receptors and adenylate cyclase towards agonists can be induced by *in vivo* or *in vitro* pretreatment with high hormone concentrations.[75-77] In this case, generally, the loss of hor-

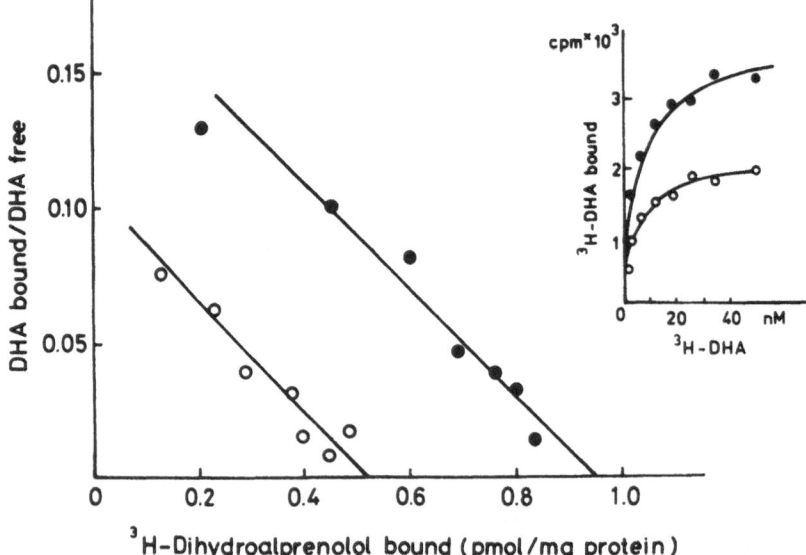

Fig. 4.   Scatchard plot of rabbit heart membrane preparation incubated with (●) or without (○) 10^{-5} M alprenolol at 33°C for 20 min, washed four times with 25 mM Tris-HCl buffer (pH 7.5) containing 5 mM MgCl$_2$, and centrifuged at room temperature at 18,000 × g. The inset shows the saturation curve of the control and alprenolol-treated membranes. The K_d for [^{3}H]DHA estimated from Scatchard and Lineweaver–Burk plots, both with and without incubation with alprenolol, was 5.7 nM.[67]

monal activation of adenylate cyclase exceeds the decrease in the number of receptors, a reaction accounted for by the fact that there is also a change in the coupling effect. Addition of guanine nucleotides might partly restore the hormonal response of adenylate cyclase,[74,78] whereas preincubation with β-blockers elevates the number of β-adrenergic receptors.[67] High concentrations of Mn^{2+} and ATP also achieve desensitization in the presence of hormone *in vitro*.[79]

Because desensitization is a general phenomenon of most receptors, it is supposed that receptors exist in two forms: the hormone is loosely bound to the receptor in the sensitized form and tightly bound in the desensitized form[80] (Fig. 6).

3.3. Pathological Changes of β-Adrenergic Receptors

Changes in β-adrenergic receptor sensitivity might also occur in pathological cases. These are mostly vegetative illnesses. In bronchial asthma, for example, a hypersensitivity of α receptors and decreased sensitivity of β receptors were claimed.[81]

In thyroid hyper- or hypofunction, a hyper- or hyposensitivity, respectively, of β-adrenergic receptors was found in heart muscle,[82,83] and a deficiency of G factor has been found in pseudohypoparathyroidism.[84]

Fig. 5. Effect of alprenolol and guanine nucleotide incubation on DHA binding. Membranes were incubated in 1-ml solutions containing 25 mM $MgCl_2$ and 75 mM Tris-HCl (pH 7.5) for 20 min at 37°C and were washed four times with 2 ml of the same buffer at 20°. For measurement of DHA binding, 190–228 μg of protein was used. Incubation was carried out at 37°C for 10 min in 0.5 ml of solution [25 mM $MgCl_2$, 75 mM Tris-HCl (pH 7.5)] with 10 nM [³H]DHA ± 2 × 10⁻⁵ M propranolol. Each value shown in the mean ± standard error of the mean of three separate experiments.[67] 1, buffer; 2, alprenolol, 10⁻⁵ M; 3, GTP, 10⁻⁴ M; 4, GTP + alprenolol; 5, Gpp(NH)p, 10⁻⁴ M; 6, Gpp(NH)p + alprenolol.[67]

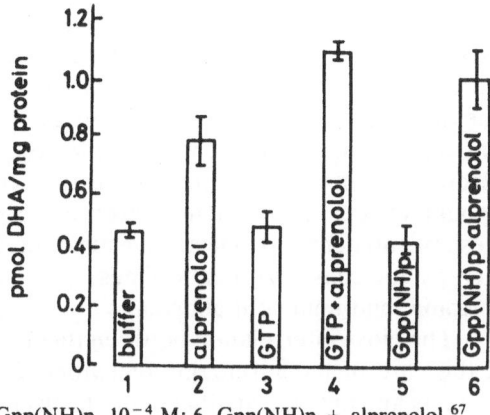

Chronic treatment with β-adrenergic receptor blockers desensitizes the heart toward β agonists, but the sudden interruption of treatment might lead to a hypersensitivity including reappearance of cardiac hypertonia and infarct.[85-88]

It was reported in rats with different forms of hypertension that adenylate cyclase from cardiac and vascular smooth muscle membranes is less sensitive to isoproterenol activation[89-91] and that the number of β-adrenoceptors is reduced.[92-95] The changes in adenylate cyclase activity are similar in this respect to withdrawal symptoms after chronic morphine treatment.[96]

The reverse is true for agonist treatments in bronchial asthma or imminent abortion, where sensitivity to agonists during chronic treatment is decreased. Because adenylate cyclase is still responsive to guanine nucleotides in the densitized state,[74] eventual treatment with them seems to offer new possibilities in desensitized cases.

4. CONCLUSIONS

The β-adrenergic receptor–adenylate cyclase enzyme complex has proved to be a very useful system for studying receptor action at the molecular level.

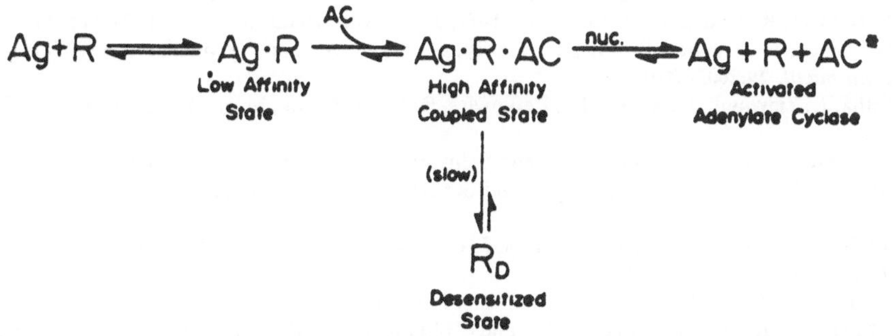

$$Ag+R \rightleftharpoons \underset{\substack{\text{Low Affinity} \\ \text{State}}}{Ag \cdot R} \xrightarrow{AC} \underset{\substack{\text{High Affinity} \\ \text{Coupled State}}}{Ag \cdot R \cdot AC} \xrightarrow{nuc.} \underset{\substack{\text{Activated} \\ \text{Adenylate Cyclase}}}{Ag + R + AC^*}$$

$$\Big\downarrow\text{(slow)}$$

$$\underset{\substack{\text{Desensitized} \\ \text{State}}}{R_D}$$

Fig. 6. Proposed model of β-adrenergic receptor activation and desensitization. (From ref. 80 courtesy of Drs. Wessels *et al.* and by kind permission of The American Society for Pharmacology and Experimental Therapeutics.

Many links that promote or inhibit the coupling of the receptor to the enzyme have been discovered in the last decade; among these, the most important are the guanine nucleotides, the G/F factor, the phospholipids, and calmodulin. The discovery of separate gene regulation for β receptor and adenylate cyclase was also a major achievement in the study of hormone action. The roles of desensitization and sensitization in the regulation of receptor response also became clearer, and certain inferences about pathology could be drawn from these phenomena too. These investigations led to an understanding of and new therapy for some autonomic illnesses such as bronchial asthma, cardiac hypertonia, and imminent abortion.

The close interrelation between theory and practice has again been proven in the case of the adrenergic receptors. From the search for new adrenergic agonists and antagonists that could be useful in therapy, the basic mechanisms of hormone action were discovered, and, finally, these new results will again be applied therapeutically.

REFERENCES

1. Ahlquist, R., 1948, *Am. J. Physiol.* **153**:586–600.
2. Dale, H. H., 1906, *J. Physiol. (Lond.)* **34**:163–206.
3. Cannon, W. B., and Rosenblueth, A., 1937, *Automatic Neuroeffector Systems*, Macmillan, New York.
4. von Euler, U. S., 1948, *Acta Physiol. Scand.* **16**:63–74.
5. Moran, N. C., and Perkins, M. E., 1958, *J. Pharmacol. Exp. Ther.* **124**:223–237.
6. Lands, A. M., Arnold, A., McAuliff, J. P., Luduena, F. P., and Brown, T. G., 1967, *Nature* **214**:597–598.
7. Berthelson, S., and Pettinger, W. A., 1977, *Life Sci.* **21**:595–606.
8. Robison, G. A., Butcher, R., and Sutherland, E. W., 1967, *Ann. N.Y. Acad. Sci.* **139**:703–723.
8a. Robison, G. A., Butcher, R., and Sutherland, E. W., 1971, *Cyclic AMP*, Academic Press, New York.
8b. Rodbard, D., 1980, *Trends Pharmacol. Sci.* **1**:222–225.
9. Lefkowitz, R. J., Sharp, G., and Haber, E., 1973, *J. Biol. Chem.* **248**:342–349.
10. Cuatrecasas, P., Tell, G. P. E., Sica, V., and Parikh, I., 1976, *Nature* **247**:92–97.
11. Lefkowitz, R. J., 1974, *Biochem. Biophys. Res. Commun.* **60**:703–709.
12. Aurbach, G. D., Fedak, S. A., Woodard, C. J., Palmer, J. S., Hauser, D., and Troxler, F., 1974, *Science* **186**:1223–1224.
13. Lefkowitz, R. J., and Williams, L. T., 1977, *Proc. Natl. Acad. Sci. U.S.A.* **74**:515–519.
14. Schmelck, H., Geynet, P., Fur, G., Hardy, C., Uzan, A., and Hanoune, J., 1979, *Biochem. Pharmacol.* **28**:2005–2010.
15. Pitha, J., Zjawiony, J., Nasrin, N., Lefkowitz, R. J., and Caron, M. G., 1980, *Life Sci.* **27**:1791–1798.
15a. Leclerc, G., Rolot, B., Velley, J., and Schwartz, J., 1981, *Trends Pharmacol. Sci.* **2**:18–20.
16. Murad, F., Chi, Y. M., Rall, T. W., and Sutherland, E. W., 1962, *J. Biol. Chem.* **237**:1233–1238.
17. Wollemann, M., Borbola, J., Jr., Papp, J. G., and Szekeres, L., 1975, *J. Mol. Cell. Cardiol.* **7**:523–533.
18. Lefkowitz, R. J., 1975, *Biochem. Pharmacol.* **24**:583–590.
19. Minneman, K. P., Hegstrand, L. R., and Molinoff, P. B., 1979, *Mol. Pharmacol.* **16**:21–33.
20. Rugg, E. L., Barnett, D. B., and Nahorski, S. R., 1978, *Mol. Pharmacol.* **14**:996–1005.
21. Caron, M. G., and Lefkowitz, R. J., 1976, *J. Biol. Chem.* **251**:2374–2384.
22. Levey, G., 1970, *Biochem. Biophys. Res. Commun.* **38**:86–92.

23. Johnson, A. R., and Sutherland, E. W., 1973, *J. Biol. Chem.* **248:**5111–5121.
24. Levey, G. S., 1971, *Ann. N.Y. Acad. Sci.* 449–457.
25. Drummond, G. J., and Dunham, J., 1978, *Arch. Biochem. Biophys.* **189:**63–75.
26. Thang, N. X., Borsodi, A., and Wollemann, M., 1980, *Biochem. Pharmacol.* **29:**2791–2797.
27. Ryan, J., and Storm, D. R., 1974, *Biochem. Biophys. Res. Commun.* **60:**304–311.
28. Thang, N. X., and Wollemann, M., 1980, *Agressologie* **21:**7–14.
29. Caron, M. G., Srinivasan, Y., Pitha, J., Kociolek, K., and Lefkowitz, R. J., 1979, *J. Biol. Chem.* **254:**2923–2927.
30. Vauquelin, G., Geynet, P., Hanoune, J., and Strosberg, A. D., 1977, *Proc. Natl. Acad. Sci. U.S.A.* **74:**3710–3714.
31. Swislocki, N. I., and Tierney, J., 1973, *Biochemistry* **12:**1862–1866.
32. Welton, A. F., Land, P. M., Newby, A. C., Yamamura, H., Nicosica, S., and Rodbell, M., 1978, *Biochim. Biophys. Acta* **522:**625–639.
33. Homcy, C., Wrenn, S., and Haber, E., 1978, *J. Biol. Chem.* **75:**59–63.
34. Stockton, J. M., and Turner, A. J., 1981, *J. Neurochem.* **36:**1722–1730.
35. Caron, M. G., and Lefkowitz, R. J., 1976, *Biochem. Biophys. Res. Commun.* **68:**315–322.
36. Atlas, D., and Levitzki, A., 1978, *Nature* **272:**370–371.
37. Neer, E. J., 1974, *J. Biol. Chem.* **249:**6527–6531.
38. Neer, E. J., Echeverria, D., and Knox, S., 1980, *J. Biol. Chem.* **255:**9782–9789.
39. Jard, S., Guillon, G., and Roy, C., 1978, *Molecules Biology and Pharmacology of Cyclic Nucleotides* (G. Folco and R. Paoletti, eds.), Elsevier/North Holland, Amsterdam, pp. 17–31.
40. Levey, G. S., 1971, *Biochem. Biophys. Res. Commun.* **43:**108–113.
41. Levey, G. S., 1971, *J. Biol. Chem.* **246:**7405–7410.
42. v. Hungen, K., and Roberts, S., 1973, *Eur. J. Biochem.* **36:**39–40.
43. Réthy, A., Tomasi, V., Trevisani, A., and Barnabei, O., 1972, *Biochim. Biophys. Acta* **290:**58–69.
44. Pohl, S. L., Krans, H. M. J., Kozyreff, V., Birnbaumer, L., and Rodbell, M., 1971, *J. Biol. Chem.* **246:**4447–4454.
45. Rubalcava, B., and Rodbell, M., 1973, *J. Biol. Chem.* **248:**3831–3837.
46. Strittmatter, W. J., Hirata, F., Axelrod, J., Mallorga, P., Tallman, J. F., and Henneberry, R. C., 1979, *Nature* **282:**857–859.
47. Hirata, F., Strittmatter, W. J., and Axelrod, J., 1979, *Proc. Natl. Acad. Sci. U.S.A.* **76:**368–372.
48. Nemecz, G., Farkas, T., and Horváth, L. I., 1981, *Arch. Biochem. Biophys.* **20:**256:263.
49. Rodbell, M., Birnbaumer, L., Pohl, S. L., and Krans, M. J., 1971, *J. Biol. Chem.* **246:**1877–1882.
50. Ho, R. L., and Sutherland, E. W., 1975, *Proc. Natl. Acad. Sci. U.S.A.* **72:**1773–1777.
51. Orly, J., and Schramm, M., 1976, *Proc. Natl. Acad. Sci. U.S.A.* **73:**4410–4414.
52. Schramm, M., Orly, J., Eimerl, S., and Korner, M., 1977, *Nature* **268:**310–313.
53. Eimerl, S., Nevfeld, G., Korner, M., and Schramm, M., 1980, *Proc. Natl. Acad. Sci. U.S.A.* **77:**760–764.
54. Brunton, L. L., Maguire, M. E., Anderson, H. J., and Gilman, A. G., 1977, *J. Biol. Chem.* **252:**1293-1302.
55. Sternweis, P. C., and Gilman, A. G., 1979, *J. Biol. Chem.* **254:**3333–3340.
56. Howlett, A. C., Sternweis, P. C., Macik, B. A., v. Arsdale, P. M., and Gilman, A. G., 1979, *J. Biol. Chem.* **253:**2287–2295.
57. Birnbaumer, L., Pohl, S. L., Krans, M. L., and Rodbell, M., 1970, *Adv. Biochem. Psychopharmacol.* **3:**185–208.
58. Ross, E. M., Howlett, A. C., Ferguson, K. M., and Gilman, A. G., 1978, *J. Biol. Chem.* **253:**6401–6412.
59. Downs, R. W., Jr., Spiegel, A. M., Signer, M., Reen, S., and Aurbach, G. D., 1980, *J. Biol. Chem.* **255:**949–954.
60. Cassel, D., and Selinger, Z., 1976, *Biochim. Biophys. Acta* **452:**538–551.
61. Cassel, D., and Selinger, Z., 1977, *Proc. Natl. Acad. Sci. U.S.A.* **74:**3307–3311.
62. Pfeuffer, T., 1977, *J. Biol. Chem.* **252:**7224–7234.
63. Pfeuffer, T., 1979, *FEBS Lett.* **101:**85–89.

64. Gill, M. D., and Meren, R., 1978, *Proc. Natl. Acad. Sci. U.S.A.* **75**:3050–3054.
65. Cassel, D., and Selinger, Z., 1978, *Proc. Natl. Acad. Sci. U.S.A.* **75**:4155–4159.
66. Kent, R. S., de Lean, A., and Lefkowitz, R. J., 1980, *Mol. Pharmacol.* **17**:14–23.
67. Tkachuk, V. A., and Wollemann, M., 1981, *Mol. Pharmacol.* **20**:224–226.
68. Axelrod, J., 1974, *Science* **184**:1341–1349.
69. Palmer, G., Spurgeon, H., and Priola, D., 1975, *Cyclic Nucleotide Res.* **1**:89–95.
70. Palmer, G., Wagner, H., and Putnam, R., 1976, *Neuropharmacology* **15**:695–702.
71. Wollemann, M., and Rózsa, K. S., 1975, *Comp. Biochem. Physiol.* **51C**:63–66.
72. Pik, K., and Wollemann, M., 1977, *Biochem. Pharmacol.* **26**:1448–1449.
73. Vetulani, J., Stawari, R. J., and Sulser, F., 1976, *J. Neurochem.* **27**:661–666.
74. Glaubiger, G., Bie Shung T., and Lefkowitz, R. J., 1978, *Nature* **273**:240–242.
75. Mukherjee, C., Caron, M., and Lefkowitz, R. J., 1975, *Proc. Natl. Acad. Sci. U.S.A.* **72**:1945–1949.
76. Mickey, J., Tate, R., and Lefkowitz, R. J., 1975, *J. Biol. Chem.* **250**:5727–5729.
77. Lefkowitz, R. J., Mullikin, D., and Williams, L. T., 1978, *Pharmacology* **14**:376–380.
78. Williams, L. T., and Lefkowitz, R. J., 1977, *J. Biol. Chem.* **252**:7207–7213.
79. Iyengar, R., Abramowitz, J., Bordelon-Riser, M., and Birnbaumer, L., 1980, *J. Biol. Chem.* **255**:3558–3564.
80. Wessels, M. R., Mullikin, D., and Lefkowitz, R. J., 1979, *Mol. Pharmacol.* **16**:10–20.
81. Barnes, P. J., Dollery, C. T., and MacDermot, J., 1980, *Nature* **285**:569–570.
82. Will-Shahab, L., Wollenberger, A., and Schulze, W., 1975, *Proc. FEBS* **37**:107–127.
83. Marshall, N. J., von Borcke, S., Shardlow, S., and Malan, P. G., 1975, *Proc. FEBS* **37**:129–137.
84. Levine, M. A., Downs, R. W., Singer, M., Marx, S. S., Aurbach, G. D., and Spiegel, A. M., 1980, *Biochem. Biophys. Res. Commun.* **94**:1319–1324.
85. Clark, B. J., 1976, *Beta Adrenoceptor Blocking Agents* (P. R. Sayena and R. P. Forsyth, eds.), North Holland, New York, pp. 45–76.
86. Foster, V. F., 1980, *Pharmacology of Antihypertensive Drugs* (A. Scriabine, ed.), Raven Press, New York, pp. 349–366.
87. Meier, M., Oriom, J., Rogg, H., and Brunner, J., 1980, *Pharmacology of Antihypertensive Drugs* (A. Schriabine, ed.), Raven Press, New York, pp. 179–197.
88. Oates, J. A., Conolly, M. E., Prichard, B. N. C., Shand, D. G., and Schapel, G., 1977, *Antihypertensive Agents* (F. Gross, ed.), Springer-Verlag, Berlin, Heidelberg, New York, pp. 571–632.
89. Bucher, B., Heitz, C., and Stoclet, J. C., 1981, *Biochem. Pharmacol.* **30**:2503–2506.
90. Bhalla, R. C., and Ashley, T., 1978, *Biochem. Pharmacol.* **27**:1967–1971.
91. Triner, L., Vuillemoz, Y., Vereshy, M., and Manger, W. M., 1975, *Biochem. Pharmacol.* **24**:743–745.
92. Bhalla, R. C., Sharma, R. V., and Ramanathan, S., 1980, *Biochim. Biophys. Acta* **632**:497–506.
93. Limas, C., and Limas, C. J., 1978, *Biochem. Biophys. Res. Commun.* **83**:710–714.
94. Limas, C. J., and Limas, C., 1979, *Biochim. Biophys. Acta* **582**:533–536.
95. Tkachuk, V. A., and Wollemann, M., 1979, *Biochem. Pharmacol.* **28**:2097–2100.
96. Sharma, S. K., Klee, A. W., and Nirenberg, M., 1975, *Proc. Natl. Acad. Sci. U.S.A.* **72**:3092–3096.

Norepinephrine

Marianne Fillenz

1. INTRODUCTION

Ramon y Cajal's neurohistological studies demonstrated that the nervous system consists of an interlacing network of nerve cells with specialized contact areas between their processes. Although the specialized area of contact, the synapse, was recognized as being the site of transmission between nerve cells, nothing was known of the mechanism of this transmission. Scott[1] in 1905 was the first to put forward a general theory of chemical transmission. He emphasized that the process of conduction to the synapse and the stimulation of the next cell were entirely different properties of the neuron. On the basis of histological resemblances between gland cells and nerve cells, he suggested that the latter are also secretory and that the arrival of the impulse at the synapse causes the discharge of a chemical substance. Furthermore, he wrote

> Since the discharge means the using up of formed material, it must be an exhaustible process, and the process of complete recovery at the synapse must depend on the integrity of the connection of the synapse with the nucleus and cell body which are the original seats of formation of the material involved in the activity.

The identity of the chemical mediator of nervous action was first investigated by Langley[2] and Elliott[3] in the sympathetic nervous system; they suggested that it was epinephrine. However, the detailed study of transmitter release was carried out by Katz[4] and colleagues on the cholinergic frog's neuromuscular junction using electrophysiological techniques. This peripheral cholinergic junction became the model for all chemically transmitting synapses, and study of the noradrenergic neuron lay dormant until the development of new techniques. The histochemical fluorescence technique first used by Eränkö[5] and further developed by Falck, Hillarp, and their colleagues[6] made possible the visualization of catecholamines and a semiquantitive mapping of their distribution in different parts of the neuron.[7] Next, the demonstration by differential centrifugation of norepinephrine storage particles[8] opened up the study of the properties and function of synaptic vesicles, whose role in the nerve terminal had previously been based on indirect evidence.

Marianne Fillenz • University Laboratory of Physiology, Oxford, Oxford OX1 3PT England.

The cholinergic and noradrenergic nerve terminals continue to be the main models for the properties of chemically operating synapses. They appear to be fundamentally different in a number of respects. This difference may be more apparent than real and be a reflection of the difference in the methods of study: the investigation of the cholinergic neuron is still largely electrophysiological, and that of the noradrenergic neuron largely biochemical; the emphasis in the cholinergic neuron is on events in the terminal, whereas in the noradrenergic neuron, the interaction and interdependence among different parts of the neuron are recognized as important influences on events in the terminal. On the other hand, it is possible that the differences will turn out to be real and that the cholinergic and noradrenergic neurons may come to serve as models of two classes of neurons with different properties and different functions. The present chapter will concentrate on the various processes within the noradrenergic neuron and their short- and long-term effect on norepinephrine release.

2. STORAGE OF NOREPINEPHRINE

Norepinephrine is stored in a number of subcellular compartments: density centrifugation reveals two populations of norepinephrine storage particles in nerve terminals but only one population in nonterminal axons.[9] Norepinephrine is also present in a soluble phase, but the exact size of this pool in the undisturbed neuron is difficult to determine: some of it is certainly the result of release from storage organelles during experimental procedures. Electron microscopy after certain specific fixatives reveals three storage organelles: large dense-cored vesicles, small dense-cored vesicles, and a third compartment, which consists of a tubular structure possibly derived from endoplasmic reticulum.[10] A cytoplasmic protein-bound fraction of norepinephrine has also been postulated on the basis of results of EM autoradiography.[11]

2.1. Large Dense-Cored Vesicles

Large dense-cored vesicles have been isolated in a high degree of purity from ox splenic nerve using an improved method of density gradient centrifugation.[12] Such experiments provide evidence for a single population of norepinephrine storage particles, which have a density of 1.178 and a mean diameter of 75 nm as measured in EMs.[13]

Biochemical analysis of this purified vesicle fraction shows that the large vesicles have a composition very similar to that of chromaffin granules: they contain proteins, cholesterol, phospholipids, catecholamines, and nucleotides.[14] The important quantitative differences, which cannot be accounted for by the difference in size and therefore surface-to-volume ratio of the two structures, are the very much lower content of chromogranin A and lysolecithin in the large axonal vesicles compared to chromaffin granules. The vesicles have a soluble core that consists of catecholamines, nucleotides, proteins, and peptides. Aldehyde fixation in the presence of dichromate without postosmication results in an electron-dense core that is reserpine sensitive and would therefore appear to be related to catecholamine content.[15] With postosmication, however, the electron density of the core no longer parallels the catecholamine content: it seems likely that after reserpine there remain the osmium-stained soluble proteins.[16,17] Of these, the most important are dopamine-β-hydroxylase (DBH) and chromogranin A. Each vesicle is estimated to contain no more than 5–12 molecules of DBH enzyme based on homospecific activity measurements.[18] Two-thirds of the enzyme is in a latent form located in the fine granular matrix of the vesicle[19]; one-third is enzymatically active and is probably associated with the inner aspect of the vesicle membrane. Recently, it has been shown that the large dense-cored vesicles from ox splenic nerve also contain opioid peptides.[20]

The exact role of the large vesicles in the nerve terminals is unknown. Their number shows great species variation: in rat noradrenergic nerve terminals, large dense-cored vesicles are only 4% of the total,[21] but in cat, dog, ox, and man, they may constitute 20–50% of the vesicle population.[17,22] Although *omega* figures involving large dense-cored vesicles suggesting exocytosis have been reported in several studies,[23-25] the stimulation of norepinephrine release has failed to produce a decrease in the large noradrenergic storage vesicles when measured either as the size of the high-density peak on sucrose density gradients[26,27] or as the number of large dense-cored vesicles in electron micrographs.[28,29] *In vitro* field stimulation of noradrenergic terminals innervating human omental veins showed an increase in large dense-cored vesicles.[25] The evidence from release of DBH is discussed below.

The formation of large dense-cored vesicles in the cell body and their movement by rapid axoplasmic transport down to the nerve terminals have been inferred from their rate of accumulation above a ligature in nerve ligation experiments.[30] Comparison of axonal vesicles with the high-density norepinephrine storage vesicles in dog spleen[31] suggested that axonal vesicles may be immature and undergo progressive changes during their transport down to the terminal. Vesicles isolated from successive proximodistal segments of the splenic nerve were compared.[32] It was found that there was a progressive increase in the norepinephrine : DBH ratio and the buoyant density of the vesicles but no change in either the ATP/protein[33] or DBH/protein[34] ratio. This yields estimated norepinephrine : ATP ratios (after allowing for postmortem changes) in the terminal large vesicles of 10–18[35,36] and suggests that variation in norepinephrine content has a significant effect on vesicle buoyant density.

The additional norepinephrine is presumably synthesized during the passage of the vesicles down the axon, but the high norepinephrine : ATP ratio raises questions about the molecular mechanisms for the amine storage. In contrast to adrenomedullary vesicles, whose catecholamines are released *in vitro* stoichiometrically with ATP with a $t_{1/2}$ of 2 h and where only 20% of the amine pool was found to be exchangeable with radioactively labeled catecholamines,[37] very little ATP is released from isolated large nerve vesicles[35,36]; the loss of norepinephrine, however, occurs with a half-time of 10 min[38] and the entire pool was found to be exchangeable.[39] More recent kinetic studies have revealed a subdivision of the amine pool in large nerve vesicles: a slowly depletable pool that resembles the storage in adrenomedullary vesicles and a rapidly depletable pool that represents the norepinephrine added to the vesicle contents during their passage towards the terminal.[14]

2.2. Small Dense-Cored Vesicles

Small dense-cored vesicles are the most conspicuous subcellular organelle in varicosities of noradrenergic terminals, although clusters of them are also found in cell bodies, dendrites, and axons.[40]

2.2.1. Biochemical Composition

Density gradient centrifugation of sympathetically innervated organs from rat, cat, rabbit, and dog all show a large proportion of the particle-bound norepinephrine in a region of the gradient corresponding to a density of 1.066. The rat vas deferens yielded a small-vesicle preparation with a norepinephrine content of 14.5 nmol/mg protein[21]; following castration, a more highly purified fraction of small dense-cored vesicles has been obtained, both in terms of norepinephrine content and the number of small dense-cored vesicles seen in electron micrographs.[41,42] This preparation has been used for the biochemical analysis of the vesicles and for comparison of their composition with that of large nerve vesicles. Small vesicles, like large vesicles, contain norepinephrine and ATP. Attempts have been made to estimate the norepinephrine content of the small vesicles. After contamination of the fraction and loss of norepinephrine during the preparative procedures were allowed for, the norepinephrine/protein ratio was calculated to be 230 nmol/mg protein; this is lower than the ratio for splenic nerve vesicles, which is estimated as being 300–400 nmol/mg.[14] The norepinephrine : ATP ratio of 20–60 in small vesicles, however, is considerably higher than that in large vesicles.

The high norepinephrine : ATP ratio found in small vesicles of rat vas deferens may not hold for all noradrenergic terminals. It has been shown, using isolated vesicle preparations, that they have a Mg^{2+}-ATP-dependent, reserpine-sensitive, high-affinity, saturable active norepinephrine uptake mechanism.[38] The apparent k_m for norepinephrine uptake in large vesicles from splenic nerve was reported to be 1.5 μM,[43] and that for small vesicles from rat vas deferens is 22 μM.[44]

Bareis and Slotkin, using a rat heart microsomal pellet, found a single linear Lineweaver–Burk plot with an apparent K_m of 6 μM.[45] From this they concluded that either the two kinds of vesicles in rat heart had the same K_m or that uptake occurred into only one vesicle population. It is possible that the large vesicles, which make up only 5% of the total, have their storage capacity saturated by the time they reach the nerve terminals and that uptake of norepinephrine occurs only into the small vesicles. Net uptake and retention of additional norepinephrine occur in vesicles after either intravenous injection of high doses of norepinephrine[46] or incubation of synaptosomes with norepinephrine.[47] When this is carried out under saturating conditions and with [³H]norepinephrine, the uptake and maximum storage capacity can be measured. The ratio of the original content : storage capacity is a measure of the degree of saturation of the vesicles with norepinephrine. The saturation of vesicles in nerve terminals of the rat vas deferens is 98%, whereas that in vesicles of the rat heart and various brain regions is only around 50%.[48,49]

The finding that after reserpine pretreatment there is no electron-dense core left in the small vesicles using either chromate–dichromate buffered aldehyde or osmium tetroxide fixation[15] suggests that in small vesicles the proteins are all contained in the membrane and that the electron-dense core represents norepinephrine. Manipulation of the norepinephrine content of isolated vesicles with various drugs yielded a close but imperfect correlation between norepinephrine content and matrix density, using aldehyde fixation with postosmication,[50] which is not the optimal method of fixation for catecholamines. On the other hand, in terminals from animals without drug treatment, there is a very close correlation between the degree of vesicle saturation and the mean core volume of small dense-cored vesicles after aldehyde–dichromate or permangate fixation.[51] Measurement of core size distribution shows a wide range of core sizes, suggesting that small vesicles are heterogeneous with respect to norepinephrine content.

To what extent the membranes of the two kinds of vesicle are similar is not yet clear: cytochrome b-561 is present in both, but chromomembrin B, present in large vesicles, has not yet been demonstrated in small vesicles.[42]

There has been considerable controversy about the DBH content of small vesicles. This is largely because of the pattern of norepinephrine and DBH distribution on gradients used for the isolation of the norepinephrine storage particles. Sucrose density gradients of dog spleen showed that whereas in the high-density region of the gradient the DBH peak coincided with the norepinephrine peak, in the low-density region of the gradient, the DBH peak occurred at a lower density than the norepinephrine peak.[52] This low-density peak was attributed to empty large vesicles or their membrane fractions, and the small vesicles were thought to lack DBH and therefore to be concerned with uptake and storage of norepinephrine and not its synthesis.

The demonstration that one-third of the DBH activity in homogenates of bovine splenic nerves appears in the region of the light-vesicle peak although this structure does not contain a significant small-vesicle population[34] has lent support to this view. There has, however, been a progressive move away from this extreme position, and the discrepancy in peak position has been explained either by the presence of two distinct populations of small vesicles, one with and the other without DBH in its membrane,[53] or, alternatively, by a small-vesicle population with a wide range of norepinephrine content, the peak of DBH with a density lower than that of norepinephrine representing vesicles with a low content of norepinephrine.

That vesicles in different nerve terminals vary in their norepinephrine content has been shown above. There are a number of reports which show a very close correspondence between the distributions of norepinephrine and DBH in the low-density region of the gradient, which contains the small vesicles,[54,55] and there is also immunocytochemical evidence for the presence of DBH in small vesicles[56]; these findings support the view that small as well as large vesicles have DBH on the inner aspect of their membrane and are therefore involved in synthesis of norepinephrine.

2.2.2. Participation in Release

There is both biochemical and morphological evidence to support the hypothesis that the small dense-cored vesicles, or some subgroup of them, constitute the releasable pool of norepinephrine.

Phenoxybenzamine administration[57] or cold stress[27] both result in a depletion of norepinephrine in the rat heart; density centrifugation shows that the depletion occurs in the peak representing the low-density norepinephrine storage particles, which correspond to the small dense-cored vesicles. There have been several reports of the ultrastructural changes seen in noradrenergic varicosities after stimulation[28,58,59]; these consist of a reduction in small dense-cored vesicles and no change in the number of large vesicles. Potassium depolarization of rat vas deferens produces a substantial norepinephrine depletion, which is paralleled by a similar reduction in the number of small dense-cored vesicles; there is an increase in small clear vesicles, but this change is much smaller.[29] There is thus a net reduction in the number of vesicles and of vesicle membrane. Similar morphological changes have been reported after transmural electrical stimulation of mouse vas deferens, although the changes in norepinephrine content were not measured. The missing vesicle membrane does not appear to be incorporated into the nerve terminal membrane as happens in the cholinergic terminal,[60] since there is no change in varicosity area, shape, or perimeter.[28,29]

Further evidence for the loss of functional vesicles from nerve terminals after transmitter release comes from measurements of changes in the vesicular norepinephrine storage capacity. Since in noradrenergic nerve terminals probably over 90% of the norepinephrine is stored in vesicles, and since in the rat 95% of these vesicles are small vesicles, changes in the number of such vesicles should be reflected in changes in norepinephrine storage capacity. This is measured as vesicular norepinephrine content after a saturating dose of exogenous norepinephrine. Such measurements show a reduction in vesicular norepinephrine storage capacity after both K^+-evoked release from hypothalamic synaptosomes and after norepinephrine depletion of rat heart by cold exposure.[47,61]

2.2.3. Origin of Small Dense-Cored Vesicles

There is at present insufficient evidence for a clear choice among the various possible origins of the small dense-cored vesicles. In early EM studies of nerve ligation experiments using osmium fixation, only large dense-cored vesicles were found to accumulate above the ligature[30]; however, the use of aldehyde in dichromate fixatives revealed that small dense-cored vesicles were also found above a ligature[62] as well as in clusters in cell body dendrites and axons of nonligated neurons,[40] although the largest number were found in the terminal. This meant that formation of small dense-cored vesicles was not confined to the nerve terminal.

Not enough is known at present about the biochemical composition of the membrane of the two classes of vesicle to exclude the formation of small vesicles from large vesicles, but the changes in their relative number after stimulation of release do not lend support to such a view. There is, on the other hand, evidence that is compatible with the separate formation of the small vesicles from the endoplasmic reticulum. This is both morphological and biochemical. There are a number of reports from electron microscopists which suggest the formation of small vesicles from endoplasmic reticulum.[63,64] It is, of course, difficult to deduce such a dynamic process from static images. The biochemical evidence comes from changes in DBH, which subject is discussed in the next section, and from the use of vesicular norepinephrine storage capacity as a measure of the number of functional vesicles. Cold exposure in rats during a 6-h period leads to fluctuations in the rate of norepinephrine release as measured by changes in plasma norepinephrine concentration[65] and to parallel changes in vesicular norepinephrine content and storage capacity in the nerve terminals of the heart.[66] The norepinephrine release shows a steady rise up to 4 h, after which it drops quite steeply. During the period of high release rate, the vesicles in the nerve terminals in the heart show an increase above the control level of storage capacity at 4 h; at 6 h, however, the vesicular norepinephrine storage capacity is below control level. This suggests that during the first 4 h, the phase of rapid release, when vesicles that have undergone exocytosis cannot refill with norepinephrine, there is an increase in the number of vesicles in the terminals; the source for the new vesicles is, however, limited, and by 6 h, it has become exhausted and so the storage capacity falls. The endoplasmic reticulum would seem a very likely candidate for the source of such newly formed vesicles in the terminal.

2.3. Dopamine-β-Hydroxylase in the Neuron

2.3.1. Subcellular Compartmentalization of DBH

The enzyme DBH catalyzes the last step in norepinephrine synthesis. Homogenization in isotonic and hypotonic media followed by differential centrifugation reveals that the subsellular distribution of DBH differs in the various parts of the neuron. In the cell body and nonterminal axon, only 50% of the DBH is particulate, the rest being sucrose soluble.[67] The latter could be DBH either in free solution in the cytoplasm or stored in a delicate structure easily disrupted by homogenization. The inhibition of norepinephrine synthesis by reserpine[68] argues against a cytoplasmic location for DBH; the presence of DBH immunoreactivity in a reticulum of interconnecting membranes suggests a location in the endoplasmic reticulum.[69] Treatment of the particulate fraction leads to a further separation of membrane-bound and osmotically releasable DBH: in axons, 20% of the DBH is osmotically releasable. These findings suggest that in nonterminal axons 50% of the DBH is in endoplasmic reticulum and the rest is in large dense-cored vesicles, 20% of whose DBH is in a soluble core.

In the terminal, only 10–15% of the DBH is sucrose soluble, and the particulate fraction includes only 5% osmotically releasable DBH.[67] This suggests an increase in vesicles at the expense of DBH-containing endoplasmic reticulum; furthermore, the reduction in the fraction of osmotically releasable

DBH means that the vesicles formed from the endoplasmic reticulum are small vesicles that lack a soluble protein core.

2.3.2. Intraneuronal Transport of DBH

The differential centrifugation experiments show the subcellular distribution of DBH but give no hint of its movement within the neuron. This has been studied in experiments using nerve ligation,[70–72] colchicine block,[67] and injection of labeled DBH antibody[73] in whole animals and the application of cold block in isolated nerves.[74,75]

In experiments interrupting axoplasmic transport by ligation, cold block, or colchicine, the time course of the redistribution of the enzyme within the nerve has been measured mostly by assaying the activity of the isolated enzyme but also by immunofluorescence measurement of the enzyme protein.[76] These experiments have revealed that DBH moves between cell body and terminals: there is bidirectional axoplasmic transport, which has a velocity of 10–12 mm/h in rabbit sciatic nerve.[74,75] Such a high rate of transport is characteristic of particle-bound substances. The anterograde movement of DBH is paralleled by transport of norepinephrine,[77] and EM evidence suggests that this represents the axoplasmic transport of large vesicles filled or filling with norepinephrine. The retrograde transport of DBH in the cold-block experiments had the same velocity as the anterograde transport, although in nerve ligation experiments using the enzyme assay, the accumulation below the ligature was only 20% of that above the ligature, suggesting a massive loss in the terminal. However, the use of immunofluorescence showed a much smaller discrepancy and suggested that a considerable proportion of the retrogradely transported enzyme had become inactive.[76]

The uptake and retrograde transport of labeled antibody[73] show that this movement of DBH is not a response to injury to the neuron but a normal process. Studies of subcellular distribution of DBH after nerve ligation and colchicine block show that it is the particulate DBH that is rapidly transported, most of the sucrose-soluble DBH being in a compartment of slowly moving or stationary enzyme. But there are also changes in osmotically releasable DBH in a ligated nerve: it decreases above the ligature but increases below the ligature.

2.3.3. Turnover of DBH

Dopamine-β-hydroxylase turnover has been calculated from its rate of accumulation above a ligature,[78] from its disappearance after protein synthesis inhibition,[79] and from its accumulation in cell bodies following colchicine injections close to a ganglion.[67] These experiments give figures ranging from 6 to 24 h for the complete turnover of DBH in the neuron. The turnover time for DBH is in very marked contrast to that of the other transmitter-specific enzyme, tyrosine hydroxylase, whose turnover time is several days.[79] The reason for the very much more rapid turnover of DBH is not clear. Release of DBH is discussed in the next section.

There is evidence that the synthesis rate of DBH is regulated by the presynaptic transmitter, which also controls the impulse traffic in and hence transmitter release from the noradrenergic neuron. A number of drugs that lead to an increase in the impulse frequency of the preganglionic fibers innervating noradrenergic neurons were found to produce a rise in the *in vitro* activity of DBH isolated from the ganglion. This increase also occurred after physiological activation of the sympathetic neurons by cold stress or swim stress.[80] Conversely, deafferentation and the administration of β-blockers caused a decrease in *in vitro* enzyme activity.[81] The increase in enzyme activity was prevented by deafferentation and could be produced by incubation of ganglia with carbamylcholine.[82]

Reserpine administration is also followed by an increase in the *in vitro* activity of DBH isolated from central noradrenergic neurons of the locus coeruleus; this is caused by an increase in the biosynthesis of DBH[84] and results in an increase in the amount of immunoprecipitable enzyme protein in the locus coeruleus. The changes in *in vitro* DBH activity in peripheral noradrenergic neurons are small: after a single dose of reserpine, DBH activity rises to 128% of control in the stellate ganglion at 3 days and to 114% of control in the heart at 4 days.[85] In the noradrenergic neurons of the locus coeruleus, DBH increases to 200% of control at 4 days, but in the frontal cortex and hypothalamus, both brain regions containing axon terminals of neurons in the locus coeruleus, there is a statistically significant decrease in the *in vitro* DBH activity to below control level.[86] The DBH in the terminals measured in terms of enzyme activity is the balance between the arrival of DBH from the cell and loss of DBH by removal, release, and inactivation. The reduction in DBH activity after reserpine in central noradrenergic terminals suggests that the rate of inactivation and/or removal exceeds the delivery of additional DBH to the terminal.

2.4. Summary

The two populations of norepinephrine storage vesicles differ in the chemical composition of their soluble core; particulate DBH is a marker for both types of vesicles, but it is not known to what extent other protein constituents are common to the two kinds of vesicle membrane. Large vesicles acquire their norepinephrine content during their transport from cell body to terminal. Small vesicles appear to be formed from endoplasmic reticulum in the nerve terminal and gradually acquire their norepinephrine as a result of local synthesis. The rapid turnover of DBH together with the reduction in the number of small vesicles following release suggests that there is little or no reuse of noradrenergic vesicles. The changes in DBH following transsynaptic induction suggest that the activation of noradrenergic neurons has long-lasting effects on vesicle turnover.

3. NOREPINEPHRINE RELEASE

3.1. Release by Exocytosis

The hypothesis for release of norepinephrine by exocytosis has strong experimental support, but much of this evidence is circumstantial. The criterion for exocytosis is the demonstration of stoichiometric release of norepinephrine and a soluble vesicle marker. Since norepinephrine is stored in two quite distinct vesicle types, whose relative contribution to release is not clear, the criterion cannot be satisfied at present. Great emphasis has been placed on the appearance of chromogranin and DBH in the perfusion medium after stimulation; these are only found in the core of the large vesicles, whose contribution to norepinephrine release is not supported by evidence from ultrastructural changes seen after release. Small dense-cored vesicles, on the other hand, whose contribution to release is supported by changes seen in both ultrastructure and norepinephrine distribution on density gradients, show heterogeneity with respect to norepinephrine content, and their only known soluble constituent in addition to norepinephrine is ATP; furthermore, most of the ATP appearing after nerve stimulation is of nonneural origin.[87]

The interpretation of the literature on DBH release is still not clear. After nerve stimulation, DBH and chromogranin were found in the perfusate, but their ratio to catecholamines was several orders of magnitude lower than in the tissue.[88] In order to resolve this discrepancy, the experiment was repeated using *in vitro* stimulation of guinea pig vas deferens.[89] These experiments, which gave a ratio of DBH:norepinephrine in the bath fluid that was just over half that in the tissue, have been variously interpreted: the authors concluded that the ratio of DBH:norepinephrine was similar to that in the tissues, whereas others have emphasized the discrepancy between the two ratios. In fact, it is surprising to find such a high DBH:norepinephrine ratio, since the soluble DBH is confined to the large vesicles, whose contribution to release, on other evidence, is small, whereas there is strong evidence for the participation of small vesicles in norepinephrine release, and they do not contain any soluble proteins.

The changes in DBH following norepinephrine release have also been studied in the nerve terminal. Subcellular fractionation of rat heart after periods of cold stress up to 30 min showed a 30% reduction in membrane-bound DBH[90,91] in the microsomal pellet containing the small dense-cored vesicles but an increase in the soluble DBH found in the final supernatant; the total DBH in the heart was reduced by 17%, assayed as enzyme activity.[91]

Since Nagatsu provided evidence for enzyme inactivation, a reduction in enzyme activity does not necessarily mean an equivalent loss of enzyme protein from the nerve terminal. There is another factor to be taken into account: it has recently been shown that drugs that lead to an increase in norepinephrine release increase the amount of labeled DBH antibody transported retrogradely from the nerve terminal to the cell body.[92] This suggests that the rapid retrograde transport represents the empty vesicle membranes after exocytosis, whose loss from the nerve terminal is manifested by the reduction in both vesicle number and norepinephrine storage capacity.

Ligation experiments suggest that the retrograde transport of DBH is not accompanied by a parallel transport of norepinephrine[70]; injection of [^{3}H]norepinephrine, however, has produced conflicting results.[93,94]

This leaves the question of the origin of DBH in plasma and cerebrospinal fluid. The parallelism between plasma norepinephrine and DBH is poor, but fluctuations of DBH in plasma and cerebrospinal fluid do occur with changes in neuronal activity.[95,96] This may be analogous to the release of the enzyme acetylcholinesterase, which is not thought to come from synaptic vesicles.[97]

The hypothesis of norepinephrine release by exocytosis has no serious challenge from an alternative theory that could account for the quantal release from a nerve terminal with a very low cytoplasmic concentration of transmitter. Measurement of quantal release in noradrenergic junctions has been complicated by multiple innervation and electrotonic coupling of smooth muscle cells. However, a recent study in the guinea pig vas deferns provides evidence for uniformly sized quanta.[98] Exocytosis is a Ca^{2+}-dependent mechanism, and Ca^{2+} dependence is widely used as a criterion for exocytosis.

3.2. Carrier-Mediated Release

In experiments using synaptosomes in a superfusion system, depolarization-induced release can be subdivided into a Ca^{2+}-dependent and a Ca^{2+}-independent fraction. The Ca^{2+}-independent release is blocked by drugs that inhibit neuronal uptake of norepinephrine and is therefore thought to be carrier mediated.[99] Calcium-independent release as a fraction of the total depolarization-induced release varies in different brain regions[47,100] and is probably an expression of the level of the cytoplasmic norepinephrine. It is uncertain what role if any it plays in the intact organism. It would provide an impulse-independent background level of release that would be increased by a rise in cytoplasmic norepinephrine as occurs during medication with monoamine oxidase inhibitors or as a result of changes in the $Na^+ : K^+$ ratio in the extracellular compartment, since this determines the direction of the carrier-mediated transport.

Exocytotic release is triggered in the intact organism by the impulse-dependent entry of Ca^{2+} into the nerve terminal. There are, however, a number of mechanisms that further regulate this release.

3.3. Regulation of Release

In *in vitro* experiments using K^+-evoked release from synaptosomes, the Ca^{2+}-dependent release of norepinephrine was found to be proportional to the size of the vesicular store.[100] Changes in the vesicular store can result either from fewer functional vesicles (demonstrable as a reduction in norepinephrine storage capacity) or from a smaller norepinephrine content (demonstrable as a reduction in vesicular saturation). Cold exposure produces a reduction in norepinephrine release rate and a parallel decrease in the vesicular norepinephrine store size.[66]

Another important mechanism for the regulation of impulse-evoked release is mediated by presynaptic receptors. Noradrenergic neurons were the first in which presynaptic inhibitory autoreceptors were demonstrated.[101] These were α receptors, and it was shown that they differed from the postsynaptic α receptors; they were therefore called $α_2$ receptors. More recent work has shown that there is no simple correlation between pharmacological properties and function of receptors, and α_2 pharmacological properties do not necessarily imply a presynaptic inhibitory role.[102] In peripheral noradrenergic neurons, a second type of autoreceptor, a β receptor, which caused enhancement of release, has also been described.[102] (See also Chapters 2 and 3.)

In addition to autoreceptors located either on nerve terminals and called presynaptic or located on cell body and dendrites and activated by recurrent collaterals or possibly by dendritic release, evidence for presynaptic actions by a number of neurotransmitters and neuromodulators has been reported. Thus, nine different presynaptic receptor mechanisms had been identified on sympathetic nerve terminals in the heart in 1979[103]; their number has undoubtedly grown by now. Although such receptors probably mediate the pharmacological effects of therapeutic agents, their role under physiological conditions is not clear; such a role requires the demonstration that the supposed agonists for these receptors are present in appropriate concentrations at the presynaptic nerve terminal. The conclusions from *in vivo* experiments concerning the normal role of presynaptic receptors are still controversial. There is also evidence for presynaptic receptors on central noradrenergic neurons. In addition to autoreceptors, the recently described effect of GABA on norepinephrine release is of interest[104,105] in view of the widespread distribution of GABA-releasing neurons in the brain.

There is little precise information about the mechanism by which presynaptic receptors modulate norepinephrine release. There is some evidence that norepinephrine acting on α_2 receptors,[106] prostaglandin,[107] and morphine[108] all depress norepinephrine release by reducing the availability of Ca^{2+} for the stimulus–secretion coupling, possibly by an action on the voltage-sensitive Ca^{2+} channel.

Chronic drug treatments and electroconvulsive shock have produced changes in the number of receptor-binding sites[109,110]; preliminary results indicate that in certain situations changes in receptor affinity also occur. One of the interesting features is that the changes in α_2 receptors differ in brain regions that receive their noradrenergic supply from the locus coeruleus. Since these neurons show extensive branching, it raises the possibility that local factors could lead to differences between nerve terminals of the same neuron.

3.4. Summary

Impulse-evoked norepinephrine release is quantal; the evidence for exocytosis is circumstantial, and the nature of DBH release is not yet clear. The amount of norepinephrine released depends on the size of the vesicular store and the activation of presynaptic receptors. In addition to Ca^{2+}-dependent release from vesicles, there is Ca^{2+}-independent carrier-mediated release from the cytoplasm.

4. NOREPINEPHRINE SYNTHESIS

4.1. Rapid Regulation

Although DBH catalyzes the last step in norepinephrine synthesis, the rate-limiting enzyme is tyrosine hydroxylase (TH), which catalyzes the first step.

This enzyme is extravesicular in localization, and it seems from both differential centrifugation and measurement of the rate of its axoplasmic transport that most of it is in solution in the cytoplasm; a small percentage, which may be functionally very important, appears to be membrane bound. The enzyme is O_2 dependent, has an absolute requirement for reduced pteridine cofactor, and is subject to feedback inhibition by the intermediate products of norepinephrine synthesis, by norepinephrine, and by its deaminated metabolites (for further details see D. M. Kuhn and W. Lovenberg, Vol. 4).

Increased release of norepinephrine is accompanied by an increase in the rate of tyrosine hydroxylation. It is now clear that the increased activity of the enzyme does not result from a direct reduction of feedback inhibition by norepinephrine release; this means that the increase in synthesis rate is not a consequence of increased release. This is further confirmed by the finding that the increases in release and synthesis have different time courses.[111] The mechanism for the rapid increase in tyrosine hydroxylation is not yet understood, and there may be more than one mechanism. A single factor, such as Ca^{2+} entry, may trigger both release and acceleration of synthesis. However, it is possible that some other mechanism may accelerate synthesis, which then, in time, increases release rate by increasing vesicle saturation and so the size of the vesicular store. If these effects are closely linked, there will be no change in norepinephrine content but a rise in norepinephrine turnover.

4.2. Slow Regulation

In addition to the rapid increase in synthesis rate, TH is subject to slow regulatory mechanisms.

All manipulations that produce induction of DBH in peripheral and central noradrenergic neurons also cause induction of TH.[80] The increase in enzyme activity in both peripheral ganglia and central neurons has the same time course as that of DBH but is much greater; this is probably because of the slower removal of the enzyme from the cell body by axoplasmic transport and the very much slower turnover of TH in the neuron. That the rise in *in vitro* TH activity in the locus coeruleus after reserpine reflects enzyme induction has been confirmed by immunoprecipitation. The rise in TH in the nucleus is followed by a rise in *in vitro* activity in the various brain regions containing axon terminals after an interval that is proportional to their distance from the cell body.[86] One cannot necessarily conclude that this represents the arrival of the additional enzyme at the nerve terminals, because there is as yet no evidence from immunoprecipitation and because the delay in the peak in the terminal region is longer than would be expected from the rate of axoplasmic transport measured in peripheral neurons. For these reasons, it has been called remote augmentation.[86] This slow remote augmentation is measured as *in vitro* activity of soluble TH. If the functionally important enzyme is the membrane-bound fraction, which amounts to only 10%, this would travel down the axon much more rapidly but would be overlooked by the assays normally used. A small functionally important fraction of enzyme may explain why in the intact neuron and in synaptosomes less than 10% of the maximum potential activity of the isolated enzyme is expressed.

The administration of muscarinic cholinergic drugs leads to an increase in the *in vitro* activity of TH from the locus coeruleus.[112] This has the same time course as enzyme induction—a delayed onset and a duration of 21 days—but immunoprecipitation shows that there is no increase in enzyme protein. This has been called delayed activation: it is not followed by remote activation and therefore can have no effect on norepinephrine synthesis in the terminals. The locus coeruleus receives a serotonergic innervation from the raphe nuclei. Neurochemical lesions of these nuclei or inhibition of serotonin synthesis by PCPA leads to an increase in the activity of TH isolated from the locus coeruleus. This is correlated with the reduction in 5-HT content of the locus coeruleus and is caused by both induction and delayed activation of the enzyme, the relative proportions of the two effects depending on which raphe nucleus is lesioned.[113]

4.3. Functional Implications

The most interesting question concerns the functional implications of enzyme induction. Evidence relevant to this question is sparse and consists mainly of the effects of drugs.

Following phenoxybenzamine administration, an increase in *in vitro* TH activity was found in the heart and was paralleled by an acceleration of norepinephrine synthesis in the intact animal.[114]

Reserpine administration results in substantial increases in activity of the soluble enzyme isolated from brain regions containing noradrenergic nerve terminals, but synaptosomes isolated from these brain regions showed no increased tyrosine hydroxylation.[115] In the adrenal medulla, reserpine also leads to an increase in tetrahydrobiopterin[116]; this may also apply to noradrenergic neurons, but its time course is not known. The expression of additional TH after reserpine may be limited by availability of reduced cofactor.

Finally, administration of 6-OH-DA into the cerebral ventricles leads to the destruction of 80% of the noradrenergic nerve terminals in the hippocampus. After a delay of 21 days, there is an increase in the *in vitro* activity of TH and an increase in synaptosomal norepinephrine synthesis[117]; in the whole animal, the metabolite level is equal to control, which is evidence of a greatly increased norepinephrine turnover in the surviving nerve terminals. Some of this is attributable to the increased firing rate that is also observed.[118] All of these drugs lead to decreased receptor activation, and the enzyme induction may represent an adaptive response enabling receptor activation to be maintained at near-normal levels. The hypothesis that enzyme induction does not have merely a homeostatic role but may be a mechanism for long-term alteration in the functional output of the noradrenergic system has as yet much less experimental support. Cold exposure leads to enzyme induction and increased norepinephrine concentration in peripheral sympathetic ganglia.[119] An increase in norepinephrine content and release in the nerve terminals has not yet been demonstrated.

5. CONCLUSION

The noradrenergic neuron differs from most other neurons in that the function of its vesicles includes not only storage and release but also a role in synthesis. The various findings are compatible with the following model of the noradrenergic neuron.

The two populations of vesicles constitute two separate systems. The large vesicles originate in the cell body, from which they are carried to the nerve terminal by rapid axoplasmic transport. Their core constituents, the soluble proteins, opioid peptides, and possibly the nucleotides, are acquired in the cell body; their norepinephrine content increases progressively as they travel towards the terminal. They appear to undergo exocytosis, during which they corelease soluble proteins, opioid peptides, and catecholamines, but it is not clear what contribution they make to impulse-evoked norepinephrine release. The hypothesis that exocytosis of large vesicles leads to their transformation into small vesicles has no experimental support but cannot at present be excluded. It would seem more likely that their empty membranes are carried back to the cell body.

Small vesicles appear to be formed from DBH-containing endoplasmic reticulum; this occurs chiefly, but not exclusively, in the nerve terminal. The newly formed vesicles contain little or no norepinephrine. Their gradual filling may be limited by both the rate of ATP uptake, which may be essential for storage, and the rate of tyrosine hydroxylation. The uniform size of norepinephrine quanta suggests that the releasable pool consists of fully saturated vesicles. Exocytosis leads to inactivation of DBH and loss of storage capacity. The empty vesicle membranes are carried to the cell body by rapid retrograde axoplasmic transport. The size of the releasable pool in the nerve terminal is therefore determined by the dynamic equilibrium among these various interacting processes. With changes in the size of the releasable pool, there will be corresponding changes in the release rate.

In the noradrenergic terminal, therefore, instead of the rapid local recycling of vesicles seen in the cholinergic nerve terminal, there is a complex sequence of events, which involves interaction among all parts of the neuron; the size of the releasable pool of norepinephrine in the nerve terminal is determined by the dynamic equilibrium among these various interacting processes. Since these processes vary in time course and in the extent to which they are regulated, the dynamic equilibrium can be set at different levels under different circumstances.

Thus, the earliest effect of an increase in impulse frequency will be an increase in both release and synthesis, with some loss of functional vesicles and replacement by new formation from endoplasmic reticulum. This will enable the terminal to maintain its norepinephrine output per impulse and to increase its output per unit time, although the increase will be limited by the negative feedback mechanism of the presynaptic receptors. At this stage, one might therefore expect a decrease in DBH enzyme activity but not in immunoprecipitable enzyme protein, an increase in vesicle saturation, and no change in vesicular storage capacity if vesicle replacement keeps up with vesicle loss.

The absence of local vesicle recycling is manifested as a progressive failure in release with more prolonged stimulation. This failure appears to result both from limited replacement of the vesicles, and thus a reduction in storage capacity, and from a failure of synthesis, since the remaining vesicles have a lower norepinephrine saturation. This second phase is followed by a third delayed phase in which induction has resulted in a greater synthetic capacity in the terminal and a delivery of a larger number of vesicles to the nerve terminal. This may automatically lead to greater release per impulse and will therefore not show up as an increase in norepinephrine in the terminal but as a higher concentration of norepinephrine metabolites–an index of turnover rate.

The hypothesis proposed here is that whereas in the cholinergic neuron the release of acetylcholine per impulse is relatively constant, and acetylcholine release is largely controlled by impulse frequency, in the noradrenergic neuron, the norepinephrine release is subject to a greater variety of regulatory mechanisms and shows slow fluctuations over long periods of time.

REFERENCES

1. Scott, F. H., 1905, *Brain* **28**:506–526.
2. Langley, J. N., 1901, *J. Physiol. (Lond.)* **27**:237–256.
3. Elliott, T. R., 1905, *J. Physiol. (Lond.)* **32**:401–467.
4. Katz, B., 1969, *The Release of Neural Transmitter Substances. The Sherrington Lectures X,* Liverpool University Press, Liverpool.
5. Eränkö, O., 1955, *Acta Endocrinol. (Kbh.)* **18**:174–179.
6. Falck, B., Hillarp, N. A., Thieme, G., and Torp, A., 1962, *J. Histochem. Cytochem.* **10**:348–354.
7. Jonsson, G., 1969, *J. Histochem. Cytochem.* **17**:714–723.
8. von Euler, V. S., and Hillarp, N.-A., 1956, *Nature* **177**:44–45.
9. Roth, R. H., Stjarne, L., Bloom, F. E., and Giarman, N. J., 1968, *J. Pharmacol. Exp. Ther.* **162**:185–194.
10. Tranzer, J. P., 1972, *Nature (New Biol.)* **237**:57–58.
11. Descarries, L., and Droz, B., 1970, *J. Cell. Biol.* **44**:385–399.
12. Lagercrantz, H., Klein, R. L., and Stjarne, L., 1970, *Life Sci.* **9**:639–650.
13. Bisby, M. A., 1971, D.Phil. Thesis, Oxford.
14. Lagercrantz, H., 1976, *Neuroscience* **1**:81–92.
15. Tranzer, J. P., and Richards, J. G., 1976, *J. Histochem. Cytochem.* **24**:1178–1193.
16. Lagercrantz, H., and Thureson-Klein, A., 1975, *Histochemistry* **43**:173–183.
17. Klein, R. L., and Thureson-Klein, A., 1974, *Fed. Proc.* **33**:2195–2206.
18. Klein, R. L., Kirksey, D. F., Rush, R. A., and Goldstein, M., 1977, *J. Neurochem.* **28**:81–86.
19. Kirksey, D. F., Klein, R. L., Baggett, J. McC., and Gasparis, M. S., 1978, *Neuroscience* **2**:621–634.
20. Wilson, S. P., Klein, R. L., Chang, K.-J., Gasparis, M. S., Viveros, O. H., and Yang, N.-H., 1980, *Nature* **288**:707–709.
21. Bisby, M. A., and Fillenz, M., 1971, *J. Physiol. (Lond.)* **215**:163–179.
22. Thureson-Klein, A., Klein, R. L., and Stjarne, L., 1976, *Neurosci. Abstr.* **2**.
23. Fillenz, M., 1971, *Phil. Trans. R. Soc. [Biol.]* **261**:319–323.
24. Thureson-Klein, A., Stjarne, L., and Brundin, J., 1976, *Neuroscience* **1**:333–337.
25. Thureson-Klein, A., Stjarne, L., and Brundin, J., 1976, *34th Ann. Proc. Electron Microscopy Soc. Amer.* (G. W. Bailey, ed.), Miami Beach, Florida, pp. 108–109.

26. Bisby, M. A., Cripps, H., and Dearneley, D. P., 1971, *J. Physiol. (Lond.)* **214**:13–14P.
27. Nelson, D. L., and Molinoff, P. B., 1976, *J. Pharmacol. Exp. Ther.* **198**:112–113.
28. Basbaum, C. B., and Heuser, J. E., 1979, *J. Cell Biol.* **80**:310–325.
29. Pollard, R. M., Fillenz, M., and Kelly, P., 1982, *Neuroscience* **7**:1623–1629.
30. Kapeller, K., and Mayor, D., 1967, *Proc. R. Soc. Lond. [Biol.]* **167**:282–292.
31. De Potter, N. P., Chubb, I. W., and De Schaepdryver, A. F., 1972, *Arch. Int. Pharmacodyn. Ther. [Suppl.]* **196**:258–287.
32. Lagercrantz, H., Kirksey, D. F., and Klein, R. L., 1974, *J. Neurochem.* **23**:769–773.
33. Yen, S. S., Klein, R. L., Chen Yen, S. H., and Thureson-Klein, A., 1976, *J. Neurobiol.* **7**:11–22.
34. Klein, R. L., Thureson-Klein, A., Chen Yen, S. H., Baggett, J. McC., Gasparis, M. S., and Kirksey, D. F., 1979, *J. Neurogiol.* **10**:291–307.
35. Yen, S. S., Klein, R. L., Chen Yen, S. H., 1973, *J. Neurocytol.* **2**:1–12.
36. Lagercrantz, H., Fried, G., and Dahlin, J., 1975, *Acta Physiol. Scand.* **94**:136–138.
37. Stjärne, L., 1972, *Handbook of Experimental Pharmacology*, Volume 33 (H. Blaschko and E. Muscholl, eds.), pp. 231–261, Springer-Verlag, Berlin, Heidelberg, New York.
38. von Euler, V. S., 1980, *Release and Uptake Functions in Adrenergic Nerve Granules. The Sherrinton Lectures XV*, Liverpool University Press, Liverpool.
39. Klein, R. L., and Lagercrantz, H., 1971, *Acta Physiol. Scand.* **83**:179–190.
40. Richards, J. G., and Da Prada, M., 1980, *Histochemistry and Cell Biology of Autonomic Neurons, SIF Cells and Paraneurons* (O. Eranko, S. Soinila and H. Paivarinta, eds.), Raven Press, New York, pp. 269–278.
41. Fried, G., Lagercrantz, H., and Hökfelt, T., 1978, *Neuroscience* **3**:1271–1291.
42. Fried, G., 1980, *Acta Physiol Scand. Suppl.* **493**:1–28.
43. von Euler, U. S., 1970, *Bayer Symposium II* (H. J. Schumann and G. Kroneberg, eds.), Springer-Verlag, Berlin, Heidelberg, New York, pp. 144–159.
44. Fried, G., 1981, *Acta Physiol. Scand.* **112**:41–46.
45. Bareis, D. L., and Slotkin, T. A., 1979, *J. Neurochem.* **32**:345–351.
46. Fillenz, M., Howe, P. R. C., and West, D. P., 1976, *J. Neurosci.* **1**:113–116.
47. West, D. P., and Fillenz, M., 1980, *J. Neurochem.* **35**:1323–1328.
48. Fillenz, M., and Stanford, S. C., 1981, *Br. J. Pharmacol.* **73**:401–404.
49. Gomez, J., and Fillenz, M., 1982, *Neurochem. Int.* **4**:135–141.
50. Fried, G., Thureson-Klein, A., and Lagercrantz, H., 1981, *Neuroscience* **6**:787–800.
51. Fillenz, M., and Pollard, R. M., 1976, *Brain Res.* **109**:443–454.
52. Chubb, I. W., De Potter, W. P., and De Schaepdryver, A. F., 1970, *Nature* **228**:1203–1204.
53. De Potter, W. P., and Chubb, I. W., 1977, *Neuroscience* **2**:167–174.
54. Nelson, D. L., and Molinoff, P. B., 1976, *J. Pharmacol. Exp. Ther.* **196**:346–359.
55. De Potter, W. P., and De Smet, F. H., 1980, *Experientia* **36**:1282–1285.
56. Rush, R. A., Millar, T. J., Chubb, I. W., and Geffen, L. B., 1978, *Catecholamines: Basic and Clinical Frontiers* (E. Usdin, I. Kopin, and J. Barchas eds.), Pergamon Press, London, pp. 331–333.
57. Fillenz, M., and Howe, P. R. C., 1975, *J. Neurochem.* **24**:683–688.
58. Coté, M. G., Palaic, D., and Panisset, J. C., 1970, *Rev. Can. Biol.* **29**:111–114.
59. Hamilton, R. C., and Robinson, P. M., 1973, *J. Neurocytol.* **2**:465–480.
60. Heuser, J., and Reese, T., 1973, *J. Cell Biol.* **57**:315–344.
61. Fillenz, M., and West, D. P., 1976, *Neurosci. Lett.* **2**:285–287.
62. Tomlinson, D., 1975, *J. Physiol. (Lond.)* **245**:727–735.
63. Holtzmann, E., 1977, *Neuroscience* **2**:327–356.
64. Fillenz, M., 1979, *The Release of Catecholamines from Adrenergic Neurons* (D. M. Paton, ed.), Pergamon Press, London, pp. 17–38.
65. Benedict, C. R., Fillenz, M., and Stanford, S. C., 1979, *Br. J. Pharmacol.* **66**:521–524.
66. Fillenz, M., Stanford, S. C., and Benedict, C. R., 1978, *Catecholamines: Basic and Clinical Frontiers* (E. Usdin, I. Kopin, and J. Barchas, eds.), Pergamon Press, London, pp. 936–938.
67. Brimijoin, S., 1974, *J. Neurochem.* **22**:347–353.
68. Stjärne, L., and Lischajko, F., 1966, *Br. J. Pharmacol.* **27**:398–404.
69. Pickel, V. M., Joh, T. H., and Reis, D. J., 1976, *J. Histochem. Cytochem.* **24**:792–806.

70. Laduron, P., and Belpaire, F., 1968, *Nature* **217**:1155–1156.
71. Coyle, J. T., and Wooten, G. F., 1972, *Brain Res.* **44**:701–705.
72. Oesch, F., Otten, V., and Thoenen, H., 1973, *J. Neurochem.* **20**:1691–1706.
73. Fillenz, M., Gagnon, C., Stoeckel, K., and Thoenen, H., 1976, *Brain Res.* **114**:293–304.
74. Brimijoin, S., 1975, *J. Neurobiol.* **6**:379–394.
75. Brimijoin, S., and Helland, L., 1976, *Brain Res.* **102**:217–228.
76. Nagatsu, I., Kondo, Y., Kato, T., and Nagatsu, T., 1976, *Brain Res.* **116**:277–285.
77. Livett, B. G., Geffen, L. B., and Austin, L., 1968, *J. Neurochem.* **15**:931–939.
78. De Potter, W. P., and Chubb, I. W., 1971, *Biochem. J.,* **125**:375–376.
79. Thoenen, H., Kettler, R., Burkard, W., and Saner, A., 1971, *Naunyn Schmiedebergs Arch. Pharmakol.* **270**:146–160.
80. Thoenen, H., 1972, *Biochem. Soc. Symp.* **36**:3–15.
81. Raine, A. E. G., and Chubb, I. W., 1977, *Nature* **267**:265–267.
82. Otten, U., and Thoenen, H., 1976, *Naunyn Schmiedebergs Arch. Pharmakol* **292**:153–159.
83. Reis, D. J., Joh, T. H., and Ross, R. A., 1975, *J. Pharmacol. Exp. Ther.* **193**:775–784.
84. Joh, T. H., Ross, R. A., and Reis, D. J., 1976, *Fed. Proc.* **35**:485.
85. Thoenen, H., and Oesch, F., 1973, *New Concepts in Neurotransmitter Regulation* (A. J. Mandell, ed.), Plenum Press, New York, pp. 33–51.
86. Reis, D. J., Ross, R. A., Pickel, V. M., and Joh, T. H., 1977, *Neurotransmitter Function: Basic and Clinical Aspects* (W. S. Fields, ed.), Symposia Specialists, Chicago, pp. 143–161.
87. Fredholm, B. B., Fried, G., and Hedqvist, P., 1981, *Eur. J. Pharmacol.* **79**:233–243.
88. Smith, A. D., De Potter, W. P., Moerman, E. J., and De Schaepdryver, A. F., 1970, *Tissue Cell* **2**:547–568.
89. Weinshilboum, R. M., Thoa, N. B., Johnson, D. G., Kopin, I. J., and Axelrod, J., 1971, *Science* **174**:1349–1351.
90. Fillenz, M., and West, D. P., 1974, *J. Neurochem.* **23**:411–416.
91. Brodde, O.-E., Hubermann, K., and Schumann, H. T., 1976, *J. Neurochem.* **27**:433–438.
92. Lees, G. J., Geffen, L. B., and Rush, R. A., 1981, *Neurosci. Lett.* **22**:115–118.
93. Geffen, L. B., Hunter, C., and Rush, R. A., 1969, *J. Neurochem.* **16**:469–474.
94. Font, C., Araneda, S., Pujol, J. F., Jouvet, M., and Bobillier, P., 1980, *Neurosci. Lett. [Suppl.]* **5**:S192.
95. Weinshilboum, R., and Axelrod, J., 1971, *Circ. Res.* **28**:307–315.
96. De Potter, W. P., Chank, C. P-H., De Smet, F., and De Schaepdryver, A. F., 1976, *Neuroscience* **1**:523–529.
97. Greenfield, S. A., and Smith, A. D., 1979, *Brain Res.* **177**:445–459.
98. Blakeley, A. G. H., and Cunnane, T. C., 1979, *J. Physiol. (Lond.)* **296**:85–96.
99. Raiteri, M., del Carmine, R., Bertollini, A., and Levi, G., 1977, *Mol. Pharmacol.* **13**:746–758.
100. West, D. P., and Fillenz, M., 1981, *J. Neurochem.* **37**:1052–1053.
101. De Potter, W. P., Chubb, I. W., Put, A., and De Schaepdryver, A. F., 1971, *Arch. Int. Pharmacodyn. Ther.* **193**:191–197.
102. Langer, S. Z., 1979, *The Release of Catecholamines from Adrenergic Neurons* (D. M. Paton, ed.), Pergamon Press, London, pp. 59–85.
103. Lokhandwala, M. F., 1979, *Life Sci.* **24**:1823–1832.
104. Arbilla, S., and Langer, S. Z., 1979, *Naunyn Schmiedebergs Arch. Pharmacol.* **306**:161–168.
105. Bowery, N. G., Hill, D. R., Hudson, A. L., Doble, A., Middlemiss, D. N., Shaw, J., and Turnbull, M., 1980, *Nature* **283**:92–94.
106. De Langen, C. D. J., and Mulder, A. H., 1980, *Brain Res.* **185**:399–408.
107. Hedqvist, P., 1976, *Br. J. Pharmacol.* **58**:599–603.
108. Gothert, M., and Wehking, E., 1980, *Experientia* **36**:239–240.
109. Reisine, T., 1981, *Neuroscience* **6**:1471–1502.
110. Stanford, S. C., and Nutt, D., 1982, *Neuroscience* **7**:1753–1757.
111. Salzman, P. M., and Roth, R. H., 1980, *J. Pharmacol. Exp. Ther.* **212**:64–73.
112. Lewander, T., Joh, T. H., and Reis, D. J., 1977, *J. Pharmacol. Exp. Ther.* **200**:523–534.
113. McRae Degueurce, A., 1980, Ph.D. Thesis, Lyon.
114. Dairman, W., and Udenfriend, S., 1970, *Mol. Pharmacol.* **6**:350–356.

115. Boarder, M. R., and Fillenz, M., 1979, *Biochem. Pharmacol.* **28**:1675–1677.
116. Abou-Donia, M. M., and Viveros, O. H., 1981, *Proc. Natl. Acad. Sci. U.S.A.* **78**:2703–2706.
117. Acheson, A. L., Zigmond, M. J., and Stricker, E. M., 1980, *Science* **207**:537–540.
118. Zigmond, M. J., Acheson, A. L., Chiodo, L. A., and Stricker, E. M., 1981, *Neurosci. Abstr.* **7**:149.
119. Costa, M., and Eränkö, O., 1974, *Histochem. J.* **6**:35–53.

Dopamine

Bertha Kalifon Madras

1. INTRODUCTION

Of all the neurotransmitter systems in the brain, dopamine (DA) has been the most widely explored. The reason becomes apparent when one considers the evidence for dopamine involvement in neurological,[1] psychiatric,[2] and endocrinological function.[3]

Dopamine research can be arbitrarily sectioned into two periods, the neurotransmitter phase (1957 to the present) and the receptor phase (1972 to the present). The first phase was prompted, in part, by the discovery of a dopamine deficit in the basal ganglia of deceased Parkinson patients[4] and the dramatic improvement in these patients following dopamine precursor administration.[5] Consequently, a broadly based exploration of the synthesis, metabolism, storage, release, reuptake, turnover, and mapping of dopaminergic pathways and levels in the brain was undertaken. Because of space limitations and the availability of excellent reviews of these topics,[6-8] only a brief overview and update of these subjects are given in Sections 2–3.

Impetus for the receptor phase came initially from indirect evidence linking the dopamine receptor with the target of antipsychotic (neuroleptic) medication for schizophrenia.[9-11] Prior to dopamine receptor identification, the site of action of these drugs was implied by neuroleptic-enhanced turnover of dopamine,[10] by neuroleptic inhibition of electrophysiological responses to dopamine,[12] and by the opposing roles of dopamine and neuroleptics on prolactin secretion.[13] In 1972, Greengard's laboratory detected a dopamine-stimulated adenylate cyclase in the brain.[14] Although it seemed an appropriate biochemical target for dopamine/neuroleptic interaction,[15] neuroleptic inhibition of the cyclase did not correlate with neuroleptic doses used to elicit clinical or behavioral responses.

Another approach proved to be more fruitful: in 1974, Dr. Philip Seeman convinced I.R.E. Belgique to tritiate the neuroleptic drug haloperidol as a probe

Bertha Kalifon Madras • Psychopharmacology Section, Clarke Institute of Psychiatry, Toronto, Ontario M5T 1R8, Canada.

for dopamine receptors. Shortly thereafter, an excellent correlation between neuroleptic drug potency for displacing [^{3}H]haloperidol bound to striatal membranes and clinical doses of neuroleptics was demonstrated.[16] After these landmark reports, [16,17] dopamine receptors became the subject of intensive investigation. As more tritiated ligands for labeling of dopamine receptors became available, discrepancies among the binding properties of these ligands emerged. To accommodate these observations, multiple types of dopamine receptors and states have been proposed. The functional relevance of the various receptor subtypes detected *in vitro* has yet to be established.

It is the receptor sites for dopamine, dopaminemimetics, and dopamine antagonists that is the focus of this chapter. Readers are also referred to a comprehensive review of dopamine receptors for additional information.[18]

2. DOPAMINE

2.1. Dopamine Distribution

Dopamine is widely distributed throughout the animal kingdom. It is found in most of the major phyla of invertebrates, where it functions as a neurotransmitter.[19–21] In lower vertebrates, high levels of dopamine are detectable in the nucleus basilus, hypothalamus, and retina. In mammals, dopamine distribution is different from that of norepinephrine and constitutes more than 50% of brain catecholamine content. It is concentrated mainly in the corpus striatum and the nucleus accumbens septi of the brain.[22] It is also found extensively in peripheral tissues, where it may function as a neurotransmitter.[23]

2.1.1. Analysis of Dopamine Levels

Detection of brain dopamine by Weil-Malherbe and Bone[24] and Carlsson *et al.*[25] was the primary step in unraveling its diversified functions. Major milestones achieved by development of a biochemical assay for dopamine included: (1) predictions of dopaminergic transmission in various brain regions; (2) a rationale for treatment of Parkinson patients[26]; (3) the feedback increase of dopamine turnover, which defined dopaminergic systems as the target of neuroleptic drugs; and (4) mapping of the major pathways of the brain by combining biochemical analysis, lesions, and histofluorescence techniques.[27]

As dopamine also serves as a precursor of norepinephrine (NE, see Section 2.4), detection of dopamine in a given region is not adequate proof of dopaminergic innervation. In general, a high ratio of dopamine/norepinephrine concentrations[23,28] combined with other functional indices (electrophysiological activity, storage, release, uptake, detection of dopamine metabolites[29]) support a neurotransmitter role.

The classical two-step procedure for assaying nanogram quantities of dopamine required an initial separation with an ion-exchange resin[25] followed by a chemical reaction to form a fluorescent product.[30] The assay was refined by automation[31] and by development of techniques to detect metabolites in the same sample.[32] It is now feasible to measure picogram to milligram quantities of dopamine and its metabolites by gas chromatography–mass spectroscopy,[33] radioenzymatic assays,[34] and electrochemical detection following high-pressure liquid chromatography.[35-38] The latter method is a recent development and is considered the current method of choice.

2.1.2. Regional Distribution

The rank order of dopamine distribution in major brain areas are the basal ganglia and nucleus accumbens > hypothalamus > amygdala > septum > mesencephalon > medulla > cortex > thalamus > hippocampus.[39,40] However, some nuclei of dopamine-poor regions have high dopamine levels. Dopamine content of the highest regions (caudate–putamen, olfactory tubercle, nucleus accumbens) is of the order of 85–100 ng/mg protein.

2.1.3. Brain Dopamine Pathways

The organization of central dopamine pathways has mainly been revealed with fluorescence histochemical techniques supplemented by other methods.[41-43] Brain dopaminergic systems can be divided into four major pathways, of which three are of long length: nigrostriatal, mesolimbic, mesocortical. The tuberoinfundibular pathway is of intermediate length, and both the retina and olfactory bulb have localized connections.

The nigrostriatal dopamine pathway originates in the substantia nigra (pars compacta) A9 and projects to the caudate–putamen.[44,45] In the striatum, the dopaminergic terminals fan out in an even distribution pattern. The dendrites of the substantia nigra also store and release dopamine, implying a neurotransmitter role for dopamine in the nigra.

Cell bodies of the mesolimbic system originate in the medial ventral tegmental area (A10). These send projections in the medial forebrain bundle that terminate primarily in the nucleus accumbens septi, olfactory tubercle, central nucleus of the amygdala, and lateral septal nucleus.[46,47]

Cells of the mesocortical system originate in the A10 region of the midbrain and send projections to discrete areas of the cortex, the cingulate gyrus, the entorhinal cortex, and the prefrontal cortex.[47] A9 and A10 cells also project to the hippocampus.

The tuberoinfundibular system projects from arcuate and periventricular nuclei of the hypothalamus to the external layer of the median eminence of the hypothalamus.[48] This projection also innervates the posterior and intermediate lobes of the pituitary. The incertohypothalamic system contains cell bodies in the dorsal and posterior hypothalamic region linked to terminals in the anterior hypothalamus[49] and lateral septal nuclei.

The retinal dopamine neurons, olfactory bulb neurons, and periaqueductal gray matter all have dopamine projections localized within them.

The identification of dopamine receptors within the major dopamine pathways of the brain is described in Section 5.2.

2.2. Synthesis and Storage

As dopamine does not readily cross the blood-brain barrier, it is synthesized within dopaminergic nerve terminals by the following reactions[50–52]:

$$\text{Phenylalanine} \xrightarrow[\text{O}_2,\ \text{Tetrahydrobiopterin}]{\text{Phenylalanine hydroxylase}} \text{Tyrosine} \tag{1}$$

$$\text{Tyrosine} \xrightarrow[\text{O}_2,\ \text{Tetrahydrobiopterin, Fe}^{2+}]{\text{Tyrosine hydroxylase}} \text{Dihydroxyphenylalanine (DOPA)} \tag{2}$$

$$\text{DOPA} \xrightarrow[\text{(AADC)}]{\text{DOPA decarboxylase}} \text{DA} + \text{CO}_2 \tag{3}$$

All enzymes necessary for the synthesis are found in the nerve terminals. There appears to be no difference in the synthetic pathway throughout brain dopaminergic regions.

Tyrosine hydroxylase is the rate-limiting enzyme for the sequence and is subject to regulation by at least two possible mechanisms[53,54]: (1) change in the activity and affinity of preformed enzyme and (2) alterations in the turnover of the enzyme (synthesis and/or degradation). Increased impulse flow changes the kinetic properties of the enzyme such that its affinity for tyrosine and for the pterin cofactor is enhanced and its affinity for dopamine decreases. This effect may be mediated by cyclic AMP.[55] The net result is an increase in dopamine synthesis. Modulation of proposed presynaptic dopamine receptors sensitive to dopamine may also regulate tyrosine hydroxylase.[56] These receptors presumably block or reverse the increase in tyrosine hydroxylase activity following cessation of impulse flow. Although transsynaptic induction of tyrosine hydroxylase activity has been shown in the adrenal medulla, it has not yet been demonstrated in brain dopaminergic neurons.[57] Tyrosine availability, however, can also regulate dopamine synthesis when dopamine turnover is accelerated[58,59] (see also references 60,61 for reviews and references 62,63 for assay method).

Aromatic amino acid decarboxylase (AADC) catalyzes the decarboxylation of DOPA,[64] 5-hydroxytryptophan (the precursor of serotonin), histidine, tyrosine, and tryptophan. It is found in a variety of peripheral tissues and brain.[65] Peripheral inhibitors of the enzyme such as carbidopa potentiate the effects of administered DOPA by elevating brain DOPA levels.[66] The enzyme is normally not rate limiting for dopamine synthesis.

In summary, brain dopamine levels can be regulated through the synthetic route by dopamine precursor availability, by induction or repression of tyrosine hydroxylase, or by inhibition of AADC. The newly synthesized dopamine is either released in response to an action potential or stored in synaptic vesicles. Although the concentration of dopamine in the nerve terminals remains fairly constant, the amount released varies considerably, probably because of feedback inhibition of tyrosine hydroxylase.

2.3. Release, Uptake, and Metabolism

2.3.1. Release

Release of dopamine both *in vivo* and *in vitro* has been extensively studied in the striatum. *In vitro* release studies, measured in slices or in synaptosomal preparations, are useful for delineating the basic mechanisms involved. *In vivo* experiments offer insight into interneuronal regulation of release. Spontaneous or calcium-dependent release of [^{3}H]dopamine from striatal slices is reduced considerably by tetrodotoxin and is enhanced by K^+ [67] or by amphetamine.[68] Dopamine agonists inhibit dopamine release, suggesting the existence of presynaptic dopamine receptors that regulate dopaminergic transmission.[69] Neuroleptics increase dopamine release.

Other neurotransmitters and/or neuromodulators in the striatum (glycine) stimulate, whereas serotonin and compounds with opiate activity inhibit dopamine release.[70–72] The interaction of dopaminergic neurons with other neurotransmitters in the striatum consequently could influence dopamine activity.

In vivo dopamine release (measured by collection of dopamine in perfusates or superfusates) is also stimulated by K^+ [73] or amphetamine (applied either locally or intravenously).

2.3.2. Uptake

Dopamine is removed from the synaptic cleft by an active uptake mechanism that transports dopamine back to the presynaptic terminals. The system has been characterized in most dopaminergic brain regions including the corpus striatum.[74] Uptake is temperature and Na^+ dependent and displays Michaelis–Menten kinetics. Differentiating between a dopamine releaser and an uptake blocker can be a challenging problem.[75]

2.3.3. Metabolism

Whether dopamine is returned to the presynaptic terminal or taken up by the postsynaptic cell or an extraneuronal site (glia, blood, capillaries), the final products of dopamine metabolism are dihydroxyphenylacetic acid (DOPAC) and homovanillic acid (HVA). The main route for dopamine metabolism is reuptake into the presynapse. Dopamine either reenters the synaptic vesicle or is metabolized by mitochondrial monoamine oxidase (MAO). Both DOPAC and HVA are detectable in brain, and HVA (CSF or plasma) is usually measured as an index of dopamine turnover.[76] 3-Methoxytyramine is increasingly recognized as a better indicator of dopamine release into the synaptic cleft than DOPAC or HVA.[77]

3. DOPAMINE FUNCTION

3.1. Psychiatric, Neurological, and Endocrinological Function

As many functions of dopamine receptors were revealed by dopamine agonist and neuroleptic drug responses, this section describes primarily dopaminergic drug effects. The results of numerous biochemical and pharmacological studies support the hypothesis that the antipsychotic, endocrinological, and extrapyramidal side effects of neuroleptic drugs are related to dopamine receptor blockade, possibly in striatal, limbic, cortical, and pituitary regions.[18]

Until binding assays were developed in 1974, the evidence was indirect but convincing. The term neuroleptic was introduced in 1955 to describe a profile of responses to chlorpromazine in schizophrenic patients. Agitated behavior, aggression, and florid symptoms of the disorder, including thought disorder, confusion, and hallucinations, were alleviated within days or weeks after onset of treatment. Neurological manifestations of drug treatment, especially a Parkinsonlike muscular rigidity and movement reduction, were also detectable. Thus, while reducing many of the psychiatric symptoms of schizophrenia, neuroleptics induced neurological side effects characteristic of dopamine underactivity. Because of the success of these drugs, a wide range of neuroleptic compounds were synthesized for clinical and research purposes.

Most of the prescribed medication for schizophrenics falls into four classes: butyrophenones (e.g., haloperidol), phenothiazines (e.g., chlorpromazine), thioxanthenes (e.g., thiothixene), and benzamides (e.g., sulpiride). This large series of neuroloptics of different chemical classes and potencies but all possessing comparable clinical efficacies provided an excellent research strategy for determining the molecular targets of neuroleptic action. Early pharmacological studies strongly suggested that dopamine receptor blockade was the primary action of neuroleptic drugs.[78] Catecholamine turnover, but more specifically dopamine turnover, was accelerated by these drugs.[78–80] Increased synthesis and utilization of dopamine were considered to be a compensatory response to dopamine receptor blockade. Furthermore, neuroleptics blocked hyperlocomotion and stereotypy induced by dopaminemimetic drugs such as apomorphine or amphetamine, again implicating the dopamine receptors as the primary target.[81,82] Neuroleptic-induced catalepsy in animals and a Parkinsonlike state in humans further consolidated the dopamine hypothesis of neuroleptic drug action. Reduced dopaminergic function consequent to either dopamine blockade or dopamine deficiency (parkinsonian brains) would predictably produce similar neurological profiles.

Two compounds that augment brain dopamine transmission, L-DOPA, the precursor of dopamine, and amphetamine, exacerbate the symptoms of schizophrenia and induce psychotic symptoms in nonschizophrenic patients.[83,84] Thus, all the indirect evidence was summarized in a dopamine hypothesis of schizophrenia (or neuroleptic action); excess dopamine activity in schizophrenic brains, is suppressed by neuroleptic blockade of dopamine receptors. With the development of the direct binding assay for dopamine receptors and the excellent correlation observed between binding potency of neuroleptics and clinical doses (see Section 4), the theory gained wide acceptance. Although no experimental data reveal increased dopamine and/or metabolites in unmedicated schizophrenic brains, increased numbers of dopamine receptors have been detected in postmortem brains (see Section 6) and may provide the rationale for excessive dopamine function.

Conversely, reduced dopaminergic function, either as a result of a dopamine deficit or neuroleptic-induced dopamine receptor blockade in the striatum, induces Parkinsonlike muscular rigidity and other extrapyramidal symptoms (tremor, akinesia). The critical role of a dopamine deficit in the development of Parkinson's disease can be regarded as established. There is a direct relationship between striatal dopamine deficiency and the major extrapyramidal symptoms of the disease. As dopamine does not readily penetrate into the brain, the precursor L-DOPA (which crosses the blood–brain barrier) is the most effective treatment for the symptoms. Other dopaminergic agonists such as bromocriptine are also used clinically.

Neurological complications of neuroleptic medication are time dependent. The initial muscular rigidity and akinesia, which resemble Parkinson's disease and respond to drug treatment, eventually diminish and are followed by the development of tardive dyskinesia in 40–50% of patients after prolonged medication. Tardive dyskinesia is characterized by abnormal repetitive movements of facial and limb musculature.

The release of several pituitary hormones (anterior, intermediate, and posterior lobes) may be regulated by the hypothalamic–pituitary dopamine pathways. Released from the median eminence and pituitary stalk, dopamine suppresses prolactin secretion in humans and animals.[85] Other agonists produce comparable results, whereas antagonists (neuroleptics) stimulate prolactin release. Dopamine agonists inhibit the release of β-endorphin and α-melanophore-stimulating hormone (α-MSH) from the intermediate lobe. Dopamine may also regulate the release of oxytocin from the posterior pituitary.

3.2. Dopaminergic Behavior in Animals

Changes in dopamine neurotransmission in animals induce a wide range of behavioral changes from akinesia to hyperactivity and stereotypy. These dopaminergic behaviors have been extensively used for drug-screening programs.

Hyperactivity can be increased by high doses of dopamine precursors and agonists apomorphine and with procedures that increase dopamine receptor numbers (see Section 6). Large doses of L-DOPA produce locomotor stimulation and involuntary oral–facial and limb movements in primates,[86] analogous to tardive dyskinesia. Both apomorphine, a widely used dopamine agonist, and amphetamine, a dopamine-releasing drug, induce a variety of characteristic behaviors in rodents including stereotypy, locomotion, sniffing, biting, licking, and gnawing. These behaviors can be attributed to the striatum (stereotypy), nucleus accumbens (locomotion), and olfactory tubercle (sniffing, repetitive movements of limbs). Apomorphine is also a powerful emetic in animals that have a chemoreceptive trigger zone (dogs), and neuroleptics are potent antiemetics, indicating that emesis is, in part, a dopamine-mediated behavior (area postrema[87]).

Unilateral intranigral injection of 6-hydroxydopamine destroys dopaminergic innervation of the striatum, causing supersensitivity of dopamine receptors on the lesioned side.[88,89] Administration of a dopamine agonist causes turning away from the lesioned side; amphetamine, which can only release dopamine from the unlesioned side, causes turning toward the lesioned side.[90]

Neuroleptics antagonize the behavioral responses to dopamine agonist drugs and induce catalepsy and decrease avoidance behavior in untreated animals.

Correlations between binding potency and drug potency for induction or reduction of dopaminergic behaviors have helped to establish the pharmacological relevance of the various dopamine receptors or binding sites detected *in vitro* (see Section 4.5).

3.3. Neurophysiological Activity

Although the precise electrophysiological role of dopamine in the CNS remains unsettled, numerous studies suggest an inhibitory role for dopamine in the striatum.[91] The response evoked by dopamine is of slow onset and long duration. Dopamine depresses spontaneous activity, excitation induced by iontophoretically applied excitatory amino acids, and stimulation-evoked activity.[92,93] Neuroleptic drugs block dopamine-induced inhibition of striatal cells.[94]

4. DOPAMINE RECEPTOR IDENTIFICATION IN VITRO

This section is a review of the literature on the use of direct binding assays to localize the molecular targets of dopamine agonist and antagonist drugs. The radioreceptor assay can be of value for identifying (1) the primary site of action of dopaminergic drugs, (2) secondary sites that may produce side effects, (3) pathological levels of receptors, and (4) long-term receptor changes induced by drugs and (5) for predicting the topography of the receptor.[18]

4.1. Terminology

Sufficient evidence has accumulated to postulate the existence of multiple dopamine receptors,[95] although an accepted classification system has yet to be devised. The electrophysiological response evoked by dopamine, dopamine linkage to adenylate cyclase, and the binding properties of dopamine receptors *in vitro* have all been suggested as a basis for receptor classification.

The electrophysiological responses evoked by dopamine are as yet controversial, and in some organs such as pituitary, dopamine is neither excitatory nor inhibitory. Linkage to adenylate cyclase stimulation was based on the observation that receptors identified in certain tissues were linked (parathyroid gland, etc.[96]), whereas receptors found in other tissues were not linked to stimulation (pituitary, a proportion of striatum receptors[95]). There is no consensus on whether dopamine-stimulated or -inhibited cyclase exists in various tissues, especially since tissue homogenization can disrupt the linkage.

Finally, *in vitro* binding characteristics of agonists and antagonists have proven to be useful for receptor classification. D_1 stimulated adenylate cyclase is responsive to micromolar concentrations of both dopamine agonists and antagonists; D_2 receptors (dopamine/neuroleptic receptors) are sensitive to nanomolar concentrations of dopamine antagonists but to variable levels of dopamine agonists; D_3 binding sites (dopamine agonist sites) are sensitive to nanomolar concentrations of dopamine agonists but to micromolar levels of dopamine antagonists. Another site, designated the D_4,[97,98] has been postulated and is sensitive to both dopamine agonists and antagonists in the nanomolar range. This site may represent a high-affinity state (D_{2H}) of the D_2 receptor.

The advantage of this classification system is its simplicity. It is based only on binding characteristics and not on complex behavioral or electrophysiological parameters. The disadvantage of the system is that certain binding affinities, especially those of agonists, are variable *in vitro* and may respond to modulators in the assay medium (e.g., GTP, Mg^{2+}, Na^+, etc.). The various types of dopamine receptors are summarized in Table I.

4.2. Tritiated Ligands for Receptors

Selection of appropriate ligands and conditions for binding is essential for acquisition of valid binding data. As outlined in the previous section, several subtypes of dopamine receptors have been proposed. Unfortunately, there are no ligands currently available that are specific for a single dopamine receptor. Rather, specificity *in vitro* is determined by the rank order of drugs that displace the [^{3}H]-labeled ligand. In general, greater selectivity is obtained with antagonists than with agonists; the latter bind to all dopamine receptors or sites, as would be expected. The ligands used for each site are outlined in individual sections.

4.3. Criteria for Receptor Identification

Both dopamine agonists and neuroleptics (antagonists) bind to a variety of specific and nonspecific sites in the brain, which may include proteins (uptake sites, other receptors) and inorganic compounds. The highly fat-soluble neuroleptics are also sequestered into lipid membranes at high concentrations. It becomes essential, therefore, to develop criteria that establish a binding site as a pharmacologically relevant receptor site.

4.3.1. Saturability

As receptors have limited capacity (the maximal number of binding sites, B^{max}, is usually less than 1000 fmol/mg protein), the specific sites should be saturable. Unfortunately, binding can be nonspecific and nonsaturable; this is usually defined as binding that occurs in the presence of an excess of (100- to 10,000-fold) nonradioactive ligand. Sites that are saturable but nonspecific are those that are labeled by [^{3}H]-labeled ligand and displaced in a saturable manner by the unlabeled ligand but not by a close congener. Thus, [^{3}H]spiperone will bind to saturable spirodecanone sites that are not displaceable by other pharmacologically active butyrophenones.[99]

Specific saturable sites are measured by exposing tissue to increasing concentrations of [^{3}H]-labeled ligand. Parallel tubes that contain an excess of unlabeled close congener are also included in the assay to measure nonspecific binding. Scatchard analysis of the data yields values for both the B^{max} and K_d.

Table I
Dopaminergic Receptors and Sites

	D_1	D_2 (low)	D_2 (high)(D_4?)	D_3 (D_1 high?)
Pharmacology				
Dopamine affinity	μM	μM	nM	nM
Butyrophenone affinity	μM	nM	nM	μM
Apomorphine	Partial agonist or antagonist	Agonist	Agonist	?
Ergots (dopaminergic)	Antagonist or partial agonist	Agonist		?
Tritiated ligands				
Antagonists	[^{3}H]*cis*-Flupenthixol[a]	[^{3}H]Spiperone[b]	[^{3}H]Spiperone	?
Agonists	[^{3}H]DA	[^{3}H]Ergots [^{3}H]DA[d] [^{3}H]NPA[e]		[^{3}H]DA[b]
Adenylate cyclase				
Stimulates	Yes	No	No	?
Inhibits	No	Yes	Yes (pituitary)	?
GTP sensitivity	Yes	Yes	Yes	?
Function	Parathyroid: Hormone release; CNS:?	CNS: DA-mediated responses	Pituitary: Hormone release	CNS: Autoreceptors?
Location				
Striatum	Intrinsic neurons	Corticostriatal tract (also D_2 high?)	Intrinsic neurons (also D_2 low?)	Intrinsic nigrostriatal tract
Substantia nigra	Striatonigral tract	Intrinsic neurons	Intrinsic neurons(?)	Intrinsic neurons
Area A10	No	Yes	?	?
Frontal cortex	Yes	Discrete areas	?	No
Retina	Yes	?	?	?
Pituitary	No(?)	Yes	Yes	No

[a] Specific if spiperone (30 nM) is used to block D_2 sites.
[b] Others include ^{3}H-sulpiride, ^{3}H-haloperidol.
[c] May label high-affinity state of D_1. Dopamine (DA).
[d] Labels D_2 only in presence of Mg^{2+}.
[e] Labels several sites. N-propylnorapomorphine (NPA).
[f] Evidence based on release of cyclic AMP by striate slices.

4.3.2. Stereoselectivity

Although neuroleptics are lipid and thus membrane soluble, the partition coefficients for lipid solubility are the same for isomeric pairs. Stereospecific displacement of [^{3}H]neuroleptic suggests that the radioligand is bound to a stereoselective site, not to a nonspecific site. The various subtypes of dopamine receptors display different stereoselectivity for ligands. The most widely used stereoisomeric pairs are ($\pm$)butaclamol and $\pm$ sulpiride. ($+$) Butaclamol is active whereas ($-$)butaclamol is not.

4.3.3. Drug Binding Profiles

Many neuroleptics are detectable in plasma water at concentrations of 1–100 nM.[18] Receptor sites detected *in vitro* should respond to neuroleptics in this range. The D_2 receptors bind neuroleptics in the appropriate concentration range. Saturation of the D_1 and D_3 sites requires higher neuroleptic concentrations than are detectable in plasma water, concentrations disproportionate to dose.

The binding profiles of the receptor should fulfill two criteria. (1) Of all the neurotransmitters identified, dopamine should be the most potent displacer of the radioligand; in brain regions where dopamine does not predominate as a neurotransmitter (frontal cortex, hippocampus), other neurotransmitters may be more potent, e.g., serotonin. (2) The drugs that displace the [^{3}H]-labeled ligand should do so in an order of potency that correlates with *in vivo* potency for eliciting clinical and behavioral responses.

4.3.4. Regional Distribution

The distribution of dopamine is uneven and highly localized in some brain regions (Section 2.1). The distribution of the receptors should correspond to those regions in the brain in which dopamine levels are high and where dopamine serves as neurotransmitter.

4.4. Properties of Dopamine Receptors

4.4.1. Dopamine-Stimulated Adenylate Cyclase

The D_1 receptor (dopamine-stimulated adenylate cyclase) was the first to be identified in the brain[14]. Because dopamine-stimulated adenylate cyclase was inhibited by neuroleptics, it was proposed to be the site of action of these drugs.[15] Subsequent detailed analysis of the binding suggested that the D_2 receptor was a more likely target of antipsychotic medication.

4.4.1a. Assay Conditions. The D_1 site is labeled dopaminergic because adenylate cyclase is activated primarily by dopamine and to a lesser extent by norepinephrine. It is identified by assaying stimulation of adenylate cyclase,[100] by measuring release of cyclic AMP by slices,[101] or by labeling the site with [^{3}H]*cis*-flupenthixol or [^{3}H]piflutixol.[102,103] The IC_{50} values for neuroleptic displacement of [^{3}H]*cis*-flupenthixol and inhibition of dopamine-stimulated adenylate cyclase correlate very highly.[104] The density of [^{3}H]flupenthixol binding sites is three to four times greater than the density detected by [^{3}H]butyrophenones (D_2 sites). Several active neuroleptics that are ineffective at inhibiting dopamine-stimulated cyclase are also ineffective at inhibiting the binding of [^{3}H]*cis*-flupenthixol, further supporting the use of this ligand to label the D_1 site. However, *cis*-flupenthixol also binds to the D_2 site (20%); blocking with spiperone improves the specificity of the D_1 binding assay.[103] High nonspecific binding is encountered with D_1^{3}H-ligands. The rate of disappearance of [^{3}H]flupenthixol binding after kainic acid lesions resembles the rate of disappearance of dopamine-stimulated adenylate cyclase.[105,106]

4.4.1b. Pharmacology of the Site. The affinity of the D_1 site for dopamine and other dopaminergic agonists is in the low micromolar range. The concentration required for half-maximal stimulation of the site is of the order of 1000–5000 nm. Other dopaminergic agonists also stimulate the site, with the rigid analogue of dopamine 2-amino-6,7-dihydroxy-1,2,3,4-tetrahydronaphthalene [(+)6,7-ADTN] being more potent than apomorphine or (−)6,7-ADTN.[107] Apomorphine is a partial agonist on the site.[108] Although the ergot alkaloid bromocriptine is a potent dopamine agonist (inhibits prolactin secretion and serves as a dopamine replacement in Parkinsonism), it inhibits adenylate cyclase in the striatum.[109,110] Other alkaloids are not potent agonists. This is additional evidence that the D_1 site linkage to dopaminergic functions in the CNS remains unknown.

Neuroleptic concentrations required to inhibit dopamine stimulation of adenylate cyclase are considerably higher than the concentrations found in the plasma water of patients. Furthermore, there is no correlation between the clinical doses of neuroleptics and the concentrations required to inhibit the D_1 site.[18] For example, the butyrophenones, which are very potent neuroleptics, are weak inhibitors of dopamine-stimulated adenylate cyclase. Others, such as the phenothiazines, which are used in considerably higher doses, are quite potent. Some of the atypical but clinically effective neuroleptics, notably sulpiride, metoclopramide, and sultopride, are practically inactive at the D_1 site.[102,104]

4.4.1c. Regional and Subcellular Distribution. Dopamine-stimulated adenylate cyclase is found in the olfactory tubercle, striatum, nucleus accumbens, substantia nigra, frontal cortex, retina, neuroblastoma cells, superior cervical ganglia, and peripheral tissues.[111–116]

Lesions of nerve tracts and cell bodies indicate that all D_1 sites are located postsynaptic to dopamine-containing neurons. All receptors in the caudate nucleus are obliterated within weeks if the cell bodies are treated with kainic acid.[105,106] In the substantia nigra, lesions of the cell bodies result in no reduction of the D_1 sites, indicating that these sites are located on terminals of neurons originating elsewhere.[117]

The subcellular distribution of the D_1 site is different from that of the D_2 site.[111] The D_1 receptor is associated mainly with particles that sediment in the mitochondrial fraction, whereas the other binding site (D_2) is also located in the microsomal pellet.

4.4.1d. Influence of GTP and Ions. GTP is usually obligatory for the regulation of adenylate cyclase, and striatal adenylate cyclase is activated by GTP and Mg^{2+}.[118] Recently, ascorbic acid has been recognized as a potent inhibitor of D_1 and D_2 receptor binding.[119–123] As ascorbate is generally used to protect dopamine from oxidation in solutions, its use has to be reassessed. The mechanism for ascorbate destruction of binding is not known but may involve lipid peroxidation.[123] D_1 sites have been extensively reviewed.[15,95,124]

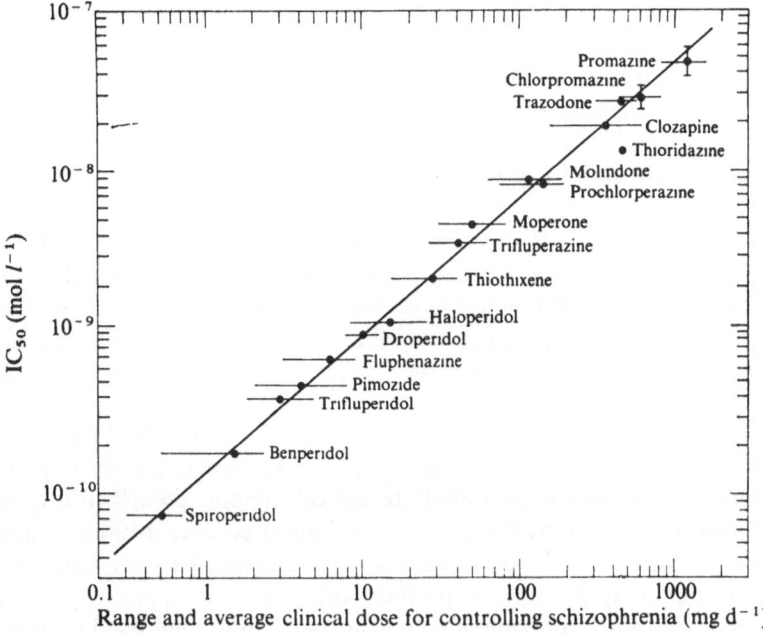

Fig. 1. The average clinical doses of neuroleptics for controlling subacute schizophrenia correlate with the IC_{50} values on D_2 dopamine receptors (neuroleptic concentrations that inhibit 50% of specific [^{3}H]haloperidol binding to membranes prepared from calf caudate). (Reprinted from Seeman *et al.*,[16] with permission.)

4.4.2. D_2 Dopamine/Neuroleptic Receptors

In contrast to the D_1 site, O_2 receptors are not linked to stimulation of adenylate cyclase. Furthermore O_2 but not O_1 sites have been identified in the pituitary, where they may inhibit hormone release by *suppressing* adenylate cyclase activity.

Using [^{3}H]haloperidol and stereoisomers of an active [(+)-butaclamol] and inactive [(−)-butaclamol] neuroleptic, Seeman's laboratory and Snyder's demonstrated stereospecific binding of [^{3}H]haloperidol to striatal tissue.[16,17] Within a year, the clinical significance of the site became apparent. All neuroleptics, regardless of chemical class or structure, displaced [^{3}H]haloperidol from binding sites with an order of potency that paralleled clinical dosage (Fig. 1). Furthermore, these receptors were elevated in postmortem schizophrenic brains.[125,126] It is increasingly apparent that although the concentrations of neuroleptics in plasma are sufficiently high for occupancy of several receptors [D_1, D_2, D_3, α-adrenergic, serotonin (5-HT_2), histamine (H_1), muscarinic cholinergic], only drug binding potency to the D_2 receptor correlates with clinical potency. There is general agreement that the D_2 receptor is the therapeutic and behavioral site of action of neuroleptics. Binding to other receptors may account in part for the side effects of these drugs.

The D_2 receptor is characterized by having low-nanomolar affinity for dopamine antagonists (neuroleptics). Affinity for dopamine agonists is variable and may range from the low-nanomolar to micromolar if the receptor converts from a high- to low-affinity state (D_{2H}, D_{2L}).

The biochemical consequences of receptor activation remain unknown. Recent reports suggest that the D_2 receptor may be linked to inhibition of adenylate cyclase.[101] The adenylate cyclase stimulated by the D_1 receptor could thereby be under inhibitory regulation of the D_2 receptor.

4.4.2a. Assay of D_2 Receptors. The receptor can be labeled by [³H]-labeled antagonists such as haloperidol, spiperone, domperidone, pimozide, sulpiride, and tiapride or by [³H]-labeled agonists (bromocriptine, pergolide, lisuride, dopamine, N-propylnorapomorphine) under appropriate conditions. Although haloperidol is a specific ligand for the D_2 site, its general use has been replaced by [³H]spiperone.

The main advantage of this "ligand of choice" is its very high affinity (K_d 0.06–0.2 nM) for the D_2 site. Although spiperone labels mainly dopamine receptors in the striatum it also binds to spirodecanone sites,[99] to α-adrenergic sites,[127] and to serotonergic sites. In the frontal cortex, it binds primarily to serotonergic sites.[128–130] The advantages of increased affinity and slow dissociation rates of spiperone are partially offset by its lipophilicity and lack of specificity. Therefore, unlabeled spiperone is generally not used to define non-specific binding. Instead, (+)-butaclamol is the traditional base line because it selects for both neuroleptic and stereospecific sites[131]; it also binds to other sites in striatal membranes.[132] Sulpiride[133] and ADTN[134] have also been suggested as base-line compounds. The binding of [³H]haloperidol (K_d 2–3 nM) and [³H]spiperone is parallel in various brain regions, although the latter is a better choice for *in vivo* studies.

Other neuroleptics have been used to label the D_2 receptor, but most present problems of low specific binding, binding to multiple sites, and rapid dissociation rates. [³H]Domperidone, which does not cross the blood–brain barrier, has been proposed as a highly specific D_2 ligand.[135] The K_d of [³H]domperidone is similar to that of [³H]spiperone, although the B^{max} is less. This lower density may reflect more selective binding of [³H]domperidone.[136]

More recently, [³H]sulpiride, a benzamide, has been introduced as a D_2 ligand.[137] Sulpiride is a unique neuroleptic: *in vivo* it is an effective antipsychotic, antiemetic, and prolactin-releasing agent; in animal models of dopamine function, sulpiride shows little of the neurological effects associated with neuroleptics.[138] Sulpiride does not inhibit dopamine-stimulated adenylate cyclase (see Section 4.4.1) in contrast to other neuroleptics, which have varying potencies on the D_1 site. Also in contrast to other neuroleptics, sulpiride and other benzamides have an absolute Na^+ requirement for displacement of [³H]spiperone.[139] Whereas *cis*-flupenthixol, a neuroleptic that preferentially labels D_1 sites, displaces [³H]sulpiride potently, molindone, which is inactive at the D_1 site but active at D_2 receptors, is a very weak displacing agent of [³H]sulpiride.[137] Until further biochemical and pharmacological studies are available, it is probably premature to assume that sulpiride labels the D_2 receptor exclusively.

Tritiated agonists can also label the D_2 site. Included in this list are [^{3}H]apomorphine, [^{3}H]N-propylnorapomorphine, and ergot alkaloids such as [^{3}H]lisuride, [^{3}H]bromocriptine, [^{3}H]pergolide, [^{3}H]dihydroergocriptine, and [^{3}H]lysergic acid diethylamide (LSD).[18] Because most of the agonists are not specific for the D_2 site, addition of drugs that block binding of the agonist to other sites improves binding specificity.

From use of [^{3}H]neuroleptics, the density of the D_2 site has been measured as 200–600 fmol/mg protein. In humans it is lower (200 fmol/mg) than in rodents (300–400) or other mammalian species. The density is calculated to be less if labeled agonists are used (100–200 fmoles/mg protein).

Assay conditions must be chosen with care. The K_d of the receptor varies with membrane concentration: the higher the protein levels, the greater the K_d, presumably because a portion of the highly fat-soluble [^{3}H]neuroleptic becomes unavailable for binding.[140] The lowest possible membrane concentration should be used to obtain accurate binding data. Receptor density and affinity may also be functions of temperature, ion concentration, and other buffer constituents (Section 4.4.2d).

4.4.2b. Pharmacology of the Site. D_2 receptors have high affinity for neuroleptics (IC_{50} 0.1–1000 nM) and variable affinity for dopamine (nM–μM) and other agonists if a simple assay medium is used. The affinity for agonist is reduced in the presence of GTP and other buffer constituents (Section 4.4.2d).[141] All neuroleptics inhibit the binding of [^{3}H]butyrophenones in the striatum regardless of chemical class or structure. The inhibition is stereoselective; active enantiomers [(+)-butaclamol] are far more potent than inactive ones.[131]

The IC_{50} values of neuroleptics for inhibition of [^{3}H]haloperidol or [^{3}H]spiperone correlate with their clinical potencies[16] and with their serum concentration in medicated schizophrenics. Several animal behaviors characteristic of dopamine receptor function are inhibited by neuroleptics in a rank order of potency that correlates with their potency in inhibiting [^{3}H]neuroleptic binding to D_2 receptors (Section 4.5).

The structural requirements for binding of a dopamine receptor agonist have been elaborated in detail by Seeman[18] and include the following: (1) a hydrogen-bonding group, which is usually OH, is most effective in the 3 position of the aromatic ring; an additional OH group on the aromatic ring increases potency; (2) the agonist possesses lipid solubility; (3) a maximum distance of 7.3 Å between the OH and the N group is optimum; and (4) the position of the N atom should be 0.6 Å from plane to ring.

The rank order of potency of neuroleptics in displacing bound [^{3}H]neuroleptic from striatal membranes varies with [^{3}H]ligand and species but generally falls into these clusters: spiperone > fluphenazine, haloperidol, (+)-butaclamol, pimozide, domperidone > chlorpromazine thioridazine, molindone, clozapine, ± sulpiride, promazine.

[^{3}H]Neuroleptics label dopaminergic sites that have variable affinities for agonists. Agonist displacement of [^{3}H]spiperone binding is at least biphasic (Hill slope < 1) (determined by computer analysis using 20 concentrations of agonist). In the presence of GTP, conversion of the high-affinity to low-affinity component occurs with consequent changes in apparent IC_{50}s. Full conversion is obtained with pituitary but not striatal receptors. The integrated rank order of potency of dopamine agonists in displacing [^{3}H]spiperone is apomorphine > ADTN > dopamine. Other catecholamines are less potent.

D_2 sites can be identified using [^{3}H-labeled agonists, but the binding is quite variable and depends on assay conditions.

[^{3}H]Dopamine binds to D_2 sites (albeit partially) only in the presence of Mg^{2+} or other divalent cations (see Section 4.4.2d).

4.4.2c. Regional, Cellular, and Subcellular Distribution. The distribution of the D_2 receptor corresponds closely to the distribution of dopaminergic neurons and to those regions in which neuroleptics are believed to produce an effect: the anterior and intermediate lobe of pituitary[141] [stimulation of prolactin and α-melanophore-stimulating hormone (α-MSH) secretion], striatum[16] (extrapyramidal movements, antipsychotic action?), nucleus accumbens[125] (reduction of hyperactivity, antipsychotic action?), the area postrema (antiemesis[142]), and discrete regions of frontal cortex (antipsychotic action?). D_2 receptors are thought to exist in postsynaptic processes in the striatum because lesions of the medial forebrain bundle invariably result in increases in receptor densities in the striatum.[143] Kainic acid destruction of the striatal cell bodies leads to a reduction but not elimination of these sites.[144] Complete elimination of striatal D_2 receptors is attained only after lesioning the corticostriatal pathways. Thus, D_2 receptors may also be located on the presynaptic axons of corticostriatal tracts.[145] Receptor densities are highest (B^{max} 200 fmol/mg protein) in dopaminergic brain regions (striatum, anterior pituitary, olfactory tubercle), lower (B^{max} 50 fmol/mg protein) in other dopamine regions (globus pallidus, cerebral cortex, hypothalamus), and lowest (B^{max} 15 fmol/mg) in thalamus, substantia nigra, hippocampus, and cerebellum.[18]

4.4.2d. Influence of Ions and Other Buffer Constituents. There is increasing evidence that D_2 receptors in the striatum, as in the pituitary, are inhibitors of cyclic AMP formation.[101] Specific D_2 agonists inhibit the efflux of cyclic AMP produced by stimulation of the D_1 receptor. The inhibition is reversed by sulpiride. GTP modulates agonist binding to the D_2 receptor in pituitary and striatum.[146–148] Micromolar concentrations of GTP, GDP, and the stable GTP analogue guanylyl-5-imidodiphosphate [Gpp(NH)p] reduce agonist potency at D_2 sites, converting high-affinity sites (D_{2H}) to low-affinity sites (D_{2L}). The Hill slope of the agonist displacement curve approaches the value of 1 as the high-affinity sites are converted to low-affinity sites (pituitary receptors). Full conversion is not attainable with striatal receptors. Heat treatment has a similar effect, possibly because of destruction of the guanine regulatory protein.[149]

[^{3}H]Spiperone binding to the D_2 receptor declines rapidly if striatal membranes are preincubated with ascorbic acid at 22°C or 38°C.[121–123] EDTA, inhibitors of lipid peroxidation, and manganese prevent the ascorbate-induced decline in binding.[121–123] Solubilized D_2 receptors are not affected by ascorbate, indicating that the effect may be indirect and on the membrane matrix of the receptor.[121,122] Peroxidation of the lipids surrounding the receptor has been proposed as the mechanism.[123,150-152] The physiological consequence of lipid peroxidation for receptor function is unknown. Administered ascorbic acid has neurolepticlike properties in rats.[123,153] Pathological states or aging, in which lipid peroxidation is enhanced (cerebrovascular accidents, high O_2 tension), may result in receptor destruction.

[^{3}H]Dopamine binding to the D_2 receptor was not successfully demonstrated until recently.[154] It is now apparent that [^{3}H]dopamine binds to at least three sites: binding to the D_1 site is both GTP and Mg^{2+} sensitive; D_2 sites can be labeled by [^{3}H]dopamine only in the presence of Mg^{2+}; D_3 sites labeled by [^{3}H]dopamine are both Mg^{2+} and GTP insensitive. Preincubation of membranes with Mg^{2+} (2 mM) enhances subsequent [^{3}H]dopamine binding to the D_2 receptors and also the D_3 sites.[154] [^{3}H]Dopamine binding to D_2 receptors, unlike the D_3 sites, is displaced stereoselectively by low concentrations of both *cis*-flupenthixol (10 nM) or (+)-butaclamol (20 nM). ([^{3}H]Dopamine bound to the D_3 sites is dislodged only by very high concentrations of neuroleptics.) GTP completely inhibits [^{3}H]dopamine binding to the D_2 receptor, further supporting the evidence that the D_2 receptor is regulated by the G protein.[154]

4.4.2e. Pituitary Receptors. The anterior and intermediate lobes of the pituitary contain a single population of D_2 receptors.[146,155-157] These receptors probably are the mediators of dopamine inhibition of pituitary hormone (prolactin, α-MSH, β-endorphin, oxytocin?) release. [^{3}H]Spiperone binding in the anterior pituitary is of high affinity, stereoselective, and saturable. The rank order of agonist and antagonist displacement of [^{3}H]neuroleptic binding correlates well with drug concentrations used to regulate prolactin release from the pituitary.[18]

The receptor may exist in two agonist states, which apparently do not select for antagonists. The proportion of the two agonist binding states (D_{2H}, D_{2L}) is determined by the level of guanine nucleotides and G regulatory protein. Agonist but not antagonist binding is completely inhibited by GTP. Higher concentrations of agonist are required for displacement of [^{3}H]spiperone binding in the presence of GTP, implying that GTP reduces the affinity of agonist for the receptor. The high-affinity state is inducible by agonist (D_{2H}) and is converted to the low-affinity state by GTP (D_{2L}).[141,147] The two states are found in similar proportions in the membrane, but the D_{2H} state is not detectable in intact cells.

4.4.2f. Assay of Receptors in Vivo. A comparison of the sites labeled by [^{3}H]haloperidol, [^{3}H]spiperone, and [^{3}H]pimozide reveals that [^{3}H-spiperone is the optimum ligand for *in vivo* labeling of the D_2 receptors.[158] [^{3}H]Haloperidol is not selectively taken up by dopamine-containing cells. [^{3}H]*n*-Propylnorapomorphine has also been shown to accumulate to saturation in dopaminergic-rich brain regions; binding is stereospecifically displaced by dopamine antagonists and agonists.[159] Administered [^{3}H]spiperone has been used for autoradiographic localization of dopamine receptors.[160]

4.4.2g. Development and Decline of D_2 Receptors. Dopamine receptors increase rapidly after birth and peak about 14–30 days later. With increasing age, the receptors decline to 60% of adult values.[161,162]

4.4.3. D_3 Binding Sites

Paradoxically, catecholamine dopamine agonists (dopamine, apomorphine, ADTN) are not the ligands of choice for labeling the D_2 receptor. Low (nanomolar) concentrations of these compounds label at least three sites in dopaminergic-rich brain regions, the most significant being a high-affinity site termed the D_3 site. The D_3 binding site is characterized by having high affinity (K_d values of 1-10 nM) for dopamine and other catecholamine agonists but low affinity and specificity for neuroleptics (K_d greater than 100 nM).[163–165] Although the physiological function of this site remains unknown, the site may function as an autoreceptor.[166] More recent evidence suggests that the sites are inducible by dopamine and thus may represent a high-affinity state of the D_1 site.[154]

4.4.3a. Assay of D_3 Sites. [^{3}H]Dopamine is probably the best ligand to use for detection of this site because at low-nanomolar concentrations it labels primarily the D_3 site. The density of this site (using apomorphine as the base line) is about 65 fmol/mg protein in rat and human striatum[164]; it is considerably higher in calf[164] (185 fmol/mg protein) and canine[167] striata. It is not detectable in the frontal cortex, hypothalamus, cerebellum, or hippocampus.[164]

The method for assaying the D_3 site is critical for obtaining reproducible data. Washing of the tissue four times (to remove endogenous dopamine) results in monophasic high-affinity binding curves for dopamine agonists.[164] Inclusion of ascorbate and EDTA is necessary in order to obtain maximum stereospecific binding.[164,168] Ascorbate is required as an antioxidant.[169] Preincubation of the receptor preparation, especially with dopamine present, increases binding, an effect prevented by EDTA and reversed by $CaCl_2$ or $MgCl_2$.[170]

4.4.3b. Pharmacology of the Site. The D_3 site is characterized by high affinity for dopamine agonists and low affinity for dopamine antagonists.[164–168] Although all neuroleptics are weak displacers of agonist binding, stereoselectivity can be demonstrated.[165] Ergot agonists, however, are less potent at these sites that at the D_2 receptors.

4.4.3c. Regional, Cellular, and Subcellular Distribution. The D_3 site is located in dopamine-rich brain regions, namely, in rat striatum and olfactory tubercle, calf caudate, and human caudate–putamen.[164] It is undetectable in frontal cortex, although both the D_1 and D_2 sites are found in this region.

Lesions of the medial forebrain bundle results in a 40–50% loss of the D_3 sites measured either by [^{3}H]dopamine or by [^{3}H]apomorphine.[171] Presynaptic depletion of dopamine rather than loss of receptor may account for the decrease.

4.4.3d. Influence of Ions and GTP. Both EDTA and ascorbate are necessary for maximum stereospecific binding measurement of the D_3 site.[164,165,169] Omission of either constituent results in an increase of total binding but a decrease of stereospecific binding. Solubilized D_3 sites respond differently to ascorbate and EDTA than do the membrane-bound preparations (B. K. Madras, unpublished data). GTP decreases agonist binding to the D_3 site, suggesting that a component of the D_3 site is a high-agonist-affinity state of the D_1 receptor.[154] Depletion of dopamine either by presynaptic lesions or by reserpine results in a loss of the D_3 site. Preincubation with dopamine restores this high-affinity binding (I. Creese, personal communication). This supports the view that the D_3 site may be a desensitized form of the D_1 site.[172]

4.4.4. Distinction among Dopamine Receptors

4.4.4a. Pharmacology. D_1 sites are characterized by having low affinity for dopamine and other agonists and for neuroleptics. Binding of butyrophenones, among the most potent group of neuroleptics, is of low affinity. There is no correlation between butyrophenone binding to the D_1 site and clinical doses of neuroleptics for controlling schizophrenia. D_2 receptors are characterized by having high affinity for neuroleptics and variable affinity for dopaminergic agonists. An excellent correlation is found between binding potency of antagonists to the D_2 receptor and potency for producing clinical and behavioral responses. D_3 sites are characterized by having high affinity for dopamine agonists and low affinity for dopamine antagonists (neuroleptics). See Table I for a summary of these points.

D_1, D_2, and D_3 sites have different structural requirements for agonists.[18]

4.4.4b. Biochemical Separation. D_1, D_2, and D_3 sites are found in different subcellular fractions.[18] D_1 and D_2 sites have reportedly been solubilized by different detergents and purified in different fractions.[173] D_2 and D_3 sites have been solubilized in the same detergent, but the recoveries are different for the two sites.[167,174] In solubilized preparations, the D_2 site is quite stable at 37°C, whereas the D_3 site declines rapidly (A. Davis, P. Seeman, and B. K. Madras, unpublished data). Phenoxybenzamine reacts with the D_2 site but leaves the D_3 site intact.[175,176]

4.4.4c. Cellular and Regional Location. D_1, D_2, and D_3 sites are not found in the same brain regions. In brain areas in which they coexist the density of the three sites is not the same. All three sites are located in the striatum; only the D_2 site is found in the anterior pituitary[95] and the ventral tegmental area (area A10).[177] The frontal cortex has considerable D_1 receptors, whereas D_2 receptors are present but localized in specific regions.[178] D_3 sites are not detectable in the frontal cortex.[164] In the retina, D_1 sites probably predominate. D_1 but not D_2 receptors are present in neuroblastoma cells.[116]

4.5. Correlation of Binding Data with Pharmacological, Clinical, and Behavioral Data

An abundance of evidence correlating direct binding assays with clinical data and behavioral models supports the concept that the primary target for dopaminergic drugs is the D_2 receptor.[18]

Although the concentrations of dopamine agonists and/or antagonists in plasma are sufficiently high for occupancy of several different receptors, including dopamine, α-adrenergic, serotonin (5-HT$_2$), histamine (H$_1$), muscarinic cholinergic, only binding potency to the D_2 receptor correlates with the "dopaminergic" action of these drugs. Dopamine agonists reduce Parkinson symptoms and induce rotation, emesis, and stereotypy with potencies that correlate with binding potency to the D_2 receptor.[18,179]

The excellent correlation of antipsychotic potency and serum concentrations of neuroleptics with binding potency to the D_2 receptor (Fig. 1) has been confirmed several times.[16,17,180,181] Several animal models of dopamine receptor activation are inhibited by neuroleptics in a rank order of potency that correlates with their rank order of binding potency to the D_2 receptor. These models include stereotypy[182] (apomorphine- or amphetamine-induced), emesis in dogs, and apomorphine-induced circling behavior in unilaterally lesioned rats.[18]

5. DOPAMINE RECEPTOR DISTRIBUTION

5.1. Occurrence in Different Species

Although interest in dopamine receptors stems from presumed receptor dysfunction in clinical syndromes, most receptor studies have been performed in animals, especially species that are convenient (rodent, calf). Dopamine receptors have been detected in rat,[183] human,[180,184,185] calf,[184] monkey, and more recently dog.[174,121,122] Human, rat, and calf caudate receptors have been most extensively explored and compared. Receptor binding conditions, which are critical, have not been standardized (buffer, ionic composition, ascorbate levels, tissue preparation, ligand and protein concentrations, GTP), resulting in different laboratories reporting disparate binding parameters (IC$_{50}$, B^{max}, and K_d values) in the same species.

Antagonist displacement of [^{3}H]spiperone binding to striatal membranes of various species is, however, similar and not subject to regulation, as is agonist binding. A major exception is the relative low potency of sulpiride, molindone, and metaclopramide for calf caudate receptors.[187] The Hill coefficient for antagonist displacement of [^{3}H]spiperone approaches 1 whereas the Hill coefficients for agonist displacement of the [^{3}H]-labeled ligand is less than 1.0.[187] Agonist binding, whether measured as displacement of [^{3}H]spiperone or using [^{3}H]-labeled agonists, is more variable in several species. Dopamine IC$_{50}$ values range from 1,200 to 30,000 in rat, although the majority are 10,000 or less.[18] These values were obtained before high and low affinity agonist states were proposed and described. Solubilization of dopamine receptors is readily achieved with canine and human brain, but calf and rat yield receptors with poor binding characteristics (see ref. 184 for summary).

5.2. Brain Regional Distribution

A correlation between the major dopaminergic regions of brain and brain dopamine receptors exists. However, not all receptor subtypes are located in all regions (Section 4).

5.2.1. Dopamine Receptors in the Pituitary

Excluding brain tissue, the most prominent location of dopamine receptors is the anterior and intermediate lobes of the pituitary gland, where dopamine receptors are considered to function as inhibitors of prolactin and α-MSH release.[188] These receptors are characterized by nanomolar affinity for neuroleptics and variable affinity for dopamine and congeners.[141] A recent model of the receptor suggests that the low-affinity state of the receptor is a binary complex, whereas the high-affinity state is a ternary complex (agonist–receptor–nucleotide-binding protein[189]). Both agonist binding and agonist displacement of antagonist binding are GTP sensitive.[190,191] GTP decreases agonist affinity of the D_2 receptor, justifying the concept of low-and high-affinity agonist states. These receptors are probably linked to inhibition of adenylate cyclase.[192,193]

5.2.2. Dopamine Receptors in the Cortex

Dopaminergic innervation of the frontal cortex has been detected by both biochemical and histochemical means. [³H]Spiperone binding to the frontal cortex, however, is mainly to serotonergic sites.[134] Serotonin is the most potent displacing agent of [³H]spiperone binding, and only spiperone, but not haloperidol, labels these sites in the frontal cerebral cortex and hippocampus. When discrete regions of the frontal cortex were analyzed for dopamine receptors and dopamine, [³H]spiperone binding predominated in the dorsal and ventral parts, the enterorhinal and cingulate cortex.[178] The pattern of dopamine-stimulated adenylate cyclase but not that of [³H]spiperone binding matched that of dopamine distribution. The B^{max} for these sites is of the order of 30 fmol/mg protein, and the K_d is 0.14 nM.

5.2.3. Dopamine Receptors in the Retina

Dopamine is the principal catecholamine in the mammalian retina where it fulfills all the criteria of a neurotransmitter.

Dopamine-stimulated adenylate cyclase has been detected in mammalian retina[194] and has been shown to be inhibited by neuroleptic drugs.[195] Although [³H]spiperone binding has also been reported in the retina,[196] [³H]domperidone, which is believed to be a highly specific label for D_2 receptors, failed to detect these receptors.[113] The neuroleptic-labeled sites have a density of 38 fmol/mg protein and a K_d of 0.3 nM; [³H]ADTN sites have a higher density (113 fmol/mg protein), of which about half are guanine nucleotide sensitive.

5.2.4. Dopamine Receptors in Other Brain Regions and Spinal Cord

Dopamine-mimetic drugs produce emesis in dogs, and neuroleptic drugs are excellent antiemetics. Recently, specific sites for [³H]spiperone binding in the area postrema of dog have been reported.[142] Agonist affinity for the site is reduced by guanine nucleotides, as observed in the striatum. Scatchard analysis of the dose–response curve revealed a density of 130 fmol/mg protein and a K_d of 0.109 nM. No dopamine-stimulated adenylate cyclase was detected, however. These presumably D_2 sites lend further support to the hypothesis that dopaminergic systems are in part responsible for emesis.

Electrophysiological, pharmacological, and microiontophoretic studies suggest that dopamine may have a specific function in the spinal cord (see ref. 197). Stereospecific binding of [³H]spiperone to discrete regions (dorsal horn, B^{max} 56 fmol/mg protein; lateral and ventral hornes, B^{max} 31–34 fmol/mg) of the spinal cord have been described.[197]

5.3. Cellular and Subcellular Distribution

The cellular localization of a receptor refers to whether it exists on neurons or other cell types and whether it is located pre- or postsynaptically. Most receptor binding studies have been performed on total membranes prepared from brain homogenates. Subcellular distribution has been reviewed.[18]

Anatomic location of brain dopamine receptors has been in part resolved using selective brain lesions. The majority of dopamine-sensitive cyclase is present on intrinsic striatal neurons. About half of all D_2 sites are found on corticostriate terminals (summarized in Table I).

5.4. Peripheral Tissue Distribution

Dopamine mediates several responses in peripheral tissues. Dopamine inhibits sympathetic function by acting on sympathetic ganglia and on the postganglionic sympathetic nerves. These may contain dopamine receptors, but controversy persists as to whether the receptors are actually dopaminergic or adrenergic. The ganglionic receptors are probably different from the vascular receptors.

Dopamine is a vasodilator in renal, mesenteric, coronary, and cerebral vascular beds.[199–202] The *in vitro* binding characteristics of the various peripheral receptors differ with respect to affinity as well as rank order of binding potency for a series of drugs. For example, (+)-sulpiride is more active than (−)-sulpiride on the postsynaptic dopamine receptor in the vascular bed, whereas on brain receptors (−)-sulpiride is more active.[201] Sulpiride is also inactive as a dopamine antagonist in splenic artery.

There are also suggestions that dopamine receptors function in the esophagus, stomach, small intestine, pancreas, and submandibular gland.[198] The receptors appear to be similar to those found in the vascular renal bed, but few or no direct binding characterization studies have been reported.

Bovine parathyroid cells secrete parathyroid hormone, a cyclic-AMP-mediated response.[203] Dopamine increases both cyclic AMP and parathyroid release in this gland. A receptor similar to the brain D_1 site has been identified in this tissue.

6. DOPAMINE RECEPTOR REGULATION

In vivo, the density of dopamine (D_2) receptors in the striatum is variable and subject to fluctuations induced by disease, lesions, or drug treatment. In general, procedures that decrease the effective availability of dopamine raise receptor densities. Conversely, procedures that increase the effective concentration of dopamine cause down-regulation of the receptor. *In vitro*, the receptor density and/or affinity is a reflection, partly, of conditions and constituents of the assay medium. The biological significance of variations in receptor binding observed *in vitro* has not been conclusively established.

6.1. Schizophrenia

The dopamine hypothesis of schizophrenia states that certain dopaminergic pathways are overactive in a subtype of schizophrenia that responds to neuroleptic medication. Evidence that brain dopamine synthesis, metabolism, or turnover is aberrant in schizophrenics remains inconclusive. With the identification of specific receptors for dopamine and neuroleptics in the brain, it became possible to examine whether dopamine receptors were abnormal in schizophrenics. Dopamine receptors in postmortem schizophrenic brains have been reported to be increased in a number of studies.[125,126,204–206] Initially, increased binding by between 50% and 75% was found in both the nucleus accumbens and the striatum of 20 schizophrenic postmortem brains using [^{3}H]haloperidol as the ligand and (+)-butaclamol as the base line.[125]

Neither the D_1 site, labeled by [^{3}H]*cis*-flupenthixol[206] nor the D_3 site labeled by [^{3}H]dopamine[18] reflected this increase.

There remains controversy as to whether the elevated density is drug induced or is a cause or consequence of the disease process.[207] Owen *et al.* and Lee and Seeman demonstrated an increase in [^{3}H]spiperone binding in brains of schizophrenics not exposed to neuroleptics for over a year.[204,205] The increased densities were less than that usually recorded for drug-treated patients. Establishing with certainty that the elevation is part of the disease process would be important for both etiological and diagnostic purposes. It would also serve as a basis for treatment strategies. The elevation of dopamine receptors in postmortem schizophrenic brains is compatible with and supportive of the dopamine hypothesis of schizophrenia. The excess receptors in the striatum, limbic, and perhaps other regions of the brain could constitute the overactive dopamine pathways that form the basis of the hypothesis. Other transmitter systems may also play a role in schizophrenia, including the GABAergic, opiate, cholinergic, and serotonergic systems.[208]

6.2. Tardive Dyskinesia and Parkinson's Disease

Tardive dyskinesia is a neurological side effect (involuntary movements of tongue, lips, facial muscles, jaw) of prolonged neuroleptic treatment. As many as 40–50% of neuroleptic-treated patients develop the characteristic involuntary motor movements.[209–212] Reversal of tardive dyskinesia is more difficult to accomplish than reversal of the initial extrapyramidal slowing observed at the onset of treatment. It can be irreversible in some cases. Both neurological entities (Parkinson's and tardive dyskinesia) coexist in about 17% of patients.[212] Dopamine receptor supersensitivity is a proposed cause of the neurological syndrome. Excess dopamine receptors induced in animals after prolonged administration of neuroleptics has been used as an experimental model of tardive dyskinesia (see Section 6.3).

Parkinson's disease is characterized by bradykinesia, tremor, muscular rigidity. Presynaptic dopamine in the striatum, is lost a consequence of nigrostriatal degeneration. Dopamine receptors (D_2), which are located postsynaptically, are elevated in untreated patients.[213] The increase has been interpreted as a compensatory response to diminished dopamine levels. D_2 receptor density of postmortem tissue from L-DOPA-treated patients is either normal or diminished.[214,215]

6.3. Neuroleptic-Induced Receptor Adaptation

As outlined in Section 6.2, the neurological side effect, tardive dyskinesia, of chronic neuroleptic treatment in humans develops after prolonged exposure. These involuntary movements may be reduced by increasing the dose of neuroleptics, but this response is usually only temporary. Cessation of therapy or treatment with dopamine agonists exacerbates the symptoms. Dopamine receptor supersensitivity may account for development of the dyskinesias.

Rats treated with antipsychotic drugs develop behavioral, electrophysiological, and pharmacological changes in dopamine neurons.

Shortly after a single injection of neuroleptics, behavioral supersensitivity to dopamine is detectable in mice.[216] The biochemical basis for the response is not an increase in dopamine receptor density.[216]

Prolonged administration of neuroleptics to animals results in behavioral supersensitivity to dopamine agonist drugs. Unlike the human response to prolonged neuroleptic medication, animals do not develop spontaneous motor movements but are behaviorally supersensitive to dopamine agonists drugs such as apomorphine or amphetamine.[217–219] Animal behavioral supersensitivity appears much sooner than that in humans. Baldessarini and Tarsy have questioned the validity of the animal model because supersensitivity develops so rapidly and is completely reversible in animals.[220] The behavioral responses are paralleled by a development of tolerance to the dopamine metabolite-elevating effects of these drugs.[221] As in humans, dopamine receptor density is increased.[219,222,223] The increase can be detected by many [³H]neuroleptics as well as labeled agonists.[219,222-225]

It is important to stress that in postmortem schizophrenic brains, there has been no reported increase in the binding of [³H]apomorphine, [³H]dopamine, or [³H]ADTN.[226] This represents yet another distinction between neuroleptic-treated schizophrenic patients and animals.

Pituitary dopamine receptor response to neuroleptics is different from that of the striatal receptors: no increase in receptor binding can be detected following prolonged exposure to neuroleptics.[227]

Lesioning the nigrostriatal tract results in supersensitive responses to dopamine agonists behaviorally,[228] electrophysiologically,[229] and neurochemically.[230] Although no change is seen in the D_1 dopamine-stimulated adenylate cyclase in homogenates of denervated tissue, D_2 receptor density increases significantly following lesions of the tract if dopamine depletion is greater than 90%.[18,219,231,232] Only the B^{max} changes, not the affinity.

Chronic lithium depresses the supersensitive responses to apomorphine in chronically treated animals; it does not repress the development of increased [³H]spiperone binding sites.[233,234]

A dopaminergic pathway has been implicated in the development of mania. As lithium is the most efficacious drug for treatment of mania, interaction of lithium and dopaminergic pathways is of interest. Other drugs increase dopamine receptor numbers.[235]

6.4. Dopamine Agonist-Induced Receptor Adaptation

Dopamine receptors are down-regulated by repeated administration of dopamine agonists. Acute and chronic amphetamine or peripheral administration of L-DOPA lower dopamine receptor density.[236,237] The ergots pergolide and bromocriptine, which have dopaminergic agonist activity, also produce dopamine receptor down-regulation.[238,239]

The mechanisms of down-regulation are unknown, but preliminary evidence, using striatal slices, indicates that down-regulation may persist following depletion of most of the agonist in the tissue.[240]

6.5. Hormone Regulation

Estrogens have a marked inhibitory effect on dopaminergic systems in both the anterior pituitary and the striatum.[241,242] Dopamine suppression of prolactin secretion is inhibited by estrogen.[243] Estrogens also inhibit apomorphine-induced rotation and potentiate neuroleptic-induced catalepsy.[244] Following estrogen administration to female (ovariectomized) or male rats, striatal dopamine receptors ([³H]spiperone) increase by about 20%.[245,246] The effects of neuroleptic and 17 β-estradiol on binding are additive.[245] As prolactin administration significantly increases the density of striatal dopamine receptors, it has been suggested that this hormone is the common mediator of estrogen- or neuroleptic-induced dopamine receptor supersensitivity.[247] However, this hypothesis has been questioned by Jenner *et al.*, who demonstrated that hypophysectomy does not prevent the onset of dopamine receptor supersensitivity induced by neuroleptic treatment.[248]

Dopamine receptors in the pituitary do not increase with chronic neuroleptic treatment.[227] Chronic treatment also does not produce tolerance to induce dopamine turnover in the hypothalamus, in contrast to the tolerance found in the striatal and limbic systems.[249] Similarly, neuroleptics do not produce tolerance to prolactin secretion in humans or animals after short- or long-term neuroleptic administration.[250]

7. CLINICAL APPLICATIONS OF DOPAMINE RECEPTORS

7.1. Assay of Neuroleptic Concentrations in Blood

There are several potential advantages to monitoring neuroleptic blood levels of patients receiving these drugs. It may provide a measure of (1) patient compliance; (2) bioavailability of the drug, which may be either too low or too high for clinical efficacy; and (3) appropriate therapeutic blood levels to minimize toxic doses. Precise assays for measuring individual neuroleptics in plasma are available. These include radioimmunoassays, gas chromatography–mass spectroscopy (GC–MS), and fluorimetry. However, these methods are not in routine use because they present technical difficulties and are restricted to a single neuroleptic (thereby exluding active metabolites), and a relationship between plasma levels of individual neuroleptics and therapeutic response has not consistently shown.[251]

An alternative to these methods is a radioreceptor assay (RRA) for plasma neuroleptics.[252] It is based on two assumptions: all neuroleptics exert their major therapeutic effect through the brain dopamine receptor, and all circulating drugs and metabolites that displace [^{3}H]spiperone bound to (calf) brain dopamine receptors *in vitro* will do so *in vivo*.

The advantages of radioreceptor assay are readily apparent:

1. It measures the sum of all circulating neuroleptics and active metabolites that bind to the clinically relevant receptor.
2. It is inexpensive, rapid, and simple.
3. It may be modified to measure plasma unbound neuroleptic concentrations, possibly the clinically more relevant concentration (unpublished observations).
4. Results from this method agree well with RIA and GC–MS methods for individual neuroleptics.
5. Metabolites active at dopamine receptors bind, whereas inactive metabolites do not.

Some of the disadvantages include the unproven assumption that all circulating compounds that bind to the dopamine receptors *in vitro* have access to brain receptors. Certain dopamine agonists (e.g., ergot derivative and a few tricyclic antidepressants) bind with high potency to the receptor and thus may contribute artificially high levels for plasma neuroleptics. Unidentified serum constituents that displace [³H]spiperone can alter control values, leading to inconsistant base lines. In spite of these drawbacks, the potential usefulness of this assay is evident. A correlation between mean plasma neuroleptic availability and clinical response to neuroleptics has been reported in some but not all studies (S. W. Tang, in preparation). Moreover, plasma prolactin levels are positively correlated with plasma neuroleptic levels measured this way.[253-257]

7.2. Lymphocyte Dopamine Receptors?

Although dopamine receptors (D_2) have been identified in various peripheral tissues and in brain, they are not readily accessible for clinical studies. Their potential value in diagnosis and monitoring of drug treatment is an incentive to find a readily available source of these receptors. LeFur and his co-workers have recently identified [³H]spiperone binding in lymphocytes of four species[258-261] including human.[261] The binding characteristics of rat lymphocytes paralleled those of striatal dopamine receptors (D_2 type). Decreased [³H]spiperone binding to lymphocytes was reported in Parkinson patients.[261] However, others using identical methods were unable to observe saturable, stereoselective, and characteristic [³H]spiperone binding to lymphocytes[262-264] (B. K. Madras and S. W. Tang, unpublished data). If lymphocyte receptors can be detected, it will be necessary to prove that they are under the same regulatory influences as the brain receptors for them to be of clinical value.

8. DOPAMINE RECEPTOR PURIFICATION

Isolated receptors will be useful for determining the morphology of the receptor, for assessing the subtypes of receptors and the organization of the receptor–effector complex, and for turnover studies. The development of immunologic probes will accelerate advances in both basic and clinical research.

8.1. Solubilization of Dopamine Receptors

The success of this step (high yield, retention of binding properties) is dependent on several factors, of which tissue source, solubilizing agent and condition, buffer, assay method, and the intrinsic stability of the receptor are the most important.[184,265-269] Solubility is established by several criteria including the following:

1. Receptor binding sites do not sediment during centrifugation at 150,000 × *g* for at least 2 h.
2. Membrane structures are not detectable in the soluble fraction by electron microscopy.
3. Passage through 0.22-μ Millipore® filters.
4. The soluble receptor should possess binding properties similar to those of the membrane receptor. Thus, for the D_2 receptor the soluble form

should retain a low dissociation constant (K_d in nanomolar or picomolar range for [³H]spiperone), saturability, stereoselectivity for (+)-butaclamol or other stereoisomeric pairs, rank order of potency for at least four classes of neuroleptics and neurotransmitters, and an appropriate brain or peripheral tissue distribution pattern.

[³H]Dopamine binding sites (D_1, adenylate cyclase activating) have been solubilized from canine tissue, as has the D_3 site.[167] Neuroleptic/dopamine receptors (D_2-type) from canine,[184,267–268] human,[185,268] calf,[184,267,270] rat,[271] and cat[271] striatum have also been solubilized with various detergents. The detergent most frequently used is digitonin, which generally yields 20–30% of the membrane receptors and 40–60% of membrane proteins. It is still unclear whether high salt concentrations release the receptor into solution or create smaller membrane fragments.[272] Both canine and human striata are excellent sources of solubilized receptors,[184] whereas calf and rat tissue yield receptors with irregular binding patterns.[184,273] Other detergents such as lysolecithin or CHAPS may prove to be effective.

The soluble receptors are measured either by Sephadex G50 chromatography or by polyethylene glycol precipitation.[274] The latter underestimates receptor densities. Other methods used for separating bound from free ligand include charcoal adsorption[271] and hydroxyapatite separation[275]; these methods have been reported without complete receptor characterizations or comparisons with other methods. Adaptation of concanavalin A–Sepharose for soluble receptor assay may prove to be valuable for dopamine receptors.[276] Although the binding properties of the solubilized and membrane receptors are very similar, the two preparations respond differently to ascorbate, EDTA, and manganese.[277]

8.2. Purification

Because the D_2 receptor is found in very low concentrations in the brain, an irreversible ligand for receptor purification would simplify the procedure. Three irreversible ligands, phenoxybenzamine,[175] vipoxin,[278] and (−)N-chloroethylapomorphine,[279] are available but are not specific for the D_2 receptor. Partial purification using conventional protein separation techniques was reported by Clement-Cormier *et al.*, but no characterization studies were included.[280] Prelabeling the receptor with [³H]spiperone, a reversible ligand, has both advantages (high affinity, reversibility) and disadvantages (nonspecificity, potential dissociation during purification). Recently, the soluble receptor from canine striatum was prelabeled with [³H]spiperone and purified by isoelectric focusing.[281,282]

Although two radioactive peaks emerged at equilibrium, only the peak with an isoelectric point of 5 retained the binding characteristics of the D_2 receptor.[282]

9. FUTURE DIRECTIONS

The first phase of dopamine receptor research focused on identifying, localizing, and characterizing the receptors in crude membrane preparations of various tissues. Major advances marked this period,[18,283] especially accumulation of evidence that the most probable target of neuroleptic action is the dopamine D_2 receptor. Yet, a host of biological and clinical questions remain. How many dopamine receptor subtypes exist, and what are their precise locations and functions? What is the morphology of the receptors? What reactions occur subsequent to receptor–ligand interaction, and how are they related to receptor structure? What is the role of the membrane in receptor function? Are the excess receptors found in schizophrenic brains of etiologic significance? What are the physiologically important determinants of receptor density and affinity? What is the evolutionary development of dopamine receptor? These questions and many others will probably be explored in the next decade of dopamine receptor research.

REFERENCES

1. Bernheimer, H., Birkmayer, W., Hornykiewicz, O., Jellinger, K., and Seitelberger, F., 1973, *J. Neurol. Sci.* **20:**415–455.
2. Seeman, P., 1977, *Biochem. Pharmacol.* **26:**1741–1748.
3. Lichtensteiger, W., 1979, *The Neurobiology of Dopamine* (A. S. Horn, J. Korf, and B. H. C. Westerink, eds.), Academic Press, New York, pp. 491–521.
4. Ehringer, H., and Hornykiewicz, O., 1960, *Klin. Wochenscher.* **38:**1236–1239.
5. Birkmayer, W., and Hornykiewicz, O., 1962, *Arch. Psychiatr. Nervekr.* **203:**560–574.
6. Horn, A. S., Korf, J., and Westerink, B. H. C. (eds.), 1979, *The Neurobiology of Dopamine*, Academic Press, New York.
7. Iversen, L. L., Iversen, S. D., and Snyder, S. H. (eds.), 1975, *Handbook of Psychopharmacology*, Volume 3, Plenum Press, New York.
8. Costa, E., and Greengard, P. (eds.), 1978, *Advances in Biochemical Psychopharmacology*, Volume 19, Raven Press, New York.
9. Van Rossum, J. M., 1966, *Arch. Int. Pharmacodyn. Ther.* **160:**492–494.
10. Matthysse, S., 1974, *J. Psychiatr. Res.* **11:**107–113.
11. Nyback, H., Sedvall, G., and Kopin, I. J., 1967, *Life Sci.* **6:**2307–2312.
12. Gallagher, J. P., Inokuchi, H., and Shinnick-Gallagher, P., 1980, *Nature* **283:**770–772.
13. Gruen, P. H., Sachar, E. H., Langer, G., Altman, N., Leifer, M., Frantz, A., and Halpern, F. S., 1978, *Arch. Gen. Psychiatry* **35:**108–116.
14. Kebabian, J. W., Petzold, G. L., and Greengard, P., 1972, *Proc. Natl. Acad. Sci. U.S.A.* **69:**2145–2149.
15. Iversen, L. L., 1975, *Science* **188:**1084–1089.
16. Seeman, P., Lee, T., Chau-Wong, M., and Wong K., 1976, *Nature* **261:**717–719.
17. Creese, I., Burt, D. R., and Snyder, S. H., 1976, *Science* **192:**481–483.
18. Seeman, P., 1980, *Pharmacol. Rev.* **32:**229–313.
19. Welsh, J., 1972, *Catecholamines* (H. Blashko and E. Muscholl, eds.), Springer-Verlag, Berlin, pp. 79–109.
20. Cottrell, G. A., Berry, M. S., and Macon, J. B., 1974, *Neuropharmacology* **13:**431–439.
21. Ginsborg, B. L., House, C. P., and Turnbull, K. W., 1976, *Br. J. Pharmacol.* **57:**133–140.
22. Holzbauer, M., and Sharman, D. F., 1979, *The Neurobiology of Dopamine* (A. S. Horn, J. Korf, and B. H. C. Westerink, eds.), Academic Press, New York, pp. 357–379.
23. Lackovic, Z., Relja, M., and Neff, N. H., 1982, *J. Neurochem.* **38:**1453–1458.
24. Weil-Malherbe, H., and Bone, A. D., 1957, *Nature* **180:**1050.
25. Carlsson, A., and Waldeck, B., 1958, *Acta Physiol. Scand.* **44:**293–298.
26. Birkmayer, W., and Hornykiewicz, O., 1961, *Wien. Klin. Wochenschr.* **73:**787–788.
27. Hokfelt, T., Fuxe, K., and Goldstein, M., 1975, *Ann. N.Y. Acad. Sci.* **254:**407–432.
28. Bell, C., and Gillespie, J. S., 1981, *J. Neurochem.* **36:**703–706.

29. Bell, C., 1982, *Neuroscience* **7**:1–8.
30. Taylor, K., and Laverty, R., 1969, *J. Neurochem.* **16**:1361–1366.
31. Westerink, B. H. C., and Korf, J., 1977, *J. Neurochem.* **29**:697–706.
32. Westerink, B. H. C., and Korf, J., 1976, *Eur. J. Pharmacol.* **38**:281–291.
33. Costa, E., Koslow, S. H., and LeFevre, H. F., 1975, *Handbook of Psychopharmacology,* Volume 1 (L. L. Iversen, S. D. Iversen, and S. H. Snyder, eds.), Plenum Press, New York, pp. 1–24.
34. Coyles, J. T., and Henry, D., 1973, *J. Neurochem.* **21**:61–67.
35. Ponzio, F., Achilli, G., and Algeri, S., 1981, *J. Neurochem.* **36**:1361–1367.
36. Westerink, B. H. C., and Mulder T. B. A., 1981, *J. Neurochem.* **36**:1449–1462.
37. Magnusson, O., Nilsson, R. B., and Westerlund, D., 1980, *J. Chromatogr.* **221**:237–247.
38. Kempf, E., and Mandel, P., 1981, *Anal. Biochem.* **112**:223–231.
39. Brownstein, M., Saavedra, J. M., and Palkovits, M., 1974, *Brain Res.* **79**:431–436.
40. Versteeg, D. H. G., van der Gugten, J., de Jong, W., and Palkovits, M., 1976, *Brain Res.* **113**:563–574.
41. Falck, B., Hillarp, N. A., Thieme, G., and Torp, A., 1962, *J. Histochem. Cytochem.* **10**:348–354.
42. Axelsson, S., Bjorklund, A., Falck, B., Lindvall, O., and Svensson, L.-A., 1973, *Acta Physiol. Scand.* **87**:57–62.
43. Lindvall, O., Bjorklund, A., Hokfelt, T., and Lungdahl, A., 1973, *Histochemie* **35**:31–38.
44. Anden, N.-E., Carlsson, A., Dahlstrom, A., Fuxe, K., Hillarp, N. A., and Larsson, N., 1964, *Life Sci.* **3**:523–530.
45. Ungerstedt, U., 1971, *Acta Physiol. Scand.* [*Suppl.*] **82**:1–93.
46. Dahlstrom, A., and Fuxe, K., 1964, *Acta Physiol. Scand.* [*Suppl.*] **232**:1–55.
47. Lindvall, O., 1975, *Brain Res.* **87**:89–95.
48. Bjorklund, A., Falck, B., Hromek, F., Owman, C., and West, K. A., 1970, *Brain Res.* **17**:1–23.
49. Moore, R. Y. and Bloom, F. E. 1978, *Ann Rev. Neurosci.* **1**:129–169.
50. Udenfriend, S., and Zaltzman-Nirenberg, P., 1963, *Science* **142**:394–396.
51. Nyback, H., Sedvall, G., Sjoquist, B., and Wiesel, F.-A., 1973, *Acta Physiol. Scand.* **87**:8A–9A.
52. Kucaenski, R. J., and Mandell, A. J., 1972, *J. Biol. Chem.* **247**:3114–3122.
53. Besson, M. J., Cheramy, A., Gauchy, C., and Musacchio, J., 1973, *Eur. J. Pharmacol.* **22**:181–186.
54. Murrin, L. C., Morgenroth, V. H. III, and Roth, R. H., 1976, *Mol. Pharmacol.* **12**:1070–1081.
55. Goldstein, M., Bronaugh, R. L., Ebstein, B., and Roberge, C., 1976, *Brain Res.* **109**:563–574.
56. Nowycky, M. C., and Roth, R. H., 1978, *Prog. Neuropsychopharmacol.* **2**:139–158.
57. Guidotti, A., and Costa, E., 1977, *Biochem. Pharmacol.* **26**:817–823.
58. Carlsson, A., and Lindqvist, M., 1978, *Naunyn Schmiedebergs Arch. Pharmacol.* **303**:157–164.
59. Sved, A. F., Fernstrom, J. D., and Wurtman, R. J., 1979, *Life Sci.* **25**:1293–1300.
60. Roth, R. H., 1979, *The Neurobiology of Dopamine* (A. S. Horn, J. Korf, and B. H. C. Westerink, eds.), Academic Press, New York, pp. 101–122.
61. Musacchio, J. M., 1975, *Handbook of Psychopharmacology* (L. L. Iversen, S. D. Iversen, and S. H. Snyder, eds.), Plenum Press, New York, pp. 4–10.
62. Nagatsu, T., Levitt, M., and Udenfriend, S., 1974, *J. Biol. Chem.* **239**:2910–2917.
63. Waymire, J. C., Bjur, R., and Weiner, N., 1971, *Anal. Biochem.* **43**:588–600.
64. Sourkes, T. L., 1955, *Rev. Can. Biol.* **14**:49–63.
65. Sourkes, T. L., 1966, *Pharmacol. Rev.* **18**:53–116.
66. Carlsson, A., and Lindqvist, M., 1973, *J. Neural. Transm.* **23**:79–91.
67. De Belleroche, J. S., Bradford, H. F., and Jones, D. G., 1976, *J. Neurochem.* **26**:561–571.
68. Liang, N. Y., and Rutledge, C. O., 1982, *Biochem. Pharmacol.* **31**:983–992.
69. Dubocovich, M. L., and Weiner, N., 1981, *J. Pharmacol Exp. Ther.* **219**:710–717.
70. Giorguieff-Chesselet, M. F., Kemel, M. L., Wandscheer, D., and Glowinski, J., 1979, *Eur. J. Pharmacol.* **60**:101–104.

71. Roberts, R. J., and Anderson, S. D., 1979, *J. Neurochem.* **32**:1539.
72. Ennis, C., Kemp, J. D., and Cox, B., 1981, *J. Neurochem.* **36**:1515–1520.
73. Bustos, G., and Roth, R. H., 1972, *Br. J. Pharmacol.* **46**:101–115.
74. Holz, R. W., and Coyle, J. T., 1974, *Mol. Pharmacol.* **10**:746–758.
75. Heikkila, R. E., Orlansky, W., and Cohen, G., 1975, *Biochem. Pharmacol.* **24**:847–852.
76. Westerink, B. H. C., 1979, *The Neurobiology of Dopamine* (A. Horn, J. Korf, and B. Westerink, eds.), Academic Press, New York, pp. 225–291.
77. Westerink, B. H. C., and Spaan, S. J., 1982, *J. Neurochem.* **38**:680–686.
78. Carlsson, A., and Lindqvist, M., 1963, *Acta Pharmacol. Toxicol.* **20**:140–144.
79. Anden, N.-E., and Stock, G., 1973, *J. Pharm. Pharmacol.* **25**:346–348.
80. Bacopoulos, N. G., Heninger, G. R., and Roth, A. H., 1978, *Life Sci.* **23**:1805–1812.
81. Costall, B., and Naylor, R. J., 1973, *Arzneim. Forsch.* **23**:674–683.
82. Ezrin-Waters, C., Muller, P., and Seeman, P., 1976, *Can. J. Physiol. Pharmacol.* **54**:516–519.
83. Snyder, S. H., 1976, *Am. J. Psychiatry* **133**:197–202.
84. Goodwin, F. K., 1972, *Psychiatric Complications of Medical Drugs,* (R. I. Shader, ed.), Raven Press, New York, pp. 149–174.
85. Smythe, G. A., 1977, *Clin. Endocrinol.* **7**:325–341.
86. Dill, R. E., Jones, D. L., Christian Gillin, J., and Murphy G., 1979, *Pharmacol. Biochem. Behav.* **20**:711–716.
87. Niemegeers, C. J. E., Heykants, J. J. P., and Janssen, P. A. J., 1981, *Psychopharmacology* **75**:240–244.
88. Ungerstedt, U., 1971, *Acta Physiol. Scand.* [*Suppl.*] **367**:69–93.
89. Ungerstedt, U., 1971, Ibid 49–68.
90. Corrodi, H., Farnebo, L., Fuxe, K., Hamberger, B., and Ungerstedt, U., *Eur. J. Pharmacol.* **20**:195–201.
91. Siggins, G. R., 1978, *Psychopharmacology: A Generation of Progress* (M. A. Lipton, A. Di Mascio, and K. F. Killam, eds.), Raven Press, New York, pp. 143–157.
92. Feltz, P., and De Champlain, J., 1972, *Brain Res.* **43**:601–605.
93. Herz, A., and Zieglgansberger, W., 1968, *Int. J. Neuropharmacol.* **7**:221–230.
94. York, D.-H., 1975, *Handbook of Psychopharmacology,* Volume 6 (L. L. Iversen, S. D. Iversen, and S. H. Snyder, eds.), Plenum Press, New York, pp. 23–61.
95. Kebabian, J. W., and Calne, D. B., 1979, *Nature* **277**:93–96.
96. Brown, E. H., Carroll, R. J., and Aurbach G. D., 1977, *Proc. Natl. Acad. Sci. U.S.A.* **74**:4210–4213.
97. Sokoloff, P., Martres, M. P., and Schwartz, J. C., 1980, *Naunyn Schmiedebergs Arch. Pharmacol.* **315**:89–102.
98. Goldberg, L. I., and Kohli, J. D., 1979, *Commun. Psychopharmacol.* **3**:447–456.
99. Leysen, J. E., and Gommeren, W., 1978, *Life Sci.* **23**:447–452.
100. Kebabian, J. W., and Greengard, P., 1971, *Science* **174**:1346–1349.
101. Stoof, J. C., Kebabian, J. W., 1981, *Nature* **294**:366–368.
102. Hyttel, J., 1980, *Psychopharmacology* **67**:107–109.
103. Cross, A. J., and Owen, F., 1980, *Eur. J. Pharmacol.* **65**:341–347.
104. Hyttel, J., 1978, *Life Sci.* **23**:551–556.
105. Garau, L., Govoni, S., Stefanini, E., Trabucchi, M., and Spano, P. F., 1978, *Life Sci.* **23**:1745–1750.
106. Minneman, K. P., Quik, M., and Emson, P. C., 1978, *Brain Res.* **151**:507–521.
107. Cannon, J. G., Costall, B., Laduron, P. M., Leysen, J. E., Naylor, R. J., 1978 *Biochem. Pharm.* **27**:1417–1420.
108. Iversen, L. L., Horn, A. S., and Miller, R. J., 1975, *Adv. Neurol.* **9**:197–212.
109. Markstein, A. M., Herrling, P. L., Burki, H. R., Asper, H., and Ruch, W., 1978, *J. Neurochem.* **31**:1163–1172.
110. Pagnini, G., Camanni, F., Crispino, A., and Portaleone, P., 1978, *J. Pharm. Pharmacol.* **30**:92–95.
111. Leysen, J., and Laduron, P., 1977, *Life Sci.* **20**:281–288.
112. Quik, M., and Iversen, L. L., 1979, *Eur. J. Pharmacol.* **56**:323–330.

113. Watling, K. J., Dowling, J. E., and Iversen, L. L., 1979, *Nature* **281**:578–580.
114. Tassin, J. P., Simon, H., Herve, D., Blanc, G., Le Moal, M., and Bockaert, J., 1982, *Nature* **295**:696–698.
115. Church, A. C., Bunney, B. S., and Krieger, N. R., 1982, *Brain Res.* **234**:369–376.
116. Hartley, E., Spuhler, K., Prasad, K. N., and Seeman, P., 1980, *Brain Res.* **190**:574–577.
117. Premont, J., Thierry, A. M., Tassin, J. P., Glowinski, J., Blanc, G., and Bockaert, J., 1976, *FEBS Lett.* **68**:99–104.
118. McSwigon, J. D. Nichol, S. E., Gottesman, I. I., Tuason V. B., and Frey, W. H. II, 1980, *J. Neurochem.* **34**:594–601.
119. Thomas, T. N., and Zemp, J. W., 1977, *J. Neurochem.* **28**:663–665.
120. Kayaalp, S. O., Rukensteen, J. S., and Neff, N. H., 1981, *Neuropharmacology* **20**:409–410.
121. Madras, B. K., Davis, A., Chan, B., and Seeman, P., 1981, *Prog. Neuropyschopharmacol.* **5**:543–548.
122. Madras, B. K., and Chan, B., 1983, *CNS Receptors, From Molecular Pharmacology to Behavior.* (P. Mandel and F. V. DeFeudis, eds.), Raven Press, pp. 275–287.
123. Heikilla, R. E., Cabbat, F. S., and Manzino, L., 1981, *Res. Commun. Chem. Pathol. Pharmacol.* **34**:409–421.
124. Kebabian, J. W., 1978, *Adv. Biochem. Psychopharmacol.* **19**:131–154.
125. Lee, T., Seeman, P., Tourtellotte, W. W., Farley, I. J., and Hornykiewicz, O., 1978, *Nature* **274**:897–900.
126. Crow, T. J., Johnstone, E. C., Longden, A. J., and Owen, F., 1978, *Life Sci.* **23**:563–568.
127. Peroutka, S. J., U'Prichard, D. C., Greenberg, D. A., and Snyder, S. H., 1977, *Neuropharmacology* **16**:549–556.
128. Leysen, J. E., Niemegeers, C. J. E., Tollenaere, J. P., and Laduron P. M., 1978, *Nature* **272**:168–171.
129. Seeman, P., Westman, K., Coscina, D., and Warsh, J. J., 1980, *Eur. J. Pharmacol.* **66**:179–191.
130. Quik, M., and Iversen, L. L., 1979, *Eur. J. Pharmacol.* **56**:323–330.
131. Seeman, P., Chau-Wong, M., Tedesco, J., and Wong, D., 1975, *Proc. Natl. Acad. Sci. U.S.A.* **72**:4376–4380.
132. Spedding, M., and Berg, C., 1982, *J. Pharm. Pharmacol.* **34**:56–58.
133. List, S., and Seeman, P., 1981, *Proc. Natl. Acad. Sci. U.S.A.* **78**:2620–2624.
134. Quik, M., Iversen, L. L., Larder, A., and Mackay, A. V. P., 1978, *Nature* **274**:513–514.
135. Baudry, M., Martres, M. P., and Schwartz, J. C., 1979, *Naunyn Schmiedebergs Arch. Pharmacol.* **308**:231–237.
136. Lazareno, S., Nahorski, S. R., 1981, *Br. J. Pharmacol.* **74**:231P–232P.
137. Freedman, S. B., Poat, J. A., and Woodruff, G. N., 1981, *Neuropharmacology* **20**:1323–1326.
138. Jenner, P., and Marsden, C. D., 1981, *Neuropharmacology* **20**:1285–1293.
139. Theodourou, A. E., Hall, M. D., Jenner, P., and Marsden, C. D., 1980, *J. Pharm. Pharmacol.* **32**:441–444.
140. Seeman, P., Ulpian, C., and Wells, J., 1982, *Soc. Neurosci. Abstr.* **8**:718.
141. Sibley, D. R., DeLean, A., and Creese, I., 1982, *J. Biol. Chem.* **257**:6351–6361.
142. Stefanini, E., and Clement-Cormier, Y. C., 1981, *Eur. J. Pharmacol.* **74**:257–260.
143. Thal, L., Mishra, R. K., Gardner, E. L., Horowitz, S. G., Varmuza, S., and Makman, M. H., 1979, *Brain Res.* **170**:381–386.
144. Murrin, L. C., Gale, K., and Kuhar, M. J., 1979, *Eur. J. Pharmacol.* **60**:229–235.
145. Schwarcz, R., Creese, I., Coyle, J. T., and Snyder, S. H., 1978, *Nature* **271**:766–768.
146. Sibley, D. R., and Creese, I., 1980, *Endocrinology* **107**:1405–1409.
147. George, S. R., Watanabe, M., and Seeman, P., 1983, in, *Dopamine Receptors* (C. Kaiser and J. Kebabian, eds.) American Chemical Society, Washington, D.C., pp 93–99.
148. Creese, I., Usdin, T., and Snyder, S. H., 1979, *Nature* **278**:557–578.
149. Hamblin, M. W., and Creese, I., 1982, *Mol. Pharmacol.* **21**:52–56.
150. Leslie, F. H., Dunlop, C. E., and Cox, B. M., 1980, *J. Neurochem.* **34**:219–221.
151. Heikilla, R. E., Cabbat, F. S., and Manzino, L., 1982, *J. Neurochem.* **38**:1000–1006.
152. Muakkassah-Kelly, S. F., Andersen, J. W., Shih, J. C., and Hochstein, P., 1982, *Biochem. Biophys. Res. Commun.* **104**:1003–1010.

153. Tolbert, L. C., Thomas, T. N., Middaugh, J. D., and Zemp, J. W., 1979, *Brain Res. Bull.* **4**:43–48.
154. Hamblin, M. W., and Creese, I., 1982, *Life Sci.* **30**:1587–1595.
155. Meltzer, H. Y., and So, R., 1979, *Life Sci.* **25**:531–536.
156. Muller, E. E., Stefanini, E., Camanni, F., Locatelli, V., Massara, F., Spano, P. F., and Cochi, D., 1979, *J. Neural Transm.* **46**:205–214.
157. Cronin, M. J., and Weiner, R. I., 1979, *Endocrinology* **104**:307–312.
158. Laduron, P. M., Janssen, P. F. M., and Leysen, J. E., 1978, *Biochem. Pharmacol.* **27**:317–321.
159. Kohler, C., Fuxe, K., and Ross, S. B., 1981, *Eur. J. Pharmacol.* **72**:397–402.
160. Klemm, N., Murrin, L. C., and Kuhar, M. J., 1979, *Brain Res.* **169**:1–9.
161. Nomura, Y., Oki, K., and Segawa, T., 1982, *J. Neurochem.* **38**:902–908.
162. Govoni, S., Spano, P. F., and Trabucchi, M., *J. Pharm. Pharmacol.* **30**:448–449.
163. Titeler, M., List, S., and Seeman, P., 1980, *Commun. Psychopharmacol.* **3**:411–420.
164. List, S. J., and Seeman, P., 1982, *J. Neurochem.* **39**:1363–1373.
165. Arana, G. W., Baldessarini, R. J., and Harding M., 1981, *Biochem. Pharmacol.* **30**:3171–3179.
166. Sokoloff, P., Martres, M.-P., and Schwartz, J.-C., 1980, *Nature* **288**:283–286.
167. Davis, A., Madras, B. K., and Seeman, P., 1982, *Biochem. Pharmacol.* **31**:1183–1187.
168. Leff, S., Sibley, D. R., Hamblin, M., and Creese, I., 1982, *Life Sci.* **29**:2081–2090.
169. Arana, G. W., Baldessarini, R. J., and Kula, N. S., 1982, *Neuropharmacology* **21**:601–604.
170. Bacopoulos, N. G., 1981, *Life Sci.* **23**:2407–2414.
171. Nagy, J. I., Lee, T., Seeman, P., and Fibiger, H. C., 1978, *Nature* **274**:278–281.
172. Leff, S. E., Hamblin, M. W., and Creese, I., 1982, *Soc. Neurosci. Abstr.* **8**:719.
173. Nishikori, K., Noshiro, O., Sano, K., and Maeno, H., 1980, *J. Biol. Chem.* **255**:10909–10915.
174. Madras, B. K., Davis, A., and Seeman, P., 1980, *Soc. Neurosci. Abstr.* **6**:240:
175. Titeler, M., 1981, *Biochem. Pharmacol.* **30**:3031–3037.
176. Hamblin, M. W., and Creese, I., 1982, *Mol. Pharmacol.* **21**:44–51.
177. Lorez, H. P., and Burkard, W. P., 1979, *Experientia* **35**:938.
178. Marchais, D., Tassin, J. P., and Bockaert, J., 1980, *Brain Res.* **183**:235–240.
179. Titeler, M., and Seeman, P., 1978, *Experientia* **34**:1490–1493.
180. Lee, T., and Seeman, P., 1979, *Soc. Neurosci. Abstr.* **5**:653.
181. Peroutka, S. J., and Snyder, S. H., 1980, *Am. J. Psychiatry* **137**:1518–1522.
182. Creese, I., Burt, D. R., and Snyder, S. H., 1978, *Handbook of Psychopharmacology*, (L. L. Iversen, S. D. Iversen, and S. H. Snyder, eds.), Plenum Press, New York, pp. 37–89.
183. Seeman, P., Wong, M., and Tedesco, J., 1975, *Soc. Neurosci. Abstr.* **1**:405.
184. Madras, B. K., Davis, A., and Seeman, P., 1982, *Eur. J. Pharmacol.* **78**:431–438.
185. Davis, A., Madras, B., and Seeman, P., 1981, *Eur. J. Pharmacol.* **70**:321–329.
186. Hartley, E. J., and Seeman, P., 1978, *Life Sci.* **23**:513–518.
187. Creese, I., Stewart, K., and Snyder, S. H., 1979, *Eur. J. Pharmacol.* **60**:55–66.
188. Caron, M. C., Beaulieu, M., Raymond, V., Gagne, B., Drouin, J., Lefkowitz, R. J., and Labrie, F., 1978, *J. Biol. Chem.* **253**:2244–2253.
189. Limbird, L. E., Gill, D. M., and Lefkowitz, R. J., 1980, *Proc. Natl. Acad. Sci.* **77**:775–779.
190. De Lean, A., Kilpatrick, B. F., and Caron, M. G., 1982, *Endocrinology* **110**:1064–1066.
191. George, S. R., Wregget, K. A., and Seeman, P., 1982, *Soc. Neurosci. Abstr.* **8**:720.
192. Onali, P., Schwartz, J. P., and Costa, E., 1981, *Proc. Natl. Acad. Sci. U.S.A.* **78**:6531–6534.
193. Cote, T. E., Greive, C. W., Tsuruta, K., Stoof, J. C., Eskay, R. L., and Kebabian, J. W., 1982, *Endocrinology* **110**:812–819.
194. Brown, J. H., and Makman, M. H., 1972, *Proc. Natl. Acad. Sci. U.S.A.* **69**:539–543.
195. Brown, J. H., and Makman, M. H., 1973, *J. Neurochem.* **21**:447–479.
196. Makman, M. H., Dvorkin, B., Horowitz, S. G., and Thal, L. J., 1980, *Brain Res.* **194**:403–418.
197. Demenge, P., Mouchet, P., Guerin, B., and Feuerstein, C., 1981, *J. Neurochem.* **37**:53–59.
198. Goldberg, L. I., Volkman, P. H., and Kohli, J. D., 1978, *Annu. Rev. Pharmacol. Toxicol.* **18**:57–79.
199. Goldberg, L. I., and Toda, N., 1975, *Circ. Res.* **36**:I97–I102.

200. Toda, N., and Goldberg, L. I., 1975, *Cardiovasc. Res.* **9:**384–89.
201. Goldberg, L. I., Kohli, J. D., Katake, A. N. and Volkman, P. H. 1978, *Fed. Proc.* **37:**2396–2402.
202. Goldberg, L. I., Kohli, J. D., Kotake, A. N., and Volkman, P. H., 1977, *Adv. Biochem. Pharmacol.* **16:**251–256.
203. Attie, M. F., Brown, E. M., Gardner, D. G., Spiegel, A. M., and Aurbach, G. D., 1980, *Endocrinology* **107:**1776–1781.
204. Lee, T., and Seeman, P., 1980, *Am. J. Psychiatry* **137:**191–197.
205. Owen, F., Crow, T. J., Poulter, M., Longden, A., and Riley, G. J., 1978, *Lancet* **2:**223–226.
206. Cross, A. J., Crow, T., and Owen, F., 1981, *Psychopharmacology* **74:**122–124.
207. Mackay, A. V. P., Bird, E. D., Spokes, E. G., Rosser, M., Iversen L. L., Creese, I., and Snyder, S. H., 1980, *Lancet* **2:**915–916.
208. Olsen, R. W., Reisine, T. D., and Yamamura, H. I., 1980, *Life Sci.* **27:**801–808.
209. Ezrin-Waters, C., Seeman, M. V., and Seeman, P., 1981, *J. Clin. Psychiatry* **42:**16–22.
210. Baldessarini, R. J., and Tarsy, D., 1981, *Annu. Rev. Neurosci.* **3:**23–41.
211. Tarsy, D., 1983, *Clin. Neuropharmacology* **6:**91–99.
212. Richardson, M. A., and Craig, T. J., 1982, *Am. J. Psychiatry* **139:**341–343.
213. Lee, T., Seeman, P., Rajput, A., Farley, I. J., and Hornykiewicz, O., 1978, *Nature* **278:**59–61.
214. Rinne, U. K., Sonninen, V., and Laaksonen, H., 1979, *Adv. Neurol.* **24:**259–274.
215. Reisine, T. D., Fields, J. Z., Yamamura, H. I., Bird, E. D., Spokes, E., Schreiner, P. S., and Enna, S. J., 1977, *Life Sci.* **21:**335–344.
216. Hyttel, J., 1979, *J. Neurochem.* **33:**641–646.
217. Tarsy, D., and Baldessarini, R. J., 1977, *Biol. Psychiatry* **12:**431–450.
218. Costall, B., Naylor, R. J., and Owen, R. T., 1978, *Eur. J. Pharmacol.* **48:**29–36.
219. Seeman, P., and List, S., 1982, *Advances in Pharmacology and Therapeutics 11*, Volume 2, (H. Yoshida, Y. Hagihara, and E. Ebashi, eds.), Pergamon Press, New York.
220. Baldessarini, R. J., and Tarsy, D., 1979, *Int. Rev. Neurobiol.* **2:**1–45.
221. Scatton, B., 1977, *Eur. J. Pharmacol.* **46:**363–369.
222. Muller, P., and Seeman, P., 1977, *Life Sci.* **21:**1751–1758.
223. Owen, F., Cross, A. J., Waddington, J. L., Poulter, M., Gamble, S. J., and Crow, T. J., 1980, *Life Sci.* **26:**55–59.
224. Hitri, A., Weiner, W. J., Borison, R. L., Diamond, B. I., Nausieda, P., and Klawans, H. L., 1978, *Ann. Neurol.* **3:**134–140.
225. Fuxe, K., Ogren, S.-O., Hall, H., Agnati, L. F., Jonsson, G., and Gustafsson, J.-A., 1980, *Adv. Biochem. Psychopharmacol.* **24:**193–206.
226. Cross, A. J., Crow, T. J., and Owen, F., 1979, *Br. J. Pharmacol.* **64:**87P–88P.
227. Friend, W. C., Brown, G. M., Jawahir, G., Lee, T., and Seeman, P., 1978, *Am. J. Psychiatry* **135:**839–841.
228. Thornburg, J. E., and Moore, K. E., 1975, *J. Pharmacol. Exp. Ther.* **192:**42–49.
229. Skirboll, L. R., and Bunney, B. S., 1979, *Life Sci.* **25:**1419–1434.
230. Ungerstedt, U., 1971, *Acta Physiol. Scand.* *[Suppl.]* **367:**69–93.
231. Creese, I., Burt, D. R., and Snyder, S. H., 1977, *Science* **197:**596–598.
232. Thal, L., Mishra, R. K., Gardner, E. L., Horowitz, S. G., Varmuza, S., and Makman, M. H., 1979, *Brain Res.* **170:**381–386.
233. Staunton, D. A., Magistretti, P. J., Shoemaker, W. J., and Bloom, F. E., 1982, *Brain Res.* **232:**391–400.
234. Staunton, D. A., Magistretti, P. J., Shoemaker, W. J., Deyo, S. N., and Bloom, F. E., 1982, *Brain Res.* **232:**401–412.
235. Creese, I., and Sibley, D. R., 1981, *Annu. Rev. Pharmacol. Toxicol.* **21:**359–391.
236. Mishra, R. K., Wong, Y.-W., Varmuza, S. L., and Tuff, L., 1978, *Life Sci.* **23:**443–46.
237. Howlett, D. R., and Nahorski, S. R., 1979, *Brain Res.* **161:**173–178.
238. Lew, J. Y., Nakamura, S., Battista, A. F., and Goldstein, M., 1979, *Commun. Psychopharmacol.* **3:**179–83.
239. Quik, M., and Iversen, L. L., 1978, *Naunyn Schmeidebergs Arch. Pharmacol.* **304:**141–145.
240. Russell, P., and Madras, B. K., 1983, submitted.

241. Labrie, F., Di Paolo, T., Raymond, V., Ferland, L., and Beaulieu, M., 1980, *Ergot Compounds and Brain Function: Neuroendocrine and Neuropsychiatric Aspects* (M. Goldstein, ed.), Raven Press, New York, p. 217.

242. Bedard, P. J., Langelier, P., Dankova, J., Villeneuve, A., Di Paolo, T., Barden, N., Labrie, F., Bossier, J. R., and Euvrard, C., 1979, *Adv. Neurol.* **24**:411.

243. Di Paolo, T., Carmichael, R., Labrie, F., and Raynaud, J.-P., 1979, *Mol. Cell. Endocrinol.* **16**:99–112.

244. Bedard, P., Dankova, J., Boucher, R., and Langelier, P., 1981, *Can. J. Physiol. Pharmacol.* **56**:538–541.

245. Di Paolo, T., Payet, P., and Labrie, F., 1981, *Eur. J. Pharmacol.* **78**:105–106.

246. Hruska, R. E., and Silbergeld, E. K., 1980, *Eur. J. Pharmacol.* **61**:397–400.

247. Hruska, R. E., Pitman, K. T., Silbergeld, E. K., and Ludmer, L. M., 1982, *Life Sci.* **30**:547–553.

248. Jenner, P., Rupniak, N. M. J., Hall, M. D., Dyer, R., Leigh, N., and Marsden, C. D., 1981, *Eur. J. Pharmacol.* **76**:31–36.

249. Bowers, M. B., and Rozitis, A. 1976, *Eur. J. Pharmacol.* **39**:109–115.

250. Chouinard, G., Annable, L., Jones, B. D., and Collu, R., 1981, *Acta Psychiatr. Scand.* **64**:353–362.

251. Curry, S. H., Davis, J. M., Marshall, H. L., and Janowsky, D. S., 1970, *Arch. Gen. Psychiatry* **22**:209–215.

252. Creese, I., and Snyder, S. H., 1977, *Nature* **270**:180–182.

253. Davis, M. M., Erickson, S., and Dekirmenjian, H., 1978, *Psychopharmacology: A Generation of Progress* (M. A. Lipton, A. DiMascio, and K. F. Killiam, eds.), Raven Press, New York, pp. 905–916.

254. Cohen, B. M., Herschel, M., Miller, E. M., Mayberg, H., and Baldessarini, R., 1980, *Neuropharmacology* **19**:663–668.

255. Cohen, B. M., Herschel, M., and Aoba, A., 1979, *Psychiatry Res.* **1**:199–208.

256. Rosenblatt, J. E., Pert, C. B., Cobson, J., Van Kammen, D. P., Scott, R., and Bunney, W. E., Jr., 1979, *Commun. Psychopharmacol.* **3**:153–168.

257. Cohen, B. M., and Lipinski, J. F., 1981, *Neuroreceptors—Basic and Clinical Aspects* (E. Usdin, W. E. Bunney, and J. M. Davis, eds.), John Wiley & Sons, New York, pp. 199–214.

258. Uzan, A., Phan, T., and Le Fur, G., 1981, *J. Pharm. Pharmacol.* **33**:102–103.

259. Le Fur, G., Phan, T., and Uzan, A., 1980, *Life Sci.* **26**:1139–1148.

260. Le Fur, G., Meininger, V., Phan, T., Gerard, A. Baulac, M., and Uzan, A., 1980, *Life Sci.* **27**:1587–1591.

261. Le Fur, G., Phan, T., Canton, T., Tur, C., and Uzan, A., 1981, *Life Sci.* **29**:2737–2749.

262. Maloteaux, J. M., Waterkein, C., and Laduron, P. M., 1982, *Arch. Int. Pharmacodyn. Ther.* **258**:174–176.

263. Fleminger, S., Jenner, P., and Marsden, C. D., 1982, *J. Pharmacol.* **34**:658–663.

264. Madras, B. K., Blaschuk, K., Scully, K., Tang, S. W., 1983, *Soc. Neurosci. Abstr.* **9**:1054.

265. Strauss, W. L., Ghai, G., Fraser, C. H., and Venter, J. C., 1979, *Arch. Biochem. Biophys.* **196**:566–573.

266. Ruegg, V. T., Hiller, J. M., and Simon, S. J., 1980, *Eur. J. Pharmacol.* **64**:367–368.

267. Gorissen, H., and Laduron, P., 1979, *Nature* **279**:72–74.

268. Madras, B. K., Davis, A., Kunashko, P., and Seeman, P., 1980, *Psychopharmacology and Biochemistry of Neurotransmitter Receptors*, (H. I. Yamamura, R. W. Olsen, and E. Usdin, eds.), Elsevier/North Holland, New York, pp. 411–419.

269. Lerner, M. H., Rosengarten, H., and Friedhoff, A. J., 1981, *Life Sci.* **29**:2367–2374.

270. Lew, J. Y., Fong, J. C., and Goldstein, M., 1981, *Eur. J. Pharmacol.* **72**:403–405.

271. Gorissen, H., Ilien, B., Aerts, G., and Laduron, P., 1980, *FEBS Lett.* **121**:133–138.

272. Clement-Cormier, Y. C., Meyerson, L. R., and McIsaac, A., 1980, *Biochem. Pharmacol.* **29**:2009–2016.

273. Gorissen, H., and Laduron, P. M., 1978, *Life Sci.* **23**:575–580.

274. Chan, B., Madras, B. K., Davis, A., and Seeman, P., 1981, *Eur. J. Pharmacol.* **74**:53–59.

275. Varmuza, S., and Mishra, R. K., 1981, *Pharmacol. Res. Commun.* **13**:587–605.

276. Lo, M. M. S., Strittmatter, S. M., and Snyder, S. H., 1982, *Proc. Natl. Acad. Sci.* **79**:680–684.

277. Chan, B., Seeman, P., Davis, A., and Madras, B. K., 1982, *Eur. J. Pharmacol.* **81**:111–116.
278. Freedman, J. E., and Snyder, S. H., 1981, *J. Biol. Chem.* **24**:13172–13179.
279. Lilly, L., Magnan, J., Davis, A., and Seeman, P., 1982, *Soc. Neurosci. Abstr.* **8**:719.
280. Clement-Cormier, Y. C., and Kendrick, P. E., 1981, *Biochem. Pharmacol.* **16**:2197–2202.
281. Lilly, L., Davis, A., Madras, B. K., Fraser, C. H., Venter, J. C., and Seeman, P., 1981, *Soc. Neurosci. Abstr.* **7**:10.
282. Madras, B. K., Davis, L., Lilly, C., Fraser, C., Venter, J. C., and Seeman, P., 1981, *Soc. Neurosci. Abstr.* **7**:10.
283. Creese, I., Morrow, A. L., Leff, S. E., Sibley, D. R., and Hamblin, M. W., 1982, *Int. Rev. Neurobiol.* **23**:255–301.

6

Central Serotonin Receptors

M. Hamon, S. Bourgoin, S. El Mestikawy, and C. Goetz

1. CURRENT VISTAS ON SEROTONIN IN THE CENTRAL NERVOUS SYSTEM

With the catecholamines (dopamine, norepinephrine, and epinephrine), serotonin (5-HT, 5-hydroxytryptamine) is one of the best known biogenic amine neurotransmitters in the CNS. During the past 20 years, an abundant literature has been devoted to this molecule, exploring its distribution, metabolism, and functions. However, investigations on the specific receptors involved in serotoninergic neurotransmission really began only in 1974 with the introduction of binding assays using tritium-labeled ligands with high specific radioactivities. The present review is an attempt to summarize the most important findings reported on central 5-HT receptors, mainly during the last 7 years.

1.1. Anatomic Distribution of 5-HT

Numerous techniques have been used for visualizing 5-HT in the CNS. They include histofluorescence from aldehyde-exposed tissues, autoradiography of [^{3}H]5-HT in brain parenchyma after its injection into ventricles or its application onto given structures, and, most recently, immunocytochemistry using a specific antibody obtained in rabbits injected with 5-HT conjugated to serum albumin. Convergent data indicate that 5-HT is selectively contained in neurons with cell bodies mainly located in the raphe area and with terminals diffusely distributed throughout the CNS.[1] Serotoninergic cell bodies that send fibers mainly to the spinal cord are distributed in posterior raphe nuclei (nucleus raphe obscurus, pallidus, and magnus), whereas those with projections invading forebrain areas, i.e., striatum, hippocampus, hypothalamus, septum, amygdala, and cerebral cortex, belong to anterior raphe nuclei (nucleus raphe dorsalis, centralis superior, and linearis rostralis).

M. Hamon, S. Bourgoin, S. El Mestikawy, and C. Goetz • Groupe NB, INSERM U.114, Collège de France, 75231 Paris, Cedex 05, France.

The density of 5-HT terminals in the CNS is generally less than 0.5%, and only a small percentage exhibit true synaptic contacts with other cells.[2] This anatomic distribution is much more consistent with 5-HT being a neuromodulator instead of an actual neurotransmitter in the CNS. Indeed, serotoninergic neurons exert an influence on several central functions (waking and sleeping, food consumption, pain transmission, neuroendocrinological functions) without directly participating in the specific neuronal organization involved in a given function.

Outside the CNS, 5-HT is also present in the pineal gland, where it is mainly the precursor of melatonin. Recently, serotoninergic neurons have been identified in the peripheral nervous system, notably in plexuses innervating the gut. These neurons seem to have metabolic properties very similar to those observed in central serotoninergic neurons.[3]

1.2. Serotonin Synthesis

Two enzymatic steps are necessary for the synthesis of 5-HT from its natural precursor, tryptophan, an essential amino acid in mammals.[4] After being taken up in serotoninergic neurons by a high-affinity carrier, tryptophan is first converted into 5-hydroxytryptophan (5-HTP) by tryptophan hydroxylase (E.C. 1.14.16.4). The 5-HTP is then decarboxylated into 5-HT by aromatic amino acid decarboxylase (E.C. 4.1.1.28) (Fig. 1). Whereas tryptophan hydroxylase is a very specific enzyme only found inside serotoninergic neurons, aromatic amino acid decarboxylase can also decarboxylate tryptophan, tyrosine, and dihydroxyphenylalanine. In addition, the latter enzyme exists in several cell types besides serotoninergic neurons, notably catecholaminergic neurons and endothelia of brain capillaries. As a consequence, 5-HT is also synthesized in these cells following the peripheral or central administration of 5-HTP.

Under normal conditions, the rate of *in vivo* tryptophan hydroxylation is much less than the V_{max} of tryptophan hydroxylase because of limited availability of substrates (tryptophan and molecular oxygen) and cofactor (tetrahydrobiopterin). In this respect, particular attention has been devoted to the problem of tryptophan transport to brain. Tryptophan is the only amino acid that is not entirely free in serum: about 80–90% is, in fact, noncovalently bound to a specific site onto serum albumin (Fig. 1). Numerous investigations have been carried out to determine whether only free or total tryptophan can cross blood–brain barrier(s) under physiological conditions. More recent studies lend support to the conclusion that total tryptophan enters the brain, since the tryptophan–serum albumin complex rapidly dissociates on reaching the blood-brain barrier.[5] However, numerous pharmacological experiments also suggest that the peripheral binding of tryptophan onto serum albumin can at least partly prevent the transport of the amino acid to brain tissues in some circumstances.[6] In all cases, tryptophan transport is limiting, since the concentration of the amino acid in serotoninergic neurons is below the K_m value of tryptophan hydroxylase for tryptophan under physiological conditions. Accordingly, any change in tryptophan transport and accumulation will affect the rate of 5-HT

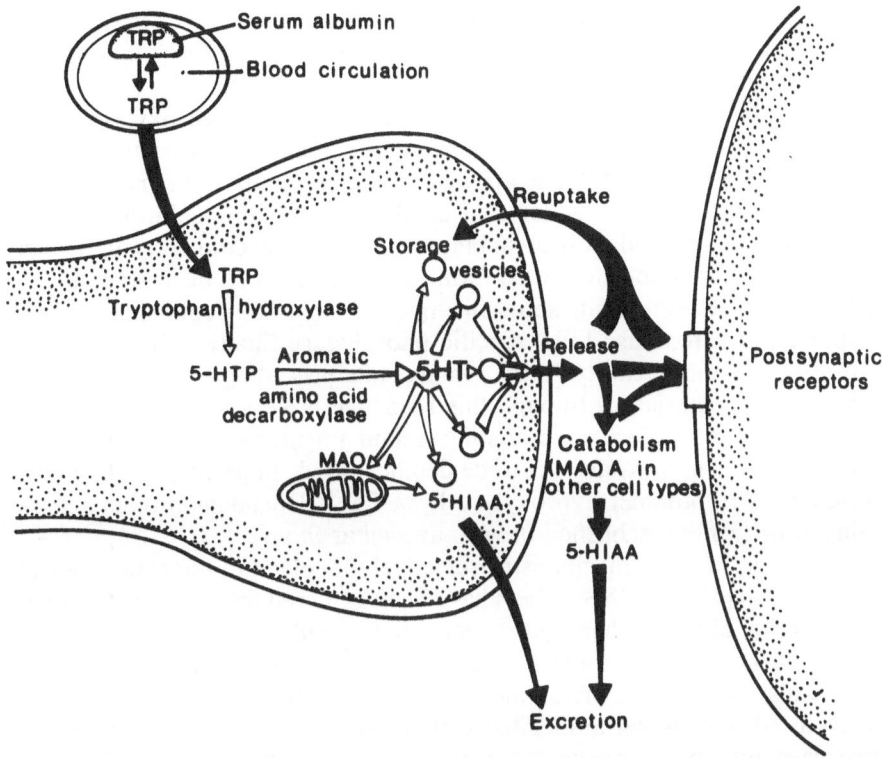

Fig. 1.   Metabolism of 5-HT in central serotoninergic neurons. Blood tryptophan is converted into 5-HT by two successive enzymatic steps: hydroxylation and decarboxylation. The inactivation of 5-HT involves a specific reuptake process and enzymatic degradation into 5-HIAA by MAO A.

synthesis. In particular, peripheral tryptophan loading induces a marked increase in brain tryptophan levels and therefore an acceleration of 5-HT synthesis in serotoninergic neurons.[4]

In numerous circumstances, changes in the rate of 5-HT synthesis occur in the absence of any significant alteration of tryptophan transport and accumulation in brain. This is notably the case during modifications of the nerve impulse flow within serotoninergic neurons. Thus, the acute lesion of serotoninergic cell bodies or fibers induces a rapid decrease in the rate of 5-HT synthesis in corresponding terminal areas.[7] Conversely, the direct electrical stimulation of raphe nuclei (containing the serotoninergic cell bodies) produces a marked elevation of 5-HT synthesis in terminal areas.[8] Furthermore, recordings of 5-HT-containing cells in raphe nuclei have shown that a positive correlation often exists between firing rate and 5-HT synthesis in various brain regions under different pharmacological conditions.[9] *In vitro* investigations using brain slices strongly suggest that these *in vivo* changes involve tryptophan hydroxylase, since the K^+-induced depolarization of serotoninergic neurons is associated with a marked activation of this enzyme, still detectable in soluble extracts of incubated tissues.[4] Calcium plays a crucial role in the depolarization-induced activation of tryptophan hydroxylase, since incubation of tissues with a Ca^{2+} chelator (EGTA) prevents the K^+ effect, whereas Ca^{2+} ionophores produce the enzyme activation even in nondepolarized tissues.[4]

Recent investigations on the regulation of tryptophan hydroxylase stress the involvement of a specific Ca^{2+} and calmodulin-dependent protein kinase in serotoninergic neurons.[4,10] The arrival of action potentials at nerve endings triggers the influx of Ca^{2+}, which first combines with calmodulin to activate this specific Ca^{2+}-dependent protein kinase. This enzyme converts tryptophan hydroxylase into its phosphorylated form, which in turn can be activated by a specific protein effector inside serotoninergic neurons. The dephosphorylation of tryptophan hydroxylase by a specific phosphatase finally permits a rapid return of the enzyme activity to the resting state preceding the arrival of additional action potentials. Although this sequence is very attractive, since it accounts for all the experimental findings relating neuronal activity and 5-HT synthesis, several points have yet to be demonstrated. In particular, the phosphorylation of tryptophan hydroxylase by a Ca^{2+}-dependent protein kinase remains to be firmly established *in vitro* as well as *in vivo*.

In addition to these mechanisms selectively involved in the acute control of 5-HT synthesis in the CNS, others have been explored that account for long-term regulation of tryptophan hydroxylase. In particular, an induction of tryptophan hydroxylase has been shown to occur following reserpine treatment in rats.[11] Similarly, the intrastriatal injection of kainic acid induces a long-lasting increase in 5-HT turnover associated with a sustained increase in tryptophan hydroxylase activity in various brain areas. Whether this effect results from an increased accumulation of tryptophan hydroxylase molecules in serotoninergic neurons or from a long-lasting activation of the existing enzyme has not yet been entirely answered.[12]

1.3. Serotonin Inactivation

In serotoninergic neurons, newly synthesized 5-HT rapidly accumulates in vesicles, where it is protected from the catabolizing enzyme, monoamine oxidase, type A (MAO A). After being released, the bulk of 5-HT is taken up by presynaptic serotoninergic terminals via a specific membrane carrier with a high affinity for the indoleamine ($K_m = 0.05$ μM). The remaining 5-HT can enter other cell types, notably glial cells, which contain MAO A activity for converting 5-HT into 5-hydroxyindole acetic acid (5-HIAA). This metabolite is then transported to blood partly from brain parenchyma and partly via its previous diffusion into cerebrospinal fluid.

Under normal conditions, only a very small fraction of 5-HT, i.e., nonvesicular-bound 5-HT, can be catabolized to 5-HIAA in central serotoninergic neurons. Accordingly, the contribution of 5-HIAA in serotoninergic neurons to total 5-HIAA in brain is likely small, and brain 5-HIAA levels can be used as an index of 5-HT release *in vivo*. Indeed, various treatments known to stimulate central serotoninergic neurons induce an elevation of 5-HIAA levels in brain.[6] However, following a tryptophan load, 5-HT synthesis is so much increased that the storage capacity of vesicles is rapidly saturated. This leads to an increase in the concentration of free 5-HT available to MAO A inside serotoninergic neurons.[13] Under such conditions, 5-HIAA levels likely reflect intraneuronal catabolism of 5-HT rather than 5-HT release. This example illustrates that changes in 5-HIAA levels or in the ratio of 5-HIAA/5-HT in brain are not always related to modifications in the activity of serotoninergic neurons. Instead, such changes can reflect modifications in the intracellular compartmentation of the indoleamine in these neurons.

2. CHARACTERISTICS OF 5-HT RECEPTORS IN THE CENTRAL NERVOUS SYSTEM

2.1. Pharmacological Observations

Pharmacological studies have shown that any change, even a minor one, in the molecular structure of an active compound can alter its properties, suggesting that this structure must be recognized by specific sites on targets to trigger the pharmacological action. This general observation also applies to 5-HT, as shown, for instance, with LSD isomers. In brain as well as in the periphery, *d*-LSD is a potent 5-HT-mimetic agent, whereas its mirror image, *l*-LSD, is devoid of any significant pharmacological action when used in concentrations even more than 100 times that of the *d* enantiomer.[14]

The first studies on 5-HT receptors concerned their pharmacological characteristics on guinea pig ileum and led Gaddum and Picarelli[15] to postulate the existence of two receptor types, M and D, based on their sensitivities to (putative) antagonists such as morphine (M) and dibenzyline (D, phenoxybenzamine). Subsequent studies have shown that this classification is too simple to give a correct picture of the actual status of 5-HT receptors even in the periphery.[16]

Development of pharmacological studies on 5-HT receptors using various experimental approaches (electrophysiology, behavioral observations, measurements of smooth muscle contractions, etc.) led to the conclusion that several types of these receptors do exist in the brain and the periphery in vertebrates and in invertebrates. However, the situation is extremely confused, since the putative receptors identified by some authors do not correspond generally to those described by others. In addition to the first classification of M and D types,[15] 5-HT receptors called A, A′, B, C, α, and β have been identified in molluscs[17]; S_1, S_2, and S_3[18] as well as 5-HT$_1$ and 5-HT$_2$[19] receptors have been characterized in the rat brain; 5-HT$_{ETMIC}$, 5-HT$_{ETMIF}$, and 5-HT$_{IL}$ have been proposed for the peripheral nervous system.[16] Undoubtedly, pharmacological analysis has been and remains a valuable approach for identifying the various kinds of serotoninergic receptors including those in the CNS. However, results obtained are extremely difficult to reconcile, indicating that additional experimental approaches are needed to better assess the actual status of 5-HT receptors in the CNS.

Table I
Tritium-Labeled Ligands for Studies on Central 5-HT Receptors

Tritium-labeled compound	*In vitro*[a] 5-HT$_1$	5-HT$_2$	*In vivo*[a]
Agonists			
[^{3}H]5-HT	Yes[b]	No[c]	No
[^{3}H]Lisuride	Yes	No	?
Antagonists			
[^{3}H]Spiperone	No	Yes	Yes
[^{3}H]Metergoline	No	Yes	Yes
[^{3}H]Mianserin	No	Yes	?
[^{3}H]Ketanserin	No	Yes	Yes
[^{3}H]Methiothepin	No	No	Yes
Mixed agonist–antagonist			
[^{3}H]LSD	Yes	Yes	Yes
Other molecule			
[^{3}H]Quipazine	No	No	?

[a] "*In vitro*" refers to binding assays with crude or synaptosomal membranes from various brain areas. "*In vivo*" refers to measurements of ^{3}H repartition in brain following i.v. administration of a tritiated compound. Studies with [^{3}H]spiperone, [^{3}H]metergoline, [^{3}H]ketanserin, [^{3}H]methiothepin, and [^{3}H]LSD have shown that these molecules labeled specific sites with pharmacological characteristics expected of typical 5-HT receptors in brain. As a rule, labeled agonists bind to 5-HT$_1$ sites, and labeled antagonists bind to 5-HT$_2$ sites.
[b] Yes: appropriate ligand.
[c] No: unsuitable ligand.

2.2. Biochemical Studies

2.2.1. Binding of Radioactive Ligands

2.2.1a. In Vitro Studies. Since the pioneer work of Marchbanks,[20] Alivisatos *et al.*,[21] and Farrow and Van Vunakis,[22] binding studies with radioactive ligands have proven to be one of the most powerful means to analyze the biochemical and pharmacological characteristics of putative 5-HT receptors in brain. In fact, the present knowledge of 5-HT receptors mainly results from binding studies carried out since 1974,[23] when rapid filtration and centrifugation techniques were introduced instead of equilibrium dialysis for measuring the proportion of radioactive ligand specifically bound to membranes.

As illustrated in Table I, several radioactive molecules—[^{3}H]5-HT itself, [^{3}H]LSD, [^{3}H]spiperone, and, more recently, [^{3}H]mianserin, [^{3}H]metergoline, [^{3}H]ketanserin, and [^{3}H]lisuride—are currently available for labeling central 5-HT receptors. Briefly, brain membranes (from lysed synaptosomes, for instance) are first washed and incubated to remove all endogenous 5-HT still bound to them.[24] They are then mixed with the radioactive ligand until the maximum binding is reached (equilibrium) and finally collected by filtration through glass fiber (GF/B or GF/C) filters. Nonspecific binding is defined as that persisting in the presence of a large excess of 5-HT itself (10 μM) or of a given agonist or antagonist (metergoline, cinanserin). In the presence of nanomolar concentrations of [^{3}H]5-HT or [^{3}H]LSD, about 70–80% of total bind-

Fig. 2. Regional distribution of [3H] in the rat brain following an intravenous injection of [3H]methiothepin. Abscissa: Total [3H] associated with crude membranes from various brain areas 24 h after an i.v. injection of [3H]methiothepin (100 μCi/rat). Ordinate: Specific binding of [3H]methiothepin to crude membranes from various brain regions 1 h after an i.v. injection of [3H]methiothepin (100 μCi/rat). Specifically bound [3H]methiothepin corresponds to the difference (△) between [3H] bound to membranes of five rats treated with [3H]methiothepin alone minus that found in five rats following coadministration of radioactive and "cold" (20 mg/kg i.p.) methiothepin. Each point is the

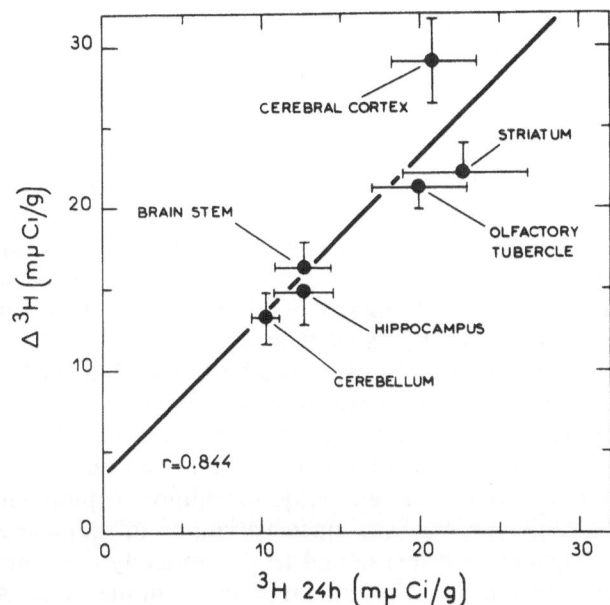

mean ± S.E.M. of five separate determinations. r = correlation coefficient between [3H] 24 h (abscissa) and △[3H] (ordinate) for the six regions analyzed.

ing is displaceable by "cold" 5-HT, agonists, and antagonists, indicating that only 20–30% of total radioactivity is associated with nonspecific sites. With [3H]spiperone, [3H]mianserin or [3H]metergoline, the specific binding maximally accounts for 50–60% of total binding even in the richest brain areas (cerebral cortex, hippocampus in the rat).

Attempts to use other radioactive molecules such as [3H]quipazine, a putative 5-HT agonist, and [3H]methiothepin, a potent 5-HT antagonist, have been largely unsuccessful. In fact, [3H]quipazine binds to a low-affinity site in brain membranes that is more or less related to the [3H]5-HT uptake carrier. Pharmacological analysis of this site clearly demonstrates that it is completely distinct from a putative low-affinity 5-HT receptor (unpublished observations). [3H]Methiothepin binds to numerous nonspecific sites, so that the relative proportion of labeling on 5-HT-related sites is too small to allow accurate quantitative studies.[25]

2.2.1b. In Vivo Studies. The heterogeneous distribution of radioactivity in brain following the peripheral administration of [3H]LSD,[26] [3H]spiperone,[27] [3H]metergoline, or [3H]methiothepin (Fig. 2) to mice or rats is considered an argument supporting the *in vivo* approach for studying central 5-HT receptors. The bound radioactivity is in fact associated with specific 5-HT sites, since pre- or coadministration of various 5-HT agonists and antagonists can prevent, at least partly, the accumulation of [3H] in several brain regions.[26,27] However, unavoidable difficulties linked to blood–brain barrier passage and catabolism of injected drugs render the measurement of *in vivo* accumulation of tritiated ligands in brain rarely sufficient for assessing the functional status of central 5-HT receptors.

2.2.2. Serotonin-Sensitive Adenylate Cyclase

Some 5-HT receptors are linked to adenylate cyclase in brain as evidenced by a more than two-fold stimulation by 5-HT of adenylate cyclase activity in colliculi homogenates of newborn rats.[28,29] Generally, brain homogenates (or membranes) are incubated under appropriate conditions with $[\alpha\text{-}^{32}P]ATP$ in the presence or the absence of 5-HT. The increase of cyclic $[^{32}P]AMP$ production in response to 5-HT is taken as a measure of 5-HT-sensitive adenylate cyclase activity in brain tissues.[29] Another method consists of using "cold" ATP as the substrate and measuring cyclic AMP formed by a specific and sensitive radioimmunoassay.[30] The latter method has obvious disadvantages because of possible artifactual interactions of unknown compounds with the tracer binding onto the cyclic AMP antibody. In addition, experimental errors are larger using the radioimmunoassay since variations in radioactivity ($[^{125}I]$succinyl-AMP-tyrosylmethyl ester) bound to the antibody are converted into exponentials, whereas those on the estimation of ^{32}P in the cyclic AMP fraction are linearly converted into picomoles of the cyclic nucleotide. These differences may explain why results on the 5-HT-sensitive adenylate cyclase using different techniques are not always confirmed by all laboratories (see ref. 31). However, all groups agree that 5-HT-sensitive adenylate cyclase corresponds to a true 5-HT receptor in brain, since several specific 5-HT agonists exert the same stimulatory effect as 5-HT.[28–32] Conversely, known 5-HT antagonists prevent the stimulatory effect of 5-HT and related agonists on adenylate cyclase activity in various brain areas.[28–32]

2.2.3. Other Biochemical Approaches

Apart from being associated with adenylate cyclase, some 5-HT receptors may well be coupled to a specific ionophore controlling the membrane permeability to a given ion. Although this has not been frequently investigated, this approach seems very promising, since Segal and Gutnick[33] have shown that 5-HT evokes an increased efflux of K^+ from hippocampal slices.

The inhibitory effect of 5-HT on the Ca^{2+}-dependent release of $[^3H]5\text{-HT}$ (triggered by K^+ or by electrical field stimulation) from brain slices apparently involves autoreceptors. Measurement of $[^3H]5\text{-HT}$ efflux from brain slices in the presence of various 5-HT agonists and antagonists is a very useful approach for characterizing the pharmacological profile of presynaptic 5-HT autoreceptors in brain.[34]

2.3. Characteristics of Specific Binding Sites for 5-HT

2.3.1. High-Affinity Binding Sites for [³H]5-HT

Scatchard plots of $[^3H]5\text{-HT}$ binding onto brain membranes allow the determination of both the apparent affinity $[K_a = K_d^{-1})$ and the total number

(B_{max}) of specific binding sites. Reported data for K_d values generally stand between 1 and 8 nM.[19,24,31,35] Using high concentrations of the labeled ligand, some authors have been able to detect another binding site with a lower affinity (K_d = 10–30 nM[35]). However, even in the case of a Scatchard plot with only one slope corresponding to an apparent K_d between 1 and 8 nM, it cannot be concluded that only one class of binding sites exists in membranes. Pedigo *et al.*[36] postulated the presence of two binding sites with the same affinity for [³H]5-HT to explain the biphasic displacement curve of bound [³H]5-HT by spiperone.

In contrast to the apparent affinity, which does not change significantly from one brain area to another, rather large differences are noted in B_{max} values. In the rat brain, higher B_{max} values are found in the hippocampus, followed by the striatum, colliculi, cerebral cortex, and hypothalamus.[24] Lowest binding capacities are observed in the spinal cord and the cerebellum. These regional differences are quite distinct from those of endogenous 5-HT levels and of [³H]5-HT reuptake sites, suggesting that high-affinity binding sites do not correspond to presynaptic serotoninergic terminals.[24] Studies on the regional distribution of [³H]5-HT binding sites in other species indicate that, even in rodents, significant interspecies differences exist. Thus, although the same distribution is observed in the guinea pig and in the rat, regional differences are generally larger in the CNS of guinea pig.[37] In contrast, no significant difference is observed among the B_{max} values determined in hippocampal, striatal, and cortical membranes in the mouse (unpublished observations). A quite different distribution is observed in the bovine brain, since maximal binding takes place in the substantia nigra, followed by the caudate nucleus, amygdala, and putamen.[37] In man, B_{max} values are highest in the frontal cortex, substantia nigra, and pallidum, whereas the hippocampus has a very low density of high-affinity [³H]5-HT binding sites.[38]

Direct visualization of [³H]5-HT high-affinity binding sites by light microscopic autoradiography of thin brain sections preincubated with 1.0 nM of [³H]5-HT confirmed their heterogeneous distribution in the rat brain.[39] Further developments of this technique using electron microscopic autoradiography would be very helpful to identify the subcellular structures that specifically bind [³H]5-HT under such conditions.[39]

Various fractionation procedures have been used to study the subcellular distribution of [³H]5-HT high-affinity binding sites in brain. In all cases, the highest density of these sites is found in synaptic membranes.[35,40,41] The specific lesion of serotoninergic neurons by an intracerebral injection of 5,7-dihydroxytryptamine (5,7-HT) does not decrease the total number of high-affinity binding sites for [³H]5-HT in any brain area, indicating that these sites are present on postsynaptic membranes. At least in the striatum, [³H]5-HT high-affinity binding sites are partly associated with membranes of intrinsic neurons (located postsynaptically with regard to serotoninergic projections), since a marked loss of these sites has been observed after the selective degeneration of neuronal cell bodies induced by kainic acid injection into this structure.[30,41]

In addition to neuronal membranes, those from glial cells have the capacity to specifically bind [³H]5-HT.[42] The apparent affinity of these glial sites for

[^{3}H]5-HT is slightly lower (K_d = 10 nM) than that of specific sites in neuronal membranes. Although [^{3}H]5-HT binding to glial membranes exhibits some pharmacological properties resembling those expected for a 5-HT receptor,[42] present knowledge of these glial sites is too limited to discuss their possible physiological function(s).

2.3.2. High-Affinity Binding Sites for [^{3}H]Spiperone

Although [^{3}H]spiperone was first proposed as a specific ligand of dopaminergic receptors, it also appears to label 5-HT-related sites in some brain areas, notably the frontal cortex.[43,44] These specific 5-HT-related sites can be separately examined by incubating membranes with [^{3}H]spiperone and a "cold" compound completely masking dopamine receptors such as 2-amino-6,7-dihydroxy-1,2,3,4-tetrahydronaphthalene (ADTN) or sulpiride.[43] Conversely, the binding of [^{3}H]spiperone onto 5-HT-related sites can be selectively inhibited by (rather) specific 5-HT antagonists such as cinanserin, metergoline, and pipamperone.[31,43]

Scatchard plots of [^{3}H]spiperone binding onto membranes of the rat frontal cortex indicate that the labeled ligand occupies only one class of noninteracting high-affinity sites characterized by a K_d equal to 0.5–1.0 nM. The apparent affinity of these sites for 5-HT (K_i = 5–10 μM) is more than three orders of magnitude lower than that of [^{3}H]5-HT high-affinity binding sites.[19,31,43,44]

As noted with [^{3}H]5-HT as the labeled ligand, [^{3}H]spiperone binding onto 5-HT-related sites exhibits marked regional differences in regard to B_{max} values. In the rat, the density of [^{3}H]spiperone binding sites (persisting when DA receptors are occupied by a "cold" ligand) is highest in the cerebral cortex (particularly the frontal cortex) and decreases in the following order: cerebral cortex > caudate nucleus > midbrain = hippocampus > hypothalamus ≥ brainstem > cerebellum. A similar regional distribution is observed in the guinea pig brain.[37] In the bovine brain, maximum binding capacity is found in the caudate nucleus, followed by the amygdala and cingulate gyrus.[37] In all species examined, the regional distribution of [^{3}H]spiperone-5-HT-related sites is therefore clearly distinct from that of [^{3}H]5-HT-labeled sites.[37]

As in the case of [^{3}H]5-HT high-affinity binding sites, those labeled by [^{3}H]spiperone are present mainly in microsomal subfractions enriched in synaptic membranes.[31] Midbrain raphe lesions[45] and intracerebral 5,7-HT treatment,[46] which both induce a marked degeneration of central serotoninergic innervation, do not reduce the total number of [^{3}H]spiperone binding sites in the cerebral cortex and in the hippocampus, indicating that they are located in postsynaptic membranes.

The large differences in densities in various brain regions[37] together with distinct pharmacological and regulatory properties have led to the conclusion that specific sites labeled with [^{3}H]5-HT or [^{3}H]spiperone in fact correspond to two separate classes of 5-HT receptors.[19,31] Peroutka and Snyder[19] proposed calling those labeled with [^{3}H]5-HT 5-HT$_1$ and those binding [^{3}H]spiperone with a high affinity 5-HT$_2$ receptors.

2.3.3. High-Affinity Binding Sites for [³H]LSD

Early studies with [³H]LSD as the labeled ligand clearly indicated that it binds to 5-HT receptors with different properties than those exhibited by [³H]5-HT high-affinity binding sites.[35,40] Recent elegant pharmacological analyses of [³H]LSD binding onto membranes from the rat frontal cerebral cortex have given a statisfactory explanation of these differences.[19] Indeed, [³H]LSD appears to bind equally to 5-HT₁ and 5-HT₂ receptors. In particular, the total number of [³H]LSD specific binding sites in the rat frontal cortex is equal to the sum of those labeled with [³H]5-HT and [³H]spiperone. The direct visualization of [³H]LSD binding sites by light microscopic autoradiography of brain slices preincubated with this labeled ligand[47] confirms that they likely correspond to both 5-HT₁ and 5-HT₂ sites.

With [³H]LSD as the ligand, 5-HT₁ and 5-HT₂ sites can be distinguished by the addition of an appropriate concentration of either spiperone (30 nM) or 5-HT (300 nM), respectively, in the assay mixture.[19] Pharmacological analyses of [³H]LSD binding sites in the presence of 30 nM spiperone closely resemble those exhibited by [³H]5-HT high-affinity binding sites (5-HT₁). Conversely, the efficacy of a given 5-HT agonist or antagonist to displace bound [³H]LSD in the presence of 300 nM 5-HT is highly correlated ($r = 0.99$) with its relative potency to inhibit [³H]spiperone binding in the frontal cortex.[19]

2.3.4. High-Affinity Binding Sites for [³H]Metergoline

Although metergoline is an ergot compound like LSD, its labeled derivative does not bind to exactly the same sites as [³H]LSD in the rat brain.[41] In contrast to LSD, which exhibits mixed agonist–antagonist properties, metergoline is a pure and rather specific 5-HT antagonist in the CNS, and the regional distribution of [³H]metergoline high-affinity binding sites is correlated ($r = 0.92$) with that of another 5-HT antagonist, [³H]spiperone, in brain.

As noted with [³H]spiperone as the labeled ligand, 5-HT antagonists are generally much more potent than 5-HT agonists to displace [³H]metergoline from its specific binding sites.[41] In particular, the apparent K_i of 5-HT against [³H]metergoline binding is only equal to 3.3 μM, a value not significantly different from the K_i of 5-HT against [³H]spiperone binding. More generally, the relative potency of a given 5-HT agonist or antagonist to inhibit [³H]metergoline binding is highly correlated ($r > 0.90$) with that determined with [³H]spiperone as the labeled ligand. These data suggest that [³H]metergoline specific binding sites in fact correspond to 5-HT₂ receptors.

2.3.5. High-Affinity Binding Sites for [³H]Mianserin

In contrast to metergoline, mianserin is not a specific 5-HT antagonist, since it also exhibits significant antihistamine effects.[48] Studies using [³H]mianserin as the ligand have confirmed that both 5-HT and histamine (H₁) specific sites are labeled.[48] However, in the presence of a saturating concentration of triptolidine to occupy all H₁ sites, [³H]mianserin selectively binds to specific sites exhibiting regional distribution and pharmacological properties similar to those of 5-HT$_2$ sites.[48] Because of its relatively high porportion of nonspecific binding, [^{3}H]mianserin can be advantageously replaced by [^{3}H]spiperone (or [^{3}H]metergoline) for studies on 5-HT$_2$ specific sites.

2.3.6. High-Affinity Binding Sites for [^{3}H]Ketanserin

Ketanserin (R 41468) is a potent 5-HT antagonist with a very poor affinity for specific sites labeled with [^{3}H]5-HT (5-HT$_1$).[49] Recent studies with [^{3}H]ketanserin as a ligand indicated that labeled binding sites correspond to 5-HT$_2$ receptors already labeled with [^{3}H]spiperone, [^{3}H]metergoline, and [^{3}H]mianserin. The regional distribution of [^{3}H]ketanserin specific binding sites in the rat brain and the relative efficacy of various 5-HT agonists and antagonists in displacing bound [^{3}H]ketanserin confirmed its usefulness for selectively studying central 5-HT$_2$ sites.[50]

2.3.7. High-Affinity Binding Sites for [^{3}H]Lisuride

Although lisuride stimulates some central 5-HT receptors, it also exhibits significant effects on adrenergic and dopaminergic receptors.[51] Binding studies with [^{3}H]lisuride as a ligand were recently reported[51] and confirmed that numerous monoamine specific sites are involved. Under certain conditions (notably in the presence of 0.1 μM clonidine to occupy adrenergic sites), [^{3}H]lisuride appears to selectively bind to 5-HT-related sites in bovine frontal cortex. Preliminary studies on the pharmacological characteristics of these sites suggest that they may correspond to 5-HT$_1$ receptors.[51]

2.3.8. Summary

In conclusion, two classes of specific binding sites called 5-HT$_1$ and 5-HT$_2$ have been characterized in the rat brain using tritium-labeled ligands. 5-HT$_1$ sites are labeled with two agonists, [^{3}H]5-HT and possibly [^{3}H]lisuride, and a mixed agonist–antagonist, [^{3}H]LSD, and exhibit higher affinities for 5-HT agonists than for antagonists. Conversely, 5-HT$_2$ sites are labeled with 5-HT antagonists including [^{3}H]spiperone, [^{3}H]metergoline, [^{3}H]ketanserin, and [^{3}H]mianserin and a mixed agonist–antagonist, [^{3}H]LSD, and have higher affinities for 5-HT antagonists than for agonists. Whether these two sites correspond in fact to two interconverting forms of the same receptor has been the subject of several reports.[31,35,40] The most recent studies stressing the regional differences between the distributions of 5-HT$_1$ and 5-HT$_2$ sites in brain and the quite distinct pharmacological properties of these two classes of sites have led to the conclusion that they in fact correspond to two separate types of 5-HT receptors.[19,31] Further studies with 5-HT agonists or antagonists binding covalently to 5-HT$_1$ or 5-HT$_2$ sites would be of great interest to demonstrate definitively that 5-HT$_1$ and 5-HT$_2$ sites are physically distinct.

2.4. Characteristics of 5-HT-Sensitive Adenylate Cyclase in the CNS

Von Hungen *et al.*[28] first reported a stimulatory effect of 5-HT on adenylate cyclase activity in brain homogenates from newborn rats. This effect requires micromolar concentrations of the indoleamine and apparently disappears during brain development. Subsequent studies[29] have shown that 5-HT also stimulates adenylate cyclase activity in the brain of adult rats. However, although basal adenylate cyclase activity increases, the absolute increment in cyclic AMP production in response to 5-HT remains constant during development, so that the relative effect of the indoleamine is markedly reduced in adults.[29] The early appearance of 5-HT-sensitive adenylate cyclase during ontogeny may suggest that the associated receptors are those involved in the possible trophic effect of the indoleamine on brain maturation.

The apparent affinity of the 5-HT receptor coupled to adenylate cyclase for 5-HT ($K_{A\mathrm{app}}$ = 0.5–1.0 μM) corresponds neither to that of 5-HT$_1$ nor 5-HT$_2$ sites. Detailed analyses of the pharmacological properties of 5-HT-sensitive adenylate cyclase indicate that this receptor is distinct from those specifically labeled with [^{3}H]5-HT or [^{3}H]spiperone (or other antagonists). In particular, several putative 5-HT agonists such as quipazine, trifluoromethylphenylpiperazine,[52] and RU 24969 [5-methoxy-3-(1,2,3,6-tetrahydro-4-pyridinyl)1H indole],[53] which displace [^{3}H]5-HT and [^{3}H]spiperone from their specific sites, do not interact with 5-HT-sensitive adenylate cyclase (Table II).

Studies on the subcellular distribution of 5-HT-sensitive adenylate cyclase indicate that it is mainly associated with synaptic membranes.[54] The failure of midbrain raphe lesions to affect the stimulatory effect of 5-HT on cyclic AMP synthesis has led to the conclusion that this enzyme is associated with a postsynaptic 5-HT receptor.[55] This has subsequently been confirmed for 5-HT-sensitive adenylate cyclase in the striatum, since the local destruction of neuronal cell bodies by coadministration of kainic acid and L-glutamic acid induces a marked reduction of this enzyme activity in newborn rats.[56]

In addition to this 5-HT-sensitive adenylate cyclase stimulated by micromolar concentrations of the indoleamine, Fillion *et al.*[30,57] have detected another 5-HT-sensitive adenylate cyclase requiring only nanomolar concentrations of 5-HT for its activity. As this latter enzyme is located in postsynaptic neuronal membranes, it might well correspond to the 5-HT$_1$ receptor, which binds [^{3}H]5-HT with a high affinity[30]. However, other groups have been unable so far to observe any stimulatory effect of nanomolar concentrations of 5-HT on adenylate cyclase activity in brain slices, homogenates, or membranes.[31,58–60]

In conclusion, a postsynaptic receptor having an apparent affinity for 5-HT in the micromolar range exists in the CNS of various species (rat[29], horse,[57] monkey,[59] and man[60]). This receptor is coupled to adenylate cyclase and is distinct from those labeled with [^{3}H]5-HT (5-HT$_1$) or [^{3}H]spiperone (5-HT$_2$).

2.5. Physicochemical Properties of Central 5-HT Receptors

Using rat cerebral cortex membranes, Bennett and Snyder[40] observed that several protein-modifying agents such as N-ethylmaleimide and 5,5'-di-

Table II
Effects of GTP and Mn^{2+} on the Apparent Affinity of 5-HT_1 Sites for Various
Agonists and Antagonists

Compound	Effect on 5-HT-sensitive adenylate cyclase[b]	IC$_{50}$ (nM)[a]			(GTP/Mn^{2+})
		Control	GTP	Mn^{2+}	
Agonists					
5-HT	+	6.0	22.3	1.5	14.9
Bufotenine	+	10.9	75.9	3.2	23.7
5-MDMT[c]	+	21.8	71.9	8.1	8.9
8-OH-NN-DPAT[c]	+	74.8	369.2	7.9	46.7
RU 24969[c]	0	20.2	130.2	5.4	24.1
TFMPP[c]	0	190.0	406.0	40.0	10.2
Antagonists					
Metergoline	—	7.8	12.4	17.4	0.7
Methysergide	—	52.0	102.6	61.7	1.7
Methiothepin	—	165.0	243.0	162.0	1.5
Pizotifen	—	334.0	366.0	516.0	0.7
Other drugs					
LSD	(+)	8.2	59.5	2.2	27.0
Quipazine	0	1120	832	1094	0.8

[a] IC$_{50}$ values are the concentrations of drugs reducing by half the specific binding of [^{3}H]5-HT onto hippocampal membranes. Binding assays were carried out with 2–3 nM [^{3}H]5-HT in the absence or the presence of 0.1 mM GTP or 1 mM MnCl$_2$. The ratio (GTP/Mn^{2+}) is obtained by dividing the IC$_{50}$ value in the presence of GTP by that determined in the presence of MnCl$_2$. Each value is the mean of at least three separate determinations.
[b] This column indicates whether a given drug stimulates (+), inhibits (−), or does not interact (0) with 5-HT-sensitive adenylate cyclase activity in colliculi homogenates from new born rats.
[c] 5-MDMT, 5-methoxy-N,N-dimethyltryptamine; TFMPP, trifluoromethylphenylpiperazine; 8-OH-NN-DPAT, 8-hydroxy-N,N-dipropyl-2-aminotetralin; RU 24969, 5-methoxy-3-(1,2,3,6-tetrahydro-4-pyridinyl)1H indole.

thiobis(2-nitrobenzoic acid) reduced the specific binding of [^{3}H]5-HT and [^{3}H]LSD, suggesting that an active SH group plays a critical role. Both proteolytic enzymes (trypsin and α-chymotrypsin) and phospholipases significantly alter [^{3}H]5-HT and [^{3}H]LSD binding, indicating that the specific sites likely involve proteolipid complexes.[40] Some differences, however, exist concerning the relative sensitivies of [^{3}H]5-HT and [^{3}H]LSD binding sites to the same agent. In particular, [^{3}H]5-HT binding appears somewhat more sensitive than [^{3}H]LSD binding to treatment of the membranes with α-chymotrypsin. Since [^{3}H]5-HT specifically labels 5-HT$_1$ sites, whereas [^{3}H]LSD binds to 5-HT$_1$ and 5-HT$_2$ sites, these observations indirectly confirm that 5-HT$_1$ and 5-HT$_2$ sites correspond to two distinct classes of receptors. This conclusion is also that of Poller et al.,[61] who have studied the heat inactivation of these two sites. They observed that [^{3}H]5-HT specific binding to membranes from the rat frontal cortex is reduced by 50% within 10 min at 45°C, whereas [^{3}H]spiperone binding is hardly affected (−10%) under the same condition. As expected for a ligand binding to both 5-HT$_1$ and 5-HT$_2$ sites, [^{3}H]LSD binding is reduced to an intermediate extent compared to [^{3}H]5-HT and [^{3}H]spiperone binding.[61]

Only few attempts have been made to further identify 5-HT receptors. Shih and Cheng[62] reported successful solubilization of 5-HT binding proteins from brain membranes using an arylazide derivative of 5-HT ([³H]nitroarylazidophenylserotonin) as a photoaffinity labeling probe. Exposure of brain membranes to UV in the presence of this compound resulted in irreversibly bound ³H at 5-HT sites. The SDS-polyacrylamide gel electrophoresis of membrane proteins revealed the presence of three distinct labeled bands with molecular weights of 80,000, 49,000, and 38,000, respectively. The proteins with molecular weights 80,000 and 38,000 may be postsynaptic, whereas that with a molecular weight of 49,000 is more likely presynaptic.[62] As the pharmacological characteristics of 5-HT binding onto these proteins have not been reported yet, their possible connection with 5-HT-sensitive adenylate cyclase and 5-HT$_1$ and/or 5-HT$_2$ sites is far from proven.

Another approach was used by Ilien *et al.*,[63] who looked for [³H]spiperone binding protein(s) in material solubilized by lysolecithin from membranes of the rat frontal cortex. A similar experimental procedure has been used successfully by Chan *et al.*[64] except that the radioactive ligand to be recognized by solubilized sites was [³H]mianserin instead of [³H]spiperone. Detailed pharmacological analyses of the solubilized [³H]spiperone and [³H]mianserin binding sites strongly suggest that they both correspond to 5-HT$_2$ sites. However, no biochemical characteristic of these solubilized sites has yet been reported.

2.6. In Vitro Modulations of 5-HT Receptors

2.6.1. Effects of Guanine Nucleotides

As observed for several receptor types, guanosine triphosphate (GTP) and its poorly metabolized analog GppNHp stimulate 5-HT-sensitive adenylate cyclase activity[57,65] and reduce the affinity of 5-HT$_1$ sites for [³H]5-HT and other 5-HT agonists but not for 5-HT antagonists.[31,66] Since this affinity change usually occurs with receptors coupled to adenylate cyclase, Peroutka *et al.*[66] concluded that 5-HT$_1$ sites are those associated with this enzyme in brain membranes. However, detailed comparison of 5-HT-sensitive adenylate cyclase to [³H]5-HT high-affinity binding sites with regard to their evolution during ontogeny, their sensitivies to 5-HT agonists and antagonists, and their fate following kainic acid-induced neuronal degeneration led to the conclusion that 5-HT$_1$ sites and 5-HT-sensitive adenylate cyclase correspond, in fact, to two distinct classes of receptors.[31,52,54,56] Some of the pharmacological evidence supporting this conclusion is given in Table II, which shows the effect of GTP on the apparent affinity of 5-HT$_1$ sites for various agonists and antagonists.

According to the coupling model (5-HT$_1$ ↔ adenylate cyclase[66]), only those 5-HT agonists stimulating adenylate cyclase should have their capacity to displace [³H]5-HT bound to 5-HT$_1$ sites significantly reduced by GTP (as a consequence of the GTP-induced decrease in the affinity of these sites for agonists). This is verified in some cases since the affinity of 5-HT$_1$ sites for agonists stimulating adenylate cyclase (5-HT, 5-methoxy-N,N-dimethyltryptamine, bufotenine, 8-hydroxy-N,N-dipropyl-2-aminotetralin) is reduced by GTP, whereas that for quipazine, a putative agonist not acting on this enzyme,[67] is not affected. Two exceptions are, however, trifluoromethylphenylpiperazine (TFMPP) and 5-methoxy-3-(1,2,3,6-tetrahydro-4-pyridinyl) 1H indole (RU 24969), which do not interact with 5-HT-sensitive adenylate cyclase although their efficacies for displacing [^{3}H]5-HT bound to brain membranes are very sensitive to GTP. Accordingly, the effect of GTP on the affinity of 5-HT$_1$ sites for 5-HT agonists cannot be considered a proof of the coupling of these sites to adenylate cyclase in brain membranes.

The ability of GTP to selectively reduce the affinity of 5-HT$_1$ sites for 5-HT agonists may be of great help in identifying such drugs. Up to now, only one putative agonist, quipazine, does not comply with the GTP effect. However, the direct agonist properties of quipazine are far from proven, and several data in the literature indicate that this drug is in fact a 5-HT antagonist.[16] Binding studies indirectly confirm this conclusion since the affinity of 5-HT$_1$ sites for quipazine is not higher than that of 5-HT$_2$ sites[31,68] in contrast to the case with all other 5-HT agonists. Therefore, reported agonist properties of quipazine likely involve indirect rather than direct interactions with 5-HT receptors. Indeed, by releasing 5-HT and blocking its reuptake and MAO A activity,[67] quipazine may increase the concentration of 5-HT in the synaptic cleft so much that its direct blocking action on postsynaptic receptors is competitively prevented by the indoleamine itself. Such a competition is markedly in favor of 5-HT, since the affinity of 5-HT$_1$ sites for the indoleamine is more than two orders of magnitude higher than that for quipazine.[24]

With [^{3}H]spiperone as the labeled ligand, contradictory results have been reported regarding the sensitivity of 5-HT$_2$ sites to guanine nucleotides. Thus, Peroutka *et al.*[66] claimed that the affinity of this receptor type for 5-HT agonists is not affected by guanine nucleotides, whereas others[31,69,70] observed significant reductions of this parameter in the presence of GTP or GppNHp. This latter finding would not allow one to conclude that 5-HT$_2$ sites are those coupled to adenylate cyclase, as GTP markedly reduces the efficacy of RU 24969 in displacing [^{3}H]spiperone bound to membranes, although, as already noted, this agonist does not interact with 5-HT-sensitive adenylate cyclase (Table II).

Similar modulations by guanine nucleotides have been recently observed using [^{3}H]metergoline as the labeled ligand. As illustrated in Fig. 3, [^{3}H]metergoline binding is not affected by GTP, but its inhibition by 5-HT decreases progressively as a function of GTP concentration in the assay mixture. Similar findings are observed with all putative 5-HT agonists (except quipazine) but not with any antagonist (unpublished observations). Since [^{3}H]metergoline very likely labels the same sites as [^{3}H]spiperone, i.e., the 5-HT$_2$ sites, these findings further suggest that not only the affinity of 5-HT$_1$ but also that of 5-HT$_2$ sites is reduced for 5-HT agonists in the presence of guanine nucleotides.

In conclusion, the specific binding of 5-HT agonists to 5-HT$_1$ and 5-HT$_2$ sites is reduced by GTP although these sites are likely not coupled to adenylate cyclase in brain membranes. Modulations by guanine nucleotides of other receptor types (angiotensin,[71] α-adrenergic,[71] benzodiazepine[72]) not coupled to adenylate cyclase further confirm that no necessary correlation exists between the GTP effect and the coupling of a given receptor to this enzyme.

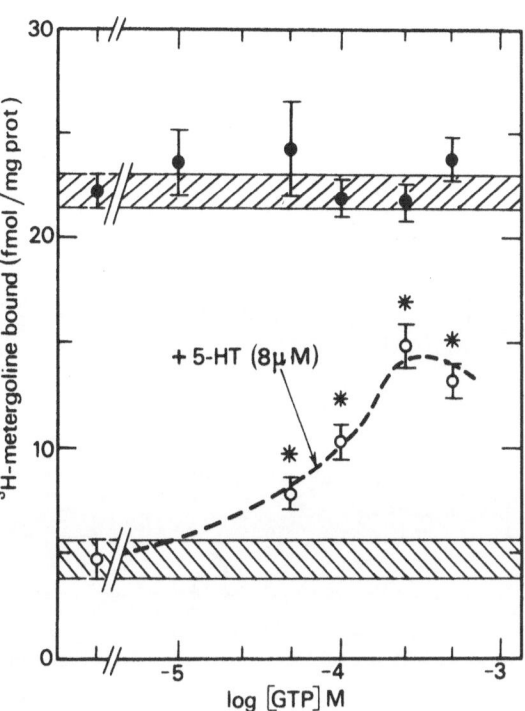

Fig. 3. Effects of GTP on the 5-HT-induced inhibition of [³H]metergoline binding onto microsomal forebrain membranes. Microsomal membranes from the rat forebrain were incubated with 0.15 nM [³H]metergoline and various concentrations of GTP in the absence (●) or the presence (○) of 8 µM 5-HT. Although GTP does not affect the specific binding of [³H]metergoline, it significantly reduces the inhibitory effect of 5-HT on this binding. *P < 0.05 when compared to [³H]metergoline specifically bound in the presence of 8 µM 5-HT alone. The ranges of [³H]metergoline specifically bound in the absence (●) and in the presence (○) of 8 µM 5-HT are indicated by hatched areas.

2.6.2. Effects of Ions

Whereas high concentrations of monovalent cations (>20 mM of Na^+ or K^+) induce a significant inhibition of [³H]5-HT and [³H]LSD binding to brain membranes,[14,40,65] millimolar concentrations of divalent cations such as Ca^{2+}, Mn^{2+}, Mg^{2+}, and Ba^{2+} and trivalent cations such as La^{3+} and Mn^{3+} consistently increase the specific binding of [³H]5-HT (Fig. 4).[40,65] In contrast, the specific binding of [³H]LSD, [³H]spiperone, or [³H]metergoline is unaffected or slightly reduced by millimolar concentrations of Mg^{2+}, Ca^{2+}, or Mn^{2+}.[14,40,41]

The effect of divalent cations results from an increased affinity of $5-HT_1$ sites for [³H]5-HT. In addition, Ca^{2+}, Ba^{2+}, and La^{3+} but not Mn^{2+} or Mn^{3+} also increase the total number of specific binding sites for [³H]5-HT in brain membranes. The effect observed with [³H]5-HT also concerns 5-HT agonists, since the ability of these drugs (but not that of 5-HT antagonists) to displace the labeled ligand bound to brain membranes is increased in the presence of Ca^{2+} or Mn^{2+}. As illustrated in Table II, this occurs not only with 5-HT agonists stimulating 5-HT-sensitive adenylate cyclase but also with other agonists such as RU 24969 and trifluoromethylphenylpiperazine, indicating that the effect of divalent cations does not depend on the possible coupling of 5-HT receptors to adenylate cyclase.

Similar effects of multivalent cations (particularly Mn^{2+} and Mn^{3+}) have been observed on the ability of 5-HT agonists to displace [³H]spiperone or [³H]metergoline from specific $5-HT_2$ sites (unpublished observations). As noted

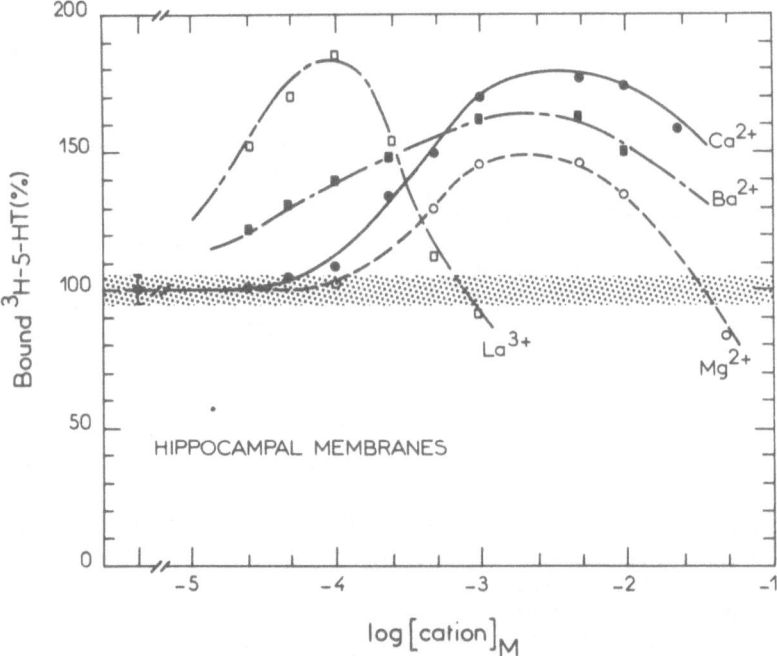

Fig. 4. Effects of various cations on the specific binding of ^{3}H-5-HT to crude membranes from the rat hippocampus. Binding assays were carried out with 1.0 nM ot the labeled ligand and various concentrations (range: 25 μM − 50 mM) of LaCl$_3$ (□), MgCl$_2$ (○), BaCl$_2$ (■) or CaCl$_2$ (●). The specific binding of ^{3}H-5-HT is expressed in percent of that found in the absence of salts. Each point is the mean of quadruplicate determinations.

for 5-HT$_1$ sites, the affinity of 5-HT$_2$ sites for antagonists is not significantly affected by divalent or trivalent cations.

In view of the opposite effects of GTP and divalent cations on the affinity of 5-HT receptors for 5-HT agonists, a simple test can be proposed to assess agonist properties of a given compound. With [^{3}H]5-HT, [^{3}H]spiperone, or [^{3}H]metergoline as the labeled ligand, comparative measurement of IC$_{50}$ values of a compound in the presence of GTP or Mn^{2+} would determine whether it exhibits agonist (or mixed agonist–antagonist) properties on postsynaptic 5-HT receptors. As illustrated in Table II, the ratio of IC$_{50}$ values determined in the presence of GTP (IC$_{50}$ GTP) or Mn^{2+} (IC$_{50}$ Mn^{2+}) is ≥10 with agonists, whereas it does not significantly differ from 1 with antagonists. This test confirms that 8-hydroxy-N,N-dipropyl-2-amino tetralin is a 5-HT agonist, whereas quipazine is likely a 5-HT antagonist (Table II).

2.6.3. Guanine Nucleotide–Divalent Cation Interactions on 5-HT Receptors

Since guanine nucleotides and divalent cations exert opposite effects on central 5-HT receptors, particular attention has been devoted to the possible interactions of these compounds at the level of GTP-binding proteins.[73] Early studies concerned the association between these proteins and receptor–adenylate cyclase complexes, as we observed that GTP stimulated whereas Ca^{2+} inhibited 5-HT-sensitive adenylate cyclase activity.[73] The specific labeling of these proteins could be achieved by choleratoxin-induced ADP-ribosylation.[74] With this technique, the maximal incorporation of $[^{32}P]$ADP-ribose into brain membranes accounted for a number of sites markedly less than that of receptors coupled to adenylate cyclase in brain membranes (unpublished observations). Furthermore, neither 5-HT nor other monoamines (dopamine and norepinephrine) altered the incorporation of $[^{32}P]$ADP-ribose into brain membranes. Therefore, choleratoxin-induced ADP-ribosylation of membrane proteins could not be used for selectively labeling GTP-binding sites modulating monoamine (notably 5-HT) receptors in brain.

Another approach was therefore selected for studying the GTP sites modulating 5-HT (and other) receptors in brain membranes. This protocol simply consists of incubating brain membranes with $[^{3}H]$GTP or $[^{3}H]$GppNHp until equilibrium is reached.[73] Membranes are then collected by centrifugation, and the amount of specifically bound $[^{3}H]$nucleotide (i.e., that displaceable by 0.1 mM unlabeled GTP) is counted. Detailed analyses of this binding indicate that it exhibits characteristics expected for the GTP sites modulating the affinity of 5-HT receptors for agonists.[73] In particular, we observed that the reduction of the GTP effect on $[^{3}H]$5-HT binding onto striatal membranes after the intrastriatal administration of kainic acid is associated with a significant reduction of the apparent affinity and number of $[^{3}H]$GTP (or $[^{3}H]$GppNHp) specific binding sites in these membranes.[73] One major discrepancy, however, concerns the total number of these specific binding sites for guanine nucleotides. In fact, this number markedly surpasses that of all monoamine receptors in brain.[73] The GTP sites coupled to 5-HT receptors therefore probably correspond to only a very small proportion of total GTP sites. This could explain why 5-HT does not significantly affect the characteristics of $[^{3}H]$GTP or $[^{3}H]$GppNHp binding onto brain membranes.[73]

Under conditions in which competitive interactions between GTP and Ca^{2+} exist for 5-HT receptors (Fig. 5), we observed that Ca^{2+}-induced changes in $[^{3}H]$GTP binding and catabolism may account for such interactions.[73] Indeed, divalent cations (notably Ca^{2+} and Mn^{2+}) markedly reduce the affinity of GTP specific sites for labeled guanine nucleotides. Furthermore, they markedly increase the activity of high-affinity GTPase(s) ($K_m = 1$–5 μM) in brain membranes. Accordingly, the stimulating effect of divalent cations on the affinity of 5-HT$_1$ and 5-HT$_2$ receptors for various agonists may be related to their ability to remove (endogenous) GTP from specific sites in membranes. As previously discussed,[73] this may also result from the formation of GTP–divalent cation complexes.

2.6.4. Effects of Membrane Viscosity

Considerable modifications in $[^{3}H]$5-HT binding onto mouse brain membranes have been reported following changes in their viscosity induced by preincubation with cholesterol or fatty acids.[75] For example, the incubation of

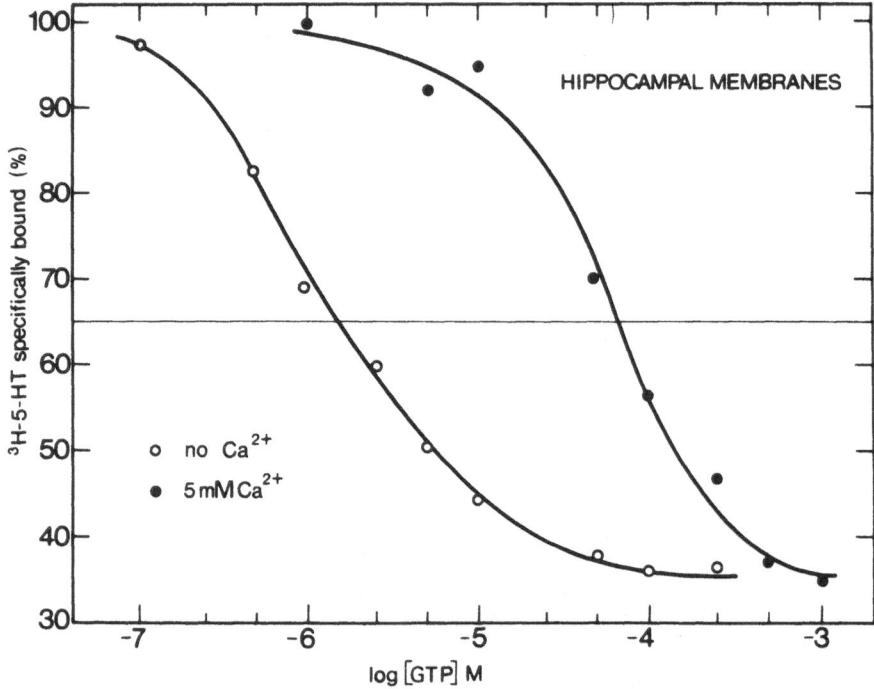

Fig. 5. Effects of Ca^{2+} on the GTP-induced inhibition of ^3H-5-HT binding to membranes from the rat hippocampus. Binding assays were performed in the presence of 1.0 nM of the labeled ligand and various concentrations of GTP in the presence (●) or the absence (○) of 5 mM $CaCl_2$. Each point corresponds to the specific binding of ^3H-5-HT expressed in percent of respective control values. The IC 35 of GTP increased from 1.5 μM to 64 μM when $CaCl_2$ is included in the assay mixture.

mouse forebrain synaptosomal membranes with cholesteryl hemisuccinate or stearic acid to increase their viscosity induces a fivefold increase in the B_{max} of 5-HT$_1$ sites and a threefold decrease in their apparent affinity for [^3H]5-HT.[75] Conversely, the incorporation of lecithin or linoleic acid into brain membranes is associated with parallel decreases in their viscosity and in their capacity to bind [^3H]5-HT onto specific 5-HT$_1$ sites.[75] This effect mainly reflects a significant reduction in the total number of 5-HT$_1$ sites in the membrane preparations.[75]

These findings suggest that changes in the membrane fluidity may be responsible for *in vivo* and/or *in vitro* modifications of [^3H]5-HT binding to brain membranes induced by various treatments.[75] In particular, the Ca^{2+}-induced increase in B_{max} of 5-HT$_1$ sites[65] may well be the consequence of the Ca^{2+}-enhanced viscosity of brain membranes. Further experiments, however, indicate that the results obtained with membranes from the mouse brain [75] may not necessarily be applicable to other membrane preparations. Using membranes from the rat hippocampus, we observed that pretreatment with *cis*-vaccenic acid (≥0.1 mM) under conditions known to decrease membrane viscosity[76] induced a marked reduction in [³H]5-HT binding capacity. However, pretreatment with palmitic or stearic acid (up to 10 mM) to raise the membrane viscosity did not induce a significant increase in [³H]5-HT binding onto rat hippocampal membranes (in contrast to the increase reported by Heron *et al.*[75] with mouse brain membranes). In addition to interspecies differences, variations in the rate of fatty acid incorporation into membranes under differing experimental conditions (those of Heron *et al.*[75] on the one hand and ours[76] on the other) may also explain these discrepancies concerning the influence of membrane viscosity on [³H]5-HT binding.

2.6.5. Effects of Other Modulating Agents

All steroids do not exert the same effect as cholesterol on [³H]5-HT binding,[75] since membrane pretreatment with low concentrations of estradiol (10^{-10}–10^{-8} M) results in a marked reduction in the total number of 5-HT₁ sites in rat brain membranes.[77] This effect may well have some functional relevance, since the B_{max} of [³H]5-HT binding sites in membranes of the basal forebrain of female rats is lower during estrus and proestrus than during diestrus.[77]

Preexposure of brain membranes to 5-HT (10 nM) or 5-HT agonists, but not 5-HT antagonists, followed by extensive washing induces a significant increase in the apparent affinity of 5-HT₁ sites for [³H]5-HT associated with a reduction of their B_{max}.[78] The same effect is evoked by nanomolar concentrations of some antidepressants.[78] The large changes in [³H]5-HT binding (-70% in the B_{max} following membrane exposure to 1 nM imipramine[78]) produced by antidepressants are, however, difficult to reconcile with consistent observations indicating that acute administration of these drugs does not alter the characteristics of [³H]5-HT binding to membranes from any region of the rat brain (Table III).

In vitro modulations of 5-HT binding sites indicate that 5-HT receptors likely exist in interconverting forms with different affinities for the agonists. Experiments with soluble receptors will be of great help in exploring the possible physicochemical changes associated with these allosteric interconversions.

2.7. In Vivo Modulations of 5-HT Binding Sites

2.7.1. Technical Comments

Although significant alterations in [³H]5-HT (Table III) and [³H]spiperone (Table IV) binding have been reported after various *in vivo* manipulations, several remarks have to be made before the possible physiological consequences of such alterations are analyzed. Endogenous 5-HT remains firmly bound to brain membranes even after extensive washing and can interfere with [³H]5-HT binding.[24] Therefore, treatments affecting 5-HT levels [*p*-chlorophenylalanine (PCPA), which inhibits 5-HT synthesis; reserpine, which depletes 5-HT stores; MAO inhibitors, which increase 5-HT levels; specific lesions of serotoninergic neurons) may well alter *in vitro* [³H]5-HT binding simply

Table III
Apparent Up- and Down-Regulation of 5-HT$_1$ Sites in Various Brain Regions Following in Vivo Treatments[a]

Treatment	Structure	K_d (%)	B_{max} (%)	Reference
Lesions				
Electrolytic lesion of the midbrain raphe 2 weeks before death	Forebrain	−52*	+6	40
Electrolytic lesion of the midbrain raphe 2 weeks before death	Hippocampus	0	(+44*)	45
	Frontal cortex	No significant change		
5,6-DHT intraventricularly 10 days before death	Whole brain	−21	−24	79
5,7-DHT intraventricularly 4 weeks before death	Cerebral cortex	(+21*)		80
5,7-DHT intracisternally 18 days before death	Spinal cord	No significant change		u.o.
5,7-DHT in forebrain serotoninergic bundles 18 days before death	Hippocampus	+17	+39*	24
	Striatum	+1	+25	
	Cerebral cortex	No significant change		
5,7-DHT in forebrain serotoninergic bundles 20 days before death	Hippocampus	−12	+31*	41
	Striatum	−18	+21	
5,7-DHT into the dorsal and the median raphe nuclei 3 weeks before death	Cerebral cortex	+16 to 17	+9 to 13	81
5,7-DHT into the dorsal raphe nucleus 1 month before death	Substantia nigra	+236*	+211*	82
	Striatum	−5	−17	
5,7-DHT into the substantia nigra 3 weeks before death	Substantia nigra	No significant change		u.o.
5-HT depletion				
Reserpine for 2 weeks	Cerebral cortex	(+54*)		40
PCPA for 2 weeks	Cerebral cortex	(+44*)		40
PCPA for 3 days	Cerebral cortex	−52*	−6	88
PCPA (3 injections within 14 days)	Forebrain	−15	+43*	89
PCPA for 12 days		−43*	+30*	
PCPA (3 injections within 7 days)	Cerebral cortex	0	+19*	80
5-HT releasers and uptake inhibitors				
d-Fenfluramine for 28 days	Dicencephalon	−40*	−30*	90
	Cerebral cortex	−40*	−25*	
Imipramine	Whole brain	−1	−36*	79
Desipramine for 3 days		+22	−29*	
Dimetacrine		+4	−36*	
Chlorimipramine for 4 weeks	Cerebral cortex	No significant change		91
Imipramine	Frontal cortex	(−34*)		92
Desipramine for 3 weeks		(−40*)		
Nisoxetine + fluoxetine		(−37*)		
Imipramine for 3 weeks	Striatum	(−67*)		92
	Hippocampus	(−63*)		

(*Continued*)

Table III   (*Continued*)

Treatment	Structure	K_d (%)	B_{max} (%)	Reference
Chlorimipramine for 4–16 days	Cerebral cortex	No significant change		93
Amitriptyline or chlorimipramine for 16 days	Hippocampus	No significant change ⎫		94
	Cerebral cortex	No significant change ⎭		
Amitriptyline or desipramine for 21 days	Cerebral cortex	No significant change		95
Imipramine for 3 weeks	Frontal cortex	(−20*)		96
Zimelidine for 2 weeks	Cerebral cortex	?	−2 ⎫	97
	Hypothalamus	?	−47* ⎭	
Zimelidine for 2 weeks	Cerebral cortex	+33	−69*	98
Desipramine for 2 weeks	Cerebral cortex	0	−18*	99
MAO inhibitors				
Nialamide for 1 week	Cerebral cortex	(−28*)		80
Clorgyline for 5 days	Cerebral cortex	0	−21*	80
Pargyline for 3 weeks	Frontal cortex	(−42*)		96
5-HT agonists				
TFMPP for 5 days	Cerebral cortex	−5	−17*	80
Quipazine for 5 days	Cerebral cortex	−9	−23*	80
*m*CPP for 4 weeks	Forebrain	−13	−24*	111
LSD for 4 days	Forebrain	−17*	−25* ⎫	112
	Brainstem + spinal cord	−23*	−30* ⎭	
5-HT antagonists				
Metergoline for 3 weeks	Cerebral cortex	No significant change		91
Metergoline for 4 weeks	Hippocampus	0	+27* ⎫	90, 111
	Striatum	0	+30* ⎬	
	Cerebral cortex	0	+47* ⎭	
Methiothepin for 3 days	Hippocampus	−6	+34*	25
Miscellaneous treatments				
Haloperidol for 3 weeks	Striatum	(+20*)	⎫	113
	Hippocampus		⎬	
	Cerebral cortex	No significant change	⎭	
Trifluoperazine for 4–6 months	Striatum	−2	+37* ⎫	114
	Cerebral cortex	+25	+43* ⎭	
ECS for 10 days	Frontal cortex	No significant change		115
Ethanol for 11–15 days	Striatum	(+63*)	⎫	116
	Brainstem	(+32*)	⎬	
	Hippocampus	(−20*)	⎭	
Lithium for 4 to 6 weeks	Hippocampus	(−27.5*)	⎫	117
	Cerebral cortex	No significant change	⎭	
Morphine for 11 days	Brainstem	−29	−42* ⎫	118
	Cerebral cortex	+20	+10 ⎭	

[a] In all cases, only chronic treatments are effective; acute treatments never significantly affect K_d or B_{max} values. K_d(%) and B_{max}(%) are the percent changes in K_d and B_{max} induced by each treatment. When binding assays were performed with only one concentration of [^{3}H]5-HT, the percent change in specific binding is given in parentheses. u.o., unpublished observations; PCPA, *para*-chlorophenylalanine; TFMPP, trifluoromethylphenylpiperazine; mCPP, *meta*-chlorophenylpiperazine; ECS, electroconvulsive shocks. *Significantly different from respective control values ($P < 0.05$).

Table IV
Apparent Up- and Down-Regulation of 5-HT$_2$ Sites in Various Brain Regions
Following in Vivo Treatments[a]

Treatment	Structure	K_d (%)	B_{max} (%)	Reference
Lesion				
Electrolytic lesion of the midbrain raphe 2 weeks before death	Hippocampus	No significant change		45
5,7-DHT intracerebrally	Dorsal hippocampus	(significant increase)		46
5-HT uptake inhibitors				
Amitriptyline ⎤		(−43)* ⎫		
Imipramine ⎬ for 3 weeks	Frontal cortex	(−40)* ⎬		96
Desipramine ⎦		(−21)* ⎭		
Amitriptyline for 3 weeks	Cerebral cortex	(−38)*		100
Amitriptyline for 6 weeks	Frontal cortex	+20	−50*	
Imipramine for 3 weeks	Cerebral cortex	(−47)* ⎫		102
	Hippocampus	(−35)* ⎭		
Amitriptyline for 3 weeks	Cerebral cortex	−8	−25* ⎫	95
Desipramine for 3 weeks	Cerebral cortex	−2	−16* ⎭	
Trazodone + phenoxy-benzamine for 4 days	Cerebral cortex	(−32)*		103
MAO inhibitors				
Tranylcypromine for 3 weeks	Cerebral cortex	(−45)*		100
Pargyline for 3 weeks	Frontal cortex	(−35)*		96
Miscellaneous treatments				
Iprindole for 3 weeks	Cerebral cortex	(−40)*		100
Iprindole for 3 weeks	Frontal cortex	(−34)*		96
Lithium for 4 to 6 weeks	Hippocampus	(−21)* ⎫		117
	Cerebral cortex	(−7) ⎭		
ECS for 10 days	Cerebral cortex	−5	+28*	100
ECS for 10 days	Frontal cortex	+35	+39*	115

[a] Data obtained with acute treatments are not presented since no significant alteration in K_d or B_{max} of [^{3}H]-spiperone binding is detected under these conditions. K_d (%) and B_{max} (%) refer to percent changes in K_d and B_{max} induced by each treatment. When binding assays were performed with only one concentration of [^{3}H]spiperone, the percent change in specific binding is given in parentheses. *Significantly different from respective control values ($P < 0.05$).

by changing the residual amount of 5-HT still bound to membranes. In particular, Bennett and Snyder[40] observed an increased binding of [^{3}H]5-HT to brain membranes from rats treated with PCPA or reserpine and concluded that this change simply reflected a decreased competition by endogenous 5-HT (still attached to membranes).

Other treatments usually involve drugs that more or less bind to membranes and interfere directly with [^{3}H]5-HT for specific binding sites. As an example, we observed that methiothepin treatment induced a long-lasting reduction in the capacity of rat brain membranes to specifically bind [^{3}H]5-HT simply because the drug remains firmly attached to 5-HT receptors for several hours (Fig. 2)[25]. Since the drug persisting in membrane preparations used for binding assays is generally not measured, results obtained following chronic

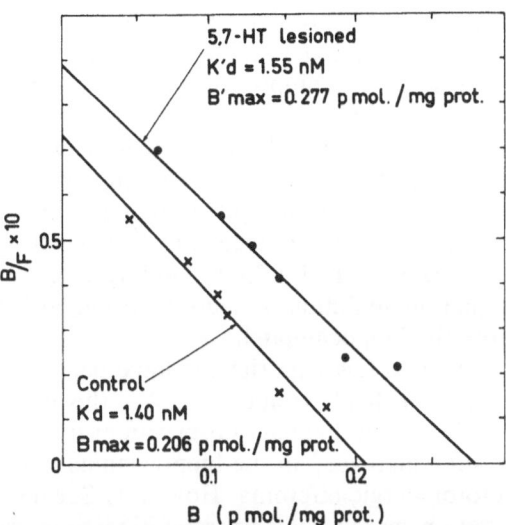

Fig. 6. Scatchard plots of specifically bound ^3H-5-HT to hippocampal membranes from control or 5,7-DHT-pretreated rats. 5,7-dihydroxytryptamine (8 μg) was administered in forebrain serotoninergic fibre bundles on the 18th day before death. Specific binding of ^3H-5-HT was measured with 0.45, 0.90, 1.35, 1.80, 4.50 and 7.40 nM of the labeled ligand. Each point is the mean of quadruplicate determinations.

treatments as compared to acute treatments must be considered with caution, since it is not known whether membranes actually contain the same amount of drug under both conditions.

2.7.2. Effects of 5-HT Depletion

Studies on [^3H]5-HT binding sites following the degeneration of serotoninergic neurons by either central 5,6-[79] or 5,7-dihydroxytryptamine[24,41,80–82] injections or electrolytic raphe lesions[40,45] have led to contradictory results (Table III). Although discrepancies may well reflect the different experimental conditions used in various laboratories, data presented in Table III also suggest that they may in fact reveal some regional differences concerning *in vivo* modulation of central 5-HT receptors. In particular, serotoninergic denervation generally has no significant effect on [^3H]5-HT binding in the spinal cord, cerebral cortex, and striatum but induces a marked increase in the capacity of hippocampal membranes to bind the tritium-labeled indoleamine (Fig. 6) (Table III). Such differences between [^3H]5-HT binding onto hippocampal, striatal, and cortical membranes cannot be explained by contaminations with variable amounts of residual endogenous 5-HT, since the same 5-HT depletion is detected in the three regions following the administration of 5,7-DHT into forebrain serotoninergic fiber bundles.[24]

Although an increase in the B_{max} of 5-HT$_1$ sites can be considered as a biochemical event associated with the well-known denervation supersensitivity to 5-HT agonists (as observed in various behavioral studies[83]), it is in contradiction with electrophysiological data obtained by the microiontophoresis technique.[84] Indeed, 5,7-DHT-induced chemical lesions of serotoninergic terminals significantly increase the sensitivity of cortical neurons to the inhibitory effect of 5-HT (in spite of the absence of any detectable change in [^3H]5-HT binding onto cortical membranes). Conversely, no change in 5-HT-evoked response of hippocampal neurons is observed following 5,7-DHT-induced degeneration of serotoninergic innervation,[85] although the density of [^{3}H]5-HT binding sites increases in the hippocampus under similar conditions (Table III). A remark must be added, however, to slightly temper this contradiction: electrophysiological recordings[85] have focused on area CA_3, which contains the lowest density of 5-HT$_1$ sites[86] in the hippocampus, whereas binding studies (Table III) likely reflect changes in dentate gyrus and CA_1. Nevertheless, such discrepancies between biochemical and electrophysiological data strongly suggest that [^{3}H]5-HT high-affinity binding sites do not correspond to 5-HT receptors triggering inhibition of neuronal firing, at least in the cerebral cortex (and possibly the hippocampus).

Only a few experiments have been specifically conducted on 5-HT$_2$ sites labeled with [^{3}H]spiperone after the destruction of serotoninergic neurons (Table IV). Preliminary observations by Azmitia[46] indicated that [^{3}H]spiperone binding increased in the hippocampus following the selective degeneration of serotoninergic afferents. However, Seeman et al.[45] failed to detect any significant change in [^{3}H]spiperone binding to membranes from this area following electrolytic raphe lesions. Since [^{3}H]5-HT binding capacity was significantly enhanced in the same membrane preparation,[45] these observations further confirm that 5-HT$_1$ and 5-HT$_2$ sites in fact correspond to two different classes of 5-HT receptors. Using another ligand of 5-HT$_2$ sites, [^{3}H]mergoline, we could detect a significant increase (+ 36%) in B_{max} in the hippocampus but not in the striatum of rats on the 20th day after intracerebral administration of 5,7-DHT.[41] Differences between the latter two studies[41,45] may be related to the fact that electrolytic raphe lesions produce nonspecific damage to several neuronal populations in addition to serotoninergic systems (selectively destroyed by 5,7-DHT treatment).

Experiments have also been carried out with [^{3}H]LSD as the ligand and show that electrolytic raphe lesions do not alter the characteristics of its specific binding onto forebrain[14,23] and cortical[87] membranes. Since [^{3}H]LSD binds equally to 5-HT$_1$ and 5-HT$_2$ sites,[19] data obtained with this ligand are difficult to interpret for 5-HT$_2$ sites. At present, data available are too limited (Table IV) to answer definitively the question of possible denervation supersensitivity with respect to 5-HT$_2$ sites in brain.

Other treatments resulting in marked depletion of 5-HT stores in brain also seem to enhance [^{3}H]5-HT binding capacity of membranes from the rat cerebral cortex (Table III). Chronic treatment with PCPA induces an increased density of 5-HT$_1$ sites and/or a higher affinity of these sites for 5-HT.[88] As already discussed by Bennett and Snyder,[40] such elevated [^{3}H]5-HT binding may well represent the reduced competitive inhibition of the exogenous ligand binding by endogenous 5-HT in membranes from PCPA-treated rats. Furthermore, the functional significance of enhanced [^{3}H]5-HT binding following PCPA treatment is questionable since the same 5-HT-evoked response was observed on neuronal firing of cortical neurons in PCPA-treated (even for 14 days) and in control rats.[84] More generally, behavioral investigations indicate that chronic PCPA administration (in contrast to intracerebral 5,7-DHT injection) does not induce supersensitivity to 5-HT agonists.[83]

2.7.3. Effects of 5-HT Releasers, MAO Inhibitors, and 5-HT Reuptake Blockers

As expected from the above findings, several groups have observed that treatments producing an increased concentration of 5-HT in synaptic clefts (MAO inhibition, reuptake blockade, releasing action) induced in contrast a decreased capacity of brain membranes to specifically bind [^{3}H]5-HT (Table III). Generally, this effect is associated with a decreased B_{max} and no significant change in the apparent affinity of 5-HT$_1$ sites for [^{3}H]5-HT. Similar findings were observed in the cerebral cortex and the hippocampus using [^{3}H]spiperone as the labeled ligand, so 5-HT$_2$ sites are also affected by these treatments (Table IV).

Since endogenous 5-HT bound to membranes used for binding assays generally accounts for less than 2.5 nM,[24] and since micromolar concentrations of the indoleamine are required to interact with 5-HT$_2$ sites,[31,43,44] modifications in [^{3}H]spiperone binding following chronic treatments with 5-HT uptake inhibitors cannot be simply explained by changes in endogenous 5-HT persisting in membranes. However, it cannot be definitively concluded that 5-HT$_2$ sites have in fact been altered as a result of chronic perturbation of serotoninergic neurotransmission, since some of the drugs used directly bind to these sites. In particular, those that, like desipramine, imipramine, and amitriptyline, induce the most significant reduction of 5-HT$_2$ sites (Table IV), are also those that directly inhibit [^{3}H]spiperone binding even at nanomolar concentrations.[104,105] Accordingly, until the concentration of these drugs persisting in the membranes used for binding assays is known, one cannot completely exclude the possibility that the *in vivo* induced changes represent in fact the persistence of tricyclic inhibitors in membranes rather than an actual decrease in the number of 5-HT$_2$ sites. This question is crucial, since Peroutka and Snyder[96] observed that chronic treatment with fluoxetine, a potent inhibitor of 5-HT reuptake with almost no effect on [^{3}H]spiperone binding,[105] did not reduce the density of 5-HT$_2$ sites in the cerebral cortex.

Comparison of these biochemical data with electrophysiological findings indicates at least some corroborations. Under conditions similar to those producing a significant reduction of 5-HT$_1$ sites,[80] Olpe[106] noted a decreased sensitivity of cortical neurons to microiontophoretically applied 5-HT following chronic clorgyline administration. However, further comparison often reveals marked discrepancies. Indeed, instead of a subsensitivity as expected from biochemical findings (Tables III and IV), De Montigny and Aghajanian[107] noted an increased response of rat forebrain neurons to iontophoretically applied 5-HT after long-term (up to 14 days) treatment with desipramine, chlorimipramine, amitriptyline, imipramine, or iprindole. Behavioral studies are in agreement with these electrophysiological data, since significant enhancements of 5-HT (and 5-HT agonist)-induced effects following chronic treatments with antidepressants have been reported in mice,[108] rats,[109] and chicks.[110]

Such discrepancies among biochemical, electrophysiological, and behavioral studies, together with those already noted following treatments inducing serotoninergic denervation or 5-HT store depletion, further suggest that the putative 5-HT receptors analyzed by biochemical techniques do not correspond to those involved in electrophysiological and behavioral responses to 5-HT and 5-HT agonists.

2.7.4. Effects of 5-HT Agonists and Antagonists

The effects of chronic treatments with agonists and antagonists acting directly on 5-HT receptors have also been investigated using [³H]5-HT as the labeled ligand (Table III). As already observed with most receptor types, chronic treatment with an agonist induces a significant decrease in the number of specific binding sites for [³H]5-HT.[80,111,112] Conversely, the chronic administration of an antagonist evokes an increased density of 5-HT_1 sites in brain (Table III). Although one group[91] failed to detect these changes, behavioral studies regularly confirm that chronic treatments with 5-HT antagonists produce an increased sensitivity of rats to 5-HT agonists.[109]

2.7.5. Effects of Miscellaneous Treatments

In addition to the above treatments, others elicited significant changes in [³H]5-HT and [³H]spiperone binding in the rat brain (Tables III and IV). Usually, a decreased binding is viewed as the consequence of an enhanced release of 5-HT in brain, whereas the reverse occurs following treatments preventing (more or less directly) serotoninergic neurotransmission. Of particular interest is the observation concerning 5-HT receptors following repeated electroconvulsive shocks.[100,115] Although this treatment approximately reproduces electroconvulsive therapy for the treatment of depression, the resulting changes in 5-HT receptors in brain are quite different from those induced by the chronic administration of antidepressant drugs. Indeed, no change in [³H]5-HT binding (Table III) and a significant increase in [³H]spiperone binding (Table IV) were observed in the cerebral cortex after chronic electroconvulsive shocks in rats. These biochemical findings corroborate behavioral observations that consistently reveal an increased sensitivity of rats to 5-HT agonists following repeated electroconvulsive shocks.[119] Accordingly, the antidepressant activity of tricyclic drugs and electroconvulsive therapy may not be related to possible alterations in 5-HT_1 and 5-HT_2 receptors.

2.8. In Vivo Modulations of 5-HT-Sensitive Adenylate Cyclase

The 5,7-DHT-induced degeneration of central serotoninergic neurons in newborn rats is associated with significant increases of 5-HT-sensitive adenylate cyclase activity in various regions, notably the hippocampus, the colliculi, and the cerebral cortex.[120] Surprisingly, a transient increase in 5-HT-sensitive adenylate cyclase activity is also observed in the brainstem, a region that exhibits a serotoninergic hyperinnervation following 5,7-DHT treatment of newborn rats.[120] In all cases, the 5,7-DHT-induced effect is selectively associated with an elevation of the $V_{\max}$ of 5-HT-sensitive adenylate cyclase (Fig. 7); the same $K_{A\text{app}}$ characterizes the enzyme in control and 5,7-DHT-treated rats (Fig.

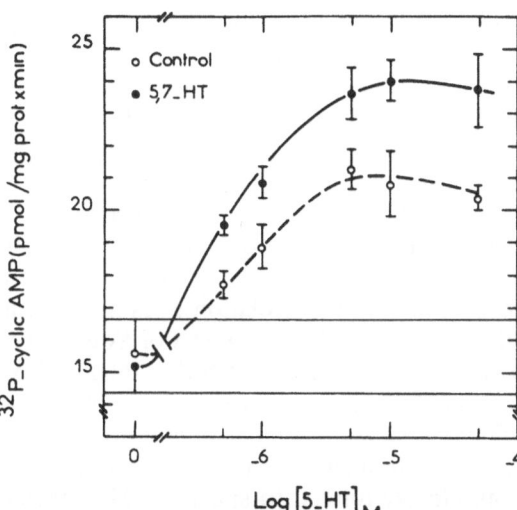

Fig. 7. 5-DHT-sensitive adenylate cyclase activity in brain-stem homogenates from control and 5,7-DHT-treated rats. 5,7-dihydroxytryptamine (2 × 100 mg/kg sc) was administered on the 1st and 2nd postnatal days. 5-HT-sensitive adenylate cyclase activity was measured in brainstem homogenates from 7-day-old control or treated rats. The two horizontal lines mark the range of basal adenylate cyclase activities in both groups of rats.

7). Since [³H]5-HT binding is not altered in any brain region following 5,7-DHT administration to newborn rats,[120] these results further confirm that 5-HT$_1$ sites do not correspond to postsynaptic 5-HT receptors coupled to adenylate cyclase in the rat brain.

Investigations on the possible down-regulation of 5-HT-sensitive adenylate cyclase have been carried out in hypotrophic rats, which exhibit an accelerated turnover of 5-HT in brain for at least the first 2 weeks following birth.[121] In spite of an increased release of 5-HT likely resulting in an overstimulation of 5-HT receptors, no significant change in 5-HT-sensitive adenylate cyclase activity could be detected in any brain area of hypotrophic rats during the first two postnatal weeks (unpublished observations).

Although the increased 5-HT-sensitive adenylate cyclase activity occurring after 5,7-DHT-induced selective degeneration of serotoninergic systems strongly suggests that this enzyme is associated with functional receptors, further investigations are necessary to explore possible adaptive changes of these receptors to experimentally induced alterations in serotoninergic neurotransmission.

2.9. Functional Correlates of 5-HT Receptors

2.9.1. Possible Significance of Correlations between in Vivo or in Vitro Effects of 5-HT and in Vitro Characteristics of 5-HT Receptors

As discussed in the previous section, attempts to find correlations between biochemical and electrophysiological or behavioral findings on 5-HT receptors have not lead to firm conclusions. Recently, Peroutka *et al.*[104] proposed that 5-HT$_2$ sites but not 5-HT$_1$ sites might correspond to those involved in the head-twitch behavior in mice. This conclusion was based on the observation that the relative efficacy of 5-HT antagonists to inhibit this behavior following 5-HTP administration is positively correlated with their ability to displace

[^{3}H]spiperone specifically bound to cortical membranes. Clearly, these data are insufficient for such a conclusion, since other observations suggest no functional relationship between 5-HT$_2$ sites and head-twitch behavior. In particular, chronic antidepressant treatment facilitates head-twitch behavior induced by 5-HT agonists[108] in spite of down-regulating 5-HT$_2$ sites in the frontal cortex and the hippocampus (Table IV).

As behavior involves complex neuronal circuits and multiple receptor types, it is undoubtedly extremely hazardous to associate any behavioral change with modifications concerning only one class of receptors in brain. That the presence of a correlation does not necessarily imply a functional relationship is illustrated by the report of Young *et al.*[122] that benzodiazepine potency *in vivo* is positively correlated to their ability to inhibit [^{3}H]strychnine binding to glycine receptors *in vitro* in spite of clear-cut evidence showing that *in vivo* effects of benzodiazepines do not involve these receptors.[123] Therefore, in this final section, instead of behavior, we focus on some other biological responses more closely related to specific 5-HT receptor stimulation.

On the basis of a possible correlation between the rank potency of several antagonists against *in vitro* [^{3}H]5-HT binding and *in vitro* 5-HT-sensitive adenylate cyclase activity, Peroutka *et al.*[104] concluded that these two markers correspond to the same receptor complex. However, a careful examination of their data justifies the elimination of LSD from the series of antagonists selected for calculating the correlation, since in fact LSD is a mixed agonist–antagonist. Of the remaining nine antagonists in their series, the ability to inhibit 5-HT-sensitive adenylate cyclase activity is in fact better correlated with the relative potency to displace [^{3}H]spiperone ($r = 0.88$) than [^{3}H]5-HT ($r = 0.80$) from 5-HT$_2$ and 5-HT$_1$ specific sites, respectively. Additionally, recent investigations on the relative efficacies of 11 selected antagonists to inhibit 5-HT-sensitive adenylate cyclase and [^{3}H]5-HT high-affinity binding indicate that no correlation can be found between the respective IC$_{50}$ values.[52] These contradictory results[52,104] further illustrate that no firm conclusion can be drawn from the existence of a possible correlation even between two *in vitro* markers of receptors. As already discussed (Section 2.4), numerous data prove in fact that the 5-HT$_1$ site, the 5-HT$_2$ site, and 5-HT-sensitive adenylate cyclase correspond to three distinct classes of postsynaptic 5-HT receptors in brain.[31,52,56]

2.9.2. Cellular Consequences of 5-HT–Receptor Interactions

In the facial nucleus in the rat[124] and also in the ganglionic nervous system of molluscs,[125] the increased cyclic AMP formation in response to 5-HT receptor stimulation triggers the activation of a specific cyclic-AMP-dependent protein kinase. This enzyme catalyzes the phosphorylation of a specific protein, protein I, associated primarily with synaptic vesicles.[124] Accordingly, it can be proposed that at least some of the postsynaptic 5-HT receptors linked to adenylate cyclase are in fact located on nonserotoninergic nerve terminals, where they likely modulate neurotransmitter release. Recent *in vitro* investigations with striatal slices have shown that 5-HT receptors can modulate the release of dopamine from dopaminergic terminals.[126] In this regard, it is interesting to notice that a decrease in [^{3}H]5-HT-high affinity binding has been observed in the putamen of parkinsonian patients,[127] therefore suggesting that some 5-HT receptors are associated with dopaminergic terminals (and may be with other neuronal systems degenerating in Parkinson's disease).

In the abdominal ganglion of *Aplysia*, Drummond *et al.*[128] have shown that the increased cyclic AMP content resulting from the stimulation of 5-HT receptors coupled to adenylate cyclase eventually evokes a hyperpolarization through an enhanced K^+ conductance. Recent *in vitro* studies on slices of the rat hippocampus suggest that a similar mechanism likely exists in the rat brain, since 5-HT induces an increase of extracellular K^+ concentration in the stratum pyramidale of area CA_1.[33] In this latter case, however, the involvement of cyclic AMP remains to be explored.

Additional investigations in molluscs have shown that not only K^+ conductance but also that of Na^+ and Cl^- may well be changed by specific 5-HT receptors.[17] Whether similar mechanisms take place in the mammalian brain is presently unknown.

Changes in ionic transport through membranes may depend not only on the phosphorylation of specific proteins (via a cyclic AMP-dependent process) or allosteric interactions between the receptor and an ionophore (as hypothesized for at least some 5-HT receptors in molluscs) but also on enzymes. In this respect, a particular role is played by Na^+/K^+-dependent ATPase, which directly controls ionic gradients through cell membranes. Serotonin has been shown to stimulate this enzyme activity. Preliminary observations suggest that this effect may directly involve a serotoninergic receptor, since it is prevented by metergoline, a specific and potent 5-HT antagonist.[129]

In light of the crucial role of Ca^{2+} and Na^+ in the control of neurotransmitter (including 5-HT[130]) release from neurons, the final result of 5-HT–receptor interactions may well be some modulation of this process via primary alterations in ionic transport through membranes. In addition to possibly explaining the direct effect of 5-HT on the release of other neurotransmitters,[126] 5-HT-induced changes in ion transport may be involved in the inhibitory effect of the indoleamine on its own release from serotoninergic terminals.[34,131–133] Detailed pharmacological analyses have confirmed that specific 5-HT receptors mediate the 5-HT-induced inhibition of [^{3}H]5-HT release from brain slices.[34,133,134] However, marked differences exist between these receptors and those specifically labeled with tritiated ligands, since, for instance, spiperone[34] and metergoline,[133] two potent 5-HT antagonists acting on 5-HT$_1$ and 5-HT$_2$ sites do not affect the depolarization-induced release of tritium from cerebral cortex slices preloaded with [^{3}H]5-HT. Since the 5-HT-induced inhibition of its own release is observed not only with slices but also with synaptosomes,[135] the receptors involved are likely located on presynaptic serotoninergic terminals. All of these observations are therefore compatible with the existence of presynaptic autoreceptors on serotoninergic terminals.

In contrast to binding studies that allow the detection of specific (postsynaptic) "receptors" with largely unknown cellular functions, in the case of presynaptic receptors, the discovery of modulations in a neuronal function, i.e., neurotransmitter release, implies their existence. However, no appropriate radioactive ligand is presently available to label these presynaptic receptors.

Table V
Principal Characteristics of Postsynaptic 5-HT Receptors in the Rat Brain

	5-HT-sensitive adenylate cyclase (newborn rats)[a]	5-HT$_1$[b]	5-HT$_2$
Richest region	Colliculi	Hippocampus	(Frontal) cerebral cortex
Selected ligand	—	[^{3}H]5-HT	[^{3}H]antagonists ([^{3}H]spiperone)
Apparent affinity for 5-HT	EC$_{50}$ = 0.5–1.0 μM	K_d = 1–8 nM	K_d = 5–10 μM
GTP effect	Stimulation	Reduction of the affinity for agonists	Reduction of the affinity for agonists
Up-regulation	After 5,7-DHT-induced serotoninergic denervation	After intracerebral 5,7-DHT or raphe lesion	After electroconvulsive shocks
Down-regulation	?	After chronic treatments with antidepressant drugs	

[a] Studies on 5-HT-sensitive adenylate cyclase are mainly carried out in newborn animals because the relative enzymic stimulation produced by 5-HT over basal activity is maximum (≥100%) during the early life period. In adult rats, the absolute stimulation by 5-HT is similar to that observed in young animals; however, as a result of a higher basal adenylate cyclase activity, the 5-HT-induced effect maximally accounts for only a 10–15% increase.
[b] At least for 5-HT$_1$ sites, the same pharmacological characteristics have been observed in newborn and in adult rats. Ontogeny is mainly associated with an increase in the density of 5-HT$_1$ sites in brain.

In addition to the various functional correlates already discussed, the stimulation of neurotransmitter receptors can also affect phospholipid methylation[126] and turnover.[137] Whether such mechanisms are also effective in the case of central 5-HT receptors has not been definitively answered.

3. CONCLUSION AND FUTURE TRENDS

Biochemical, electrophysiological, and behavioral studies are compatible with the hypothesis that several classes of 5-HT receptors exist in the CNS.

At least three receptor types are present at the postsynaptic level (Table V, Fig. 8):that coupled to adenylate cyclase, that having a high affinity for 5-HT (5-HT$_1$ site), and that labeled with [^{3}H]spiperone (and other tritium-labeled 5-HT antagonists) (5-HT$_2$ site).

In addition, two receptors exist on serotoninergic neurons themselves and can be called presynaptic autoreceptors (Fig. 8). The stimulation of receptor A$_1$ on raphe cell bodies and/or dendrites (Fig. 8) triggers a negative control on neuronal firing[18] associated with a significant reduction of 5-HT release from serotoninergic terminals in forebrain regions.[138] Autoreceptors of the A$_2$ type (Fig. 8) are located on presynaptic terminals and modulate the Ca^{2+}-dependent 5-HT release (and possibly 5-HT synthesis[134]).

Pharmacological characteristics of these various receptors have been well explored and clearly confirm that they are distinct entities. Apparent changes

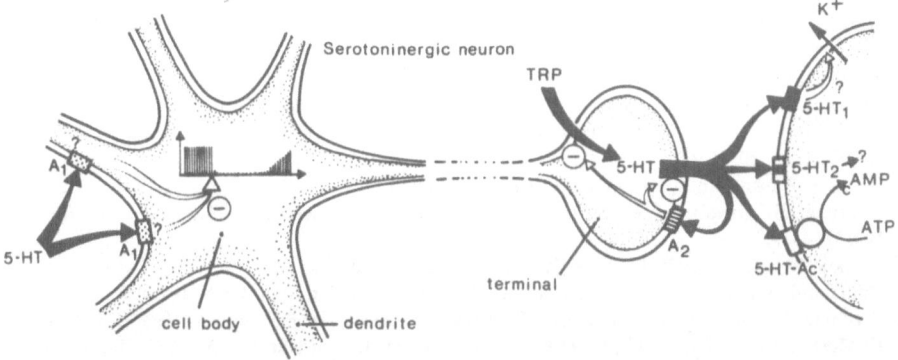

Fig. 8. Schematic representation of the heterogeneity of central 5-HT receptors. The stimulation of autoreceptors A_1 on cell bodies and/or dendrites induces a transient inhibition of nerve impulse flow within serotoninergic neurons. Presynaptic autoreceptors A_2 are involved in negative feedback mechanisms controlling 5-HT release (and possibly synthesis). Three classes of postsynaptic receptors can be identified by biochemical techniques: 5-HT$_1$ which is selectively labeled by ^{3}H-5-HT, 5-HT$_2$ which is selectively labeled by ^{3}H-antagonists and the receptor coupled to adenylate cyclase. Indirect evidence suggests that the 5-HT$_1$ receptor can influence the K^+ conductance.

in 5-HT receptor characteristics following chronic alterations of serotoninergic neurotransmission correspond to the up- and down-regulation expected of functional receptors. However, except in the case of the receptor coupled to adenylate cyclase, the cellular events triggered by the binding of 5-HT onto these various types of receptors are poorly known, particularly in the mammalian brain. Investigations in this direction for a better knowledge of the functional correlates of 5-HT receptors would be very fruitful.

Regarding the biochemistry of central 5-HT receptors, much effort also remains to be devoted to their solubilization and purification. It must be stressed that only the physical separation of the various classes of 5-HT receptors will be the final proof of the heterogeneity postulated presently.

Finally, more precise characterization of the various types of 5-HT receptors would be of great help in synthesizing new drugs that would interact selectively with only one receptor type. Such drugs would allow the further exploration of the role of various classes of 5-HT receptors in central functions. One expected outcome of such studies is to develop new therapies of disease states related to these functions (sleep disorders, defective pain control, etc.).

ACKNOWLEDGMENTS. This research has been supported by grants from CNRS (ATP 4149), INSERM, DRET, and Rhône-Poulenc S.A.

REFERENCES

1. Azmitia, E. C., 1978, *Handbook of Psychopharmacology*, Volume 9 (L. L. Iversen, S. D. Iversen, and S. H. Snyder, eds.), Plenum Press, New York, pp. 233–314.
2. Beaudet, A., and Descarries, L., 1981, *J. Physiol. (Paris)* 77:193–203.
3. Gershon, M. D., 1981, *J. Physiol. (Paris)* 77:257–265.

4. Hamon, M., Bourgoin, S., Artaud, F., and El Mestikawy, S., 1981, *J. Physiol. (Paris)* **77:**269–279.

5. Pardridge, W. M., 1979, *J. Neural Transm. [Suppl.]* **15:**43–54.

6. Gessa, G. L., and Tagliamonte, A., 1974, *Aromatic Amino Acids in the Brain* (G. E. W. Wolstenholme and D. W. Fitzsimons, eds.), Elsevier-North Holland-Excerpta Medica, Amsterdam, pp. 207–216.

7. Herr, B. E., and Roth, R. H., 1976, *Brain Res.* **110:**189–193.

8. Bourgoin, S., Oliveras, J. L., Bruxelle, J., Hamon, M., and Besson, J. M., 1980, *Brain Res.* **194:**377–389.

9. Hamon, M., and Glowinski, J., 1974, *Life Sci.* **15:**1533–1548.

10. Yamauchi, T., Nakata, H., and Fujisawa, H., 1981, *J. Biol. Chem.* **256:**5404–5409.

11. Zivcovic, B., Guidotti, A., and Costa, E., 1974, *Adv. Biochem. Psychopharmacol.* **11:**19–30.

12. El Mestikawy, S., Goetz, C., Pasquier, A., Glowinski, J., and Hamon, M., 1982, *Brain Res.* **244:**319–329.

13. Bourgoin, S., Faivre-Bauman, A., Benda, P., Glowinski, J., and Hamon, M., 1974, *J. Neurochem.* **23:**319–327.

14. Bennett, J. P., and Snyder, S. H., 1975, *Brain Res.* **94:**523–544.

15. Gaddum, J. H., and Picarelli, Z. P., 1957, *Brit. J. Pharmacol.* **12:**323–328.

16. Wallis, D., 1981, *Life Sci.* **29:**2345–2355.

17. Gerschenfeld, H. M., and Paupardin-Tritsch, D., 1974, *J. Physiol. (Lond.)* **243:**427–457.

18. Rogawski, M. A., and Aghajanian, G. K., 1981, *J. Neurosci.* **1:**1148–1154.

19. Peroutka, S. J., and Snyder, S. H., 1979, *Mol. Pharmacol.* **16:**687–699.

20. Marchbanks, R. M., 1966, *J. Neurochem.* **13:**1481–1493.

21. Alivisatos, S. G. A., Ungar, F., Seth, P. K., Levitt, L. P., Geroulis, A. J., and Meyer, T., 1971, *Science* **171:**809–812.

22. Farrow, J. T., and Van Vunakis, H., 1973, *Biochem. Pharmacol.* **22:**1103–1113.

23. Bennett, J. L., and Aghajanian, G. K., 1974, *Life Sci.* **15:**1934–1944.

24. Nelson, D. L., Herbet, A., Bourgoin, S., Glowinski, J., and Hamon, M., 1978, *Mol. Pharmacol.* **14:**983–995.

25. Nelson, D. L., Herbet, A., Pichat, L., Glowinski, J., and Hamon, M., 1979, *Naunyn-Schmiedeberg's Arch. Pharmacol.* **310:**25–33.

26. Duchemin, A.-M., Quach, T. T., Rose, C., and Schwartz, J. C., 1979, *Life Sci.* **24:**401–410.

27. Clements-Jewery, S., and Robson, P. A., 1980, *Neuropharmacology* **19:**657–661.

28. Von Hungen, K., Roberts, S., and Hill, D. F., 1975, *Brain Res.* **84:**257–267.

29. Enjalbert, A., Bourgoin, S., Hamon, M., Adrien, J., and Bockaert, J., 1978, *Mol. Pharmacol.* **14:**2–10.

30. Fillion, G., Beaudoin, D., Rousselle, J. C., Deniau, J. M., Fillion, M. P., Dray, F., and Jacob, J., 1979, *J. Neurochem.* **33:**567–570.

31. Hamon, M., Nelson, D. L., Herbet, A., and Glowinski, J., 1980, *Adv. Biochem. Psychopharmacol.* **21:**223–233.

32. Enjalbert, A., Hamon, M., Bourgoin, S., and Bockaert, J., 1978, *Mol. Pharmacol.* **14:**11–23.

33. Segal, M., and Gutnick, M. J., 1980, *Brain Res.* **195:**389–401.

34. Baumann, P. A., and Waldemeier, P. C., 1981, *Naunyn-Schmiedeberg's Arch. Pharmacol.* **317:**36–43.

35. Fillion, G. M. B., Rousselle, J.-C., Fillion, M.-P., Beaudoin, D. M., Goiny, M. R., Deniau, J.-M., and Jacob, J. J., 1978, *Mol. Pharmacol.* **14:**50–59.

36. Pedigo, N. W., Yamamura, H. I., and Nelson, D. L., 1981, *J. Neurochem.* **36:**220–226.

37. Peroutka, S. J., and Snyder, S. H., 1981, *Brain Res.* **208:**339–347.

38. Heltzel, J. A., Boehme, D. H., and Vogel, W. H., 1981, *Brain Res.* **204:**451–454.

39. Young, W. S. III, and Kuhar, M. J., 1980, *Eur. J. Pharmacol.* **62:**237–239.

40. Bennett, J. P., Jr., and Snyder, S. H., 1976, *Mol. Pharmacol.* **12:**373–389.

41. Hamon, M., Mallat, M., Herbet, A., Nelson, D. L., Audinot, M., Pichat, L., and Glowinski, J., 1981, *J. Neurochem.* **36:**613–626.

42. Fillion, G., Beaudoin, D., Rousselle, J. C., and Jacob, J., 1980, *Brain Res.* **198:**361–374.

43. Leysen, J. E., Niemegeers, C. J. E., Tollenaere, J. P., and Laduron, P. M., 1978, *Nature* **272:**168–171.

44. Creese, I., and Snyder, S. H., 1978, *Eur. J. Pharmacol.* **49**:201–202.
45. Seeman, P., Westman, K., Coscina, D., and Warsh, J. J., 1980, *Eur. J. Pharmacol.* **66**:179–191.
46. Azmitia, E., 1978, *Trends Neurosci.* August 45–48.
47. Meibach, R. C., Maayani, S., and Green, J. P., 1980, *Eur. J. Pharmacol.* **67**:371–382.
48. Peroutka, S. J., and Snyder, S. H., 1981, *J. Pharmacol. Exp. Ther.* **216**:142–148.
49. Leysen, J. E., Awouters, F., Kennis, L., Laduron, P. M., Vandenberk, J., and Janssen, P. A. J., 1981, *Life Sci.* **28**:1015–1022.
50. Leysen, J. E., Verwimp, M., Geerts, R., Gommeren, W., and Laudron, P. M., 1981, *Neurosci. Abstract* 6.3, p. 7.
51. Battaglia, G., and Titeler, M., 1981, *Life Sci.*, **29**:909–916.
52. Nelson, D. L., Herbet, A., Enjalbert, A., Bockaert, J., and Hamon, M., 1980, *Biochem. Pharmacol.* **29**:2445–2453.
53. Euvrard, C., and Boissier, J. R., 1980, *Eur. J. Pharmacol.* **63**:65–72.
54. Hamon, M., Nelson, D. L., Herbet, A., Bockaert, J., and Glowinski, J., 1980, *Monogr. Neural Sci.* **7**:161–175.
55. Bourgoin, S., Enjalbert, A., Adrien, J., Héry, F., and Hamon, M., 1977, *Brain Res.* **127**:111–126.
56. Nelson, D. L., Herbet, A., Adrien, J., Bockaert, J., and Hamon, M., 1980, *Biochem. Pharmacol.* **29**:2455–2463.
57. Fillion, G., Rousselle, J. C., Beaudoin, D., Pradelles, P., Goiny, M., Dray, F., and Jacob, J., 1979, *Life Sci.* **24**:1813–1822.
58. Daszuta, A., Pons, F., and Cadilhac, J., 1979, *Eur. J. Pharmacol.* **56**:397–401.
59. Ahn, H. S., and Makman, M. H., 1978, *Brain Res.* **153**:636–640.
60. Tsang, D., and Lal, S., 1977, *Can. J. Physiol. Pharmacol.* **55**:1263–1269.
61. Poller, D., Carroll, J. A., and Middlemiss, D. N., 1981, *Eur. J. Pharmacol.* **72**:121–123.
62. Shih, J. C., and Cheng, S. H., 1981, *Adv. Exp. Med. Biol.* **133**:319–325.
63. Ilien, B., Gorissen, H., and Laduron, P., 1980, *Biochem. Pharmacol.* **29**:3341–3344.
64. Chan, B., Seeman, P., Dumbrille-Ross, A., and Madras, B. K., 1981, *Neurosci. Abstract* 6.4, p. 7.
65. Mallat, M., and Hamon, M., 1982, *J. Neurochem.* **38**:151–161.
66. Peroutka, S. J., Lebovitz, R. M., and Snyder, S. H., 1979, *Mol. Pharmacol.* **16**:700–708.
67. Hamon, M., Bourgoin, S., Enjalbert, A., Bockaert, J., Héry, F., Ternaux, J. P., and Glowinski, J., 1976, *Naunyn-Schmiedeberg's Arch. Pharmacol.* **294**:99–108.
68. Middlemiss, D. N., Carroll, J. A., Fisher, R. W., and Mounsey, I. J., 1980, *Eur. J. Pharmacol.* **66**:253–254.
69. Marchais, D., and Bockaert, J., 1980, *Biochem. Pharmacol.* **29**:1331–1336.
70. Rosenfeld, M. R., and Makman, M. H., 1981, *J. Pharmacol. Exp. Ther.* **216**:526–531.
71. Rodbell, M., 1980, *Nature* **284**:17–22.
72. Fong, H. M. C., Okada, K., and Goldstein, M., 1982, *Eur. J. Pharmacol.* **77**:57–59.
73. Hamon, M., Mallat, M., El Mestikawy, S., and Pasquier, A., 1982, *J. Neurochem.* **38**:162–172.
74. Schleifer, L. S., Garrison, J. C., Sternweis, P. C., Northup, J. K., and Gilman, A. G., 1980, *J. Biol. Chem.* **225**:2641–2644.
75. Heron, D. S., Shinitzky, M., Hershkowitz, M., and Samuel, D., 1980, *Proc. Natl. Acad. Sci. U.S.A.* **77**:7463–7467.
76. Hanski, E., Rimon, G., and Levitzki, A., 1979, *Biochemistry* **18**:846–853.
77. Biegon, A., and Samuel, D., 1980, *Neurotransmitters and Their Receptors* (U. Z. Littauer, Y. Dudai, I. Silman, V. I. Teichberg, and Z. Vogel, eds.), John Wiley & Sons, New York, Chichester, pp. 119–124.
78. Fillion, G., and Fillion, M. P., 1981, *Nature* **292**:349–351.
79. Segawa, T., Mizuta, T., and Nomura, Y., 1979, *Eur. J. Pharmacol.* **58**:75–83.
80. Savage, D. D., Mendels, J., and Frazer, A., 1980, *Neuropharmacol.* **19**:1063–1070.
81. Whitaker, P. M., and Deakin, J. F. W., 1981, *Eur. J. Pharmacol.* **73**:349–351.
82. Blackburn, T. P., Cox, B., Heapy, C. G., Lee, T. F., and Middlemiss, D. N., 1981, *Eur. J. Pharmacol.* **71**:343–346.

83. Trulson, M. E., Eubanks, E. E., and Jacobs, B. L., 1976, *J. Pharmacol. Exp. Ther.* **198**:23–32.
84. Ferron, A., Descarries, L., and Reader, T. A., 1982, *Brain Res.* **231**:93–108.
85. De Montigny, C., Wang, R. Y., Reader, T. A. and Aghajanian, G. K., 1980, *Brain Res.* **200**:363–376.
86. Segal, M., 1981, *Adv. Exp. Med. Biol.* **133**:375–390.
87. Blackshear, M. A., Steranka, L. R., and Sanders-Bush, E., 1981, *Eur. J. Pharmacol.* **76**:325–334.
88. Fleisher, L. N., Simon, J. R., and Aprison, M. H., 1979, *J. Neurochem.* **32**:1613–1619.
89. Steigrad, P., Tobler, I., Waser, P. G., and Borbély, A. A., 1978, *Naunyn-Schmiedeberg's Arch. Pharmacol.* **305**:143–148.
90. Samanin, R., Mennini, T., Ferraris, A., Bendotti, C., and Borsini, F., 1980, *Brain Res.* **189**:449–457.
91. Wirz-Justice, A., Krauchi, K., Lichsteiner, M., and Feer, H., 1978, *Life Sci.* **23**:1249–1254.
92. Maggi, A., U'Pritchard, D. C., and Enna, S. J., 1980, *Eur. J. Pharmacol.* **61**:91–98.
93. Savage, D. D., Frazer, A., and Mendels, J., 1979, *Eur. J. Pharmacol.* **58**:87–88.
94. Savage, D. D., Mendels, J., and Frazer, A., 1980, *J. Pharmacol. Exp. Ther.* **212**:259–263.
95. Tang, S. W., Seeman, P., and Kwan, S., 1981, *Psychiatr. Res.* **4**:129–138.
96. Peroutka, S. J., and Snyder, S. H., 1980, *Science* **210**:88–90.
97. Fuxe, K., Ögren, S.-O., and Agnati, L. F., 1979, *Neurosci. Lett.* **13**:307–312.
98. Fuxe, K., Ögren, S.-O., Agnati, L. F., Eneroth, P., Holm, A. C., and Andersson, K., 1981, *Neurosci. Lett.* **21**:57–62.
99. Bergstrom, D. A., and Kellar, K. J., 1979, *J. Pharmacol. Exp. Ther.* **209**:256–261.
100. Kellar, K. J., Cascio, C. S., Butler, J. A., and Kurtzke, R. N., 1981, *Eur. J. Pharmacol.* **69**:515–518.
101. Peroutka, S. J., and Snyder, S. H., 1980, *J. Pharmacol. Exp. Ther.* **215**:582–587.
102. Kendall, D. A., Stancel, G. M., and Enna, S. J., 1981, *Science* **211**:1183–1185.
103. Taylor, D. P., Allen, L. E., Ashworth, E. M., Becker, J. A., Hyslop, D. K., and Riblet, L. A., 1981, *Neuropharmacology* **20**:513–516.
104. Peroutka, S. J., Lebovitz, R. M., and Snyder, S. H., 1981, *Science* **212**:827–829.
105. Dumbrille-Ross, A., Tang, S. W., and Coscina, D. V., 1981, *Life Sci.* **29**:2049–2058.
106. Olpe, H.-R., 1981, *Eur. J. Pharmacol.* **69**:375–377.
107. De Montigny, C., and Aghajanian, G. K., 1978, *Science* **202**:1303–1306.
108. Friedman, E., and Dallob, A., 1979, *Commun. Psychopharmacol.* **3**:89–92.
109. Buxton, D. A., Marsden, C. A., and Stolz, J. F., 1981, *Br. J. Pharmacol.* **74**:215p.
110. Jones, R. S. G., 1980, *Psychopharmacology* **69**:307–311.
111. Mennini, T., Poggesi, E., Caccia, S., Bendotti, C., Borsini, F., and Samanin, R., 1980, *Neurotransmitters and their Receptors* (U. Z. Littauer, Y. Dudai, I. Silman, V. I. Teichberg, and Z. Vogel, eds.), John Wiley & Sons, New York, Chichester, pp. 101–118.
112. Trulson, M. E., and Jacobs, B. L., 1979, *Life Sci.* **24**:2053–2062.
113. Muller, P., and Seeman, P., 1977, *Life Sci.* **21**:1751–1758.
114. Dawbarn, D., Long, S. K., and Pycock, C. J., 1981, *Br. J. Pharmacol.* **73**:149–156.
115. Vetulani, J., Lebrecht, U., and Pilc, A., 1981, *Eur. J. Pharmacol.* **76**:81–85.
116. Muller, P., Britton, R. S., and Seeman, P., 1980, *Eur. J. Pharmacol.* **65**:31–37.
117. Treiser, S. L., Cascio, C. S., O'Donohue, T. L., Thoa, N. B., Jacobowitz, D. M., and Kellar, K. J., 1981, *Science* **213**:1529–1531.
118. Samanin, R., Cervo, L., Rochat, C., Poggesi, E., and Mennini, T., 1980, *Life Sci.* **27**:1141–1146.
119. Green, A. R., Heal, D. J., and Grahame-Smith, D. G., 1977, *Psychopharmacology* **52**:195–200.
120. Hamon, M., Nelson, D. L., Mallat, M., and Bourgoin, S., 1981, *Neurochem. Int.* **3**:69–79.
121. Chanez, C., Priam, M., Flexor, M.-A., Hamon, M., Bourgoin, S., Kordon, C., and Minkowski, A., 1981, *Brain Res.* **207**:397–408.
122. Young, A. B., Zukin, S. R., and Snyder, S. H., 1974, *Proc. Natl. Acad. Sci. U.S.A.* **71**:2246–2250.
123. Dray, A., and Straughan, D. W., 1976, *J. Pharm. Pharmacol.* **28**:314–315.

124. Dolphin, A. C., and Greengard, P., 1981, *J. Neurosci.* **1**:192–203.
125. Kandel, E. R., 1981, *Nature* **293**:697–700.
126. Ennis, C., Kemp. J. D., and Cox, B., 1981, *J. Neurochem.* **36**:1515–1520.
127. Enna, S. J., 1981, *Serotonin—Current Aspects of Neurochemistry and Function* (B. Haber, S. Gabay, M. R. Issidorides, and S. G. A. Alivisatos, eds.), Plenum Press, New York, pp. 347–357.
128. Drummond, A. H., Benson, J. A., and Levitan, I. B., 1980, *Proc. Natl. Acad. Sci. U.S.A.* **77**:5013–5017.
129. Wu, P. H., and Phillis, J. W., 1979, *J. Pharm. Pharmacol.* **31**:782–784.
130. Maura, G., Gemignani, A., Versace, P., Martire, M., and Raiteri, M., 1982, *Neurochem. Int.* **4**:219–224.
131. Göthert, M., and Weinheimer, G., 1979, *Naunyn-Schmiedeberg's Arch. Pharmacol.* **310**:93–96.
132. Mitchell, R., and Fleetwood-Walker, S., 1981, *Eur. J. Pharmacol.* **76**:119–120.
133. Mounsey, I., Brady, K. A., Carroll, J., Fisher, R., and Middlemiss, D. N., 1982, *Biochem. Pharmacol.* **31**:49–53.
134. Bourgoin, S., Artaud, F., Enjalbert, A., Héry, F., Glowinski, J., and Hamon, M., 1977, *J. Pharmacol. Exp. Ther.* **202**:519–531.
135. Cerrito, F., and Raiteri, M., 1979, *Eur. J. Pharmacol.* **57**:427–430.
136. Hirata, F., and Axelrod, J., 1980, *Science* **209**:1082–1090.
137. Berridge, M. J., and Heslop, J. P., 1981, *Br. J. Pharmacol.* **73**:729–738.
138. Bourgoin, S., Soubrié, P., Artaud, F., Reisine, T. D., and Glowinski, J., 1981, *J. Physiol. (Paris)* **77**:303–307.

Histamine and Its Receptors in the Nervous System

Lindsay B. Hough and Jack Peter Green

1. INTRODUCTION

Before much was known about histamine in the brain—even before its concentration in brain was accurately measured—histamine by chance sparked the development of modern psychopharmacology and therefore modern biological psychiatry. Phenothiazines were antihistamines (i.e., H_1 antagonists) that were observed to produce a "euphoric quietude"[1,2]; and from the phenothiazines developed the antidepressant drugs, which had also been designed as H_1 antagonists.[3] Histamine was dismissed as having a role in either the pharmacological or therapeutic effects of either neuroleptic or antidepressant drugs. As summarized below (Sections 12.2 and 12.3), these drugs do in fact interact with histamine receptors at drug concentrations that are found in plasma of patients and in brain of experimental animals. Other psychotropic drugs affect brain histamine (Section 12). Neither the pharmacological nor the therapeutic consequences of these effects are clear because the functions of histamine in brain are not certain.

The main obstacles to understanding the functions of histamine in brain have been the insensitivity of methods to show putative histaminergic fibers *in situ*, the lack of suitable methods to measure histamine and its catabolites, and problems in the characterization of histamine receptors in the nervous system. Much progress has been made in surmounting these impediments, yielding evidence[4-11] supporting the early hypothesis[12] that histamine functions in neural processes.

2. MEASUREMENT OF HISTAMINE IN NEURAL TISSUES

Some methods used to measure histamine in nervous tissue have produced spuriously high values and confusion,[4,13] as they have when applied to plasma,[14]

Lindsay B. Hough and Jack Peter Green • Department of Pharmacology, Mount Sinai School of Medicine of the City University of New York, New York, New York, 10029.

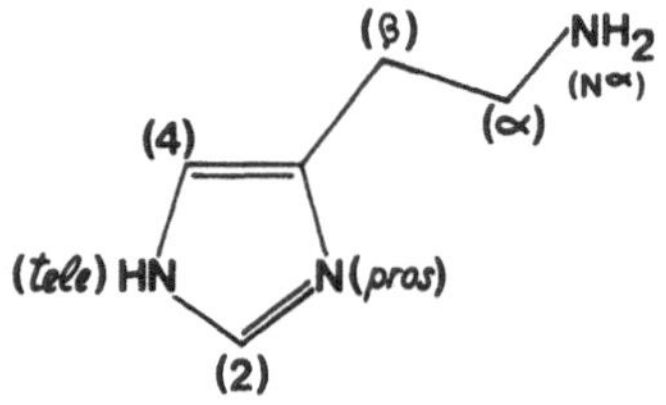

Fig. 1. Nomenclature for histamine derivatives.[20]

and some methodological problems remain. If histamine concentrations in mature whole mammalian brain appear to be higher than 75 ng/g,[4] one can suspect that histamine is not being specifically measured.

Histamine can be accurately measured by bioassay on the guinea pig ileum if interfering substances are removed and if specificity is shown by antagonism of responses by H_1 blockers.[4,15] Without these steps, the method yields brain histamine levels 10- to 50-fold too high.[13]

The fluorometric determination of histamine[16] yields accurate histamine levels in tissues except for blood[14] and brain.[4] Attempts to remove the interfering polyamines, e.g., spermidine, by solvent extraction[17] still yield brain histamine levels higher than those obtained by other methods,[4,17] but purification of brain extracts by ion-exchange chromatography[18] before fluorometric assay gives valid levels. Measuring brain histamine by fluorimetry requires removing all interfering materials, for some of them, like spermidine, are in a 500-fold excess over histamine in brain.[4] Another way to attain specificity is the combined enzymatic–fluorometric method[14] in which the fluorescence measured from assays of biological samples is attributed to histamine only when it is destroyed by incubation with histamine methyltranferase before derivatization. Though specific, these methods lack the sensitivity to measure histamine in small areas of brain.

The radioenzymatic method for measuring histamine has had a major impact on studies of histamine in brain. The method[19] exploits the specificity of histamine methyltransferase to place a labeled methyl group on the ring nitrogen of histamine. Unpurified tissue homogenates are incubated with isotopically labeled S-adenosylmethionine (the methylating cofactor) and partially purified histamine methyltransferase to produce labeled *tele*-methylhistamine (Fig. 1), which is extracted and counted. Modifications have been described[21-24] varying the type and amount of isotope, the incubation volume, and the conditions to extract *tele*-methylhistamine. The advantages of the method are high sensitivity, permitting assays of very small brain regions, and the ability to analyze unpurified homogenates. The method has room for error, as shown by the large variation in the results of analyses of the same plasma samples in different laboratories.[25] Care is especially needed when the single isotope modification of the method is used, since no internal standard is present. Tissue that has not been sufficiently diluted or that contains drugs can inhibit histamine methylation and thereby give falsely low histamine values. There is argument over whether tissue homogenates should be boiled before analysis by this method.

Some[21,26] recommend boiling to denature protein and to destroy endogenous S-adenosylmethionine, thus increasing the efficiency of methylation; others[22,27] suggest that boiling may destroy endogenous histamine. Details of the method have been described.[27]

The limitations inherent in the enzymatic assay—e.g., necessity of routine recovery measurements, inhibition of histamine methyltransferase by drugs— have given incentive to the search for chemical methods. Histamine has been derivatized to compounds suitable for analysis by gas chromatography[28-30] and gas chromatography–mass spectrometry,[31] but these derivatives have not been used to assay histamine in biological samples. In an attempt to circumvent the gas chromatographic problems associated with the piperidine-like nitrogen of histamine, e.g., tailing, it was derivatized in a novel way, and a gas chroma- tographic–mass spectrometric method was developed to measure histamine in biological samples.[32] Determinations of histamine release from antigen-chal- lenged leukocytes by this method agree with parallel determinations by the radioenzymatic method. The specificity of this method is unquestioned, but its sensitivity does not appear sufficient to assay brain histamine.

Some success in measuring histamine by high-pressure liquid chromatog- raphy (HPLC) has been reported. Histamine was measured by fluorometry after ion-exchange chromatography, dansylation, and HPLC separation.[33] The method was sensitive enough to measure whole mouse brain histamine; HPLC was used to separate and detect the o-phthalaldehyde derivative of histam- ine.[34-36] Although the sensitivity of these methods does not exceed that of the radioenzymatic assay, regional brain assays may be feasible, as 0.5 ng was reported to be detectable in rat brain.[36]

One aspect of the measurement of brain histamine that has received little attention is the method of killing the animal. It was reported that measurements of brain histamine were higher and more reproducible if animals were immersed in liquid nitrogen before decapitation and that the addition of Triton X-100 to the homogenizing buffer improved the extraction of histamine from brain.[26] Others[37] reported no difference in rat whole brain histamine levels after these procedures.

When rats were killed by focused microwave irradiation, whole brain his- tamine levels were threefold higher than after decapitation.[37,38] The increase was not caused by the rapid inactivation of histamine catabolism, and no ev- idence was obtained that these higher histamine levels have greater validity than those found after decapitation.[37] The radioactivity produced by radiome- thylation of homogenates of microwave-treated brain migrated as *tele*-meth- ylhistamine in three different solvent systems in thin-layer chromatography, and the high values persisted after the brain extract was purified by ion-ex- change chromatography before assay.[37] Others[39] confirmed the microwave ef- fect, but their thin-layer chromatograms indicated a microwave-induced artifact of the radioenzymatic assay. The different findings need to be resolved.

Unless essential, anesthetics (e.g., barbiturates) should be avoided when histamine is to be measured in brain. Mouse brain histamine content is in- creased,[40] and rat brain histamine turnover is decreased,[41] by these agents.

3. FORMATION OF HISTAMINE

Because histamine poorly penetrates the brain from blood,[4,42,43] brain histamine levels arise from histamine synthesis *in situ* from histidine. Active transport of histidine by brain slices has been shown,[44] but little is known about the process. Since histidine loading elevates brain histamine,[40,43,45] histidine transport might be a controlling factor in brain histamine synthesis. The characteristics of histidine uptake and its relationship to histaminergic mechanisms remain poorly understood.

Two enzymes are capable of decarboxylating histidine *in vitro*,[46] the L-aromatic amino acid decarboxylase (i.e., DOPA decarboxylase, E.C. 4.1.1.26) as well as the "specific" histidine decarboxylase (E.C. 4.1.1.22). The pH optimum, affinity for histidine, effects of selective inhibitors, and regional distribution of histamine-synthesizing activity indicate that the specific histidine decarboxylase is responsible for histamine biosynthesis in brain.[47,48] The enzyme in human brain is similar to that in rat brain.[49] The instability of histidine decarboxylase and its low activity in brain precluded purification of the enzyme. But rat fetal histidine decarboxylase was purified to near homogeneity.[50-52] Studies of its pH optimum, cofactor requirements, inhibitor sensitivity, and antigenic properties showed that the brain and fetal enzymes are very similar, clearly different from both the mammalian DOPA decarboxylase and from the bacterial histidine decarboxylase.[51,52] Immunohistochemical visualization of this enzyme in stomach and brain has begun[52] (Section 5.2.2).

Estimates of the K_m of histidine for brain histidine decarboxylase are much higher than the concentration of histidine in plasma or brain,[6] implying that the enzyme is not saturated *in vivo*; consequently, administration of L- but not D-histidine increased brain histamine levels.[40,43,45,53] Although these studies help to reveal the dynamics of histamine in brain, histidine loading is not a method to increase specifically brain histamine, since histidine may alter other brain transmitter systems; e.g., it lowers mouse brain 5-hydroxytryptamine levels.[40]

There are surprisingly few selective inhibitors of histidine decarboxylase. Since pyridoxal phosphate is a cofactor for this enzyme, any inhibitor acting at the cofactor site (e.g., semicarbazide) inhibits histidine decarboxylase and thus histamine formation.[54,55] Such nonselective inhibitors have limited value in the elucidation of histaminergic mechanisms. A series of benzyloxyamines, the most studied of which is brocresine, are potent histidine decarboxylase inhibitors, but they also inhibit other pyridoxal-requiring enzymes.[56] Other less potent inhibitors (e.g., α-methylhistidine, α-hydrazinohistidine, 4-imidazolyl-3-amino-2-butanone) are reported to be more selective.[54-56] Although these drugs inhibit histidine decarboxylase in brain homogenates,[47,57] they vary in ability to penetrate and act on brain.[58,59] Recently, α-fluoromethylhistidine was characterized as an irreversible inhibitor of brain histidine decarboxylase.[60] This compound almost completely abolished cortical and hypothalamic histidine decarboxylase activity in mouse brain homogenates at concentrations lacking effects on histamine methyltransferase, DOPA decarboxylase, or glutamate decarboxylase. This inhibitor decreased endogenous histamine levels in some regions of mouse brain.[60] Further studies are needed to evaluate the specificity and usefulness of this compound.

4. STORAGE AND FATE OF HISTAMINE

4.1. Storage

Subcellular fractionation of brain supports the hypothesis that a portion of brain histamine is stored in neurons (Table I). Early studies[13,61] showed that microsomal fractions contain nearly half of the histamine content of rat brain (Table I). Purification of this fraction by sucrose gradients and examination by electron microscopy revealed the presence of synaptosomes free of myelin and mitochondria.[61] Subsequent work (Table I) confirmed that a particulate fraction of brain contains appreciable amounts of histamine, the fraction containing the highest levels depending on the conditions used. Crude mitochondrial fractions of rat hypothalamus contained 46% of the histamine and 25% of the histidine decarboxylase activity. Osmotic lysis of particulate fractions resulted in the retention of histamine by synaptic vesicles and membranes,[61,63] whereas histidine decarboxylase was partially released,[63,66] suggesting a differential localization within nerve endings. By incomplete equilibrium sedimentation techniques, the distribution of histamine was reported to be like that of norepinephrine and other synaptosomal markers.[62] All workers have reported a significant portion, about 20%, of adult rat brain histamine to be in the nuclear (i.e., P_1) subcellular fraction (Table I). Several kinds of studies have led to the suggestion that the histamine in this fraction is in mast cells (Section 5.2.1 and 6) or in other cells associated with blood vessels (Section 11.1).

Consistent with the idea that a portion of histamine is stored in neuronal vesicles is that reserpine released histamine from rat hypothalamic slices.[67,68] *In vivo*, reserpine had no effect on rat whole brain histamine levels,[26,69] but it increased the rate of disappearance of labeled histamine from rat brain.[70] Reserpine decreased histamine content in mouse whole brain[40] and in cat hypothalamus and medial thalamus but not in cat hypophysis, where most of the histamine may be in mast cells.[15]

If histamine is stored in neuronal vesicles, the compound or compounds to which it is bound is not known. Histamine is known to form complexes with acidic lipids and other acidic substances, including sulfomucopolysaccharides, which may bind histamine in mast cells.[4,71-74] Brain sulfomucopolysaccharides (Section 5.2.1) are found in the same subcellular fraction as histamine[75] and in the same fraction that sequesters histamine both *in vitro*[76] and during intraventricular perfusion.[77] The affinity of the brain sulfomucopolysaccharides to complex histamine is not known; it is known that heparins from different species vary in their capacities to bind histamine and other biogenic amines[72] and that rat brain contains a lipid-soluble sulfomucopolysaccharide not found in other rat organs[75] (Section 5.2.1).

Table I

Subcellular Distribution of Histamine, tele-Methylhistamine, and Related Enzymes in Rat Brain

Substance	Tissue	Method	Percent of total			
			Nuclear (P_1)	Mitochondrial (P_2)	Microsomal (P_3)	Supernatant (S)
Histamine[13]	Whole brain	Biological	17.6	18.2	44.3	19.9
Histamine[61]	Cortex	Biological	7.7	21.0	51.4	19.9
Histamine[62]	Hypothalamus[a]	Enzymatic	25.7	37.8	—	36.5[b]
Histamine[63]	Hypothalamus	Enzymatic	20.2	46.4	15.6	17.8
Histamine[64]	Cortex	Fluorometric	33.0	34.0	17.0	16.0
Histamine[65]	Whole brain	Enzymatic	25.4	26.6	14.7	33.3
tele-Methylhistamine[65]	Whole brain	GC–MS	10.0	27.7	6.6	55.7
Histidine decarboxylase[66]	Cortex	Enzymatic[c]	9.0	46.0	—	45.0[b]
Histidine decarboxylase[63]	Hypothalamus	Enzymatic	10.4	24.7	5.7	61.1
Histamine methyltransferase[63]	Hypothalamus	Isotopic	5.5	13.7	0.9	79.9

[a] Five regions were studied.
[b] Including P_3.
[c] Results were very similar with a radiochromatographic method.

4.2. Release

Almost all work on histamine release has been done on preparations *in vitro*. Endogenous histamine is released from rat hypothalamic slices by potassium by a temperature-sensitive process that is inhibited by EDTA.[67] Similar characteristics were reported for the release of labeled histamine from rat hypothalamic slices after preincubation with labeled histidine.[78] When rat brain slices from several regions were incubated with labeled histamine and superfused, potassium evoked a rapid, calcium-dependent release of radioactivity despite the fact that there was little net uptake of radioactivity.[79] After guinea pig hypothalamic slices were labeled with histamine and superfused, electrical field stimulation caused a calcium-dependent release of radioactive histamine, the magnitude of which was related to the frequency of stimulation.[80] The cells from which this release occurs remain unknown. Since potassium does not release histamine from peritoneal mast cells, it has been argued that the histamine release from brain slices is not from brain mast cells.[78,79] Even so, other brain cells should be considered. For example, glia accumulate γ-aminobutyric acid and release it in response to depolarizing stimuli.[81]

The release of endogenous histamine from brain after electrical stimulation of discrete brain regions *in vivo* has not been rigorously demonstrated. Attempts to show this may have been hampered by inadequate methods for measuring histamine. Monnier[82] showed that electrical stimulation of rabbit reticular formation activated the EEG and increased the histamine-like activity in the cranial venous sinus effluent, whereas stimulation of thalamic nuclei that produced "relaxation" was associated with fivefold lower effluent histamine levels. These findings are especially noteworthy, for they are consistent with recent work (Section 5.2.2) suggesting that histaminergic fibers originating in the reticular formation innervate most of the telencephalon and might mediate wakefulness (Section 11.2). Although the finding that the effluent histamine levels increase during this stimulation may be accurate, histamine was measured in this study by bioassay without the necessary precautions (Section 2). The lowest histamine content found in 1 ml of effluent was about five times the amount in the entire rabbit brain.[43] This study[82] merits reinvestigation.

4.3. Uptake

One circumstance that has prevented the biochemical and morphological characterization of histaminergic neurons is the apparent absence of a high-affinity uptake system for histamine in brain. Several investigators have reported the inability of rat brain slices or synaptosomes to accumulate labeled histamine against a concentration gradient.[44,83,84] *tele*-Methylhistamine is also not taken up by crude or purified rat cortical synaptosomes (L. B. Hough and E. Domino, unpublished data). From these studies it has been concluded that, unlike catecholamine- and 5-hydroxytryptamine-synthesizing nerve endings, histamine-containing neurons do not use an uptake system for inactivation of histamine but rather rely on metabolism.[85]

Some studies suggest that this conclusion may not apply to all species. Synaptic vesicles from pig striatum exhibited a concentration-dependent ac-

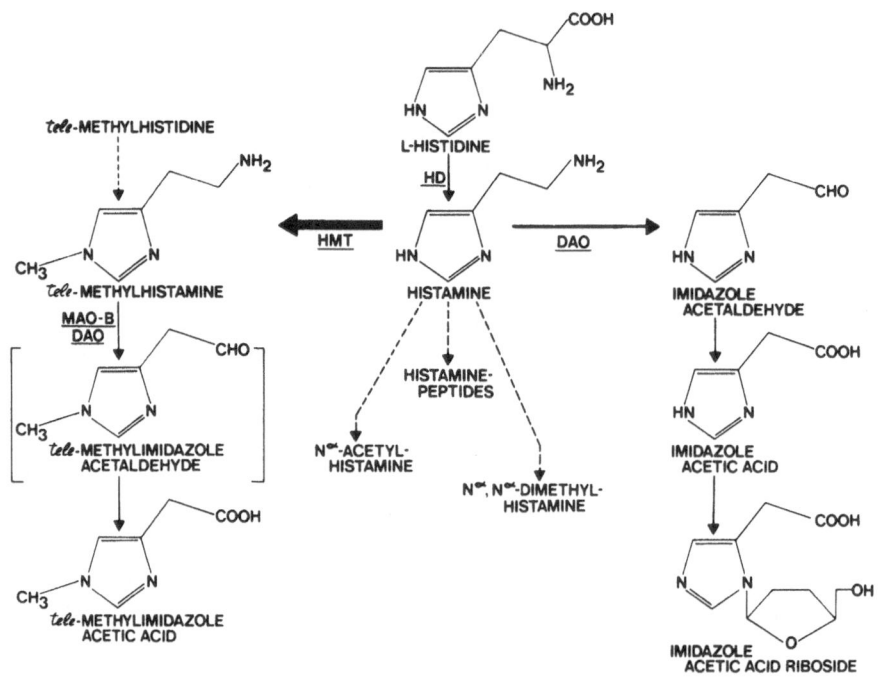

Fig. 2. Formation and major metabolic pathways of histamine. Abbreviations: DAO, diamine oxidase; HD, histidine decarboxylase; HMT, histamine methyltransferase; MAO-B, monoamine oxidase B.

cumulation of histamine that was unaffected by ATP and magnesium; the saturability and energy requirements for the process were not established.[86] A temperature- and sodium-dependent accumulation of labeled histamine was observed in rabbit hypothalamic slices, and tissue/medium ratios were 2 to 3, which could suggest accumulation against a concentration gradient.[87] The saturability and kinetics of this uptake were not evaluated, but the accumulation was inhibited by 5-hydroxytryptamine and desipramine.[87] Incubation of labeled histamine with guinea pig hypothalamic slices for 90 min resulted in a tissue/medium ratio of 2.95[80]; this uptake was not further characterized. Incubations of rabbit iris with labeled histamine produced an autoradiographic pattern resembling the distribution of sympathetic nerve fibers[88] (Section 8). An uptake system for histamine has been characterized in the invertebrate nervous system (Section 7).

4.4. Catabolism

Histamine is mainly metabolized by two distinct enzymatic systems in mammals (Fig. 2). It is oxidized by diamine oxidase (E.C. 1.4.3.6) to imidazoleacetaldehyde and then imidazoleacetic acid[89] as well as methylated by histamine methyltransferase (E.C. 2.1.1.8) to produce *tele*-methylhistamine (i.e., 1-methyl-4-imidazolylethylamine). Mammalian brain lacks the ability to oxidize histamine[90-92] and nearly quantitatively methylates it.[91,92] Histamine methyltransferase is highly selective for histamine; neither histidine nor other endogenous imidazoles with the possible exception of N^α-methylhistamines are known to be ring-methylated by this enzyme.[93,94] Histamine methyltransferase uses as methyl donor S-adenosyl-L-methionine, the cofactor for many methylation reactions including those of other biogenic amines.

The mechanism of enzymatic histamine methylation remains unclear. Studies of purified histamine methyltransferase from guinea pig brain suggested a ping–pong mechanism.[95] Kinetic studies of histamine methyltransferase in human skin suggested the ordered sequential Bi–Bi mechanism.[96] The mechanisms could be different in different species or tissues, especially in view of the likely presence of isozymes of histamine methyltransferase.[97] The products of the reaction, both *tele*-methylhistamine (see Fig. 2) and S-adenosylhomocysteine, are inhibitors of enzyme activity.[92,98]

Except for the possible side-chain N^α-methylation of histamine in stomach,[94] endogenous histamine is methylated only in the *tele*- position.[94,99] Methylation of histamine in the *pros*- position of the ring was inferred after the administration of large amounts of exogenous histamine,[100] but no endogenous *pros*-methylhistamine could be detected in rat brain by chromatographic[101] and gas chromatographic–mass spectrometric methods.[102,103] Although there are species differences in the importance of histamine methylation in peripheral tissues, no other major pathways of histamine metabolism in mammalian brain have been discovered. An early study describing histamine oxidation by rat brain[77] has not been confirmed.[91,92] Formation of γ-glutamylhistamine, a quantitatively minor pathway in mammalian brain,[105] has been reported to be a major metabolic mechanism in the histaminergic neuron of *Aplysia* (Section 7).

Compounds of several chemical classes are inhibitors of histamine methyltransferase. The aminoisoquinolines—amodiaquine and chloroquine—are competitive inhibitors of the enzyme *in vitro*,[106] as are the diaminopyrimidines, e.g., pyrimethamine, metoprine, and etoprine.[107] The latter are effective *in vivo*, elevating endogenous rat brain histamine content.[107]

It has become clear that many histamine agonists and antagonists are potent inhibitors of histamine metabolism (see Table II). H_1-receptor antagonists can either potentiate or inhibit histamine methyltransferase, depending on substrate, cofactor, and inhibitor concentrations.[109] Dimaprit, the highly selective H_2 agonist, is a noncompetitive inhibitor of histamine methyltransferase with a K_i similar to its EC_{50} on the H_2 receptor.[110,111] This finding led to studies of dimaprit analogues on histamine methyltransferase and to the discovery that SKF 91488 is a potent inhibitor of this enzyme and devoid of agonist activity.[110] Another highly selective H_2 agonist, impromidine, inhibits histamine methyltransferase in the micromolar range, somewhat lower than its EC_{50} on H_2 receptors, but inhibits diamine oxidase in the nanomolar range,[110] showing more affinity than aminoguanidine for this enzyme.[110]

tele-Methylhistamine undergoes oxidative deamination by either diamine oxidase or monoamine oxidase.[114-118] The affinity is higher for diamine oxidase.[114] In brain, which lacks diamine oxidase, *tele*-methylhistamine is oxidized

Table II
Inhibitors of Histamine Methyltransferase

Compound	Pharmacological classification	Type of inhibition[a]	K_i (μM)
Aminoisoquinolines			
Amodiaquine[108]	Antifolate[b]	Competitive	2.5[g]
Quinacrine[109]	Antifolate[b]	Competitive	0.1
Diaminopyrimidines[107]			
Pyrimethamine	Antifolate[b]	Competitive	0.9
Metoprine	Antifolate[b]	Competitive	0.1
Etoprine	Antifolate[b]	Competitive	0.8
Thioureas			
Dimaprit[110,111]	Histamine agonist[c]	Noncompetitive	8[f]
SKF 91488[110]	None	Noncompetitive	0.9–1.6
Imidazoles			
2-Methylhistamine[110]	Histamine agonist[d]	Uncompetitive	14[e]
4-Methylhistamine[110]	Histamine agonist[c]	Uncompetitive	18[e]
tele-Methylhistamine[110]	None	Uncompetitive	300
Impromidine[112]	Histamine agonist[c]	Competitive	0.5
Burimamide[113]	H_2 antagonist	Mixed	—[f]
Metiamide[113]	H_2 antagonist	Competitive	—[f]
Cimetidine[113]	H_2 antagonist	Competitive	—[f]
Alkylamines[109]			
d-Chlorpheniramine	H_1 antagonist	Competitive	0.7
l-Chlorpheniramine	H_1 antagonist	Competitive	9
Diphenhydramine	H_1 antagonist	Competitive	40
Phenothiazines[109]			
Chlorpromazine	Neuroleptic	—	14
Promethazine	H_1 antagonist	—	9
Other heterocycles[110]			
Pyridylethylamine	Histamine agonist[d]		300
Thiazolylethylamine	Histamine agonist[d]		100

[a] Type of inhibition with respect to substrate i.e., histamine. Nearly all inhibitors activate the enzyme at higher substrate concentration.
[b] Inhibitor of dihydrofolate reductase.
[c] Agonist affinity greater for H_2 receptor.
[d] Agonist affinity greater for H_1 receptor.
[e] IC_{50} at 1 μM histamine.
[f] Determination of K_i complicated by enzyme activation.
[g] IC_{50} at 100 μM histamine.

by monoamine oxidase, classified as type B.[115,116] The presumed aldehyde intermediate (never isolated) is rapidly oxidized to 1-methyl-4-imidazoleacetic acid,[92,115,116] i.e., *tele*-methylimidazoleacetic acid. The possibility that brain forms imidazoleacetic acid can not be completely dismissed, since labeled imidazoleacetic acid ribotide and riboside were detected in rat brain after peripheral administration of labeled histidine.[119] Imidazoleacetic acid could arise from transamination and decarboxylation of histidine as well as by histamine oxidation[89]; or it may enter brain from plasma, as implied by the central effects it produces in animals after peripheral administration (Section 4.6).

The concentrations of histamine metabolites in brain and spinal fluid may reflect the utilization of histamine by brain cells. Early reports of the presence of endogenous brain *tele*-methylhistamine[120] have been confirmed and extended by use of newer methods. Hough *et al.* developed[102] and then improved[103] a gas chromatographic–mass spectrometric method for detection and measurement of nanogram quantities of *tele*-methylhistamine in brain as well as in other tissues and fluids.[102,121] Recently, other laboratories have used this method or a modification of it.[122,123]

Histamine methylation may occur in a cellular location different from its synthesis or storage, since histamine methyltransferase and *tele*-methylhistamine are found mostly in the supernatant subcellular fraction, in contrast to histamine and histidine decarboxylase (Table I). Furthermore, osmotic lysis of the particulate fractions of hypothalamus released histamine methyltransferase but not histamine.[63] Lesions of medial forebrain bundle (Section 5.2.2) that reduce forebrain histidine decarboxylase have less effect on histamine methyltransferase,[124] suggesting that histamine could be methylated after its release from neurons. Perhaps consistent with this is that histamine methyltransferase has been found in cultured glial cells.[125] In contrast to rat peritoneal cells, which contain extremely small amounts of *tele*-methylhistamine[126] (in accord with earlier work showing little or no metabolism of histamine by these cells[127]), rat brain contains similar amounts of histamine and *tele*-methylhistamine (Table III), suggesting that brain *tele*-methylhistamine levels may be indicative of the activity of non-mast cell stores of histamine such as those in neurons (but see Section 5.2.1). Consistent with this idea, the regional distribution of brain *tele*-methylhistamine is similar to that of histamine[103,128] (Table III).

However, *tele*-methylhistamine may arise from a source other than histamine. Administration of labeled *tele*-methylhistidine resulted in the formation of labeled *tele*-methylhistamine, presumably by direct decarboxylation.[129] Although the highest levels of *tele*-methylhistidine are in muscle, brain levels are about 100 times that of *tele*-methylhistamine at least after protein hydrolysis.[130] Furthermore, although *tele*-methylhistidine is very nearly quantitatively excreted (either free or as the N-acetyl derivative), a trace amount of labeled *tele*-methylimidazoleacetic acid was identified in urine after administration of *tele*-methylhistidine.[131] The suggestion that brain *tele*-methylhistamine derives from histamine because of their similar regional distributions[103,128] (Table III) does not argue against formation of *tele*-methylhistamine by decarboxylation, because histidine decarboxylase, also distributed like these amines (Table III), could directly decarboxylate histidine and *tele*-methylhistidine. The importance of this pathway remains to be established.

Although labeled *tele*-methylimidazoleacetic acid had been detected in brain after incubation or administration of labeled histamine or histidine,[92,115,116] endogenous unlabeled *tele*-methylimidazoleacetic acid was detected in brain or tissues only recently. Khandelwal *et al.*[132] described the detection and measurement of *tele*- and *pros*-methylimidazoleacetic acids in brain, plasma, CSF, and urine. Previously described methods to quantify these acids[133] lacked the sensitivity for tissue measurements and were applicable only to analysis of urine. *tele*-Methylimidazoleacetic acid is probably the end product of histamine metabolism in brain, and its measurement may be useful in assessing brain histaminergic activity, although, as noted above, it could

Table III

Regional Distribution of Histamine, tele-Methylhistamine, and Related Enzymes in Rat Brain

Region	Histamine (ng/g)		Histidine decarboxylase (nmol/g · h)		Histamine methyltransferase (nmol/g · h)		tele-Methylhistamine (ng/g)	Monoamine oxidase (nmol/mg · h)[e]
	Ref. 21	Ref. 10	Ref. 21	Ref. 10	Ref. 21	Ref. 10	Ref. 128	Ref. 116
Cerebellum	26	22	3.3	0.7	370	5	11	5.2
Medulla–pons	25	29[a]	4.7	3.9[a]	470	13[a]	10	5.6
Thalamus–midbrain	75	52[b]	12.6	15[b]	910	18[b]	35[c]	5.8
Hypothalamus	209	196	28.8	61.8	1,050	17	179	10.8
Striatum	56	77	10.4	13.6	920	19	36[d]	6.9
Hippocampus	50	35	6.8	12	830	15	24	5.5
Cerebral cortex	39	42	6.3	12	820	18	30	6.5
Spinal cord	—	59	—	1.4	—	6	—	—

[a] Medulla oblongata only.
[b] "Midbrain" includes mesencephalon, thalamus, and subthalamus.
[c] Mean of mesencephalon (23 ng/g) and thalamus (47 ng/g).
[d] Anterior caudate nucleus and globus pallidus.
[e] Determined with tele-methylhistamine as substrate; this activity is similar to type B MAO.

arise from *tele*-methylhistidine. *pros*-Methylimidazoleacetic acid, recently detected in plasma, CSF, and brain,[132,134] is probably not a metabolite of histamine but rather of *pros*-methylhistidine.[91] With this method, no imidazoleacetic acid was found in brain.[132]

Histamine metabolites have only recently been identified and measured in human cerebrospinal fluid.[121,132,134] Mean levels of *tele*-methylhistamine in fresh spinal fluid were 2.18 pmol/ml,[121] somewhat lower than 9.12 pmol/ml, the mean value (obtained by similar methods) of samples stored for several years.[122] Regression analysis[121,134] of the values from the latter study showed a significant positive correlation ($r = 0.835$, $n = 12$) of *tele*-methylhistamine levels with storage time of the 12 adult samples. The intercept of the line, predictive of *tele*-methylhistamine levels in fresh samples, was 1.76 pmol/ml, in close agreement with the other value for fresh samples.[121] When the regression analysis was done on all 18 bloodless samples,[122] six of which were from children, the correlation fell ($r = 0.65$), but it was still significant ($p < 0.005$). It is not clear how levels of this amine could increase in samples stored frozen. Spinal fluid may contain histamine methyltransferase, but it was not detectable in 0.2 ml after a 15-min incubation,[121] and it was very low in goat spinal fluid.[135] The apparent increase over time might be related to the number of times the sample was thawed rather than to how long it was stored. Repeated freezing and thawing of human spinal fluid increased γ-aminobutyric acid levels.[136] Human spinal fluid levels of *tele*-methylimidazoleacetic acid were 22.8 pmol/ml.[134] High levels of histamine were found in human spinal fluid, concentrations high enough to activate brain histamine receptors.[121]

4.5. Turnover

The turnover rate of histamine could be changed by stimuli that affect processes in which histamine may function. The half-life of labeled histamine formed in brain after intracisternal administration of labeled histidine was estimated to be 0.5 to 3 h in different rat brain regions, but no correction was made for synthesis occurring during catabolism.[91] The formation of labeled histamine after intraventricular administration of labeled histidine obeyed a precursor–product relationship and permitted the determination of a conversion index, i.e., the ratio of labeled histamine formed over the precursor specific activity.[41] With the assumption of a one-compartment open model and steady-state histamine levels, the conversion index estimates the brain histamine turnover labeled by the precursor. This method gave a turnover rate of 34 ng/g·h for whole rat brain, a half-life of 46 min.[41] The histamine in brain seems not to be within a single compartment, so refinements in the model are needed that may require the ability to measure the different cell contributions to brain histamine. The conversion index differed among the rat brain regions. This distribution is similar to that of histidine decarboxylase[41] but may have been influenced by the intraventricular route of administration.

The rate of decline in rat brain histamine content after large doses of inhibitors of histidine decarboxylase led to an estimate[59] of less than 1 min for the half-life of brain histamine. Since even lethal doses of these drugs depleted no more than 30% of the histamine in any region, it is likely that not all the histamine in brain undergoes so rapid a turnover.[59] These estimates also presume that the turnover rate is not changed by inhibition of the enzyme. In the same study, labeled histidine was given intraventricularly, and the specific activities of brain histidine and histamine were measured.[59] Two studies with these methods[41,59] found nearly identical time courses of specific activity. One group[41] estimated the half-life of hypothalamic histamine to be about 10 min, and the other,[59] using a different kinetic model, estimated less than 1 min. Accurate estimates of brain histamine turnover require validation of some of the assumptions of the kinetic models. The rate of accumulation of endogenous rat brain *tele*-methylhistamine after a large dose of the monoamine oxidase inhibitor pargyline—a novel means to estimate histamine turnover—showed the half-life of whole rat brain histamine to be 50 min,[104] remarkably similar to the 46 min obtained by isotopic methods.[41]

4.6. *Pharmacological Activity of Histamine Catabolites*

The proposition[4] that some of the metabolites of histamine may have pharmacological activities has support, but the physiological implications are not clear. Imidazoles, which have been known for a long time to catalyze the hydrolysis of esters, enhance cyclic 3′,5′-nucleotide phosphodiesterase in whole homogenates or dialyzed homogenates of mouse brain. Among the active imidazoles were *tele*-methylimidazoleacetic acid, imidazoleacetic acid, and histamine itself.[137] This effect is provocative because, in principle, it could offer a novel means of arresting an effect of an endogenous substance; i.e., histamine increases cyclic AMP and cyclic GMP in some brain regions (Section 10), after which histamine is metabolized to *tele*-methylimidazoleacetic acid, which could then activate destruction of the cyclic nucleotides. The concentration[137] (5 mM) of the acid used to activate was, however, far beyond the levels found in, at least, rat brain, which are less than 0.4 nmol/g[132]; hence, the physiological significance of the activation is unclear.

Parenteral injections of large amounts of imidazoleacetic acid protected mice against anaphylaxis,[138] produced hypothermia,[139] and caused progressive hyperactivity, ataxia, catalepsy, and a loss of righting reflex.[140] Imidazoleacetic acid enhanced the lethal effects of strychnine, picrotoxin, and pentylenetetrazol.[140] The rat showed similar responses, which were like those produced by γ-hydroxybutyrate[140,141] and drugs such as D-LSD and phencyclidine.[140] Administration of imidazoleacetic acid into the cerebral ventricle of cat reduced arterial pressure,[142,143] heart rate, and renal sympathetic nerve discharge, all of which were reduced by bicuculline.[143]

In contrast with imidazoleacetic acid, *tele*-methylimidazoleacetic acid was inactive on both crayfish stretch receptor neurons[144] and cat cortical neurons.[145] On microiontophoresis, imidazoleacetic acid depressed the firing rate of neurons in the cat cortex[145,146] and medulla oblongata[147] with potency similar to that of γ-aminobutyric acid. Bicuculline, but not strychnine, antagonized the effects of both substances.[146,147] In both mouse and rat brain, imidazoleacetic acid increased the pool of γ-aminobutyric acid.[148,149] Experiments on nerve endings from rat brain showed the imidazoleacetic acid blocked the uptake system for γ-aminobutyric acid and for β-alanine.[150] Imidazoleacetic acid (1 mM) also inhibited uptake of γ-aminobutyric acid (0.2 mM) by rat superior cervical ganglia.[151] Both acids depolarized dorsal root neurons in frog spinal cord, but the effect of imidazoleacetic acid was blocked by either strychnine or picrotoxin, whereas the effect of γ-aminobutyric acid was blocked only by picrotoxin.[152] On the stretch receptor neuron of some crayfish, imidazoleacetic acid and γ-aminobutyric acid are about equipotent,[144,153] and on others, the latter is slightly more potent.[153] The dose–response curve of imidazoleacetic acid in depolarizing rat superior cervical ganglia paralleled that of γ-aminobutyric acid, and the effects of both were blocked by bicuculline.[151] Further evidence that they act on the same or similar receptors is that binding of γ-aminobutyric acid by rat brain membranes, a bicuculline-sensitive site, was reduced by imidazoleacetic acid[150]; the apparent affinity of the latter was about ten times less than that of γ-aminobutyric acid. Imidazoleacetic acid appears not to be formed by brain (Section 4.4), but it may enter brain, as shown by the pharmacological effects it elicited after parenteral administration of large doses (200–500 mg/kg)[138–143] and by the presence of labeled imidazoleacetic acid riboside and ribotide in the brain of rats that had received a peripheral injection of labeled histidine.[119]

It is puzzling that *tele*-methylhistamine, which has less than 2% of the activity of histamine on either H_1 or H_2 receptors,[154] on microiontophoresis onto rat and cat neurons has activity about equal to that of histamine.[147,155–157] Microinjection of *tele*-methylhistamine into the caudal hippocampus increased the locomotor, rearing, sniffing, and grooming activities of isolated rats; histamine, in contrast, decreased these activities.[158] It is possible that *tele*-methylhistamine is acting to release another substance, such as norepinephrine. Precedent for an indirect action not mediated by a known histamine receptor is the demonstration that impromidine, a histamine agonist, releases norepinephrine from atrial tissue by a process that was not blocked by histamine antagonists.[159]

Two other derivatives of histamine that may be catabolites, at least in peripheral tissues, N-acetylhistamine and N,N-dimethylhistamine, produce an arousal action (as does histamine, Section 11.2) in pentobarbital-treated rabbits.[160] The receptor associated with this effect has not been examined.

5. LOCALIZATION OF HISTAMINE

5.1. Regional Localization

When specific methods are applied to brain, all mammals studied have whole brain levels of histamine between 45 and 75 ng/g.[4] There are interesting variations in the levels of histamine and its metabolic enzymes in brains of lower vertebrates, some of which contain high histamine levels.[161]

The regional distribution in brain of dog, cat, rabbit, monkey, and frog has been tabulated[4]; there have since been additional descriptions of dog,[162] rabbit[53,163] rat,[163] monkey,[164,165] man,[166] and goat.[135] In all mammalian species, the distribution is very similar (in contrast with the differences among species in the distribution of H_1 receptors; Section 10.1.2). Table III shows the distribution in rat brain. All mammalian species studied have highest concentrations of histamine in hypothalamus and lowest in brainstem and cerebellum. Large variations in the histamine content within a particular brain region have have also been described; e.g., a 25-fold variation in the histamine content was found among hypothalamic nuclei of monkey,[165] mammillary bodies and ventromedial nuclei having the highest concentration. In contrast to the other nuclei, the histamine content of the supraoptic nucleus varied as much as tenfold among the animals studied, an observation perhaps relevant to the proposed function of histamine in antidiuretic hormone secretion (Section 11.5). Heterogeneity in histamine distribution within rat hypothalamus has also been shown.[167] Highest levels were found in the median eminence, ventral premammilary, arcuate, and suprachiasmatic nuclei, a distribution different from that of norepinephrine, dopamine, 5-hydroxytryptamine, and choline acetyltransferase.[167]

Histidine decarboxylase exhibits a regional distribution similar to that of histamine in rat (Table III) and human brain,[49] in accord with the suggestion that brain histamine content is controlled by its endogenous biosynthesis. Recent lesion and immunofluorescence studies are in accord with the suggestion that rat brain histidine decarboxylase is in neurons (Section 5.2.2).

In contrast to histidine decarboxylase, histamine methyltransferase activity is more nearly evenly distributed in brain regions, although still higher in forebrain areas (Table III), similar to the distribution of S-adenosine-L-methionine.[168] The enzyme was identified in cultured glia,[125] and, hence, it is probably not only a neuronal enzyme. Lesions produced less of a fall in histamine methyltransferase than in histidine decarboxylase.[124,169]

tele-Methylhistamine levels in whole brain are similar to those of histamine,[65,103] and the regional distributions of tele-methylhistamine and histamine in rat brain (Table III) are highly correlated ($r = 0.99$, $P < 0.01$).[65] When these levels are plotted against the regional histamine methyltransferase activity (Table III), the correlation is not significant. These findings suggest that histamine methyltransferase may not be saturated in vivo, but rather that tele-methylhistamine formation depends on substrate availability, i.e., accessibility of histamine to the enzyme.

5.2. Cellular Localization

Since histamine is stored in high concentrations in mast cells, i.e., 20 pg/cell,[170] it is necessary to consider that at least some of the histamine in brain is stored in mast cells. Until recently, this consideration was dismissed, since early studies reported the absence of mast cells in brain.[171] It is now certain that mast cells are present in mammalian brain, and it seems likely that histamine in brain is present in mast cells, perhaps also in the related neurolipomastocytoid cells, and in nerves as well, as pointed out below.

The distinction between neurons and nonneural cells is becoming increasingly less complete. Indeed, cells having many combinations of characteristics of both nerves and endocrine cells have been identified and termed paraneurons.[172] The mast cell has been classified as a paraneuron.[172] Perhaps heralding the proposed continuum between neural and nonneural cells are early studies of "clasmatocytes," described as arborizing mast cells in amphibian (and maybe mammalian) tissues containing metachromatic granules but easily confused with nerves.[170] Similarly, there may be a heterogeneity of mast cells in mammalian brain, as described below.

5.2.1. Mast Cells and the Nervous System

Interest in histamine as a neurotransmitter has also brought about renewed interest in mast cells. There is now considerable interest in mast cells in the nervous system because, if for no other reason, their contribution to histamine stores must be known before a neuronal localization can be invoked. And they may function in brain (Section 11.1).

A clear distinction must be made between the definition of a mast cell and the characteristics of a mast cell. In many studies attempting to identify the cellular localization of histamine in brain, the contribution of mast cells has been dismissed on the basis of the kinetics, subcellular distribution, and ontogeny of brain histamine, or on the effect of mast cell degranulators on brain histamine content. None of these characteristics is an invariable, exclusive criterion of mast cells. Mast cells differ not only among species[173] but within the same animal: mast cells from different tissues show great variation in immunoreactivity[174] and response to drugs.[175]

Until unequivocal criteria are found that embrace all mast cells, the presence of mast cells in tissues must still be judged by the original definition[170]: "A mast cell is a connective-tissue element which possesses cytoplasmic granules that stain metachromatically under ordinary circumstances." By this definition, mast cells are numerous in peripheral motor nerves of many species.[176] The histamine content of these nerves correlates with their mast cell numbers; and histochemical studies also support the presence of 5-hydroxytryptamine and dopamine in the mast cells of peripheral nerves of some species.[176]

Detailed studies of the exact localization of mast cells within nerve bundles, as well as their amine contents, embryogenesis, and reactivity to 48/80 have been reviewed.[4,176] The increase in mast cell numbers and histamine content after experimental or pathological lesions of peripheral nerve[176,177] suggests that these cells may respond to neuronal injury, perhaps as a mediator of inflammation, demyelinization, or tissue repair.[176] Little is known about changes in mast cell numbers after experimental brain lesions, although mast cell numbers were greater in the area postrema in monkeys receiving systemic 6-hydroxydopamine.[178] Brain histamine levels were increased after some surgical lesions (Section 5.2.2). Mast cells appear around syphilitic lesions and multiple sclerosis plaques in human brain[176] but were not found in greater numbers in human brains after stroke or lobotomy.[179]

Mast cells had been considered absent from brain[171] except for structures containing connective tissues, e.g., meninges, choroid plexus, area postrema, pineal body, and pituitary[176]. But detailed studies show mast cells within the CNS of rat,[180–183] hamster,[180,183,184] gerbil,[181] cat,[180,183] hedgehog,[185] and numerous other species.[180,186] Interestingly, mast cells are reported to be absent in brains of guinea pig, rabbit, dog, monkey, and man.[180] Absence of mast cells from rabbit and guinea pig brain has been disputed,[183] but absence from human[179] and *Macaca mulatta* (but not other) monkey brain[186] has support. Neonatal rat brain has more mast cells than adult brain (Section 6). A detailed review of the anatomy and possible functions of mast cells in brain has appeared.[187]

Not only do the numbers of mast cells in brain differ among species, but even within the same species, there is extraordinary variability in their numbers. Among ten hooded rats,[186] the number of mast cells in brain varied from 333 to 4358, and in cerebral leptomeninges from 18 to 3321. Similar variations were found by others.[181,188] It is not likely that the variability is a histological artifact, since similar variations have been found in numerous variations in methods. In one study of possible sources of variability,[189] no differences in brain mast cells were found after different fixatives, or in the presence or absence of perfusion or anesthesia. The action of the microtome blade can cause the denucleation of 1–2% of brain mast cells, but this cannot account for the large variation in brain mast cell numbers.[190]

The heterogeneity in staining capacity of brain mast cells,[188] as of mast cells in other organs, is indicative of the heterogeneity in mast cell granulation, perhaps reflecting their tendency to degranulate.[188] This variability in staining properties, which is reflective of the sulfomucopolysaccharide (e.g., heparin) content of the granules, could account for the variability in the counts of mast cells. Thus, different brains may in fact have similar numbers of mast cells, but differences in the degree of their granulation or sulfomucopolysaccharide content could result in poor quantification.

Despite this variation, many investigators found mast cells to be highest within diencephalic areas of adult brains, particularly in rodent thalamus.[180–182,184,185,191] Noted to contain especially high mast cell numbers were the dorsomedial, lateral, ventral, and paraventricular nuclei and the medial and lateral geniculate bodies. The significance of mast cells in these nuclei is obscure. However, in hamster brain, the greatest mast cell density was found around the large thalamic vein between the anterior dorsal and lateral thalamic nuclei.[184] It is not certain that mast cells in a thalamic nucleus can always be accounted for by the proximity of that nucleus to major blood vessels, but nearly all studies of mast cells in brain report that mast cells are often found in the vicinity of arterioles and venules.

Based on the histamine content of peritoneal mast cells, 20 pg/cell,[170] and mean mast cell numbers in brain, estimates of the degree to which rat brain mast cells contribute to brain histamine content have been made: estimates in the studies cited range from 10 to 100%. In light of the large variability of brain mast cell numbers, correlative studies of mast cell numbers and histamine content might improve these estimates. Large sample sizes must be surveyed;

studies of small numbers of brains sometimes report means with small variances that may be in error.[182,183] Thus, the variability in brain mast cell numbers and problems with the measurement of brain histamine (Section 2) have hampered the determination of the extent to which mast cells contribute to brain histamine, at least in rat. Although histochemical studies indicate that brain mast cells do contain histamine,[183] the histamine concentration in these cells is not known. If the concentrations of histamine in mast cells of rat brain resemble those of rat peritoneal mast cells, then it appears likely that a considerable amount of rat brain histamine could reside in mast cells. If so, then the great variability of brain mast cell numbers could be artifactual, since far less variability has been noted in brain histamine levels (Section 5.1).

The results of studies of a mutant mouse deficient in mast cells make the existence of non-mast-cell histamine in brain very likely. Mice of the genotype W/W^v (a pleiotropic mutant exhibiting defective hematopoiesis, melanogenesis, and gonadogenesis[192]) have only moderate erythrocytopenia but less than 1% of the normal number of skin mast cells and no detectable mast cells in brain or other organs.[36,193,194] The whole brain histamine content of this mutant, measured by a new HPLC method,[36] was about one-half that of the littermate controls, implying that one-half of brain histamine is in cells other than mast cells, perhaps neurons. This inference rests on the assumption that the non-mast cell histamine pool is the same in the mutant as in controls. Since this mutant is pleiotropic, it differs from its littermate in many ways[192,193]; hence, differences, found in this genotype may not be simply attributable to an absence of mast cells. It is worth noting too that mouse brain in general has fewer mast cells than has rat brain.[180]

A further confusion is the postulated existence of two types of brain mast cells. According to this classification,[180] type I cells, clearly the well-defined mast cell observed by others, are round to oval cells with round nuclei and uniform, metachromatic granules; these were concentrated around blood vessels and in dorsal thalamic areas of small rodents and cat, but they were not found in other mammalian brains. Type II cells are oval or spindle shaped, containing lipid-rich granules more heterogeneous in size. These cells are associated with blood vessels, distributed parallel to regional blood flow, and present in all species examined. Histochemical examination of type II cells gave equivocal results for the presence of histamine.[180] This classification has not been unanimously accepted[195] because the type II mast cell was initially described as lacking metachromasia.[180] In subsequent papers, the cell was characterized as metachromatic or normochromatic.[196] Reluctance by others to accept this cell as a mast cell may have prompted its being renamed the "neurolipomastocytoid cell,"[196] a designation that has received some acceptance.[197]

Brain neurolipomastocytoid cells may lack heparin, the source of metachromasia in mast cells, and may or may not contain histamine—both important questions to resolve. These cells need to be noted because they resemble true mast cells. These cells could be confused with the "faded blue" color of brain mast cells described in some studies.[188] The lack of metachromasia may be the basis of the controversy about the presence of mast cells in rabbit and guinea pig brain.[180,183] In further resemblance to the mast cell, the neurolipomastocytoid cell degranulates in response to compound 48/80 and radiation.[198] Some studies *in vitro* suggest that brain macrophages, mast cells, and neurolipomastocytoid cells share a common precursor.[199] It was proposed that the neurolipomastocytoid cells may function in place of mast cells in the brains of animals higher in evolution.[196] Whatever that function may be, these cells may be another example of the proposed continuum between neural and nonneural cells.

Relevant to metachromatic cells in brain is the characterization of sulfomucopolysaccharides. Sulfomucopolysaccharides from rat brain[75] containing glucosamine, galactosamine, hexuronic acid, and ester-sulfate were separated into two fractions, one a liposulfomucopolysaccharide not found in any other organ examined and clearly distinct from heparin. Brain sulfomucopolysaccharides have a far longer half-life than those in other rat organs or in murine neoplastic mast cells, and sediment mostly in the microsomal, i.e., P_3, fraction, different from heparin in mast cells, which sediments with the nuclei, i.e., the P_1 fraction. Autoradiography showed that brain sulfomucopolysaccharides were higher in gray matter than in white, as are the neurolipomastocytoid cells (i.e., type II mast cells) of brain. The solubility of the liposulfomucopolysaccharide in chloroform–methanol may relate to the loss of granules from the neurolipomastocytoid cell when extracted with the same solvent mixture.[180]

These sulfomucopolysaccharides[75] may function to store brain histamine (Section 4.1) and perhaps other amines.

5.2.2. Neurons

Notwithstanding the sparse information about mast cells and mast cell-like cells in brain, evidence is persuasive that at least some of the histamine in brain resides in cells other than mast cells: (1) brain mast cells have an uneven distribution within brain, but, as noted above, this distribution appears not to be the same as the regional distribution of histamine (Table III); (2) among species there are clear differences in brain mast cell numbers (Section 5.2.1), but these species have similar whole- and regional brain histamine levels (Section 5.1); (3) rat brain exhibits a postnatal decline in mast cell numbers but increasing activity of the histamine-synthesizing enzyme[64] (Section 6); (4) the subcellular distribution of histamine in brain (Section 4.1) is different from that of histamine in peripheral mast cells[64]; (5) brain lesions decrease histamine synthesis in regions distal to the lesion[7]; mast cell numbers show either no change or increase after nerve degeneration, at least in peripheral nerves[176,177]; (6) the W/W^v mutant mouse, which is reported to contain no brain mast cells, has about 60% of the brain histamine content of littermate controls.[36]

A major obstacle in characterizing histamine as a brain transmitter has been the inability to identify histamine in neurons by histochemical techniques that revealed histamine in brain mast cells.[183] Despite this failure, lesion experiments in rats have yielded evidence for the presence and distribution of histaminergic fibers. Lesions of lateral hypothalamus[200] or fornix,[201] or surgical isolation of hypothalamus[202] either had no effect or increased brain histamine content. Lesions of the medial forebrain bundle at the level of the lateral hypothalamus decreased ipsilateral forebrain histamine levels by 30% and reduced histidine decarboxylase by 36–60%.[203,240] The time course of the effect was consistent with that of degenerating nerve fibers, and the reduction in cortical histamine content was caused by decreases in the mitochondrial and supernatant subcellular fractions.[204] Surgical isolation of cortex[205] or deafferentation of hippocampus[206] or amygdala[207] caused large decreases in histidine decarboxylase. The results of many lesion experiments[7,10] suggest the presence of a diffuse ascending histaminergic system joining the medial forebrain bundle in caudal hypothalamus and projecting to ipsilateral cortex, hippocampus, striatum, and anterior hypothalamus. Experiments with colchicine are consistent with such a system.[10]

The anatomy of the ascending histaminergic system shares many of the features of the other diffuse monoaminergic systems. In further analogy to these systems, a search was made for the discrete nuclei containing the cell bodies of the ascending histaminergic pathway.[7,10] Selective lesions of several anterior hypothalamic or upper midbrain areas failed to identify discrete anatomical areas responsible for the forebrain histidine decarboxylase, but permitted the determination of the most "efficient lesion areas."[7,10] These studies suggest that mammillary bodies contain histaminergic cell bodies projecting to striatum, whereas the mesencephalic reticular formation contains histamine-synthesizing cells projecting to ipsilateral cerebral cortex.[7,10] Such a distribution is consistent with the suggested function of histamine in arousal (Section 11.2). Reduction in hindbrain histidine decarboxylase after hypothalamic lesions suggests a descending histaminergic pathway.[208]

Brain lesions decrease histidine decarboxylase activity more than they decrease histamine levels,[204] suggesting that the enzyme may be a better marker for histaminergic neurons than is histamine content, perhaps because histamine is in brain mast cells as well. Peripheral mast cells contain high histamine levels and low histidine decarboxylase.[204] Brain lesions may simultaneously destroy histamine-synthesizing neurons and stimulate brain mast cell infiltration or growth, which could explain the increases in brain histamine content after some lesions.[201,202] Mast cell numbers increase after surgical lesions of peripheral nerve,[176,177] but this has not been shown in brain.

The close similarity in the distributions of the proposed histaminergic fibers and of catecholamine- and 5-hydroxytryptamine-containing fibers raised the question of whether histidine decarboxylase was present in neurons containing catecholamines or 5-hydroxytryptamine. Administration of 6-hydroxydopamine or 5,6-dihydroxytryptamine reduced DOPA and 5-hydroxytryptophan decarboxylation in some brain regions without reducing histidine decarboxylase, suggesting that the histamine-synthesizing fibers are distinct from those containing catecholamines and 5-hydroxytryptamine.[204] But these experiments do not completely rule out localization of histidine decarboxylase in noradrenergic or serotonergic fibers, since neither neurotoxin abolished more than 54% of DOPA or 5-hydroxytryptophan decarboxylation, and neither had effects in other regions known to contain these fibers. It is necessary to consider that histidine decarboxylase activity increased in the remaining fibers in response to the lesion—as has been demonstrated for dopaminergic parameters after incomplete lesions of nigrostriatal fibers[209]—thereby obscuring a decline. In fact, histidine decarboxylase was elevated in cortex after 6-hydroxydopamine.[204] However, other studies[7,10] that determined the most caudal lesion that was effective in reducing forebrain histidine decarboxylase suggest that the histaminergic cell bodies are probably distinct from the locus coeruleus and raphe nuclei.

The inability to abolish forebrain histidine decarboxylase with lesions of discrete brain areas is consistent with a diffuse localization of histaminergic cell bodies. This localization awaits validation by a specific and sensitive histochemical method for visualizing these cells. Antibodies to purified histidine decarboxylase may offer a method, as preliminary immunohistochemical studies suggest.[52]

6. ONTOGENY OF HISTAMINE IN BRAIN

In contrast with other brain biogenic amines (which are found in low levels at birth, mainly in the P_2 subcellular fraction, and increase with the biosynthetic enzymes during synaptogenesis[210]), rat brain histamine concentration is highest at birth and is found in the P_1 (i.e., nuclear) subcellular fraction.[65,211–215] A rapid postnatal decline occurs in whole brain histamine levels, accompanied by a "shift" from the nuclear to the mitochondrial (P_2) fraction.[65,213,216] Paradoxically, histidine decarboxylase activity increases during this period.[214,215]

Although the histamine concentration (per gram tissue) decreases from birth to day 10 because of the rapid growth, it is clear that the 10-day-old rat brain has the highest amount of histamine (per brain).[65,211,212] The levels of *tele*-methylhistamine[65] and histamine methyltransferase[65,217] are also maximal at the tenth postnatal day, indicating that histamine methylation is a major metabolic mechanism in neonatal brain, as it is in adult brain. Correlations[65] of histamine and *tele*-methylhistamine levels in each rat brain of various postnatal ages showed that histamine methylation, perhaps turnover, is highest on the tenth postnatal day. *tele*-Methylhistamine in neonatal brain (as in adults, Table III) is found mostly in supernatant fractions.[65]

The cellular origin of neonatal brain histamine remains unknown. Several studies are consistent with a mast cell localization. Mast cells (Section 5.2.1) are numerous in the newborn rat brain,[186,218] and in one study the postnatal decline in brain mast cell numbers was similar to the postnatal disappearance of histamine (measured fluorometrically) from the nuclear subcellular fraction of rat brain.[218] However, other studies measuring histamine by a radioenzymatic method[65,213] found somewhat different postnatal changes in the histamine content of the subcellular fractions of rat brain. An interesting finding[218] in agreement with others[186] is that meninges account for 80% of the mast cells in a 1-day-old rat brain. Histamine assays of 1-day-old rat brain with and without meninges should help test the hypothesis that mast cells account for most of the neonatal histamine.

Some results implicating mast cells in the storage of neonatal brain histamine are based on the parallel findings in neonatal brain and adult peritoneal mast cells. Histamine in the neonatal brain is recovered mostly in the nuclear subcellular fraction, as is the histamine in peritoneal mast cells added to neonatal brain homogenates.[215,216] Like peritoneal mast cells, neonatal rat brain exhibits a low ratio of histidine decarboxylase/histamine.[215] Finally, the half-lives of labeled histamine in mast cells and in neonatal brain after labeled histidine administration were said to be similar.[215]

It is known that several neonatal rat tissues besides brain have an increased capacity to synthesize histamine.[54] Mast cells were dismissed as the cellular origin for this histamine because it was believed that most of the tissues lacked mast cells. However, a recent report[219] showed that the high levels of whole body histamine in neonatal mice was not evident in W/W (mast cell-deficient) mice, strongly suggesting that the neonatal histamine levels of at least some tissues represent mast cells. Analogous studies are needed on brain.

Neonatal brain histamine may be in mast cells, but several findings prevent ready acceptance of this conclusion. Purified neonatal brain nuclei retained 90% of the P_1 histamine, and this fraction was free of mast cell granules by microscopic examination.[213] It could be argued, however, that only a small number of mast cell granules are needed to account for the neonatal P_1 histamine content if their histamine content resembles that of peritoneal mast cell granules.[220] A finding that might be inconsistent with the idea that neonatal brain histamine can be ascribed to mast cells emerged from measurements of *tele*-methylhistamine. Neonatal brain histamine is methylated, and neonatal brain, like adult brain, has similar amounts of histamine and *tele*-methylhistamine.[65] In contrast, peritoneal mast cells—both crude and purified—have a concentration of *tele*-methylhistamine of about 0.2% of that of histamine, and histamine methyltransferase activity is undetectable in these cells.[126] These results agree with earlier ones showing that rat peritoneal mast cells[127] or mouse neoplastic mast cells[221] do not metabolize histamine. These results cannot be reconciled with the localization of neonatal brain histamine in mast cells unless neonatal brain mast cells release histamine at a much higher rate than do adult peritoneal mast cells. It is known that young peritoneal mast cells synthesize histamine more rapidly than do adult mast cells.[222]

The striking ontogenic changes in rat brain histamine suggest that it might function in development, as has been suggested for other transmitter substances.[223] Many experiments suggest a role of histamine in rapidly growing tissue, including fetal tissue.[54] The finding[213] that purified nuclei from rat neonatal brain retained 90% of their histamine suggests that neonatal histamine might be a regulator of brain growth, as has been postulated for polyamines such as spermidine.[224] The ontogeny of spermidine in rat brain bears remarkable resemblance to that of histamine.[211] Many critical developmental changes are occurring in brain during the time that neonatal histamine content peaks, and histamine (and/or mast cells) may serve as a modulator of neuronal differentiation, myelinization, or brain angiogenesis. The stimulation of mast cell growth by nerve growth factor[225] (Section 11.1) and the close relationship of mast cells to blood vessels (Section 5.2.1) could be clues to the elucidation of a developmental role.

Other ontogenic studies of brain histamine support a neurotransmitter role. Neonatal rat hypothalamic histamine levels increase[226] from days 5 to 20, similar to the ontogenic patterns of other transmitters. Comparable studies are needed of hippocampus and cortex, where there is more evidence of histaminergic transmission. In contrast to rat brain, nearly all regions of fetal guinea pig brain increase in histamine concentration during the last one-third of gestation, and no postnatal decline in histamine levels was found.[227] These prenatal increases parallel synaptogenesis, since the guinea pig, in contrast with rat, demonstrates precocial brain development.[228] A mast cell origin for neonatal rat brain histamine might be consistent with the different ontogeny found in guinea pig brain, the latter being said to have few or no mast cells, at least in adult brain.[180]

7. HISTAMINE IN THE INVERTEBRATE NERVOUS SYSTEM

Progress has been made in examining the hypothesis that histamine is a neurotransmitter in invertebrates. Assay of individual neurons of the cerebral ganglion of the snail *Aplysia californica* showed that in cell bodies of the LC2 and RC2 neurons histamine levels were 50 to 100 times higher than in other areas.[229] These neurons also contain much higher levels of histidine decarboxylase[230]; the histidine levels were like those in other cells.[231] High histamine levels were also found in the giant visceral neuron of the pond snail, *Lymnaea stagnalis*.[232]

Neurophysiological responses to histamine have been studied in invertebrates. Three classes of cells within *Aplysia* cerebral ganglia responded to the stimulation of C2 neurons with monosynaptic latencies; all responded similarly to the iontophoretic application of histamine.[233] Two distinct hyperpolarizing responses were described after the iontophoretic application of histamine to *Aplysia* neurons within the A cluster.[234] These responses were a fast, chloride-mediated hyperpolarization and a slower hyperpolarization mediated by increased potassium conductance. Pharmacological characterization of these responses indicated that the potassium-mediated response is probably mediated by H_2 receptors,[234,235] whereas the chloride-mediated response does not resemble an event mediated by either the H_1 or H_2 receptor.[234] Studies reporting the presence of an H_1-receptor-mediated response in *Aplysia* neurons[236] used concentrations of pyrilamine six orders of magnitude higher than its apparent dissociation constant for the H_1 receptor, precluding classification of the receptor associated with the response (Section 9). The iontophoretic effects of histamine on other invertebrate nervous systems have been reported.[237,238]

The inactivation mechanism for histamine in invertebrate neurons remains unclear; evidence suggests that both uptake and metabolism occur. Autoradiography of *Lymnaea* neurons after incubation with labeled histamine showed a selective accumulation of radioactivity in a large visceral neuron containing very high endogenous histamine levels.[232] The ganglia of a snail (*Helix pomatia*) showed a saturable sodium- and energy-dependent transport system for histamine.[239] In this system, the histamine taken up was metabolized rapidly.[239] It is not known if the uptake system is associated with histaminergic transmission.

In *Aplysia*, histamine is neither oxidized nor methylated, but instead metabolized to the peptide, γ-glutamylhistamine.[240] Other invertebrates also lacked the histamine metabolites seen in mammals and showed an unknown metabolite, possibly γ-glutamylhistamine.[239,241] The enzyme mediating this reaction has been characterized in *Aplysia*.[242] As this compound lacks neurophysiological effects,[242] it may be a mechanism for inactivation of histamine in the invertebrate nervous system. In mammals, this is a minor pathway of histamine metabolism; it has been described in rat brain.[105]

After labeled histamine was injected into cell bodies located on the anterior dorsal surface of the visceral ganglion of *Lymnaea*, axonal ramifications were seen in the neuropils of several ganglia.[243] Electron microscopic examination of the perikaryon of the neuron showed the large aggregations of granulated vesicles.[243] Ultrastructural studies of the C2 (presumably histaminergic) neuron of *Aplysia* with horseradish peroxidase found two morphologically distinct populations of vesicles within axonal varicosities of this neuron, similar to the pattern noted in other neurons of this invertebrate.[244] Further studies could likely reveal that these vesicles contain high concentrations of histamine.

8. HISTAMINE IN THE PERIPHERAL NERVOUS SYSTEM

The presence of high concentrations of histamine in peripheral nerves has been known for a long time,[4] but its localization and function remain speculative. Suggestive of the localization is that the histamine content of several motor neurons correlates with their mast cell numbers.[4,176]

Sympathetic nerves also contain histamine. As measured by bioassay, the superior cervical ganglion of many species appeared to contain high concentrations of histamine,[245] but by radioenzymatic assay the histamine content of rat superior cervical ganglion was only one-tenth of that found by bioassay.[246] The cellular origin of this histamine remains uncertain. Sympathetic denervation decreases norepinephrine levels while producing little change in the histamine content,[247,248] implying that the histamine is not within adrenergic neurons. As the histamine content of sympathetic nerve was not reduced by treatment with compound 48/80, it was concluded that the histamine was not in mast cells,[249] but it is now clear that not all mast cells in nerves are degranulated by compound 48/80.[176] Exogenous histamine can modulate ganglionic and postganglionic sympathetic transmission; the receptors subserving some of these effects have been identified as reviewed.[250] Evidence also exists that histamine is released after sympathetic activity and that histamine functions as a sympathetic nonadrenergic transmitter mediating active vasodilatation.[249,251] Whether the source of this histamine is the mast cell or the neuron, the process would be no less neural, as emphasized previously.[4]

Other provocative experiments show a relationship between histamine and sympathetic nerves. After rabbit iris was incubated with labeled histamine, radioactivity appeared in a pattern similar to that of the fluorescence of adrenergic nerves, the latter demonstrated concomitantly.[88] Decentralization of the superior cervical ganglion as well as sympathetic denervation abolished appearance of radioactivity, but decentralization did not abolish the adrenergic fluorescence, indicating that the radioactivity was not within adrenergic neurons.[88] The pattern of radioactivity was not attributable to mast cells, which are reportedly absent from rabbit iris.[88] Labeled histidine did not produce such a pattern, nor did incubation of labeled histamine with isolated ganglia. These very interesting findings are consistent with the presence in iris of sympathetic fibers that take up histamine but whose origin is not in ganglia.[88]

9. CONSIDERATIONS IN THE USE OF AGONISTS AND ANTAGONISTS TO CLASSIFY HISTAMINE RECEPTORS AND TO IMPUTE FUNCTIONS TO HISTAMINE

The prudent use of histamine agonists and antagonists classified the histamine receptors, leading directly to the discovery of histamine H_2 antagonists,[154,252] and could help to reveal functions associated with histamine receptors in brain. Following the rules of "pharmacological taxonomy"[253] facilitates attaining either goal. When these rules are ignored in work designed to classify receptors and to learn functions, the results are often ambiguous.[11,254–256]

One mistake is to attribute specificity to agonists that possess only relative selectivity. With the exception of dimaprit[257,258] and impromidine,[259] which are specific for the H_2 receptor, the agonists used to define histamine receptors are full agonists on both histamine H_1 and H_2 receptors,[154] and at high enough concentrations, they will stimulate both receptors. For example, at 0.1 mM, a concentration often used to classify histamine receptors, 2-methylhistamine, which has relative selectivity for the H_1 receptor, produces 80% of the maximum response on the H_2 receptor while producing 97% of the maximum response on the H_1 receptor.[255] Even when cautions are observed, agonists may produce misleading results in classification.[255] The apparent H_2 agonist activity of tolazoline and clonidine may rest on their ability to release endogenous histamine.[260]

Histamine agonists and antagonists influence the activity of histamine methyltransferase (Table II). For example, although dimaprit does not activate the H_1 receptor, it might result in activation of the H_1 receptor by inhibiting histamine methyltransferase, thereby increasing endogenous histamine levels. Indirect effects of histamine need to be considered, for stimulation of a histamine receptor can release another endogenous substance or substances. For example, stimulation of the H_2 receptor in the guinea pig myenteric plexus appears to release acetylcholine, a substance P-like compound, 5-hydroxytryptamine, and a compound(s) derived from arachidonic acid.[261] Histamine has long been known to release catecholamines.[71] In addition, agonists may have effects that are nonspecific or, at least, effects that elude simple classification. Both impromidine and the relatively selective H_1 agonist pyridylethylamine increased stimulation-induced norepinephrine efflux from guinea pig atria, but the use of antagonists failed to reveal the specific prejunctional receptors associated with the release.[159] Analogously, the analeptic effect of dimaprit after intraventricular administration was neither mimicked by impromidine nor antagonized by metiamide, results suggesting an effect not attributable to the H_2 receptor.[262]

Competitive antagonists provide less ambiguity than agonists in the classification of receptors if the antagonists are used at concentrations at which they express their selectivity, i.e., near their K_B values, for at high concentrations they block other receptors and sites. Sometimes the K_B values have been ignored; e.g., H_1 and H_2 antagonists have been used at the same concentrations, commonly at 10^{-6} M, which is more than 100 times the K_B of most H_1 antagonists and near the K_B of most H_2 antagonists. As H_1 antagonists have high affinity for the H_1 receptor when used *in vitro* at concentrations closer to their K_B values for this receptor, they unambiguously showed processes associated with the H_1 receptor and, by implication, histamine-sensitive processes that were likely attributable to another histamine receptor.[252] For example, mepyramine, though a competitive antagonist at both H_1 and muscarinic receptors, has 40,000 times the affinity for the H_1 receptor as for the muscarinic receptor[263]; other H_1 antagonists show less ability to discriminate between these two sites.[264–266] A new H_1 antagonist, astemizole, lacks affinity for the muscarinic receptor but has affinities for α-adrenergic and 5-hydroxytryptamine receptors.[266] At a concentration of 10^{-6} M, mepyramine and other H_1 antagonists block the H_2 receptor.[254,255,267–269]

More subtle and therefore more likely to mislead are effects of drugs that obfuscate their effects on receptors. After antidepressant drugs were observed to block competitively the histamine H_2 receptor linked to adenylate cyclase in brain homogenates,[269–272] Angus and Black[273] tested amitriptyline, the antidepressant with the highest affinity, on the H_2 receptor in guinea pig atrium, rat uterus, and mouse stomach. On these preparations, amitriptyline did not antagonize histamine, but on the H_2 receptor associated with contraction of the guinea pig papillary muscle, amitriptyline competitively antagonized the effect of histamine.[273] The authors suggested that tissue factors rather than heterogeneity of receptors explain the differences between atrium and papillary muscle in the response to the drug. One factor could be phosphodiesterase activity. An inhibitor of phosphodiesterase activity had greater stimulant activity on atrium than on papillary muscle.[273] As amitriptyline is a competitive inhibitor of phosphodiesterase,[274,275] its block of both the enzyme and the H_2 receptor in the atrium may have, as Angus and Black state,[273] "self-cancelled." Compatible with this idea is that the combination of phosphodiesterase inhibitor and metiamide, an H_2 antagonist, produced on the atrium an effect very similar to that of amitriptyline alone.[273] Analogous work showed that propantheline derivatives block both the H_2 receptor linked to adenylate cyclase in brain homogenates and the H_2 receptor in guinea pig papillary muscle but not the H_2 receptor in guinea pig atria.[276] An additional site that amitriptyline (and many other drugs) interacts with that could obfuscate interaction with the H_2 receptor is calmodulin.[277]

The H_2 antagonists are selective for the H_2 receptor, but at high concentrations they too affect other sites and other receptors. The affinity of cimetidine for the H_2 receptor, $K_B = 8 \times 10^{-7}$ M, is nearly 1000 times its affinities for the H_1 and muscarinic receptors,[278,279] where it appears to block competitively.[279] At the postjunctional α-adrenergic receptor, where cimetidine is a noncompetitive antagonist, a concentration of 10^{-6} M slightly but measurably shifted the dose–response curve to norepinephrine.[280] Burimamide was a competitive antagonist at the α-adrenergic receptor, for which its affinity[280] is only about 1/50 that for the H_2 receptor.[281]

H_2 antagonists have presynaptic effects as well. Cimetidine appears to release acetylcholine from the guinea pig ileum, an effect seen at a concentration of 8 μM with an EC_{50} of about 15 μM.[282] Ranitidine also appeared to release acetylcholine from some sites with an ED_{50} of 5×10^{-6} M on the most sensitive preparation.[283] Cimetidine, at a concentration of 3 μM, enhanced stimulation-induced release of norepinephrine although having, at a concentration of 10^{-5} M, a very slight effect in enhancing resting release and no effect on uptake of norepinephrine.[280] Burimamide and metiamide were even more potent than cimetidine in increasing stimulation-induced release of norepinephrine.[278,280]

Another source of error in classifying histamine receptors is in the use of binding techniques (Section 10.2) that do not consider artifacts, which have been well documented.[284-287]

When experiments are done *in vivo* to identify the receptor and/or associated functions, the complications are multiplied, such as by the numerous ways that a gross physiological phenomenon (e.g., sedation) can be brought about, by the numerous interactions an administered drug undergoes before reacting with a receptor, and by the (usually) unknown concentration of the drug in the system being appraised. Another consideration is that the number of receptors in an organ may vary among individuals, as is suggested for H_1 and H_2 receptors in human bronchi in studies *in vivo*.[288] In three strains of mice, the EC_{50} of histamine and the K_B of cimetidine on the isolated vas deferens were like those on the H_2 receptor in other tissues, but in six other strains of mice, the EC_{50} and K_B were more than five times higher.[289] Another concern is that treatment with an antagonist can alter levels of the agonist; e.g., an H_1 antagonist changed levels of histamine in some parts of brain,[290] part of which may stem from inhibition of histamine methyltransferase (Table II). Whatever the reasons, the altered levels of histamine could also influence events associated with the H_2 receptor.

Another pervasive issue in appraising the effect of a drug *in vivo* is its interaction with other drugs. Cimetidine inhibits metabolism of drugs by the mixed-function oxidase system, thereby exaggerating the effects of some drugs, including those that act on the central nervous system.[291,292] Other specific effects *in vivo* could lead to false attribution. Mepyramine *in vivo* antagonized effects of apomorphine and tryptamine as well as physostigmine.[293] The effects of chlorpheniramine *in vivo* were quantitatively like those of imipramine and amitriptyline on, respectively, 5-hydroxytryptamine and norepinephrine sites associated with neural functions.[294,295] The effect of H_1 antagonists in potentiating the central actions of catecholamines echoes the potentiating effects on peripheral systems, e.g., blood pressure.[296,297] These actions on 5-hydroxytryptamine and norepinephrine are attributable to blockade of the sites of uptake for these amines (as of dopamine as well).[294,295,298,299] Mepyramine, for example, competitively antagonized 5-hydroxytryptamine uptake by synaptosomes with a K_B similar to that of fluoxetine, the drug commonly used to block 5-hydroxytryptamine uptake.[299] In the high doses usually used in studies on whole animal including man, there is reason to suspect that H_1 antagonists block some of these sites in addition to the H_1 receptor.

The diverse effects of substances that produce muddled inferences are not confined to xenobiotics. Infusions of histidine not only raise levels of histamine in brain,[40,43,45,53] but lower levels of 5-hydroxytryptamine,[40] and, *pari passu*, infusions of DOPA and 5-hydroxytryptophan raise histamine levels.[53] All this work on histamine and its antagonists may simply exemplify the restraint needed before a receptor or a function can be imputed. There is no reason to presume that these cautions do not apply to other drugs. Histamine and drugs that block it are probably distinct only in having been studied for a long time in many systems and with many techniques.

Procedures used to classify receptors are often applicable to studies of function. Using agonists and antagonists of different chemical classes at concentrations appropriate for their EC_{50} and K_B values is likely to yield less equivocal results. In experiments on isolated systems, e.g., some electrophysiological studies, the use of agonists and antagonists in tandem has been revealing. In whole animals and man, testing several antagonists, including enantiomers, that differ sufficiently in their host of known pharmacological affinities and at different doses can yield useful information.

10. HISTAMINE RECEPTORS

At this time only two types of histamine receptors have been unambiguously shown in brain or in any other organ or tissue.

10.1. The Histamine H_1 Receptor

10.1.1. The H_1 Receptor Characterized by Binding Studies

Brain, of all organs and tissues examined, has the highest density of H_1 receptors as revealed by [^{3}H]mepyramine binding in the species studied.[300,301] It is surprising that the K_B of mepyramine and some other H_1 antagonists for the receptor in whole brain homogenates differed between guinea pig and other species. The binding of labeled mepyramine in the presence of triprolidine to the H_1 receptor in membranes from whole brain of guinea pig[301] has a Hill coefficient not different from 1.0, and Scatchard analysis indicates a homogeneous population of binding sites with $K_B = 0.5$ nM. The comparable K_B values of rat, rabbit, and mouse brains were between 3 and 4 nM, and that of human frontal cortex was 1.0 nM; all of these also showed saturable and homogeneous binding sites.[301] The rate of association of mepyramine is faster and the rate of dissociation slower in guinea pig membranes than in rat membranes.[301] The K_B values of other H_1 antagonists differ among species.[301,302]

All membranes studied discriminate between the stereoisomers of chlorpheniramine, but the guinea pig brain membranes showed higher affinity (as much as tenfold) than either the rat or human preparations for d-chlorpheniramine, triprolidine, chlorpromazine, and promazine. Other organs and tissues differ among species in their affinities for H_1 antagonists; within the same species, tissues varied in these affinities.[300] The affinities of H_1 antagonists for the H_1 receptor in rat membranes are better correlated with affinities for the H_1 receptor linked to histamine contraction of the guinea pig ileum than for the H_1 receptor in guinea pig brain membranes.[301]

Another group using [^{3}H]mepyramine and promethazine to label the H_1 receptors employed different methods, notably using membranes that sedimented at 6000 $\times$ g rather than at 50,000 $\times$ g[301]; the H_1 receptors are predominantly in the synaptosomal membranes.[303,304] Another salient difference is that the membranes, after incubation with the ligand, were collected by centrifugation rather than by filtration. The proportion of promethazine-insensitive binding sites was higher in this system[303] than the proportion of triprolidine-insensitive binding sites found by the other group,[301] a difference that is more likely attributable to the membrane fraction than to the masking ligand. On these guinea pig brain membranes, a second mepyramine binding site with lower affinity was revealed when the concentration of labeled mepyramine was increased.[302]

The K_B of the high affinity site in brain membranes, 0.83 nM,[302] was almost identical to that in guinea pig intestinal membranes and slightly higher than that, 0.50 nM, obtained by Chang et al.[301]; the B_{max}[303] was considerably higher than that found by Chang et al.[301] Binding of antagonists showed stereoselectivity. Most antagonists showed Hill coefficients differing from unity,[303] in contrast with the results of Chang et al.[301] The affinities (actually IC_{50} values in absence of simple competition) of these antagonists for the binding sites in guinea pig membranes correlated with their affinities for binding to membranes from guinea pig small intestine and for blocking histamine-induced contractions of the guinea pig ileum.[303] With these membranes, too, rat brain showed less affinity for H_1 antagonists than did guinea pig brain.[302]

The H_1 receptors are sensitive to proteolytic enzymes and to phospholipases A and C.[301] Binding of H_1 agonists to the receptor has sensitivity reminiscent of other receptors.[305] Sodium ions, especially, decrease the affinity of histamine for the H_1 receptor tenfold as measured by mepyramine binding; GTP, its analogue, and GDP decrease affinity. The effects of sodium and GTP are additive. Affinities of H_1 agonists are enhanced by manganese, to a lesser extent by magnesium, but not by calcium. None of these chemicals influenced the K_B values of antagonists.[305] A soluble digitonin extract of guinea pig brain retained sensitivities to sodium and divalent cations but not to GTP.[306,307]

Binding studies revealed that the H_1 receptor density in rat brain at birth is about 10–15% of that in the adult, which is attained 15–20 days after birth.[226,304] The K_B remains the same. In all regions studied, the ontogeny is similar.[304] H_1 receptor development occurs earlier than development of histidine decarboxylase and differs from the ontogeny of histamine content (Section 6). The development of the H_1 receptor paralleled the development of the ca-

Table IV
Regional Distribution of the Histamine H_1 Receptor and the H_2-Linked Adenylate
Cyclase Activity in Brain

Region	H_1 Receptor[310]		H_2-linked adenylate cyclase activity[b] (guinea pig)
	Rat[a]	Guinea pig[a]	
Telencephalon			
Neocortex	3.0	3.4	117[c]
Corpus striatum	2.2	1.7	28[c]
Hippocampus	2.2	4.5	104[c]
Diencephalon			
Thalamus	2.9	5.7	46[c]
Hypothalamus	4.2	3.8	15[c]
Mesencephalon	3.0	3.8	
Central gray			8[c]
Cerebellum	1.7	9.2	0[d]
Pons and medulla		2.9	0[d]
Other			
Optic chiasma			0[d]

[a] [^{3}H]Mepyramine binding, pmol/g wet weight.[310]
[b] Not necessarily a measure of H_2-receptor density.
[c] Formation of cyclic AMP by membranes. Percent increase over basal activity by 10^{-3} M histamine.[254]
[d] Formation of cyclic AMP by homogenates. Percent increase over basal activity by 10^{-4} M histamine.[311]

pacity to incorporate inorganic phosphate (administered by intracisternal injection) into phospholipids 5 min after intracisternal injection of histamine,[308] which enhances incorporation by acting on the H_1 receptor.[308,309]

10.1.2. Regional Distribution of the H_1 Receptor

In whole brain membranes of guinea pig, rat, and mouse and in membranes of human frontal cortex, the B_{max} values were similar, 3–10 pmol/g wet weight; rabbit preparations showed somewhat less density.[301] But there are distinct differences among species in the regional distribution of the H_1 receptor in brain.[301–303] Guinea pigs showed highest density in the cerebellum and lowest in the corpus striatum, whereas in rat the hypothalamus was highest and the cerebellum lowest (Table IV); in calf, the parietal and occipital cortices were richest, the pons, medulla and midbrain lowest; and in man, the frontal, parietal, cingulate, and temporal cortices, and amygdala were high, and the pons, medulla, and cerebellum low.[301] With the exception of hypothalamus, the regional distribution of the H_1 receptor was generally not related to the distribution of histamine or of histidine decarboxylase (cf. Tables III and IV), an independence common to other systems.[301]

The distribution of labeled mepyramine in brain after intravenous injections in mice was similar to the B_{max} of different brain regions, and the potency (ED_{50}) of H_1 antagonists in competing for labeled mepyramine levels in brain did not differ greatly from their affinities for the H_1 receptor.[312,313] In similar experiments *in vivo*, the H_1 antagonists reduced the amount of labeled mepyramine in the particulate fraction, the potency of some H_1 antagonists in this system again reflecting, though inexactly, their affinities for the H_1 receptor.[314]

Radioautography of slices of the guinea pig brain[315] showed a high density of H_1 receptors in the molecular layer of the cerebellum; few densities were seen in this layer in the rat cerebellum. In hippocampal formation of the guinea pig, the dentate gyrus was especially rich in H_1 receptors, and CA4 had greater densities than CA1, CA2, or CA3[315]; in the rat hippocampal formation, high densities were seen in CA3 and in the subiculum.[316] In rat hypothalamus, high densities were observed in the supraoptic and suprachiasmatic nuclei, ventromedial nucleus, and nucleus premammilaris,[316] which are also rich in histamine[167] and which may be related to the function of histamine (Section 11). Possibly related to the ability of histamine to activate afferents (Section 11.2) is the autoradiographic demonstration[317] in monkey of H_1 receptors in layers I and II of the dorsal horn, in the ventral horn, and in dorsal root ganglia; a large proportion of the cells within the ganglia had both H_1 and opiate receptors. Binding studies had showed H_1 receptors in guinea pig spinal cord.[303]

Binding studies of brain homogenates do not support the idea that all H_1 receptors are on neurons. The H_1 receptor, of all receptors studied, best sustained lesions in rats.[318] H_1 binding by striatal homogenates was not altered by injection of 6-hydroxydopamine into the substantia nigra, by hemisection of brainstem, or ablation of the cerebral cortex. Binding by hippocampal homogenates was unaffected by medial forebrain lesions or transection of the fimbria and fornix. Kainate injection into the hippocampus was followed in 4 days by a fall in H_1 binding, but at 15 days or longer no difference from untreated rats was seen.[318] It is difficult to attribute the localization of H_1 receptors to blood vessels, as the receptor distribution in brain regions is discrete; the receptor distribution is also distinct from that described for mast cells (Section 5.2.1). Some of the H_1 receptors in brain may be associated with glia, as suggested.[318] It should be emphasized, however, that in guinea pig, injections of kainate into the cerebellum produced a decrease in H_1 receptors in the molecular layer, observations suggesting that the receptors there are associated with neurons.[319] Dorsal root section reduced the density of H_1 receptors by 50% in layers I and II, as observed by autoradiography.[317]

10.1.3. Cyclic GMP Formation and the H_1 Receptor

In cultured mouse neuroblastoma, histamine stimulates cyclic GMP formation.[320] The activation, which peaks around 30 s after exposure to histamine, 0.1 mM, increases cyclic GMP formation as much as 50-fold. The ED_{50} of histamine was that for the H_1 receptor in peripheral tissue. Mepyramine, at appropriate concentrations, 3 and 6 nM, shifted the dose–response curve to the right without altering the slope. The affinities of other H_1 antagonists for this receptor were like those for the H_1 receptor in guinea pig ileum.[321]

There is evidence that in mammalian neural tissue, histamine enhances cyclic GMP formation, but the receptor mediating the effect remains unclassified. Histamine increases the cyclic GMP content of cerebral cortical slices of rabbit[322] and of guinea pig[323] but not of cerebellar slices of rabbit,[322] guinea pig,[324] or mouse.[325] Histamine also increases the cyclic GMP content of rat pineal body[326] and bovine superior cervical ganglion.[327] This effect of histamine requires calcium ions[323] (in contrast with the effect of histamine in stimulating cyclic AMP formation[328]). It may be pertinent that suggestive evidence has been presented that histamine may increase calcium influx in membranes of smooth muscle containing the H_1 receptor but not in membranes from smooth muscle lacking this receptor.[329] Calcium ions may then activate guanylate cyclase.[323]

10.1.4. Cyclic AMP Formation and the H_1 Receptor

One of the first systems used to show histamine receptors in brain was the stimulation of cyclic AMP formation by histamine in brain slices of different species.[330,331] This work was not primarily designed to define rigorously the histamine receptor(s) associated with cyclic AMP formation in brain slices. In brain homogenates in Tris buffer, adenylate cyclase activity was clearly shown to be associated with the H_2 receptor, and no H_1 receptor linked to adenylate cyclase activity could be demonstrated in these homogenates (Section 10.2.1). The finding that only the H_2 receptor is coupled to adenylate cyclase in brain *homogenates* does not rule out the presence of an H_1 receptor linked to cyclic AMP formation in brain *slices*. Conditions used to evoke it in homogenates may be inadequate, or homogenizing conditions could destroy the coupling. A homogenate from guinea pig cerebral cortex, prepared in Krebs–Ringer solution and containing large vesicular sacs, showed histamine-enhanced formation of cyclic AMP from adenine-labeled ATP, likely associated with an H_1 receptor.[332] In experiments showing only the H_2-linked adenylate cyclase, the homogenates were prepared in Tris buffer or sucrose, and exogenous ATP was used as substrate[267,311]; under these conditions, no histamine-stimulated cyclic AMP formation could be evinced in slices or in vesicular sacs, which require Krebs–Ringer solution and ATP stores labeled endogenously with adenine.[333]

Early studies showed evidence of an H_1-linked adenylate cyclase in slices, as previously summarized.[256] The histamine stimulation of the cyclic AMP formation in slices is about fivefold,[331] whereas in homogenates the maximal stimulation is about onefold.[267,311] The activity could be elicited in guinea pig cerebellar slices but not in cerebellar homogenates.[333] The relative potencies of a series of drugs in blocking histamine-stimulated adenylate cyclase in rabbit cerebral cortical slices differed from their potencies in blocking histamine-stimulated adenylate cyclase in comparable homogenates.[334] In the vesicular sacs or slices of guinea pig cortex,[335-338] H_1 antagonists reduced histamine-stimulated cyclic AMP formation at concentrations that could implicate a role of the H_1 receptor. In hippocampal slices of the same species, this reduction by the H_1 antagonists was much less than in cerebral cortical slices,[335] a difference that could imply that the histamine-stimulated increase in cyclic AMP formation in the hippocampal slice is linked to both receptors, whereas in the cortical slice it is linked to the H_1 receptor. Perhaps supporting this suggestion is that the activity of H_2 linked adenylate cyclase activity in hippocampal and cerebral cortical homogenates of the guinea pig is about the same,[267,311] whereas in slices histamine stimulated much more cyclic AMP formation in hippocampus than in the cerebral cortex.[331,335]

Evidence was presented that in hippocampal slices of the guinea pig, stimulation of both the H_1 receptor and the H_2 receptor was associated with increased cyclic AMP formation.[339] H_1 antagonists in appropriate concentrations and over a range sufficient to do a Schild plot showed competitive antagonism. On cerebral cortical slices, the K_B values of five H_1 antagonists in blocking histamine potentiation of adenosine-stimulated cyclic AMP formation and the K_B values in blocking mepyramine binding were similar.[340]

From work on slices and on vesicular sacs, it was suggested that H_2 or adenosine receptors are needed for H_1-stimulated cyclic AMP formation.[337] A study of the dose–response curves to histamine agonists in the presence and absence of adenosine showed that adenosine increased the affinity of the H_1 agonist thiazolylethylamine, but not that of 4-methylhistamine, an agonist more selective for H_2 receptors.[341] It was concluded that the H_1 receptor is not directly coupled to cyclic AMP formation in brain but rather facilitates activation by H_2 and adenosine receptors.[337]

These results showing differences between (1) slices and vesicular sacs in Krebs–Ringer solution with endogenously labeled ATP on the one hand and (2) the conventionally prepared lysed homogenates in Tris buffer with exogenously labeled ATP on the other are reminiscent of, though not paralleled by, results on the β-adrenergic receptor in bovine cerebellum.[342] Here, the intact sacs showed characteristics of having a β_1 receptor linked to the cyclase when assayed with adenine-labeled ATP and of a β_2 receptor linked to cyclase when assayed with exogenous ATP. The lysed synaptosomes showed two populations of β-adrenergic-linked cyclase.[342]

An ontogeny study on rabbit cerebral cortical slices showed histamine-stimulated cyclic AMP formation to be evident at fetal day 25, to peak at 8 days after birth, and to fall to adult levels at about 20 days.[343] Tripelennamine reduced the histamine-stimulated cyclic AMP formation; it is likely that at least part of the measurement was attributable to the H_1 receptor.[343] The ontogeny of H_1 binding in rats was different[226] (Section 10.1.1).

10.1.5. Other Effector Mechanisms Linked to the H_1 Receptor

Histamine enhances incorporation of inorganic phosphate into phospholipids in rat brain[308,309] and stimulates glycogenolysis in mouse brain,[344] both effects apparently linked to the H_1 receptor. In chick brain, the increased glycogenolysis was concomitant with a conversion of phosphorylase *b* to *a*.[345]

10.1.6. Desensitization of the H_1 Receptor

In neuroblastoma, the H_1 receptor linked to cyclic GMP formation was desensitized by histamine.[320] Mepyramine, 50 nM, but not metiamide, 1.5×10^{-5}M, prevented the desensitization. The rate of desensitization increased with increased histamine concentration. Desensitization was not accompanied by a change in the EC_{50} of histamine but rather in the reduction of the maximal response. Also indicating that the receptor itself is desensitized is that omission of calcium from the preincubation medium, which reduces cyclic GMP formation, did not affect development of desensitization.[320]

Early studies of histamine-stimulated cyclic AMP formation in guinea pig cerebral cortical slices, probably mediated by both histamine receptors, showed desensitization to histamine that was partly attributable to enhanced phosphodiesterase activity.[346] Histamine stimulation of glycogenolysis in mouse cerebral cortical slices was desensitized, as manifest not in a reduction in the maximal response but rather in a change in the EC_{50} of the agonist.[344]

Treating rat pups with diphenhydramine or mepyramine, each at 10 mg/kg, increased the density of H_1 binding sites in whole brain and hypothalamus.[347] But prolonged treatment of mature guinea pigs with mepyramine did not alter the B_{max} or K_B of mepyramine binding by membranes from different parts of the brain or from smooth muscle.[348] Tolerance to an H_1 antagonist in adult rats and mice[349] may not be related to an effect on the H_1 receptor.

10.2. The Histamine H_2 Receptor

All biochemical studies on the H_2 receptor in brain have been done on the H_2 linked adenylate cyclase. Binding studies with H_2 antagonists have been unsuccessful in specifically labeling the H_2 receptor. Cimetidine binding to sites that had been inferred to be H_2 receptors was shown to be resistant to many H_2 antagonists.[350,351] In one of the published studies that fostered inferences on H_2 binding sites, cited by these authors,[350,351] radioactive cimetidine was used not at a concentration near its K_B (6.03×10^{-7} M) but at 6.25×10^{-9} M, a concentration that could occupy no more than a fraction of a percent of the H_2 receptors, as calculated from percent occupancy = $[D/(K_B + D)] \times 100$. Availability of labeled tiotidine, the thiazolyl derivative that has far greater H_2 affinity than cimetidine, presented hope of specifically labeling the receptor, but the nonspecific binding of tiotidine was, like that of cimetidine, so high that the interaction with the H_2 receptor was obscured: the discernible H_2 binding by labeled tiotidine was insufficient to characterize the H_2 receptor.[269] Methods are needed to reduce the nonspecific binding of these ligands; or ligands with less nonspecific binding need to be discovered.

10.2.1. The H_2 Linked Adenylate Cyclase

In guinea pig brain homogenates (prepared in the conventional way), as contrasted with slices and vesicular sacs (Section 10.4.1), only an H_2 receptor linked to adenylate cyclase could be revealed.[267,311,352,353] Modeling the system showed that if homogenates contain an H_1 receptor linked to the cyclase, it could contribute no more than 15% to the activation.[256] That the receptor is H_2 was suggested by the relative EC_{50} values of a series of agonists on this preparation,[267,311] which were similar to the EC_{50} values on well-defined peripheral H_2 receptors. More persuasive is that seven antagonists with affinities ranging over four orders of magnitude for the H_2 receptor in peripheral tissues had affinities for the histamine-linked adenylate cyclase in homogenates[267,269]

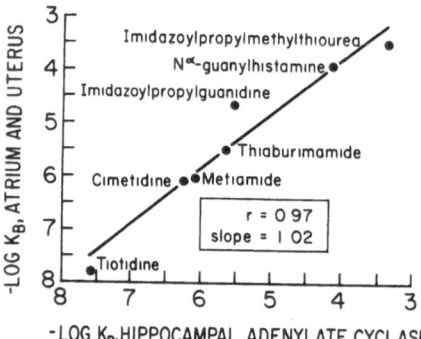

Fig. 3. The K_B values of histamine H_2 receptor antagonists on peripheral H_2 receptors and on the histamine receptor linked to adenylate cyclase in homogenates of the guinea pig hippocampus. The values on the peripheral receptors have been tabulated by Ganellin[281] with the exception of tiotidine, which was reported by Yellin *et al.*[354] The values on the hippocampal homogenates were obtained by Green et al.[267] except for tiotidine, which was reported by Maayani *et al.*[269]

similar to those on the peripheral H_2 receptor[281,354] (Fig. 3). Criteria for competitive antagonism were met. All compounds caused a parallel dextral shift in the dose–respnse curve. Some of the values shown in Fig. 3 were obtained from Schild plots[263,355] (i.e., log [dose ratio − 1] *vs.* log [antagonist]) in which the slopes did not significantly differ from 1.0 as predicted from simple competitive antagonism. Also as expected of simple competitive antagonism, the same K_B values were obtained with different agonists, including dimaprit,[267] which activates the H_2 receptor with negligible or no activation of the H_1 receptor.[257,258] The same Schild plot resulted when cimetidine was tested on either hippocampal or cortical membranes,[267] implying that the receptors in the two regions are not different.

For H_2 activation of adenylate cyclase,[356] GTP (or an analog) is very nearly obligatory. The guanine nucleotides increase V_{max} for histamine (at 10^{-4} M histamine). The ED_{50} for histamine is decreased by the guanine nucleotides. Implicit in this work[356] is suggestive evidence that just as the guanine nucleotides influence the action of histamine on the H_2-linked cyclase (and the binding of agonists to the receptor), histamine may influence the action of the guanine nucleotides on the cyclase: a Michaelis–Menten plot of these data[356] suggests that the EC_{50} of the GTP analogue [i.e., p(NH)ppG] is decreased threefold (from 3.5×10^{-6} M to 1×10^{-6}) in the presence of histamine, 10^{-4} M.

Another consideration in the H_2-linked adenylate cyclase is the magnesium ion: Mg^{2+} complexes with ATP^{4-} to form magnesium ATP (MgATP), which is recognized as the substrate for the enzyme. Magnesium ions may also stimulate by chelating free ATP, which inhibits the enzyme. Because more Mg^{2+} is required than is necessary for stoichiometric formation of MgATP, it was suggested that magnesium ions may act directly on an allosteric site on the enzyme.[356] Analysis of published data[356] shows that increasing magnesium ions increased the affinity of the cyclase for its substrate, MgATP (the apparent K_m value falls). This plot also suggests that the cyclase is inhibited at high concentrations of MgATP. In brain homogenates, the H_2-linked adenylate cyclase has a sensitivity toward magnesium ions in the presence of histamine, 10^{-3} M, that is tenfold greater than in its absence. Analogously, in the presence of histamine, ATP is a more potent inhibitor of the cyclase than in the absence of histamine (K_i = 0.16 mM *vs*. 0.33 mM).[356]

In guinea pig brain, the cortex and hippocampus showed greatest activity of the H_2-stimulated adenylate cyclase (Table IV). It needs to be emphasized that the distribution of H_2-stimulated adenylate cyclase activity is not necessarily the distribution of the H_2 receptor, as adenylate cyclase stimulation is a function not only of receptor density but of the coupling system and of the factors that control activity of the enzyme. The H_2 receptor-linked adenylate cyclase is also found in homogenates of monkey brain, where cortex and hippocampus also show greatest activity.[357] Within the hippocampus, the subiculum was most responsive.[357]

Experiments on chicks also showed adenylate cyclase to be linked to the H_2 receptor.[358,359] Subcutaneous injection of histamine (10 mg/kg body weight) to neonates, which lack a blood–brain barrier to histamine, increased the cyclic AMP content of cerebral hemispheres by three- to fourfold. Mepyramine, in doses as high as 20 mg/kg, did not block this rise, but the H_2 antagonists burimamide, 10 and 20 mg/kg, and metiamide, 2 and 5 mg/kg, produced dose-related reductions in the response. *In vitro*, the histamine-stimulated adenylate cyclase of chick brain slices (more responsive than brain slices of any other species reported) was antagonized by metiamide, 1 and 5 μM, and burimamide, 10 and 50 μM, but not by mepyramine, 50 μM. The cerebral hemispheres showed more activity than did the cerebellum, medulla–diencephalon, or optic lobes.[360]

Histamine stimulated adenylate cyclase activity in homogenates of rabbit frontal cortex, anterior limbic cortex, hypothalamus—in all of which activity diminished with age[361]—and in rat neocortex[311] and hippocampus,[267] but the receptor with which the activity is associated was not defined. Histamine evoked no stimulation of adenylate cyclase in homogenates of whole brain of gerbil or hamster,[362] rat corpus striatum or cerebellum,[311] and either monkey[357,363] or rat hypothalamus.[362,363] Rat brain homogenates are far less responsive in this regard than are guinea pig brain homogenates,[267,311] at least under the assay conditions used. A study that included Triton in the medium reported histamine-stimulated cyclic AMP formation in rat hypothalamic homogenates.[267,311] Failure to show histamine-stimulated cyclic AMP formation in homogenates does not imply that the H_2 receptor is not present. Electrophysiological studies of rat brain strongly suggest that the H_2 receptor is there and responding (Section 10.3).

The subcellular fraction of guinea pig cerebral cortex containing synaptic membranes was especially rich in the H_2 receptor.[356] From this finding and the nonuniform regional distribution, it was suggested that the receptor is associated with neurons rather than glia.[356] Others noted that the H_2 receptor is associated with glycogenolysis in chick brain slices,[359] that glycogen is associated with glial cells,[359] and that histamine stimulates cyclic AMP formation in astrocytoma cells,[365] all leading to the proposal that the H_2 receptor may be located on the surface of glial cells. If so, this localization may not be exclusive,

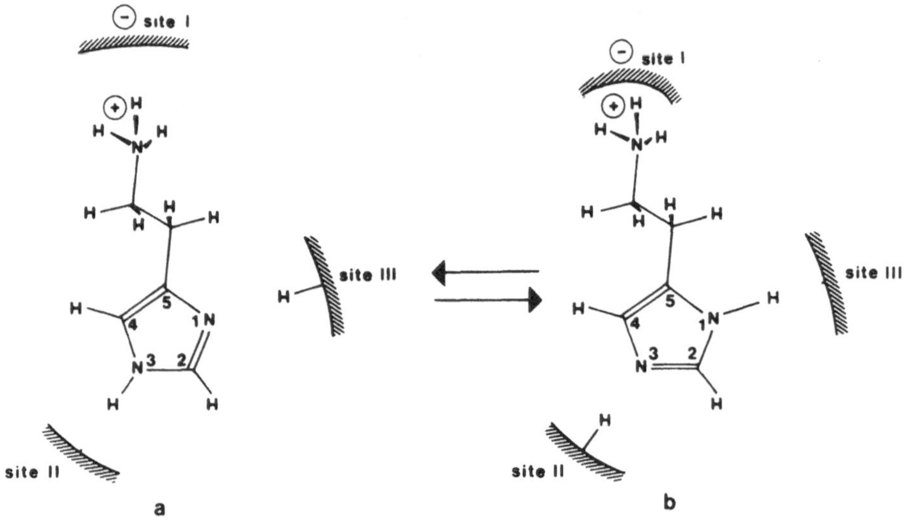

Fig. 4.   Charge-relay mechanism proposed for histamine activation of the H_2 receptor.[9,368] Histamine reacts with the receptor through its cationic head (i.e., the protonated amino group at site I and through the N(3)H and N(I) groups at sites II and III, respectively. Neutralization of the cationic head triggers transition of a proton from N(3)H to site II and transition of a proton from site III of the receptor to N(1). (Reprinted by permission from Yellin.[368a])

as kainate lesions in the guinea pig hippocampus almost completely eliminated histamine-stimulated cyclic AMP formation in slices.[366]

10.2.2. Activation of the H_2 Receptor

Tautomerism of the imidazole ring is a determinant of the action of histamine agonists on the H_2 receptor.[367] To test a mechanism by which tautomerism could be involved in receptor activation, quantum chemical calculations were used[368] to evaluate the changes that occur in the electronic structure of histamine when the protonated side chain amine of the monocation interacts with a negatively charged group serving as a model for an anionic receptor site. These changes in electronic structure were compared with those that occur when the monocation side chain is deprotonated to give the free base, since it had been established by both experiments and theoretical calculations that the tautomeric preference in histamine changes with ionization: in the free base, the more stable tautomer is N(1)H, whereas in the monocation, the more stable tautomer is N(3)H (Fig. 4). The results showed that neutralization of histamine monocation, either by interaction of the protonated amine with a simulated anionic site or by deprotonation of the side chain amine, produces a nearly identical effect on the electronic charge distribution in the structurally remote region of the imidazole ring. With the interaction of the cation with the anionic group, N(1) becomes more nucleophilic than N(3). These changes in reactivity parallel the shift in predominant tautomeric forms found both experimentally and theoretically in the transition from monocation to free base.

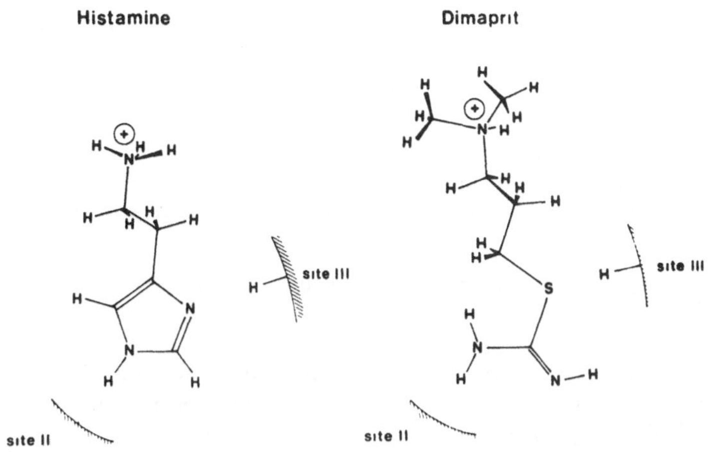

Fig. 5. Analogy proposed between the charge-relay models for histamine and dimaprit.[9] (Reprinted by permission from Yellin.[368a])

A mechanism for the direct involvement of histamine tautomerism in receptor activation was proposed (Fig. 4) in which histamine interacts through its cationic head with an anionic site I while the interaction of the N(3)H group with a matching site II and that of the imine nitrogen N(1) at site III provide the additional points of attachment. As the protonated amine on the side chain approaches the anionic site I, the interaction with this site causes N(1) to become protonated and N(3) to release its proton to site II. These results present a mechanism of drug–receptor interaction that induces a "charge relay system" to trigger a biological response.

Other H_2 agonists were then studied. Dimaprit (Fig. 5) is an H_2 agonist with no H_1 activity. Because of π-electron conjugation, the isothiourea group in dimaprit is planar and sterically congruent with the imidazole ring of histamine. The interaction scheme for dimaprit is analogous to that for histamine.[9] The N(1) of histamine is replaced by the sulfur atom, which, calculations show, generates in its surroundings an electric field that would be strongly attractive to the proton at the proposed site III in the interaction scheme of histamine. Because of conjugation, the interaction at site III will result in a decrease in the proton affinity of the nitrogen corresponding to N(3) of histamine. This change in reactivity may cause a temporary shift of a proton from the dimaprit molecule to site II. The process is reversible and entirely mimics in its effect the charge relay caused by the tautomeric shift in histamine as a result of the interaction with an anionic site.

4-Methylhistamine, which has relatively high affinity for the H_2 receptor, was compared with 2-methylhistamine, which has less.[154] The two compounds differed in the proton affinities of the imidazole nitrogens. In the 4-methylhistamine monocation, the proton affinity of N(1) in the N(3)H tautomer is only slightly increased compared with histamine. But in the 2-methylhistamine cation, the proton affinity is considerably higher. As a consequence, the dication species, in which both ring nitrogens are bound to hydrogens, will be much more abundant in 2-methylhistamine than in either histamine or 4-methylhistamine. Since only the monocation and the free base can undergo tautomerism, the charge relay is not feasible for the dication. Also supporting this proposal is that 4-methylhistamine is less potent than histamine on the H_2 receptor, and the comparable pK_a values (6.76 and 6.07)[369] imply that at physiological pH, a greater portion of 4-methylhistamine than of histamine exists as a dication.

10.3. Electrophysiological Effects Associated with Histamine Receptors

Neurons in rat hypothalamus show either increased or decreased firing rate when exposed to histamine.[370,371] Observations on perfused rat hypothalamic cultures showed that excitation by histamine is blocked by promethazine, an H_1 antagonist lacking local anesthetic activity in low doses, but not by the H_2 antagonist metiamide.[372] Promethazine, but not metiamide, also blocked the slow depolarization produced by histamine on rat hippocampal CA3 pyramidal cells *in vitro*.[373]

On the longitudinal muscle of the guinea pig ileum, histamine, probably by acting on the H_1 receptor, increased the frequency of action potentials and, with increasing concentration, produced depolarizations, increased conductance, and increased potassium efflux by opening a channel shared by the muscarinic receptor.[374] The H_1-receptor-mediated increase in contractile force in guinea pig ventricle was inferred to be associated with release of intracellular calcium, not to increased calcium influx.[375]

More intensively studied have been the electrophysiological effects associated with the H_2 receptor. The EC_{50} of histamine in stimulating the H_2-linked adenylate cyclase in guinea pig hippocampal homogenates was very similar to the EC_{50} in increasing the firing rate of CA3 cells in guinea pig hippocampal slices, an effect seen with impromidine and blocked by tiotidine at a concentration that blocks the peripheral H_2 receptor.[353] The increase in firing rate is not commonly seen in other species. Parallel studies like this one would be difficult to perform on the rat hippocampus. Although the histamine-stimulated adenylate cyclase is present in homogenates of rat hippocampus, the maximum cyclic AMP formed is far less than in homogenates of guinea pig hippocampus and not even sufficient to characterize the receptor (Section 10.2.1). But electrophysiological experiments suggest that the H_2 receptor is in rat hippocampus. Histamine depresses the firing rate of rat hippocampus,[370] as it does that of many neurons in mammalian brain [and molluscan neurons as well (Section 7)], an effect that is mediated by H_2 receptors (in contrast with the H_1 mediated increase in firing rate mentioned above). The depression is blocked by H_2 antagonists.[370,371]

Further, in slices of rat hippocampus, histamine and impromidine hyperpolarize about half of the CA1 pyramidal cells and dentate granule cells when applied by iontophoresis or pressure, with a concomitant reduction in excitability and spontaneous firing rate.[376,377] Some cells are depolarized by histamine and impromidine, the probability being higher on application by microdrops.[377] Although depolarization is usually accompanied by an increase in spontaneous firing rate, decreases were also observed.[376,377] In a calcium-deficient or magnesium-enriched medium, which depresses synaptic transmission, the hyperpolarizations were not blocked, but the depolarizations were. The hyperpolarizations were attributed to a postsynaptic action, possibly to increased K^+ conductance in dendrites. Histamine also reduces homocysteate-evoked firing.[376] Compared with the action of homocysteate, the hyperpolarizations produced by histamine (as by 5-hydroxytryptamine) are slow in onset and long in duration, properties that may imply modulation rather than transmission.[376] The depolarizations are probably indirect actions of histamine on other cells.[373,376,377] The depolarizations coupled with reduced firing rate, Haas suggested,[377] are consistent with dendritic inhibition and somatic disinhibition. It should be noted that all of these electrophysiological effects of histamine were blocked by appropriate concentrations of metiamide.[377] In rat hypothalamic neurons in culture, the H_2-mediated depression also persists in calcium-free medium, again suggesting that this H_2 effect is postsynaptic.[372]

Electrical stimulation of the rat medial forebrain bundle depresses firing of cortical neurons,[378] and stimulation of fornix depresses hippocampal CA1 pyramidal cells.[379] Of special interest is that metiamide reduced these electrically induced depressions,[379] in accord with the proposed histaminergic pathways (Section 5.2.2). It seems clear that this depressant effect of histamine on neurons is mediated by the H_2 receptor; however, it is not certain that this H_2 effect is linked to adenylate cyclase in rat. Destruction of the medial forebrain bundle in guinea pigs produced hypersensitivity to iontophoretically applied histamine without altering the EC_{50} or V_{max} of histamine-stimulated adenylate cyclase activity in slices from cortex or hippocampus or in hippocampal homogenates, results implying that in these neurons, the H_2 cyclase is not related to the electrophysiological response.[380,381] Also in guinea pigs, chronic drug treatment enhanced the electrophysiological response of the hippocampus to histamine[353] but not the adenylate cyclase response in hippocampal homogenates.[269,353] But in rat, the same lesion resulted in hypersensitivity of histamine-stimulated adenylate cyclase in cortical and hippocampal slices.[382] In rat hypothalamic cultures, Geller[383] found that the H_2-linked depressant response was enhanced by phosphodiesterase inhibitors. Kainic acid injections into guinea pig hippocampus abolished histamine-stimulated cyclic AMP synthesis in hippocampal slices,[366] where the stimulation is partly linked to the H_2 receptor.[339]

The monosynaptically excitatory field responses of the rabbit nucleus accumbens, driven by electrical stimulation of the hippocampal fimbria, were reduced by iontophoretic application of 4-methylhistamine, the relatively selective H_2 agonist, but not by 2-methylhistamine, the relatively selective H_1 agonist.[384] This effect of 4-methylhistamine was diminished by metiamide. Most interesting, the effect was diminished by bicuculline, the antagonist of γ-aminobutyric acid.[384]

11. FUNCTIONS OF HISTAMINE IN BRAIN

Some functions of histamine may be related to brain mast cells. Other experiments suggest a function for histamine without regard to the cellular source.

Section 9 reviews the cautions needed before inferring functions from observations after injecting histamine agonists and antagonists.

11.1. Functions that May Be Associated with Mast Cells

As mast cells contain vasoactive substances and are associated with blood vessels, they may function in microvascular regulation. The diencephalic localization of mast cells in hibernating animals[184] and the increase in brain mast cell numbers during hibernation[385] suggested that mast cells may regulate the microvasculature during hibernation. However, the similar localization of mast cells in nonhibernating animals and the increased mast cells in peripheral tissues during hibernation [184] may question this specific role.

Other studies support a role for histamine and/or mast cells in brain microcirculation. Local application of 48/80 caused pial vasoconstriction,[386] suggestive of a role of mast cells and/or neurolipomastocytoid cells. The presence of mast cells in human meninges and the relaxation of human pial artery by histamine, probably by stimulation of the H_2 receptor, were shown.[387] Recent studies have shown the presence of histamine[388,389] and histamine H_2 receptors[390] in purified fractions of brain blood vessels. Intraarterial histamine increased the brain permeability of labeled sucrose, an effect antagonized by metiamide but not mepyramine, suggesting that stimulation of cerebral endothelial H_2 receptors increases the permeability of the blood–brain barrier.[391] Perhaps relevant to the control of cerebrospinal fluid is the high concentration of mast cells in the choroid plexus and cells lining the subarachnoid space (Section 5.2.1).

Other suggested functions for mast cells in brain are even more speculative. Brain mast cells, like peripheral mast cells, may have IgE bound to their surfaces, but this has not been determined, nor has their participation in any immunologic processes been revealed. If immunocompetent, then brain mast cells may be the only cell in brain capable of reacting to immune stimuli with the release of neurally active substances. Unfortunately, little is known about such processes in brain.[392]

Brain mast cells may respond to other stimuli. Five-day-old rats handled daily for 10 or 15 days had fewer than one-half the number of brain mast cells found in rats that were not handled.[393] Newborn rats exposed to complete darkness for the first 3 months had more than twice the mast cell numbers in the dorsal part of the lateral geniculate body and in the overlying leptomeninges than did controls exposed to normal light–dark cycles.[394] As discussed by these authors, a hormonal mechanism might explain these findings.

Another observation that suggests a possible relationship between mast cells and neurons is that injections of nerve growth factor to neonatal rats resulted in a ten-fold increase in mast cell numbers in sympathetically innervated tissues, an effect not produced in adults.[225] The question of whether or not the mast cell response was secondary to the growth of the fibers was answered: administration of nerve growth factor to animals treated with 6-hydroxydopamine was not accompanied by sympathetic innervation but did produce the increase in mast cell numbers. Thus, nerve growth factor, not sympathetic neurons, stimulates the proliferation of mast cells. That mast cells are needed for the proper development of sympathetic fibers is not known but could be tested by studying adrenergic ontogeny in the W/W^v mouse, which lacks mast cells (Section 5.2.1).

11.2. Behavior

Exogenous histamine produces behavioral changes. Injected into the lateral hypothalamus, histamine caused a dose-dependent inhibition of rat hypothalamic self-stimulation, an effect antagonized by microinjections of H$_1$ blockers.[395] This antagonism is clearly not simple, since qualitatively different effects of different H$_1$ antagonists on self-stimulation have been described.[396] As the dose of mepyramine was increased, self-stimulation decreased, whereas chlorpheniramine produced, at a low dose, a decrease in self-stimulation and with higher doses a dose-related increase in rate of self-stimulation.[396] The effects of neither drug on self-stimulation can clearly be attributed to H$_1$ antagonism (Section 9). The effect of mepyramine in decreasing self-stimulation was seen after a subcutaneous dose, 2.5 mg/kg, that was below the ED$_{50}$ and confidence limits of H$_1$ antagonism (5.39 and 3.34–8.69 mg/kg), as measured by blockade of the effects of compound 48/80. Similarly, the comparable doses of chlorpheniramine were (mg/kg) 0.63 to decrease self-stimulation and 2.5 and greater to increase self-stimulation; for H$_1$-antagonism, the ED$_{50}$ was 14.2 with confidence limits of 8.26–24.4.[396]

Intraventricular injections of histamine caused a dose-dependent anorexia in cats that was blocked by H$_1$ antagonists.[397] Several hypothalamic nuclei that contain histamine and its receptors responded to histamine with increased drinking in rats[398]; H$_1$ blockers suppress thirst-induced drinking by an unknown mechanism.[399] Consistent with its inability to penetrate brain, histamine causes drinking by at least two peripheral mechanisms.[400] On intraperitoneal injection, histamine produced drinking that was severely attenuated by bilateral vagotomy.[401] Subsequent studies showed the effect to be mediated by stimulation of gastric vagal afferents.[400] Because vagotomy antagonizes but does not abolish the effect,[400] an additional mechanism was suspected. The vagotomy-resistant stimulation of drinking by histamine was completely blocked by inhibition of angiotensin-converting enzyme, suggesting that peripheral histamine also stimulates drinking through angiotensin, a hypothesis consistent with the hypotensive effects of these doses of histamine, known to release angiotensin.[400] Metiamide and cimetidine, given intraperitoneally, antagonize histamine-activated drinking,[401,402] suggesting that at least one of these peripheral mechanisms is mediated by H$_2$ receptors. Some evidence suggests that peritoneal mast cells respond to changes in osmolality by histamine release, with the subsequent activation of these dipsogenic mechanisms.[402]

Intravenously, histamine was an effective punisher of schedule-controlled behavior in squirrel monkeys; the effect, unrelated to cardiovascular actions, was blocked by diphenhydramine but not cimetidine.[403]

These studies, showing that histamine can produce behavioral effects by acting on sites outside of brain, are reminiscent of earlier work showing the activation of visceral afferent nerves by histamine.[82] The effects of histamine on afferent fibers were shown on the rabbit nodose ganglia.[404] Histamine, applied at micromolar concentrations by superfusion, elicited a slow-onset, long-lasting depolarization accompanied by decreased membrane conductance, probably to potassium, in some type C neurons. Histamine had no effect on type A neurons, the myelinated and more rapidly conducting ones. The histamine receptor associated with the response was not examined, but it may be relevant that H_1 receptor binding sites are found in the dorsal ganglia of monkey[317] (Section 10.1.2).

Indirect evidence supports the suggestion that histamine functions as a mediator of arousal. Histamine given peripherally or into the brain activates the EEG and enhances reticulocortical evoked potentials,[82] perhaps consistent with the suggested site of the cell bodies of the ascending histaminergic system (Section 5.2.2). Dimaprit had an analeptic effect during urethane anesthesia,[262] but the effect was not antagonized by H_2 blockers or mimicked by other H_2 agonists, suggesting a mechanism independent of the H_2 receptor (for similar examples, see Section 9). The rat hypothalamic histamine content fluctuates over the light–dark cycle, perhaps reflecting different stages of arousal.[405] The ability of CNS depressants to decrease the estimated turnover of rat brain histamine (Section 4.5) may support this hypothesis, but the mechanism of this effect remains unknown. Although the method for measuring histamine was suspect, the suggestion (Section 4.2) that histamine release was increased by stimulation of waking centers supports a role of histamine in arousal. Also in accord with a role of histamine in arousal is the sedative nature of H_1 antagonists in man, an effect not readily shown in laboratory animals. In man, no simple, consistent relationship has been found between peripheral response to H_1 blockers and central effects including sedation,[406,407] a lack of concordance that could rest partly on pharmacokinetics and individual differences.[408] Drugs that produce sedation in man compete *in vivo* for the binding of labeled mepyramine in rat brain.[312]

It is interesting that in man, H_1 antagonists impaired psychomotor performance, and there was no temporal concordance in the effects on sedation and on psychomotor performance, the two effects appearing to be independent.[408] It may be relevant that autoradiography showed a high density of H_1 antagonist binding to structures of the rat brain that are associated with the auditory system,[316] which may relate to the impairment of auditory (but not all) vigilance occurring with H_1 blockers in man.[406] Similarities in the pharmacological effects of H_1 antagonists and antidepressant drugs are observed in laboratory behavioral tests for antidepressant drugs,[409,410] the two classes of drugs producing responses so alike (also see Section 9) that the H_1 antagonists are characterized as "false positives"[411] in the search for antidepressant drugs. Both classes of drugs antagonize reserpinelike drugs, potentiate amphetaminelike drugs, and block muricide behavior and isolation-induced fighting, but these effects were not correlated with H_1 antagonist activity.[409,410] Potency of H_1 antagonists in antagonizing isolation-induced fighting of mice more nearly paralleled antimuscarinic potency than potency for the H_1 receptor.[410]

Treatment of depressed patients with high doses of diphenhydramine yielded ambiguous results.[412] Even if the clinical results were more successful[409] than the authors assessed,[412] they cannot be confidently attributed to H_1 antagonism, for 69% of the patients showed side effects, most of which can be attributed to muscarinic blockade.[412] An intriguing observation is that a low dose of diphenhydramine, 50 mg, was reported to heighten mood in healthy people while decreasing psychomotor activity; with increased dose, no effect on mood was seen.[413] At a dose of 100 mg, diphenhydramine altered time perception in man (subjects thought time passed more slowly), an uncommon effect of a drug and like that of cannabis; mood and cognitive processes were unchanged.[414] The heightened mood seen in healthy people at the low dose of diphenhydramine cannot be regarded as an effect of an antidepressant drug, for the antidepressants used clinically do not heighten the mood of healthy people but rather, like most H_1 antagonists, sedate them.

Cimetidine has had extensive clinical use, and adverse reactions are uncommon, but when they occur they are sometimes referable to the central nervous system. It is now clear that cimetidine can enter the brain, despite early work indicating otherwise. Measurable levels were present in all of seven postmortem brains examined from patients that had been treated with cimetidine, the brain: serum ratio averaging about 1.[415,416] In patients with advanced renal or hepatic disease, the blood levels rise, and in patients with diseases of both organs, the half-time of cimetidine is prolonged to ten times the normal value.[415,416] Although uncommon, the most frequently reported adverse CNS reactions to cimetidine are dose-related, reversible mental symptoms which, in order of appearance, are restlessness, confusion, disorientation, agitation, and visual and auditory hallucinations.[415-418] Mental changes have been reported with serum trough concentrations as low as 0.25 to 0.5 µg/ml,[415] probably because of alterations in the blood–brain barrier with hepatic disease. The usual cimetidine cerebrospinal: serum ratio is 0.29,[416] but one agitated, confused, and paranoid patient with hepatic failure but with normal serum levels had a ratio of 0.51.[417]

Cerebrospinal fluid levels of cimetidine were measured in five patients.[415,416] In the three patients with unchanged mental status, the concentrations of cimetidine were 0.7, 1.0, and 1.0 µM; the cerebrospinal fluid of the two patients with mental symptoms contained concentrations of cimetidine of 3.4 and 5.5 µM. With the assumption that spinal fluid levels are in equilibrium with brain receptors and that most histamine in adult brain is held in nerve endings, we calculate (from the equation in Section 10.2) that in the three patients without mental changes, 48 and 57% of the H_2 receptors were occupied by cimetidine, whereas in the two patients with mental changes, 82 and 88% of the H_2 receptors were occupied by cimetidine. These concentrations of cimetidine are also enough (Section 9) to reduce, but only slightly, postjunctional adrenergic activity and to increase stimulation-induced release of norepinephrine.

There are no comparable clinical studies of ranitidine, another H_2 antagonist used in man. But in four people without renal or hepatic disease who received two oral doses of 200 mg of ranitidine, CSF levels were no higher than 1×10^{-7} M,[419] which, with the assumptions stated above, could occupy 62% of the H_2 receptors in brain. It remains to be seen whether on a therapeutic regimen in patients with renal and hepatic disease, ranitidine produces greater CSF levels than those found in this study. Additional work is needed to test the hypothesis that the mental signs associated with high cimetidine concentrations in cerebrospinal fluid result from H_2 blockade, but if verified, it would suggest the importance of H_2 receptor activation for maintenance of normal mentation.

11.3. Central Sympathetic Activity

Microinjections of histamine into anterior or posterior hypothalamus produce hypertension, effects blocked by H_1 antagonists.[420] Suggestive of a role for endogenous histamine in central cardiovascular mechanisms are the findings that intraventricular histamine or SKF 91488 (the selective histamine methyltransferase inhibitor, Table II) caused dose-related increases in blood pressure.[421] The effect of SKF 91488 was associated with increases in hypothalamic histamine content. Although several studies show that the antihypertensive agent clonidine can stimulate brain H_2 receptors, studies suggesting that this mechanism contributes to the lowering of blood pressure[422] cannot be obtained in all species.[423] The potentiation of the cardiotoxicity of oubain by either histamine or the H_2 agonist dimaprit, and the antagonism of oubain by intraventricular cimetidine suggest that part of the centrally mediated cardiotoxicity of oubain may stem from a central histamine-related mechanism.[424]

11.4. Temperature Regulation

Histamine may function in thermoregulation.[425] Intrahypothalamic injections of histamine lower body temperature. This may result from a lowering in the thermoregulatory set point, and it is an effect blocked by peripheral H_1 antagonists.[425,426] Other studies indicate that an H_2 receptor, possibly in the wall of the third ventricle, may mediate a heat-dissipating mechanism.[425] This or yet another hypothermic effect of H_2 stimulation may be mediated by 5-hydroxytryptamine.[427] In man, an uncommon adverse effect of cimetidine is fever,[8] which may be caused by blockade of the heat-dissipating mechanism.

11.5. Hormone Release

Histamine may mediate the release of antidiuretic hormone. The supraoptic nucleus, known to mediate this release, contains a highly variable concentration of histamine.[167] Histamine microinjections near this structure decreased urine volume, an effect not produced by other vasodilators and blocked by mepyramine as well as by lesions of the median eminence.[428] Radioimmunoassay has confirmed that intraperitoneal as well as intracerebral histamine injections increase plasma arginine vasopressin.[429]

Intraventricular histamine elevates plasma prolactin and luteinizing hormone but not growth hormone.[430] The histamine-induced release of prolactin seems not to be through direct action on the pituitary gland.[431] Both histamine receptors have been implicated in prolactin release, for H_1 antagonists as well as H_2 agonists inhibit suckling-induced prolactin release in rats.[432] H_1 antagonists also lower serum prolactin levels in men and women.[433] These observations on H_2 agonists, suggestive of a tonic H_2-receptor inhibition of prolactin release, are consistent with the prolactin-releasing effects of cimetidine in rats[432] and man.[434,435] This cimetidine effect in man seems to be centrally mediated, may be independent of dopamine, and is more pronounced in women than men.[433] However, other H_2 antagonists do not induce prolactin release,[435,436] and the prolactin release by cimetidine was not reversed by the H_2 agonist impromidine.[435] Further studies are needed to determine whether these differences are pharmacodynamic (perhaps related to a differential penetration of the antagonists into brain) or whether they demonstrate a property of cimetidine unrelated to the H_2 receptor. Ovarian venous progestin was increased by histamine injected into the tuber cinereum but not when given into the pituitary.[437] Oophorectomy and castration increase hypothalamic histidine decarboxylase[438] and histamine.[405] Other studies on women show that histamine metabolism is influenced by estrogen and progestins.[439,440]

Plasma luteinizing hormone levels are increased not only by intraventricular histamine in animals[430] but by intravenous histamine in man.[433] H_1 antagonists increased these levels in men and lowered them in women. Although many studies show that histamine raises adrenocorticotrophic hormone and cortisol, the site and mechanism of action are unclear.[8]

12. EFFECTS OF PSYCHOTROPIC DRUGS ON BRAIN HISTAMINE AND ITS RECEPTORS

Several classes of centrally acting drugs have effects on brain histamine. Barbiturates have been reported to increase histamine levels in rat midbrain and caudate nucleus,[441] in cat hypothalamus and medial thalamus,[15] and in mouse whole brain.[41] Barbiturates and other CNS depressants decreased the estimated rate of whole brain histamine turnover in rats[41] (Section 4.5). The drugs may be acting by impairing release of histamine. Ethanol also increases rat[442,443] and mouse[444] brain histamine levels and inhibits its release from hypothalamus.[442,445] Reserpine reduced brain histamine levels in some species (Section 4.1). Bulbocapnine increased histamine levels in the cat hypothalamus.[15]

12.1. Opiates

Whole brain histamine levels were changed by acute morphine administration in mice[446] but not in rats[447] or cats.[15] The mouse study reported that the locomotor but not the analgesic effect of morphine was inversely correlated with mouse whole brain histamine levels, high and low doses producing opposite effects.[446] Other studies suggest a role of the H_2 receptor and brain histamine in the development of tolerance or physical dependence to opiates. The agonist 4-methylhistamine (used as a specific H_2 agonist, but see Section 9) on intracisternal injection inhibited naloxone-induced jumping in morphine-dependent mice, an effect reversed by intracisternal metiamide.[448] Chronic morphine administration lowered rat hypothalamic and cortical histamine levels and increased histidine decarboxylase in these regions.[447] These results may relate to the finding[449] that injections of histamine or dimaprit into rat dorsal hippocampus induced teeth chattering, wet dog shakes, and other signs of opiate abstinence, all of which were blocked by cimetidine.[449] Perhaps relevant to the acute effects of morphine are the findings that microinjections of histamine or dimaprit into the dorsal raphe produced analgesia, an effect blocked by cimetidine.[449]

Morphine and other (but not all) opiates release histamine from mast cells in many species, including man, in whom the fall in systemic vascular resistance produced by morphine infusions correlated well with the rise in plasma histamine.[450] In dog, the peripheral effects of morphine were mimicked by compound 48/80, which also releases histamine from mast cells, and was prevented by pretreatment with compound 48/80 and H_1 antagonists.[451,452] Histamine release is not restricted to peripheral tissues as shown by release of histamine when the lateral ventricle of the rabbit was perfused with morphine.[453] The source of this histamine may be the mast cells that line the subarachnoid space and other structures in contact with cerebrospinal fluid.[121] A rise in histamine levels in spinal fluid influences brain activity, for intracisternal injections of histamine in rats increased adenylate cyclase activity in cerebral cortex[454] and in rabbits raised the levels of cyclic AMP in cerebrospinal fluid.[455]

12.2. Antidepressant Drugs

Monoamine oxidase inhibitors have acute and chronic effects on brain histamine metabolism. Endogenous brain *tele*-methylhistamine, oxidized by type B MAO (Section 4.4), increases after treatment with pargyline[104,116] or deprenyl.[116] This treatment also caused a slight initial increase in brain histamine content, which quickly returned to control levels,[104] suggestive of mechanisms regulating brain histamine levels. The rise in histamine may be caused by inhibition of histamine methyltransferase by the increased levels of *tele*-methylhistamine (Section 4.4). Iproniazid raised histamine concentration in the cat hypothalamus.[15] The persistent increase in *tele*-methylhistamine levels after pargyline may imply that MAO inhibitors may have long-term effects on histaminergic dynamics. Patients receiving an MAO inhibitor had increased urinary levels of *tele*-methylhistamine.[456]

The tricyclic and tetracyclic antidepressant drugs competitively antagonize histamine at the H_2 receptor as measured on the H_2-linked adenylate cyclase.[269–272,457] Angus and Black[273] showed that amitriptyline competitively blocks the H_2 receptor associated with contraction of the guinea pig papillary muscle. For those antidepressant drugs that have been studied, their affinities for the H_2 receptor are about the same as or greater than their affinities for the muscarinic receptor; and the drugs are more potent in blocking the H_2 receptor than in inhibiting uptake of 5-hydroxytryptamine or norepinephrine.[270] Mianserin blocked the H_2-linked stimulation of the firing rate of CA3 cells in guinea pig hippocampal slices.[353]

That H_2 receptor blockade by antidepressants occurs in brain *in vivo* was clearly shown in measurements on two different functions. Amitryptyline reduced the H_2 receptor-linked depression of firing rate of rat cortical neurons.[458] This effect, similar to that of metiamide, was seen after direct application of amitriptyline as well as after treating rats with amitriptyline, as little as 3 mg/kg body weight.[458] The hypothermia produced in rats by intraventricular injection of dimaprit and other histamine agonists was significantly reduced by pretreatment with imipramine or amitriptyline.[459] The blockade occurred after intraperitoneal injection, 10 and 20 mg/kg, or after intraventricular injection.[459] Chronic treatment with mianserin or imipramine increased the sensitivity to histamine of CA3 cells in guinea pig hippocampal slices.[353]

The H_1 receptor is also blocked by antidepressant drugs. The blockade was demonstrated to be competitive by Schild plots in antagonism to histamine on the H_1 receptor linked to contraction of the guinea pig ileum,[460] by parallel shifts in the dose–response curve in the H_1-linked guanylate cyclase system of mouse neuroblastoma cells,[321] and by blockade of mepyramine binding to brain homogenates of guinea pig[457,461] and rat.[461,462] These drugs also block histamine-stimulated cyclic AMP formation in vesicular sacs of guinea pig brain.[463] Furthermore, labeled antidepressant drugs bind to H_1 sites in rat and guinea pig brain homogenates.[464,465]

Blockade of the histamine receptors may influence histamine metabolism. Acute administration of imipramine to rats increased histamine concentrations in cortex and midbrain but not in hypothalamus; only in midbrain was there an increase in histidine decarboxylase activity.[466] After treatment for 20 days, histamine levels fell in cortex and midbrain, and histidine decarboxylase activity increased in hypothalamus and decreased in midbrain; body temperature fell too.[466]

Table V shows that the rank order of affinities of these drugs for the H_2 receptor is similar to that for the H_1 receptor, doxepin being a notable exception. Doxepin has a greater affinity for the H_1 receptor than have those drugs that are conventionally classified as H_1 antagonists. The affinities of some antidepressants, e.g., amitriptyline, mianserin, and doxepin, for the H_1 receptor are high enough for them to be used as labeling ligands for the H_1 receptor in binding studies.[464–468] Treating mice with antidepressant drugs reduced the amount of [^{3}H]mepyramine retained by whole brain[313] and brain particulate fractions.[314] Pharmacological effects shared by H_1 antagonists and antidepressant drugs, both *in vitro* and *in vivo*, have been pointed out (Sections 9 and 11.2).

It is noteworthy that almost every antidepressant drug that has been studied has a higher affinity for the H_1 receptor than for the muscarinic receptor or the postjunctional α-adrenergic receptor,[461] and all have much higher affin-

Table V
Apparent Dissociation Constants, $K_b(nM)$, of Antidepressant Drugs for Histamine
Receptors

| Drug | H₁ receptor | | | H₂ receptor |
	Guinea pig ileum[460]	Mouse neuroblastoma cells[321]	Rat brain [³H]mepyramine binding[461]	Guinea pig hippocampus[270,272]
Doxepin	0.06	0.03	0.03	150
Mianserin				65
Amitriptyline	0.08	0.13	0.10	45
Clomipramine				200
Nortriptyline	3.2	7	13	760
Imipramine	13	10	27	240
Protriptyline	69	35	66	2000
Desipramine	150	260	230	1100
Dibenzepin				1500
Iprindole				3200

ities for the H₁ receptor than for the H₂ receptor (Table V). Since at ordinary doses in rats, amitriptyline and imipramine blocked the H₂ receptor *in vivo*,[458,459] and since the concentrations of antidepressant drugs in the plasma of patients and in the brain of rats[469] are sufficient to occupy the H₂ receptor,[270] it follows that *in vivo* most antidepressant drugs block H₂, H₁, and α-adrenergic receptors; some block the muscarinic receptor and the presynaptic sites of uptake of 5-hydroxytryptamine and/or norepinephrine and/or dopamine; and they react with other sites, some of which have been delineated, such as phosphodiesterase and calmodulin.[277]

12.3. Neuroleptic Drugs

The neuroleptic drugs block both histamine receptors. For the H₁ receptor, blockade was shown on histamine-linked cyclic GMP formation in mouse neuroblastoma cells[331] and in binding of labeled mepyramine to brain homogenates of guinea pig, rat, and man.[461] On smooth muscle preparations, blockade of the H₁ receptor by at least some neuroleptics is noncompetitive.[470,471] The rank order of potencies on all H₁ receptor systems is similar. Many of these drugs competed *in vivo* for [³H]mepyramine binding.[312,313]

These drugs block histamine-stimulated adenylate cyclase activity in rabbit and guinea pig brain slices and homogenates[271,334,472] and the H₂-linked adenylate cyclase activity in homogenates of guinea pig hippocampus or cortex.[269,272,457,473] Stereoselectivity was exhibited with both flupenthixol and butaclamol, the clinically active isomer showing much greater potency than the other isomer.[473] Of all neuroleptic drugs tested, only chlorpromazine showed, like the antidepressant drugs, competitive antagonism at the H₂ receptor.[269] All other neuroleptic agents tested shifted the dose–response curve to histamine in a nonparallel manner. The rank order of potencies of these drugs for the H₂ receptor in two independent studies were in agreement.[269,272,473]

The affinities of these drugs for the H_1 and H_2 receptors were inversely related; i.e., $r = -0.95$,[269] in contrast with the antidepressant drugs (Table V), in which the rank orders of affinities for the two histamine receptors are not very disparate. This inverse relationship suggests that judicious choices of these neuroleptic drugs in combination with relatively selective histamine agonists can be used to reveal the effect of these drugs that may be attributable to a specific histamine receptor. The inverse relationship may also be exploited to help design selective histamine antagonists. The affinities of the neuroleptic agents for the histamine receptors appear in general to be less than their affinities for muscarinic, dopamine, 5-hydroxytryptamine, and α-adrenergic receptors,[11,321,461] but at clinical doses, it would be expected that all of these receptors would be at least partially occupied by the drugs.

High doses of chlorpromazine raised histamine levels in whole rat brain[69] and in the hypothalamus and medial thalamus of cat,[15] perhaps by inhibiting histamine methyltransferase.[98] The effect of haloperidol on histamine and histidine decarboxylase activity in rat brain has been described.[466] Acute administration of haloperidol decreased histidine decarboxylase activity in hypothalamus and cortex without influencing histamine levels, but not in other regions of brain; treatment for 20 days decreased histamine levels in both hypothalamus and cortex. Acute, not chronic, treatment was accompanied by a fall in body temperature.[466]

12.4. Hallucinogens

D-LSD is a competitive antagonist of histamine at both H_1[47] and H_2 receptors.[267] The inactive stereoisomer, L-LSD, showed less activity than D-LSD on the H_1 receptor[474] and no measurable activity on the H_2 receptor.[267] Other hallucinogens, e.g., psilocin and mescaline, showed affinities for the H_1 receptor but not for the H_2 receptor.[267] (Also inactive on the H_2 receptor were many other types of drugs such as opiates, anticonvulsants, minor tranquilizers, Δ^9-THC, and barbiturates.[269]) For the H_1 receptor, 2-bromo-LSD and D-LSD have about the same affinity, but 2-bromo-LSD has about ten times greater affinity than D-LSD for the H_2 receptor.[267] Angus and Black[273] showed that 2-bromo-LSD was a competitive antagonist of histamine on the H_2 receptor in guinea pig papillary muscle.

2-Bromo-LSD also has greater affinity than D-LSD for haloperidol binding sites,[475] and it is more potent that D-LSD in increasing levels of DOPA[476] and DOPAC[477] in rat striatum. 2-Bromo-LSD blocks head twitches induced in mice by D-LSD.[478] 2-Bromo-LSD lacks the agonist activity of D-LSD, but its affinity for postsynaptic sites is similar to or greater than that of D-LSD.[254,267] Whatever may be the group of receptors that give rise to the hallucinogenic effect of D-LSD, it is most likely that 2-bromo-LSD shares them, for 2-bromo-LSD blocks the hallucinogenic effect of D-LSD in man[479] at a dosage interval before cross tolerance is seen. The effect of a dose of D-LSD of 1 µg/kg was blocked by a dose of 2-bromo-LSD of 32 to 654 µg/kg.[480]

Of greater interest is that 2-bromo-LSD causes behavioral changes in laboratory animals and in man. Larger doses of 2-bromo-LSD than LSD are needed to show changes in animal behavior, but they occur.[481] In clinical studies, 2-bromo-LSD produced all the central effects of D-LSD except hallucinations.[480,482] What contribution blockade of the histamine receptor makes to the behavioral effects of D-LSD is not known. The hallucinations observed in toxic reactions to H_1 antagonists may be caused by muscarinic blockade.[483] Adverse reactions to cimetidine in man include hallucinations and hyperpyrexia (Section 11.2), which D-LSD produces as well. From measurements of D-LSD levels in the brain of laboratory animals, it appears that its concentration may be sufficient to interact with both histamine receptors.[267]

12.5. Significance of the Interaction of Psychotropic Drugs with Histamine Receptors

It appears to be enigmatic that three classes of drugs with very different behavioral effects—antidepressant, neuroleptic, and hallucinogenic drugs (D-LSD)—all block histamine receptors. The histamine receptors are not unique in this regard, for dopamine and 5-hydroxytryptamine receptors also have high affinity for both neuroleptic agents and D-LSD. As the behavioral effects of these classes of drugs are very different, it follows that their behavioral effects cannot be ascribed to interaction with one receptor alone.

Since, at least for antidepressant drugs, usual plasma and brain levels are sufficient to block even the receptor that has least affinity for the drugs, namely, the H_2 receptor, it follows that many receptors are blocked after the usual dosing. The behavioral effects of these drugs may rest on their affinities for so many varied sites, the aggregations of these interactions producing the behavioral and therapeutic response.[254,269,270,484,485] It may be useful to recall that the way in which the visual cortex processes a spot of light of known wavelength, intensity, and duration focused directly on a receptor has been described as "hypercomplex," for the "response properties of most neurons depend on subtle interactions within an intricate system of more and less specific excitatory and inhibitory influences, whose cooperative details determine most aspects of neuronal behaviours."[486] It is hard to believe that the events by which psychotropic drugs alter human behavior are less complex.

Whether any of the known sites with which the psychotropic drugs interact accounts for, or even contributes to, the behavioral effects of any of these classes of drugs is not known. None of the relatively selective antagonists of any one biogenic amine can produce the behavioral effects of any one of these classes of psychotropic drugs. It is tempting to attribute a behavioral effect to the site at which the drug acts at lowest concentration; this could be misleading. With antidepressant and neuroleptic drugs, high and prolonged dosing is required to produce a therapeutic effect, at which time, at least for the antidepressant drugs, numerous receptors are blocked. The lowest dose of D-LSD that produced psychological effects in man (7 μg/man) elicited a shift in affect that the subjects were unaware of but that was discernible in a questionnaire and in interview, which also revealed hypomania, increased psychomotor activity, tension, and irritability; 20 μg/man caused changes in body image and thought processes; only above a dose of 30 μg/man did visual illusions and hallucinations evolve.[487] It would be interesting to know what receptors are marshaled as the dose and effects progress in man. In brain of treated laboratory animals, the concentration of D-LSD is probably sufficient to react with receptors for dopamine, 5-hydroxytryptamine, and histamine.[267] It was suggested[267] that neither blockade of the H_2 receptor nor any other single action of the drug can alone account for all the behavioral or numerous other pharmacological effects of D-LSD but that blockade of the H_2 receptor may account for the effect of D-LSD on perseveration and habituation.

Perhaps specific and tractable questions should be asked about the pharmacology of psychotropic drugs before asking how they alter mood and behavior. As emphasized in the initial report[270] describing competitive blockade of the H_2 receptor by antidepressant drugs,

> the information available cannot support an assertion that blockade of the H_2 receptor (or any other single action of these drugs) contributes to their therapeutic effect. But some of the pharmacological effects of the antidepressant drugs may rest on blockade of H_2 receptors. For example, histamine increases fluid consumption in rats by a central mechanism, and tricyclic antidepressant drugs diminish fluid consumption in rats.[270]

Thus, there may be a pharmacological effect common to the three different classes of psychotropic drugs that is attributable to blockade of the H_1 receptor and another pharmacological effect attributable to blockade of the H_2 receptor. Learning what effect is associated with which receptor may depend on learning the specific parts of the brain the psychotropic drugs act on, on what receptors in these discrete places, and what specific change(s) in the receptor-coupled responses ensues. Then perhaps the sum of all the pharmacological effects will account for the behavioral effects. Whether blockade of a histamine receptor or receptors contributes to the behavioral effect or to a pharmacological effect (or to a side effect), awaits further work.

The most compelling question may be the molecular basis by which psychotropic drugs seem to block so many receptors: each class of psychotropic drugs appears to have affinity for a plethora of receptors. And even within a single class of psychotropic drugs, e.g., the antidepressant or the neuroleptic agents, there are extraordinarily varied chemical structures that not only elicit similar behavioral effects but block the same receptors. Not all of these chemical structures bear resemblance to any of the endogenous agonists—certainly not to all the agonists—with which they compete for receptors. Possibly each class of psychotropic drugs alters an event(s) in membranes, which is then manifest as interactions with specific receptors, even sometimes expressed as competitive antagonism.[485] The fundamental membrane event(s) is unknown.

13. SUMMARY

Evidence is compelling that histamine serves as a mediator in brain. Histamine, the specific histidine decarboxylase, and *tele*-methylhistamine (its major metabolite in brain) all have similar uneven regional distributions in brain (Section 5.1). Highest levels are in hypothalamus, with lowest in cerebellum and brainstem. Synaptosomes have histamine and histidine decarboxylase, suggesting localization in nerve endings. Histidine decarboxylase and *tele*-methylhistamine may be better indices of histaminergic neurons than is histamine (Sections 4 and 5). Histamine is released from brain slices by potassium or electrical stimulation by a calcium-dependent mechanism (Section 4.2). Estimates of the turnover rate of brain histamine by different methods suggest a half-life of less than 1 h (Section 4.5), similar to estimates of other transmitters. Biochemical (Sections 10.1 and 10.2) and electrophysiological (Section 10.3) studies show that histamine H_1 and H_2 receptors are present in brain, and some studies indicate that these receptors may function in synaptic transmission. Most persuasively, lesions of midbrain or caudal hypothalamus cause a progressive loss of forebrain histidine decarboxylase, implying the existence of an ascending histamine-synthesizing system that joins the medial forebrain bundle and projects ipsilaterally to cortex, hippocampus, and other forebrain areas (Section 5.2.2).

Additional work is needed to study the transmitter role of histamine. The histamine-synthesizing fibers need to be directly visualized; the lesion studies imply a diffuse distribution of cell bodies. Electrical stimulation of these cell bodies should release histamine from these fibers *in vivo*; this remains to be unequivocally shown. Also lacking is a secure understanding of the mechanism of inactivating histamine: in some species, metabolism of histamine appears to be the means of inactivation (Sections 4.3 and 4.4) [although some histamine metabolites have pharmacological activity (Section 4.6)]. In others, including invertebrates (Section 7), there may be a reuptake process for histamine (Section 4.3). Newly developed chemical methods for measuring histamine catabolites (Section 4.4) will help us to understand the roles of histamine in brain. Analogous chemical methods to measure histamine still lack sufficient sensitivity; the radioenzymatic assay is sensitive but may give spurious results (Section 2).

Unlike the distribution of histamine and its metabolites, the distribution of the H_1 receptor, shown by binding studies and autoradiography, differs among species (Section 10.1). The H_1 receptor is associated with ion fluxes, cyclic nucleotide synthesis, phosphorylation, and glycogenolysis (Section 10.1.3–10.1.5 and 10.3). The distribution of the H_2 receptor is not known because an acceptable binding method is lacking (Section 10.2). The presence of the H_2 receptor has been shown in some species by its coupling to adenylate cyclase (Section 10.2.1) and in other species by electrophysiological techniques (Section 10.3). Both histamine receptors are probably on neurons as well as other cells. No conclusive evidence has shown more than the two histamine receptors.

A portion of the histamine in some mammalian brains is likely to be present in cells other than neurons (Section 5.2). Although previously regarded as absent from brain, mast cells have been clearly shown in brains of some species (Section 5.2.1). These cells, commonly observed in rodent and cat brain, are most prominent in dorsal thalamus and exhibit a distribution different from that of histamine. Animals of the same species show great variation in mast cell numbers in brain. Histochemical studies show histamine in these mast cells, but the variability in numbers and uncertainty of the cellular concentration of histamine make an assessment of their contribution to brain histamine levels difficult. The contribution may be substantial, as inferred from observations of a mutant mouse deficient in mast cells. Granular perivascular cells, termed neurolipomastocytoid cells, may be mistaken for mast cells in brain; their amine content is unknown (Section 5.2.1).

Functions of histamine in brain are still speculative. The high levels of histamine in hypothalamus (Section 5.1), the presence of histamine receptors in hypothalamus (Section 10), and the ability of histamine to alter water and food intake, thermoregulation, autonomic activity, and hormone release (Section 11) implicate histamine in some of these vegetative functions. A role of histamine in arousal is suggested by the proposed ascending histaminergic fibers emanating from reticular formation (Section 5.2.2), as well as the activating effect of histamine on brain and the sedative nature of H_1 antagonists (Section 11.2). The H_2 receptor may also be involved in brain function, as suggested by the adverse effects of cimetidine (Section 11). But some functions have been incorrectly attributed to histamine because of imprecise use of agonists and antagonists (Section 9). Histamine may also have a function in brain development (Section 6), as suspected from its high levels and nuclear distribution in neonatal brain. Neonatal brain is also rich in mast cells.

Other functions of histamine in brain may relate to mast cells (Section 11.1). They may play roles in brain analogous to their suspected roles in the periphery, e.g., inflammation. Their perivascular location and the vascular actions of histamine favor a role in vascular regulation. Mast cells might also mediate immune responses in brain. Their circumventricular location (Section 5.2.1) and the potent effects of histamine on neurons (Section 10.3) suggest that brain mast cells are a likely target for drugs, such as some opiates (Section 12.1), which cause histamine release.

Various classes of psychotropic drugs affect histamine and its receptors in brain (Section 12). Drugs that produce sedation decrease the rate of histamine synthesis. Monoamine oxidase inhibitors alter the levels of histamine and its metabolites. Antidepressant and neuroleptic agents and hallucinogens block both histamine receptors; the contribution of histaminergic mechanisms to the behavioral and/or pharmacological effects of these drugs is not known, in part because these drugs interact with numerous other sites.

ACKNOWLEDGMENTS. This work was supported by research grants from the National Institute of Mental Health (MH-31805) and the National Institute on Drug Abuse (DA-01875). L.B.H. is recipient of a faculty development award in pharmacology from the Pharmaceutical Manufacturers Association. Plotting and statistical analyses were done on the PROPHET system, a national computer resource sponsored by the National Institutes of Health through the Chemical/Biological Information-Handling Program, Division of Research. We are grateful to Mr. David Allen for patience and sensible and excellent typing, to Mr. Jay Cohen for reading proof. Dr. Joseph Goldfarb was most helpful in interpreting the electrophysiology.

REFERENCES

1. Swazey, J. P., 1974, *Chlorpromazine in Psychiatry: A Study of Therapeutic Innovation*, MIT Press, Cambridge.
2. Kety, S. S., 1978, *Psychopharmacology: A Generation of Progress* (M. A. Lipton, A. DiMascio, and K. F. Killam, eds.), Raven Press, New York, pp. 7–11.
3. Kuhn, R., 1970, *Discoveries in Biological Psychiatry* (F. J. Ayd, Jr. and B. Blackwell, eds.), E. B. Lippincott, Philadelphia, pp. 205–217.
4. Green, J. P., 1970, *Handbook of Neurochemistry*, Volume 4 (A. Lajtha, ed.), Plenum Press, New York, pp. 221–250.
5. Boissier, J.-R., 1971, *Acta Pharmacol.* **24**:52–91.
6. Schwartz, J. C., 1977, *Annu. Rev. Pharmacol. Toxicol.* **17**:325–339.
7. Garbarg, M., Barbin, G., Llorens, C., Palacios, J. M., Pollard, H., and Schwartz, J. C., 1980, *Neurotransmitters, Receptors and Drug Action* (W. B. Essman, ed.), Spectrum, New York, pp. 179–202.
8. Hough, L. B., and Green, J. P., 1980, *Psychopharmacol. Bull.* **16**:42–44.
9. Green, J. P., Johnson, C. L., and Weinstein, H., 1978, *Psychopharmacology: A Generation of Progress* (M. A. Lipton, A. DiMascio, and K. F. Killam, eds.), Raven Press, New York, pp. 319–332.
10. Schwartz, J. C., Barbin, G., Duchemin, A. M., Garbarg, M., Pollard, H., and Quach, T. T., 1981, *Neuropharmacology of Central Nervous System and Behavioral Disorders* (C. G. Palmer, ed.), Academic Press, New York, pp. 539–570.
11. Green, J. P., 1983, *Biochemical Studies of CNS Receptors, Handbook of Psychopharmacology*, New Series, Volume 17 (L. L. Iversen, S. D. Iversen, and S. H. Snyder, eds.), Plenum Press, New York, pp. 385–420.
12. Green, J. P., 1964, *Fed. Proc.* **23**:1095–1102.
13. Carlini, E. A., and Green, J. P., 1963, *Br. J. Pharmacol.* **20**:264–277.
14. Neugebauer, E., and Lorenz, W., 1981, *Behring Inst. Mitt.* **68**:102–133.
15. Adam, H. M., and Hye, H. K. A., 1966, *Br. J. Pharmacol.* **28**:137–152.
16. Shore, P. A., 1971, *Methods Enzymol.* **17B**:842–845.
17. Anton, A. H., and Sayre, D. F., 1969, *J. Pharmacol. Exp. Ther.* **166**:285–292.
18. Endo, Y., and Ogura, Y., 1973, *Eur. J. Pharmacol.* **21**:293–298.
19. Snyder, S. H., Baldessarini, R. J., and Axelrod, J., 1966, *J. Pharmacol. Exp. Ther.* **153**:544–549.
20. Black, J. W., and Ganellin, C. R., 1974, *Experientia* **30**:111–113.
21. Taylor, K. M., and Snyder, S. H., 1972, *J. Neurochem.* **19**:1343–1358.
22. Beaven, M. A., Jacobsen, S., and Horakova, Z., 1972, *Clin. Chim. Acta* **37**:91–103.
23. Salberg, D. J., Hough, L. B., Kaplan, D. E., and Domino, E. F., 1977, *Life Sci.* **21**:1439–1446.
24. Kobayashi, Y., and Maudsley, D. V., 1972, *Anal. Biochem.* **46**:85–90.
25. Gleich, G. J., and Hull, W. M., 1980, *J. Allergy Clin. Immunol.* **66**:295–298.
26. Taylor, K. M., and Snyder, S. H., 1972, *J. Pharmacol. Exp. Ther.* **173**:619–633.
27. Beaven, M. A., and Horakova, Z., 1978, *Handbook of Experimental Pharmacology*, Volume XVIII/2 (A. Rocha e Silva, ed.), Springer-Verlag, Berlin, Heidelberg, New York, pp. 151–173.
28. Navert, H., 1975, *J. Chromatogr.* **106**:218–224.
29. Mahy, N., and Gelpi, E., 1977, *J. Chromatogr.* **130**:237–242.
30. Doshi, P. S., and Edwards, D. J., 1979, *J. Chromatogr.* **176**:359–366.
31. Mahy, N., and Gelpi, E., 1978, *Chromatographia* **11**:573–577.
32. Mita, H., Yasueda, H., and Shida, T., 1980, *J. Chromatogr.* **181**:153–159.
33. Yamatodani, A., Seki, T., Taneda, M., and Wada, H., 1977, *J. Chromatogr.* **144**:141–145.
34. Tsuruta, Y., Kohashi, K., and Ohkura, Y., 1978, *J. Chromatogr.* **146**:490–493.
35. Tsuruta, Y., Kohashi, K., and Ohkura, Y., 1981, *J. Chromatogr.* **224**:105–110.
36. Yamatodani, A., Maeyama, K., Watanabe, T., Wada, H., and Kitamura, Y., 1982, *Biochem. Pharmacol.* **31**:305–309.

37. Hough, L. B., and Domino, E. F., 1977, *J. Neurochem.* **29**:199–204.
38. Orr, E. L., and Eichelman, B., 1979, *J. Neurochem.* **33**:303–308.
39. Subramanian, N., Schinzel, W., Mitznegg, P., and Estler, C. J., 1978, *Agents Actions* **8**:488–490.
40. Taylor, K. M., and Snyder, S. H., 1972, *J. Neurochem.* **19**:341–354.
41. Pollard, H., Bischoff, S., and Schwartz, J. C., 1974, *J. Pharmacol. Exp. Ther.* **190**:88–99.
42. Schayer, B. W., and Reilly, M. A., 1970, *J. Neurochem.* **17**:1649–1655.
43. Bulfield, G., and Kacser, H., 1975, *J. Neurochem.* **24**:403–405.
44. Neame, K. D., 1964, *J. Neurochem.* **11**:655–662.
45. Schwartz, J. C., Lampart, C., and Rose, C., 1972, *J. Neurochem.* **19**:801–810.
46. Aures, D., Hakanson, R., and Clark, W. G., 1970, *Handbook of Neurochemistry*, Volume 4 (A. Lajtha, ed.), Plenum Press, New York, pp. 165–196.
47. Schwartz, J. C., Lampart, C., and Rose, C., 1970, *J. Neurochem.* **17**:1527–1534.
48. Palacios, J. M., Mengod, G., Picatoste, F., Grau, M., and Blanco, I., 1976, *J. Neurochem.* **27**:1455–1460.
49. Barbin, G., Palacios, J. M., Garbarg, M., Schwartz, J. C., Gaspar, P., Javoy-Agid, F., and Agid, Y., 1980, *J. Neurochem.* **35**:400–406.
50. Watanabe, T., Nakamura, H., Liang, L. Y., Yamatodani, A., and Wada, H., 1979, *Biochem. Pharmacol.* **28**:1149–1155.
51. Fukui, H., Watanabe, T., and Wada, H., 1980, *Biochem. Biophys. Res. Commun.* **93**:333–339.
52. Tran, V. T., and Snyder, S. H., 1981, *J. Biol. Chem.* **256**:680–686.
53. Abou, Y. Z., Adam, H. M., Stephen, W. R. G., 1973, *Br. J. Pharmacol.* **48**:577–589.
54. Kahlson, G., and Rosengren, E., 1971, *Biogenesis and Physiology of Histamine*, Williams & Wilkins, Baltimore.
55. Maslinski, C., 1975, *Agents Actions* **5**:89–107.
56. Taylor, R. J., Leinweber, F. J., and Braun, G. A., 1973, *Biochem. Pharmacol.* **22**:2299–2310.
57. Bielkiewicz, B., and Maslinski, C., 1978, *Biochem. Pharmacol.* **27**:2977–2978.
58. Schayer, R. W., and Reilly, M. A., 1974, *Agents Actions* **4**:133–138.
59. Dismukes, K., and Snyder, S. H., 1974, *Brain Res.* **78**:467–481.
60. Garbarg, M., Barbin, G., Rodergas, E., and Schwartz, J. C., 1980, *J. Neurochem.* **35**:1045–1052.
61. Kataoka, K., and De Robertis, E., 1967, *J. Pharmacol. Exp. Ther.* **156**:114–125.
62. Kuhar, M. J., Taylor, K. M., and Snyder, S. H., 1971, *J. Neurochem.* **18**:1515–1527.
63. Snyder, S. H., Brown, B., and Kuhar, M. J., 1974, *J. Neurochem.* **23**:37–45.
64. Schwartz, J. C., 1975, *Life Sci.* **17**:503–518.
65. Hough, L. B., Khandelwal, J. K., and Green, J. P., 1982, *J. Neurochem.* **38**:1593–1599.
66. Baudry, M., Martes, M. P., and Schwartz, J. C., 1973, *J. Neurochem.* **21**:1301–1309.
67. Taylor, K. M., and Snyder, S. H., 1973, *J. Neurochem.* **21**:1215–1223.
68. Verdiere, M., Rose, C., and Schwartz, J. C., 1974, *Agents Actions* **4**:184–185.
69. Green, H., and Erickson, R. W., 1964, *Neuropharmacology* **3**:315–320.
70. Pollard, H., Bischoff, S., and Schwartz, J. C., 1973, *Eur. J. Pharmacol.* **24**:399–401.
71. Green, J. P., 1962, *Adv. Pharmacol.* **1**:349–422.
72. Green, J. P., and Day, M., 1963, *Ann. N.Y. Acad. Sci.* **103**:334–350.
73. Green, J. P., 1966, *Fed. Proc.* **26**:211–218.
74. Green, J. P., 1967, *The Molecular Basis of Some Aspects of Mental Activity*, Volume 2 (O. Walaas, ed.), Academic Press, New York, pp. 96–111.
75. Robinson, J. D., Jr., and Green, J. P., 1962, *Yale J. Biol. Med.* **35**:248–257.
76. Robinson, J. D., Anderson, J. H., and Green, J. P., 1965, *J. Pharmacol. Exp. Ther.* **147**:236–243.
77. Snyder, S. H., Glowinski, J., and Axelrod, J., 1966, *J. Pharmacol. Exp. Ther.* **153**:8–14.
78. Verdiere, M., Rose, C., and Schwartz, J. C., 1975, *Eur. J. Pharmacol.* **34**:157–168.
79. Subramanian, N., and Mulder, A. H., 1976, *Eur. J. Pharmacol.* **35**:203–206.
80. Biggs, M. J., and Johnson, E. S., 1980, *Br. J. Pharmacol.* **70**:555–560.
81. Minchin, M. C. W., and Iversen, L. L., 1974, *J. Neurochem.* **23**:533–540.
82. Monnier, M., Sauer, R., and Hatt, A. M., 1970, *Int. Rev. Neurobiol.* **12**:265–305.

83. Snyder, S. H., and Taylor, K. M., 1972, *Perspectives in Neuropharmacology, A Tribute to Julius Axelrod* (S. H. Snyder, ed.), Oxford University Press, New York, pp. 43–75.
84. Honegger, C. G., Krepelka, L. M., Steinmann, V., and Von Hahn, H. P., 1974, *Eur. Neurol.* **12**:236–252.
85. Schwartz, J. C., Baudry, M., Chast, F., Pollard, H., Bischoff, S., and Krishnamoorthy, M. S., 1974, *CNS—Studies on Metabolic Regulation and Function* (E. Genazzani and H. Herkin, eds.), Springer-Verlag, Berlin, Heidelberg, New York, pp. 172–184.
86. Phillipu, A., and Matthaei, H., 1975, *Naunyn Schmiedebergs Arch. Pharmacol.* **287**:191–204.
87. Tuomisto, L., Tuomisto, J., and Walaszek, E. J., 1975, *Med. Biol.* **53**:40–46.
88. Ehinger, B., 1974, *Acta Physiol. Scand.* **90**:218–225.
89. Maslinski, C., 1975, *Agents Actions* **5**:183–225.
90. Burkhard, W. P., Gey, K. F., and Pletscher, R., 1963, *J. Neurochem.* **10**:183–186.
91. Schayer, R. W., and Reilly, M. A., 1973, *J. Pharmacol. Exp. Ther.* **184**:33–40.
92. Schwartz, J. C., Pollard, H., Bischoff, S., Rehault, M. C., and Verdiere-Sahuque, M., 1971, *Eur. J. Pharmacol.* **16**:326–335.
93. Barth, H., Lorenz, W., and Niemeyer, I., 1973, *Hoppe Seylers Z. Physiol. Chem.* **354**:1021–1026.
94. Barth, H., Crombach, M., Schunack, W., and Lorenz, W., 1980, *Biochem. Pharmacol.* **29**:1399–1407.
95. Thithapandha, A., and Cohn, V. H., 1978, *Biochem. Pharmacol.* **27**:263–271.
96. Francois, D. M., Thompson, M. F., and Greaves, M. W., 1980, *Biochem. J.* **187**:819–828.
97. Axelrod, J., and Vesell, E. S., 1970, *Mol. Pharmacol.* **6**:78–84.
98. Brown, D. D., Tomchick, R., and Axelrod, J., 1959, *J. Biol. Chem.* **234**:2948–2950.
99. Hough, L. B., Khandelwal, J. K., and Mittag, T. W., 1981, *Agents Actions* **11**:427–430.
100. Karjala, S. A., and Turnquest, B. W., 1955, *J. Am. Chem. Soc.* **77**:6358–6363.
101. Nakajima, T., Wolfgram, F., and Clark, W. G., 1967, *J. Neurochem.* **14**:1113–1118.
102. Hough, L. B., Stetson, P. L., and Domino, E. F., 1979, *Anal. Biochem.* **96**:56–63.
103. Hough, L. B., Khandelwal, J. K., Morrishow, A., and Green, J. P., 1981, *J. Pharmacol. Methods* **5**:143–148.
104. Hough, L. B., Khandelwal, J., and Green, J. P., 1982, *Biochem. Pharmacol.* **31**:4074–4076.
105. Konishi, H., and Kakimoto, Y., 1976, *J. Neurochem.* **27**:1461–1463.
106. Cohn, V. H., 1965, *Biochem. Pharmacol.* **14**:1686–1688.
107. Duch, D. S., Edelstein, M. P., and Nichol, C. A., 1980, *Mol. Pharmacol.* **18**:100–104.
108. Barth, H., Lorenz, W., and Troidl, H., 1975, *Br. J. Pharmacol.* **55**:321–327.
109. Taylor, K. M., and Snyder, S. H., 1972, *Mol. Pharmacol.* **8**:300–310.
110. Beaven, M. A., and Shaff, R. E., 1979, *Biochem. Pharmacol.* **28**:183–188.
111. Barth, H., and Lorenz, W., 1978, *Agents Actions* **8**:359–365.
112. Beaven, M. A., and Roderick, N. B., 1980, *Biochem. Pharmacol.* **29**:2897–2900.
113. Barth, H., Niemeyer, I., and Lorenz, W., 1973, *Agents Actions* **3**:138–147.
114. Lindell, S. E., and Westling, H., 1957, *Acta Physiol. Scand.* **39**:370–384.
115. Waldmeier, P. C., Feldtrauer, J. J., and Maitre, L., 1977, *J. Neurochem.* **29**:785–790.
116. Hough, L. B., and Domino, E. F., 1979, *J. Pharmacol. Exp. Ther.* **208**:422–428.
117. Suzuki, O., Katsumata, Y., and Oya, M., 1979, *Life Sci.* **24**:2227–2230.
118. Elsworth, J. D., Glover, V., and Sandler, M., 1980, *Psychopharmacology* **69**:287–290.
119. Robinson, J. D., and Green, J. P., 1964, *Nature* **203**:1178–1179.
120. Fram, D. H., and Green, J. P., 1968, *J. Neurochem.* **15**:597–602.
121. Khandelwal, J. K., Hough, L. B., Morrishow, A. M., Green, J. P., 1982, *Agents Actions* **12**:583–590.
122. Swahn, C. G., and Sedvall, G., 1981, *J. Neurochem.* **37**:461–466.
123. Keyzer, J. J., Wolthers, B. G., Muskiet, F. A. J., Kauffman, H. F., and Groen, A., 1981, *Clin. Chim. Acta* **113**:165–173.
124. Bischoff, S., and Korf, J., 1978, *Brain Res.* **141**:375–379.
125. Garbarg, M., Baudry, M., Brenda, P., and Schwartz, J. C., 1975, *Brain Res.* **83**:583–591.
126. Goldschmidt, R. C., Khandelwal, J. K., and Hough, L. B., 1983, *Agents Actions* (in press).
127. Furano, A. V., and Green, J. R., 1964, *J. Physiol.* (*Lond.*) **170**:263–271.
128. Hough, L. B., and Domino, E. F., 1979, *J. Neurochem.* **32**:1865–1866.

129. Schwartz, J. C., Rose, C., and Caillens, H., 1973, *J. Pharmacol. Exp. Ther.* **184**:766–779.
130. Haverberg, L. N., Omstedt, P. T., Munro, H. N., and Young, V. R., 1975, *Biochim. Biophys. Acta* **405**:67–71.
131. Young, V. R., Alexis, S. D., Baliga, B. S., and Munro, H. N., 1972, *J. Biol. Chem.* **247**:3592–3600.
132. Khandelwal, J. K., Hough, L. B., Pazhenchevsky, B., Morrishow, A. M., and Green, J. P., 1982, *J. Biol. Chem.* **257**:12815–12819.
133. Tham, R., and Holmstedt, B., 1965, *J. Chromatogr.* **19**:286–295.
134. Khandelwal, J. K., Hough, L. B., and Green, J. P., 1982, *Klin. Wochenschr.* **60**:914–918.
135. Tuomisto, L., and Eriksson, L., 1982, *Agents Actions* **12**:142–145.
136. Abbott, R. J., Keidan, J., Pye, I. F., and Nahorski, S. R., 1981, *J. Neurochem.* **37**:1042–1044.
137. Roberts, E., and Simonsen, D. G., 1970, *Brain Res.* **24**:91–111.
138. Fox, C. L., and Lasker, S. E., 1962, *Am. J. Physiol.* **202**:111–113.
139. Tunnicliff, G., Wein, J., and Roberts, E., 1972, *J. Neurochem.* **19**:2017–2023.
140. Marcus, R. J., Winters, W. D., Roberts, E., and Simonsen, D. G., 1971, *Neuropharmacology* **10**:203–215.
141. Benton, D., Kyriacou, P., Rick, J. T., and Taberner, P. V., 1974, *Eur. J. Pharmacol.* **27**:288–293.
142. Clipsham, P. J., Hamilton, T. C., Hunt, A. E., and Poyser, R. H., 1980, *Eur. J. Pharmacol.* **65**:193–200.
143. Antonaccio, M. J., and Snyder, D. W., 1981, *J. Pharmacol. Exp. Ther.* **218**:200–205.
144. McGeer, E. G., McGeer, P. L., and McLennan, H., 1961, *J. Neurochem.* **8**:36–49.
145. Godfraind, J. M., Krnjevic, K., Maretic, H., and Pumain, R., 1973, *Can. J. Physiol. Pharmacol.* **51**:790–797.
146. Curtis, D. R., and Felix, D., 1971, *Brain Res.* **34**:301–321.
147. Haas, H. L., Anderson, E. G., and Hosli, L., 1973, *Brain Res.* **51**:269–278.
148. Clifford, J. M., Taberner, P. V., Tunnicliff, G., Rick, J. T., and Kerkut, G. A., 1973, *Biochem. Pharmacol.* **22**:535–542.
149. Tunnicliff, G., 1976, *Gen. Pharmacol.* **7**:259–262.
150. Hitzemann, R. J., and Loh, H. H., 1978, *J. Neurochem.* **30**:471–477.
151. Bowery, N. G., and Jones, G. P., 1976, *Br. J. Pharmacol.* **56**:323–330.
152. Nistri, A., and Corradetti, R., 1978, *Neuropharmacology* **17**:13–19.
153. Swagel, M. W., Ikeda, K., and Roberts, E., 1973, *Nature* [*New Biol.*] **246**:91–92.
154. Black, J. W., Duncan, W. A. M., Durant, C. J., Ganellin, C. R., and Parsons, E. M., 1972, *Nature* [*New Biol.*] **236**:385–390.
155. Phillis, J. W., Tebecis, A. K., and York, D. H., 1968, *Br. J. Pharmacol. Chemother.* **33**:426–440.
156. Haas, H. L., and Wolf, P., 1977, *Brain Res.* **122**:269–276.
157. Roberts, F., 1981, *Neuropharmacology* **20**:711–714.
158. Alvarez, E., and Guerra, F., 1982, *Physiol. Behav.* **28**:1035–1040.
159. Rand, M. J., Story, D. F., and Wong-Dusting, H., 1982, *Br. J. Pharmacol.* **76**:305–311.
160. Goldstein, L., Pfeiffer, C. C., and Munoz, C., 1963, *Fed. Proc.* **22**:424.
161. Almeida, A. P., and Beaven, M. A., 1981, *Brain Res.* **208**:244–250.
162. Watanabe, H. Y., Tsuriya, Y., and Kasuya, Y., 1974, *Chem. Pharm. Bull. (Tokyo)* **22**:950–952.
163. Blanco, I., Blanco, M., Grau, M., Palacios, J. M., Picatoste, F., and Scherk, G., 1973, *Experientia* **29**:791–793.
164. Michaelson, I. A., Coffman, P. Z., and Vedral, D. F., 1968, *Biochem. Pharmacol.* **17**:2435–2441.
165. Taylor, K. M., Gfeller, E., and Snyder, S. H., 1972, *Brain Res.* **41**:171–179.
166. Lipinski, J. F., Schaumberg, H. H., and Baldessarini, R. J., 1973, *Brain Res.* **52**:403–408.
167. Brownstein, M. J., Saavedra, J. M., Palkovits, M., and Axelrod, J., 1974, *Brain Res.* **77**:151–156.
168. Gharib, A., Sarda, N., Chabannes, B., Cronenberger, L., and Pacheco, H., 1982, *J. Neurochem.* **38**:810–815.

169. Sperk, G., Hortnagl, H., Reither, H., and Hornykiewicz, O., 1981, *J. Neurochem.* **37**:525–526.
170. Selye, H., 1965, *The Mast Cells*, Butterworths, London.
171. Riley, J. F., 1959, *The Mast Cells*, Livingstone, Edinburgh.
172. Fujita, T., and Kobayashi, S., 1979, *Trends Neurosci.* **2**:27–30.
173. Ennis, M., 1982, *Agents Actions* **12**:60–63.
174. Ennis, M., and Pearce, F. L., 1980, *Eur. J. Pharmacol.* **66**:339–345.
175. Barrett, K. E., and Pearce, F. L., 1982, *Agents Actions* **12**:186–188.
176. Olsson, Y., 1968, *Int. Rev. Cytol.* **24**:27–70.
177. MacDonald, S. M., Mezei, M., and Mezei, C., 1981, *J. Neurochem.* **36**:9–16.
178. Brizee, K. R., Palazoo, M. C., Klara, P. M., and Hofer, H., 1978, *Cell Tissue Res.* **187**:115–127.
179. Dropp, J. J., 1979, *Acta Anat.* **105**:505–513.
180. Ibrahim, M. Z. M., 1974, *J. Neurol. Sci.* **21**:431–478.
181. Dropp, J. J., 1972, *Anat. Rec.* **174**:227–238.
182. Kruger, P. G., 1974, *Experientia* **30**:810–811.
183. Edvinsson, L., Cervos-Navarro, J., Larsson, L. I., Owman, C., and Ronnberg, A. L., 1977, *Neurology (Minneap.)* **27**:878–883.
184. Kelsall, M. A., 1966, *Anat. Rec.* **154**:727–740.
185. Campbell, D. J., and Kiernan, J. A., 1966, *Nature [New Biol.]* **210**:756–757.
186. Dropp, J. J., 1976, *Acta Anat.* **94**:1–21.
187. Persinger, M. A., 1977, *Physiol. Psychol.* **5**:166–176.
188. Cammermeyer, J., 1972, *Z. Anat. Entwick.* **139**:71–92.
189. Persinger, M. A., 1979, *Behav. Neural Biol.* **25**:380–386.
190. Cammermeyer, J., 1976, *Acta Anat.* **96**:459–468.
191. Persinger, M. A., 1977, *Behav. Neural Biol.* **21**:426–431.
192. Green, E., 1966, *Biology of the Laboratory Mouse*, Dover, New York,
193. Kitamura, Y., Go, S., and Hatanaka, K., 1978, *Blood* **52**:447–452.
194. Watanabe, T., Maeyama, K., Yamatodani, A., Yamada, M., Kitamura, Y., and Wada, H., 1980, *Life Sci.* **26**:1569–1574.
195. Kiernan, J. A., 1976, *J. Anat.* **121**:303–311.
196. Ibrahim, M. Z. M., Munib, E. A., and Bahuth, N., 1979, *Acta Anat.* **104**:134–154.
197. Sturrock, R. R., 1980, *Neuropathol. Appl. Neurobiol.* **6**:211–219.
198. Ibrahim, M. Z. M., 1970, *Brain Res.* **17**:348–350.
199. Ibrahim, M. Z. M., Waziri, R., and Kamath, S., 1979, *Cell Tissue Res.* **204**:217–232.
200. Garbarg, M., Krishnamoorthy, M. S., Feger, J., and Schwartz, J. C., 1973, *Brain Res.* **50**:361–367.
201. Krishnamoorthy, M. S., Garbarg, M., Feger, J., and Schwartz, J. C., 1973, *Agents Actions* **3**:181.
202. Dismukes, K., Kuhar, M. J., and Snyder, S. H., 1974, *Brain Res.* **78**:144–151.
203. Garbarg, M., Barbin, G., Feger, J., and Schwartz, J. C., 1974, *Science* **186**:833–835.
204. Garbarg, M., Barbin, G., Bischoff, S., Pollard, H., and Schwartz, J. C., 1976, *Brain Res.* **106**:333–348.
205. Barbin, G., Hirsch, J. C., Garbarg, M., and Schwartz, J. C., 1975, *Brain Res.* **92**:170–174.
206. Barbin, G., Garbarg, M., Schwartz, J. C., and Storm-Mathisen, J., 1976, *J. Neurochem.* **26**:259–263.
207. Ben-Ari, Y., La Salle, G. L., Barbin, G., Schwartz, J. C., and Garbarg, M., 1977, *Brain Res.* **138**:285–294.
208. Pollard, H., Llorens-Cortes, C., Barbin, G., Garbarg, M., and Schwartz, J. C., 1978, *Brain Res.* **157**:178–181.
209. Agid, Y., Javoy, F., and Glowinski, J., 1973, *Nature [New Biol.]* **245**:150–151.
210. Coyle, J. T., 1977, *Int. Rev. Neurobiol.* **20**:65–103.
211. Pearce, L. A., and Schanberg, S. M., 1969, *Science* **166**:1301–1303.
212. Tillement, J. P., Guernet, M., Diehl, A., Blanco, I., Blanco, M., and Boissier, J. R., 1971, *J. Pharmacol. (Paris)* **2**:1–10.
213. Young, A. G., Pert, C. D., Brown, D. G., Taylor, K. M., and Snyder, S. H., 1971, *Science* **173**:247–249.

214. Schwartz, J. C., Lampart, C., Rose, C., Rehault, M. C., Bischoff, S., and Pollard, H., 1971, *J. Neurochem.* **18**:1787–1789.
215. Martres, M. P., Baudry, M., and Schwartz, J. C., 1975, *Brain Res.* **83**:265–275.
216. Picatoste, F., Blanco, I., and Palacios, J. M., 1977, *J. Neurochem.* **29**:735–737.
217. Kouvelas, E. D., Savakis, C. E., Tzebelikos, E. T., Bonatsos, G., and Mitrossilis, S., 1976, *Experientia* **32**:1136–1138.
218. Ferrer, I., Picatoste, F., Rodergas, I., Garcia, A., Sabria, J., and Blanco, I., 1979, *J. Neurochem.* **32**:587–592.
219. Watanabe, T., Kitamura, Y., Maeyama, K., Go, S., Yamatodani, A., and Wada, H., 1981, *Proc. Natl. Acad. Sci. U.S.A.* **78**:4209–4212.
220. Kruger, P. G., Lagunoff, D., and Wan, H., 1980, *Exp. Cell Res.* **129**:83–93.
221. Green, J. P., 1968, *Eur. J. Pharmacol.* **3**:68–73.
222. Beaven, M. A., Aiken, D., and Soll, A. H., 1982, *Fed. Proc.* **41**:1709.
223. Hamon, M., and Bourgoin, S., 1981, *Physiological and Biochemical Basis for Perinatal Medicine* (M. Monset-Couchord and A. Minkowski, eds.), S. Karger, Basel, pp. 286–295.
224. Shaw, G. G., 1979, *Biochem. Pharmacol.* **28**:1–6.
225. Aloe, L., and Levi-Montalcini, R., 1977, *Brain Res.* **133**:358–366.
226. Subramanian, N., Whitmore, W. L., Seidler, F. J., and Slotkin, T. A., 1981, *J. Neurochem.* **36**:1137–1141.
227. Tuomisto, L., 1977, *J. Neurochem.* **28**:271–276.
228. Booth, R. F. G., Patel, T. B., and Clark, J. B., 1980, *J. Neurochem.* **34**:17–25.
229. Weinrich, D., Weiner, C., and McCaman, R., 1975, *Brain Res.* **84**:341–345.
230. Weinrich, D., and Yu, Y., 1977, *J. Neurochem.* **28**:361–369.
231. Weinrich, D., and Weinrich, C. A., 1977, *Comp. Biochem. Physiol.* **56C**:1–4.
232. Turner, J. D., and Cottrell, G. A., 1977, *Nature* **267**:447–449.
233. Weinreich, D., 1977, *Nature* **267**:854–857.
234. Groul, D. L., and Weinrich, D., 1979, *Brain Res.* **162**:281–301.
235. Gruol, D. L., and Weinrich, D., 1979, *Neuropharmacology* **18**:415–424.
236. Carpenter, D. O., and Gaubatz, G. L., 1975, *Nature* **254**:343–345.
237. Gotow, T., Kirkpatrick, C. T., and Tomita, T., 1980, *Brain Res.* **196**:151–167.
238. Gotow, T., Kirkpatrick, C. T., and Tomita, T., 1980, *Brain Res.* **196**:169–182.
239. Osborne, N. N., Wolter, K. D., and Neuhoff, V., 1979, *Biochem. Pharmacol.* **28**:2799–2805.
240. Weinrich, D., 1979, *J. Neurochem.* **32**:363–369.
241. Huggins, A. K., and Woodruff, G. N., 1968, *Comp. Biochem. Physiol.* **26**:1107–1111.
242. Stein, C., and Weinrich, D., 1982, *J. Neurochem.* **38**:204–214.
243. Turner, J. D., Powell, B., and Cottrell, G. A., 1980, *J. Neurocytol.* **9**:1–14.
244. Bailey, C. H., Chen, M. C., Weiss, K. R., and Kupfermann, I., 1982, *Brain Res.* **238**:205–210.
245. Euler, U. S. v., 1966, *Handbook of Experimental Pharmacology*, Volume 18/1 (M. Roche e Silva, ed.), Springer-Verlag, Berlin, pp. 318–333.
246. Lindl, T., Behrendt, H., Henl-Sawaja, M. C. B., Teufel, E., and Cramer, H., 1974, *Naunyn Schmiedebergs Arch. Pharmacol.* **286**:283–296.
247. Euler, U. S. v., and Purkhold, A., 1951, *Acta Pysiol. Scand.* **24**:218–224.
248. Ryan, M. J., and Brody, M. J., 1972, *J. Pharmacol. Exp. Ther.* **181**:83–91.
249. Ryan, M. J., and Brody, M. J., 1970, *J. Pharmacol. Exp. Ther.* **174**:123–132.
250. Marshall, I., 1981, *J. Autonom. Pharmacol.* **1**:235–250.
251. Heitz, D. C., and Brody, M. J., 1975, *Am. J. Physiol.* **228**:1351–1357.
252. Schild, H. O., 1981, *Agents Actions* **11**:12–19.
253. Black, J. W., 1976, *Proceedings of the Sixth International Congress of Pharmacology*, Volume I (E. Klinge, ed.), Plenum Press, New York, pp. 3–16.
254. Green, J. P., Weinstein, H., and Maayani, S., 1978, *Quantitative Structure Activity Relationships of Analgesics, Narcotic Antagonists and Hallucinogens* (G. Barnett, M. Trsic, and R. Willette, eds.), "QuaSAR" Research Monograph 22, National Institute on Drug Abuse, United States Government Printing Office, Washington, pp. 38–58.
255. Green, J. P., and Hough, L. B., 1980, *Cellular Receptor for Hormones and Neurotransmitters* (D. Schulster and A. Levitzki, eds.), John Wiley & Sons, New York, pp. 287–305.

256. Hough, L. B., Weinstein, H., and Green, J. P., 1980, *Adv. Biochem. Psychopharmacol.* **21**:183–192.
257. Parsons, M. E., Owen, D. A. A., Ganellin, C. R., and Durant, C. J., 1977, *Agents Actions* **7**:31–38.
258. Bertaccini, G., Molina, E., Zappia, L., and Zseli, J., 1979, *Naunyn Schiedebergs Arch. Pharmacol.* **309**:65–68.
259. Durant, G. J., Duncan, W. A. M., Ganellin, C. R., Parsons, M. E., Blakemore, R. C., and Rasmussen, A. C., 1978, *Nature* **276**:403–405.
260. Kenakin, T. P., and Angus, J. A., 1981, *J. Pharmacol. Exp. Ther.* **219**:474–480.
261. Barker, L. A., and Ebersole, B. J., 1982, *J. Pharmacol. Exp. Ther.* **221**:69–75.
262. Pakkari, I., and Karppanen, H., 1982, *Neuropharmacology* **21**:171–178.
263. Schild, H. O., 1947, *Br. J. Pharmacol.* **2**:189–206.
264. Van den Brink, F. G., and Lien, E. J., 1978, *Handbook of Experimental Pharmacology*, Volume 18/2 (M. Rocha e Silva, ed.), Springer-Verlag, Berlin, pp. 333–367.
265. Roche e Silva, M., and Antonio, A., 1978, *Handbook of Experimental Pharmacology*, Volume 18/2 (M. Rocha e Silva, ed.), Springer-Verlag, Berlin, pp. 381–437.
266. Laduron, P. M., Janssen, P. F. M., Gommeren, W., and Leysen, J. E., 1982, *Mol. Pharmacol.* **21**:294–300.
267. Green, J. P., Johnson, C. L., Weinstein, H., and Maayani, S., 1977, *Proc. Natl. Acad. Sci. U.S.A.* **74**:5697–5701.
268. Kanof, P. D., and Greengard, P., 1979, *J. Pharmacol. Exp. Ther.* **209**:87–96.
269. Maayani, S., Hough, L. B., Weinstein, H., and Green, J. P., 1982, *Adv. Biochem. Psychopharmacol.* **31**:133–147.
270. Green, J. P., and Maayani, S., 1977, *Nature* **269**:163–165.
271. Palmer, G. C., Wagner, H. R., Palmer, S. J., and Manian, A. A., 1977, *Commun. Pharmacol.* **1**:61–69.
272. Kanof, P. D., and Greengard, P., 1978, *Nature* **272**:329–333.
273. Angus, J. A., and Black, J. W., 1980, *Circ. Res.* **46**(Suppl. I):64–69.
274. Berndt, S., and Schwabe, U., 1973, *Brain Res.* **63**:303–312.
275. Levin, R. M., and Weiss, B., 1976, *Mol. Pharmacol.* **12**:581–589.
276. Hough, L. B., and Barker, L. A., 1981, *J. Pharmacol. Exp. Ther.* **219**:453–458.
277. Weiss, B., Prozialeck, W., and Cimimo, M., 1980, *Advances in Cyclic Nucleotide Research* Volume 12 (P. Hanet and H. Sands, eds.), Raven Press, New York, pp. 213–225.
278. Brimblecombe, R. W., Duncan, W. A. M., Durant, G. J., Emmett, J. C., Ganellin, C. R., and Parsons, M. E., 1975, *J. Int. Med. Res.* **3**:86–92.
279. Barker, L. A., 1981, *Agents Actions* **11**:699–705.
280. McCulloch, M. W., Medgett, I. C., and Rand, M. J., 1979, *Br. J. Pharmacol.* **67**:535–543.
281. Ganellin, C. R., 1978, *Handbook of Experimental Pharmacology*, Volume 18/2 (M. Rocha e Silva, ed.), Springer-Verlag, Berlin, pp. 251–294.
282. Vyas, S., and Verma, S., 1981, *Agents Actions* **11**:193–195.
283. Bertaccini, G., and Coruzzi, G., 1982, *Agents Actions* **12**:168–171.
284. Hollenberg, M. D., and Cuatrecasas, P., 1975, *Handbook of Psychopharmacology*, Section I: *Basic Pharmacology*, Volume 2 (L. L. Iversen, S. D. Iversen, and S. H. Snyder, eds.), Plenum Press, New York, pp. 129–177.
285. Bennett, J. P., Jr., 1978, *Neurotransmitter Receptor Binding* (H. I. Yamamura, S. J. Enna, and M. J. Kuhar, eds.), Raven Press, New York, pp. 57–90.
286. Burt, D. R., 1978, *Neurotransmitter Receptor Binding* (H. I. Yamamura, S. J. Enna, and M. J. Kuhar, eds.), Raven Press, New York, pp. 41–55.
287. Hollenberg, M. D., and Cuatrecasas, P., 1979, *The Receptors, A Comprehensive Treatise.* Volume I, *General Principles and Procedures* (R. D. O'Brien, ed.), Plenum Press, New York, pp. 193–214.
288. Michoud, M. C., Lelorier, J., and Amyot, R., 1981, *Bull. Eur. Physiopathol. Resp.* **17**:807–821.
289. Lush, I. E., Sloan, T. P., and Smith, R. L., 1982, *Br. J. Pharmacol.* **76**:196P.
290. Boissier, J. R., Guernet, M., Tillement, J. P., Blanco, I., and Blanco, M., 1970, *Life Sci.* **9**:249–256.

291. Henry, D. A., and Langman, M. J. S., 1981, *Drugs* **21**:444–459.
292. Bauman, J. H., and Kimelblatt, B. J., 1982, *Drug Intell. Clin. Pharm.* **16**:380–386.
293. Wauquier, A., Van den Broeck, W. A. E., Awouters, F., and Janssen, P. A. J., 1981, *Neuropharmacology* **20**:853–859.
294. Carlsson, A., and Lindqvist, M., 1969, *J. Pharm. Pharmacol.* **21**:460–464.
295. Lidbrink, P., Jonsson, G., and Fuxe, K., 1971, *Neuropharmacology* **10**:521–536.
296. Johnson, G. L., and Kahn, J. B., Jr., 1966, *J. Pharmacol. Exp. Ther.* **152**:458–468.
297. Isaac, L., and Goth, A., 1967, *J. Pharmacol. Exp. Ther.* **156**:463–468.
298. Coyle, J. T., and Snyder, S. H., 1969, *Science* **166**:899–901.
299. Brown, P. A., and Vernikos, J., 1980, *Eur. J. Pharmacol.* **65**:89–92.
300. Chang, R. S. L., Tran, V. T., and Snyder, S. H., 1979, *J. Pharmacol. Exp. Ther.* **209**:437–442.
301. Chang, R. S. L., Tran, V. T., and Snyder, S. H., 1979, *J. Neurochem.* **32**:1653–1663.
302. Hill, S. J., and Young, J. M., 1980, *Br. J. Pharmacol.* **68**:687–696.
303. Hill, S. J., Emson, P. C., and Young, J. M., 1978, *J. Neurochem.* **31**:997–1004.
304. Tan-Tran, V., Freeman, A. D., Chang, R. S. L., and Snyder, S. H., 1980, *J. Neurochem.* **34**:1609–1613.
305. Chang, R. S. L., and Snyder, S. H., 1980, *J. Neurochem.* **34**:916–922.
306. Gavish, M., Chang, R. S. L., and Snyder, S. H., 1979, *Life Sci.* **25**:783–790.
307. Toll, L., Tran, V. T., Gavish, M., and Snyder, S. H., 1980, *Psychopharmacology and Biochemistry of Neurotransmitter Receptors* (H. I. Yamamura, R. W. Olsen, E. Usdin, eds.), Elsevier/North Holland, New York, pp. 301–311.
308. Subramanian, N., Seidler, F. J., Whitmore, W. L., and Slotkin, T. A., 1980, *Life Sci.* **27**:1315–1319.
309. Friedel, R. O., and Schanberg, S. M., 1975, *J. Neurochem.* **24**:819–820.
310. Tan-Tran, V., Chang, R. S. L., and Snyder, S. H., 1978, *Proc. Natl. Acad. Sci. U.S.A.* **75**:6290–6294.
311. Hegstrand, L. R., Kanof, P. D., and Greengard, P., 1976, *Nature* **260**:163–165.
312. Quach, T. T., Duchemin, A. M., Rose, C., and Schwartz, J. C., 1979, *Eur. J. Pharmacol.* **60**:391–392.
313. Quach, T. T., Duchemin, A. M., Rose, C., and Schwartz, J. C., 1980, *Neurosci. Lett.* **17**:49–54.
314. Diffley, D., Tran, V. T., and Snyder, S. H., 1980, *Eur. J. Pharmacol.* **64**:177–181.
315. Palacios, J. M., Young, W. S. III, and Kuhar, M. J., 1979, *Eur. J. Pharmacol.* **58**:295–304.
316. Palacios, J. M., Wamsley, J. K., and Kuhar, M. J., 1981, *J. Neurosci.* **6**:15–37.
317. Ninkovic, M., Hunt, S. P., and Gleave, J. R. W., 1982, *Brain Res.* **241**:197–206.
318. Chang, R. S. L., Tran, V. T., and Snyder, S. H., 1980, *Brain Res.* **190**:95–110.
319. Palacios, J. M., Wamsley, J. K., and Kuhar, M. J., 1981, *Brain Res.* **214**:155–162.
320. Taylor, J. E., and Richelson, E., 1979, *Mol. Pharmacol.* **15**:462–471.
321. Richelson, E., 1980, *Psychopharmacology and Biochemistry of Neurotransmitter Receptors* (H. I. Yamamura, R. W. Olsen, and E. Usdin, eds.), Elsevier/North Holland, New York, pp. 263–277.
322. Kuo, J.-F., Lee, T.-P., Reyes, P. L., Walton, K. G., Donnelly, T. E., Jr., and Greengard, P., 1972, *J. Biol. Chem.* **247**:16–22.
323. Schwabe, U., Ohga, Y., and Daly, J. W., 1978, *Naunyn Schmiedebergs Arch. Pharmacol.* **302**:141–151.
324. Ohga, Y., and Daly, J. W., 1977, *Biochim. Biophys. Acta* **498**:46–60.
325. Ferrendelli, J. A., Kinscherf, D. A., and Chang, M.-M., 1975, *Brain Res.* **84**:63–73.
326. O'Dea, R. F., and Zatz, M., 1976, *Proc. Natl. Acad. Sci. U.S.A.* **73**:3398–3402.
327. Study, R. E., and Greengard, P., 1978, *J. Pharmacol. Exp. Ther.* **207**:767–778.
328. Shimizu, H., Creveling, C. R., and Daly, J. W., 1970, *Mol. Pharmacol.* **6**:184–188.
329. Uchida, M., 1980, *Eur. J. Pharmacol.* **64**:357–360.
330. Kakiuchi, S., and Rall, T. W., 1968, *Mol. Pharmacol.* **4**:367–378.
331. Daly, J., 1977, *Cyclic Nuceotides in the Nervous System*, Plenum Press, New York, pp. 97–179.
332. McNeal, E. T., Creveling, C. R., and Daly, J. W., 1980, *J. Neurochem.* **35**:338–342.

333. Chasin, M., Mamrak, F., and Samaneigo, S. G., 1974, *J. Neurochem.* **22**:1031–1038.
334. Spiker, M. D., Palmer, G. C., and Manian, A. A., 1976, *Brain Res.* **104**:401–406.
335. Chasin, M., Mamrak, F., Samaniego, S. G., and Hess, S. M., 1973, *J. Neurochem.* **21**:1415–1427.
336. Psychoyos, S., 1978, *Life Sci.* **23**:2155–2162.
337. Daly, J. W., McNeal, E., Partington, C., Neuwirth, M., and Creveling, C. R., 1980, *J. Neurochem.* **35**:326–337.
338. Psychoyos, S., Dove, J., Stowbridge, B., and Nusynowite, I., 1982, *J. Neurochem.* **38**:1437–1445.
339. Palacios, J. M., Garbarg, M., Barbin, G., and Schwartz, J. C., 1978, *Mol. Pharmacol.* **14**:971–982.
340. Hill, S. J., Daum, P., and Young, J. M., 1981, *J. Neurochem.* **37**:1357–1360.
341. Dismukes, R. K., Rogers, M., and Daly, J. W., 1976, *J. Neurochem.* **26**:785–790.
342. Novak-Hofer, I., and Malnoë, A., 1981, *Biochim. Biophys. Acta* **677**:160–162.
343. Palmer, G. C., Schmidt, M. J., and Robison, G. A., 1972, *J. Neurochem.* **19**:2251–2256.
344. Quach, T. T., Duchemin, A., Rose, C., and Schwartz, J., 1981, *Mol. Pharmacol.* **20**:331–338.
345. Edwards, C., Nahorski, S. R., and Rogers, K. J., 1974, *J. Neurochem.* **22**:565–572.
346. Schultz, J., and Daly, J. W., 1973, *J. Biol. Chem.* **248**:860–866.
347. Subramanian, N., Whitmore, W. L., and Slotkin, T. A., 1981, *J. Neurosci.* **1**:674–678.
348. Hill, S. J., Hiley, C. R., and Young, J. M., 1981, *Eur. J. Pharmacol.* **71**:421–428.
349. Heinrich, M. A., 1953, *Arch. Int. Pharmacodyn.* **92**:444–463.
350. Smith, I. R., Cleverley, M. T., Ganellin, C. R., and Metters, K. M., 1980, *Agents Actions* **10**:422–426.
351. Rising, T. J., Norris, D. B., Warrander, S. E., and Wood, T. P., 1980, *Life Sci.* **27**:199–206.
352. Black, J. W., Gerskowitch, V. P., Randall, P. J., and Trist, D. G., 1981, *Br. J. Pharmacol.* **74**:978P–979P.
353. Olianas, M., Oliver, A. P., and Neff, N. H., 1982, *Adv. Biochem. Psychopharmacol.* **31**:149–156.
354. Yellin, T. O., Buck, S. H., Gilman, D., Jones, D. F., and Wardleworth, J. M., 1979, *Life Sci.* **25**:2001–2009.
355. Arunlakshana, O., and Schiild, H. O., 1959, *Br. J. Pharmacol.* **14**:48–59.
356. Kanof, P. D., Hegstrand, L. R., and Greengard, P., 1977, *Arch. Biochem. Biophys.* **182**:321–334.
357. Newton, M. V., Hough, L. B., and Azmitia, E. C., 1982, *Brain Res.* **239**:639–643.
358. Nahorski, S. R., Rogers, K. J., and Smith, B. M., 1974, *Life Sci.* **15**:1887–1894.
359. Nahorski, S. R., Rogers, K. J., and Smith, B. M., 1977, *Brain Res.* **126**:387–390.
360. Nahorski, S. R., and Smith, B. M., 1976, *Eur. J. Pharmacol.* **40**:273–278.
361. Makman, M. H., Ahn, H. S., Thal, L. G., Sharpless, N. S., Dvorkin, B., Horowitz, S. G., and Rosenfeld, M., 1980, *Brain Res.* **192**:177–183.
362. Hough, L. B., and Green, J. P., 1981, *Brain Res.* **219**:363–370.
363. Ahn, H. S., and Makman, M. H., 1977, *Brain Res.* **138**:125–138.
364. Huszti, Z., 1981, *Agents Actions* **11**:135–142.
365. Clark, R. B., and Perkins, J. P., 1971, *Proc. Natl. Acad. Sci. U.S.A.* **68**:2757–2760.
366. Garbarg, M., Barbin, G., Palacios, J. M., and Schwartz, J. C., 1978, *Brain Res.* **150**:638–641.
367. Durant, G. J., Ganellin, C. R., and Parson, M. E., 1975, *J. Med. Chem.* **18**:905–909.
368. Weinstein, H., Chou, D., Johnson, C. L., Kang, S., and Green, J. P., 1976, *Mol. Pharmacol.* **12**:738–745.
368a. Yellin, T. O., 1979, *Histamine Receptors*, Spectrum, New York, pp. 201–202.
369. Ganellin, C. R., 1974, *Molecular and Quantum Pharmacology* (E. G. Bergman and B. Pullman, eds.), D. Reidel, Dordrecht, pp. 43–53.
370. Haas, H. L., 1974, *Brain Res.* **76**:363–366.
371. Renaud, L. P., 1976, *Brain Res.* **115**:339–344.
372. Geller, H. M., 1981, *Dev. Brain Res.* **1**:89–101.
373. Segal, M., 1981, *Brain Res.* **213**:443–448.

374. Bolton, T. B., Clark, J. P., Kitamura, K., and Lang, R. J., 1981, *J. Physiol.* (*Lond.*) **320**:363–379.
375. Mantelli, L., Amerini, S., Picchi, A., Mugelli, A., and Ledda, F., 1982, *Agents Actions* **12**:122–130.
376. Haas, H. L., 1981, *Agents Actions* **11**:125–128.
377. Haas, H. L., 1981, *Neurosci. Lett.* **22**:75–78.
378. Sastry, B. S. R., and Phillis, J. W., 1976, *Can. J. Physiol. Pharmacol.* **54**:782–786.
379. Haas, H. L., and Wolf, P., 1977, *Brain Res.* **122**:269–279.
380. Dismukes, R. K., Ghosh, P., Creveling, C. R., and Daly, J. W., 1976, *Exp. Neurol.* **52**:206–215.
381. Haas, H. L., Wolf, P., Palacios, J. M., Garbarg, M., Barbin, G., and Schwartz, J. C., 1978, *Brain Res.* **156**:275–291.
382. Dismukes, R. K., Ghosh, P., Creveling, C. R., and Daly, J. W., 1975, *Exp. Neurol.* **49**:725–735.
383. Geller, H. M., 1979, *Neurosci. Lett.* **14**:49–53.
384. Chronister, R. B., Palmer, G. C., Defrance, J. F., Sikes, R. W., and Hubbard, J. I., 1982, *J. Neurobiol.* **13**:23–37.
385. Kruger, P. G., 1970, *Acta Zool.* **51**:85–93.
386. Rosenblum, W. I., 1973, *Brain Res.* **49**:75–82.
387. Edvinsson, L., Owman, C., and Sjoberg, N., 1976, *Brain Res.* **115**:377–393.
388. Jarrot, B., Hjelle, J. T., and Spector, S., 1979, *Brain Res.* **168**:323–330.
389. Joo, F., Dux, E., Karnushina, I. L., Halasz, N., Gecse, A., Ottlecz, A., and Mezei, Z., 1981, *Agents Actions* **11**:129–134.
390. Karnushina, I. L., Palacios, J. M., Barbin, G., Dux, E., Joo, F., and Schwartz, J. C., 1980, *J. Neurochem.* **34**:1201–1208.
391. Gross, P. M., Teasdale, G. M., Angerson, W. J., and Harper, A. M., 1981, *Brain Res.* **210**:396–400.
392. Solomon, G. F., 1981, *Psychoneuroimmunology* (R. Ader, ed.), Academic Press, New York, pp. 259–280.
393. Persinger, M. A., 1980, *Behav. Neural Biol.* **30**:448–459.
394. Mares, V., Bruckner, G., and Biesold, D., 1979, *Exp. Neurol.* **65**:278–283.
395. Cohn, C. K., Ball, C. G., and Hirsch, J., 1973, *Science* **180**:757–759.
396. Wauquier, A., and Niemegeers, C. J. E., 1981, *Eur. J. Pharmacol.* **72**:245–248.
397. Clineschmidt, B. V., and Lotti, V. J., 1973, *Arch. Int. Pharmacodyn.* **206**:288–298.
398. Leibowitz, S. F., 1973, *Brain Res.* **63**:440–444.
399. Gerald, M. C., and Maickel, R. P., 1972, *Br. J. Pharmacol.* **44**:462–471.
400. Kraly, F. S., and Miller, L. A., 1982, *Physiol. Behav.* **28**:841–846.
401. Kraly, F. S., and June, K. R., 1982, *J. Comp. Physiol. Psychol.* **96**:89–104.
402. Goldstein, D. J., and Halperin, J. A., 1977, *Nature* **267**:250–252.
403. Goldberg, S. R., 1980, *J. Pharmacol. Exp. Ther.* **214**:726–736.
404. Higashi, H., Ueda, N., and Nishi, S., 1982, *Brain Res. Bull.* **8**:23–32.
405. Orr, E. L., and Quay, W. B., 1975, *Endocrinology* **97**:481–484.
406. Peck, A. W., Fowle, A. S. E., and Bye, C., 1975, *Eur. J. Clin. Pharmacol.* **8**:455–463.
407. Carruthers, S. G., Shoeman, D. W., Hignite, C. E., and Azarnoff, D. L., 1978, *Clin. Pharmacol. Ther.* **23**:375–382.
408. Seppala, T., Nuotto, E., and Kortilla, K., 1981, *Br. J. Clin. Pharmacol.* **12**:179–188.
409. Barnett, A., Taber, R. I., and Roth, F. E., 1969, *Int. J. Neuropharmacol.* **8**:373–379.
410. Barnett, A., Makick, J. B., and Taber, R. I., 1971, *Psychopharmacologia* **19**:359–365.
411. Wallach, M. B., and Hedley, L. R., 1979, *Commun. Psychopharmacol.* **3**:35–39.
412. Hankoff, L. D., Gundlach, R. H., Paley, H. M., and Rudorfer, L., 1964, *Dis. Nerv. Syst.* **25**:547–553.
413. Jäätelä, A. Mannisto, P., Paatero, H., and Tuomisto, J., 1971, *Psychopharmacologia* **21**:202–211.
414. Mohs, R. C., Tinklenberg, J. R., Roth, W. T., and Kopell, B. S., 1978, *Psychopharmacology* **59**:13–19.
415. Schentag, J. J., Cerra, F. B., Calleri, G., DeGlopper, E., Rose, J. G., and Bernhard, H., 1979, *Lancet* **1**:177–181.

416. Schentag, J. J., Cerra, F. B., Calleri, G. M., Leising, M. E., French, M. A., and Bernhard, H., 1981, *Clin. Pharmacol. Ther.* **29:**737–743.
417. Kimelblatt, B. J., Cerra, F. B., Caller, G., Berg, M. J., McMillen, M. A., and Schentag, J. J., 1980, *Gastroenterology* **78:**791–795.
418. Schentag, J. J., 1980, *Ther. Drug Monit.* **2:**133–142.
419. Walt, R. P., LaBrooy, S. J., Avgerinos, A., Oehr, T., Riley, A., and Miseiwicz, J. J., 1981, *Scand. J. Gastroenterol.* **16**(Suppl. 69):19–23.
420. Finch, L., and Hicks, P. E., 1977, *Neuropharmacology* **16:**211–218.
421. Klein, M. C., and Gertner, S. B., 1981, *J. Pharmacol. Exp. Ther.* **216:**315–320.
422. Karppanen, H., Paakkari, I., Paakkari, P., Huotari, R., and Orma, A.-L., 1976, *Nature* **259:**587–588.
423. Finch, L., and Hicks, P. E., 1976, *Eur. J. Pharmacol.* **40:**365–368.
424. Tackett, R. L., and Holl, J. E., 1980, *J. Pharmacol. Exp. Ther.* **215:**552–556.
425. Lomax, P., and Green, M. D., 1981, *Fed. Proc.* **40:**2741–2745.
426. Bugajski, J., and Zacny, E., 1981, *Agents Actions* **11:**442–447.
427. Pilc, A., Rogoz, Z., and Byrska, B., 1980, *Neuropharmacology* **19:**947–950.
428. Bennett, C. T., and Pert, A., 1974, *Brain Res.* **78:**151–156.
429. Dogterom, J., van Wimersma-Greidanus, T. B., and DeWied, D., 1976, *Experientia* **32:**659–660.
430. Libertun, C., and McCann, S. M., 1976, *Neuroendocrinology (Basel)* **20:**110–120.
431. Rivier, C., and Vale, W., 1977, *Endocrinology* **101:**506–511.
432. Arakelian, M. C., and Libertun, C., 1977, *Endocrinology* **100:**890–895.
433. Pontiroli, A. E., and Pozza, G., 1978, *Acta Endocrinol. (Kbh.)* **88:**23–28.
434. Carlson, H. E., and Ippoliti, A. F., 1977, *Clin. Endocrinol. Metab.* **45:**367–370.
435. Sharpe, P. C., Melvin, M. A., Mills, J. G., Burland, W. L., and Groom, G. V., 1980, *Acta Endocrinol. (Kbh.)* **95:**308–313.
436. Robins, A. H., and McFadyen, M. L., 1981, *J. Pharm. Pharmacol.* **33:**615–616.
437. Endroczi, E., and Hilliard, J., 1965, *Endocrinology* **77:**667–673.
438. Bjorklund, A., Hakanson, R., Nobin, A., and Sjoberg, N., 1972, *Experientia* **28:**1232–1233.
439. Fram, D. H., and Green, J. P., 1965, *J. Biol. Chem.* **240:**2036–2042.
440. Green, J. P., Fram, D. H., and Kase, N., 1964, *Nature* **204:**1165–1168.
441. Friedman, A. H., and Walker, C. A., 1969, *J. Physiol. (Lond.)* **202:**133–146.
442. Subramanian, N., Mitznegg, P., and Estler, C. J., 1978, *Naunyn Schmiedebergs Arch. Pharmacol.* **302:**119–121.
443. Rawat, A. K., 1980, *Res. Commun. Chem. Pathol. Pharmacol.* **27:**91–103.
444. Papanicolaou, J., and Fennessy, M. R., 1980, *Psychopharmacologia* **72:**73–77.
445. Subramanian, N., Schinzel, W., Mitznegg, P., and Estler, C. J., 1980, *Pharmacology* **20:**42–45.
446. Lee, J. R., and Fennessy, M. R., 1976, *Clin. Exp. Pharmacol. Physiol.* **3:**179–189.
447. Mazurkiewicz-Kwilecki, I., and Henwood, R. W., 1976, *Agents Actions* **6:**402–408.
448. Wong, C. L., and Roberts, M. B., 1976, *Agents Actions* **6:**569–576.
449. Glick, S. D., and Crane, L., 1978, *Nature* **273:**547–549.
450. Rosow, C. E., Moss, J., Philbin, D. M., and Savarese, J. J., 1982, *Anesthesiology* **56:**93–96.
451. Gershon, S., and Shaw, F. H., 1958, *J. Pharm. Pharmacol.* **10:**22–29.
452. Akcasu, A., and Unna, K. R., 1970, *Eur. J. Pharmacol.* **13:**103–107.
453. Tanaka, K., and Lin, Y., 1969, *Jpn. J. Pharmacol.* **19:**510–514.
454. Chou, W. S., Ho, A. K. S., and Loh, H. H., 1971, *Nature [New Biol.]* **233:**280–281.
455. Sebens, J. B., and Korf, J., 1975, *Exp. Neurol.* **46:**333–344.
456. Fram, D. H., and Green, J. P., 1968, *Clin. Pharmacol. Ther.* **9:**355–357.
457. Coupet, J., and Szuchs-Meyers, V. A., 1981, *Eur. J. Pharmacol.* **74:**149–155.
458. Haas, H. J., 1979, *Agents Actions* **9:**83–84.
459. Nowak, J. Z., Bielkiewicz, B., and Lebrecht, U., 1979, *Neuropharmacology* **18:**783–789.
460. Figge, J., Leonard, P., and Richelson, E., 1979, *Eur. J. Pharmacol.* **58:**479–483.
461. Snyder, S. H., 1980, *Pharmakopsychiatry* **13:**62–67.
462. Taylor, J. E., and Richelson, E., 1980, *Eur. J. Pharmacol.* **67:**41–46.
463. Psychoyos, S., 1981, *Biochem. Pharmacol.* **30:**2182–2185.

464. Tan-Tran, V., Lebovitz, R., Toll, L., and Snyder, S. H., 1981, *Eur. J. Pharmacol.* **70**:501–509.
465. Taylor, J. E., and Richelson, E., 1982, *Eur. J. Pharmacol.* **78**:279–285.
466. Mazurkiewicz-Kwilecki, I. M., and Bielkiewicz, B., 1977, *Prog. Neuro psychopharmacol.* **1**:115–124.
467. Maayani, S., Weinstein, and Green, J. P., 1979, **38**:376.
468. Rehavi, M., and Sokolovsky, M., 1978, *Brain Res.* **149**:525–529.
469. Nagy, A., 1977, *J. Pharm. Pharmacol.* **29**:104–107.
470. Fjalland, B., and Boeck, V., 1978, *Acta Pharmacol. Toxicol.* **42**:206–211.
471. Fontaine, J., and Reuse, J., 1978, *Arch. Int. Pharmacodyn.* **235**:51–61.
472. Free, C. A., Paik, V. S., and Shada, J. D., 1974, *The Phenothiazines and Structurally Related Drugs* (I. S. Forrest, C. J. Carr, and E. Usdin, eds.), Raven Press, New York, pp. 739–748.
473. Maayani, S., Green, J. P., and Weinstein, H., 1978, *Fed. Proc.* **37**:612.
474. Frederickson, P. A., and Richelson, E., 1979, *Eur. J. Pharmacol.* **56**:261–264.
475. Creese, I., Burt, D. R., and Snyder, S. H., 1976, *Life Sci.* **17**:1715–1720.
476. Persson, S.-A., 1977, *Eur. J. Pharmacol.* **43**:73–83.
477. Persson, S.-A., 1977, *Life Sci.* **20**:1199–1206.
478. Corne, S. J., and Pickering, R. W., 1967, *Psychopharmacologia* **11**:65–78.
479. Ginzel, K. H., and Mayer-Gross, W., 1956, *Nature [New Biol.]* **178**:210.
480. Bertino, J. R., Klee, G. D., and Weintraub, M. D., 1959, *J. Clin. Exp. Psychopathol.* **20**:218–227.
481. Uyeno, E. T., and Mitoma, C., 1969, *Psychopharmacologia* **16**:73–80.
482. Schneckloth, R., Page, I. H., del Greco, F., and Corcoran, A. C., 1957, *Circulation* **16**:523–532.
483. Nigro, S. A., 1968, *J.A.M.A.* **203**:301–302.
484. Green, J. P., Maayani, S., Weinstein, H., and Hough, L. B., 1980, *Psychopharmacol. Bull.* **16**:36–38.
485. Green, J. P., 1981, *Trends Pharmacol. Sci.* **2**:VII–IX.
486. Movshon, J. A., 1978, *Nature* **272**:305–306.
487. Greiner, T., Burch, N. R., and Edelburg, R., 1958, *Arch. Neurol.* **79**:208–210.

GABA Receptors

G. A. R. Johnston, R. D. Allan, and J. H. Skerritt

1. INTRODUCTION

The simple amino acid GABA (γ-aminobutyric acid, 4-aminobutanoic acid) is a transmitter of major significance in the mammalian central nervous system. Receptors for GABA are found at both presynaptic ("axoaxonic") and postsynaptic ("axosomatic" and "axodendritic") sites and also on nerve fibers ("nonsynaptic" or "fiber" receptors). There is strong evidence for a multiplicity of GABA receptors. Activation of GABA receptors may be studied *in vivo* using electrophysiological or behavioral methods and *in vitro* using isolated tissue preparations. Such studies of receptor activation are complemented by neurochemical investigation of the binding of GABA and related ligands to a variety of CNS membrane preparations. These studies provide direct information on the kinetics of binding interactions but only rather indirect information on the relevance of these interactions to physiological processes. Binding to the recognition sites on GABA receptors is merely the initial stage in producing the changes in ionic permeability that influence membrane potential and neuronal excitability. GABA agonist binding is considered to change receptor conformation with the activated receptor influencing an associated ionophore.

Several classes of centrally acting drugs act on GABA receptors. Barbiturates and benzodiazepines appear to modulate the activation of some GABA receptors by potentiating the coupling between these receptors and ionophores. Many convulsants are GABA antagonists acting on GABA recognition sites (e.g., bicuculline) and/or on associated ionophores (e.g., picrotoxinin).

GABA receptors might be considered as part of complex domains consisting of receptors, modulator sites, and ionophores, with all three influencing the final outcome of the initial activation of the GABA receptors. In general, activation of GABA receptors results in an increased membrane permeability

G. A. R. Johnston, R. D. Allan, and J. H. Skerritt • Department of Pharmacology, University of Sydney, Sydney, NSW 2006, Australia.

to chloride ions, which is associated with a hyperpolarizing response on nerve cell bodies and a depolarizing response on presynaptic terminals and nerve fibers.

2. GABA ANALOGUES

A variety of analogues of GABA are known that have relatively selective actions as GABA agonists.[1] In general, these are conformationally restricted analogues of GABA representing particular shapes of the GABA molecule. GABA itself has considerable flexibility as a result of free rotation around the single bonds in its carbon chain. This flexibility can be reduced by incorporation of unsaturated and/or ring structures into the GABA molecule, and a systematic study of GABA analogues of restricted conformation has shown that GABA interacts in different shapes with its various receptors, transport carriers, and enzymes.[2] Extensive information about the properties of GABA analogues is essential to study the multiplicity of GABA receptors, since receptors may be differentiated on the basis of their agonist and antagonist selectivity.

Some GABA analogues are being investigated as possible therapeutic agents.[3] Since GABA does not pass the blood–brain barriers on systemic administration, considerable effort has been devoted to developing lipophilic analogues of GABA that might act directly in the CNS after systemic administration or as prodrugs liberating the active drug within the CNS.[1] Baclofen [β-(*p*-chlorophenyl)-GABA] was developed for this purpose, and recent evidence indicates that it is a selective agonist for a novel class of GABA receptors.[4] Muscimol (1-aminomethylisoxazol-3-ol), a cyclic analogue of GABA from the mushroom *Amanita muscaria*, does produce central effects on systemic administration, but these may be caused by metabolites rather than the parent isoxazole.[3] Tetrahydroisoxazol[4,5-c]pyridin-3-ol (THIP), a cyclic analogue of muscimol and thus a bicyclic analogue of GABA, shows particular promise as a systemically active GABA agonist.[1] Other promising GABA analogues include kojic amine (2-aminomethyl-5-hydroxy-4H-pyran-4-one),[5] SL-76002 [α-(4'-chlorophenyl)-5-fluoro-2-hydroxybenzylideneamino-4-butyramide],[6] esters of GABA with long-chain alcohols, e.g., cetyl-GABA,[7] and certain amides of GABA, e.g., N-pivaloyl-GABA.[8]

Although most GABA analogues contain an amino group and a carboxyl group or their equivalents (e.g., the imidazole residue in imidazole-4-acetic acid and the 3-hydroxyisoxazole structure in muscimol), agents lacking one of these functional groups do appear to be able to act on GABA receptors. Such agents include ethylenediamine,[9] SL-75102 [sodium α-(4'-chlorophenyl)-5-fluoro-2-hydroxybenzylideneamino-4-butyrate],[10] pentobarbital[11],[12] and chloralose,[11] which may directly activate GABA receptors in a bicuculline-sensitive manner. Furthermore, bicuculline may be regarded as a GABA analogue[2] that acts directly on certain GABA receptors to antagonize their activation by GABA in a competitive manner.

3. ELECTROPHYSIOLOGICAL STUDIES

3.1. GABA Antagonists

Sensitivity to the GABA antagonists bicuculline and picrotoxinin is used extensively in electrophysiological studies to characterize the activation of GABA receptors. This represents a gross oversimplification for the following reasons: (1) there is now extensive evidence for bicuculline-insensitive GABA receptors selectively activated by baclofen; (2) there is evidence accumulating for a multiplicity of bicuculline-sensitive receptors; (3) bicuculline may antagonize the action of other neuroactive substances such as taurine; and (4) picrotoxinin may also antagonize the action of other transmitters such as serotonin.[13,14] The development of more selective GABA antagonists will be essential to further progress in the characterization of GABA receptors.

3.1.1. Bicuculline

The most widely used GABA antagonist is the phthalide isoquinoline alkaloid bicuculline.[13] It is a potent convulsant on systemic administration and antagonizes synaptic inhibitions in most, if not all, areas of the CNS, consistent with the widespread function of GABA as an inhibitory neurotransmitter. Much more information is needed, however, than bicuculline antagonism to characterize any particular synaptic inhibitory process as involving GABA as transmitter. Other bicuculline-sensitive neuronal inhibitors, including imidazole-4-acetic acid, δ-aminolevulinic acid, ethylenediamine, β-alanine, and taurine, are found in the brain, and thus not all bicuculline-sensitive synaptic inhibitions may necessarily involve GABA. A variety of neurochemical data does indicate that GABA is involved in most bicuculline-sensitive inhibitions.[14]

Bicuculline is a competitive antagonist of the action of GABA with a pA_2 of 5.35 measured against a presynaptic action of GABA in cuneate slices.[15] This is consistent with the proposal, based on structural similarities between bicuculline and certain GABA agonists, that GABA and bicuculline share some common receptors.[16]

Bicuculline is unstable under physiological conditions, having a half-life of only a few minutes and being hydrolyzed to the relatively inactive bicucine. At pH 2–3, protonated bicuculline appears to be relatively stable, and most electrophysiological studies use solutions at this pH in microelectrodes for local administration by iontophoresis. Quaternary salts such as bicuculline methochloride and bicuculline methiodide (incorrectly called "N-methylbicuculline") are not only more water soluble (0.25 M aqueous solutions of bicuculline methochloride can be prepared) but are stable in the pH range 2–8, making them preferable to bicuculline for many experiments. It is essential to use bicuculline itself to study CNS effects following systemic administration, since blood–brain barriers exclude the quaternary salts much more efficiently than the free base that is in an equilibrium with the simple salts. The exclusion of the quaternary salts from the CNS can be useful if it is the peripheral effects of bicuculline that are to be studied without complications from central effects.

Administered by iontophoresis, bicuculline and its quaternary salts antagonize the action of GABA and of many GABA-like substances on most CNS neurons and also on neurons in the periphery. Some regional and species variation in the susceptibility of agonists to bicuculline antagonism is apparent, indicative of multiple bicuculline-sensitive receptors showing regional and species differences. An important regional difference, in view of the possible transmitter role of taurine, is that bicuculline antagonizes the inhibitory action of taurine in the cerebral and cerebellar cortices but not in the spinal cord and medulla.[14]

Bicuculline and its quaternary salts have other actions in addition to antagonizing the action of GABA and related amino acids. They are weak competitive inhibitors of acetylcholinesterase and, particularly the quaternary salts, have direct neuronal excitant actions.

The optical enantiomer of bicuculline, (-)-bicuculline, is inactive as a GABA antagonist, and only phthalide isoquinoline alkaloids having the $1S,9R$ absolute configuration, (+)-bicuculline and corlumine, have GABA antagonist properties. Many isoquinoline alkaloids are convulsants instead because of a weak strychninelike antagonism of the inhibitory action of glycine.[13,17]

3.1.2. Picrotoxinin

Picrotoxinin is the active principle of picrotoxin (a 1 : 1 mixture of picrotoxinin and picrotin). It is a potent convulsant used extensively in investigations of synaptic inhibitions *in vivo* by systemic administration and *in vitro* by bath application, but its use in iontophoretic experiments is limited by low aqueous solubility and the lack of readily ionizable groups.[17]

Picrotoxinin acts as a noncompetitive inhibitor of the action of GABA, acting at different sites from bicuculline.[18] Binding studies suggest that picrotoxin interacts with chloride ionophores rather than with GABA recognition sites.[19] Picrotoxinin also antagonizes certain actions of serotonin, indicating that GABA and serotonin may activate ionophores with common pharmacological properties.[20]

Many compounds structurally related to picrotoxinin, including anisatin, coriamyrtin, shikimin, and tutin, are also convulsants, and most have GABA antagonist properties, but none is selective for GABA.[17] Dihydropicrotoxinin has similar properties to picrotoxinin and has been used as a radioactive ligand for binding studies related to GABA-activated chloride ionophores.[19]

3.1.3. Other GABA Antagonists

GABA antagonism may be an important factor in the epileptogenic properties of penicillins. Benzylpenicillin and ampicillin are weak but relatively selective GABA antagonists.[13] Benzylpenicillin may act on ionophores rather than on GABA receptors, since it is not a competitive antagonist.[20]

Many convulsants, including tetramethyldisulfotetramine, bicyclophosphate, and bicyclocarboxylate ortho esters, have a cagelike structure and appear to act as GABA antagonists as the result of interaction with ionophores.[20]

This may also be true of other convulsants such as substituted caprolactams and thiolactams.[17] The polyacetylenic alcohol cunaniol is a potent convulsant that shows some selectivity as a GABA antagonist.[15] A convulsant benzodiazepine, Ro5-3663, antagonizes GABA-mediated synaptic inhibition.[22] Bemegride and metrazole are weak noncompetitive antagonists of the presynaptic action of GAB in cuneate slices.[15]

Certain drugs used as selective antagonists of other neurotransmitters at specific receptors can act as GABA antagonists under appropriate conditions. Tubocurarine, the classic nicotinic antagonist, is a relatively potent GABA antagonist but is unspecific in that it also acts as a glycine antagonist.[13] The opiate antagonist naloxone is a weak competitive GABA antagonist, acting in a nonstereospecific manner.[23] The selective histamine H_2 receptor antagonists metiamide and cimetidine may also act as GABA antagonists.[24] The bicyclic GABA analogues isoTHIP (tetrahydroisxazolo[3,4-c]pyridin-3-ol) and iso-THAZ (tetrahydro-4H-isoxazolo[3,4-d]azepin-3-ol) have GABA antagonist properties,[1] as does γ-acetylenic-GABA.[25]

In view of the multiplicity of GABA receptors and the value of selective antagonists with which to distinguish among different populations of GABA receptors, it would seem very worthwhile to investigate some of the above antagonists and related compounds in more detail, particularly with respect to antagonist ability distinguishing among various classes of GABA agonists.

3.2. GABA Agonists

GABA agonist activity may be assessed by a variety of electrophysiological procedures, e.g., by measuring extracellularly changes in neuronal firing frequency or by measuring intracellularly changes in membrane conductance, excitability, potential, or current following local extracellular administration by microelectrophoresis from a micropipette *in vivo* or bath application to a neuronal culture or tissue slice *in vitro*. Since a variety of receptors for GABA and related substances are likely to exist, the particular receptors involved in any agonist response need to be defined on the basis of agonist and antagonist specificity. In very broad terms, GABA agonist activity may be divided initially on the basis of sensitivity to antagonism by bicuculline.

It may be necessary to take into account various GABA transport processes when studying potential GABA agonists. GABA is known to be rapidly removed from the synaptic environment by uptake into presynaptic terminals, glial elements, and neuronal cell bodies, and it can be released from these sites under appropriate conditions.[26] Thus, it may be important to distinguish between direct and indirect agonist actions, indirect effects arising from a substance releasing GABA from an intracellular store into the extracellular environment, where it can activate GABA receptors. Furthermore, the observed potency of a GABA agonist may be limited by uptake processes, and it is necessary to make some allowance for this when attempting to correlate structure–activity data. Ideally, it is best to compare GABA receptor agonists that do not interact with GABA transport systems.

Sigmoidal dose–response curves have been noted in various preparations for the action of GABA and several GABA agonists, indicative of cooperative interactions in the activation of GABA receptors.[17,27]

3.2.1. Bicuculline-Sensititive GABA Agonists

In most if not all regions of the mammalian CNS, local administration of GABA results in a depression of neuronal firing, and this depression can be antagonized by local or systemic administration of bicuculline.

The most extensive studies of bicuculline-sensitive GABA agonists *in vivo* have been made on postsynaptic receptors of spinal interneurons of anesthetized cats. A wide variety of GABA analogues have been observed to act as bicuculline-sensitive depressants in these studies, of particular interest being the analogues of restricted conformation: these studies have led to a pattern emerging of the likely range of "active conformations" of GABA acting on bicuculline-sensitive feline spinal postsynaptic receptors[16] consistent with the proposal that bicuculline and GABA agonists such as muscimol share common receptors. Extensive structure–activity studies on GABA led to the concept of GABA interacting with these receptors in partially extended and almost planar conformations.[28]

Model building and interactive computer graphics techniques indicate that GABA could interact in a range of "active conformations" from the "bicuculline conformation" (defined on the basis of the structure of the antagonist bicuculline) to the "muscimol conformation" (defined on the basis of the most potent known bicuculline-sensitive GABA agonist, muscimol).[16] These two conformations of GABA may define the limits of the range of "active conformations" at bicuculline-sensitive receptors. Within this range, the agonists muscimol, THIP, *trans*-4-aminocrotonic acid, and (+)-*cis*-3R-aminocyclopentane-1S-carboxylic acid may adopt conformations close to the "muscimol conformation," whereas 4,5-dihydromuscimol and (+)-*trans*-3S-aminocyclopentane-1S-carboxylic acid may adopt the "bicuculline conformation." These GABA agonists may form the basis for distinguishing among different classes of bicuculline-sensitive GABA receptors. A key compound in these tests is THIP, a bicyclic GABA analogue that has a semirigid structure that appears to endow particular specificity on its interaction with certain receptors[1,28]; THIP does not interact with GABA uptake systems.

GABA receptors sensitive to antagonism by bicuculline may occur as widely on nerve terminals ("presynaptic") as they do on cell bodies ("postsynaptic").[29] Presynaptic GABA receptors on primary afferent terminals in feline spinal cord appear to be very similar to postsynaptic GABA receptors on feline spinal interneurons,[30] although all of the key compounds listed above have yet to be tested. Structure–activity studies on bicuculline-sensitive GABA receptors in the feline neurohypophysis suggest some differences from spinal receptors.[31] Furthermore, the relative potency of GABA and muscimol as bicuculline-sensitive depressants differs in the feline cerebral cortex and spinal cord,[32] and glycine appears to be able to activate bicuculline-sensitive postsynaptic receptors in the rat cerebral cortex but not in the rat spinal cord.[33]

Glycine responses are somewhat more sensitive to antagonism by bicuculline on spinal neurons of the mouse than in the rat or cat.[34] In these examples, the glycine responses can also be antagonized by strychnine and thus can be distinguished from the GABA responses. Taurine has a strychnine-sensitive, bicuculline-insensitive depressant action in the cat spinal cord, a strychnine-sensitive, bicuculline-sensitive action in the cat cerebral cortex, and a bicuculline-sensitive, strychnine-insensitive action in the rat cerebellum.[32,35] Clearly, a number of bicuculline-sensitive receptors exist for depressant amino acids with regional and species differences as revealed by electrophysiological studies in the mammalian CNS.

GABA also has "nonsynaptic" actions.[36] In a study of the relative potencies of a variety of conformationally restricted GABA analogues as depressants of neuronal activity in the hemisected spinal cord from immature rats (a "synaptic" action) and as depolarizers of dorsal root fibers from immature rats (a "nonsynaptic" action), there were clear differences observed in the relative potencies of the analogues,[37] indicating that different populations of bicuculline-sensitive receptors were involved. The nonsynaptic receptors present on the spinal root fibers are probably very similar to the receptors widely distributed on unmyelinated peripheral nerve fibers.[36] It seems likely that both "synaptic" and "nonsynaptic" classes of bicuculline-sensitive receptors occur in the CNS; this would explain the biphasic action of GABA on CA1 pyramidal cells in the rat hippocampus, producing depolarization and hyperpolarization at different sites with different agonist and antagonist profiles.[38] The hyperpolarizing responses appear to arise primarily from receptors on the cell soma, whereas the depolarizing responses arise from dendritic "nonsynaptic" receptors. THIP appears to be relatively much more potent as a GABA agonist on "synaptic" receptors than on "nonsynaptic" receptors, both in the above spinal cord/spinal root studies[37] and in hippocampal slice experiments.[38]

The fluctuations in membrane current ("membrane noise") produced by GABA in neurons cultured from mouse spinal cord and dorsal root ganglia can be interpreted as reflecting the kinetic behavior of a single population of two-state chloride ion channels.[39] Fluctuation analyses of current responses to a number of GABA analogues indicate that they activate channels of very similar conductance but significantly different duration, and thus the differences in potency can be attributed to differences in the lifetime of the activated channels.[40] It is not clear whether the GABA receptors studied in these dissociated neuronal cultures are "synaptic," "nonsynaptic," or a mixture thereof, though the relatively short duration of the action of THIP (13 ms) compared to that of GABA (30 ms) and muscimol (76 ms) suggests that predominantly "nonsynaptic" receptors are involved. These observations may have to be revised, however, if studies using patch-clamp recording with an improved resolution reveal that what previously appeared to be single channel openings are interspersed with brief (<300 μs) closings, as has been found for glutamate-activated channels.[41] Some evidence for complex kinetic behavior of GABA-activated channels has been obtained using low agonist concentrations.[42]

Ethylenediamine has a potent bicuculline-sensitive depressant action of the firing of neurons in rat cerebral cortex and globus pallidus and a bicuculline-sensitive depolarizing action on rat dorsal root ganglia.[9,43] These actions may well represent direct activation of bicuculline-sensitive receptors by ethylenediamine, since ethylenediamine can weakly displace GABA in ligand-binding studies. Ethylenediamine, however, does interact with neuronal and glial GABA transport systems, and its effects may include an indirect action via the release of GABA under certain conditions.[9,44] Ethylenediamine does not appear to show any particular preference for "synaptic" or "nonsynaptic" receptors.

The Schiff base SL-75102 can act as a direct bicuculline-sensitive GABA agonist on dorsal root ganglia, whereas the corresponding amide, SL-76002 (progabide) acts as a prodrug, breaking down to directly acting GABA agonists in the CNS after systemic administration.[6,10]

Pentobarbital and related barbiturates have direct bicuculline-sensitive GABA agonist actions on motoneurons in the hemisected frob spinal cord[11] and on cultured neurons from mouse spinal cord.[45] The GABA agonist action requires approximately fivefold higher concentrations of barbiturates than the GABA-enhancing action discussed in Section 2.3. GABA and (−)-pentobarbital appear to open chloride ion channels of similar conductance, but the lifetime of the channels opened by (−)-pentobarbitone is much greater (129 ms) than that for GABA-activated channels.[45]

Chloralose has been reported to have a bicuculline-sensitive GABA agonist action similar to pentobarbital on frog spinal motoneurons.[11]

3.2.2. Bicuculline-Insensitive GABA Agonists

The GABA analogue baclofen (Lioresal®) was prepared as a lipophillic derivative likely to penetrate the blood–brain barriers more effectively than GABA. On local administration, baclofen depressed the firing frequency of neurons in most areas of the mammalian CNS, but unlike the depressant action of GABA, that induced by baclofen was insensitive to antagonism by bicuculline.[46,47] Subsequently, a number of conformationally restricted analogues of GABA were found to have similar bicuculline-insensitive depressant actions: cis-4-aminocrotonic acid,[48] trans-2- and trans-3-aminocyclohexane carboxylic acid,[49] and cis-2-(aminomethyl)-cyclopropane carboxylic acid.[50] The structural similarities among these compounds prompted the proposal that they constitute another class of GABA agonist acting on bicuculline-insensitive receptors.[17,49] The possibility of bicuculline-insensitive GABA receptors in the CNS had also been proposed on the basis of other considerations by Ryall.[51] From structure–activity considerations, it seemed that GABA might interact with bicuculline-insensitive receptors in folded conformations, whereas extended conformations of GABA were necessary for interaction with the "classical" bicuculline-sensitive receptors.[17]

These speculations were linked to physiological events by Bowery and colleagues, whose studies on baclofen led them to propose that baclofen is a selective agonist for a novel class of receptors, which are insensitive to antagonism by bicuculline and which modulate monoamine release in the CNS and also in the periphery at autonomic nerve terminals.[4,52] The agonist profile of these bicuculline-insensitive receptors was different from that for bicuculline-sensitive receptors; in particular, 3-aminopropanesulfonic acid and THIP were essentially inactive. Furthermore, these receptors were not linked to chloride channels, their activation appearing to involve changes in calcium fluxes. The action of baclofen was stereoselective, with the ($-$) isomer being some 100 times more active than the ($+$) isomer.[52] The receptors involved appeared to be GABA receptors, since GABA was able to mimic the actions of baclofen in a bicuculline-insensitive manner. Studies on embryonic sensory neurons in tissue culture have provided evidence for bicuculline-sensitive GABA receptors selectively activated by muscimol, which produces a decrease in membrane resistance, and bicuculline-insensitive GABA receptors selectively activated by baclofen, which produces a decrease in action potential duration on the same neurons.[53] The two classes of GABA receptors exhibit different time courses of desensitization and different developmental patterns.

Unfortunately, not all actions of baclofen can be mimicked by GABA, and it is apparent that baclofen can activate receptors other than GABA receptors. The most potent action of baclofen in many instances is a selective reduction of monosynaptic excitation in the spinal cord. This action is stereoselective for the ($-$) isomer and is likely to result from the activation of presynaptic receptors that are not activated by GABA.[54–56] Similar observations have been made in the hippocampus.[57]

It is possible that some of the conformationally restricted analogues of GABA in folded conformations may be more selective than baclofen for the activation of bicuculline-insensitive GABA receptors. Furthermore, conformational restriction of baclofen by methods similar to those developed for GABA may yield baclofen analogues with more selective actions than those of the parent compound.[58] The development of selective antagonists for bicuculline-insensitive GABA receptors is an urgent need.

3.3. GABA Modulators

Barbiturates and benzodiazepines potentiate GABA-mediated synaptic transmission in the mammalian CNS.[59–63] These drugs appear to act by modulating GABA–receptor–ionophore complexes.[19,64] Specific receptor sites for these modulating drugs distinct from GABA recognition sites, appear to exist such that barbiturates and benzodiazepines can act in different ways to potentiate GABA responses. Not all GABA receptors are associated with barbiturate and/or benzodiazepine receptors.

It is often found that barbiturates, but not benzodiazepines, potentiate the effects of locally administered GABA on CNS neurons *in vivo*, whereas both classes of drugs potentiate GABA-mediated synaptic transmission. A number of studies indicate that benzodiazepines act presynaptically to increase GABA release,[65,66] whereas postsynaptic actions to potentiate GABA responses are less commonly seen.[67] Barbiturates have readily observable presynaptic and postsynaptic actions at GABA synapses.[59,60,68] Barbiturates and benzodiazepines show some selectivity with respect to different populations of GABA receptors in hippocampal slices in that pentobarbital enhanced the depolarizing ("nonsynaptic") responses to GABA to a much greater extent that the hyperpolarizing ("synaptic") responses, whereas diazepam showed the opposite selectivity.[38]

The depolarizing action of GABA on afferent nerve fibers in rat cuneate slices can be potentiated by flurazepam (Dalmane®). Flurazepam preferentially attenuated the GABA antagonist action of picrotoxin rather than that of bicuculline, indicating that this benzodiazepine potentiated the effects of GABA by acting at sites closer to the chloride ionophores than the GABA recognition sites.[69] Flurazepam potentiates GABA responses in the rat *in vivo* on cerebellar cortical neurons more readily than on cerebral cortical neurons, suggesting some regional selectivity in the action of benzodiazepines.[67]

Diazepam (Valium®) and pentobarbital (Nembutal®) potentiate the increase in chloride conductance induced by GABA in cultured mouse spinal neurons, but fluctuation analysis indicated that the drugs act in different ways.[70] Diazepam increased the frequency of opening of the chloride channels with little effect on open-channel lifetime, whereas pentobarbital decreased the frequency and increased average open-channel lifetime. The benzodiazapine antagonist Ro15-1788 blocks the potentiation of GABA-induced depolarization in rat sympathetic ganglia produced by chlordiazepoxide but not that produced by phenobarbital.[71]

Diphenylhydantoin (phenytoin, Dilantin®, DPH) has been shown to potentiate the action of both locally applied and synaptically released GABA on the crayfish stretch receptor.[72] In the rat dorsal raphe nucleus, diphenylhydantoin does not potentiate the action of GABA but appears to unmask a depressant action of benzodiazepines in that following systemic administration of diphenylhydantoin, similarly administered diazepam produced a picrotoxin-sensitive depression of neuronal firing.[73] This suggests that diphenylhydantoin and diazepam in combination can activate picrotoxin-sensitive ionophores and may indicate that endogenous benzodiazepinelike agents play a role in the anticonvulsant action of diphenylhydantoin.

Sodium di-*n*-propylacetate (sodium valproate, DPA), on local application to neurons in rat cerebral cortex by microelectrophoresis, potentiates the depression of neuronal firing induced by GABA and muscimol but not by glycine,[74] supporting previous observations made on cultured neurons.[75] This potentiation of GABA action may contribute to the anticonvulsant effects of sodium di-*n*-propylacetate, which are apparent very soon after systemic administration of the drug, before changes in GABA metabolism are observed.

4. BEHAVIORAL STUDIES

Animal models have been developed for studying drug effects on GABA receptors *in vivo*.[76,77] For example, unilateral intranigral injection of certain GABA agonists induces contralateral turning in rats, whereas injection of GABA antagonists induces ipsilateral turning. Barbiturates and benzodiazepines potentitate many of the behavioral effects of GABA agonists in these animal models. The results of such studies with GABA agonists, antagonists, and modulators, which reflect the interactions of many neuronal systems, are usually more difficult to interpret than electrophysiological studies on single neurons. Thus, intranigral GABA potentiates dopamine-dependent stereotyped behavior but decreases dopamine-dependent locomotor activity.

Both bicuculline-sensitive and bicuculline-insensitive GABA receptors may be studied in animal models. Thus, baclofen and muscimol have different behavioral effects on intranigral injection.[77,78] Baclofen may have a partial agonist action on bicuculline-insensitive GABA receptors in the substantia nigra related to modulation of dopamine release. Although single-cell studies indicate that activation of GABA receptors by baclofen is not antagonized by picrotoxin or potentiated by benzodiazepines, the behavioral effects of baclofen following intranigral injection can be blocked by picrotoxin and potentiated by flurazepam,[79] indicating that other receptors are involved than those directly activated by baclofen.

There is evidence from behavioral studies for denervation supersensitivity of GABA receptors in that an increased response to intranigral muscimol is observed following lesion of the striatonigral tract and degeneration of GABA terminals in the substantia nigra.[79]

There is considerable evidence to implicate GABA receptors in analgesia. The GABA agonists muscimol, THIP, and baclofen produce analgesia on systemic administration.[80] This effect is insensitive to the opiate antagonist naloxone and thus independent of enkephalin-related analgesia. A variety of experimental stressors produce analgesia that has naloxone-insensitive components,[81] and thus GABA-related drugs may have an important future in the pharmacology of stress-related behavior and its consequences.

GABA systems are involved in the regulation of blood pressure, with receptors in the central and peripheral nervous systems being involved.[82] GABA agonists exert a hypotensive action that can be antagonized by bicuculline and picrotoxin and potentiated by benzodiazepines. In view of the involvement of GABA systems in stress-related phenomena and the link between stress and heart disease, it is clear that a more detailed study of GABA receptors and their activation could be of vital importance to one of the current major health problems.

5. BINDING STUDIES

There are many studies on the binding of radioactive GABA and related compounds to a variety of membrane preparations. Most claim to be studies of GABA receptors relevant to the function of GABA as a synaptic transmitter, but this is doubtful, since the binding as usually studied is more characteristic of what is known of "nonsynaptic" than "synaptic" GABA receptors on the basis of structure–activity studies using conformationally restricted analogues of GABA.[37] It may be that the organization of GABA receptors in synaptic structures changes receptor properties, e.g., cooperativity and structural specificity, compared with those of receptors on nerve fibers and that the preparation of membranes for binding experiments changes this organization such that most isolated GABA receptors behave as if they were "nonsynaptic" in binding experiments. Furthermore, as discussed in Section 6, endogenous in-hibitors of GABA receptors are present in neuronal membranes in association with the receptors, modifying their availability and affinity. Different proce-dures for receptor purification influence these associations in different ways. Thus, the choice of preparation is particularly important in studying GABA binding and attempting to relate it to the function of synaptic GABA receptors under physiological conditions. Structure–activity studies show that different membrane preparations can favor different classes of GABA binding sites.[83]

Fractionation studies show that most GABA binding is associated with synaptic membranes, and structure–activity studies show a rough correlation between the ability of analogues to displace bound GABA and their electro-physiological potencies as GABA receptor agonists.[3,19] Furthermore, lesioning studies and experiments on cultured neurons and glia indicate that most GABA binding is associated with neurons, although GABA binding sites have been reported on glial cells.[84] High-affinity GABA binding sites are particularly en-riched in postsynaptic densities isolated from rat brain,[85] but detailed structure–activity studies have yet to be carried out with such relatively highly purified membrane structures in order to determine the "synaptic" or "nonsynaptic" nature of the binding.

5.1. Ligands

A variety of radioactive ligands in addition to GABA have been used in binding studies relevant to GABA receptors. Muscimol binding has been widely studied[86]; there are differences between muscimol and GABA binding, sug-gesting that it is likely that GABA does not interact with all muscimol binding sites and *vice versa*.[2,16,86] Muscimol binding has been used for the autoradi-ographic localization of "high-affinity GABA receptors."[87] Isoguvacine,[88] pi-peridine-4-sulfonic acid,[89] and THIP[90] have been used to study bicuculline-sensitive GABA-binding sites, but these ligands show a greater degree of non-saturable binding (and thus less precise measurement of saturable binding) than either muscimol or GABA. Baclofen[91] binds to brain membranes in a bicu-culline-insensitive, calcium-dependent manner.

In contrast to the numerous studies on agonist binding, relatively few stud-ies have been carried out with antagonists. Tritium-labeled bicuculline meth-iodide appears to bind to only a single population of GABA-sensitive sites with relatively low affinity.[92] Bicuculline binding has also been studied using a flu-orescent antibody technique.[93] Dihydropicrotoxinin, a derivative of picrotox-inin readily labeled with tritium, binds to GABA-insensitive sites that appear to be associated with a GABA-activated chloride ionophore.[94] There are far fewer dihydropicrotoxinin binding sites than GABA binding sites on rat brain membranes, suggesting that GABA binding sites are not associated with ion-ophores on a 1 : 1 basis, but this may be misleading because of the compli-cations of endogenous inhibitors of both GABA binding (see Section 6) and of dihydropicrotoxinin binding.[95]

5.2. Detergents, Inorganic Ions, and Multiple Binding Sites

There are many macromolecules in brain tissue in addition to GABA receptors that can bind GABA, e.g., those associated with specific uptake carriers in neuronal and glial elements and those associated with enzymes involved in the synthesis and degradation of GABA. These macromolecules have differing structural specificities with respect to GABA agonists (substrates) and antagonists (inhibitors), differing subcellular localizations, and differing physical properties, which generally enable them to be distinguished from one another.

In a crude homogenate of rat brain, most of the GABA bound to membranes incubated in a "physiological" medium appears to be bound to uptake carriers.[3] Such binding is sodium dependent and can be antagonized by uptake inhibitors such as nipecotic acid but not by the receptor antagonist bicuculline. It can be minimized by incubation in sodium-free media, by addition of known selective uptake inhibitors, or by treating the membranes in such a way as to remove or inactivate the uptake carriers, e.g., extraction with Triton X-100. Alternatively, one can use binding ligands such as muscimol,[96] isoguvacine,[88] piperidine-4-sulfonic acid,[89] or THIP,[90] which have little or no interaction with uptake carriers under appropriate conditions but are relatively selective agonists for certain classes of GABA receptors.

Sodium-dependent binding of GABA to brain membranes can usually be reduced, and sodium-independent binding apparently enhanced, by freezing and then thawing the preparation. Repeated freeze-thawing and several washes of the membranes yield preparations with which sodium-independent binding of GABA can be readily studied with a high ratio of specific to nonspecific binding.[97] Extraction of the membranes with the nonionic detergent Triton X-100 (e.g., 0.5% Triton X-100 in buffer at 37°C for 30 min) further enhances sodium-independent GABA binding, increasing both the apparent affinity of binding and the apparent density of binding sites.[98-100] These changes in binding properties appear to be the result of the removal of endogenous inhibitors (see Section 6); these substances are present in the supernatant washings from the frozen-thawed and detergent-extracted membrane preparations and potently inhibit GABA binding when added back to these preparations.[100]

Sodium-independent GABA binding sites appear to be relatively resistant to inactivation or solubilization by Triton treatment compared to binding sites for many other neurotransmitters and centrally active drugs. Triton dissolves the limiting membranes of nerve endings, leaving the junctional complex with attached subsynaptic structures relatively intact.[101] Repeated extractions of junctional complexes with 0.5% Triton X-100 do not influence the apparent affinity or number of binding sites of sodium-independent GABA binding.[102] Incubation of retinal membranes with 100 mM sodium perchlorate seems to have a similar but not identical action to Triton X-100 on GABA binding,[103] possibly indicating a differential effect of Triton X-100 and sodium perchlorate on certain endogenous inhibitors. Whereas Triton extraction enhances agonist binding, binding of the GABA antagonist bicuculline is reduced by such treatment.[92] GABA binding sites can be solubilized by treatment with 1% Triton X-100[104] or sodium deoxycholate.[19] Optimal solubilization of sodium-independent, bicuculline-sensitive GABA binding sites has been reported to be obtained with 2% sodium deoxycholate, yielding a preparation devoid of benzodiazepine binding sites.[105]

Although most GABA binding studies to date have been carried out in Tris–citrate incubation media, it is known that certain inorganic ions can significantly influence the characteristics of GABA binding.[98] Chloride ions, in particular, appear to play an important role in the interactions between GABA and barbiturate and benzodiazepine receptors, as discussed in Section 8. To relate observed binding phenomena to physiological processes, it would seem reasonable to study, where possible, the binding of GABA in the presence of "physiologically relevant" concentrations of inorganic ions. Of course, many binding studies aim to study molecular properties of receptors that are difficult to investigate in any other way. They are not planned simply to mimic what we already know about physiological processes and what can be better investigated by other techniques.

Bicuculline-displaceable binding of GABA[106] and muscimol[96] to brain membranes has been reported in the presence of physiologically relevant concentrations of sodium, potassium, calcium, magnesium, and chloride ions. Baclofen-sensitive GABA binding has also been reported under these conditions, indicative of the activity of bicuculline-insensitive GABA binding sites,[107] and such binding is known to be calcium dependent.[91] The presence of inorganic ions can alter the apparent substrate specificity of binding; thus, ethylenediamine only weakly inhibits the binding of GABA, muscimol, or baclofen in Tris–citrate or Tris–HC1 medium but inhibits the binding of all three ligands in Krebs-Henseleit medium, producing the same maximum displacement as GABA.[107] Several "good" buffers, including HEPES (N-2-hydroxyethylpiperazine-N'-2-ethanesulfonic acid), are competitive inhibitors of GABA binding with little effect on muscimol binding to rat brain membranes, providing further evidence of differences between the sodium-independent binding of GABA and muscimol.[108]

Anions may have a differential effect on agonist and antagonist binding to certain sites. Thus, whereas many anions decrease the sodium-independent, bicuculline-sensitive binding of GABA and muscimol to brain membranes,[98] they increase the binding of bicuculline and the potency of bicuculline as an inhibitor of GABA binding.[92] These results suggest that certain GABA receptors may exist in interconvertible forms with selectively higher affinity for either agonists or antagonists.

The kinetics of the sodium-independent, bicuculline-sensitive binding of GABA and related agonists to brain membranes are usually biphasic, being analyzed in terms of high- and low-affinity binding sites.[19] The absolute values for the dissociation constants characterizing these two populations of binding sites vary over orders of magnitude depending on the membrane preparation, and there may well be many more than two populations involved. The apparent kinetic parameters of GABA binding are determined to a large extent by the presence of absence or various endogenous inhibitors,[100] including GABA itself. [109]

Hill plots show no evidence of cooperativity between high- and low-affinity GABA binding sites,[99] unlike physiological observations on the activation of

GABA receptors. These kinetically different sites are also functionally distinct, although they show a remarkably similar structural specificity with respect to GABA analogues that can displace bound GABA.[110] They differ in rates of heat inactivation and sensitivity to changes in pH, temperature, and ionic environment. Ammonium thiocyanate, 50 mM, selectively destroys high-affinity GABA binding without influencing low-affinity binding or the ability of GABA to stimulate benzodiazepine binding[111]; this chaotropic agent significantly increases the binding of radioactive bicuculline to brain membranes[92] and the ability of bicuculline to displace bound GABA.[111] Furthermore, certain drugs have differential actions on the two populations; thus, barbiturates increase the affinity of GABA binding to high-affinity sites,[64] whereas benzodiazepines increase the affinity of GABA binding to low-affinity sites.[112]

6. ENDOGENOUS INHIBITORS (GABARINS)

The concept of endogenous inhibitors or modulators of GABA receptors arose from studies on the binding of radioactive GABA to rat brain membranes. As noted in Section 5, the methods used to prepare the membranes for binding studies greatly influenced the apparent affinity of GABA binding and the number of apparent binding sites. The exposure of these latent GABA binding sites appears to require the removal of substances that inhibit GABA binding and which are normally incorporated in the membranes. The freeze-thaw regimens and Triton extraction procedures that were developed to optimize GABA binding[97–99] remove these inhibitors from their association with GABA binding sites on the membranes, and their presence in the supernatant washes of such membrane preparations is readily demonstrated.[100] A variety of such endogenous inhibitors appears to exist, and they have been called collectively GABARINS (GABA Receptor INhibitorS).

GABARINS may function to control the affinity and availability of certain GABA receptors, thus altering the properties of GABA–receptor–ionophore complexes. The interactions between GABA receptors and GABARINS appear to be important sites of drug action.[81] GABARINS may play an important role in synaptic development and in synaptic changes involved in learning and memory.[81] They may be involved in desensitization and supersensitivity phenomena; e.g., removal of GABARINS could underlie denervation supersensitivity. Thus, following striatal lesions that lead to degeneration of GABA-releasing terminals in the nigra, an increased behavioral response to intranigral injections of muscimol and an increased apparent density of GABA binding sites have been noted.[113]

GABARIN activity has been associated with bicuculline-sensitive CNS GABA binding sites and in the modulation of these sites by barbiturates and benzodiazepines. It is not yet known whether peripheral bicuculline-sensitive sites or any of the bicuculline-insensitive GABA binding sites are influenced by GABARINS.

6.1. Occluded GABA

A substantial proportion, but certainly not all,[100] of the GABARIN activity in detergent extracts of rat brain membranes can be attributed to competitive inhibition by endogenous GABA released from mitochondria and other sites.[114] This "free" GABA competes with exogenous radiolabeled GABA for GABA receptors unless it is removed from the membrane preparation by thorough washing. Free GABA can appreciably influence the apparent kinetic properties of GABA binding, and failure to take into account even very low amounts of GABA leads to misinterpretation of binding data.[109]

GABA is also occluded by the membranes; such GABA in the "occult form"[115] may represent a storage or inactivated form of GABA and make up some 5% of total brain GABA. It is in nonexchangeable, osmotically insensitive compartments possibly associated with "cryptic" GABA receptors buried in the membranes[16] (G.A. R. Johnston and S. M. E. Kennedy, unpublished observations). Occluded GABA could be the basis of the apparent desensitization of GABA responses seen *in vivo*, GABA combining with the receptor, and both being buried in the membrane. Occluded GABA may be liberated from thoroughly washed brain membranes by detergent extraction or acid treatment. Such tightly bound GABA does not exchange with radiolabeled GABA added in the initial homogenization and thus does not represent redistributed GABA.[16] Even though the interaction between liberated occluded GABA and its receptors is necessarily competitive, the occlusion of GABA results in a reduction of the number of available GABA binding sites rather than an alteration in their apparent affinity. Changes in the amount of occluded GABA could effectively result in up- or down-regulation of available GABA receptors. This phenomenon could underlie the rapid changes seen in the number of available GABA receptors observed under certain conditions; e.g., following a 3-min swim stress of mice, there is an apparent 70% increase in the number of low-affinity GABA binding sites in the forebrain.[116]

6.2. Phospholipids

Membrane phospholipids are recognized to be important in the receptor interactions of several hormones and neurotransmitters.[117] Lipids may regulate receptor binding affinity or modulate subsequent molecular events. Chemical similarities between the structures of GABA and phosphatidylethanolamine were first noted by Watkins,[118] who suggested that GABA may induce chloride permeability changes by acting on a phosphatidylethanolamine–receptor complex, with GABA and the polar head group of phosphatidylethanolamine competing for similar portions of the receptor complex. There is now a considerable amount of evidence to support a role for phosphatidylethanolamine in GABA receptor function.

Phospholipids are abundant in detergent extracts of brain membranes, and derivatives of phosphatidylethanolamine in particular show potent GABARIN activity.[100] Treatment of crude synaptic membranes with phospholipase C, which removes polar head groups, increases the availability and affinity of

GABA receptors,[119] as does enzymic base exchange with ethanolamine (G. A. R. Johnston and S. M. E. Kennedy, unpublished observations).

Abnormalities in the interactions of phospholipids and GABA receptors may be involved in Huntington's disease, in which there is widespread degeneration of GABA neurons. Increased GABA binding is observed in some brain areas, together with increased levels of free glycerophosphoethanolamine.[120] The lower endogenous GABA levels in the brain membranes of Huntington's disease victims may explain some of the apparent alterations in receptor binding.[121] It is possible, however, that part of the increase in GABA binding results from a decreased association of phosphatidylethanolamine with the receptors, since it has been found that detergent extraction or phospholipase C treatment had no effect on GABA binding in postmortem Huntington's disease cerebellar membranes and that this binding was initially of greater affinity than that in control membranes but was similar to that in control membranes after extraction with detergent or enzyme treatment.[120]

The increases in GABA binding following Triton extraction of brain homogenates are severalfold higher than those obtained by treatment with phospholipase C producing similar depletion of membrane phospholipids.[122] In reconstitution experiments, treatment of detergent-extracted membranes with phosphatidylethanolamine yielded membranes with GABA binding properties intermediate between those of well-washed membranes and detergent-extracted membranes.[100,122] (G. A. R. Johnston and S. M. E. Kennedy, unpublished observations). These results are consistent with phosphatidylethanolamine being responsible for only that part of the GABARIN activity concerned with the regulation of GABA receptors.

Lipid-soluble drugs might influence GABA receptors by interactions with membrane phospholipids. Barbiturates, in particular, are likely to act in this way, since the enhancement of GABA binding to brain membranes by these drugs is destroyed by detergent extraction, and the structure–activity profile in such enhancement parallels their lipophillic properties, consistent with studies on the binding of phenobarbital.[64] The enhancement of GABA binding by benzodiazepines also appears to be dependent on the presence of lipids that can be extracted by detergents, which, interestingly, does not prevent the enhancement of benzodiazepine binding by GABA, a phenomenon than may even be observed in solubilized receptor complexes.[112]

Sulfolipids may be associated with GABA receptors, since incubation with arylsulfatase A decreases the number of available binding sites for GABA and muscimol on bovine cerebellar membranes without their affinity being altered.[123]

6.3. Peptides

An acidic peptide, GABA-modulin, of molecular weight 15,000, has been reported to noncompetitively inhibit the binding of GABA to brain membranes.[124] This peptide appears to be very difficult to isolate and characterize.[19] A reliable preparation would enable its role in modulation of GABA receptor function to be objectively assessed. A recent report indicates that GABA-modulin has "metamorphosed" into a basic protein of molecular weight 18,500.[125]

Smaller peptides have been reported to be able to inhibit the binding of GABA to brain membranes. Thus, a peptidelike thermostable substance of molecular weight 3000 from pig brain inhibits GABA binding to rat brain membranes; this material antagonizes the anticonvulsant action of diazepam, which suggests that it may act as an endogenous antagonist of GABA or of the coupling between GABA and benzodiazepine receptors.[126] Peptides of less than 1000 daltons that inhibit GABA binding have been reported in extracts of rat brain.[127]

It seems likely that a rich variety of endogenous peptides have GABARIN-like activity, similar to the many endogenous peptides with opiatelike activity.

6.4. Other Endogenous Inhibitors

Yoneda and Kuriyama[128] have described a low-molecular-weight nonpeptide inhibitor (GRIF; GABA Receptor Inhibitory Factor) of muscimol binding in extracts of rat brain membranes. GRIF can be readily separated from GABA by cation-exchange chromatography since it lacks a primary amino function and is stable to heat, trypsin, collagenase, peroxidase, and phospholipase C treatment. It has a molecular weight of less than 500 based on gel and ultrafiltration studies.

Purines such as adenosine, hypoxanthine, and inosine may act as endogenous modulators of GABA receptors. These purines are relatively weak noncompetitive inhibitors of GABA binding[129] and thus distinct from GRIF, which is a competitive inhibitor of GABA binding.[128]

7. DEVELOPMENTAL AND PHYLOGENETIC TRENDS

GABA receptors are present in the brain at birth and, in common with most other neurotransmitter receptors, increase in number appreciably during postnatal development.[130,131] Endogenous inhibitors of GABA binding are also present in brain extracts early in development, before most synapses have been formed.[81,130,131] However, unlike GABA binding sites, these endogenous inhibitors appear to be in greater concentration in neonatal brain than in adult brain. Decreases in endogenous inhibitors during maturation may be related to the development of synaptic activity, and GABA receptors may be up-regulated on synapse formation.

Benzodiazepine binding sites, like the endogenous inhibitors of GABA binding, are found in brain in high levels at birth, and these levels decrease with increasing age.[132] Furthermore, the ability of GABA to enhance benzodiazepine binding decreases with increasing age,[133] suggesting a link between GABARINS and benzodiazepine receptors.

Detergent extraction has no influence on GABA binding in nervous tissue from primitive vertebrates such as the hagfish and produces only slight enhancement of GABA binding in amphibian brain extracts compared with that observed on similar extraction of avian or mammalian brain tissue.[134] Hagfish

GABA receptors are of higher affinity than mammalian sites, suggesting that GABARINS may evolve in vertebrates as the central nervous system increases in complexity. Hagfish brain also lacks benzodiazepine receptors, providing further evidence that the coupling of GABA and benzodiazepine receptors involves certain of the GABARINS.[135]

8. DRUG EFFECTS ON GABA BINDING

There is much electrophysiological evidence for the modulation of GABA receptor activation by drugs such as barbiturates and benzodiazepines (see Section 3.3). Under appropriate circumstances, these drugs can be shown to modulate the binding of GABA to brain membranes. Membrane integrity is vital to these studies, as the coupling between the sites of drug action and the GABA binding sites is relatively fragile and can be rendered inoperative by procedures commonly used to "optimize" GABA binding.

The acute effects of barbiturates and benzodiazepines are discussed in more detail below. Many drugs have chronic effects on GABA binding. Thus, during withdrawal from chronic ethanol administration, the affinity of low-affinity GABA binding in rat brain was significantly lower than in pair-fed control animals.[136] Chronic elevation of brain GABA levels induced by GABA-T inhibitors such as isonicotinic acid hydrazide reduced GABA binding.[137] Naloxone-precipitated withdrawal in morphine-dependent rats resulted in a decrease in the number of low-affinity GABA binding sites in the cerebellum and striatum.[138] Chronic administration of phenobarbital or diazepam decreased the number of high-affinity GABA binding sites in rat striatum.[139]

8.1. Enhancement of GABA Binding by Barbiturates

Anesthetic, anticonvulsant, and convulsant barbiturates enhance the binding of GABA to extensively washed crude synaptic membrane preparations from rat brain.[64,140] For pentobarbital, half-maximal enhancement of GABA binding was produced at 38 μM, which is well within the range of concentrations found in the brains of anesthetized laboratory animals. This enhancement involved an increased affinity of GABA for the high-affinity binding sites and resulted from a decreased rate of dissociation of GABA from these sites.[141] An increased lifetime of barbiturate-coupled GABA–receptor complexes would be consistent with the observed increased lifetime of GABA-activated chloride channels in cultured neurons in the presence of pentobarbital[70] discussed in Section 3.3.

This enhancement of GABA binding by barbiturates appears to involve a close association with GABA-activated chloride ionophores, since the enhancement is blocked by the ionophore antagonist picrotoxinin. Since picrotoxinin does not influence GABA binding in chloride-free media, barbiturates are not acting directly on GABA receptors to enhance GABA binding. At high concentrations (>0.5 mM), certain barbiturates do exhibit a direct GABAmimetic effect, which is not influenced by detergent extraction of the membranes.[142] On the other hand, the ability of barbiturates to enhance GABA binding is lost on detergent extraction, indicating the involvement of certain endogenous inhibitors of GABA binding.

Barbiturate binding sites have been studied using radiolabeled phenobarbital, which binds with relatively low affinity to rat brain membranes.[143] These binding sites are abolished on detergent extraction, and barbiturate binding is not influenced by GABA, bicuculline, or picrotoxinin. Structure–activity studies for a range of substituted barbiturates showed an excellent direct correlation between ability to enhance GABA binding and ability to displace phenobarbital binding in that the concentrations required for half-maximal enhancement and half-maximal displacement were almost identical. This indicates that both phenomena are directly linked and that detergent-sensitive components are responsible for the coupling between the two classes of sites.

Furthermore, there is a close correlation between the octanol–water partition coefficients for these barbiturates and their ability to enhance GABA binding, suggesting that lipids may be important components of barbiturate–GABA-receptor complexes. Barbiturates may interact with specific phospholipids to enhance GABA binding via changes in membrane fluidity.[64]

Barbiturates also enhance GABA binding by an anion-dependent mechanism involving an apparent increase in the density of GABA sites[19] and requiring a fivefold higher concentration of barbiturate than the affinity change discussed above. This density effect is not so dependent on membrane integrity as the affinity change and would not account for the increased lifetime of GABA-activated chloride channels found in cultured neurons.[70] The apparent increase in density could, however, be caused by an increase in affinity of normally undetectable low-affinity GABA binding sites.[19]

8.2. Enhancement of GABA Binding by Benzodiazepines

Many groups have reported on the stimulation of benzodiazepine binding by GABA,[144–147] citing this phenomenon as biochemical support for electrophysiological evidence of a specific interaction between benzodiazepines and GABA-mediated synaptic transmission. The electrophysiological observations pertain to the enhancement of the action of GABA induced by benzodiazepines as discussed in Section 3.3, and any enhancement of benzodiazepine binding by GABA is not necessarily relevant to these electrophysiological observations. The relevant phenomenon of stimulation of GABA binding by benzodiazepines is quite difficult to observe and, as for barbiturate enhancement of GABA binding, is greatly dependent on the integrity of the membrane preparation.

Benzodiazepines enhance the binding of GABA to fresh, well-washed synaptosomal membranes from rat brain by increasing the affinity of the low-affinity site.[112] How this relates to the observed increase in the frequency of opening of GABA-activated chloride channels induced by benzodiazepines is unclear.[70] The ability of benzodiazepines to enhance GABA binding is halved on freezing the membranes and lost on detergent extraction, whereas these procedures did not influence the ability of GABA to enhance benzodiazepine binding (J. H. Skerritt, S. Chen Chow, and G. A. R. Johnston, unpublished observations). GABA enhancement of benzodiazepine binding appears to be a relatively robust phenomenon, whereas benzodiazepine enhancement of GABA binding is quite sensitive to membrane perturbations.

Enhancement of GABA binding by benzodiazepines and enhancement of benzodiazepine binding of GABA are different phenomena. When the nature of the endogenous ligands for benzodiazepine receptors is sorted out,[148] the physiological relevance of the ability of GABA to enhance benzodiazepine binding may become clearer.

9. CONCLUSIONS

The multimolecular units of receptors, modulator sites, and ionophores in which GABA receptors exist present an exceedingly complex problem to neuroscientists. The very simple molecule GABA appears to interact with a multiplicity of receptors, and these interactions may be modulated by a wide variety of endogenous agents including peptides, phospholipids, and purines. Electrophysiological studies remain the optimal way in which to study the activation of GABA receptors, whereas binding studies have led to a greater understanding of the nature of GABA receptors and how they may be influenced by various endogenous and exogenous sustances including widely prescribed drugs. Neurochemists should continue to develop agents of increasing selectivity with which to investigate GABA receptors. Particular attention should be given to the development of selective agonists and antagonists for bicuculline-insensitive GABA receptors and selective agonists and antagonists for subclasses of bicuculline-sensitive GABA receptors.

ACKNOWLEDGMENTS. The authors are grateful to Dr. J. L. Barker, Dr. N. G. Bowery, Dr. P. Krogsgaard-Larsen, Dr. R. W. Olsen, Dr. R. A. Nicoll, and Dr. S. J. Enna for sending manuscripts prior to publication and to Mrs. C. Drew, Dr. L. P. Davies, Dr. J. Hambley, Dr. I. Spence, and Dr. M. Willow for helpful discussions and to the National Health and Medical Research Council for support.

REFERENCES

1. Krogsgaard-Larsen, P., and Falch, E., 1981, *Mol. Cell. Biochem.* **38**:129–146.
2. Johnston, G. A. R., Allan, R. D., Kennedy, S. M. E., and Twitchin, B., 1979, *GABA-Neurotransmitters* (P. Krogsgaard-Larsen, J. Scheel-Kruger, and H. Kofod, eds.), Munksgaard, Copenhagen, pp. 149–164.
3. Enna, S. J., 1981, *Biochem. Pharmacol.* **30**:907–913.
4. Bowery, N. G., Hill, D. R., Hudson, A. L., Doble, A., Middlemiss, D. N., Shaw, J., and Turnbull, M., 1980, *Nature* **283**:92–94.
5. Atkinson, J. G., Girard, Y., Rokack, J., Rooney, C. S., McFarlane, C. S., Rackham, A., and Share, N. N., 1979, *J. Med. Chem.* **22**:99–106.
6. Lloyd, K. G., Worms, P., Deportere, H., and Bartholini, G., 1979, *GABA-Neurotransmitters* (P. Krogsgaard-Larsen, J. Scheel-Kruger, and H. Kofod, eds.), Munksgaard, Copenhagen, pp. 308–325.
7. Sytinsky, I. A., Soldatenkov, A. T., and Lajtha, A., 1978, *Prog. Neurobiol.* **10**:89–133.

8. Gakzigna, L., Garbin, L., Bianchi, M., and Marzotto, A., 1978, *Arch. Int. Pharmacodyn. Ther.* **235**:73–85.

9. Perkins, M. N., Bowery, N. G., Hill, D. R., and Stone, T. W., 1981, *Neurosci. Lett.* **23**:325–327.

10. Desarmenien, M., Feltz, P., Headley, P. M., and Santangelo, F., 1981, *Br. J. Pharm.* **72**:355–364.

11. Nicoll, R. A., and Wojtowicz, J. M., 1980, *Brain Res.* **191**:225–237.

12. Willow, M., and Johnston, G. A. R., 1981, *J. Neurochem.* **37**:1291–1294.

13. Curtis, D. R., and Johnston, G. A. R., 1974, *Neuropoisons, Their Pathophysiological Actions*, Volume 2 (L. L. Simpson and D. R. Curtis, eds.), Plenum Press, New York, pp. 207–248.

14. Johnston, G. A. R., 1978, *Annu. Rev. Pharmacol. Toxicol.* **18**:269–289.

15. Simmonds, M. A., 1978, *Br. J. Pharm.* **63**:495–502.

16. Andrews, P. R., and Johnston, G. A. R., 1979, *Biochem. Pharmacol.* **28**:2697–2702.

17. Johnston, G. A. R., 1978, *Receptors in Pharmacology* (J. R. Smythies and R. J. Bradley, eds.), Marcel Dekker, New York, pp. 295–333.

18. Simmonds, M. A., 1980, *Neuropharmacology* **19**:39–45.

19. Olsen, R. W., 1981, *J. Neurochem.* **37**:1–13.

20. Segal, M., 1976, *Brain Res.* **103**:161–166.

21. Pickles, H. G., and Simmonds, M. A., 1980, *Neuropharmacology* **19**:35–38.

22. Schlosser W., and Franco, S., 1979, *J. Pharmacol. Exp. Ther.* **211**:290–295.

23. Gruol, D. L., Barker, J. L., and Smith, T. G., 1980, *Brain Res.* **198**:323–332.

24. Antonaccio, M. J., Asaad, M., and Boccagno, J., 1981, *Eur. J. Pharmacol.* **72**:369–372.

25. Gent, J. P., and Normanton, J. R., 1978, *Br. J. Pharmacol.* **64**:383P–384P.

26. Johnston, G. A. R., 1978, *Proc. Aust. Physiol. Pharmacol. Soc.* **9**:94–98.

27. Krause, D. N., Ikeda, K., and Roberts, E., 1981, *Brain Res.* **255**:319–332.

28. Krogsgaard-Larsen, P., Johnston, G. A. R., Lodge, D., and Curtis, D. R., 1977, *Nature* **268**:53–55.

29. Pickles, H. G., 1979, *Br. J. Pharmacol.* **65**:223–228.

30. Curtis, D. R., Bornstein, J. C., and Lodge, D., 1980, *Brain Res.* **194**:255–258.

31. Mathison, R. D., and Dreifuss, J. J., 1980, *Brain Res.* **187**:476–480.

32. Curtis, D. R., Duggan, A. W., Felix, D., Johnston, G. A. R., and McLennan, H., 1971, *Brain Res.* **70**:493–499.

33. Biscoe, T. J., Duggan, A. W., and Lodge, D., 1972, *Comp. Gen. Pharmacol.* **3**:423–433.

34. Martin, M. R., McHanwell, S., and Biscoe, T. J., 1978, *Brain Res.* **151**:225–233.

35. Frederickson, R. C. A., Neuss, M., Morzorati, S. L., and McBride, W. J., 1978, *Brain Res.* **145**:117–126.

36. Brown, D. A., 1979, *Trends Neurosci.* **2**:271–273.

37. Allan, R. D., Evans, R. H., and Johnston, G. A. R., 1980, *Br. J. Pharmacol.* **70**:609–615.

38. Alger, B. E., and Nicoll, R. A., 1982, *J. Physiol.* (*Lond.*) **328**:125–141.

39. Barker, J. L., McBurney, R. N., and MacDonald, J. F., 1982, *J. Physiol.* **322**:365–387.

40. Barker, J. L., and Mathers, D. A., 1981, *Science* **212**:358–361.

41. Cull-Candy, S. G., and Parker, I., 1982, *Nature* **295**:410–412.

42. Mathers, D. A., and Barker, J. L., 1981, *Proc. Soc. Neurosci.* **6**:726.

43. Perkins, M. N., and Stone, T. W., 1982, *Br. J. Pharmacol.* **75**:93–99.

44. Davies, L. P., Hambley, J. W., and Johnston, G. A. R., 1982, *Neurosci. Lett.* **29**:57–61.

45. Barker, J. L., and Mathers, D. A., 1981, *Trends Neurosci.* **4**:10–13.

46. Davies, J., and Watkins, J. C., 1974, *Brain Res.* **70**:501–505.

47. Curtis, D. R., Game, C. J. A., Johnston, G. A. R., and McCulloch, R. M., 1974, *Brain Res.* **70**:493–499.

48. Johnston, G. A. R., Curtis, D. R., Beart, P. M., Game, C. G. A., McCulloch, R. M., and Twitchin, B., 1975, *J. Neurochem.* **24**:157–160.

49. Johnston, G. A. R., 1976, *GABA in Nervous System Function* (E. Roberts, T. N. Chase, and D. B. Tower, eds.), Raven Press, New York, pp. 395–411.

50. Allan, R. D., Curtis, D. R., Headley, P. M., Johnston, G. A. R., Lodge, D., and Twitchin, B., 1980, *J. Neurochem.* **34**:652–654.

51. Ryall, R. W., 1975, *Handbook Psychopharmacol.* **4**:83–128.

52. Bowery, N. G. Doble, A., Hill, D. R., Hudson, A. L., Shaw, J. S., Turnbull, M. G., and Warrington, R., 1981, *Eur. J. Pharmacol.* **71**:53–70.
53. Dunlap, K., 1981, *Br. J. Pharmacol.* **74**:579–585.
54. Ault, B., and Evans, R. H., 1981, *Eur. J. Pharmacol.* **71**:357–364.
55. Curtis, D. R., Lodge, D., Bornstein, J. C., and Peet, M. J., 1981, *Exp. Brain Res.* **42**:158–170.
56. Davies, J., 1981, *Br. J. Pharmacol.* **72**:373–384.
57. Lanthorn, T. H., and Cotman, C. W., 1981, *Brain Res.* **225**:171–178.
58. Allan, R. D., Tran, H., and Skerritt, J. H., 1982, *Neurosci. Lett.* **S8**:30.
59. Ransom, B. R., and Barker, J. L., 1976, *Brain Res.* **114**:530–535.
60. Polc, P., and Haefely, W., 1976, *Naunyn Schmiedebergs Arch. Pharmacol.* **294**:121–131.
61. Nicoll, R. A., 1978, *Psychopharmacology: A Generation of Progress* (M. A. Lipton, A. Dimascio, and K. F. Killiam, eds.), Raven Press, New York, pp. 1337–1348.
62. MacDonald, R. L., and Barker, J. L., 1978, *Nature* **271**:563–564.
63. Evans, R. H., 1979, *Brain Res.* **171**:113–120.
64. Johnston, G. A. R., and Willow, M., 1982, *Trends Pharmacol.* **3**:328–833.
65. Polc, P., Mohler, H., and Haefely, W., 1974, *Naunyn Schmiedebergs Arch. Pharmacol.* **284**:319–337.
66. Curtis, D. R., Lodge, D., Johnston, G. A. R., and Brand, S. J., 1976, *Brain Res.* **118**:344–347.
67. Jiang, Z.-G., 1981, *Can. J. Physiol. Pharmacol.* **59**:595–598.
68. Lodge, D., and Curtis, D. R., 1978, *Neurosci. Lett.* **8**:125–129.
69. Simmonds, M. A., 1980, *Nature* **284**:558–560.
70. Study, R. E., and Barker, J. L., 1981, *Proc. Natl. Acad. Sci. U.S.A.* **78**:7180–7184.
71. Nutt, D. J., and Cowen, P. J., 1982, *Nature* **295**:436–438.
72. Deisz, R. A., and Lux, H. D., 1977, *Neurosci. Lett.* **5**:199–203.
73. Gallagher, D. W., Mallorga, P., and Tallman, J. F., 1980, *Brain Res.* **189**:209–220.
74. Kerwin, R. W., Olpe, H.-R., and Schmutz, M., 1980, *Br. J. Pharmacol.* **71**:545–551.
75. MacDonald, R. L., and Bergey, K. B., 1979, *Brain Res.* **170**:558–562.
76. Scheel-Kruger, J., Arnt, J., Braestrup, C., Christensen, A. V., and Magelund, G., 1979, *GABA-Neurotransmitters* (P. Krogsgaard-Larsen, J. Scheel-Kruger, and H. Kofod, eds.), Munksgaard, Copenhagen, pp. 447–464.
77. Waddington, J. L., and Cross, A. J., 1979, *Naunyn Schmiedebergs Arch. Pharmacol.* **306**:275–280.
78. Delini-Stula, A., 1979, *GABA-Neurotransmitters* (P. Krogsgaard-Larsen, J. Scheel-Kruger, and H. Kofod, eds.), Munksgaard, Copenhagen, pp. 482–499.
79. Waddington, J. L., 1978, *Eur. J. Pharmacol.* **51**:417–422.
80. Hill, R. C., Maurer, R., Buescher, H.-H., and Roemer, D., 1981, *Eur. J. Pharmacol.* **69**:221–224.
81. Johnston, G. A. R., Skerritt, J. H., and Willow, M., 1982, *Problems in GABA Research* (Y. Okada and E. Roberts, eds.), Excerpta Medica, Amsterdam, pp. 293–301.
82. DeFeudis, F. V., 1981, *Neurochem. Int.* **3**:113–122.
83. Krogsgaard-Larsen, P., Hjeds, H., Curtis, D. R., Lodge, D., and Johnston, G. A. R., 1979, *J. Neurochem.* **32**:1717–1724.
84. Spano, P. F., Riccardi, F., Kobayashi, H., Memo, M., and Trabucchi, M., 1981, *Abstracts, Eighth International Congress of Pharmacology*, Japanese Pharmacological Society, Tokyo, p. 307.
85. Matus, A., Pehling, G., and Wilkinson, D., 1981, *J. Neurobiol.* **12**:67–73.
86. DeFeudis, F. V., 1980, *Neuroscience* **5**:675–688.
87. Palacios, J. M., Wamsley, J. K., and Kuhar, M. J., 1981, *Brain Res.* **222**:285–307.
88. Morin, A. M., and Wasterlain, C. G., 1980, *Life Sci.* **26**:1239–1245.
89. Krogsgaard-Larsen, P., Snowman, A., Lummis, S. C., and Olsen, R. W., 1981, *J. Neurochem.* **37**:401–409.
90. Falch, E., and Krogsgaard-Larsen, P., 1982, *J. Neurochem.* **38**:1123–1129.
91. Hill, D. R., and Bowery, N. G., 1981, *Nature* **290**:149–152.
92. Molher, H., and Okada, T., 1978, *Mol. Pharmacol.* **14**:256–265.

93. Bhatacharyya, A., Madyastha, K. M., Bhattacharyya, P. K., and Devanandan, M. S., 1981, *Biochim. Biophys. Acta* **98**:520–526.
94. Olsen, R. W., Ticku, M. K., Greenlee, D., and Van Ness, P., 1979, *GABA-Neurotransmitters* (P. Krogsgaard-Larsen, J. Scheel-Kruger, and H. Kofod, eds.), Munksgaard, Copenhagen, pp. 165–178.
95. Olsen, R. W., and Leep-Lundberg, F. L., 1980, *Eur. J. Pharmacol.* **65**:101–104.
96. Johnston, G. A. R., Kennedy, S. M. E., and Lodge, D., 1978, *J. Neurochem.* **31**:1519–1523.
97. Enna, S. J., and Snyder, S. H., 1975, *Brain Res.* **100**:81–97.
98. Enna, S. J., and Snyder, S. H., 1977, *Mol. Pharmac.* **13**:442–453.
99. Horng, J. S., and Wong, D. T., 1979, *J. Neurochem.* **32**:1379–1386.
100. Johnston, G. A. R., and Kennedy, S. M. E., 1978, *Amino Acids as Chemical Transmitters* (F. Fonnum, ed.), Plenum Press, New York, pp. 507–516.
101. Fiszer, S., and DeRobertis, E., 1967, *Brain Res.* **5**:31–44.
102. Lester, B. R., Miller, A. L., and Peck, E. J., 1981, *J. Neurochem.* **36**:154–164.
103. Redburn, D. A., and Mitchell, C. K., 1981, *Life Sci.* **28**:541–549.
104. Mazzari, S., Massotti, A., Guidotti, A., and Costa, E., 1981, *GABA and Benzodiazepine Receptors* (E. Costa, G. DiChiara, and G. L. Gessa, eds.), Raven Press, New York, pp. 1–8.
105. Greenlee, D. V., and Olsen, R. W., 1979, *Biochem. Biophys, Res. Commun.* **88**:380–387.
106. DeFeudis, F. V., and Somoza, E., 1977, *Gen. Pharmacol.* **8**:181–187.
107. Bowery, N. G., Hill, D. R., Hudson, A. L., Perkins, M. N., and Stone, T. W., 1982, *Br. J. Pharmacol.* **75**:47P.
108. Tunnicliff, G., and Smith, J. A., 1981, *J. Neurochem.* **36**:1122–1126.
109. Gardner, C. R., Klein, J., and Grove, J., 1981, *Eur. J. Pharmacol.* **75**:83–92.
110. Olsen, R. W., Bergman, M. O., Van Ness, P. C., Lummis, S. C., Watkins, A. E., Napias, C., and Greenlee, D. V., 1981, *Mol. Pharmacol.* **19**:217–227.
111. Browner, M., Ferkany, J. W., and Enna, S. J., 1981, *J. Neurosci.* **1**:514–518.
112. Skerritt, J. H., Willow, M., and Johnston, G. A. R., 1982, *Neurosci. Lett.* **29**:63–66.
113. Waddington, J. L., and Cross, A. J., 1978, *Nature* **276**:618–620.
114. Napias, C., Bergman, M. O., Van Ness, P. C., Greenlee, D. V., and Olsen, R. W., 1980, *Life Sci.* **27**:1001–1011.
115. Elliott, K. A. C., and Van Gelder, N. M., 1958, *J. Neurochem.* **3**:28–40.
116. Skerritt, J. H., Trisdikoon, P., and Johnston, G. A. R., 1981, *Brain Res.* **215**:398–401.
117. Loh, H. H., and Low, P. Y., 1980, *Annu. Rev. Pharmacol.* **20**:201–234.
118. Watkins, J. C., 1965, *J. Theor. Biol.* **9**:37–50.
119. Giambalvo, C., and Rosenberg, P., 1976, *Biochim. Biophys. Acta* **436**:741–756.
120. Lloyd, K. G., and Davidson, L., 1979, *Science* **205**:1147–1149.
121. Olsen, R. W., Van Ness, P., Napias, C., Bergman, M., and Tourtellotte, W. W., 1980, *Receptors for Neurotransmitters and Peptide Hormones* (G. Pepeu, M. J. Kuhar, and S. J. Enna, eds.), Raven Press, New York, pp. 451–460.
122. Toffano, G., Aldino, C., Balzano, M., Leon, A., and Savoini, G., 1981, *Brain Res.* **222**:95–102.
123. Ebadi, M., and Chweh, A., 1980, *Neuropharmacology* **19**:1105–1111.
124. Guidotti, A., Toffano, G., and Costa, E., 1978, *Nature* **275**:553–555.
125. Ebstein, B., Guidotti, A., and Costa, E., 1982, *Problems in GABA Research* (Y. Okada and E. Roberts, eds.), Excerpta Medica, Amsterdam, pp. 348–354.
126. Nagy, J., Kardos, J., Maksay, G., and Simonyi, M., 1981, *Neuropharmacology* **20**:529–533.
127. Johnston, G. A. R., and Kennedy, S. M. E., 1978, *Clin. Exp. Pharmacol. Physiol.* **6**:686–687.
128. Yoneda, Y., and Kuriyama, K., 1980, *Nature* **285**:670–673.
129. Ticku, M. K., and Burch, T., 1980, *Biochem. Pharmacol.* **29**:1217–1220.
130. Skerritt, J. H., and Johnston, G. A. R., 1982, *Dev. Neurosci.* **5**:189–197.
131. Aldinio, C., Balzano, M. A., and Toffano, G., 1980, *Pharmacol. Res. Commun.* **12**:495–500.
132. Candy, J. M., and Martin, I. L., 1979, *J. Neurochem.* **32**:655–658.
133. Mallorga, P., Hamburg, M., Tallman, J. F., and Gallager, D. W., 1980, *Neuropharmacology* **19**:405–408.

134. Mann, E., and Enna, S. J., 1980, *Brain Res.* **184**:367–373.
135. Braestrup, C., and Nielsen, M., 1978, *Brain Res.* **147**:170–173.
136. Ticku, M. J., 1980, *Br. J. Pharmacol.* **70**:403–410.
137. Enna, S. J., Ferkany, J. W., and Strong, R., 1980, *Receptors for Neurotransmitters and Peptide Hormones* (G. Pepeu, M. J., Kuhar, and S. J. Enna, ed.), Raven Press, New York, pp. 253–263.
138. Ticku, M. J., and Huffman, R. D., 1980, *Eur. J. Pharmacol.* **68**:97–106.
139. Mohler, H., Okada, T., and Enna, S. J., 1978, *Brain Res.* **156**:391–395.
140. Willow, M., and Johnston, G. A. R., 1981, *J. Neurosci.* **1**:364–367.
141. Willow, M., and Johnston, G. A. R., 1981, *Neurosci. Lett.* **23**:71–74.
142. Willow, M., and Johnston, G. A. R., 1981, *J. Neurochem.* **37**:1291–1294.
143. Willow, M., Morgan, I. G., and Johnston, G. A. R., 1981, *Neurosci. Lett.* **24**:301–306.
144. Tallman, J. F., Thomas, J. W., and Gallager, D. W., 1978, *Nature* **274**:383–385.
145. Braestrup, C., Neilson, M., Krogsgaard-Larsen, P., and Falch, E., 1980, *Nature* **280**:331–333.
146. Garvish, M., and Snyder, S. H., 1979, *Life Sci.* **26**:579–582.
147. Massotti, M., and Guidotti, A., 1980, *Life Sci.* **27**:847–854.
148. Mohler, H., 1981, *Trends Pharmacol. Sci.* **2**:116–119.

Excitatory Amino Acid Receptors

Najam A. Sharif

1. INTRODUCTION

Despite limitations imposed by their ubiquitous occurrence, universal neuronal excitant properties, and involvement in metabolic functions, glutamic (Glu) and aspartic (Asp) acids now satisfy many of the basic criteria employed to assess and ascribe a transmitter role to neuroactive substances.[1-4] The dicarboxylate amino acids (AAs) exhibit a heterogeneous distribution in the mammalian central nervous system (CNS), being enriched, together with their synthetic enzymes, within synaptosomal fractions. Release of endogenous and exogenous AAs from CNS preparations by depolarizing stimuli has been demonstrated. Iontophoresed Glu, Asp, and their analogs produce neuronal excitation, mimicking the responses evoked by stimulation of excitatory pathways, and their actions are terminated by specific uptake mechanisms. Furthermore, specific synaptic binding proteins and receptor-mediated ionic fluxes and cyclic nucleotide generation associated with Glu and Asp have been detected and extensively evaluated.[2-4]

The neuroactivity of putative transmitter agents was previously assessed by iontophoretic procedures. Because of inherent methodological problems, inability to predict concentration and precise locus (pre- or postsynaptic) of applied drugs, their usefulness has been limited to yielding only qualitative information. However, the concept of multiple excitatory AA receptors has primarily originated from these types of studies. More recently, the advent and commercial availability of radioactive derivatives of putative neurotransmitters has allowed a more quantitative appraisal of the identity, specificity, and localization of central receptors by *in vitro* ligand-binding techniques. This chapter reviews the status of such radioreceptor studies for Glu, Asp, cysteine sulfinate, and kainic acid, attempting to reconcile binding properties with criteria employed for receptor identification.[5] Some other aspects of AA physiology relevant to receptor function are also discussed.

Najam A. Sharif • Department of Pharmacology and Experimental Therapeutics, University of Maryland School of Medicine, Baltimore, Maryland 21201. *Present address:* Department of Biochemistry, University of Nottingham, Queen's Medical Centre, Nottingham, NG7 2UH England.

2. RECEPTOR-LABELING STUDIES

2.1. Binding Methodology and Interpretation

At a very simplistic level, the technique of radioreceptor labeling involves incubation of radioisotopic derivatives of the putative neurotransmitter or its agonist/antagonist analogues with cellular fractions of target organs known to receive the specific transmitter innervation. To limit binding occurring to non-receptor sites, high-affinity ligands labeled to high specific activity are utilized. The "nonspecific" component, which often represents interaction of the ligand with membrane surfaces by hydrophobic attractions, adsorption, etc., can be resolved from the "specific" receptor binding by addition of excess unlabeled ligand to some assay tubes ("blanks") and the assay taken to equilibrium and terminated.

The specific binding of the radioligand must undergo rigorous assessment according to certain criteria[5] before one can equate a binding site with a physiological receptor. For instance, the receptor population must display high specific binding and be associated with a finite number of sites of affinity in accord with the likely synaptic concentration of the ligand. The binding should be located and concentrated in subcellular fractions containing synaptic elements and above all be displaced readily and reversibly by ligand analogues with a potency and pharmacological specificity similar to that observed in physiological responses.

Binding studies yield kinetic parameters for ligand–receptor interactions such as affinity constant (K_d) and receptor number (B_{max}). Further manipulation of the saturation isotherms allows the detection of dynamic aspects of ligand–protein binding such as cooperativity and heterogeneity by Hill and Scatchard plots, respectively.[6,7] In addition, quantification of analogue affinity for binding in competition experiments (K_i, IC_{50} values) can help clarify the structural requirements of binding sites and can eventually lead to the development of therapeutically useful drugs.

However, despite these advantages of binding studies, one must be wary of some inherent problems, such as artifactual binding to nonprotein materials.[8] Secondly, it is not often possible to distinguish ligand binding occuring to pre- and postsynaptic sites unless specific deafferentation or neuronal depletion studies are conducted concurrently. Moreover, binding assays cannot determine whether a ligand–site interaction represents activation of a specific agonist or antagonist (or combination) conformation of the receptor unless labeled derivatives of both types of analogues are employed. This is an obvious drawback for excitatory AA studies since "selective" antagonists[4,9] are not yet readily available in radioactive form.

2.2. Optimization of Conditions for Excitatory Amino Acid Binding

Since the synaptic properties of excitatory AAs differ vastly from those of other transmitters, conditions for demonstration of their receptor binding must be outlined briefly.

Although degradative and converting enzymes exist for the inactivation of numerous transmitters, the actions of the AAs Glu, Asp, and γ-aminobutyric acid (GABA) are terminated by Na$^+$-dependent uptake systems.[10] Therefore, to surmount the problems posed by possible interference from binding to enzyme and transport sites, binding assays for radiolabeled AAs are conducted in Na$^+$-free media buffered to physiological pH. However, routine addition of peptidase, Glu uptake, and Glu-metabolizing enzyme inhibitors to incubation buffer is not recommended as a means for eliminating[^{3}H]Glu binding to non-receptor sites because these purported blockers possess some agonistic activity and thus reduce the overall receptor binding (N. A. Sharif and P. J. Roberts, unpublished data).

In vitro preparations for study have included tissue slices, slide-mounted tissue sections,[11] synaptic plasma membranes,[12-22] synaptic junctional complexes (SJC),[13] and postsynaptic densities (PSD).[14] To eliminate possible interference from the high endogenous concentrations of AAs, membrane preparations are subjected to vigorous washing procedures. Thus, specific [^{3}H]GABA binding has been maximized by mild detergent treatment,[15] and by repeated freeze-thawing and washings.[16,17] To optimize excitatory AA receptor binding, mild sonication, preincubation (37°C for 30 min), and extensive washings of membranes were necessary to remove inhibitory endogenous AAs, cyclic nucleotides, and other unidentified substances.[18,20] Osmotic lysis by freeze-thawing was found to be detrimental, resulting in drastically reduced binding capacity.[18-21]

Since acidic AA binding attains equilibrium within 5–20 min at 37°C and is reversible ($t_{1/2} \simeq 30$ s),[19] a rapid centrifugation method is employed to separate the bound from free radioactivity. Recently, filtration under vacuum through Millipore® cellulose or Whatman GF/B® filters has been successfully and reproducibly utilized to harvest receptor-bound label.[22,23] Superficial rinsing of the pelleted and filtered membranes helps to remove adsorbed radioactivity and thus reduces nonspecific binding.

Such optimization of conditions for tissue preparation and for performing radioreceptor labeling experiments ensures maximization of sensitivity of the binding assay and allows a more accurate determination of excitatory AA receptor properties. Further details regarding relevant methodological problems and assay procedures are discussed exhaustively in a recent review.[21]

3. CEREBRAL GLUTAMATE RECEPTORS

3.1. Binding Characteristics

Glutamate may exert its multitudinous neurotransmitter functions by a combination of enhancement of cationic permeability[2,24] and elevation of intracellular nucleotide levels.[25-27] The expression of these physiological responses, manifested as depolarization and neuronal excitation,[2-4,28] must therefore be mediated via specific membrane-bound Glu receptors located on neuronal perikarya. The biochemical demonstration of these putative receptor sites has become one of the most prominent goals of excitatory AA neuro-chemistry ever since the initial characterization of acidic AA electrophysiology.[2-4]

Receptor labeling is usually best accomplished by employment of radiol-abeled antagonists, because of their greater affinity. In the absence of potent and specific pharmacological antagonists for excitatory AAs (a situation par-tially remedied recently[4,9]), the detection and characterization of acidic AA binding sites have relied exclusively on agonist binding. Problems associated with agonist-binding include the faster dissociation of rates of agonists and their possible enzymic metabolism.

Early studies of Glu binding sites in rat brain employed low-specific-ac-tivity L-[^{14}C]Glu (250–300 mCi/mmol). Roberts[29] reported a single class of low-affinity [^{14}C]Glu recognition sites on rat cortical membranes (K_d = 0.8 μM, B_{max} = 30 pmol/mg protein). Michaelis et al.,[30] however, observed multiple components of stereospecific [^{14}C]Glu binding in whole brain membranes (K_d = 0.2 μM, B_{max} = 1.8 nmol/mg; K_d = 4.4 μM, B_{max} = 8.8 nmol/mg) and subsequently showed these properties to be retained by an isolated glycopro-tein.[31] Rat-cortex-derived proteolipid fractions exhibited triphasic binding is-otherms (K_ds ranged from 0.3 to 55 μM, B_{max} ranged from 0.5 to 166 nmol/mg protein); the high density of binding sites probably represented sequestering of the label within lipids.[32]

The presence of stereoselective sites for L-[^{3}H]Glu (K_d = 0.74 μM, B_{max} = 73 pmol/mg) on fresh cerebellar membranes was first described in 1978.[33] Following optimization of membrane preparation, which involved sonication, preincubation, and thorough washing, Sharif and Roberts[18,19] demonstrated a substantial enhancement of binding affinity and capacity (K_d = 0.36 μM, B_{max} = 117 pmol/mg protein). Sodium-independent binding to cerebellar membranes attained equilibrium within 10 min and was optimal at pH 7.1 and 37°C in buffered Tris medium. Subcellular fractionation revealed maximal binding ac-tivity to be associated with the synaptosomal elements. This activity repre-sented interaction with a homogeneous class of noncooperative binding sites, which displayed a heterogeneous distribution. Greatest enrichment of sites was within central regions, specifically, the cerebellum, cortex, and striatum, con-sonant with known glutamatergic innervation of these structures.

Striatal homogenates prepared from previously frozen tissue exhibited [^{3}H]Glu binding components of surprisingly high affinity (K_ds = 11–80 nM)[34]; and Na$^+$-independent assays failed to reveal any specific binding following freeze-thawing of the preparation.[35] These two latter studies emphasize the danger of utilizing frozen preparations for studying excitatory AA receptors, and this aspect is discussed in detail in Section 7.3.

Two distinct [^{3}H]Glu binding sites have been detected in hippocampal membranes[22,36] and differentiated by their requirement for Na$^+$ and differing kinetic and pharmacological properties. The Na$^+$-independent component, rep-resenting postsynaptic sites, possessed slower rates of association (K_1 = 0.5 vs. 1.27 μM^{-1} min^{-1}) and dissociation (K_{-1} = 0.65 vs. 1.3 min^{-1}) than the Na$^+$-dependent binding to uptake sites. The latter premise was further sub-stantiated pharmacologically, since inhibitors of receptor binding were poor displacers of Na^+-dependent binding and *vice versa*.[22,29] Most recently, Werling and Nadler,[37] in contrast to previous studies,[22,36] found two [^{3}H]Glu binding sites in the absence of Na^+ on hippocampal membranes (K_{d1} = 0.011 μM, B_{max1} = 2.5 pmol/mg protein; K_{d2} = 0.57 μM, B_{max2} = 47 pmol/mg protein). In addition, they reported that binding activity was potentiated by low concentrations of Glu analogues (e.g., 1–50 nM L-homocysteate). Such cooperative interactions between excitatory AA analogues at the Glu receptor are analogous to the situation reported for the invertebrate neuromuscular junction[38] and therefore merit further investigation, especially since the biochemical correlates of long-term potentiation (LTP)[39,40] may involve such a phenomenon.

Although extensive studies of excitatory AA receptor binding to neuronal elements of mammalian species other than the rat have not yet appeared, apparent high-affinity, Na^+-independent [^{3}H]Glu binding sites responsive to antagonism by 2-aminophosphonobutyrate (2-APB) have been identified in bovine retina (K_ds ≈ 0.01 and 0.8 μM)[41] and feline cerebellar membranes (K_ds ≈ 0.33 and 1.8 μM).[42]

Any ligand binding detected in crude synaptic membranes must ultimately be related to a specific interaction with synaptic receptor proteins in order to attach physiological significance to the binding. Accordingly, Cotman *et al.*[13] have demonstrated a substantial enrichment of labeled acidic AA binding sites on synaptic junctional complexes (SJC) isolated by detergent solubilization of plasma membranes. A subsequent analysis of the binding kinetics using double-label techniques has revealed a distinct linear Scatchard plot for [^{3}H]Glu, a finding compatible with the prevalance of one class of high-affinity Glu receptors on SJC derived from rat forebrain (K_d = 0.45 μM, B_{max} = 91 pmol/mg protein).[43] Although these equilibrium data are in excellent agreement with those obtained from brain synaptic membranes[18–22,37] and thus provide compelling evidence for labeling of the same Glu binding sites in these preparations, other functional aspects, such as tissue and pharmacological specificity and reconstitution into liposomes, have not yet been explored. Of course the low yield of purified SJC is an extremely limiting factor here.

In conclusion, numerous reports describing binding of radioactive Glu to *in vitro* mammalian CNS preparations have been published since the mid-1970s. Unfortunately, by virtue of the adoption of dissimilar preparative (fresh *vs.* frozen; homogenates *vs.* synaptic membranes *vs.* SJC or PSD) and assay (microcentrifugation *vs.* filtration) procedures, quite complex and variable equilibrium kinetics have been documented, which have posed numerous interpretative problems. Thus, the erroneously high-affinity (K_d = 10–90 nM)[34] [^{3}H]Glu binding sites detected seem nonphysiological, since these components would be permanently saturated under normal *in vivo* conditions, where synaptic Glu levels may reach as high as 1 mM following release from the presynaptic terminals. This possible artifact could have resulted from the use of frozen tissue homogenates and employment of a narrow range of [^{3}H]Glu concentrations in saturation experiments. Similarly, binding of [^{14}C]Glu to lipid moieties and uptake sites could have accounted for the exceedingly high density of binding sites found on organically extracted proteolipids.[32]

Incidentally, it must also be borne in mind that because of the liberation of substantial quantities of endogenous L-Glu and other acidic AAs into the homogenization media, binding studies conducted on homogenates and/or inadequately washed crude membranes (perhaps not even sonicated or preincubated)[18] may be measuring a somewhat "saturated" form of the Glu receptor; hence, the documentation of relatively low-affinity binding sites ($K_d \simeq 0.8$ μM) in early studies (Table I).[29-33] Invariably, this treatment would affect the masking of high-affinity binding sites and, because of the presence of high levels of extracellular sodium, would also tend to favor Glu binding to uptake sites. This would result in an artifactual overestimate of binding site capacity and a reduced affinity. Therefore, future studies of excitatory AA binding would greatly benefit from application of common, optimized tissue preparation and assay methodologies.

3.2. Pharmacology of [^{3}H]Glutamate Binding

The involvement of specific receptors in binding measurements can usually be inferred (in part) providing putative neurotransmitter analogues displace ligand binding with a profile of activity similar to that required for eliciting physiological responses.

The pharmacological specificity of [^{3}H]Glu binding is essentially uniform throughout the different CNS regions, suggesting a possible homogeneity of binding sites. Of all the neuroexcitants tested,[33] L-Glu was the most potent displacer of Na$^+$-independent binding (IC$_{50}$ = 4.8 μM) followed closely by its rigid analogues ibotenate and DL-quisqualate (Quis) (IC$_{50}$ = 8.1, 8.4 μM, respectively) and then DL-homocysteate. Curiously, D-Glu was six times less potent than L-Glu, and kainate (KA) was devoid of any activity, even though both of these compounds are at least equipotent with L-Glu at depolarizing neurons.[46] However, increasing evidence (Section 4) suggests autonomous sites of action of these molecules (KA and L-Glu). Antagonists suspected of differential activity at the "glutamate-preferring" receptors, such as 2-amino-4-phosphonobutyrate (2-APB),[47] and (±)2-amino-7-phosphonoheptanoate (APH),[9] were good displacers although some 7–8 times less potent than L-Glu,[48] whereas Glu-diethylester (GDEE) and α-aminoadipate showed minimal affinity for the binding. Enhancement of antagonist (2-APB) potency in the presence of 2.5 mM Ca^{2+} in electrophysiological[49] and binding systems[50] (N. A. Sharif and P. J. Roberts, unpublished data) is extremely intriguing and may involve the modulatory properties of calmodulin[51] at the receptor site. Interestingly, the phosphonate analogues probably provide the best correlation among *in vivo* field potential inhibition,[49] *in vitro* cyclic GMP production, and receptor binding displacement activities.[48,50] The inactivity of enzyme and uptake blockers[52] against [^{3}H]Glu binding has supported the involvement of receptors in these studies.

Among the recently synthesized ibotenate derivates,[53] only homoibotenate exhibited any inhibitory activity (IC$_{50}$ = 7.5 μM) at competing for cerebellar [^{3}H]Glu binding.[54] The 5-methyl-4-isoxazol proprionate analogue (AMPA), a powerful neuronal excitant, failed to displace the binding.[54] These apparent disparities between the electrophysiological and *in vitro* binding studies have been difficult to resolve and may arise as a consequence of dissimilarities in

Table I
Selected Glutamate Binding Studies

Rat brain region	Tissue preparation	Assay conditions	Radioligand employed	K_d (μM)	B_{max} (pmol/mg protein)	Reference
Whole brain	Crude membranes	Na$^+$-free, filtration	L-[^{14}C]Glu (0.1–30 μM)	(a) 0.18 (b) 2.10	(a) 4,440 (b) —	30
Cerebral cortex	Crude membranes	Na$^+$-free, filtration	L-[^{14}C]Glu (0.4–8.7 μM)	(a) 4.0 (b) 8.3	(a) 200 (b) 28	29
Cerebral cortex	Proteolipid fractions	Sephadex chromatography	L-[^{14}C]Glu (0.06–250 μM)	(a) 0.3 (b) 5.0 (c) 55.0	(a) 530 (b) 32,000 (c) 166,000	32
Cerebral cortex	Crude membranes	±Na$^+$, centrifugation	L-[^{3}H]Glu (0.005–1.6 μM)	(a) 1.34 (b) 0.37	(a) 210 (+Na$^+$) (b) 8.4 (−Na$^+$)	44
Cerebellum	Synaptic membranes	Na$^+$-free, centrifugation	L-[^{3}H]Glu (0.001–1.8 μM)	(a) 0.74	(a) 73	33
Cerebellum	Synaptic membranes	Na$^+$-free, centrifugation (sonicated + preincubated)	L-[^{3}H]Glu (0.001–1.8 μM)	(a) 0.36	(a) 117	18
Striatum	Crude membranes	Na$^+$-free, centrifugation	L-[^{3}H]Glu (0.001–1 μM)	(a) 0.68	(a) 70	45
Hippocampus	Synaptic membranes	±Na$^+$, filtration	L][^{3}H]Glu (0.05–10 μM)	(a) 2.40 (b) 0.77	(a) 75 (+Na$^+$) (b) 6.5 (−Na$^+$)	22
Hippocampus	Synaptic membranes	Na$^+$-free, centrifugation	L-[^{3}H]Glu (0.001–1 μM)	(a) 0.011 (b) 0.57	(a) 2.5 (b) 47.0	37
Forebrain	SJCs	Na$^+$-free, centrifugation	L-[^{3}H]Glu (0.01–1 μM)	(a) 0.45	(a) 91	43

preparations, ionic environment, uptake resistance, and even lipid solubilities of compounds. These areas therefore invite further investigation and other suitable explanations. However, the general pharmacological specificity (barring a few anomalies) has strengthened the contention that under Na^+-free conditions [^{3}H]Glu interacts with a postsynaptic receptor protein.

Excitatory AAs are envisaged to interact with the receptor recognition template by a three-point attachment via two anionic and a single cationic group.[4,55] However, although it was previously thought that the "glutamate-preferring" receptor mainly accepts molecules in an extended conformation,[4,55] recent iontophoretic evidence[56] points to a converse conclusion. Thus, compounds capable of assuming a folded state (e.g., L-Glu, Quis, AMPA) preferrentially bind to this Glu receptor conformation; i.e., the determining factor for receptor preference is the distance between the cationic amino group and the distal anionic residue of the molecule. It is hoped that more detailed structure–activity studies will allow resolution of this dilemma and help to define the steric requirements of these receptor sites.

The underlying mechanism behind the remarkable actions of $CaCl_2$ on excitatory AA receptor systems[40,43,49,50] presents an extremely interesting puzzle. A feasible explanation may be as follows: partial cyclization of acidic AAs, accomplished by chelation of Ca^{2+} ions by the anionic groups of the compounds, would theoretically prime the ligands for receptor binding.[56] Depletion of Ca^{2+} from the ligand–Ca^{2+} complexes by high-affinity calmodulin[43,51] would activate the latter[51] and help to concentrate the ligands at the postsynaptic site. In addition, the Ca^{2+}-induced activation of calmodulin may subsequently lead to favorable conformational changes, perhaps even proteolysis,[57–59] and result in the exposure of new binding sites. Thus, if the ligand is [^{3}H]Glu, an enhanced specific [^{3}H]Glu binding would be observed; if the ligand is a competitive antagonist (e.g., 2-APB), then a greater inhibition of radioligand binding[50] and suppression of synaptically or AA-evoked excitations[49] *in vivo* would ensue. A possible alternative may be that Ca^{2+} complexes with the endogenous inhibitor agents,[18,60,61] disinhibits the receptor, and consequently allows a greater access of the ligands to the previously occluded recognition site. Obviously, experimental pursuit of other suitable explanations for these Ca^{2+} phenomena is desirable, and it is hoped that speculations of this type may promote further research in this area and help to answer some of these outstanding questions.

3.3. Regulation of Glutamate Receptors

The possible modulation of ligand binding by a variety of exogenous agents and/or treatments can often reveal regulatory mechanisms that may operate *in vivo* and thus further underline the possible physiological relevance of that receptor system.

In numerous cyclase–receptor coupled systems, guanine nucleotides (GTP, GDP) stimulate cyclase activity but attenuate agonist–receptor interaction.[25,62] In support of this seemingly universal phenomenon, we have recently demonstrated a similar "guanine-specific" inhibition of cerebellar [^{3}H]Glu binding.[20,63] Sodium chloride displayed a biphasic effect: low concentrations (0.1–10 mM) produced 23–50% inhibition, whereas higher levels markedly stimulated binding (215% at 100 mM). The inhibitory action of Na^+ and guanine nucleotides was mediated by an apparent 12-fold reduction in binding affinity, with receptor density remaining relatively unchanged.[20] The sodium effect is in accord with previous observations for hippocampal[36] and cerebellar binding[18] and probably represents protein folding and/or interaction at the receptor–ionophore complex.[24,36] The guanine nucleotide modulation is likely to be directed through a regulatory protein linked to the receptor and a cyclase moiety,[62] and their influence exerted via acceleration of the dissociation of the AA from its recognition site, as evidenced by a transient saturability of the binding.[20] The dramatic elevation of cyclic GMP levels by exogenous glutamate and their reduction by purported antagonists[64] imply the prevalence of a guanylate-cyclase-coupled Glu receptor system in the cerebellum and lends credibility to the guanine nucleotide modulation of [³H]Glu binding. Although more definitive data are required to make this hypothesis even more convincing, such nucleotide modulation of ligand binding may represent negative feedback mechanisms *in vivo*.

Physiological concentrations of Ca^{2+} (0.01–1.25 mM) and other divalent cations enhanced hippocampal [³H]Glu binding[36] and cerebellar [³H]Asp binding,[19] with a progressive inhibition occurring at higher levels. Interestingly, potentiated synaptic activity following repetitive stimulation (LTP) has led to a duplication of the Ca^{2+}-induced effect and greatly increased capacity of [³H]Glu binding sites in membranes prepared from hippocampal slices exposed to LTP[39,65] has been observed. The proposal that a specific Ca^{2+}-dependent thiol protease activated by endogenously released Ca^{2+} may be responsible for the unmasking of latent receptors[57,58,66] has been supported somewhat by the finding that the Ca^{2+}-dependent proteolysis and the enhanced binding were competitively antagonized by leupeptin, a cysteine proteinase inhibitor. In addition, the increased binding sites detected following preincubation,[18] their cold lability,[18] and the sharp pH and temperature optimi of binding all implicate a possible enzyme-mediated stimulation of [³H]Glu binding and/or the inactivation of an endogenous inhibitor. The possible existence of the latter has been demonstrated by a 67% reduction in [³H]Glu binding on readdition of postincubation supernatant (devoid of acidic AAs) to control cerebellar membranes.[18,20] The pronounced augmentation of binding following this "heat activation" has now been confirmed in the cerebellum[54] and hippocampus[37,59] and may involve the removal of inhibitory nucleotides, phospholipids, and perhaps a protein, as has been claimed for the GABA system.[16] Evidently Michaelis *et al.*[60] have extracted a thermostable, nondialyzable binding inhibitor by cholate treatment of membranes and subsequently showed its properties to be mimicked by exogenously added bovine gangliosides. Therefore, the unidentified endogenous inhibitor[18] may be a phospholipid.

It must be mentioned in passing that although LTP has not been detected in the cerebellum, the enhanced [³H]Glu binding induced by ultrasonication and lyophilization[18,21] may reflect a similar type of receptor/membrane modification as that produced by high-frequency stimulation of hippocampal pathways.[39,40,57] Therefore, investigations of possible neuronal plasticity in the cerebellum may also be warranted, especially since hyperactivity of intrinsic and/or extrinsic excitatory projections[67] of this structure may be able to induce sustained neuronal activity and cause convulsions.

In vivo regulation of receptors can be manifested as an increase or decrease in their availability and affinity. Changes in normal cerebral function are often affected by chronic drug administration[68] and/or by alteration of the status of presynaptic elements.[69] In this context, we have described the development of Glu receptor supersensitivity in the rat striatum following deafferentation of the glutamatergic corticostriatal projection[45]: a 34% increase in [^{3}H]Glu binding was observed.

The age-related ontogeny of cerebellar [^{3}H]Glu binding sites has been correlated with the concurrent development of glutamatergic terminals and the onset of kainate toxicity,[70] thus highlighting another facet of *in vivo* Glu receptor regulation.

3.4. Chemical Modification Studies

The interaction of [^{3}H]Glu with a protein moiety is most likely in view of the marked pH and temperature sensitivity of the binding. To date, however, little information is available as to the functional groups involved and the potential importance of lipid–protein interactions at the Glu recognition site. To this end we have examined the effect of covalent modifications and membrane perturbations of cerebellar synaptic membranes on their ability to bind [^{3}H]Glu.[71,72] Dithiothreitol-induced (DTT) reduction of disulfide bonds and alkylation of thiol residues by N-ethylmaleimide (NEM) or iodoacetamide and their oxidation by 5,5'-dithio-bis(2-nitrobenzoic acid) (DTNB) all produced substantial inactivation of [^{3}H]Glu binding in a time-, temperature-, and concentration-dependent manner.[71,73] These inhibitions reflected decrements in receptor density with affinity parameters remaining relatively unchanged, as has been demonstrated for muscarinic[74] and dopamine receptors.[75] The reductive effects of DTT could be reversed by sequential treatment of membranes with stoichiometric amounts of NEM and DTNB, with consequential regeneration of disulfide bridges with the latter agent.[71] Pretreatment of membranes with low concentrations of L-Glu before exposure to sulfhydryl reagents afforded a dose-related protection against modification (e.g., 17 and 69% at 50 μM and 1 mM L-Glu against 10 mM DTT).[73] Protection against DTNB was relatively more difficult to achieve because of its rather harsh oxidizing properties.

Treatment of membranes with proteolytic and lipolytic enzymes, nondegradative lipid perturbants, and solubilization agents all resulted in drastically reduced [^{3}H]Glu binding, indicating the importance of lipids here.[59,71]

Studies of this type have indicated the presence of essential disulfide bonds and thiol groups at or in very close proximity to the Glu binding protein (as evidenced by the protective effects of L-Glu) and emphasized the importance of protein–lipid cooperation for maintenance of a stable conformation of this receptor complex. Recently, Michaelis[76] proposed the existence of an iron–sulfur center associated with the Glu binding site on the basis of azide-induced inhibition of binding. The physiological implications and relevance of this discovery are unclear at present, but it may have some bearing on the possible purification of this moiety.

Specific [³H]Glu binding is adversely effected by detergent treatments, as documented for cerebellar,[59] hippocampal,[59] and brain membranes.[31] Binding could be enhanced, however, provided the treated membranes were thoroughly washed[59] or if soluble fractions[31] were employed in assays. With affinity chromatography, a detergent-solubilized brain glycoprotein has been isolated and partially purified[31] and shown to be responsive to azides and gangliosides. Attempts to reconstitute this glycoprotein into lipid bilayers have been unsuccessful so far, resulting in altered specificity and lowered [³H]Glu binding capacity.[77] However, future experiments with purified SJCs[13] and these glycoproteins ought to be fruitful in unraveling the chemical composition and structure of the Glu binding protein.

4. KAINATE BINDING

Kainic acid (KA) is a rigid analogue of glutamate. Although very early iontophoretic studies implicated a common excitatory receptor system for Glu, quisqualate (Quis), and KA, overwhelming multidisciplinary evidence now argues against this hypothesis. For instance, Quis- (or Glu-) but not KA-induced depolarizations of feline spinal neurons are blocked by GDEE,[78,79] with the depeptide γ-D-glutamylglycine displaying the opposite specificity.[4] Separate loci of action for KA and Glu have been further invoked by the potentiation effects of KA on Glu-induced responses in cortex[38] and at the invertebrate neuromuscular junction,[80] at promoting striatal [³H]dopamine release,[81] and at stimulating cerebellar cyclic GMP synthesis.[26] The ontogeny of cerebellar KA and Glu receptors also differs markedly.[82] Additional information concerning the autonomous nature of dicarboxylate receptors has emerged from [³H]KA binding studies.

Although specific, saturable, and reversible Na^+-independent binding of [³H]KA to neuronal membranes of several species has been demonstrated,[35,83–86] the assay conditions employed (2–4°C for 1 h) were quite dissimilar to those for [³H]Glu binding (see Sections 2 and 3). Two binding sites were labeled by [³H]KA. The low-affinity population ($K_d = 50$ nM, $B_{max} = 1$ pmol/mg protein) had a heterogeneous distribution, whereas the high-affinity component ($K_d = 5$ nM) was specifically localized in diencephalon regions (striatum > hippocampus > cortex > cerebellum). These kinetic parameters and receptor distribution (and the pharmacology) bear only a slight resemblance to those pertaining to [³H]Glu binding sites. It is possible, however, the [³H]KA interacts with a small fraction of a subtype of Glu receptors able to accept L-Glu (and KA) in an extended conformation.[83]

Other characteristics of [³H]KA binding, such as synaptosomal enrichment[83] and postsynaptic junctional localization,[84,87,88] suggest a putative synaptic function. This could be vestigial in higher animals, since a phylogenetic study has revealed the presence of an exceedingly high density of [³H]KA sites in lower vertebrates and nonchordates.[86] However, their physiological signif-

icance is emphasized by the recent correlation of the time course of KA neurotoxicity with the development of striatal [³H]KA binding sites and the etiology
of Huntington's disease (HD).[89] Thus, the biochemical and behavioral correlates of intrastriatal KA injections characterized by neuronal degeneration and
loss of [³H]KA binding closely resemble the symptoms and caudate neuronal
alterations observed in HD.[90,91]

In comparision with cerebellar [³H]Glu binding,[33] specific [³H]KA binding[92] was inhibited by excitant analogues in the following order of potency:
domoate (K_i = 6 nM) > KA (K_i = 23 nM) > L-Glu (K_i = 0.4 μM) >> ibotenate
and DL-homocysteate. The purported Glu antagonists GDEE and 2-APB were
ineffective here. Furthermore, the negative cooperativity seen for L-Glu inhibition of [³H]KA binding has indicated allosteric interactions and also the
prevalence of separate receptors for these molecules. Further proof for this
contention has been provided by a recent differentiation of KA sites (reduced)
and [³H]Glu sites (enhanced) by Na⁺-cholate treatment of brain membranes[60]
and the development of relatively specific KA antagonists based on KA[4,9,93]
and γ-glutamyl dipeptides.[4,9,94]

In a recent search for a natural, endogenous ligand for KA binding sites,
only high concentrations of Glu were detected in isolated brain synaptic vesicles.[95] Of course, a synaptic equivalent, perhaps a small peptide, may be elud-
ing discovery at the present time.

5. L-ASPARTATE RECEPTORS

Because of the structural similarity of L-Glu to L-aspartic acid (Asp), the
central actions of these AAs could be predicted to occur at a common site.
Specific "aspartate-preferring" receptors, however, were postulated following
observations of the differential sensitivity of feline and rat spinal neurons to
iontophoretic application of L-Glu and L-Asp,[4,55,96,97] and the subsequent excitation of cells previously desensitized to Glu by L-Asp.[98] Further evidence
supporting these results came from the specific antagonism of N-methyl-D-Asp
(NMDA)-evoked and synaptically induced depolarizations by Mg^{2+} and certain
organic compounds.[55] However, until recently, there was a relative paucity of
detailed knowledge concerning the biochemical demonstration of distinct receptors for L-Asp and L-Glu.

The first report of dissimilar binding[32] described little more than kinetic
properties. Three distinct L-[¹⁴C]Asp recognition sites ($K_d \simeq$ 0.2–50 μM) were
detected on hydrophobic cortical proteins.[99] The high-affinity component interacted selectively with unlabeled NMDA and L-Asp but least with L-Glu.
However, the high density of sites ($B_{max} \simeq$ 3–617 nmol/mg protein) seemed
nonphysiological and may have represented binding of [³H]KA to lipid components and to transport sites. A similar paradoxical report of high-affinity (K_d
= 10 nM) [³H]NMDA (a potent L-Asp derivative) binding[100] has not been
confirmed despite many efforts to repeat this work.

Using L-[³H]Asp and fresh cerebellar synaptic membranes, we have demonstrated a single class of specific, saturable (K_d = 0.87 μM, B_{max} = 44

pmol/mg protein) and reversible ($t_{1/2}$ dissociation = 32 s) binding sites for this ligand.[19] Ligand–receptor interaction equilibrated within 10 min, exhibited pH and temperature optimi,[19] and was enriched within synaptosomal fractions (unpublished data). By comparison, [³H]Asp binding displayed about half the affinity and a third of the capacity of [³H]Glu binding in the same preparation and a pharmacological sensitivity that was also different. Therefore, though L-Glu and L-Asp were the most potent displacers (IC_{50} = 2, 5 μM, respectively), and ibotenate and L-cysteine sulfinate were effective inhibitors, other neuroexcitants such as Quis, 4-F-Glu, and DL-homocysteate, which showed much affinity for [³H]Glu binding, were almost inactive in this system. However, DL-α-aminoadipate (α-AA), DL-α-aminosuberate, HA-966,[19] and the phosphonate derivatives of α-AA, e.g., (−)-2-amino-5-phosphonovalerate (2-APV), hexanoate, and the heptanoate analogues,[48] all inhibited [³H]Asp binding in a profile closely resembling their *in vivo* antagonistic potency at suppressing L-Asp (and NMDA) and synaptically induced excitations.[4,55] The inactivity of KA and NMDA in the two systems was striking. Although the latter compound ought to have displaced some [³H]Asp binding, its low purity and high racemicity could have accounted for its lack of efficacy.

In common with the influence of cations on [³H]Glu binding,[36] [³H]Asp binding was inhibited by monovalent cations and stimulated by low levels (1–10 mM) of Ca^{2+} and Mg^{2+} ions.[19] However, the differential mechanism of stabilization of cold-labile [³H]Glu and [³H]Asp binding,[19] their varied sensitivity to detergents,[101] together with the abovementioned kinetic and pharmacological dissimilarities all suggest separate sites of action of L-Glu and L-Asp. Supportive evidence for this inference is the detection of kinetically different binding sites for these AAs on isolated SJCs.[43]

Many of the features described for [³H]Asp binding[19] have recently been confirmed and extended to show that the [³H]Asp receptors are mainly located on postsynaptic elements.[102]

Of late, specific NMDA-type receptors have been identified on the basis of agonist-stimulated cerebellar cyclic GMP production[64] and striatal[93,103] ²²Na⁺ efflux and [¹⁴C]acetylcholine release. These responses have been envisaged to be receptor mediated since their blockade was affected by putative aspartergic (NMDA) antagonists. However, the question of whether NMDA (or a related AA) represents the natural transmitter of the previously termed "aspartergic synapses" remains to be resolved, especially since exogenously administered NMDA exhibits neurotoxic properties.[104]

6. CYSTEINE SULFINATE BINDING

The potent neuroexcitant properties of sulfur-containing AAs[28] have invoked a tentative synaptic role. Cysteine sulfinate (CSA), a structural analogue of Glu and Asp, fulfills some of the properties suggesting neurotransmitter identity,[3,105] one of which is the demonstration of specific binding sites for this ligand.

Radioreceptor labeling studies for CSA are still in their infancy, and to date, only two reports have described CSA binding properties.[23,105] The Na^+-independent binding of [³H]CSA to previously frozen brain membranes displayed high affinity and low capacity (K_d = 100 nM, B_{max} = 2.4 pmol/mg protein); these characteristics were the converse of the biphasic Na^+-dependent uptake system (K_d = 27 and 398 μM).[105] Whereas CSA was the most potent binding inhibitor (K_i = 0.09 μM) (3-fold greater than L-Glu; 80-fold greater than L-Asp), L-Asp best antagonized the [³H]CSA transport. Binding was greatest in the cerebellum and lowest in retina, whereas the striatum and cortex displayed the highest density of uptake sites. Hence, pre- and postsynaptic binding are easily distinguished.

Many of the basic properties of the [³H]CSA binding reported,[23,105] are consistent with interaction of the AA with receptors, most probably a subpopulation of Glu and/or Asp recognition sites. However, it is quite probable that CSA or a related sulfur-containing acidic AA may be operative as the unidentified excitatory neurotransmitter at cerebellar climbing fiber terminals and at some synapses of the corticostriatal and corticonigral projections. Therefore, research in this area should be productive in the future, providing common assay procedures are adopted by the various researchers in order to reconcile some of the methodological problems noted previously.[18,106]

7. ADDITIONAL TOPICS

7.1. Ontogeny and Localization of Receptors

Further synaptic relevance of excitatory AAs is corroborated by the age-dependent proliferation of their relevant postsynaptic receptors and presynaptic elements in rat brain.[40,70,83,107]

The profiles of maturation of [³H]Glu binding sites in the cerebellum,[70,82] hippocampus,[40] striatum,[89] and cerebral cortex[44] are essentially very similar. Binding site affinity and density increase with time in parallel with protein and DNA deposition[107] and maximal synaptogenesis[67] and plateaus from 25 days post-partum to adulthood when represented per region. In contrast, the Na^+-dependent uptake sites are more numerous from birth than Na^+-independent receptor sites and are also present before the latter but progressively decline in density with age. The first demonstrable release of endogenous Glu occurs around postnatal day 12[70] (PND), coincides with the peak of Na^+-dependent [³H]Glu uptake,[70] and has been correlated with the first observation of the onset of KA neurotoxicity in the striatum[89] and cerebellum[70] and with LTP in the hippocampus.[108] These results substantiate the premise that KA may exert its toxic effects by inducing Glu release (see ref. 90) and by activating receptors distinct from those pertaining to Glu. Further evidence for the latter probability is the slower pattern of ontogenetic development of [³H]KA binding sites than of those corresponding to [³H]Glu.[82]

Of particular importance are the observations of correlative endogenous Glu release[70] and cyclic GMP synthesis[27] in cerebellar slices and the first detection of hippocampal LTP and Ca^{2+}-stimulated [^{3}H]Glu binding.[40] The profile of development of these responses, peaking at PNDs 8–12, suggests an intimate cooperativity between pre- and postsynaptic elements of glutamatergic innervation, long suspected as a widespread ontogenetic phenomenon.[67]

Postsynaptic localization of [^{3}H]Glu binding sites seems most likely in the cerebellum on Purkinje cells, as evidenced by cyclic GMP[27] and iontophoretic responses,[109] on pyramidal and granule cells in the hippocampus, and on interneurons in the striatum,[45] as evidenced by ibotenate injections. However, the reduced [^{3}H]Glu binding following nigral 6-hydroxydopamine lesions[45] and Glu-stimulated [^{3}H]dopamine release from striatal slices[81] both suggest the existence of presynaptic Glu binding sites on dopaminergic terminals. It is obvious, however, that studies involving neurotoxin-induced lesions will be of utmost importance for the elucidation of excitatory AA pathways and in the neuronal localization of AA binding sites in other brain regions and that multiple-lesion experiments may be necessary to discern the identity of the [^{3}H]Glu binding sites that become supersensitive following hemidecortication.[45]

7.2. Neuropathology of Excitatory Amino Acids

In view of their putative transmitter role in the retina, cortex, hippocampus, striatum, cerebellum, and spinal cord,[13] Glu and Asp have been implicated in such diverse physiological functions as vision, olfaction, memory, and locomotion. Therefore, it would seem likely that a number of pathological disorders of the mammalian CNS may involve dysfunction of excitatory AA function.

In this respect, a chronic overstimulation of striatal Glu receptors has been suggested as a factor in the development of Huntington's disease[92] and in the etiology of hippocampal status epilepticus.[110] In contrast, lowered synaptic Glu receptor activation in the mesolimbic system has been proposed as an alternative precipitating factor for schizophrenia.[111]

High dietary intake (and elevated endogenous levels) of excitotoxic AAs has been linked to the so-called "Chinese restaurant syndrome," epilepsy, senile dementia, hippocampal stroke,[112] and amytotropic lateral sclerosis (irreversible paralysis).[113]

Studies of this type emphasize that endogenous AAs under certain conditions (e.g., neuronal hyperactivity, receptor sub- or supersensitivity) are likely to alter neuronal properties and produce clinically relevant CNS disorders. Future goals of excitatory AA research will be the development of suitable drugs to help alleviate some of these neuropathological disorders, especially since the excitant and neurotoxic potency of Glu, Asp, and their agonist analogues are closely related.

7.3. Stability of Receptor Sites

To avoid having to perform binding assays immediately after membrane preparation, often a long and laborious procedure,[12,18] a number of investigators have resorted to utilizing previously frozen tissue preparations. How-

ever, the possible deleterious effects of cold storage on the stability and properties of receptors have not always been recognized in the past. Therefore, a
brief appraisal of AA receptor sensitivity to such conditions may be relevant
here, in order that appropriate attention be drawn to the possible dangers of
using frozen-thawed membranes for characterization of receptor properties.

Purified brain synaptic membranes maintained as buffered suspensions or
pellets at sub-zero temperatures display progressive loss of binding capacity
for [^{3}H]Glu and [^{3}H]Asp[18,19] with minimal effects on binding affinity. This cold
lability of dicarboxylate AA receptors has since been repeatedly confirmed for
[^{3}H]Glu binding to membranes of rat striatum,[34,35] hippocampus,[37] neuroblastoma cells,[114] and forebrain SJCs.[43] A similar sensitivity to low-temperature
storage has been exhibited by binding sites for [^{3}H]Asp,[19,102] kainate,[95] and
[^{3}H]CSA.[23,105] Protein folding and/or aggregations of synaptic vesicles may be
responsible for these effects, since warming of freeze-thawed suspensions
raised [^{3}H]Glu binding and thus allowed a partial recovery from this cold-
lability.[18,73] Even though "cold denaturation" appears not to be irreversible
as originally thought, it could not be retarded either by cryoprotectants[18] or
by peptidase inhibitors (unpublished data).

In recent studies we have found that lyophilization of synaptic membranes
not only afforded protection but greatly enhanced the binding properties for
[^{3}H]Glu and [^{3}H]Asp.[18,19,106] This technique has subsequently been employed
to stabilize [^{3}H]GABA receptors[115] and [^{3}H]kainate binding.[95] Since the induced stability extends over several weeks, lyophilization offers a unique
means of membrane storage without loss of receptor binding properties.[18,19]

8. CONCLUDING REMARKS

There is now overwhelming electrophysiological, anatomic, and neurochemical evidence supporting the contention of acidic AA involvement in neurotransmission in the mammalian brain via multiple excitatory receptors.

The task of differentiating among the various binding sites, at a neurophysiological level at least, has become less ardous as a result of the development of potent and relatively specific organic antagonists. Future studies to
label and characterize these distinct sites would be further facilitated by employment of radioactive derivatives of these putative antagonists.

Isolation and purification of the receptor proteins, especially from cultured
homogeneous cells, will be immensely facilitated by the discovery and application of potent and specific irreversible analogues and/or natural toxins, as
has been the case for the nicotinic cholinergic receptor.[116] Indeed, δ-philan-
thotoxin, a constituent of wasp venom, already shows considerable promise
as a blocker of excitatory synaptic activity at the glutamatergic insect neuromuscular junction.[117] Notably, some progress has already been made towards
isolation of SJCs[43] and a partially purified glycoprotein[31] related to a [^{3}H]Glu
binding protein by conventional biochemical techniques. Therefore, the pros-
pects of direct localization and visualization of excitatory AA receptors by
autoradiography (ARG) and immunohistochemical methods appear bright. I

feel that the development of monoclonal antibodies to the isolated protein moieties and their subsequent utilization for the latter goal will provide a great impetus towards the delineation of excitatory pathways and help differentiate the whole spectrum of acidic AA receptors.

New approaches to study ligand–receptor binding by nuclear magnetic resonance[118] will also find application towards understanding cooperative interactions at receptors[37] and may help to elucidate the role played by Ca^{2+}-binding proteins[51] in neuronal plasticity,[39,40] in stimulating [^{3}H]Glu binding,[57] and in altering antagonist potency.[50] Excitatory AA neurochemistry awaits these developments.

Note Added in Proof

A multitude of exciting and important reports on various aspects of excitatory AA receptors have appeared since this chapter was submitted for publication. Substantial evidence has accrued supporting the concept that KA neurotoxicity may result from pronounced activation of presynaptic KA receptors to enhance release of endogenous (and labeled) Glu and Asp from CNS preparations[119,120,121] and by augmenting Ca^{2+} and Na^+ uptake.[122] While two phosphonate-based antagonists, ($-$)APH and ($-$)APV, appear to block ibotenate- and NMDA-induced neuronal depletions,[123] and sound- and NMDA-initiated seizures,[124] no such compounds have emerged that selectively prevent the neurotoxic effects of KA. In the same context a tryptophan metabolite, quinolinic acid, has been shown to mimick the neurodegenerative properties of cyclic excitotoxins[125] suggesting its possible involvement in the etiology of HD.

Significant progress in the visualization of excitatory AA receptors by ARG has occurred for [^{3}H]KA[126] and L-[^{3}H]Glu.[127,128] Similarly, the anticipated existence of an endogenous ligand for Glu receptors (aside from L-Glu itself) has been strengthened by the discovery of N-acetylaspartyl-glutamate in rat brain which potently competes for L-[^{3}H]Glu binding.[129] Evidently neuropeptides containing acidic AA residues may play a role in excitatory neurotransmission.

Results of pioneering studies on the receptor binding of putative excitant AA antagonists have now been reported. DL-[^{3}H]APB interaction with rat brain membranes exhibited a relatively low affinity (K_d = 1.3 µM) but Glu analogs competed potently for the binding (IC$_{50}$s (µM): Quis = 0.3; L-homocysteate = 1.1; L-Glu = 2.1; (L)APB = 1.4, (D)APB = 41.4; DL-APV = 45).[130] This specificity is apparently in accord with the drug selectivity of a (Quis-type) Glu receptor (Ref. 73). Receptors for NMDA appear to have been labeled with D-[^{3}H]APV.[131] Here the most potent inhibitors included L-Glu, DL-APH and L-homocysteate. Recent evidence for Glu receptor heterogeneity derives from shallow inhibition curves for phosphonate analogues in their displacement of cerebellar L-[^{3}H]Glu binding.[48,132] The apparent isolation and purification of a Glu-binding protein (M_r = 14,300) from rat brain (K_ds = 0.17 and 0.8 µM) has been redocumented.[133] Detailed pharmacological specificity of this protein should be undertaken next.

ACKNOWLEDGMENTS. I sincerely thank Dr. Peter J. Roberts (Southampton, U.K.) for his support and encouragement during my early research career. I am also indebted to Dr. David R. Burt for useful advice during preparation of this manuscript.

REFERENCES

1. Werman, R., 1966, *Comp. Biochem. Physiol.* **18**:745–766.
2. Curtis, D. R., and Johnston, G. A. R., 1974, *Ergeb. Physiol.* **69**:97–188.
3. Johnson, J. L., 1978, *Prog. Neurobiol.* **10**:155–202.
4. Watkins, J. C., and Evans, R. H., 1981, *Annu. Rev. Pharmacol. Toxicol.* **21**:165–204.
5. Burt, D. R., 1978, *Neurotransmitter Receptor Binding* (H. I. Yamamura, S. J., Enna, and M. J. Kuhar, eds.), Raven Press, New York, pp. 41–55.
6. Weiland, G. A., and Molinoff, P. B., 1981, *Life Sci.* **29**:313–330.
7. Molinoff, P. B., Wolfe, B. B., and Weiland, G. A., 1981, *Life Sci.* **29**:427–443.
8. Cuatracasas, P., and Hollenberg, M. D., 1975, *Biochem. Biophys. Res. Commun.* **62**:31–41.
9. Evans, R. H., and Watkins, J. C., 1981, *Life Sci.* **28**:1303–1308.
10. Logan, W. J., and Snyder, S. H., 1972, *Brain Res.* **11**:199–212.
11. Palacios, J. M., Niehoff, D. L., and Kuhar, M. J., 1981, *Neurosci. Lett.* **25**:101–105.
12. Enna, S. J., and Snyder, S. H., 1975, *Brain Res.* **100**:81–97.
13. Cotman, C. W., Foster, A. C., and Lanthorn, T., 1981, *Glutamate as a Neurotransmitter* (G. DiChiara and G. L. Gessa, eds.), Raven Press, New York, pp. 1–27.
14. Matus, A., Pehling, G., and Wilkinson, D., 1981, *J. Neurobiol.* **12**:67–73.
15. Horng, J. S., and Wong, D. T., 1979, *J. Neurochem.* **32**:1379–1386.
16. Toffano, G., Guidotti, A., and Costa, E., 1978, *Proc. Natl. Acad. Sci. U.S.A.* **75**:4024–4028.
17. Napias, C., Bergmann, M. O., Van Ness, P. C., Greenlee, D. V., and Olsen, R. W., 1980, *Life Sci.* **27**:1001–1011.
18. Sharif, N. A., and Roberts, P. J., 1980, *J. Neurochem.* **34**:779–784.
19. Sharif, N. A., and Roberts, P. J., 1981, *Brain Res.* **211**:293–303.
20. Sharif, N. A., and Roberts, P. J., 1981, *Biochem. Pharm.* **30**:3019–3022.
21. Roberts, P. J., and Sharif, N. A., 1983, *Methods Neurobiol.* (P. J. Marangos, I. Campbell, and R. M. Cohen, eds.), Academic Press, New York (in press).
22. Baudry, M., and Lynch, G., 1981, *J. Neurochem.* **36**:811–820.
23. Iwata, H., Yamagami, S., and Baba, A., 1982, *J. Neurochem.* **38**:1275–1279.
24. Hosli, J. S., Andres, P. F., and Hosli, E., 1976, *Pflugers Arch.* **363**:43–48.
25. Greengard, P., 1979, *Fed. Proc.* **38**:2208–2217.
26. Foster, G. A., and Roberts, P. J., 1980, *Life Sci.* **27**:215–221.
27. Garthwaite, J., and Balázs, R., 1978, *Nature* **275**:328–329.
28. Curtis, D. R., and Watkins, J. C., 1963, *J. Physiol. (Lond.)* **166**:1–14.
29. Roberts, P. J., 1974, *Nature* **252**:399–401.
30. Michaelis, E. K., Michaelis, M. K., and Boyarsky, L. L., 1974, *Biochim. Biophys. Acta* **367**:338–348.
31. Michaelis, E. K., 1975, *Biochem. Biophys. Res. Commun.* **87**:106–112.
32. DeRobertis, E., and Fiszer de Plazas, S., 1976, *J. Neurochem.* **26**:1237–1243.
33. Foster, A. C., and Roberts, P. J., 1978, *J. Neurochem.* **31**:1467–1477.
34. Biziere, K., Thompson, H., and Coyle, J. T., 1980, *Brain Res.* **183**:421–433.
35. Vincent, S. R., and McGeer, E. G., 1980, *Brain Res.* **184**:99–108.
36. Baudry, M., and Lynch, G., 1979, *Eur. J. Pharmacol.* **57**:283–285.
37. Werling, L. L., and Nadler, J. V., 1982, *J. Neurochem.* **38**:1050–1062.
38. Shinozaki, H., and Konishi, S., 1970, *Brain Res.* **24**:368–371.
39. Baudry, M., Oliver, M., Creager, R., Wieraszko, A., and Lynch, G., 1980, *Life Sci.* **27**:325–330.
40. Baudry, M., Arst, D., Oliver, M., and Lynch, G., 1981, *Dev. Brain Res.* **1**:37–48.
41. Mitchell, C. K., and Redburn, D. A., 1982, *Neurosci. Lett.* **28**:241–246.

42. Head, R. A., Tunnicliff, G., and Matheson, G. K., 1980, *Can. J. Biochem.* **58**:534–538.
43. Foster, A. C., Mena, E. E., Fagg, G. E., and Cotman, C. W., 1981, *J. Neurosci.* **1**:620–625.
44. Sanderson, C., and Murphy, S., 1982, *Dev. Brain Res.* **2**:329–339.
45. Roberts, P. J., McBean, G. J., Sharif, N. A., and Thomas, E. M., 1982, *Brain Res.* **235**:83–91.
46. Johnston, G. A. R., Curtis, D. R., Davies, J., and McCulloch, R. M., 1974, *Nature* **248**:804.
47. Cull-Candy, S. G., Donnellan, J. F., James, R. W., and Lunt, G. G., 1976, *Nature* **262**:408–409.
48. Roberts, P. J., Foster, G. A., Sharif, N. A., and Collins, J. F., 1982, *Brain Res.* **238**:475–479.
49. Koerner, J. F., and Cotman, C. W., 1981, *Brain Res.* **216**:192–198.
50. Fagg, G. E., Foster, A. C., Mena, E. E., Koerner, J. F., and Cotman, C. W., 1981, *Trans. Am. Soc. Neurochem.* **12**:21.
51. Cheung, W. Y., 1982, *Sci. Am.* **246**:62–71.
52. Roberts, P. J., and Watkins, J. C., 1975, *Brain Res.* **85**:120–125.
53. Krogsgaard-Larsen, P., and Honoré, T., 1983, *TIPS* **1**:31–33.
54. Honoré, T., Lauridsen, J., and Krogsgaard-Larsen, P., 1981, *J. Neurochem.* **36**:1302–1304.
55. Watkins, J. C., 1981, *Glutamate: Transmitter in the CNS* (P. J. Roberts, J. Storm-Mathisen, and G. A. R. Johnston, eds.), John Wiley & Sons, New York, pp. 1–24.
56. McLennan, H., Hicks, T. P., and Liu, J. R., 1982, *Neuropharmacology* **21**:549–554.
57. Baudry, M., Bundman, M. C., Smith, E. K., and Lynch, G., 1981, *Science* **212**:937–938.
58. Vargas, F., Greenbaum, L., and Costa, E., 1981, *Neuropharmacology* **19**:791–794.
59. Baudry, M., Smith, E., and Lynch, G., 1981, *Mol. Pharmacol.* **20**:280–286.
60. Michaelis, E. K., Michaelis, M. L., Chang, H. H., Grubbs, R. D., and Kuonen, D. R. 1981, *Mol. Cell. Biochem.* **38**:163–179.
61. Foster, A. C., Fagg, G. E., Harris, E. W., and Cotman, C. W., 1982, *Brain Res.* **242**:374–377.
62. Rodbell, M., 1980, *Nature* **284**:17–22.
63. Sharif, N. A., and Roberts, P. J., 1980, *Eur. J. Pharmacol.* **61**:213–214.
64. Foster, G. A., and Roberts, P. J., 1981, *Br. J. Pharm.* **74**:723–729.
65. Baudry, M., and Lynch, G., 1980, *Exp. Neurol.* **68**:202–204.
66. Baudry, M., and Lynch, G., 1980, *Proc. Natl. Acad. Sci. U.S.A.* **77**:2298–2302.
67. Altman, J., 1972, *J. Comp. Neurol.* **145**:353–514.
68. Rosenblatt, J. E., Shore, D., Neckers, L. M., Perlow, M. J., Freed, W. J., and Wyatt, R. J., 1979, *Eur. J. Pharmacol.* **60**:387–388.
69. Møller Nielsen, I., Christensen, A. V., and Hyttel, J., 1978, *Advances in Biochemical Psychopharmacology*, Volume 19, *Dopamine*, (P. J. Roberts, G. N. Woodruff, and L. L. Iversen, eds.) Raven Press, New York, pp. 267–274.
70. Foster, G. A., Roberts, P. J., Rowlands, G. J., and Sharif, N. A., 1981, *Br. J. Pharm.* **73**:235.
71. Sharif, N. A., and Roberts, P. J., 1980 *Brain Res.* **194**:594–597.
72. Sharif, N. A., and Roberts, P. J., 1980, *Neurotransmitters and their Receptors* (U. Z. Littauer, Y. Dudai, I. Silman, V. I. Teichberg, and Z. Vogel, eds.), John Wiley & Sons, pp. 369–372.
73. Sharif, N. A., and Roberts, P. J., 1983, *Neurochem. Res.* (in press).
74. Hedlund, B., and Bartfai, T., 1979, *Mol. Pharmacol.* **15**:531–544.
75. Hamblin, M. W., and Creese, I., 1981, *Mol. Pharmacol.* **21**:44–51.
76. Michaelis, E. K., 1979, *Biochem. Biophys. Res. Commun.* **87**:106–112.
77. Grubbs, R. D., and Michaelis, E. K., 1979, *Soc. Neurosci Abstr.* **5**:304.
78. McLennan, H., and Lodge, D., 1979, *Brain Res.* **169**:83–90.
79. McLennan, H., and Wheal, H. V., 1976, *Nueropharmacology* **15**:709–712.
80. Shinozaki, H., and Shibuya, I., 1974, *Neuropharmacology* **13**:1057–1065.
81. Roberts, P. J., and Sharif, N. A., 1978, *Brain Res.* **157**:391–395.
82. Slevin, J. T., and Coyle, J. T., 1981, *J. Neurochem.* **37**:531–533.
83. Simon, J. R., Contrera, J. F., and Kuhar, M. J., 1976, *J. Neurochem.* **26**:141–147.
84. London, E. D., and Coyle, J. T., 1979, *Eur. J. Pharmacol.* **56**:287–290.
85. London, E. D., and Coyle, J. T., 1979, *Mol. Pharmacol.* **15**:492–505.
86. Henke, H., 1980, *Amino Acid Transmitters* (F. V. DeFeudis and P. Mandel, eds.), Raven Press, New York, pp. 30–38.

87. Schwarcz, R., and Fuxe, K., 1979, *Life Sci.* **24:**1471–1480.
88. Foster, A. C., Mena, E. E., Monaghan, D. T., and Cotman, C. W., 1981, *Nature* **289:**73–75.
89. Campochiaro, P., and Coyle, J. T., 1978, *Proc. Natl. Acad. Sci. U.S.A.* **75:**2025–2029.
90. Coyle, J. T., McGeer, E. G., McGeer, P. L., and Schwarcz, R., 1978, *Kainic Acid as a Tool in Neurobiology* (E. G. McGeer, J. W. Olney, and P. L. McGeer, eds.), Raven Press, New York, pp. 139–159.
91. Beaumont, K., Maurin, Y., Reisine, T. D., Fields, J. Z., Spokes, E., Bird, E. D., and Yamamura, H. I., 1979, *Life Sci.* **24:**809–816.
92. Coyle, J. T., Zaczek, R., Slevin, J., and Collins, J., 1981, *Glutamate as a Neurotransmitter* (G. Di Chiara and G. L. Gessa, eds.), Raven Press, New York, pp. 337–346.
93. Luini, A., Goldberg, O., and Teichberg, V. I., 1981, *Proc. Natl. Acad. Sci. U.S.A.* **78:**3250–3254.
94. Tal, N., Goldberg, O., Luini, A., and Teichberg, V. I., 1982, *J. Neurochem.* **39:**574–576.
95. Riveros, N., and Orrego, F., 1982, *Brain Res.* **236:**492–496.
96. Duggan, A. W., 1974, *Exp. Brain Res.* **19:**522–525.
97. Biscoe, T. J., Evans, R. H., Headley, P. M. Martin, P. M., and Watkins, J. C., 1976, *Br. J. Pharmacol.* **58:**373–383.
98. Dostrovsky, J. O., and Pomeranz, B., 1977, *Neurosci. Lett.* **4:**315–319.
99. Fiszer de Plazas, S., and De Robertis, E., 1976, *J. Neurochem.* **27:**889–894.
100. Snodgrass, S. R., 1979, *Soc. Neurosci. Abstr.* **5:**572.
101. Foster, A. C., Fagg, G. E., Mena, E. E., and Cotman, C. W., 1981, *Brain Res.* **229:**246–250.
102. Di Lauro, A., Meek, J. L., and Costa, E., 1982, *J. Neurochem.* **38:**1261–1267.
103. Scatton, B., and Leymann, J., 1982, *Nature* **297:**422–424.
104. Olney, J. W., Degubareff, T., and Labruyere, J., 1979, *Life Sci.* **25:**537–540.
105. Recasen, M., Varga, V., Nanopoulous, D., Saadoun, F., Vincendon, G., and Benavides, J., 1982, *Brain Res.* **239:**153–173.
106. Sharif, N. A., and Roberts, P. J., 1983, *J. Neurosci. Methods* submitted.
107. De Barry, J., Vincendon, G., and Gombos, G., 1980, *FEBS. Lett.* **109:**175–179.
108. Skrede, K. K., and Malthe-Sørenssen, D., 1981, *Brain Res.* **208:**436–441.
109. Chujo, T., Yamada, Y., and Yamamoto, C., 1975, *Exp. Brain Res.* **23:**293–300.
110. Nadler, J. V., Evensen, D. A., and Smith, E. M., 1981, *Brain Res.* **205:**405–410.
111. Kim, J. S., Kornhuber, H. H., Schmid-Burgk, and Holzmuller, B., 1980, *Neuroscience* **20:**379.
112. Olney, J. W., Fuller, T. W., and Degubareff, T., 1979, *Brain Res.* **176:**91–100.
113. Barrow, M. V., Simpson, C. F., and Miller, E. J., 1974, *Q. Rev. Biol.* **49:**101–128.
114. Prasad, K. N., Nayak, M., Prasad, E. J., Cummings, S., and Pattisapu, K., 1980, *Life Sci.* **27:**2251–2259.
115. Chang, L. R., Barnard, E. A., Lo, M. M. S., and Dolly, J. O., 1981, *FEBS Lett.* **126:**309–312.
116. Heidmann, T., and Changeaux, J.-P., 1978, *Annu. Rev. Biochem.* **47:**317–357.
117. Clark, R. B., Donaldson, P. L., Gration, K. A. F., Lambert, J. J., Piek, T., Ramsey, R., Spanuer, W., and Usherwood, P. N. R., 1982, *Brain Res.* **241:**105–114.
118. Brown, C. E., 1981, *J. Neurosci. Methods* **3:**339–363.
119. Ferkany, J. W., Zaczek, R., and Coyle, J. T., 1982, *Nature,* **298:**757–759.
120. Potashner, S. J., and Gerard, D., 1983, *J. Neurochem.* **40:**1548–1557.
121. Collins, G. S., Anson, J., and Surtes, L., 1983, *Brain Res.* **265:**157–159.
122. Berdichevsky, E., Riveros, N., Sanchez-Armass, S., and Orrego, F., 1983, *Neurosci. Lett.* **36:**75–80.
123. Schwarcz, R., Collins, J. F., and Parks, D. A., 1982, *Neurosci. Lett.* **33:**85–90.
124. Croucher, R., Collins, J. F., and Meldrum, B. S. 1982, *Science,* **216:**899–901.
125. Schwarcz, R., Whetsell, W. O., and Mangano, R. M., 1983, *Science,* **219:**316–318.
126. Monaghan, D. T., and Cotman, C. W., 1982, *Brain Res.* **252:**91–100.
127. Halpain, S., Parsons, B., and Rainbow, T. C., 1983, *Eur. J. Pharmacol.* **86:**313–314.
128. Greenamyre, J. T., Young, A. B., and Penney, J. B., 1983, *Neurosci. Lett.* **37:**155–160.
129. Zaczek, R., Koller, K., Cotter, R., Heller, D., and Coyle, J. T., 1983, *Proc. Natl. Acad. Sci. U.S.A.* **80:**1116–1119.

130. Butcher, S. P., Roberts, P. J., and Collins, J. F., 1983, *IRCS Med. Sci.* **11**:42–43.
131. Jones, A. W. Olverman, H. J., and Watkins, J. C., 1983, *J. Physiol.* **340**:45P.
132. Slevin, J., Collins, J., Lindsley, K., and Coyle, J. T., 1982, *Brain Res.* **249**:353–360.
133. Michaelis, E. K., Michaelis, M. L., Stormann, T. M., Chittenden, W. L., and Grubbs, R. D., 1983 *J. Neurochem.* **40**:1742–1753.

Benzodiazepine Receptors

Richard F. Squires

1. INTRODUCTION

The author's interest in benzodiazepine (BZ) receptors was aroused by a report[1] that a series of BZs displaced [³H]strychnine from specific binding sites on membranes from rat brainstem and spinal cord with good correlations between the potencies of the BZs in displacing [³H]strychnine *in vitro*, and in *in vivo* pharmacological tests predictive of clinical efficacy. Candace Pert introduced me to affinity binding methodology in 1972. In December, 1974, at a benzo-diazepine symposium,[2] strong biochemical[3] and electrophysiological evidence[4] for an involvement of GABA in the action of BZs was presented. The decisive event leading to the discovery of the BZ receptors was a visit to Hoffmann-LaRoche in Basel on July 20, 1976. During this visit, Willy Haefely generously offered to provide me with a sample of [³H]diazepam ([³H]DZP) synthesized at Roche for radiohistochemical studies. On October 21, 1976, I received a gift of 100 mCi [³H]DZP (specific activity 14.5 Ci/mmol) from Hoffman-LaRoche, and a day or two later I did my first experiment with it, which was a complete success.

It is now clear that this success was obtained because I incubated the [³H]DZP (2–3 nM) with rat brain membranes first at 37°C and then at 0°C, a procedure that was inspired by a paper by Bennett and Snyder[5] on the binding of [³H]LSD to rat brain membranes. They found that the association and dissociation rates for [³H]LSD were rapid at 37°C but extremely slow at 4°C, so that [³H]LSD that had bound to its receptor at 37°C was effectively "locked" onto the receptor by the subsequent reduction in temperature to 4°C.[5] [³H]Diazepam and especially [³H]flunitrazepam (FLU)[272] were later shown to resemble [³H]LSD somewhat in this respect.

For the next month, together with my three technicians, Helle Carstensen, Pia Jacobsen, and Anne Mette Larsen, I worked very intensively with [³H]DZP binding and, in a letter to Willy Haefely (at Hoffmann-LaRoche, Basel) dated November 26, 1976, I was able to report:

Richard F. Squires • Nathan S. Kline Institute for Psychiatric Research, Orangeburg, New York 10962

For the past few weeks we have been working quite intensively on an affinity binding method using the tritiated Valium which you so kindly sent us. At concentrations of 2–3 nM ^{3}H-Valium binds very well to a crude P_2 fraction from rat whole brain, and more than 90% of this binding can be displaced with low concentrations (0.01–0.1 μM) of cold Valium and many other benzodiazepines which are pharmacologically and/or clinically active. A group of benzodiazepines, including Ro-4933, -3636, -5807, -4864, -3785, and -4556, are essentially inactive at concentrations near 0.3 μM. We have also tested a large number of diverse reference substances and have so far found nothing (aside from the active benzodiazepines) which can displace ^{3}H-Valium in concentrations up to 0.1 mM. Several supposed neurotransmitter substances, including GABA, glycine, L-glutamate, acetylcholine, noradrenaline, dopamine, and 5-HT were inactive at 1 mM. Several neuroleptics, thymoleptics, opiates, convulsants (picrotoxin, bicuculline, metrazol, strychnine), barbiturates, anticonvulsants of other types, meprobamate, phosphodiesterase inhibitors, alpha and beta-adrenergic blockers, and Baclofen are also inactive. . . . The Valium receptors are unevenly distributed in brain with low concentrations in cerebellum and, especially, pons–medulla. There are high concentrations in all cortical regions. There is some specific binding in P_2 fractions from liver and kidney, but not from small or large intestine. These are, of course, very preliminary results which will have to be repeated. However, so far the picture looks very exciting.

Most of the data for our first benzodiazepine papers[6–8] were generated by my three technicians at Ferrosan during our first month's work with [^{3}H]DZP.

Claus Braestrup became interested in my BZ project, and, in December, 1976, requested my permission to write a Ph.D. dissertation on the BZ work, which I was pleased to help him with.[10–15] I hired Claus Braestrup on August 16, 1976 to do other work.[9]

Two techniques that I had learned earlier as a graduate student at the California Institute of Technology proved to be of unusual value later in characterizing the BZ receptors:

1. Heat inactivation, used to differentiate allelic variants of the enzyme tyrosinase in *Neurospora crassa*.[16] I later used this technique successfully to differentiate MAO-A and MAO-B in certain species[17] and to characterize different subpopulations of BZ receptors.[18–21]
2. Hill analysis of concentration–response curves,[22] which I used to describe the interactions between Na^+ and K^+ with the $(Na^+ + K^+)$-dependent ATPase from rat brain[23] and the interactions of various ligands with the BZ receptors.[8,11,19,21,24]

2. GENERAL PROPERTIES OF BENZODIAZEPINE RECEPTORS

Initial work on the binding of [^{3}H]DZP to membranes from the brain of several mammalian species led to the following conclusions.

[^{3}H]Diazepam binds with high affinity (K_d near 3 nM at 0°C) to a single class of receptors located on membranes in mammalian brain.[6,11,25–28] In addition to linear Scatchard plots for [^{3}H]DZP and [^{3}H]FLU binding, a single class of binding sites was indicated by apparently monophasic inactivation of specific [^{3}H]DZP binding sites at 60°C in 50 mM Tris HCl, pH 7.4[7] and apparently monophasic dissociation of both [^{3}H]DZP and [^{3}H]FLU from specific receptors in brain.[11] Later studies, however, revealed the existence of multiple specific BZ receptors in brain (see Section 9).

Specific binding of [³H]DZP and [³H]FLU to specific brain receptors is temperature dependent, with greatest binding at lowest temperatures (0°C–4°C)[7-11,26,29]; [³H]DZP and [³H]FLU bind to the same receptors in brain,[11,29-31] but [³H]FLU has a higher affinity for BZ receptors than [³H]DZP and a slower rate of dissociation from them.[11,29-31]

There was a good correlation between the potencies of a large number of BZs in displacing [³H]DZP from specific brain sites *in vitro* and the potencies of the same BZs (ED_{50} values) in several *in vivo* pharmacological tests thought to be predictive of clinical efficacy.[6,7] Binding to specific brain BZ receptors is stereospecific.[25]

[³H]Diazepam was not displaced from specific brain sites by any known neurotransmitter or hormone, nor by numerous drugs representing 22 distinct pharmacological classes,[6,8,25] leading to the conclusion that the BZ receptor might be a unique receptor, perhaps for an undiscovered neurotransmitter[6,8] analogous to the enkephalins[32,33] and endorphins,[34,35] the "endogenous ligands" for the opiate receptors[36] (but see Section 12).

Benzodiazepine receptors are unevenly distributed in the brain, and their distribution does not parallel the distribution of any known neurotransmitters or receptors for them.[6,7,10,25,28,31] The greatest densities of BZ receptors are in all cortical areas, with decreasing levels in the limbic structures > cerebellum > pons-medulla > spinal cord.[6,7,10,25,28,31]

The localization of specific brain BZ receptors is probably neuronal, since they were not detectable on primary mouse astrocytes in tissue culture,[14] and locally injected kainic acid reduced the density of BZ receptors in rat substantia nigra,[11] striatum,[37] and cerebellum.[38] Neuronal localization is also consistent with the known electrophysiological effects of the benzodiazepines.[4,42-50]

Phylogenetically, high-affinity [³H]DZP binding sites were not found in the nervous tissue of five invertebrate species[12] or in the brains of cyclostomes (hag fish, lamprey) or elasmobranchs (sharks and rabbit fish),[39] but they have been found in the central nervous system (CNS) of all vertebrate species studied beginning with the higher bony fishes (osteichthys).[12,39]

Another type of BZ receptor (called "peripheral") was found in high concentrations in mammalian kidney and in lower concentrations in liver and lung.[7] Binding of [³H]DZP to the kidney receptor is displaced by nanomolar concentrations of Ro 5-4864 (*p*-chlorodiazepam) but not by micromolar concentrations of clonazepam. The relative potencies of these two BZs in displacing [³H]DZP from specific brain binding sites is just reversed.[7] Peripheral or kidney-type BZ receptors are discussed in more detail in Section 4.

3. LOCALIZATION OF BENZODIAZEPINE RECEPTORS IN THE CENTRAL NERVOUS SYSTEM

The high densities and uneven distribution of characteristic BZ receptors in brain and their apparent absence from all other organs studied suggest that these receptors might be located mainly on neurons. A large body of electro-physiological evidence[4,40–50] showing the BZs selectively potentiate the inhibitory actions of GABA on various kinds of neurons in many parts of the CNS greatly strengthens the idea of neuronal localization.

Further support for neuronal localization of BZ binding sites comes from radiohistochemical studies,[51–55] kainic acid lesion studies,[11,37,38,55,56] and decreases in BZ receptor densities associated with other types of neuronal degeneration.[57–60]

In rat brain, radiohistochemical studies[51,52] indicate high densities of [³H]FLU binding sites in olfactory bulb (molecular and mitral cell layer), nucleus olfactorius, rostal medial forbrain bundle, hippocampus, dentate gyrus, pyriform cortex, septum, amygdala, and cortex (especially lamina I, IV, and VI). There are low BZ receptor densities in the caudate–putamen and the basal ganglia in general, with the exception of the nucleus entopeduncularis. In the diencephalon, the lateral preoptic nucleus, the ventromedial hypothalamic nuclei, the lateral mamillary nuclei, the periventricular rotundocellular nuclei, the subthalamic nuclei, and the zona incerta exhibited high receptor densities. In general, the hindbrain has very low BZ receptor densities with the exception of nucleus ventralis rostralis lemnisci lateralis, the inferior and superior colliculi, and the substantia nigra (pars lateralis). In the cerebellum, high densities are found in the molecular layer and dentate nuclei. The substatia gelatinosa of the spinal trigeminal nucleus and laminae II, III, IV, and X of the spinal cord exhibit high BZ receptor densities. The inner plexiform layer of the retina was also labelled.[52] White matter areas of the brain are essentially devoid of BZ receptors.

These light microscopic autoradiographic studies are not able to determine the exact location of the receptors. For example, they might be on pre- or postsynaptic elements in a highly labeled layer or nucleus. However, a kainic acid lesion study in the guinea pig cerebellum indicates that high densities of BZ receptors in the molecular layer are located on the dendrites of Purkinje cells.[55] GABA receptors in the granule cell layer immediately adjacent to the kainic acid lesion site are unaffected.[55] Selective decreases in BZ receptor densities in the cerebella of "nervous" mutant mice, with known degeneration of cerebellar Purkinje cells,[61] provide further evidence for the presence of BZ receptors on these neurons.[59,60,62] A comparison of the respective radiohistochemical localizations of BZ and muscimol binding sites suggests that all BZ receptors may be coupled to GABA receptors but that many GABA receptors may not be coupled to BZ receptors.[51] The respective patterns of [³H]muscimol and [³H]FlU binding in brain are distinctly different, although there is considerable overlapping.[51] The lack of correspondence between [³H]muscimol and [³H]FLU labeling seems to result in part from the failure of the radiohistochemical method to detect low-affinity [³H]muscimol binding sites. [³H]Flunitrazepam, however, binds with uniformly high affinity to all brain specific BZ receptors.

Even in brain areas where [³H]FLU seems to bind alone, with little or no detectable [³H]muscimol binding, the addition of exogenous GABA increases [³H]FLU binding.[51,54] However, the magnitude of the increase in [³H]FLU binding caused by the addition of exogenous GABA varies dramatically from about 50% in cortex (lamina IV) to 230% in the molecular layer of the cerebellum.[54] This differential response to GABA may reflect the existence of several types of BZ/ion/GABA receptor complexes. Certain substances such as CL 218,872[19,63] and ethyl-β-carboline-3-carboxylate, (ethyl-BCC)[64] exhibit higher affinity for a subpopulation of BZ receptor complexes (called BZ_1 receptors), which constitute most of the BZ receptors in cerebellum, than for another subpopulation (called BZ_2 receptors), which constitute about 50% of the BZ receptors in the hippocampus (see Section 9). Autoradiographic studies clearly show that [³H]FLU binding in some brain regions was substantially reduced by the triazolopyridine CL 218,872,[53] whereas binding in other regions was affected less.[53] CL 218,872 reduces [³H]FLU binding selectively in the cerebellum, globus pallidus, and parts of the cerebral cortex.[53] Areas in which [³H]FLU binding is less affected by CL 218,872 include the superficial layer of the superior colliculus, the caudate–putamen, and parts of the dentate gyrus.[53] Methyl-BCC produces a selective pattern of [³H]FLU displacement in brain similar to that produced by CL 218,872 (M. Kuhar, D. L. Niehoff, and J. R. Unnerstall, personal communication). However, CL 218,872 exhibits a greater degree of discrimination between BZ receptor types than the BCC esters.

Benzodiazepine receptors are probably located on denrites, soma, and nerve terminals. The evidence for this is partly electrophysiological and partly derived from kainic acid lesion studies. It has been known for at least 16 years that BZs potentiate and prolong presynaptic inhibition,[4,40–42] which is mediated by GABA and antagonized by picrotoxin. As mentioned above, a large body of electrophysiological data demonstrates that BZs selectively potentiate the inhibitory action of GABA on neuronal cell bodies (soma).[44–50] A novel, partially depolarizing, response to GABA applied to the dendrites of hippocampal pyramidal cells[65–67] is potentiated by BZs.[68] High concentrations of brain-type BZ receptors were found in rat,[69,70] bovine,[69,71] and chicken[72] retina.

The survival of some BZ receptors near the site of kainic acid injections into several brain regions[11,37,38,56] suggests that some of these receptors may be located on terminals that, in general, are resistant to kainic acid.

4. "PERIPHERAL" OR "KIDNEY-TYPE" BENZODIAZEPINE RECEPTORS

I discovered the "peripheral" BZ receptor during my first month's work with [³H]DZP,[7] shortly after the discovery of the brain specific receptor. Clonazepam was the most, and Ro 5-4864 the least, potent in displacing [³H]DZP from the binding sites in brain. Thus, when I found binding of [³H]DZP to rat kidney P_2 membranes, I tested these two BZs on the "kidney" binding to determine whether these binding sites were different from those in brain. The results demonstrated clearly that the kidney BZ receptor was quite different from the brain receptor,[7] with the relative affinities of clonazepam and Ro 5-4864 for the kidney receptor just the reverse of their relative affinities in brain.[7]

The BZ receptor in rat kidney also proved to be more thermolabile and to have a slightly lower affinity for [³H]DZP than the BZ receptor in rat brain.[7] Peripheral BZ receptors cannot be photolabeled with [³H]FLU (see Section 11), nor can they be solubilized in active form (see Section 10).[73] The binding of [³H]Ro5-4864 to specific kidney sites is not ion dependent (M. Reilly and R. F. Squires, unpublished data), in contrast to [³H]FLU or [³H]CGS-8216 binding to brain specific sites (see Sections 6 and 7). Peripheral BZ receptors have been identified on primary cultures of mouse cerebral astrocytes,[14,74] six astrocytoma cell lines,[75,76] B16/C3 melanoma cells,[77] rat blood platelets,[78] rat peritoneal mast cells,[79] rat and guinea pig heart,[80,264] as well as rat and mouse glia cells.[81–84] The number of peripheral BZ receptors is reduced in kidney and increased in blood platelets of spontaneous hypertensive rats.[85] Renal BZ binding sites are increased in deoxycorticosterone/salt hypertensive rats.[86]

The physiological roles of peripheral BZ receptors are at present unknown. It has been reported that BZs, including Ro5-4864, stimulate the methylation of phospholipids in the membranes of C6 astrocytoma cells[76] and induce melanogenesis in B16/C3 melanoma cells.[77] Radiohistochemical localization using [³H]DZP revealed high densities of these receptors in collecting ducts and cortex of the rat kidney, suggesting that they may play some role in transport of substances from urine to blood (J. Tallman, personal communication).

Inactivation of peripheral BZ receptors by bombardment with high-energy electrons indicates an average monomeric subunit molecular weight (mol. wt.) near 32,000 in contrast to an average mol. wt. of 57,000 for the brain-specific receptor.[87] In a recent study high densities of ³H-Ro5-4864 binding sites were found in kidney, heart, brain and adrenal gland of the rat. Specific ³H-Ro5-4864 binding was potently displaced by a novel isoquinoline derivative PK11195[265,266] and less potently by diazepam. Interestingly, PK11195 was 5 to 10 times more effective in displacing ³H-Ro5-4864 from sites in cerebral cortex and striatum than from sites in the peripheral organs, suggesting the existence of several types of ³H-Ro5-4864 binding sites. Radiohistochemical studies using ³H-PK-11195 reveal high densities of binding sites in the choreoid plexus and olfactory bulb as well as in the expected peripheral organs with highest densities in the adrenal gland. So far no pharmacological effects of PK11195 have been detected in experimental animals, in doses up to 100 mg/Kg (G. Le Fur, personal communication).

Recently, Ro5-4864 has been reported to be a potent convulsant, especially in guinea pigs,[267] and to have anxiogenic properties in male hooded rats which can be reversed by phenytoin but not by PK 11195.[268] Evidence has been obtained that Ro5-4864 acts as a potent GABA antagonist on a small subpopulation of ³⁵S-TBPS-coupled GABA-A receptors (R. Squires and E. Saederup, in preparation).

5. ASSOCIATION OF GABA RECEPTORS WITH BENZODIAZEPINE RECEPTORS

The first indication of coupling between GABA and BZ receptors came from electrophysiological work[4,40–50] and from indirect biochemical evidence.[3,88–91]

In early experiments, GABA, GABAmimetics, and GABA antagonists did not seem to significantly affect specific [³H]DZP binding to rat brain membranes.[6-8,25,26] However, in 1978, Tallman, Thomas, and Gallager[92] reported that GABA or muscimol increased the binding of [³H]DZP to twice-washed rat brain membranes in 50 mM Tris HCl, pH 7.5. These results were rapidly confirmed and extended by several other investigators.[72,93-98] GABA and a group of GABAmimetics were found to increase the affinity of [³H]DZP or [³H]FLU for their receptors with only small or no increases in the total number of binding sites (B_{max} values).[72,92-98]

In January, 1979, I made two findings that convinced me of a direct coupling between GABA and at least some BZ receptors: (1) GABA, muscimol, or β-guanidinopropionate (BGPA), in the presence of 25 mM Tris HCl, pH 7.5, provided great protection against heat inactivation (60°C for 30 min)[20]; (2) two piperidine-derived GABAmimetics, 4,5,6,7-tetrahydroisoxazolo-4,5c-pyridine-3-ol (THIP) and isoguvacine (IGV),[99] gave plateau inhibition of [³H]FLU binding that was dependent on the kind and concentration of buffers (ions) in the incubation medium.[20,21,24] The plateau inhibitions by THIP and isoguvacine were reversible by GABA.[20] In the presence of high salt concentrations (100 mM Tris citrate, pH 7.4), two piperidine-derived GABAmimetics, THIP[99] and piperidine-4-sulfonate,[100] were found to have little or no effect alone on [³H]DZP binding[98,101] but effectively antagonized the ability of GABA or muscimol to enhance [³H]DZP binding in a competitive way.[101,20,21,24]

Continued investigations of the plateau inhibitions caused by THIP and IGV confirmed that they were highly reproducible under a variety of conditions, reversible by GABA and muscimol, and dependent on the kind and concentration of ions in the incubation medium.[21,24] In particular, the plateau inhibitions were obtained under conditions (prolonged incubations) in which all the ligands involved ([³H]FLU, THIP, and/or GABA, and ions) were almost certainly at equilibrium with their respective receptors. I also found that the binding of [³H]FLU to brain specific receptors is almost entirely dependent on the addition of ions[21,24]: in the absence of added ions, specific [³H]FLU binding is reduced to less than 10% of the maximum binding, and a large variety of Tris salts increased binding in a concentration-dependent, saturable way, with 50% of maximum binding obtained at concentrations ranging from less than 1 mM for phosphate and maleate and near 4 mM for the halides to 14 mM for glutamate and aspartate. The piperidine-derived GABAmimetics THIP, piperidine-4-sulfonate (P4S), and IGV decreased the affinity of chloride and several other anions that enhanced binding about eight- to tenfold.[24,21]

At a concentration of 200 μM, THIP decreased the affinity for the activating Tris salts by factors ranging from about 3 to about 12, depending on the anion.[21] These effects of THIP are reversed by GABA or muscimol without affecting the maximum binding.[24] Scatchard analysis of [³H]FLU binding in low Tris HCl concentrations (1 and 10 mM) shows that the reduction of [³H]FLU binding reflects only a reduction in B_{max} values with no change in the affinity (K_d) for [³H]FLU.[63] In the presence of 25 mM Tris HCl, THIP (100 μM) both increased the K_d for [³H]FLU and decreased the B_{max}.[63] Both THIP and P4S, in the presence of 150 mM NaCl and 50 mM Tris citrate, pH

7.1, potently enhance [³H]FLU (0.25 nM) binding at 30°C but not at 0°C.[102] This enhancement was also dependent on the presence of chloride ion[102] and was not observed using 50 mM Tris citrate alone.

With EDTA/water-dialyzed membranes in the presence of 50 mM Tris HC1 pH 7.5 plus 200 mM NaC1, P4S enhanced [³H]FLU (0.2 nM) binding even at 0°C. However, the concentration of chloride ion required to give 50% of the maximum [³H]FLU binding is 10–20 times greater in the presence of P4S (50 μM) than in the presence of GABA (200 μM), both at 0°C and 30°C.[272] Thus, ions alone, but not GABA alone, are able to promote BZ binding.

The pervasive effects of GABA (or muscimol) and THIP (or P4S) in regulating (in opposite directions) the affinities of ions required for binding suggest that all BZ receptors may be indirectly coupled to GABA receptors through ion recognition sites.[21,24] This conclusion is also supported by the radiohistochemical findings.[51,54]

Heat inactivation studies[19,21,103,104,272] also indicate that ions (e.g., NaC1) alone can protect BZ receptors against heat inactivation. GABA and a number of GABAmimetics, which alone have no protective effect, dramatically decrease the concentration of NaC1 required to provide 50% of its maximum protective effect.[103,104] Interestingly, the concentration–response curves for NaC1 in the heat protection paradigm are characterized by Hill coefficients near 2, strongly suggesting cooperative interactions between interacting binding sites in the BZ/ion/GABA receptor complexes.[103,104] Evidence reviewed below raises the possiblity that both cation and anion recognition sites are coupled to BZ receptors and that separate but interacting Na^+ and $C1^-$ ion recognition sites may be involved in the heat protection effects (Section 7). Sodium chloride KC1, and CsC1 produce similar protection patterns alone and with GABA (500 μM), whereas sodium or potassium phosphate (pH 7.5), with and without GABA, produce clearly different patterns.

All GABAmimetics tested, with the possible exception of THIP, can potentiate the protective effects of NaC1, but to different degrees depending on the GABAmimetic.[103] Gavish and Snyder[105] confirmed that, in the presence of 50 mM Tris citrate buffer, pH 7.1, muscimol, GABA, imidazoleacetate (ImAA), and THIP, in decreasing order of potency, could significantly protect BZ ([³H]FLU) receptors in calf brain membranes against heat inactivation (30 min at 55°C). I found that the time courses of heat inactivation of BZ receptors from rat brain at 60°C in the presence of 50 mM NaC1 and saturating concentrations of GABAmimetics are all polyphasic, exhibiting "fast" components with half-lives of a few minutes and "homogeneous" slow components, in both forebrain and cerebellum, with half-lives in the 70- to 100-min range.[103] When extrapolated back to time zero, the B_0 values for the slow components ranged from about 30% in the presence of saturating P4S to 80% in the presence of saturating GABA (expressed as percent of unheated control), indicating extensive multiplicity of BZ/ion/GABA receptor complexes[103,104,272] (see Section 9).

Picrotoxin, which is known to block the inhibitory effects of GABA on neuronal firing in a way reversible by BZs,[106] potentiates the protective effects of high (>100 mM) but not low (<50 mM) NaC1 concentrations,[104,272] also suggesting that the protective effects of picrotoxin could be dependent on both Na^+ and Cl^- recognition sites (see Section 8 below).

Dialysis of rat brain P_2 membranes against 1 mM EDTA, then water, removes large amounts of GABA that cannot be removed by dialysis against water alone.[272] Apparently, much GABA is bound to particulate matter in brain in a divalent-cation-dependent way.[104]

In such EDTA- and water-dialyzed membranes, $CaCl_2$ (5 mM) plus GABA (500 μM), but neither substance alone, provided about 50% protection of [^{3}H]FLU binding sites against heat inactivation (60°C for 30 min); Mg^{2+}, Ba^{2+}, Sr^{2+}, and Mn^{2+} can substitute for Ca^{2+} in the presence of GABA, and all GABA-A receptor agonists tested, including muscimol, 3-aminopropane sulfonate (APSA), BGPA, ImAA, IGV, P4S, THIP, *trans*-4-aminocrotonic acid, dihydromuscimol, and *d,l*-homo-β-proline, can fully or partially substitute for GABA in the presence of Ca^{2+} ion. Baclofen, a GABA-B receptor agonist, is inactive alone or in the presence of Ca^{2+}. Picrotoxin, which has no protective effect alone or together with low concentrations of NaCl (50 mM), potentiated the protective effect of Ca^{2+} ion in a way that is further enhanced synergistically by 50 mM NaCl.[104]

Close coupling between BZ and GABA receptors is also indicated by experiments showing that either BZs or GABAmimetics can protect both BZ and GABA receptors simultaneously from heat inactivation,[105] that GABA and BZ receptors from rat brain, solubilized with sodium desoxycholate, both have the same sedimentation coefficient (11.3 S) using sucrose density gradient centrifugation,[107,108] that binding of [^{3}H]FLU to BZ receptors solubilized with either sodium desoxycholate[108–110] or Triton X-100[111,112] is enhanced by GABA and GABA-A receptor agonists, and that such receptors retain the ability to bind [^{3}H]GABA[107,110] and [^{3}H]muscimol.[112] However, binding of [^{3}H]DZP to BZ receptors solubilized from rat brain membranes with 0.5% Lubrol-PX does not appear to be enhanced by GABA.[113] Lower concentrations of Triton X-100 (e.g., 0.05%) appear to preferentially solubilize BZ receptors from rat brain membranes, leaving at least some GABA receptors in the membrane.[114,115] These, together with the results of some kainic acid lesion[38] and radiohistochemical[51] experiments, support the conclusion that whereas all BZ receptors are probably coupled to GABA receptors, not all GABA receptors are coupled to BZ receptors. GABA autoreceptors, which regulate the release of GABA from GABA neurons, are not coupled to BZ receptors.[116]

It now appears likely that all of the central actions of the BZs, including their anticonflict and anxiolytic activities, are GABA mediated.[117,118]

6. ASSOCIATION OF ANION RECOGNITION SITES WITH BENZODIAZEPINE RECEPTORS

Since all known inhibitory actions of GABA on vertebrate neurons[119–123] and the crustacean neuromuscular junction[124,125] are associated with large increases in anion (chloride) conductances across the neuronal (or muscle) membrane, and all "brain specific" BZ receptors in higher vertebrates are probably coupled to GABA receptors, it is reasonable to assume that BZ receptors might be coupled to chloride recognition sites.

A number of biochemical studies do indicate coupling between BZ and chloride (anion) recognition sites. T. Costa *et al.*[126] reported that in the presence of 50 mM Tris maleate buffer, pH 7.6, several of the "Eccles" anions,[119] especially iodide, bromide, chloride, nitrate, and thiocyanate, increase the binding of 0.3 nM [³H]DZP. The enhancing effect of iodide ion was caused by an increase in the affinity for [³H]DZP with no change in the total number of binding sites (B_{max}).[126] Martin and Candy[127] also reported an enhancement of [³H]DZP binding (3 nM) to rat brain membranes by several anions in the presence of 100 mM Tris citrate buffer, pH 7.1. Of nine anions tested, all except isethionate produced significant enhancement of [³H]DZP binding, but to different degrees ($Br^- > I^- > NO_3^- > CL^- > F^- > IO_3^- > SCN^- > ClO_3^-$). Again, Scatchard analysis of [³H]DZP binding revealed that Br^- ion increased the affinity of [³H]DZP without changing the total number of sites available.[127] The "background" buffers used in these two studies (50 mM Tris maleate[126] and 100 mM Tris citrate,[127] respectively) probably modify the response to the other ions added.[21,24]

The piperidine-derived GABAmimetics[99,100] THIP, IGV, and P4S were found to produce plateau inhibitions of [³H]FLU binding to brain specific sites the magnitude of which was dependent on the kind and concentration of the buffer (ions) used.[20,21,24] In 25 mM Tris HCl (pH 7.5), THIP (100 μM) reduced both the affinity of [³H]FLU and the total number of binding sites available.[63] The binding of [³H]FLU proved to be almost entirely dependent on the presence of ions,[21,24] with 50% of maximum binding obtained at a concentration near 4 mM Tris HCl. The reduction in the binding of [³H]FLU in the presence of low salt (buffer) concentrations results entirely from a reduction in the number of available binding sites (i.e., a reduction in B_{max}) with no change in the affinity of these binding sites for [³H]FLU[63] (R. Squires, unpublished observation). THIP, IGV, and P4S all reduce the affinity of the BZ/ion/GABA receptor complexes for the anion required for [³H]FLU binding,[21,24] effects that can be reversed by GABA or muscimol.[24] Similarly, the binding of [³H]CGS-8216, a potent BZ antagonist,[251] is also ion dependent and regulated by GABA receptors in qualitatively the same way as [³H]FLU binding.[272]

The effects of several salts in enhancing [³H]FLU binding seem to reside more in the anion than in the cation, since the concentrations of various Tris salts required for 50% of maximum stimulation of [³H]FLU binding ranged from less than 1 mM for maleate and phosphate to about 14–15 mM for glutamate and aspartate,[24] and the magnitude of the increase in the concentration of a particular Tris salt required for 50% of maximum binding produced by THIP (200 μM) ranged from a factor of about 5 (Tris benzoate) to 12 (Tris acetate), again suggesting an important role for the anion.

Clearly, the ion requirement for BZ binding to brain specific sites is not restricted to the "Eccles" anions.[119] Every anion tested, with Tris or Na^+ as the counterion, was able to greatly enhance [³H]FLU binding.[21,24]

Various salts can protect BZ receptors against heat inactivation, and again the anions appear to play an important role in the protective effect.[103] Lithium chloride, NaCl, KCl, and CsCl alone can provide almost complete protection against heat inactivation, with 50% of maximum protection near 280 mM and Hill numbers greater than 1. Sodium and potassium phosphates (pH 7.5) alone seem to provide only about 70% of the maximum protection, with half-maximum effects near 50 mM and Hill numbers near 1. GABA (200 μM) reduces the concentration of NaCl, KCl, or CsCl required for 50% protection from near 280 mM to about 40 mM and increases the Hill number to about 2.[103,104] In contrast, GABA reduces the concentration of phosphate ion required for 50% of maximum protection from about 50 to 14 mM and decreases the Hill number from about 1 to 0.8.[103]

It is now established that independent sites that bind picrotoxin (picrotoxinin), related ("cage") convulsants, barbiturates, and certain pyrazolopyridines (SQ 20,009, SQ 65,396) are coupled to, and interact allosterically with, separate BZ, GABA, and ion recognition sites[103,104,106,128-130] (see Section 8). The enhancement of [³H]FLU or [³H]DZP binding to brain specific receptors by the pyrazolopyridines SQ 20,009 and SQ 65,396[129-132] and by certain barbiturates[133] is dependent on chloride or some other "Eccles" anions (but not fluoride, sulfate, acetate, and phosphate) and is antagonized by picrotoxin or isopropyl bicyclophosphate (IPTBO). The enhancement of [³H]DZP binding by pentobarbital in the presence of 200 mM NaCl is competitively inhibited by picrotoxin with an apparent K_i near 10 μM.[133] Picrotoxin and IPTBO in the absence of GABA enhance [³H]DZP[134] and [³H]FLU[131] binding in a chloride-ion-dependent manner but inhibit the increase in [³H]BZ binding in response to GABA, also in a chloride-ion-dependent way.[131,135] These actions of picrotoxin and IPTBO may be much more potent when [³H]FLU binding is carried out at 35°C[131] rather than 0°C (see Section 8).

7. ASSOCIATION OF CATION RECOGNITION SITES WITH BENZODIAZEPINE RECEPTORS

In addition to the evidence for anion recognition sites in BZ receptor complexes, presented above, there is now also an extensive accumulation of evidence for the presence of independent, physiologically relevant cation recognition sites associated with at least some BZ receptors.

Mackerer and Kochman[136] discovered that several divalent cations, most notably Ni^{2+}, enhance [³H]DZP binding to brain specific sites by increasing its affinity. Nickel ion also dramatically antagonizes the inhibition of [³H]FLU binding by BCC methyl and ethyl esters.[21]

Silver ion has recently been reported to selectively increase the affinity of 3H-muscimol for GABA-A receptors coupled to BZ receptors.[137]

[³H]DZP binding sites in rat cortex, striatum, cerebellum, and retina were increased 27–60% 3 days after an acute oral dose of methyl mercury (MeHg). This treatment had no effect on the affinity (K_d) of [³H]DZP except in the retina, where MeHg treatment increased both the number of binding sites and the affinity of [³H]DZP for them. Methyl mercury did not affect [³H]GABA or

[³H]spiperone binding in the same areas of the rat CNS but did decrease the content of cyclic GMP in cerebellar cortex.[138]

The binding of [³H]baclofen or [³H]GABA to GABA-B receptors is Ca^{2+} or Mg^{2+} dependent,[139] and GABA (acting on a GABA-B like receptor) was found to decrease Ca^{2+} "spikes" in tetrodotoxin-blocked neurons from chick dorsal root ganglia grown in tissue culture.[140,141] Although BZs do not interact significantly with these GABA-B receptors, their interaction with Ca^{2+} ions raises the possibility that coupling to divalent cation sites may be a common feature of several types of GABA receptors, including those coupled to BZ receptors.

Benzodiazepines have been reported to modify, in a complex way, Ca^{2+} fluxes in rat skeletal muscle,[142] frog skeletal muscle,[143] and synaptosomes from mouse brain.[144] Diazepam at 50 μM significantly decreased the concentration of Na^+ in frog sartorius muscle.[143] Benzodiazepines in high concentrations (10–70 μM) inhibit a Ca^{2+}, calmodulin-stimulated protein kinase in rat brain membrane.[145]

Barbiturates, which interact with a picrotoxin binding site coupled to BZ receptors,[128,273] inhibit Ca^{2+} ion uptake into synaptosomes from rat,[146–150] rabbit, and mouse[147] brains and depress K^+-facilitated, Ca^{2+}-dependent release of [³H]norepinephrine and [³H]GABA from mouse forebrain synaptosomes.[151] GABA and DZP individually block the Ca^{2+}-dependent, K^+-evoked release of [³H]serotonin from rat hippocampal synaptosomes, and DZP enhanced the inhibitory effect of GABA on K^+-evoked release.[152] Flurazepam and diazepam were both reported to inhibit K^+ and Ca^{2+} induced contractures of the guinea-pig taenia coli and it was suggested that these BZs inhibit the transmembrane influx of Ca^{2+}.[269] Flurazepam was also reported to block both Na^+ and K^+ currents across the membranes of voltage-clamped squid giant axons with the K^+ current being more potently inhibited than the Na^+ current.[270] Experimental evidence suggested that flurazepam acts from inside the axon and blocks K^+-channels only after they have opened.[270] Similar effects of flurazepam on voltage-clamped myelinated nerve fibres of the frog (*Rana esculenta*) have been reported.[271] [³H]DZP binds to a "peripheral" type binding site in longitudinal muscle-myenteric plexus strips of the guinea pig ileum (Kd = 43 nM). Several BZs were reported to decrease, in a dose-dependent manner, the electrically induced contractions of the longitudinal muscle strip but their potencies in this test did not correlate with their binding affinities.[299] Diazepam antagonized the contractions of the longitudinal muscle strip caused by K^+ and Ca^{2+} and this effect of diazepam was reversed by increasing the Ca^{2+} concentration in the medium.[299]

Ethanol, an anxiolytic that also selectively potentiates GABA-mediated neurotransmission in brain,[153] has been reported to augment Ca^{2+}-mediated mechanisms both pre- and postsynaptically in the CAl region of rat hippocampus.[154] Ethanol and pentobarbital were reported to inhibit intrasynaptosomal sequestration of Ca^{2+}.[155]

GABA is a neurotransmitter in many invertebrate species.[156] Five responses to GABA have been identified on neurons of the marine mollusc *Apylsia*.[157] Two of these are inhibitory, and three excitatory. The excitatory responses to GABA are associated with conductance increases to Na^+ and/or conductance decreases to K^+.[157,158] One of the inhibitory responses to GABA may be associated with a conductance increase to K^+.[157] Although the nervous tissue of invertebrate species does not seem to contain high-affinity [³H]DZP binding sites that can be detected using the filtration assay,[12] BZs do affect certain invertebrate neurons and muscle.[159,160] In *Aplysia*, chlordiazepoxide and flurazepam were found to produce a GABA-like presynaptic inhibition of acetylcholine (ACh) release without, however, potentiating the action of GABA.[159] Ro 11-3128, a BZ that produces spastic paralysis in schistosomes, stimulated the influx of Ca^{2+} and Na^+ into the schistosomes while decreasing the influx of K^+.[160] Removal of Ca^{2+} from, or addition of Mg^{2+} to, the incubation medium blocked the spastic paralysis induced by Ro 11-3128.[160]

In guinea pig cerebellar slices, chlordiazepoxide potentiated the inhibitory action of GABA on spontaneous spike discharges and reversed the GABA antagonistic effect of picrotoxin in this system.[161] The experimental evidence suggested that chlordiazepoxide potentiated GABA by increasing membrane permeability to K^+ ion but not to chloride ion.[161] Earlier work by Okamoto[162] using this system suggested that inhibition of spontaneous spikes by GABA was associated with increases in K^+ as well as Cl^- permeability.[162]

GABA depolarizes the dorsal roots of the frog spinal cord and reduces dorsal root potentials in a way antagonized by bicuculline and picrotoxin.[163] This depolarizing response to GABA remained in the absence of external chloride ion but was abolished by removing external Na^+.[163] Primary afferent depolarization in dorsal roots (presynaptic inhibition) is potentiated by BZs[40,41] and antagonized by picrotoxin and bicuculline.[4,163]

GABA also partially depolarizes and inhibits hippocampal pyramidal cells when applied to the dendrites, as opposed to hyperpolarizing (also inhibitory) effects when applied to the soma.[65-68] The depolarizing response to GABA is also potentiated by BZs and involves increases in the conductance of both chloride and a cation, probably sodium ion.[65-68] Nanomolar concentrations of midazolam, when applied to CA_1 hippocampal neurons in vitro, cause a long-lasting hyperpolarization and moderate conductance increase which was not blocked by tetrodotoxin or intra-cellular injections of Cl^- ion, but which was blocked by omitting Ca^{2+} from the medium. Calcium spikes, in the presence of tetrodotoxin, were enhanced by midazolam at concentrations which did not potentiate the effects of GABA. It was concluded that these low nanomolar concentrations of midazolam inhibited the firing of CA_1 neurones by augmenting a Ca^{2+}-mediated K^+-conductance.[288]

In EDTA/water-dialyzed rat brain P_2 preparation, Ca^{2+}, Mg^{2+}, Ba^{2+}, Sr^{2+}, and Mn^{2+}, in concentrations up to 5 mM, have little effect on [³H]FLU binding. The BZ receptors in this preparation are extremely sensitive to inactivation by heat, with an overall half-life of less than 2 min at 60°C. More than 97% of the receptors are destroyed by heating for 30 min 60°C.[104] GABA (500 μM) or $CaCl_2$ (5 mM) individually provided little (<5%) protection, but the combination of GABA plus $CaCl_2$ provided about 50% protection (30 min, 60°C). Concentration–response curves for $CaCl_2$ in the presence of 500 μM GABA revealed a concentration-dependent, saturable protective effect with

50% of maximum protection near 500 μM $CaCl_2$ and a Hill number near or slightly less than 1. Conversely, a GABA concentration–response curve in the presence of 5 mM $CaCl_2$ produced a similar concentration-dependent, saturable protective effect with 50% of the maximum protective effect near 17 μM GABA.

Magnesium, Ba^{2+}, Sr^{2+} and Mn^{2+} can substitute for Ca^{2+} in this system, but all other divalent and trivalent cations tested (Be^{2+}, Ni^{2+}, Zn^{2+}, Cu^{2+}, Pb^{2+}, Co^{2+}, Hg^{2+}, Cd^{2+}, and La^{3+}) were inactive and antagonized Ca^{2+} plus GABA in essentially noncompetitive ways.[104]

A number of "classical" GABA-A receptor antagonists could substitute for GABA in the presence of 5 mM $CaCl_2$. These included APSA, BGPA, ImAA, muscimol, *trans*-4-aminocrotonate, dihydromuscimol, (S)-methylmus-cimol, d,l-homo-β-proline, IGV, P4S, and THIP. Muscimol is the most potent (EC_{50} = 5 μM), and ImAA the least (EC_{50} = 130 μM) potent. The magnitude of the protective effect, however, varies with the GABAmimetic, THIP providing the least (16%) and GABA the most (55%) protection (30 min at 60°C). Time courses of heat inactivation at 60°C in the presence of saturating concentrations of $CaCl_2$ (5 mM) and GABAmimetics were all bi- or polyphasic, with fast components having half-lives of less than 5 min and slow components with half-lives ranging from 28 min (THIP) to 95 min (GABA). These slow components, extrapolated back to zero time, yielded values (B_0 values) ranging from 42% (for THIP) to 69% (for GABA) expressed as percent of unheated control.[104]

Addition of 50 mM NaCl to saturating $CaCl_2$ + GABA increased the half-life of the slow component from 102 to 173 min and the B_0 value from 63 to 90%.[104]

Picrotoxin acts somewhat like GABA in the heat inactivation system: alone, picrotoxin has no protective effect, but together with $CaCl_2$ (5 mM), it protects about 22%, with 50% of its maximum protective effect at 154 μM and a Hill number near unity. In the presence of NaCl (50 mM), which alone has little (4%) protective effect, picrotoxin plus $CaCl_2$ now protect 39%, and picrotoxin produces 50% of its maximum effect at 50 μM (Hill number near unity). Picrotoxin does not affect the concentration–response curve for GABA in the presence of 5 mM $CaCl_2$. The results with picrotoxin suggest that picrotoxin, chloride, and Ca^{2+}, acting on separate but interacting receptors, synergistically protect BZ receptors from heat inactivation.[104]

In the EDTA- and water-dialyzed membrane preparation, NaCl alone can also provide almost complete protection against heat inactivation, with 50% of its maximum protective effect near 250 mM and a Hill number near 2. GABA (500 μM), which has no protective effect alone, shifts the NaCl concentration–response curve to the left and decreases the 50% effective concentration of NaCl to near 30 mM. Picrotoxin (100 μM) also shifts the concentration–response curve for NaCl slightly to the left (EC_{50} = 128 μM) but dramatically increases the Hill number to 3. Picrotoxin plus $CaCl_2$ further reduce the EC_{50} for NaCl to 66 μM and decrease the Hill number to 1.8.[104]

The Hill number (>2) for NaCl strongly suggest positive cooperative interactions between separate receptors in the BZ receptor complex. Positive cooperativity between separate recognition sites for Na^+ and Cl^- seems possible. The relatively low Hill number for NaCl in the presence of 5 mM $CaCl_2$ (1.2) tends to support this conclusion, since Ca^{2+} may have high affinity for the same site occupied by Na^+ ion, and when the cation site is nearly saturated with Ca^{2+}, the NaCl concentration–response curve would reflect the effect of chloride ion alone (EC_{50} near 211 mM).

Picrotoxinin, but not picrotin, can fully substitute for picrotoxin, together with NaCl and $CaCl_2$, in protecting BZ receptors against heat inactivation.[104,272]

Picrotoxin and related convulsants are most potent in potentiating the protective effect of 200 mM NaCl against heat inactivation (30 min, 60°C). Picrotoxin doubles the number of [^{3}H]FLU binding sites protected in the presence of NaCl from 35 to 70% of the unheated control with an EC_{50} near 15 μM, whereas picrotoxinin, anisatin, *t*-butylbicyclophosphate, IPTBO, and isopropylbicyclophosphotionate have EC_{50} values near 25, 3, 31, 12, and 3 μM, respectively.[104,272] Picrotin was inactive at 100 μM. The ability of picrotoxin (picrotoxinin), anisatin, and several "cage" convulsants to protect BZ receptors with increased potency in high but not low NaCl concentrations also indicates positive cooperativity between independent but interacting binding sites for cations, anions, and picrotoxin.

Taken together, these results indicate the existence of separate but interacting sites for BZs, cations, anions, picrotoxin, and GABA and are consistent with the idea that BZ-associated GABA receptors may regulate cation as well as anion (chloride) channels.[67,68,161]

It is fascinating that several (but not all) GABA-containing neurons (e.g., cerebellar Purkinje cells) also contain calcium-binding proteins.[164–167] It has been pointed out that three types of neurons known to produce voltage-dependent calcium spikes within their dendritic trees (cerebellar Purkinje cells, the neurons of the inferior olive, and CAl pyramidal cells of the guinea pig hippocampus) contain vitamin-D-dependent calcium-binding proteins.[166] Benzodiazepine receptors are probably located on the dendrites of cerebellar Purkinje cells[55,56,59,60,62] and of hippocampal CAl pyramidal cells.[68] There is a long-term reduction in the concentration of a calcium-binding protein in the hippocampus following kindling-induced epilepsy.[168]

8. ASSOCIATION OF PICROTOXIN–BARBITURATE SITES WITH BENZODIAZEPINE RECEPTORS

Four purine analogues, etazolate (SQ 20,009),[129–132,169,170] cartazolate (SQ 65,396),[132,170,171] tracazolate (ICI 136,753),[172] and EMD 28,422,[173,174] are able to restore behavior suppressed by punishment and to enhance binding of [^{3}H)DZP and [^{3}H]FLU to brain specific receptors.

Supavilai and Karobath[129,130,132] demonstrated that the enhancement of BZ binding by SQ 20,009 is dependent on chloride (or certain other "Eccles" anions) and antagonized by picrotoxin. This was confirmed by Leeb-Lundberg *et al.*,[169,175,176] who also found that [^{3}H]dihydropicrotoxinin ([^{3}H]DHP) binding to brain specific sites was competitively displaced by SQ 20,009 and SQ 65,396, with 50% displacement in the low micromolar range.[128,169,176]

Similarly, hypnotic barbiturates (e.g., pentobarbital) enhance BZ binding in a chloride-ion-dependent and picrotoxin-reversible way.[128,133,177,178] The (+) isomer of etomidate, a potent nonbarbiturate hypnotic, enhances [^{3}H]DZP binding to sites in rat forebrain membranes but not to sites in cerebellum.[179] Barbiturates that enhance the binding of BZ do so with effective concentrations similar to those that displace [^{3}H]DHP.[128,133] The antagonism of this barbiturate effect is achieved by concentrations of picrotoxin (1 μM) near the K_d for [^{3}H]DHP. Other convulsants that displace [^{3}H)DHP from brain specific sites in the low micromolar range [*t*-butylbicyclophosphate, tetramethylene disulfotetramine (TETS), and Ro 5-3663] also reverse the enhancing effects of barbiturates on BZ binding.[128,169]

SQ 20,009,[132] pentobarbital,[133] and (+)-etomidate[179] enhance BZ binding to brain specific sites by increasing BZ affinity without changing the total number of BZ binding sites (B_{max}). The enhancing effects of the purine analogue EMD 28,442 on [^{3}H]DZP binding may be more an effect on B_{max} than K_d.[173,174]

The potency of several hypnotic barbiturates in enhancing BZ binding correlates well with their anesthetic/hypnotic activity or potency in reversing GABA antagonists electrophysiologically.[128,133] However, some anticonvulsant barbiturates (phenobarbital and metharbital) as well as chlormethiazole[178] fail to enhance [^{3}H]DZP binding but, instead, antagonize the enhancement produced by pentobarbital and SQ 20,009 in concentrations similar to those required to displace [^{3}H]DHP binding.[178]

Barbiturate and SQ 20,009 enhancements of BZ binding are not additive, whereas enhancement of BZ binding by GABA is additive with the effect of SQ 20,009 or pentobarbital.[169,176] The enhancement of BZ binding by GABA, which is not chloride ion dependent, is selectively blocked by THIP or imidazole acetate, whereas the enhancement by pentobarbital or SQ 20,009 (which is chloride ion dependent) is selectively blocked by picrotoxin.[169]

Taken together, the above findings support the conclusion that picrotoxin, barbiturates, and the pyrazolopyridines (SQ 20,009 and SQ 65,396) bind to the same or overlapping sites in the BZ receptor complex, which is different from independent (but interacting) sites for BZs, GABA, anions, and cations (see also Section 7).

"Picrotoxin" binding sites (like GABA receptors) may exhibit a rather high degree of heterogeneity. The displacement of [^{3}H]DHP binding by certain drugs yields "flat" curves with Hill numbers less than 1, indicating possible heterogeneity of DHP binding sites.[128] Further, the very potent convulsants *t*-butylbicyclophosphate (which produces convulsions in 100% of adult mice at about 25 μg/kg)[180] and TETS (LD_{50} about 200 μg/kg in mice) are surprisingly weak in displacing [^{3}H]DHP binding, with IC_{50} values near 2 and 4 μM, respectively.[176] This suggests that these two convulsants, especially *t*-butylbicyclophosphate, bind with much higher affinity to another, perhaps similar, site, which is not significantly labelled by [^{3}H]DHP. Phenobarbital seems to be an effective antidote to bicyclophosphate poisoning,[181,182] whereas DZP does not appear to be as effective.[91] Recently, picrotoxin and its active component picrotoxinin, as well as the related convulsant anisatin, isopropylbicyclophosphate, isopropylbicyclophosphorothionate, TETS, and *t*-butylbicyclophosphate were reported to protect ^{3}H-flunitrazepam binding sites, on rat brain membranes, from heat inactivation (30 min at 60°C) in the presence of 200 mM NaCl (but not 50 mM NaCl), Picrotin was inactive in this test.[104,272]

^{35}S-t-butylbicylophosphorothionate (TBPS) has proved to be an improved ligand for picrotoxinin binding sites.[273] ^{35}S-TBPS binding (Kd = 17 nM at 25°C in 200 mM KBr) can be measured using a conventional filtration assay and non-specific binding is typically 10–15% of the total binding.[273] The optimum temperature for ^{35}S-TBPS binding is near 21°C with none at 0°C. Binding is entirely ion dependent and 50% of maximum binding is obtained near 80 mM KBr or NaBr or 110 mM NaCl or KCl. GABA and all GABA-A receptor agonists tested, potently inhibit ^{35}S-TBPS in the presence of 200 mM KBr at 25°C, but GABA fails to inhibit in the presence of Na_2SO_4, $NaPO_4$ (pH 7.5) or NaF (non-Eccles anions). In 200 mM KBr at 25°C the inhibitory effect of GABA is competitively reversed by an ultrapotent, bicuculline-like competitive GABA antagonist, R5135,[252] as well as by bicuculline itself, strychnine, brucine, d-tubocurarine, amoxapine and a series of N-aryl-piperazine GABA antagonists (R. Squires and E. Saederup, in preparation). The GABA-B receptor agonist, baclofen, has no effect on ^{35}S-TBPS binding. The complete inhibition of ^{35}S-TBPS binding by GABA and the GABA-A receptor agonists, demonstrates that all TBPS binding sites in brain are coupled to GABA-A receptors. Negligible specific TBPS binding is found outside the brain. ^{35}S-TBPS binding is also inhibited potently by a large number of "cage" convulsants, bemegride and a series of tetrazole convulsants the most potent of which is 8-t-butylpentamethylene tetrazole (IC_{50} = 3.7 μM) (R. Squires, E. Saederup, J. Crawley, P. Skolnick, and Paul in preparation). The pyrazolopyridines SQ 65,396 (IC_{50} = 190 nM) and SQ 20,009 (880 nM) (+) etomidate (2.8 μM), (−) secobarbital (31 μM), methaqualone (43 μM), ethaqualone (13 μM), nisobamate (22 μM), carisoprodol (120 μM), (mebrobamate analogues) the cannabinoids Δ^9-THC (11 μM), dimethylheptylpyran (6.5 μM), androsterone (1.4 μM), diethylstilbestrol (3.2 μM), cyproheptadine (59 μM) and pizotifen (90 μM) are among the substances recently found to potently inhibit ^{35}S-TBPS binding. The inhibitory effects of all anxiolytics, sedatives, hypnotics and anticonvulsants tested were at least partially reversed by R5135 (10 nM) while the inhibitory effects of all the convulsants tested, with the exception of anisatin, were not reversed by 10 nM R5135.[273]

TBPS binding sites appear to be heterogeneous since ^{35}S-TBPS dissociates from specific bindings sites in a polyphasic way (slow component constitutes about 71% of all binding sites), and the binding sites are inactivated by 60°C in a polyphasic way in 200 mM NaCl, also with a slow component constituting about 66% of all sites.[273] Further, inhibition of ^{35}S-TBPS binding by 5 μM GABA is partially reversed (66% to 75%) by Loxapine, pipazethate, thebaine, emetine and 1-(m-chlorophenyl)-piperazine (R. Squires and E. Saederup, in preparation).

9. BENZODIAZEPINE RECEPTOR MULTIPLICITY

The extraordinary degree of GABA receptor multiplicity in invertebrate species, illustrated by five known types of GABA receptors in *Aplysia*[157] and four known types in the crayfish (*Astacus fluviatilis*),[183] suggest that the degree of GABA receptor multiplicity in vertebrate species could be even greater. There is already considerable evidence for extensive GABA receptor multiplicity in vertebrate species, with BZ receptors representing a subpopulation of GABA receptors, which are themselves heterogeneous. Radiohistochemical studies indicate that some GABA (muscimol) receptors are probably not associated with BZ receptors.[51,184,185] In addition to these non-BZ-coupled [^{3}H]muscimol binding sites, there are Ca^{2+}-dependent GABA-B receptors that exhibit high affinity for GABA and (−)-baclofen but very low affinity for most GABA-A receptor agonists, including muscimol, and the GABA-A receptor blocker bicuculline methobromide.[139,186] A GABA receptor exhibiting several of the properties of GABA-B receptors acts to decrease calcium-dependent action potentials in sensory neurons from embryonic chick dorsal root ganglia in tissue culture.[140,141]

On the basis of early studies on [^{3}H]DZP binding, it was incorrectly concluded that BZ receptors were homogeneous.[6,7,11,25-29]

The first evidence for BZ receptor heterogeneity came from heat inactivation experiments in sodium phosphate buffer.[18,19] Benzodiazepine receptors in baboon (*Papio papio*) brain are inactivated at 65°C in 50 mM $NaPO_4$, pH 7.4, in a bi- or polyphasic manner,[18] in contrast to apparent monophasic inactivation of BZ receptors from rat brain when heated at 60°C in 50 mM Tris HCl, pH 7.5.[7] When the heat inactivation experiments were repeated using rat brain membranes heated at 60°C in 50 mM Tris HCl and 50 mM $NaPO_4$ buffers, pH 7.5, respectively, it was found that the BZ receptors in rat brain also disappeared in an apparently biphasic fashion in sodium phosphate but almost monophasically in Tris HCl.[19]

Early studies of [^{3}H]DZP and [^{3}H]FLU dissociation from receptors in rat brain P_2 membranes indicated monophasic dissociation of both ligands.[11] A more careful reinvestigation of the dissociation of these two ligands revealed at least two components for both, regardless of the buffer use.[19] Costa *et al.*[126] also reported a polyphasic dissociation of [^{3}H]DZP from rat brain membranes.

A group of triazolopyridazines (e.g., CL 218,872) was found to displace [^{3}H]FLU from brain specific binding sites with Hill numbers less than 1, a finding that is consistant with (but does not prove) multiple binding sites with different affinities for CL 218,872.[19]

The idea of multiple BZ receptors received further support from the finding that CL 218,872 displaces [^{3}H]FLU from sites on rat cerebellar and hippocampal membranes with IC_{50} values near 37 nM and 330 nM and Hill numbers of 0.9 and 0.6, respectively.[63] The selective displacement of [^{3}H]FLU from certain well-defined brain regions by CL 218,872 was confirmed by radiohistochemical studies.[53] Light microscopic autoradiography revealed that CL 218,872 preferentially displaces [^{3}H]FLU from a subclass of binding sites (so-called BZ_1 sites) present in the highest proportion in cerebellum but also present in relatively high concentrations in the globus pallidus and parts of the cerebral cortex.[53] Brain areas having high densities of [^{3}H]FLU binding sites with relatively low affinity for CL 218,872 (BZ$_2$ sites) include superior colliculus, caudate–putamen, and parts of the dentate gyrus.[53] Like CL 218,872, methyl- and ethyl-BCC were also found to displace [^{3}H]FLU from specific binding sites on hippocampal membranes with IC$_{50}$ value about six- to eightfold higher than those obtained in cerebellum.[21,64] Hill numbers for ethyl-BCC were 0.95 and 0.75 in rat cerebellum and hippocampus, respectively,[64] whereas in calf brain, the corresponding Hill numbers for methyl-BCC were 0.74 and 0.64.[21] The ability of the BCC esters to discriminate between BZ receptors in cerebellum and hippocampus have been demonstrated in rat, mouse, guinea pig,[187] bovine,[21,187] and human brains (R. Squires, unpublished data).

Radiohistochemical studies similar to the CL 218,872 studies,[53] indicate the ethyl-BCC exhibits a pattern of selective [^{3}H]FLU displacement in brain similar to the pattern produced by CL 218,872, although the latter substance achieves a greater degree of discrimination (M. Kuhar, D. Niehoff, and J. Unnerstall, personal communication). Also, with a quantitative light microscopic radiohistochemical method, it was found that in more than 200 brain regions, the addition of GABA (10 μM) to the incubation medium increased [^{3}H]FLU binding (compared to the no-GABA control).[54] However, the magnitude of the increase in [^{3}H]FLU binding brought about by GABA varied dramatically from one brain region to another: in lamina IV of the cortex, the head of the caudate, the CA3 region of the hippocampus, the mediolateral thalamus, and the entire dentate gyrus, the increase in [^{3}H]FLU binding induced by GABA was less than 100%, whereas in the molecular layer of the cerebellum and the globus pallidus, the increase was more than 200%.[54] These results indicate that there are different types of BZ/GABA receptor complexes, which can be activated by GABA to different degrees.

In preincubated brain slices, bicuculline (100 μM) reduced [^{3}H]FLU binding in lamina IV of the cerebral cortex and molecular layer of the dentate gyrus by 40–50% but had no effect on [^{3}H]FLU binding in the molecular layer of the cerebellum or the globus pallidus (two regions in which [^{3}H]FLU binding is enhanced more than 200% by the addition of GABA).[54] Bicuculline also inhibits the binding of [^{3}H]FLU to rat brain P$_2$ membranes in a bi- or polyphasic way, supporting the idea that bicuculline discriminates between different kinds of BZ receptors.[20]

Time courses of heat inactivation of BZ receptors at 60°C in the presence of 50 mM NaCl plus saturating concentrations of GABAmimetics were found to be polyphasic in all cases[103] in both forebrain and cerebellum, with fast components (half-lives less than 5 min) and relatively homogeneous slow components with half-lives ranging from about 50 to 100 min, depending on the GABAmimetic used. When these slow components are extrapolated back to zero time, B_0 values, expressed as percent of unheated control binding, ranged from about 30% for P4S to 75% for GABA; ImAA yielded a B_0 value near 50%, whereas THIP was inactive in this experimental situation. In rat cerebellum, the B_0 values tended to be somewhat less than in forebrain.[103] It is clear, therefore, that by the criterion of polyphasic heat inactivation, BZ receptors in rat cerebellum as well as forebrain exhibit extensive heterogeneity, with presumptive evidence for at least four BZ receptor complexes in both brain regions.[103]

Time courses of BZ receptor inactivation by heat in the presence of saturating concentrations of $CaCl_2$ and GABAmimetics yield a somewhat different picture.[104] In this system also, all time courses were polyphasic in both cerebellum and forebrain, but THIP, IGV, and P4S yielded B_0 values near 50% of the unheated control, whereas GABA, APSA, BGPA, ImAA, and muscimol yielded B_0 values near 65%, providing evidence for at least three BZ receptor complexes in both cerebellum and forebrain.[104]

Scatchard and competition analysis of [³H]propyl-BCC (PrBCC) binding revealed the presence of three classes of binding sites with superhigh, high, and low affinities.[188] The superhigh-affinity PrBCC sites, however, constitute only 3–6% of the total sites.[188]

In the filtration assay, a low concentration (0.3 nM) of [³H]PrBCC selectively labeled BZ_1 receptors, since [³H]PrBCC is washed off BZ_2 receptors during filtration.[187] The ratio of specifically bound [³H]PrBCC to [³H]FLU varies from one brain region to another, being highest in cerebellum ("pure BZ_1 sites") and lowest in hippocampus and globus pallidus.[187]

Kenazepine, an alkylating benzodiazepine, reacts noncompetitively and irreversibly with some BZ receptors and competitively (reversibly) with others.[189] Cerebellum contains the largest proportion (80%) of the noncompetitive type, whereas hippocampus and cortex contain a preponderance of competitive-type receptors (about 80 and 50%, respectively). Hill numbers for kenazepine are about 0.7 in cortex and cerebellum but near 1 in dorsal hippocampus.[189] These, and the heat inactivation experiments described above, clearly indicate a more complex pattern of receptor heterogeneity than that described by differential interactions with CL 218,872 or BCC esters (BZ_1 and BZ_2 receptors).

In rabbit cerebral cortex, two [³H]DZP binding sites have been reported, with K_d values of 5 nM and 84 nM, constituting 30 and 70% of the total binding sites, respectively.[190] Similarly, in rat hippocampus, two [³H]DZP binding sites have been reported, with K_d values of 3 nM and 900 nM, constituting 5% and 95% of the total binding sites, respectively. The low-affinity [³H]DZP binding site was not detected in cortex or cerebellum or in any of the three brain regions using [³H]FLU as the radioligand.[191]

Diethyl pyrocarbonate was reported to selectively inhibit, in an irreversible manner, about 40% of high-affinity [³H]DZP binding sites in whole-rat-brain membranes.[192] The pyrocarbonate-resistant fraction of [³H]DZP binding was also enhanced by muscimol or by pentobarbital to the same extent as the untreated sites.[192] Because of the great instability of diethyl pyrocarbonate in aqueous media (half-life about 30 min at 0°C in water), it is not possible to obtain meaningful time courses of BZ receptor inactivation in the presence of this reagent. At 1 mM, diethyl pyrocarbonate (20 min at 0°C) inhibits roughly to the same extent in all brain regions examined (cortex, cerebellum, hippocampus, and others) (R. Ticku, personal communication).

Under appropriate conditions, membranes prepared from rat cerebral cortex, bulbus olfactorius, striatum, and hippocampus can be covalently photolabeled with [³H]FLU (see Section 11). Following solubilization, four labeled proteins with mol. wts. 51,000, 53,000, 55,000, and 59,000 can be separated by SDS-polyacrylamide gel electrophoresis.[274,300] In contrast, membranes from rat cerebellum photolabeled with [³H]FLU in the same way yielded mainly only one labeled protein with mol. wt. of 51,000.[274,300] CL 218,872 inhibited the irreversible photolabeling of the 51k, 53k, 55k, and 59k proteins, by ³H-FLU, with IC_{50} values of 126 nM, 520 nM, 1,778 nM and 1,000 nM, respectively, while ethyl BCC inhibited photolabeling with IC_{50} values of 0.63 nM, 3.7 nM, 8.6 nM, and 32 nM, respectively.[300] With [³H]FLU at 0.4 nM, GABA (10 μM) enhanced, and bicuculline (10 μM) reduced, specific photolabeling to both 51,000-dalton and 55,000-dalton proteins.[193]

Several groups of investigators have noted that not all BZ receptors can be solubilized from brain membranes with detergents, even after repeated extractions[108,109,113,196] (see Section 10). Lo, Strittmatter, and Snyder[196] found that 2% Triton X-100 could only extract a limited fraction of BZ receptors from calf brain membranes (40°C at 1 h) even after repeated extractions. This fraction ranged from 7–9% in corpus striatum and cerebellum to 45% in hippocampus. The solubilized receptor fraction had lower affinity for CL 218,872 and methyl-BCC than the membrane-bound receptors that resist solubilization, indicating that Triton selectively solubilized BZ_2 receptors while leaving BZ_1 receptors in the membrane.[196] However, earlier work by Gavish and Snyder[112] showed that the fraction of BZ receptors selectively solubilized from whole calf brain by 1% Triton X-100 (presumably BZ_2 receptors) is itself heterogeneous, since it yields two labeled proteins (55,000 and 62,000 daltons) after photolabeling with [³H]FLU and SDS-polyacrylamide gel electrophoresis.[112]

Similarly, the solubilization-resistant (BZ_1) fraction is also heterogeneous, since it contains both high- and low-affinity CL-218,872 binding sites in cortex, striatum, hypothalamus, and colliculi.[197] The same differential solubilization has been obtained with other detergents, including sodium desoxycholate (DOC), digitonin, Nonidet P40, and Lubrol-PX.[196] A fraction of BZ receptors selectively solubilized from rat cerebral cortex by 0.5% Lubrol PX was found to be heterogeneous with respect to heat inactivation.[113] About half of the fraction resistant to solubilization with 2% Triton X-100 (the BZ_1 fraction) can be solubilized by a combination of Triton plus 1 M NaCl.[196] After photolabeling the Triton-soluble and Triton-resistant fractions, respectively, and selective degradation with trypsin, chymotrypsin, or cyanogen bromide, the two fractions yield distinctly different patterns of labeled peptides following HPLC separation (S. H. Snyder and M. M. S. Lo, personal communication).

Native BZ receptors are probably pentamers or higher polymers built up of monomers with mol. wts. in the 50–60,000 range (see Section 10). Obviously, several types of monomers, which might combine in different proportions in the polymeric forms, could result in rather extensive BZ receptor heterogeneity. The evidence presented above for BZ receptor heterogeneity in rat cerebellum, a region apparently containing only 51,000-dalton subunits, suggests that the 51,000-dalton fraction itself may be heterogeneous, as pointed out by Sieghart Karobath.[193] Thus, both "BZ_1" and "BZ_2" receptors appear to represent heterogeneous subpopulations. Recently, we have photolabelled membranes from rat forebrain and cerebellum with [³H]FLU in 200 mM NaCl containing 500 μM GABA, washed the membranes by centrifugation at 37°C to remove free ³H-flunitrazepam, denatured the washed membrane fraction by heating in a boiling waterbath for 10 min, followed by another centrifugation. The pellet was resuspended in 50 mM Tris-HCl, pH 8.5 containing 20 mM CaCl₂ and 1.5 mg/ml TPCK-treated bovine pancreatic trypsin and incubated at 37°C for 30, 60 and 90 min. The trypsin was inactivated by placing the reaction mixture in a boiling waterbath for 10 min, then centrifuging. The supernatant, which now contained most of the radioactivity, was put onto a Bondapak C18 reverse phase column (10 μ particle size) and eluted using a 0 to 60% acetonitrile gradient containing 0.25% tri-fluoro acetic acid. Using rat forebrain, five major peaks of radioactivity emerge from the column, only two of which are present in the cerebellum. The first three peaks are probably tetra or penta-peptides, the fourth peak contains about 20 amino-acid residues and the fifth about 25. Peaks 3 and 5 are the ones found in the cerebellum. Photolabeling of peaks 3 and 5 is selectively blocked by methyl BCC while all five peaks are blocked by CGS 8216. These peaks are apparently not precursors or products of each other since the relative amounts of the peaks do not change significantly during the time course of trypsin digestion indicating that the five peaks are "core" tryptic peptides (K. -S. Hui, M. Hui, E. Saederup and R. Squires, in preparation). These results are consistent with, but do not prove, the assumption that there are five different genes, coding for five different but homologous proteins which can be photolabelled with ³H-flunitrazepam. This assumption is consistent with the findings of Sieghart et al.[274,300] who reported the presence of four proteins with slightly different molecular weights (51 to 59k) which are specifically photolabelled with [³H]FLU, and separated by SDS-poly acrylamide gel electrophoresis. Cerebellum was reported to contain almost exclusively the 51k species and it is tempting to speculate that this fraction is, in fact, heterogeneous consisting of two major proteins which yield tryptic peptides 3 and 5, respectively. In the nerve cell membrane, BZ receptors consist of aggregates, probably pentamers, or larger, of 50 to 60k subunits. The acetylcholine nicotinic receptor model, a pentamer consisting of four homologous sub units,[275,285] each coded for by separate genes ($\alpha_2 \beta \gamma \delta$), may be similar in some respects to the arrangements of sub units in the BZ/GABA receptor complexes.

10. SOLUBILIZED BENZODIAZEPINE RECEPTORS

Benzodiazepine receptors that retain many of the properties of the membrane-bound receptors can easily be solubilized with a variety of detergents.

The first two reports of solubilized BZ receptors were published almost simultaneously in 1979 by Lang et al.[198] and by Yousufi et al.[113] and were followed rapidly by a third (Gavish et al.[199]).

The detergents used successfully to solubilize BZ receptors include Triton X-100,[111,112,114,115,196,198–200] Lubrol-PX,[113,201] sodium DOC,[107,109,110] and digitonin.[199] Triton X-100 was usually used at a concentration near 1% with an extraction time of 60 min near 0°C.[111,112,196,199,200] In two reports,[114,115] 0.05%

Triton X-100 selectively solubilized BZ receptors without GABA receptors when membranes were incubated for 30 min at 37°C. Lubrol-PX was used at a concentration of 0.5% or 1%, and brain membranes were extracted for 30 min at 0 to 4°C.[113,201] Digitonin was used at a concentration of 1%, and extraction was for 30 min at 4°C.[199]

Solubilized BZ receptors are usually assayed by incubating with [³H]DZP or [³H]FLU, usually 15 to 60 min at 0–4°C,[107,113,115] sometimes preceded by a preincubation for 15 min at room temperature[198] or 37°C[200] then precipitating the receptor-bound [³H]ligand with PEG[111,113,200,201] or with 20% to 50% saturated ammonium sulfate.[109,199] [³H]Flunitrazepam bound to the solubilized receptors can also be separated from free [³H]FLU by passing the assay mixture through a column of Amberlite® XAD-2 and eluting with cold 50 mM Tris HCl, pH 7.4. Free [³H]FLU was adsorbed to the XAD-2, whereas [³H]FLU bound to the soluble receptors passed through the column.[108] Since BZ receptors seem to be glycoproteins,[286] solubilized receptors can be immobilized on concanavalin A-Sepharose beads.[196] Soluble receptors with bound [³H]ligand can also be adsorbed directly onto Whatman DEAE cellulose filters[198,200] or onto Whatman GF/B glass fiber filters.[107] Benzodiazepine receptors solubilized with Trition X-100, Lubrol-PX, digitonin, or sodium desoxycholate (DOC) bind [³H]DZP and [³H]FLU with K_d values somewhat higher than those found using membrane-bound BZ receptors. However, several BZs were reported to displace bound [³H]BZs from soluble receptors with K_i values similar to those found using intact brain membranes.[108,109,113,199–201] Like membrane-bound BZ receptors,[7] soluble BZ receptors are destroyed by proteolytic enzymes,[108,198] 1 mM urea, or 2 M guanidinium chloride.[113]

The mol. wt. of soluble BZ receptors was reported to be near 200–250,000[107,108,110,113,198,200,279] determined by sucrose density gradient centrifugation or gel filtration. This value is approximately four to five times the value obtained by SDS-poly acrylamide gel electrophoresis after covalently photolabeling with [³H]FLU.[73,112,193,194,202,203,278] However, under certain conditions, different mol. wts. for soluble BZ receptors have been obtained: a MW near 670,000, determined by Sepharose gel filtration, was reported in the presence of 50 mM Tris HCl, pH 7.4, 0.5% Lubrol-PX, 0.15 M KCl, and 20% glycerol.[110,277] A mol. wt. near 61,000 for Lubrol-PX-solubilized BZ receptor was reported by gel filtration on Sephadex G-200, eluting with 200 mM NaCl plus 10 mM sodium phosphate, pH 7.0.[201] A mol. wt. near 111–116,000 for photolabeled receptors was reported using gel filtration on Ultrogel® AcA-34 in the presence of 0.1% Triton X-100, 150 mM KCl, and 35 mM potassium phosphate.[202] The presence of higher salt concentrations may favor dissociation of polymeric BZ receptors into dimers[202] or monomers.[201]

It has been noted by several groups of investigators that detergents are only able to solubilize part of the BZ receptors from membranes.[108,109,113,115,196] The fraction of BZ receptors preferentially solubilized by detergents was reported to have some of the properties of "BZ₂" receptors[196]: they exhibit lower affinities for CL 218,872 and β-carboline-3-carboxylate esters, and very little BZ binding material can be extracted from bovine cerebellum, an area thought to contain almost exclusively "BZ₁" receptors. However, two findings indicate that detergent-solubilized BZ receptors are not homogeneous: (1) Lubrol-PX-solubilized receptors are inactivated by heat in a bi- or polyphasic manner,[113] and (2) Triton X-100-solubilized BZ receptors, after covalent photolabeling with [³H]FLU and SDS-polyacrylamide gel electrophoresis, yield two proteins with mol. wts. of 55,000 and 62,000.[112]

Binding of [³H]BZs to soluble BZ receptors has, in most cases, been reported to be enhanced by GABA and GABA-A receptor agonists,[108,109,111,112,200] and [³H]GABA or [³H]muscimol appear to bind to the same material that binds [³H]BZs[107,108,110,112,277,278,279] even after extensive purification.[110,112,278,301] About two to four molecules of [³H]GABA or [³H]muscimol were bound to soluble BZ receptors for every [³H]FLU molecule bound.[110,197,277,278] Some groups of investigators, however, did not find enhancement by GABA of [³H]BZ binding to detergent-solubilized BZ receptors,[113,115,201] and three groups of investigators[114,115,201] have reported that, using appropriate extraction procedures, [³H]muscimol (or[³H]GABA) binding material can be physically separated from [³H]BZ binding material. It cannot be excluded, however, that GABA receptors are covalently linked to BZ receptors but selectively inactivated during some solubilization procedures.

Effective purification of soluble BZ receptors is achieved by affinity chromatography on Sepharose 4B.[112,200] Purification, 5200-fold, to apparently homogeneity was achieved on an affinity column made of Sepharose 4B coupled to Ro 5-3027 (delorazepam).[200] After the column was extensively washed to remove all extraneous protein, the BZ receptors were eluted with 6 mM chlorazepate. The purified receptor had an apparent mol. wt. near 200,000 when determined by gel filtration on Ultrogel® AcA 22 in 20 mM Tris HCl, pH 7.4, containing 0.1% Triton X-100. The same material, when subjected to SDS-urea-polyacrylamide gel electrophoresis, yielded a single protein band with an apparent mol. wt. of 60,000.[200] Sigel et al.[278] purified BZ/GABA receptor complexes 1,800-fold by BZ-agarose and ion exchange chromatography. [³H]muscimol and [³H]FLU were reported to bind to the same physical structure in a ratio between 3 and 4. This binding material had a Stokes radius of 7.3 nm and a sedimentation coefficient, of 11 S. Polyacrylamide gel electrophoresis of this unlabelled material in the presence of SDS and dithiothreitol revealed two major proteins with MWs of 57k and 53k. Both proteins could be photolabeled with [³H]FLU.[278]

11. PHOTOLABELING OF BENZODIAZEPINE RECEPTORS

The ability of [³H]FLU to form a covalent bond with brain specific BZ receptors when irradiated with near-UV light was discovered by Hanns Mohler and associates.[194,203] Such photolabeling with FLU requires low temperatures,[194] is blocked by other BZs with potencies corresponding to their affinities for the receptor,[203] and has a time course corresponding to the time course of irreversible inhibition of BZ receptors as measured by subsequent [³H]DZP binding.[195,203] Half-maximum incorporation of [³H]FLU occurred at a concentration similar to the apparent dissociation constant for [³H]FLU. After about 50% inactivation of BZ receptors by UV irradiation in the presence of 3 nM nonradioactive FLU, Scatchard analysis revealed that the B_{max} was decreased at least 50% and the K_d was increased about 30%.[203] When BZ receptors were photolabeled with [³H]FLU, solubilized, and subjected to sodium dodecyl sulfate (SDS)-polyacrylamide gel electrophoresis, a tritium-labeled protein with a molecular weight near 50,000 was detected.[194,203] Electron microscopic autoradiography of brain slices photolabeled with [³H]FLU revealed a predominant localization of labeling near synapses.[203] One-third of the photolabeled BZ receptors were found to be associated with nerve endings that were stained with antibodies to glutamate decarboxylase, a marker of GABA neurons.[204]

The nitro group in the 7 position of FLU appears to be essential for photolabeling of BZ receptors, since two other BZs with nitro groups in the same position also irreversibly block BZ receptors when irradiated with UV light, whereas three other BZs lacking the 7-nitro group do not.[195] Photolabeling of BZ receptors with [³H]FLU can be blocked by low concentrations of active BZs but not by several ligands having low affinity for BZ receptors but high affinity for other receptors (e.g., spiperone, atropine, naltrexone, propranolol, and GABA).[195] About 15–20% of all BZ receptors in rat cortex[194] or whole bovine brain[195] are not photolabeled by [³H]FLU even after 24 h of UV irradiation.[195] Benzodiazepine receptors first solubilized from bovine brain membranes using Triton X-100[112] or from rat brain membranes using Lubrol-PX[73] can be subsequently photolabeled. When subjected to SDS-polyacrylamide gel electrophoresis, the Triton X-100-extracted, photolabeled receptors from bovine brain yielded two labeled protein (55,000 and 62,000 daltons[112]), whereas the Lubrol-PX-extracted receptors yielded only one band (48,000 daltons[73]).

Although both soluble and membrane-bound BZ receptors can be covalently photolabeled with [³H]FLU, there is some apparent disagreement regarding the number of proteins labeled. Three groups of investigators found only one protein specifically photolabeled with [³H]FLU,[73,194,202,203] whereas two others reported that two[112] or three[193] or four[274,300] proteins could be labeled. The mol. wts. of all the labeled proteins, determined by PAG-SDS electrophoresis, are similar and range from 48,000[73] to 62,000.[112] The labeling procedures were similar and consisted of a preincubation of brain membranes with [³H]FLU (3 to 10 nM) at 0–4°C for 60 to 90 min before UV irradiation for 4 to 10 min. However, the ionic composition of the incubation media varied significantly: 50 mM Tris HCl, pH 7.4[73]; Krebs–Tris (containing about 120 mM NaCl)[112,203];150 mM NaCl plus 50 mM Tris citrate, pH 7.1[193]; and 100 mM Tris citrate.[202] Where two, three or four bands were labeled, the incubation mixture contained high (>100 mM) chloride ion concentrations.[112,193,274,300] Although a high chloride concentration does not seem to be a sufficient condition for multiple band labeling,[194] it may be necessary. Sieghart and Karobath,[193] who reported the photolabeling of three bands with [³H]FLU using membranes from cortex, striatum, olfactory bulb, and hippocampus (but only one band in cerebellum), used the highest NaCl concentration, the longest preincubation time (90 min), the highest [³H]FLU concentration (10 nM), and the longest UV irradiation time (10 min). There seems to be a tendency, in all cases, for the lower-mol.-wt. bands (near 50,000) to be preferentially labeled.

I have recently found that [³H]FLU in concentrations of 0.2 or 1 nM reached equilibrium very slowly at 0°C in the presence of 50 mM Tris HCl and 200 mM NaCl with or without GABA. Only about 50% of equilibrium binding is reached after a 3-h incubation at 0°C without preincubation.[272] Furthermore, the association of [³H]FLU with BZ receptors is polyphasic, with a slow component having a half-life of several hours at 0°C.[272] Equilibrium binding of [³H]FLU (0.2 nM) at 0° is achieved in 60–90 min after a 10- to 15-min preincubation at 37°C.[272] It therefore seems possible that the lower-mol.-wt. proteins (48–51,000) are preferentially labeled under the conditions used in all published reports.[73,112,193–195,202,203]

It would be of interest to repeat these photolabeling experiments starting with a 10- to 15-min preincubation at 37° in 100–200 mM NaCl followed by an additional 60- to 90-min incubation at 0° before UV irradiation.

Using a Lubrol-PX-solubilized BZ receptor preparation, Thomas and Tallman[73] found that about one site was blocked and unavailable for reversible binding for each site photolabeled. In contrast, when membrane-bound sites were photolabeled, about four sites were inactivated for each site photolabeled.[73,203] Sieghart and Drexler[274] examined photolabeling of BZ receptors, using ³H-FLU, in 14 regions of the rat brain and found that only 12.5% to 18% of the reversible ³H-FLU binding sites can become irreversibly photolabeled, depending on the brain region. In other words, in rat brain, only one out of 5 to 8 FLU binding sites can be irreversibly photolabeled and the remaining binding sites have much reduced affinity for many diazepam-like ligands (for example, the affinity for diazepam is reduced by a factor of 50 by prephotolabeling with FLU), but there are unchanged affinities for several GABA-neutral or GABA-negative ligands for BZ receptors[283] and increased affinities for strongly GABA-negative convulsant ligands such as methyl-BCC and DMCM.[276,287] However, this differential "photo shift" in the affinities for BZ receptor ligands is not a universal discriminator between GABA-positive and GABA-negative ligands for BZ receptors since the GABA-positive ligands CGS-9896[284] and Suriclone[302] (J.-C. Blanchard and L. Julou, personal communication) retain their high affinities for BZ receptors after photolabeling with FLU. Peripheral (kidney-type) BZ receptors (see Section 4) are not photolabeled by [³H]FLU.[73]

12. THE ELUSIVE "ENDOGENOUS LIGAND"

On the basis of early work with BZ receptors, [6–8] Braestrup and I, together with others working in the field,[29,205–207] seriously considered the possibility that there might be hitherto undiscovered "endogenous ligands" acting on the receptor, perhaps analogous to the enkephalins[32,33] and endorphins,[34,35] the endogenous ligands for the opiate receptors.[36] After the discovery of coupling between BZ and GABA receptors, (See section 5 above) this hypothesis became much less probable. In 1978, a noncompetitive protein inhibitor of [³H]GABA binding extracted from rat brain membranes was proposed as an endogenous regulatory of GABA function at the receptor level.[208] This substance ("GABA-modulin"), an acidic protein with a mol. wt. near 15,000 was reported to be a thermostable (95°C for 15 min) competitive inhibitor of [³H]DZP binding, but only after preincubation with BZ receptors for at least 15 min.[208,209] Further, several BZs were reported to partially reverse the inhibitory effects of "GABA-modulin" on [³H]GABA binding.[209] It was speculated that this endogenous protein inhibitor might be an endogenous ligand for the entire BZ/GABA receptor complex[206,209] and play a role in determining the level of anxiety.[209] These results have not been independently confirmed, and the original "endogenous inhibitory protein" proved to be a mixture of proteins or peptides.

Further purification separated the [³H]GABA-inhibitory from the [³H]DZP-inhibitory activity.[115] The "new" GABA-modulin[115] has a molecular weight of 15,000 is inactivated at 95°C, is destroyed by proteolytic enzymes, is precipitated by 30–60% saturated ammonium sulfate, binds neither [³H]GABA nor [³H]DZP, inhibits [³H]GABA binding by a noncompetitive mechanism, and does not inhibit [³H]DZP binding.[115] Another protein (or peptide) called an "endogenous inhibitor of [³H]DZP binding" (DBI) is a competitive inhibitor of [³H]DZP binding with a mol. wt. near 11,000, has a K_d for the BZ receptor near 1 μM, is not destroyed by heating at 95°C for 15 min, is destroyed by proteolytic enzymes, does not bind [³H]GABA or [³H]DZP, and does not inhibit [³H]GABA binding. DBI has a tyrosine residue in the C-terminal position while the amino-terminus was blocked. It contains 105 amino acids, of which lysine (19), glutamate/glutamine (12), asparagine/aspartate (10) and serine (10) were most abundant.[280] Cysteine was absent but there are two methionine residues, and cyanogen bromide cleavage of DBI at these residues yielded three peptides, two of which have been sequenced. DBI is a competitive inhibitor of four BZ receptor ligands, two GABA-positive, two GABA-neutral (or GABA-negative). Concentrations of DBI which inhibit the binding of these ligands 50% fail to displace specifically bound [³H]etorphine, [³H]GABA, [³H]QNB, [³H]dihydroalprenolol, [³H]adenosine or [³H]imipramine. Injected intraventricularly, DBI reversed an anticonflict effect of diazepam and facilitated shock-induced suppression of drinking in rats.[280]

If there are peptide (protein) endogenous ligands for the BZ/GABA receptor complex, it seems possible that they could "coexist" with GABA and be coreleased together with GABA from at least some GABA nerve terminals. There are at least 11 known examples of peptides thought to be neurotransmitters present in neurons containing "classical" neurotransmitters such as the catecholamines, serotonin, and acetylcholine.[210] In fact, a subpopulation of serotonin neurons in the rat medulla oblongata also contains substance P as well as TRH.[210] The physiological significance of such coexistence is not yet clear, but it is fascinating to speculate that one or both of the above mentioned peptides (GABA-modulin and/or DBI) might coexist with GABA in certain neurons.

Other peptide inhibitors of BZ binding have been reported. Several substances extracted from porcine cerebral cortex with mol. wts. ranging from 500 to 100,000 were reported to inhibit [³H]DZP binding.[211] One fraction containing material in the 30- to 70-kilodalton range competitively inhibited [³H]FLU binding.[211] This inhibitory material was destroyed by trypsin and acid hydrolysis (100°C in 6 N HCl overnight) but not by heat (100°C for 15 min). It was suggested that the 30- to 70-kilodalton material might represent the "proforms" of lower-mol.-wt. physiologically relevant ligands for the BZ receptor.[211] This report has not been followed up or confirmed by others. Kenessey *et al.*[212] obtained four fractions by chromatography on a Sephadex G-75 column from an aqueous extract of bovine brain, with mol. wts. ranging from 10,000 to 67,000 that inhibited [³H]DZP binding to rat brain membranes.[212] The 42- to 67-kilodalton fraction inhibits [³H]DZP binding in a noncompetitive way.[212]

Davis and Cohen[213] isolated a peptide from bovine brain P₂ fractions that competitively inhibited [³H]DZP binding. This peptide was reported to have a mol. wt. near 3000, to be heat stable (no loss of activity after heating at 100°C for 15 min), but to be destroyed by papain. It apparently contained alanine, histidine, isoleucine, leucine, lysine, methionine, tyrosine, valine, and four other minor amino acids.[213] After intracerebral-ventricular injection, the BZ receptor active peptide was reported to produce effects similar to those produced by DZP, including EEG changes and ability to restore behavior suppressed by electric shock (thirsty rat conflict test).[213,214]

Chiu and Rosenberg[215] reported the solubilization with 1% digitonin of a heat-stable, trypsin-sensitive substance from rat brain membranes that noncompetitively inhibited [³H]FLU binding.

Nagy *et al.*[216] isolated a peptide from porcine cerebral cortex homogenates. After centrifugation, the supernatant was heated at 95°C for 15 min at pH 7.1. Recentrifugation and ultrafiltration of the second supernatant yielded a substance with mol. wt. near 3000 that inhibited both [³H]GABA and, less potently, [³H]DZP binding and antagonized the anticonvulsant effect of DZP when both were injected into the amygdala of rats.[216]

The numerous reports of protein and peptide inhibitors of [³H]DZP and [³H]FLU binding is striking. It seems possible that at least some of these may be fragments of BZ receptors or parts of BZ/GABA receptor complexes. It is probable that BZ receptors are proteins folded in characteristic ways. Such folding could be due in part to complementary amino acid sequences in different parts of the protein which have high affinity for each other. Thus, a peptide containing one such amino acid sequence could disrupt a BZ receptor by preventing its normal folding.

Karobath *et al.*[217] extracted material from rat brain with acetone that inhibited [³H]DZP binding, was resistent to proteolytic enzymes, and had an apparent mol. wt. of 500. The substance(s) was apparently unevenly distributed in brain and was present in even higher concentration in several peripheral organs including heart, striated muscle, spleen, lung, kidney, and liver. This preliminary work was not followed up.

Skolnick *et al.*[218] extracted dialyzable, heat stable competitive inhibitors of [³H]DZP binding from bovine brain that were resistant to degradation by proteolytic enzymes and were subsequently identified as inosine and hypoxanthine.[218] They had K_i values near 1 mM. Subsequent studies revealed that many purines inhibit [³H]DZP binding competitively with K_i values ranging from about 20 μM to 5 mM.[219,220] 3-Isobutyl-1-methylxanthine,[220] caffeine, 2′-deoxyinosine, and 2′-deoxyguanosine are among the more potent of the purines tested, with K_i values near 200–300 μM. However, the purine exhibiting the highest affinity for BZ receptors so far reported in the literature is 1-methylisoguanosine, a marine natural product with muscle-relaxant properties, isolated from the sponge *Tedania digitata*.[221] It has a K_i value near 20 μM. Inosine, one of the substances found in brain that might be considered an "endogenous ligand" for BZ receptor, does have some of the pharmacological and electrophysiological properties of the BZs.[222,223] However, the high concentrations of inosine required to significantly occupy BZ receptors raises doubts concerning the physiological relevance of this, and other naturally occurring purines, as endogenous ligands for BZ receptors. Nicotinamide is another substance found in brain with low affinity for BZ receptors (IC_{50} in displacing [³H]DZP near 4 mM) that produces several BZ-like effects, including ability to restore behavior in rats that had been suppressed by punishment ("conflict test") as well as anticonvulsant, antiaggressive, muscle relaxant, hypnotic, and BZ-like actions on the electrical activity of the spinal cord.[224] Although very high doses (>500 mg/kg) were required to produce these effects, nicotinamide was more potent than hypoxanthine or inosine in protecting mice against 3-mercaptopropionate-induced seizures.[224]

In summary, although much work has been devoted to isolating and characterizing possible "endogenous ligands" for BZ receptors, there is no compelling evidence that any of the candidates proposed to date is a physiologically relevant "endogenous ligand."[225] In fact, the existence of an endogenous ligand for the BZ receptors analogous to the enkephalins and endorphins now appears highly improbable. Tallman *et al.*[225] concluded their review of presumptive endogenous ligands for BZ receptors in January, 1980, with the statement: "Thus, the existence of an additional transmitter substance or endogenous ligand may not be required to explain the pharmacological properties of the benzodiazepines.[225]" This conclusion still appears to be valid. There is still no consensus regarding the nature of a postulated "endogenous ligand" and no compelling reason to believe that one should exist.

13. THE β-CARBOLINE STORY

Early in 1977, I initiated a search for BZ active substances in human urine (see ref. 11, p. 258). Working alone, I found that much BZ receptor active material in human or bovine urine was adsorbed onto Amerlite® XAD-2 and could be selectively eluted with 50% ethanol–water.[11] This active material proved to be heterogeneous and could be quantitatively and irreversibly adsorbed onto activated charcoal. Braestrup and I hypothesized that some of the "endogenous ligand" might be excreted in urine in conjugated form, and to test this hypothesis, we proposed experiments to determine whether BZ receptor activity would increase after glusulase treatment or acid hydrolysis to "deconjugate" the "endogenous ligand."

Heating the 50% ethanol eluate from Amberlite® XAD-2 (Chromosorb®) at 80°C for 24 h in 0.1 N HCl resulted in a large increase in BZ receptor activity,[226] which was subsequently shown to be due to the formation of ethyl-BCC.[227] Braestrup and Nielsen published the results of this project as their own.[226] In spite of much speculation that BCC might be related to an "endogenous ligand" for the BZ receptor,[64,226–229] no evidence has been obtained to support this speculation. On the contrary, there is much evidence indicating that BCC is a chemical artifact.

1. The β-carbolines are only formed by prolonged heating of urine or brain extracts in acid, conditions known to favor the formation of BCC from tryptophan-containing proteins and peptides by a Bischler–Napieralski-type reaction.[230]
2. Several protein-containing materials of diverse origin, including casein, bacto-tryptose, dried yeast extract, dried pituitary powder, chicken feathers, and sheep's wool (as sources of keratin, a tryptophan-containing protein), when heated in 90% ethanol containing 0.3 N HCl at 80°C for several days as described by Braestrup et al.,[227] yielded large amounts of BZ receptor active material with properties of BCC esters.[21]
3. Ethyl-BCC was identified in HCl–ethanol extracts of liver and kidney, after heating at 80°C for 20–24 h, in amounts comparable to those obtained from brain using the same procedure[227] (M. Nielsen, personal communication).
4. Rechromatography of the 50% ethanol eluate from Amberlite® XAD-2 on lipophilic Sephadex (LH-20) columns revealed three or four major peaks of BZ receptor activity, probably of fairly high mol. wt. judging from their appearance near the void volume of the LH-20 column. When the fractions from this column are heated in HCl–ethanol, several new peaks appear, also near the void volume, but which do not correspond to the original peaks of BZ receptor active material (R. Squires, and E. Saederup, unpublished observations). These results show that the β-carboline precursors in urine are heterogeneous, do not correspond to the original inhibitory activity, and could well be tryptophan-containing peptides, which yield ethyl-BCC when heated in HCl–ethanol by a Bischler–Napieralski-type reaction as described by Uphaus et al.[230]

The physiological roles of these BCC derivatives, if any, and the mechanisms by which they might be formed have been the subject of much speculation.[64,226–229,231–240] The present consensus is that the BCC derivatives are chemical artifacts (C. Braestrup, personal communications).

Although these BCCs are chemical artifacts not formed in brain or elsewhere in the mammalian organism in significant amounts, they have interesting pharmacological and physiological properties. Pharmacologically, ethyl-BCC has been shown to be a GABA-negative ligand for BZ receptors* and proconvulsant in rats and mice[231,232,234,236-238] and baboons.[240] Methyl-BCC is a convulsant in several mammalian species including mice, rats, cats, and monkeys (B. Jones, N. Oakley, and P. Skolnick, personal communications).[229] Propyl-BCC has some weak anticonvulsant activity and is most effective against convulsions produced by the methyl-BCC (B. Jones, personal communication).

Several beta carbolines electronegatively substituted in the 3-position, have been reported to be anxiogenic in a variety of animal tests as well as in humans. File *et al.*[290] reported that ethyl-BCC had a potent anxiogenic action in a rat social interaction test. The GABA-neutral BZ receptor ligand Ro 15-1788 had little or no anxiogenic or anxiolytic effect in this test but counteracted the anxiogenic effect of ethyl-BCC.[290] In rhesus monkeys, ethyl-BCC produces an acute "anxiety" syndrome characterized by dramatic elevations in heart rate, blood pressure, plasma cortisol and catecholamines, effects which were also blocked by Ro 15-1788.[291] Another test for anxiogenic substances involves enhancing the shock-induced suppression of drinking in rats.[292] A current of 0.35 mA, delivered through the drinking tube, has little suppressive effect on drinking in thirsty rats. However, this current becomes aversive and suppresses drinking when the thirsty rats are pre-treated with ethyl-BCC, methyl-BCC, DMCM, or FG 7142 (beta carboline-3-carboxylate-N-methyl amide). The suppressive effects of these beta carbolines were antagonized by Ro 15-1788, CGS-8216, and diazepam. Pentylene tetrazole (metrazol, PTZ) produced suppressive effects similar to those produced by the beta carbolines, but, in contrast to the beta carbolines, the suppressive effect of PTZ was not reversed by Ro 15-1788. Strychnine was not anxiogenic in this system.[292] Methyl-BCC, in sub-convulsant doses lowered the rate of lever pressing to obtain a food reward during periods when the mice also received painful electric shocks to their feet (modified Geller conflict procedure).[293] Diazepam increased lever pressing during conflict periods and was antagonized by both Ro 15-1788 and methyl-BCC which had no analgesic effects at the doses used. Methyl-BCC also reversed diazepam-induced ataxia (rotarod) in mice.[293] Intravenous administration of ethyl-BCC to rats was reported to produce multiple electroencephalographic bursts, characteristic of seizures, which could be prevented by CGS 8216.[294] Seizures produced by methyl-BCC in mice were blocked by diazepam, Ro 15-

---

* The terms "agonist", "antagonist", and "inverse agonist" to describe the various ligands for BZ receptors does not now seem appropriate since the term "agonist" has traditionally been used to denote a physiological effect of a substance similar to that produced by a neurotransmitter or hormone. It is now clear that BZ and picrotoxinin binding sites are sites which allosterically modify GABA neurotransmission. Therefore, terms GABA-positive, GABA-neutral and GABA-negative ligands for both BZ and picrotoxinin receptors would appear more appropriate. Thus, diazepam-like BZs would be GABA-positive ligands for BZ receptors since they facilitate GABA neurotransmission. Substances like Ro 15-1788 and propyl-BCC (see below) are almost GABA-neutral while methyl-BCC and DMCM are GABA-negative. Similarly, barbiturates, etomidate and methaqualone are GABA-positive ligands for picrotoxinin (TBPS) sites, etazolate and cartazolate are closer to GABA-neutral, while metrazol, bemegride and the cage convulsants are GABA-negative ligands for picrotoxinin (TBPS) sites.

It is also clear that there is a continuum of activities of ligands for both BZ and picrotoxinin (TBPS) sites, ranging from extreme positive, through neutral to extreme negative.

1788 and CGS 8216.[295] 3-hydroxymethyl-beta-carboline (3-HMC) a GABA-negative ligand for the BZ receptors, causes insomnia in rats and, in doses which do not affect sleep, blocks the sleep-inducing effect of a large dose of flurazepam.[296] Discrimination of the interoceptive stimulus produced by PTZ serves as the basis for another animal model of anxiety.[297] Rats are provided with two levers to press to obtain a food reward. One lever rewards the animal after it has received a saline injection, the other provides a reward after the animal has received PTZ. Rats learn to discriminate between saline and the interoceptive PTZ stimulus and press the appropriate lever for reward. Bemegride, harmane, Ro 5-3663, cocaine and the GABA-negative beta-carbolines generalize with the PTZ stimulus while GABA-positive BZs, barbiturates, meprobamate and valproate antagonize the interoceptive PTZ stimulus.[297] FG 7142, like PTZ, has been reported to cause intense anxiety in humans (R. Dorow, CINP abstract, 1982; C. Braestrup personal communication).

Biochemically, the BCC esters bind with higher affinity to BZ receptors in cerebellum ("BZ_1-receptors") than in hippocampus[21,64,187,229] in all mammalian species examined including man (R. Squires, unpublished) and displace [^{3}H]BZs from receptors in forebrain membranes with Hill number less than 1[21,64,227,228] reflecting the existence of both BZ_1 and BZ_2 receptors in forebrain, as opposed to cerebellum which appears to contain almost entirely "BZ_1" receptors.[63,64,187,229] Thus, the BCC esters exhibit a pattern of selective affinity for BZ_1 receptors qualitatively similar to that of CL 218,872.[19,53,63] The ethyl and propyl esters exhibit a high degree of selectivity for brain specific BZ receptors. They have low affinity for all other receptor types investigated including peripheral BZ receptors.[227,228]

With the filtration binding assay, low concentrations (0.3–0.4 nM) of the tritiated propyl ester ([^{3}H]PrCC), "BZ_1" receptors are selectively labelled[187] and range from about 57% of all [^{3}H]FLU binding sites in hippocampus to 100% in cerebellum. Similar results are obtained using [^{3}H]methyl-BCC.[229] The dissociation rate constant for [^{3}H]PrBCC varies with the brain region (hippocampus > cerebral cortex > cerebellum).[187]

The BCC esters are chemically unstable and hydrolyze readily in aqueous media. The tritiated BCC esters are not among the best ligands for measuring BZ receptors because of the relatively large nonspecific binding, ranging from 14% in cortex to 68% in pons–medulla.[228] Preliminary radiohistochemical results reveal that the tritiated BCC esters are not well suited for light microscopic autoradiography because of high nonspecific background binding (M. Kuhar and D. Niehoff, personal communication). Other chemically more stable ligands having lower nonspecific binding and exhibiting different patterns of differential affinity among the various BZ receptor complexes can be expected.[241–243]

GABA and muscimol affect the binding of the tritiated BCC esters differently depending on the ester: in 100 mM Tris citrate buffer, pH 7.1, the affinity of the convulsants methyl-BCC[229] and DMCM[276,287] for BZ receptors is reduced by GABA and muscimol in a bicuculline-reversible way; the affinity of ethyl-BCC is unaffected by GABA or muscimol[229,244]; the affinity of propyl-BCC is slightly increased by GABA or muscimol.[228,229] The failure of Ehlert [233] to detect an effect of GABA on the affinity of the [³H]PrBCC for BZ receptors may have resulted from the use of sodium/potassium phosphate buffer instead of Tris citrate buffer. The binding of the strongly convulsant beta carboline DMCM[276,287] is inhibited by eight GABA-A receptor agonists, the most potent being muscimol followed by GABA. The inhibitory effects of GABA and muscimol on [³H]DMCM binding were reversed by the potent bicuculline-like competitive GABA antagonist R 5135. Although [³H]DMCM and [³H]propyl-BCC are potently displaced by a series of ligands having high affinity for BZ receptors, the ratio of IC$_{50}$ values, using the two ligands, varied by almost a factor of 60 with ethyl-BCC being 6-fold more potent in displacing [³H]propyl BCC than [³H]DMCM. Conversely, DMCM was 8-fold more potent in displacing [³H]DMCM than [³H]propyl BCC.[287] Pre-photolabelling with FLU greatly increases the number of ³H-DMCM binding sites while decreasing [³H]FLU binding. Treatment of BZ receptors with high energy electrons or Ag$^+$ also increased [³H]DMCM binding.[287] DMCM exhibits somewhat higher affinity for the "BZ$_2$" than for the "BZ$_1$" fraction, just the opposite selectivity of methyl,ethyl- and propyl-BCC and CL 218,872. Association and dissociation of ³H-DMCM to (and from) specific sites in rat hippocampus were both polyphasic.[287]

Ethyl-BCC also antagonizes the action of DZP on a presynaptic GABA receptor.[238] Potassium-ion-induced release of [³H]glutamate from striatal prisms is enhanced by muscimol in a way that is potentiated by DZP. Ethyl-BCC antagonizes the potentiation by DZP but has no effect on [³H]glutamate release alone.[238] Hydroxymethyl-β-carboline (3-HMC), which has a lower affinity (K_d = 1.5 μM) but greater *in vivo* stability than the BCC esters, was also reported to antagonize the anticonvulsant effects of DZP as well as to increase exploratory behavior in mice.[244] Convulsions induced in baboons by methyl BCC were antagonized by two GABA-neutral ligands for BZ receptors Ro 15-1788 and propyl-BCC.[281] Ethyl-BCC alone had no effect on the concentration of cyclic GMP in mouse cerebellum, but inhibited the diazepam-induced decrease in cyclic GMP.[282] On the other hand Koe and Lebel[298] reported that ethyl-BCC increased the basal level of cyclic-GMP in mouse cerebellum, and augmented the increases in cycylic GMP caused by isoniazide. Diazepam blocked the increases produced by both ethyl-BCC and isoniazide, and alone reduced the level of cyclic GMP. Ro 15-1788 blocked both the decrease in cyclic GMP produced by diazepam and the increases produced by ethyl-BCC.[298]

Hydroxymethyl-β-carboline and 3-acetyl-β-carboline (K_d = 50 nM)[243] both reduce overall sleep time and REM sleep and increase latency to sleep in rats.[242]

14. OTHER NOVEL SUBSTANCES THAT INTERACT WITH BENZODIAZEPINE RECEPTOR COMPLEXES

After the discovery of BZ receptors, many pharmaceutical firms started systematic screening of large numbers of substances on hand for BZ activity.

Many active nonbenzodiazepine structures, some having very high affinity for BZ receptors, have been found. Because of the secrecy required for effective commercial development of new drugs, most of these active structures have not yet been published. Companies known to have found novel BZ receptor active substances include Hoffmann-LaRoche, Ciba-Geigy, Searle, Rhone-Poulenc, Hoechst-Roussel, Synthelabo, Lederle, Merck Darmstadt, Merck (USA), Eli Lilly, Thomae, Sandoz, Pharmuka, Glaxo, Ferrosan A/S, and Schering AG. Because of a hasty publicity campaign and unwarranted speculations that they might be related to physiologically relevant "endogenous ligands" for BZ receptors, the BCC esters have received disproportionate attention.

Other BZ receptor active substances that are at least as interesting and important tools for studying the properties and physiological functions of BZ receptors were developed in the same period of time. Some of the properties of the triazolopyridazines (e.g., CL 218,872)[245] have already been described above.[19,53,63]

The properties of an imidazodiazepine (a BZ derivative), Ro 15-1788, which is a potent and highly selective BZ antagonist, have recently been described[246-248] (for structure see ref. 246). Ro 15-1788 has a high affinity for BZ receptors ($K_d = 0.86$ nM at 0°C) but does not produce any of the behavioral or neurological effects typical of the BZs. It has very low toxicity and, in contrast to ethyl-BCC, is not a proconvulsant (e.g., it does not sensitize animals to convulsants such as pentylenetetrazol) or a central stimulant.

Ro 15-1788 effectively and selectively antagonizes the anticonvulsant, ataxia, and sedation-producing effects of the BZs. Further, it reverses the ability of DZP to increase behavior suppressed by punishment (anticonflict test) and reverses the ability of BZs to potentiate the electrophysiological effects of GABA. Biochemically, Ro 15-1788 reverses the ability of BZs to attentuate the increase in the concentration of homovanillic acid produced in brain by neuroleptics. By itself, Ro 15-1788 has no effect on the concentration of cyclic GMP in rat cerebellum but antagonizes the cyclic-GMP-reducing effect of DZP in a dose-dependent way. Ro 15-1788 does not antagonize the behavioral or biochemical effects of barbiturates, meprobamate, ethanol, or muscimol but does antagonize the behavioral effects of zopiclone, chemically a non-BZ that has a BZ pharmacological profile and displaces [³H]DZP and [³H]FLU in the 40–60 nM range[249] (for structure of zopiclone, see ref. 249). Ro 15-1788 exhibits very low affinity for peripheral type BZ receptors and does not antagonize the antischistosomal effect of Ro 11-3128, a "classical" BZ, but does antagonize its centrally mediated effects in mammals.[246]

Radiohistochemical studies with [³H]Ro 15-1788 revealed the same regional distribution pattern in brain as with [³H]FLU.[247] The K_d values for [³H]Ro 15-1788 in several brain regions are similar, as are the K_d values for [³H]clonazepam.[248] The B_{max} values obtained for the two ligands are almost identical in six brain regions and are about three times higher in cortex than in pons–medulla.[248] A variety of structurally different ligands for BZ receptors showed very similar potencies in displacing either [³H]clonazepam or [³H]Ro 15-1788. GABA (10 μM), presumably in the presence of 50 mM Tris HCl, provided similar protection against heat inactivation (60°C for 30 min) of [³H]clonazepam or [³H]Ro 15-1788 binding sites. However, [³H]Ro 15-1788 binding is not enhanced by GABA, pentobarbital, SQ 20,009, $NiCl_2$, or high concentrations of iodide, bromide, or chloride ion, as is [³H]DZP binding. With extensively washed rat cerebral cortex membranes and [³H]Ro 15-1788 as the radioligand, the IC_{50} values for several GABA-positive ligands BZ receptors including FLU, DZP, zopiclone, oxazepam, and CL 218,872, were decreased (by a factor of 3 for FLU and DZP) when the binding was performed at 35°C in the presence of GABA (100 μM). The IC_{50} value for methyl-BCC eas significantly increased by GABA at 30°C but not at 0°C. Van't Hoff analysis of the dissociation constants (K_d) for [³H]Ro 15-1788 and [³H]clonazepam, respectively, as a function of temperature showed an inclination point at 21°C for [³H]clonazepam but not for [³H]Ro 15-1788, which yielded a linear Van't Hoff plot.[248]

It is clear that different ligands with different structures occupy BZ receptors and surrounding domains in different ways and produce different effects (both biochemical and behavioral), as would be expected.[250] Ligands for all five known independent receptors in the GABA/cation/picrotoxin/anion/BZ receptor complexes behave in analogous ways (e.g., THIP or P4S versus GABA, chloride versus citrate, picrotoxin versus barbiturates, etc.).

The BZ receptor screening program at Hoffmann-LaRoche has already resulted in the "identification of many compounds possessing either pure antagonistic or antagonistic properties combined with weak agonistic activity and has revealed interesting and unexpected structure–activity relationships."[246]

CGS-8216, a pyrazoloquinoline (for structure, see ref. 251), is another extremely potent GABA-neutral or GABA-negative ligand for BZ with a novel chemical structure quite different from the BZs.[251] CGS-8216 probably has the highest affinity for BZ receptors of any substance so far reported in the literature (K_d near 0.04 nM at 0°C).

Pharmacologically and behaviorally, CGS-8216 alone produces little or no effect, but potently antagonizes the action of DZP *in vivo* with a long (6-h) duration of action. CGS-8216 does not bind significantly to α- or β-adrenergic, H_1-histamine, or GABA receptors but does weakly inhibit the activation by adenosine of cyclic AMP formation in vesicles from guinea pig brain (IC_{50} = 5 μM).[251] CGS-8216, like ethyl-BCC, antagonizes sedative barbiturates and potentiates the convulsant effects of pentylenetetrazol. [³H]CGS-8216 was displaced from brain specific sites by BZ receptor active substances with an order of potencies similar to that observed when [³H]DZP or [³H]FLU were used as radioligands. The regional distribution of [³H]CGS-8216 and [³H]FLU binding sites in rat brain are similar, and [³H]CGS-8216 binds with the same high affinity to BZ receptors in all parts of the brain examined. However, the total number of binding sites (B_{max} values) was greater for [³H]FLU than for [³H]CGS-8216. CGS-8216 exhibited mixed-type inhibition of [³H]FLU binding; that is, Scatchard analysis of [³H]FLU binding in the presence of CGS-8216 revealed both a decrease in B_{max} and an increase in K_d for [³H]FLU. These results suggest that [³H]CGS-8216 may bind selectively to a subpopulation of [³H]FLU binding sites as found with the filtration assay system, perhaps analogous to the selective binding of [³H]PrBCC to BZ_1 receptors when the filtration assay is used.[187]

As with [³H]FLU,[21,24] the binding of [³H]CGS-8216 is almost totally dependent on ions, the affinities of which are regulated by GABA receptors.[272] In the presence of GABA (200 µM), Tris HCl produces 50% of its maximum enhancing effect on the binding of [³H]CGS-8216 at a concentration near 1 mM, whereas in the presence of P4S (50 µM), the EC_{50} for Tris HCl is increased to 8 mM. GABA and P4S are mutually antagonistic, in a competitive way, with respect to their effects on ion affinities with either [³H]FLU or [³H]CGS-8216 binding. The Hill numbers for the chloride ion (Tris HCl) enhancement of [³H]CGS-8216 binding are near 0.7.

[³H]CGS-8216 binding sites, like [³H]FLU binding sites, are almost completely protected from heat inactivation (60°C for 30 min) by NaCl alone, with 50% of maximum protection near 240 mM. In the presence of GABA (200 µM) or picrotoxin (100 µM), the concentrations of NaCl required for 50% of maximum protection of [³H]CGS-8216 binding sites are reduced to 30 mM and 130 mM, respectively. The Hill numbers for NaCl, alone or in the presence of GABA, are near 2, and in the presence of picrotoxin, near 3. [³H]CGS-8216 binding sites, like [³H]FLU binding sites, are also protected against heat inactivation (30 min at 60°C) by $CaCl_2$ plus GABA. However, [³H]CGS-8216 binding sites are protected to a significantly greater extent than [³H]FLU sites.[272]

Dose–response curves for GABA in the presence of 5 mM $CaCl_2$ or $CaCl_2$ in the presence of 500 µM GABA are consistently higher (i.e., greater protection expressed as percent of unheated control) with [³H]CGS-8216 compared to [³H]FLU as ligand. The time course of heat inactivation at 60°C in the presence of saturating GABA (500 µM) and $CaCl_2$ (5 mM) is almost monophasic with [³H]CGS-8216, with a B_0 value near 92% in both rat forebrain and cerebellum.[272] In contrast, the B_0 values obtained using [³H]FLU as ligand are near 70%.[104,272] These results suggest that [³H]CGS-8216 binds preferentially to a subpopulation of [³H]FLU binding sites that is selectively protected by $CaCl_2$ plus GABA in both rat forebrain and cerebellum.

A closely related substance, CGS-9896, which differs from GCS-8216 only by the presence of a *para*-chloro atom on the phenyl ring, is a highly potent (K_d near 0.1–0.2 nM) DZP-like compound. The IC_{50} value for CGS-9896 in displacing [³H]CGS-8216 is not affected by GABA, which reduces the IC_{50} values for several GABA-positive ligands for BZ receptors (H. Kalinsky, B. Petrack, and W. Cash, personal communication).

A steroid convulsant, R-5135 (for structure, see ref. 252), was reported to potently reverse the enhancement of [³H)DZP binding by GABA. The inhibition by R-5135 of [³H]DZP binding to unwashed rat brain membrane preparations was caused mainly by a decrease in the affinity of [³H]DZP with little effect on the B_{max}. In extensively washed membranes in 100 mM Tris citrate, pH 7.7, R-5135 did not inhibit below 1 µM but effectively antagonized in the 10–100 nM range the increase in [³H]DZP binding resulting from the addition of GABA. R-5135 appears to be a bicuculline-like competitive GABA antagonist about 500 times more potent than bicuculline.[252] R-5135 does not significantly displace [³H]CGS-8216 at 10 μM (R. Squires, unpublished data). R-5135 is a potent inhibitor of [³H]GABA and [³H]strychnine binding (IC$_{50}$ values near 20 nM and 7 nM, respectively.[252]

Fominoben (see *Merck Index*, 9th edition, for structure) was reported by Antoniadis *et al.*[253] to displace [³H]FLU from brain specific sites with an IC$_{50}$ value near 1.5 μM. I have confirmed this. Geller and Baldino[254] reported that fominoben antagonized the depression in firing of hypothalamic neurons, growing in tissue culture, induced by both GABA and glycine, again indicating common structural features in GABA and glycine receptors. Pharmacologically, fominoben exhibits both antitussive and respiratory stimulant properties, an unusual combination (see *Merck Index*, 9th edition). Unexpectedly, fominoben was found to antagonize pentylenetetrazol convulsions, an effect that was reversed by the BZ antagonist Ro 15-1788 (H. Geller, personal communication). Fominoben also antagonized the convulsions produced by 3-mercaptopropionic acid, a GABA synthesis inhibitor (H. Geller, personal communication).

Avermectin B1a (AVM) is a macrocyclic lactone (for structure, see ref. 255) derived from mycelia of *Streptomyes avermitilis* that has potent broad-spectrum antihelmintic activity.[255] Studies in invertebrate species (ascaris and lobster)[255,256] indicate that it binds irreversibly to neuronal membranes and opens chloride channels, causing permanent inhibition of the neurons, which can, however, be transiently reversed by picrotoxin.[255,256] [³H]Avermectin binds with high affinity to membranes from dog and rat brain.[257] The [³H]AVM binding sites are unevenly distributed in brain, with highest concentrations in cerebellum, followed in decreasing order by midbrain, hypothalamus, and frontal cortex. No specific [³H]AVM binding was detected in any white matter or in the pons–medulla.[257] Avermectin caused a concentration-dependent increase in [³H]DZP[258] and [³H]FLU[259] binding to brain specific receptors *in vitro*. The increases in binding mainly reflect increases in affinity for the radioligands and a smaller increase (12 to 25%) in the number of binding sites (B_{max}).[258,259] The increases in [³H]DZP[260] and [³H]FLU[259] binding caused by AVM are not reversed by extensive washing of the membrane preparations. Pretreatment of cerebellar membranes with Triton X-100 completely abolishes the action of AVM on [³H]FLU binding.[259] The enhancing effects of AVM were also diminished by THIP and P4S as well as by SQ 20,009 and SQ 65,396.[259] The AVM-antagonistic effects of P4S and THIP were attenuated by chloride ion, whereas the antagonistic effects of SQ 20,009 and SQ 65,396 were potentiated by chloride ion.[259] When [³H]DZP (0.5 nM) binding to extensively washed membranes from rat cerebral cortex was maximally enhanced by GABA (30 μM) plus NaCl (200 mM) no further enhancement could be obtained with AVM.[260] It was suggested that AVM enhanced BZ binding by acting on a site different from GABA receptors or picrotoxin–pyrazolopyridine receptors.[260]

The highest concentration of [³H]AVM binding sites are in cerebellum,[257] where BZ receptors have been shown to interact with AVM,[260] raising the possibility that AVM selectively interacts with BZ$_1$ receptors.

15. TENTATIVE CONCLUSIONS

Several new, or hitherto not sufficiently appreciated, concepts have emerged from the study of BZ/GABA receptor complexes: 1) BZ and picro-toxinin (TBPS) binding sites are separate allosteric sites which modify GABA neurotransmission. There may be other hitherto unidentified allosteric sites in the BZ/GABA receptor complex, possibly for cannabinoids and for silatranes. Similar allosteric sites may be coupled to receptors for other neurotransmitters such as dopamine, serotonin and acetylcholine. It is clear that these sites can best be studied using specific high-affinity ligands. 2) A "receptor" or "binding site" is defined by its ligand. It is now clear that a ligand can bind to its receptor or binding site through a fairly large number of "contract points." In general, the greater the size of the ligand and the greater its affinity for the binding site the larger the number of "contact points" is likely to be. Thus, binding sites for two ligands with different sizes and shapes may only partially overlap. Some ligands may overlap two or more separate but allosterically interacting sites. For example, a series of tetrazole convulsant appear to overlap both BZ and TBPS (picrotoxinin) sites, usually with somewhat higher affinity for TBPS sites (Squires, Saederup, Paul and Skolnick, in preparation). An exception is un-decamethylene tetrazole which exhibits higher affinity for the BZ than for the TBPS site. 3) One of the many functions of GABA is to dampen anxiety, and the potentiation of this function by the BZs is the main reason for their wide-spread use in medicine. It seems probable that a sub-population of BZ/GABA complexes is selectively involved in reducing anxiety. A variety of substances are now known to increase anxiety by inhibiting GABA neurotransmission. Such substances can act directly on GABA receptors (Amoxapine), on the picrotoxinin (TBPS) site (metrazole), or on BZ receptors (FG 7142). While it is clear that GABA supresses anxiety, the mechanism which produces or "drives" anxiety is still unknown. It seems reasonable to assume that anxiety, like convulsions, is driven by an excitatory neurotransmitter with glutamate and aspartate being prime candidates. The recent finding that the glutamate antagonist 2-amino-7-phosphono-heptanoic acid protects against the seizures induced by diverse chemical convulsants, including the GABA-negative BZ receptor ligand DMCM,[289] strengthens this view. Thus, selective glutamate or aspartate antagonists may prove to be effective anxiolytics.

Brain specific BZ receptors mediate most, if not all, of the pharmacolog-ical, behavioral, and therapeutic effects of the BZs. Although high-affinity bind-ing sites for [^{3}H]DZP and [^{3}H]FLU have so far not been detected in invertebrate species, there is evidence for the existence of physiologically relevant BZ re-ceptors in several invertebrate species. The brain specific BZ receptors found in vertebrate species are restricted to neurons and seem to be located on den-trites, cell bodies, and nerve terminals. All brain specific receptors are coupled to GABA receptors, and all of the centrally mediated effects of the BZs involve GABA. There are independent recognition sites (binding sites) for BZs, anions, cations, picrotoxin (barbiturates), GABA, and possibly AVM in BZ receptor complexes. There is a high degree of BZ receptor heterogeneity, with at least four, possibly five, different receptor complexes.

There are no "endogenous ligands" for the BZ receptors analogous to the enkephalins and endorphins for the opiate receptors. New drugs are being developed that discriminate among the different BZ receptor complexes. Several GABA-neutral and GABA-negative ligands for BZ receptors have been discovered, and some new drugs appear to be GABA-positive at some subpopulations of receptors and simultaneously GABA-neutral or GABA-negative at others. Some new drugs exhibit different affinities for the various subpopulations of BZ receptor complexes. In the neuronal membrane, BZ receptors exist as polymers consisting of four to 12 subunits with mol. wts. in the 50–60,000 range. Apparently four or five subunits, differing slightly in mol.wt., can be photolabeled with [^{3}H]FLU. The peripheral BZ receptors are fundamentally different from the brain specific receptors. The physiological function of peripheral BZ receptors is, at present, unknown.

REFERENCES

1. Young, A. B., Zukin, S. R., and Snyder, S. H., 1974, *Proc. Natl. Acad. Sci. U.S.A.* **71**:2246–2250.
2. Costa, E., and Greengard, P. (eds.), 1975, *Advances in Biochemical Psychopharmacology*, Volume 14, Raven Press, New York.
3. Costa, E., Guidotti, A., and Mao, C. C., 1975, *Adv. Biochem. Psychopharmacol.* **14**:113–130.
4. Haefely, W., Kulcsar, A., Mohler, H., Pieri, L., Polc, P., and Schaffner, R., 1975, *Adv. Biochem. Psychopharmacol.* **14**:131–151.
5. Bennett, J. P., and Snyder, S. H., 1975, *Brain Res.* **94**:523–544.
6. Squires, R. F., and Braestrup, C., 1977, *Nature* **266**:732–734.
7. Braestrup, C., and Squires, R. F., 1977, *Proc. Natl. Acad. Sci. U.S.A.* **74**:3805–3809.
8. Braestrup, C., and Squires, R. F., 1978, *Eur. J. Pharmacol.* **48**:263–270.
9. Squires, R. F., and Braestrup, C., 1978, *J. Neurochem.* **30**:231–236.
10. Braestrup, C., Albrechtsen, R., and Squires, R. F., 1977, *Nature* **269**:702–704.
11. Braestrup, C., and Squires, R. F., 1978, *Br. J. Psychiatry* **133**:249–260.
12. Nielsen, M., Braestrup, C., and Squires, R. F., 1978, *Brain Res.* **141**:342–346.
13. Braestrup, C., Nielsen, M., Squires, R. F., and Laurberg, S., 1978, *Acta Psychiatr. Scand.* **274**:27–32.
14. Braestrup, C., Nissen, C., Squires, R. F., and Schousboe, A., 1978, *Neurosci. Lett.* **9**:45–49.
15. Braestrup, C., Squires, R. F., Bock, E., Pedersen, C. T., and Nielsen, M., 1978, *Adv. Pharmacol. Ther.* **7**:173–185.
16. Horowitz, N. H., Fling, M., MacLeod, H., and Sueoka, N., 1961, *Genetics* **46**:1015–1024.
17. Squires, R. F., 1972, *Adv. Biochem. Psychopharmacol.* **5**:355–370.
18. Squires, R. F., Naquet, R., Riche, N., and Braestrup, C., 1979, *Epilepsia* **20**:215–221.
19. Squires, R. F., Benson, D. I., Braestrup, C., Coupet, J., Klepner, C. A., Myers, V., and Beer, B., 1979, *Pharmacol. Biochem. Behav.* **10**:825–830.
20. Squires, R. F., Klepner, C. A., and Benson, D. I., 1980, *Adv. Biochem. Psychopharmacol.* **21**:285–293.
21. Squires, R. F., 1981, *Adv Biochem. Psychopharmacol.* **26**:129–138.
22. Monod, J., Changeux, J.-P., and Jacob, F., 1963, *J. Mol. Biol.* **6**:306–329.
23. Squires, R. F., 1965, *Biochem. Biophys. Res. Commun.* **19**:27–32.
24. Squires, R. F., 1981, *Drug Dev. Res*, **1**:211–221.
25. Mohler, H., and Okada, T., 1977, *Science* **198**:849–851.
26. Mohler, H., and Okada, T., 1977, *Life Sci.* **20**:2101–2110.
27. Bosmann, H. B., Case, K. R., and DiStefano, P., 1977, *FEBS Lett.* **82**:368–372.

28. Mackerer, C. R., Kochman, R. L. Bierschenk, B. A., and Bremner, S. S., 1978, *J. Pharmacol. Exp. Ther.* **206**:405–413.
29. Speth, R. C., Wastek, G. J., and Yamamura, H. I., 1979, *Life Sci.* **24**:351–358.
30. Damm, H. W., Muller, W. E., Schlafer, U., and Woller, U., 1978, *Res. Commun. Chem. Pathol. Pharmac.* **22**:597–600.
31. Speth, R. C., Wastek, G. J., Johnson, P. C., and Yamamura, H. I., 1978, *Life Sci.* **22**:859–866.
32. Hughes, J., 1975, *Brain Res.* **88**:295–308.
33. Hughes, J., Smith, T. W., Kosterlitz, H. W., Fothergill, L. A., Morgan, B. A., and Morris, H. R., 1975, *Nature* **258**:577–579.
34. Teschemacher, H., Opheim, K. E., Cox, B. M., and Goldstein, A., 1975, *Life Sci.* **16**:1771–1776.
35. Li, C. H., and Chung, D., 1976, *Proc. Natl. Acad. Sci. U.S.A.* **73**:1145–1148.
36. Pert, C. B., and Snyder, S. J., 1973, *Science* **179**:1011–1014.
37. Sperk, G., and Schlogl, E., 1979, *Brain Res.* **170**:563–567.
38. Biggio, G., Corda, M. G., De Montis, G., Stefanini, E., and Gessa, G. L., 1980, *Brain Res.* **193**:589–593.
39. Fernholm, B., Nielsen, M., and Braestrup, C., 1979, *Comp. Biochem. Physiol.* **62C**:209–211.
40. Schmidt, R. F., Vogel, M. E., and Zimmerman, M., 1967, *Naunyn Schmiedebergs Arch. Pharmacol.* **285**:69–82.
41. Stratten, W. P., and Barnes, C. D., 1971, *Neuropharmacology* **10**:685–696.
42. Polc, P., Mohler, H., and Haefely, W., 1974, *Naunyn Schmiedebergs Arch. Pharmacol.* **284**:319–337.
43. Wolf, P., and Haas, H. L., 1977, *Naunyn Schmiedebergs Arch. Pharmacol.* **299**:211–218.
44. Kozhechkin, S. N., and Ostrovskaya, R. U., 1977, *Nature* **269**:72–73.
45. Choi, D. W., Farb, D. H., and Fischbach, G. D., 1977, *Nature* **269**:342–344.
46. Geller, H. M., Taylor, D. A., and Hoffer, B. J., 1978, *Naunyn Schmiedebergs Arch. Pharmacol.* **304**:81–88.
47. Geller, H. M., Hoffer, B. J., and Taylor, D. A., 1980, *Fed. Proc.* **39**:3016–3023.
48. MacDonald, R., and Barker, J. L., 1978, *Nature* **271**:563–564.
49. Tsuchiya, T., and Fukushima, H., 1978, *Eur. J. Pharmacol.* **48**:421–424.
50. Gallager, D. W., 1978, *Eur. J. Pharmacol.* **49**:133–143.
51. Young, W. S. III, and Kuhar, M. J., 1979, *Nature* **280**:393–395.
52. Young, W. S. III, and Kuhar, M. J., 1980, *J. Pharmacol. Exp. Ther.* **212**:337–346.
53. Young, W. S. III, Niehoff, D. L., Kuhar, M. J., Beer, B., and Lippa, A. S., 1981, *J. Pharmacol. Exp. Ther.* **216**:425–430.
54. Unnerstall J. R., Kuhar, M. J., Niehoff, D. L., and Palacios, J. M., 1981, *J. Pharmacol. Exp. Ther.* **218**:797–804.
55. Palacios, J. M., Wamsley, J. K., and Kuhar, M. J., 1981, *Brain Res.* **214**:155–162.
56. Chang, R. S., Tran, V. T., and Snyder, S. H., 1980, *Brain Res.* **190**:95–110.
57. Mohler, H., and Okada, T., 1978, *Br. J. Psychiatry* **133**:261–268.
58. Reisine, T. D., Wastek, G. J., Speth, R. C., Bird, E. D., and Yamamura, H. I., 1979, *Brain Res.* **165**:183–187.
59. Speth, R. C., and Yamamura, H. I., 1979, *Eur. J. Pharmacol.* **54**:397–399.
60. Skolnick, P., Syapin, P. J., Paugh, B. A., and Paul, S. M., 1979, *Nature* **277**:397–398.
61. Mao, C. C., Guidotti, A., and Landis, S., 1975, *Brain Res.* **90**:335–339.
62. Lippa, A. S., Sano, M. C., Coupet, J., Klepner, C. A., and Beer, B., 1978, *Life Sci.* **23**:2213–2218.
63. Klepner, C. A., Lippa, A. S., Benson, D. I., Sano, M. C., and Beer, B., 1979, *Pharmacol. Biochem. Behav.* **11**:457–462.
64. Nielsen, M., and Braestrup, C., 1980, *Nature* **286**:606–607.
65. Alger, B. E., and Nicoll, R. A., 1979, *Nature* **281**:315–317.
66. Andersen, P., Dingledine, R., Gjerstad, L., Langmoen, I. A., and Laursen, A. M., 1980, *J. Physiol. (Lond.)* **305**:279–296.
67. Djorup, A., Jahnsen, H., and Laursen, A. M., 1981, *Brain Res.* **219**:196–201.
68. Jahnsen, H., and Laursen, A. M., 1981, *Brain Res.* **207**:214–217.

69. Howells, R. D., Hiller, J. M., and Simon, E. J., 1979, *Life Sci.* 25:2131–2136.
70. Paul, S. M. Zatz, M., and Skolnick, P., 1980, *Brain Res.* 187:243–246.
71. Syapin, P. J., and Skolnick, P., 1979, *J. Neurochem.* 32:1047–1051.
72. Howells, R. D., and Simon, E. J., 1980, *Eur. J. Pharmacol.* 67:133–137.
73. Thomas, J. W., and Tallman, J. F., 1981, *J. Biol. Chem.* 256:9838–9842.
74. Tardy, M., Costa, M. F., Rolland, B., Fages, C., and Gonnard, P., 1981, *J. Neurochem.* 36:1587–1589.
75. Syapin, P. J., and Skolnick, P., 1979, *J. Neurochem.* 32:1047–1051.
76. Strittmatter, W. J., Hirata, F., Axelrod, J., Mallorga, P., Tallman, J. F., and Henneberry, R. C., 1979, *Nature* 282:857–859.
77. Matthew, E., Laskin, J. D., Zimmerman, E. A., Weinstein, I. B., Hsu, K. C., and Englehardt, D. L., 1981, *Proc. Natl. Acad. Sci. U.S.A.* 78:3935–3939.
78. Wang, J. K. T., Taniguchi, T., and Spector, S., 1980, *Life Sci.* 27:1881–1888.
79. Taniguchi, T., Wang, J. K. T., and Spector, S., 1980, *Life Sci.* 27:171–178.
80. Davies, L. P., and Huston, V., 1981, *Eur. J. Pharmacol.* 73:209–211.
81. McCarthy, K. D., and Harden, T. K., 1981, *J. Pharmacol. Exp. Ther.* 216:183–191.
82. Gallager, D. W., Mallorga, P., Oertel, W.,Henneberry, R., and Tallman, J., 1981, *J. Neurosci.* 1:218–225.
83. Shibla, D. B., Gardell, M. A., and Neale, J. H., 1981, *Brain Res.* 210:471–474.
84. Schoemaker, H., Bliss, M., and Yamamura, H. I., 1981, *Eur. J. Pharmacol.* 71:173–175.
85. Taniguchi, T., Wang, J. K. T., and Spector, S., 1981, *Eur. J. Pharmacol.* 70:587–588.
86. Regan, J. W., Yamamura, H. I., Yamada, S., and Roeske, W. R., 1980, *Eur. J. Pharmacol.* 67:167–168.
87. Paul, S. M., Kempner, E. S., and Skolnick, P., 1982, *Eur. J. Pharmacol.* 1981, 76:465–466.
88. Costa, E., Guidotti, A., Mao, C. C., and Suria, A., 1975, *Life Sci.* 17:167–186.
89. Mao, C. C., Guidotti, A., and Costa, E., 1975, *Brain Res.* 83:516–519.
90. Biggio, G., Brodie, B. B., Costa, E., and Guidotti, A., 1977, *Proc. Natl. Acad. Sci. U.S.A.* 74:3592–3596.
91. Mattsson, H., Brandt, K., and Heilbronn, E., 1977, *Nature* 268:52–53.
92. Tallman, J. F., Thomas, J. W., and Gallager, D. W., 1978, *Nature* 274:383–385.
93. Wastek, G. J., Speth, R. C., Reisine, T. D., and Yamamura, H. I., 1978, *Eur. J. Pharmacol.* 50:445–447.
94. Briley, M. S., and Langer, S. Z., 1978, *Eur. J. Pharmacol.* 52:129–132.
95. Martin, I. L., and Candy, J. M., 1978, *Neuropharmacology* 17:993–998.
96. Dudai, Y., 1979, *Brain Res.* 167:422–425.
97. Chiu, T. H., and Rosenberg, H. C., 1979, *Eur. J. Pharmacol.* 56:337–345.
98. Karobath, M., and Sperk, G., 1979, *Proc. Natl. Acad. Sci. U.S.A.* 76:1004–1006.
99. Krogsgaard-Larsen, P., Johnston, G. A. R., Lodge, D., and Curtis, D. R., 1977, *Nature* 268:53–55.
100. Krogsgaard-Larsen, P., Falch, E., Schousboe, A., Curtis, D. R., and Lodge, D., 1980, *J. Neurochem.* 34:756–759.
101. Braestrup, C., Nielsen, M., Krogsgaard-Larsen, P., and Falch, E., 1979, *Nature* 280:331–333.
102. Supavilai, P., and Karobath, M., 1980, *Neurosci. Lett.* 19:337–341.
103. Squires, R. F., 1982, *Brain Peptides and Hormones* (R. Collu, J.-R., Ducharme, A. Barbeau, and G. Tolis, eds.), Raven Press, New York, pp. 93–106.
104. Squires, R. F. and Saederup, E., 1982 *Mol. Pharm.* 22:327–334.
105. Gavish, M., and Snyder, S. H., 1980, *Nature* 287:651–652.
106. Simmonds, M. A., 1980, *Nature* 284:558–560.
107. Asano, T., and Ogasawara, N., 1980, *Life Sci.* 26:1131–1137.
108. Asano, T., and Ogasawara, N., 1980, *Life Sci.* 26:607–613.
109. Sherman-Gold, R., and Dudai, Y., 1980, *Brain Res.* 198:485–490.
110. Asano, T., and Ogasawara, N., 1981, *Life Sci.* 29:193–200.
111. Gavish, M., and Snyder, S. H., 1980, *Life Sci.* 26:579–582.
112. Gavish, M., and Snyder, S. H., 1981, *Proc. Natl. Acad. Sci. U.S.A.* 78:1939–1942.
113. Yousufi, M. A. K., Thomas, J. W., and Tallman, J. F., 1979, *Life Sci.* 25:463–470.

114. Chiu, T. H., and Rosenberg, H. C., 1979, *Eur. J. Pharmacol.* **58**:335–338.
115. Massotti, M., Guidotti, A., and Costa, E., 1981, *J. Neurosci.* **1**:409–418.
116. Brennan, M. J. W., 1982, *J. Neurochem.* **38**:264–266.
117. Cananzi, A. R. Costa, E., and Guidotti, A., 1980, *Brain Res.* **196**:447–453.
118. Lal, H., Shearman, G. T., Fielding, S., Dunn, R., Kruse, H., and Theurer, K., 1980, *Neuropharmacology* **19**:785–789.
119. Eccles, J. C., 1964, *Science* **145**:1140–1147.
120. Krnjevic, K., 1976, *GABA in Nervous System Function* (E. Roberts, T. N. Chase, and D. B. Tower, eds.), Raven Press, New York, pp. 269–281.
121. Barker, J. L., and Ransom, B. R., 1978, *J. Physiol. (Lond.)* **280**:331–354.
122. Gallagher, J. P., Higashi, H., and Nishi, S., 1978, *J. Physiol. (Lond.)* **275**:263–282.
123. Study, R. E., and Barker, J. L., 1981, *Proc. Natl. Acad. Sci. U.S.A.* **78**:7180–7184.
124. Takeuchi, A., and Takeuchi, N., 1967, *J. Physiol. (Lond.)* **191**:575–590.
125. Constanti, A., and Nistri, A., 1976, *Br. J. Pharmacol.* **57**:347–358.
126. Costa, T., Rodbard, D., and Pert, C. B., 1979, *Nature* **277**:315–317.
127. Martin, I. L., and Candy, J. M., 1980, *Neuropharmacology* **19**:175–179.
128. Olsen, R. W., 1981, *J. Neurochem.* **37**:1–13.
129. Supavilai, P., and Karobath, M., 1979, *Eur. J. Pharmacol.* **60**:111–113.
130. Supavilai, P., and Karobath, M., 1980, *Eur. J. Pharmacol.* **62**:229–233.
131. Karobath, M., Drexler, G., and Supavilai, P., 1981, *Life Sci.* **28**:307–313.
132. Supavilai, P., and Karobath, M., 1981, *Eur. J. Pharmacol.* **70**:183–193.
133. Leeb-Lundberg, F., Snowman, A., and Olsen, R. W., 1980, *Proc. Natl. Acad. Sci. U.S.A.* **77**:7468–7472.
134. Placheta, P., and Karobath, M., 1980, *Eur. J. Pharmacol.* **62**:225–228.
135. Fujimoto, M., and Okabayashi, T., 1981, *Life Sci.* **28**:895–901.
136. Mackerer, C. R., and Kochman, R. L., 1978, *Proc. Soc.Exp. Biol. Med.* **158**:393–397.
137. Supavilai, P., Mannonen, A., and Karobath, M., 1982, *Neurochem. Int.* **4**:259–268.
138. Corda, M. G., Concas, A., Rossetti, Z., Guarneri, P., Corongiu, F. P., and Biggio, G., 1981, *Brain Res.* **229**:264–269.
139. Hill, D. R., and Bowery, N. G., 1981, *Nature* **290**:149–152.
140. Dunlap, K., and Fischbach, G. D., 1978, *Nature* **276**:837–839.
141. Dunlap, K., 1981, *Br. J. Pharmacol.* **74**:579–585.
142. Bihler, I., and Sawh, P. C., 1978, *Mol. Pharmacol.* **14**:879–883.
143. Degroof, R. C., Bianchi, C. P., and Narayan, S., 1980, *Eur. J. Pharmacol.* **66**:193–199.
144. Leslie, S. W., Friedman, M. B., and Coleman, R. R., 1980, *Biochem. Pharmacol.* **29**:2439–2443.
145. DeLorenzo, R. J., Burdette, S., and Holderness, J., 1981, *Science* **213**:546–549.
146. Blaustein, M. P., and Ector, A. C., 1975, *Mol. Pharmacol.* **11**:369–378.
147. Friedman, M. B., Coleman, R., and Leslie, S. W., 1979, *Life Sci.* **25**:735–738.
148. Leslie, S. W., Friedman, S. W., Wilcox, R. E., and Elrod, S. V., 1980, *Brain Res.* **185**:409–417.
149. Elrod, S. V., and Leslie, S. W., 1980, *J. Pharmacol. Exp. Ther.* **212**:131–136.
150. Pincus, J. H., and Hsiao, K., 1981, *Brain Res.* **217**:119–127.
151. Haycock, J. W., Levy, W. B., and Cotman, C. W., 1977, *Biochem. Pharmacol.* **26**:159–161.
152. Balfour, D. J. K., 1980, *Eur. J. Pharmacol.* **68**:11–16.
153. Nestoros, J. N., 1979, *Science* **209**:708–710.
154. Carlen, P. L., Gurevich, N., and Durand, D., 1982, *Science* **215**:306–309.
155. Harris, R. A., 1981, *Biochem. Pharmacol.* **30**:3209–3215.
156. Gerschenfeld, H. M., 1973, *Physiol. Rev.* **53**:1–119.
157. Yarowsky, P. J., and Carpenter, D. O., 1978, *Brain Res.* **144**:75–94.
158. Yarowsky, P. J., and Carpenter, D. O., 1977, *Life Sci.* **20**:1441–1448.
159. Tremblay, J. P., and Grenon, G., 1980, *Life Sci.* **27**:491–496.
160. Pax, R., Bennett, J. L., and Fetterer, R., 1978, *Naunyn Schmiedebergs Arch. Pharmacol.* **304**:309–315.
161. Okamoto, K., and Sakai, Y., 1979, *Br. J. Pharmacol.* **65**:277–285.
162. Okamoto, K., Quastel, D. M. J., and Quastel, J. H., 1976, *Brain Res.* **113**:147–158.

163. Barker, J. L., and Nicoll, R. A., 1973, *J. Physiol. (Lond.)* **228**:259–277.
164. Celio, M. R., Heizmann, C. W., 1981, *Nature* **293**:300–302.
165. Roth, J., Baetens, D., Norman, A. W., and Garcia-Segura, L.-M., 1981, *Brain Res.* **222**:452–457.
166. Jande, S. S., Maler, L., and Lawson, D. E. M., 1981, *Nature* **294**:765–767.
167. Bainbridge, K. G., and Miller, J. J., 1981, *Soc. Neurosci. Abstr.* **7**:588.
168. Miller, J. J., and Bainbridge, K. G., 1981, *Soc. Neurosci. Abstr.* **7**:588.
169. Leeb-Lundberg, F., Snowman, A., and Olsen, R. W., 1981, *J. Neurosci.* **1**:471–477.
170. Williams, M., and Risley, E. A., 1979, *Life Sci.* **24**:833–842.
171. Beer, B., Klepner, C. A., Lippa, A. S., and Squires, R. F., 1978, *Pharmacol. Biochem. Behav.* **9**:849–851.
172. Meiners, B. A., and Salama, A. I., 1982, *Europ. J. Pharmac.* **78**:315–322.
173. Skolnick, P., Lock, K.-L., Paugh, B., Marangos, P., Windsor, R., and Paul, S., 1980, *Pharmacol. Biochem. Behav.* **12**:685–689.
174. Skolnick, P., Lock, K.-L., Paul, S. M., Marangos, P. J., Jones, R., and Irmscher, K., 1980, *Eur. J. Pharmacol.* **67**:179–186.
175. Leeb-Lundberg, F., and Olsen, R. W., 1980, *Psychopharmacology and Biochemistry of Neurotransmitter Receptors* (H. I. Yamamura, R. W. Olsen, and E. Usdin, eds.), Elsevier, New York, pp. 593–606.
176. Olsen, R. W., and Leeb-Lundberg, F., 1981, *Adv. Biochem. Psychopharmacol.* **26**:93–102.
177. Skolnick, P., Paul, S. M., and Barker, J. L., 1980, *Eur. J. Pharmacol.* **65**:125–127.
178. Leeb-Lundberg, F., Snowman, A., and Olsen, R. W., 1981, *Eur. J. Pharmacol.* **72**:125–129.
179. Ashton, D., Geerts, R., Waterkeyn, C., and Leysen, J. E., 1981, *Life Sci.* **29**:2631–2636.
180. Bowery, N. G., Collins, J. F., Hill, R. G., and Pearson, S., 1977, *Br. J. Pharmacol.* **60**:275–276P.
181. Bellet, E. M., and Casida, J. E., 1973, *Science* **182**:1135–1136.
182. Casida, J. E., Eto, M., Moscioni, A. D., Engel, J. L., Milbrath, D. S., and Verkade, J. G., 1976, *Pharm. Appl. Pharmacol.* **36**:261–279.
183. Dudel, J., and Hatt, H., 1976, *Pflugers Arch.* **364**:217–222.
184. Palacios, J. M., Wamsley, J. K., and Kuhar, M. J., 1981, *Brain Res.* **222**:285–307.
185. Palacios, J. M., Unnerstall, J. R., Young, W. S., and Kuhar, M. J., 1981, *Adv. Biochem. Psychopharmacol.* **26**:53–60.
186. Bowery, N. G., Hill, D. R., Hudson, A. L., Doble, A., Middlemiss, D. N., Shaw, J., and Turnbull, M., 1980, *Nature* **283**:92–94.
187. Braestrup, C., and Nielsen, M., 1981, *J. Neurochem.* **37**:333–341.
188. Ehlert, F. J., Roeske, W. R., and Yamamura, H. I., 1981, *Life Sci.* **29**:235–248.
189. Williams, E. F., Rice, K. C., Paul, S. M., and Skolnick, P., 1980, *J. Neurochem.* **35**:591–597.
190. Yokoi, I., Rose, S. E., and Yanagihara, T., 1981, *Life Sci.* **28**:1591–1595.
191. Volicer, L., and Biagioni, T. M., 1982, *J. Neurochem.* **38**:591–593.
192. Burch, T. P., and Ticku, M. K., 1981, *Proc. Natl. Acad. Sci. U.S.A.* **78**:3945–3949.
193. Sieghart, W., and Karobath, M., 1980, *Nature* **286**:285–287.
194. Battersby, M. K., Richards, J. G., and Mohler, H., 1979, *Eur. J. Pharmacol.* **57**:277–278.
195. Johnson, R. W., and Yamamura, H. I., 1979, *Life Sci.* **25**:1613–1620.
196. Lo, M. M. S., Strittmatter, S. M., and Snyder, S. H., 1982, *Proc. Natl. Acad. Sci. U.S.A.* **79**:680–684.
197. Lo, M. M. S., and Snyder, S. H., 1981, *Soc. Neurosci. Abstr.* **7**:613.
198. Lang, B., Barnard, E. A., Chang, L.-R., and Dolly, J. O., 1979, *FEBS Lett.* **104**:149–153.
199. Gavish, M., Chang, R. S. L., and Snyder, S. H., 1979, *Life Sci.* **25**:783–790.
200. Martini, C., Lucacchini, A., Ronca, G., Hrelia, S., and Rossi, C. A., 1982, *J. Neurochem.* **38**:15–19.
201. Davis, W. C., and Ticku, M. K., 1981, *J. Neurosci.* **1**:1036–1042.
202. Braestrup, C., Nielsen, M., Skovjberg, H., and Gredal, O., 1981, *Adv. Biochem. Psychopharmacol.* **26**:147–155.
203. Mohler, H., Battersby, M. K., and Richards, J. G., 1980, *Proc. Natl. Acad. Sci. U.S.A.* **77**:1666–1670.

204. Mohler, H., Richards, J. G., and Wu, J.-Y., 1981, *Proc. Natl. Acad. Sci. U.S.A.* **78:**1935–1938.
205. Iversen, L., 1977, *Nature* **266:**678.
206. Iversen, L., 1978, *Nature* **275:**477.
207. Iversen, L., 1980, *Nature* **285:**285–286.
208. Toffano, G., Guidotti, A., and Costa, E., 1978, *Proc. Natl. Acad. Sci. U.S.A.* **75:**4024–4028.
209. Guidotti, A., Toffano, G., and Costa, E., 1978, *Nature* **275:**553–555.
210. Hokfelt, T., Johansson, O., Ljungdahl, A., Lundberg, J. M., and Schultzberg, M., 1980, *Nature* **284:**515–521.
211. Colello, G. D., Hockenbery, D. M., Bosmann, H. B., Fuchs, S., an Folkers, K., 1978, *Proc. Natl. Acad. Sci. U.S.A.* **75:**6319–6323.
212. Kenessey, A., Lang, T., and Graf, L., 1981, *Int. J. Peptide Protein Res.* **18:**103–106.
213. Davis, L. G., and Cohen, R. K., 1980, *Biochem. Biophys. Res. Commun.* **92:**141–148.
214. Davis, L. G., McIntosh, H., and Reker, D., 1981, *Pharmacol. Biochem. Behav.* **14:**839–844.
215. Chiu, T. H., and Rosenberg, H. C., 1981, *J. Neurochem.* **36:**336–338.
216. Nagy, J., Kardos, J., Maksay, G., and Simonyi, M., 1981, *Neuropharmacology* **20:**529–533.
217. Karobath, M., Sperk, G., and Schonbeck, G., 1978, *Eur. J. Pharmacol.* **49:**323–326.
218. Skolnick, P., Marangos, P. J., Goodwin, F. K., Edwards, M., and Paul, S., 1978, *Life Sci.* **23:**1473–1480.
219. Marangos, P. J., Paul, S. M., Parma, A. M., Goodwin, F. K., Syapin, P., and Skolnick, P., 1979, *Life Sci.* **24:**851–858.
220. Asano, T., and Spector, S., 1979, *Proc. Natl. Acad. Sci. U.S.A.* **76:**977–981.
221. Davies, L. P., Cook, A. F., Poonian, M., and Taylor, K. M., 1980, *Life Sci.* **26:**1089–1095.
222. Skolnick, P., Paul, S. M., and Marangos, P. J., 1980, *Fed. Proc.* **39:**3050–3055.
223. MacDonald, J. F., Barker, J. L., Paul, S. M., Marangos, P. J., and Skolnick, P., 1979, *Science* **205:**715–717.
224. Mohler, H., Polc, P., Cumin, R., Pieri, L., and Kettler, R., 1979, *Nature* **278:**563–565.
225. Tallman, J. F., Paul, S. M., Skolnick, P., and Gallager, D. W., 1980, *Science* **207:**274–281.
226. Nielsen, M., Gredal, O., and Braestrup, C., 1979, *Life Sci.* **25:**679–686.
227. Braestrup, C., Nielsen, M., and Olsen, C. E., 1980, *Proc. Natl. Acad. Sci. U.S.A.* **77:**2288–2292.
228. Nielsen, M., Schou, H., and Braestrup, C., 1981, *J. Neurochem.* **36:**276–285.
229. Braestrup, C., and Nielsen, M., 1981, *Nature* **294:**472–474.
230. Uphaus, R. A., Grossweiner, L. I., Katz, J. J., and Kopple, K. D., 1959, *Science* **129:**641–643.
231. O'Brien, R. A., Schlosser, W., Spirt, N. M., Franco, S., Horst, W. D., Polc, P., and Bonetti, E. P., 1981, *Life Sci.* **29:**75–82.
232. Cowen, P. J., Green, A. R., and Nutt, D. J., and Martin, I. L., 1981, *Nature* **290:**54–55.
233. Ehlert, F. J., Roeske, W. R., Braestrup, C., Yamamura, S. H., and Yamamura, H. I., 1981, *Eur. J. Pharmacol.* **70:**593–596.
234. Polc, P., Ropert, N., and Wright, D. M., 1981, *Brain Res.* **217:**216–220.
235. Muller, W. E., Fehske, K. J., Borbe, H. O., Wollert, U., Nanz, C., and Rommelspacher, H., 1981, *Pharmacol. Biochem. Behav.* **14:**693–699.
236. Rommelspacher, H., Nanz, C., Borbe, H. O., Fehske, K. J., Muller, W. E., and Wollert, U., 1981, *Eur. J. Pharmacol.* **70:**409–416.
237. Tenen, S. S., and Hirsch, J. D., 1980, *Nature* **288:**609–610.
238. Mitchell, R., and Martin, I., 1980, *Eur. J Pharmacol.* **68:**513–514.
239. Glover, V., Bhattacharya, S. K., and Sandler, M., 1981, *Nature* **292:**347–349.
240. Cepeda, C., Tanaka, T., Besselievre, R., Potier, P., Naquet, R., and Rossier, J., 1981, *Neurosci. Lett.* **24:**53–57.
241. Skolnick, P., Williams, E. F., Cook, J. M., Cain, M., Rice, K. C., Mendelson, W. B., Crawley, J., and Paul, S. M., 1982, *Prog. Clin. Biol. Res.* **90:**233–252.
242. Mendelson, W. B., Cain, M., Cook, J. M., Paul, S. M., and Skolnick, P., 1982, *Prog. Clin. Biol. Res.* **90:**253–261.
243. Cain, M., Weber, R. W., Guzman, F., Cook, J. N., Paul, S. M., and Skolnick, P., 1982, *J. Med. Chem.* **25:**1081–1091.

244. Skolnick, P., Paul, S., Crawley, J., Rice, K., Barker, S., Weber, R., Cain, M., and Cook, J., 1981, *Eur. J. Pharmacol.* **69**:525–527.
245. Lippa, A. S., Coupet, J., Greenblatt, E. N., Klepner, C. A., and Beer, B., 1979, *Pharmacol. Biochem. Behav.* **11**:99–106.
246. Hunkeler, W., Mohler, H., Pieri, L., Polc, P., Bonetti, E. P., Cumin, R., Schaffner, R., and Haefely, W., 1981, *Nature* **290**:514–516.
247. Mohler, H., Burkard, W. P., Keller, H. H., Richards, J. G., and Haefely, W., 1981, *J. Neurochem.* **37**:714–722.
248. Mohler, H., and Richards. J. G., 1981, *Nature* **294**:763–765.
249. Blanchard, J. C., Boireau, A., Garret, C., and Julou, L., 1979, *Life Sci.* **24**:2417–2420.
250. Fujimoto, M., Hirai, K., and Okabayashi, T., 1981, *Life Sci.* **30**:51–57.
251. Czernik, A. J., Petrack, B., Kalinsky, H. J., Psychoyos, S., Cash, W. D., Tsai, C., Rinehart, R. K., Granat, F. R., Lovell, R. A., Brundish, D. E., and Wade, R., 1982, *Life Sci.* **30**:363–372.
252. Hunt, P., and Clements-Jewery, S., 1981, *Neuropharmacology* **20**:357–361.
253. Antoniadis, A., Muller, W. E., and Wollert, U., 1980, *Neuropharmacology* **19**:121–124.
254. Geller, H. M., and Baldino, F., 1981, *Soc. Neurosci. Abstr.* **7**:110.
255. Fritz, L. C., Wang, C. C., and Gorio, A., 1979, *Proc. Natl. Acad. Sci. U.S.A.* **76**:2062–2066.
256. Kass, I. S., Wang, C. C., Walrond, J. P., and Stretton, A. O. W., 1980, *Proc. Natl. Acad. Sci. U.S.A.* **77**:6211–6215.
257. Pong, S. S., and Wang, C. C., 1980, *Neuropharmacology* **19**:311–317.
258. Williams, M., and Yarbrough, G. G., 1979, *Eur. J. Pharmacol.* **56**:273–276.
259. Supavilai, P., and Karobath, M., 1981, *J. Neurochem.* **36**:798–803.
260. Paul, S. M., Skolnick, P., and Zatz, M., 1980, *Biochem. Biophys. Res. Commun.* **96**:632–638.
261. Karobath, M., Placheta, P., Lippitsch, M., and Krogsgaard-Larsen, P., 1979, *Nature* **278**:748–749.
262. Maurer, R., 1979, *Neurosci. Lett.* **12**:65–68.
263. Karobath, M., and Lippitsch, M., 1979, *Eur. J. Pharmacol.* **58**:485–488.
264. Taniguchi, T., Wang, J. K. T., and Spector, S., 1982, *Biochem. Pharmacol.* **31**:589–590.
265. Le Fur, G., Perrier, M. L., Vaucher, N., Imbault, F., Flamier, A., Benavides, J., Uzan, A., Renault, C., Dubroeucq, M. C., and Gueremy, C., 1983, *Life Sci.* **32**:1839–1847.
266. Le Fur, G., Guilloux, F., Rufat, P., Benavides, J., Uzan, A., Renault, C., Dubroeucq, M. C., and Gueremy, C., 1983, *Life Sci.* **32**:1849–1856.
267. Weissman, B. A., Cott, J., Paul, S. M., and Skolnick, P., 1983, *Eur. J. Pharmacol.* **90**:149–150.
268. File, S. E., and Lister, R. G., 1983, *Neurosci. Let.*, **35**:93–96.
269. Ishii, K., Takashi, K., Akutagawa, M., Makino, M., Tanaka, T., and Ando, J., 1982, *Eur. J. Pharmacol.* **83**:329–333.
270. Swenson, Jr., R. P., 1982, *Brain Res.* **241**:317–322.
271. Schwarz, J. R., and Spielmann, R. P., 1983, *Eur. J. Pharmacol.* **90**:359–366.
272. Squires, R. F., and Saederup, E., 1982, *Pharmacol. of Benzodiazepines*, ed. E. Usdin et al. MacMillan Press (London), pp. 567–582.
273. Squires, R. F., Casida, J. E., Richardson, M., and Saederup, E., 1982, *Mol. Pharm.* **23**:326–336.
274. Sieghart, W., and Drexler, G., 1983, *J. Neurochem.* **41**:47–55.
275. Raftery, M. A., Hunkapiller, M. W., Strader, C. D., and Hood, L. E., 1980, *Science* **208**:1454–1457.
276. Braestrup, C., Schmiechen, R., Neef, G., Nielsen, M., and Petersen, E. N., 1982, *Science*, **216**:1241–1243.
277. Asano, T., Yamada, Y., and Ogasawara, N., 1983, *J. Neurochem.* **40**:209–214.
278. Sigel, E., Stephenson, F. A., Mamalaki, C., and Barnard, E. A., 1983, *J. Biol. Chem.* **258**:2965–2971.
279. Chang, L.-R., and Barnard, E. A., 1982, *J. Neurochem.* **39**:1507–1518.
280. Guidotti, A., Forchetti, C. M., Corda, M. G., Konkel, D., Bennett, C. D., and Costa, E., 1983, *Proc. Natl. Acad. Sci. U.S.A.* **80**:3531–3535.

281. Valin, A., Dodd, R. H., Liston, D. R., Potier, P., and Rossier, J., 1982, *Eur. J. Pharmacol.* **85**:93–97.
282. Fujimoto, M., Kawasaki, K., Matsushita, A., and Okabayashi, T., 1982, *Eur. J. Pharmacol.* **80**:259–262.
283. Mohler, H., 1982, *Eur. J. Pharmacol.* **80**:435–436.
284. Brown, C., and Martin, I. L., 1983, *Neurosci. Letters* **35**:37–40.
285. Noda, M., Takahashi, H., Tanabe, T., Toyosato, M., Kikyotani, S., Furutani, Y., Hirose, T., Takashima, H., Inayama, S., Miyata, T., and Numa, S., 1983, *Nature* **302**:528–532.
286. Sherman-Gold, R., and Dudai, Y., 1983, *J. Neurosci. Res.* **10**:27–33.
287. Braestrup, C., Nielsen, M., and Honore, T., 1983, *J. Neurochem.* **41**:454–465.
288. Carlen, P. L., Gurevich, N., and Polc, P., 1983, *Brain Res.* **271**:358–364.
289. Czuczwar, S. J., and Meldrum, B., 1982, *Eur. J. Pharmacol.* **83**:335–338.
290. File, S. E., Lister, R. G., and Nutt, D. J., 1982, *Neuropharmacology* **21**:1033–1037.
291. Ninan, P. T., Insel, T. M., Cohen, R. M., Cook, J. M., Skolnick, P., and Paul, S. M., 1982, *Science* **218**:1332–1334.
292. Corda, M. G., Blaker, W. D., Mendelsen, W. B., Guidotti, A., and Costa, E., 1983, *Proc. Natl. Acad. Sci. U.S.A.* **80**:2072–2076.
293. de Carvalho, L. P., Grecksch, G., Chapouthier, G., and Rossier, J., 1983, *Nature* **301**:64–66.
294. Skolnick, P., Schweri, M. M., Paul, S. M., Martin, J. V., Wagner, R. L., and Mendelson, W. B., 1983, *Life Sci.* **32**:2439–2445.
295. Schweri, M., Cain, M., Cook, Paul, S., and Skolnick, P., 1982, *Pharmac. Biochem. Behav.* **17**:457–460.
296. Mendelson, W. B., Cain, M., Cook, J. M., Paul, S. M., and Skolnick, P., 1983, *Science* **219**:414–416.
297. Lal, H., and Emmett-Oglesby, M. W., *Neuropharmacology*, in press 1983.
298. Koe, B. K., and Lebel, L. A., 1983, *Eur. J. Pharmacol.* **90**:97–102.
299. Hullihan, J. P., Spector, S., Taniguchi, T., and Wang, J. K. T., 1983, *Br. J. Pharmac.* **78**:321–327.
300. Sieghart, W., Mayer, A., and Drexler, G., 1983, *Eur. J. Pharmacol.*, **88**:291–299.
301. Martini, C., Rigacci, T., and Lucacchini, A., 1983, *J. Neurochem.*, **41**:1183–1185.
302. Blanchard, J.-C., and Julou, L., 1983, *J. Neurochem.*, **40**:601–607.

Recognition Sites for Antidepressant Drugs

De-Maw Chuang and E. Costa

1. INTRODUCTION

It has been almost two decades since the discovery that tricyclic antidepressants and monoamine oxidase (MAO) inhibitors were clinically effective in the treatment of depression. In spite of intensive investigations, neither the molecular mechanisms whereby antidepressants operate nor the etiology of this disease is well understood. The knowledge that reserpine, which depletes monoamine stores,[1,2] can induce a depressed mood [3] has stimulated speculation on a causal relationship between a defect in monoaminergic transmission and endogenous depression.

This speculation was indirectly supported by the finding that drugs that increase the availability of norepinephrine (NE) and/or serotonin (5-HT) at receptor recognition sites through an inhibition of monoamine uptake[4-9] can ameliorate the symptoms of depression. At its peak of credibility, the monoamine hypothesis on the etiology of affective disorders stated that the symptoms of these disorders reflect either a decreased availability of catecholamines at specific synaptic receptors or a change in the balance between the function of 5-HT and NE neurons.[10-16] However, the popularity of the theory began to decline when it was reported that the ability to increase monoamine availability through an inhibition of their uptake is not a common feature of all clinically effective antidepressants. Among others, mianserin[17,18] iprindole,[19] and bupropion[20] fail to inhibit monoamine uptake, and because of this deviation from the norm, these drugs were termed atypical antidepressants. Conversely, a potent blocker of catecholamine uptake such as cocaine[21] fails to relieve the symptoms of depression. This discrepancy began to undermine the validity of the monoamine hypothesis, and its impact was further increased when various investigators began to compare the short latency time required by the anti-

De-Maw Chuang and E. Costa • Laboratory of Preclinical Pharmacology, National Institute of Mental Health, Saint Elizabeths Hospital, Washington, D. C. 20032.

depressant to elicit monoamine uptake inhibition with the protracted time delay required by these drugs to elicit a beneficial effect in depression. Hence, it is extremely doubtful that the inhibition of monoamine uptake is an explanation for their therapeutic action. This realization became a serious challenge to the monoamine theory in the etiology of endogenous depressions.

With the realization that the beneficial effects of antidepressants have a delayed onset, investigations on the mode of action of antidepressants have been focused on those molecular events that are elicited after 2 to 3 weeks of treatment with antidepressants. Therefore, in this chapter, we have attempted to present relevant information on molecular events elicited in brain by long-term treatment with antidepressants and MAO inhibitors. This information is given in the perspective of present knowledge on the presence of specific recognition sites for certain antidepressants located in brain synaptic membrane. The present understanding of the function of such specific high-affinity binding sites for typical and atypical antidepressants and the role of the adaptive changes that occur in these sites following long-term administration of these drugs are discussed in terms of their significance in mediating the therapeutic action of antidepressants. The chemical structures of several antidepressant drugs discussed in this chapter are shown in Fig. 1.

2. PRESYNAPTIC ADAPTIVE CHANGES FOLLOWING REPEATED INJECTIONS OF ANTIDEPRESSANT DRUGS

2.1. Changes in Monoamines

Although a single dose of desipramine and of other tricyclic antidepressants with a secondary amine group decreases the NE turnover rate in rat brain,[22–25] repeated daily doses of typical antidepressants accelerate brain NE turnover rate.[19,23,25,26–28] Long-term daily doses of imipramine or protriptyline, which are tricyclic antidepressants with a tertiary amine group, also increase NE turnover in rat brain.[29] In contrast, repeated daily injections of amitriptyline, another antidepressant with a tertiary amine group, fail to change the NE turnover in rat brain.[23,25] It has been reported that repeated daily doses[30] or a single dose of chlorimipramine reduces the 5-HT turnover[30–33] in rat brain. However, another report shows that the inhibition of brain 5-HT turnover by chlorimipramine or imipramine shows adaptation after repeated daily doses.[34] It is important to keep in mind that the turnover rate of dopamine is unaffected following protracted treatment with daily doses of desipramine,[19,26,28,35] protriptyline,[26] and chlorimipramine.[28]

Mianserin, a tetracyclic atypical antidepressant, has little or no effect on central NE uptake *in vivo*,[17] although it displays moderate activity in blocking NE uptake *in vitro*.[18,36] Repeated daily injections as well as a single dose of mianserin increase the turnover of NE in rat brain.[37] Mianserin fails to block 5-HT uptake *in vivo*,[18] and it is very weak in blocking 5-HT uptake *in vitro*.[18,36] Although mianserin is reported to be a 5-HT receptor antagonist *in vivo*,[38,39] it fails to alter 5-HT turnover in rat brain following a single dose[37] or repeated daily injections.[28,37]

SOME TYPICAL ANTIDEPRESSANTS

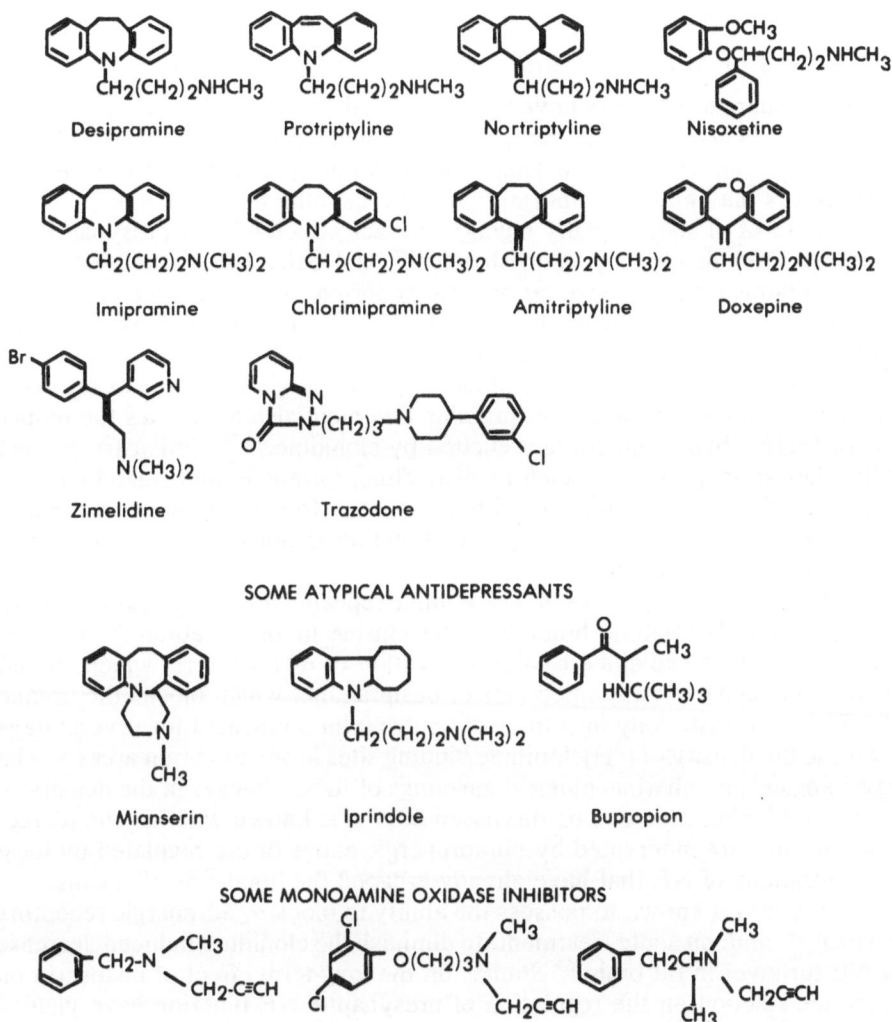

Fig. 1. Chemical structures of some antidepressants.

Iprindole, another atypical antidepressant with little or no effect on monoamine uptake,[19] does not change NE turnover following repeated daily injections.[19,40] After such a treatment, iprindole can increase 5-HT turnover in rat neocortex[41] but not in mouse brain.[42] Zimelidine and fluoxetine, two antidepressants that cause a selective and potent inhibition of 5-HT uptake, [43]–[45] reduce the turnover of 5-HT in rat and mouse brain after a single dose or repeated daily injections.[46,47] Thus, the effects of long-term daily injections of typical or atypical antidepressants on monoamine turnover differ. These discrepancies may be related to a number of factors, including drug dosage, frequency of administration, and time intervals between the last injection and the measurement of amine turnover.

2.2. Effects of Antidepressants on α_2-Adrenergic Receptors

α-Adrenergic receptors have been classified into two subtypes, termed α_1 and α_2.[48–50] Whereas the α_1 receptor site is located postsynaptically, the α_2 receptor site can be found presynaptically and postsynaptically.[51] Presynaptic α_2 receptors may play an important role in regulating the release of NE.[52–54] A low dose of clonidine (<0.1 mg/kg) may act selectively on presynaptic α_2 receptors, as indicated by neurochemical[55] and behavioral[56,57] studies. Repeated administration of desipramine was reported to desensitize α_2 receptors of rat heart, as assessed from the [^{3}H]NE efflux from prelabeled tissue during field stimulation.[57] Electrophysiological studies confirm that after daily doses of desipramine repeated for several days, α_2 adrenergic receptors of rat brain are down-regulated[58]; such a desipramine treatment also attenuates the reduction of MHPG-SO$_4$ brain content elicited by clonidine.[58,59] Similar treatments with other antidepressants, such as nisoxetine, iprindole, and trazodone, fail to change the clonidine-induced diminution of MHPG-SO$_4$ brain content.[60] Hence, down-regulation of α_2 receptor function is not a long-term response elicited by every antidepressant.

Interestingly, daily doses of desipramine repeated for several days increase the B_{max} of [^{3}H]clonidine binding in the murine limbic forebrain.[61] This response cannot be explained by desensitization of α_2 receptors, which should occur after repeated daily injections of desipramine, which blocks the uptake of NE. In contrast, daily injections of amitriptyline, repeated for several days decrease the density of [^{3}H]clonidine binding sites in several brain areas.[62] The physiological and pharmacological meanings of these changes in the density of clonidine binding sites can be discussed after it is known whether the α_2 recognition sites are innervated by noradrenergic axons or are regulated by local concentrations of NE that have already escaped the uptake mechanisms.

Mianserin is known to possess the ability to block α_2 adrenergic receptors *in vitro*[63,64] and, on acute treatment, to diminish the clonidine-induced decrease of NE turnover in rat brain.[65] Studies on the long-term effect of mianserin on clonidine's action on the regulation of presynaptic NE function have yielded contradictory results. Two-week treatments with mianserin have no effect on clonidine (25 μg/kg)-induced reduction of the MHPG-SO$_4$ content of rat brain.[66] Other studies indicate that repeated daily injections of mianserin block the ability of clonidine (350 μg/kg) to lower the brain content of MHPG.[67] These results may be attributed at least in part to different sites of action of clonidine given in doses that have a more than tenfold difference.

3. POSTSYNAPTIC ADAPTIVE CHANGES FOLLOWING REPEATED DAILY INJECTIONS OF ANTIDEPRESSANTS

3.1. β-Adrenergic Receptors

A long-term treatment with daily injections of antidepressants, but not a single injection of a typical or atypical antidepressant, decreases the number of β-adrenergic receptor recognition sites and the signal amplification elicited by NE on the adenylate cyclase of rat brain. In this section, up-to-date information regarding the modification of β-adrenergic receptors elicited by long-term treatment with antidepressants is summarized, and its significance to the interpretation of the therapeutic action of antidepressants is discussed.

3.1.1. Recognition Sites for β-Adrenergic Receptors

Banerjee *et al.*[68] first reported that long-term daily injections, but not a single injection, of a tricyclic antidepressant (desipramine, iprindole, and doxepine) decrease the number of recognition sites for β-adrenergic ligands located in synaptic membranes prepared from rat brain, but that the affinity of the β-adrenergic receptor for the specific ligands remains substantially unchanged. Such a down-regulation of β-adrenergic receptor binding sites by repeated daily doses of desipramine was confirmed by successive studies.[64,69–74] A similar response is elicited following long-term treatment with imipramine,[64,73,75,76] amitriptyline,[73,74] nortriptyline, chlorimipramine,[74] bupropion,[74] and trazodone.[77] The down-regulation of β-adrenergic receptor recognition sites occurs in various structures of the rat brain, but the duration of the treatment required to down-regulate the receptor response differs in various brain areas. Whereas in the cortex and cerebellum the decrease of B_{max} for ligand binding can be detected within 7–10 days after the desipramine or imipramine treatments,[71,75] in the hippocampus, such a down-regulation requires daily injections of these drugs for 3 weeks or longer.[71,75] Though rat brain contains both β_1 and β_2 receptors,[78] the down-regulation elicited by antidepressants involves only β_1 receptors.[79] The time characteristics to elicit changes in β_1 receptor recognition sites and in the signal amplification elicited by isoproterenol on adenylate cyclase caused by long-term treatment with antidepressants are similar.[70] Although repeated daily doses of desipramine decrease the density of β-adrenergic receptor recognition sites and the extent of the signal amplifications elicited by NE on pineal adenylate cyclase,[80] such treatment fails to change β-adrenergic receptor function in rat heart.[70]

The molecular mechanisms underlying the loss of β-adrenergic receptor recognition sites are presently undefined. Since desipramine has a very weak affinity for the adrenergic receptor recognition sites,[63,73] it is unlikely that the down-regulation of β-adrenergic receptors represents an irreversible binding of the antidepressant to β-adrenergic receptor recognition sites. Experiments using plasma membranes incubated with GTP have ruled out the possibility that a persistent NE binding to β-adrenergic receptor site is responsible for the down-regulation of β-adrenergic receptor induced by desipramine.[72] Other molecular events that may explain the down-regulation of the β-adrenergic receptor elicited by antidepressants may include receptor inactivation, decreased synthesis or enhanced degradation of the receptor protein, or internalization of receptor recognition site. The last event is particularly intriguing, because we have shown that internalization of β-adrenergic receptor recognition sites occurs in frog erythrocytes during desensitization induced by persistent occupancy of the recognition site by isoproterenol.[81–84]

A loss in the number of β-adrenergic receptor recognition sites located in crude synaptic membranes prepared from rat brain has also been reported following repeated treatments with MAO inhibitors[70,73,74] and electroconvulsive therapy,[85-87] which are two additional procedures used in the clinic to relieve the symptoms of depression. Long-term administration of lithium, which is used to treat patients with bipolar depression, also evokes a small but statistically significant reduction of the B_{max} of β-adrenergic receptors located in membranes prepared from rat brain cortex.[76] However, no change in the binding characteristics was found in cerebellum after chronic lithium treatment.[88] Long-term daily injections of mianserin,[89] nisoxetine,[89] and possibly zimelidine[89,90] fail to modify the number of central β-adrenergic receptor binding sites. Despite the inability of these drugs to decrease the number of β-adrenergic receptor recognition sites, some of these drugs reduce the signal amplification elicited by NE on brain adenylate cyclase[89,90] (see Section 3.1.2).

The desipramine-induced loss of β-adrenergic receptor recognition sites can be accelerated by the concomitant administration of α-receptor blockers such as yomhibine[61] or phenoxybenzamine.[91] This facilitation was tentatively explained by the suggestion that a blockade of presynaptic α_2 receptors facilitates the release of NE, thereby accelerating the down-regulation of β receptors. However, evidence is lacking that these inhibitors are acting exclusively on presynaptic α_2 receptors. Moreover, it is an oversimplification to assume that increased NE availability is the mechanism involved in the down-regulation of β-adrenergic receptors elicited by antidepressants. This is particularly true when one considers either the action of mianserin, which down-regulate β-adrenergic receptor function without changing the B_{max} for the ligand binding to β-adrenergic receptor recognition sites,[89] or that of iprindole, which fails to change NE uptake *in vivo*[19] but down-regulates β-adrenergic receptors in rat brain.[68,70,75,92] Conversely, cocaine, which is a potent NE uptake inhibitor,[21] fails to decrease the number of central β-adrenergic receptor recognition sites following repeated injections[74] or prolonged intravenous infusion.[92] Furthermore, several lines of independent investigation clearly indicate that the down-regulation of β-adrenergic receptors in the rat brain requires the function of serotonergic neurons. This finding brings up the possibility that an interneuronal system connecting 5-HT and NE neurons is operative in bringing about the desensitization of β-adrenergic receptors elicited by antidepressants (for details, see Section 5.4).

3.1.2. Norepinephrine-Sensitive Adenylate Cyclase

Sulser and co-workers[93,94] reported that the signal amplification elicited by NE on adenylate cyclase in brain slices can be attenuated following repeated daily injections of antidepressants. Experiments using rats treated repeatedly with various typical or atypical antidepressants, MAO inhibitors, or electroshock have shown that each therapeutic procedure that relieves the symptoms of depression is able to decrease the signal amplification elicited by NE on the adenylate cyclase of rat brain.[86,89,93,94] The delayed onset of the drug-elicited attenuation of adenylate cyclase responsiveness to NE stimulation can be correlated with the appearance of their clinical effects. Undoubtedly, the decrease in the density of membrane-bound β-adrenergic receptor recognition sites is not essential to cause the decrease of the signal amplification elicited by NE on brain adenylate cyclase. In fact, mianserin,[95] nisoxetine,[96] and possibly zimelidine[95] can reduce the signal amplification elicited by NE on cyclase without decreasing the B_{max} for the binding of specific ligands to β-adrenergic receptors. Since these receptors function as a supramolecular entity that includes a detector (recognition site), a coupling device [G/F protein(s)], and a transducer (adenylate cyclase),[81,82] one may speculate that the desensitization of NE-coupled adenylate cyclase elicited by mianserin and other drugs may involve an alteration of G/F protein(s) function.

The desipramine-induced decrease in the signal amplification elicited by NE on cortical adenylate cyclase is abolished by destruction of NE axonal terminals with 6-OH-DA[70,97] or lesion of the locus coeruleus.[98] In this condition, the signal is maximally amplified because of the absence of innervation, and presynaptic modulation is paralyzed because of the denervation. This suggests that with the innervation comes a modulatory influence that may not reside exclusively in the amount of NE reaching the receptor but may involve an additional chemical signal generated by the noradrenergic axon. This possibility is in keeping with current evidence suggesting that synaptic communication may involve more than one chemical signal (see Section 5.4).

3.2. α-Adrenergic Receptors

Long-term administration of daily doses of tricyclics, such as desipramine,[73,85] imipramine, amitriptyline,[73,76] and MAO inhibitors[73] fails to change the binding characteristics of α_1-adrenergic receptor recognition sites in rat brain. The recent report that long-term amitriptyline treatment increases the number of α_1-receptor recognition sites and of cholinergic muscarinic receptor recognition sites[99] represents an exception and may be related to the unusual high affinity of amitriptyline for the recognition sites for α_1-adrenergic receptors[63,100] and cholinergic muscarinic receptors.[101] It has been reported that in rat brain, the binding of [^{3}H]clonidine, an α_2-receptor ligand, is increased after repeated daily injections of desipramine[61] or iprindole.[102] However, the time course of this increase can be dissociated from that of the down-regulation of β-adrenergic receptor recognition sites. Conversely, after 2 weeks of twice-daily administration of amitriptyline to rats, the density of α_2 clonidine binding site is decreased in several brain areas.[103] As discussed in Section 2.2, neither the exact location of these drug-modified α_2 recognition sites nor the pharmacological implication of these changes in α_2 recognition sites is clear at this time.

3.3. Serotonergic Receptors

Two classes of 5-HT recognition sites are thought to be located in crude synaptic membranes; they have been termed 5-HT$_1$ and 5-HT$_2$. The 5-HT$_1$ recognition site binds 5-HT and other 5-HT receptor agonists with high affinity, and some of these sites are linked to adenylate cyclase.[104-106] The 5-HT receptors in rat hippocampus show supersensitivity following denervation[106] and are down-regulated following persistent occupancy by the endogenous ligands, as is the case with pretreatment with pargyline[73] and MAO type A inhibitors.[107,108] The 5-HT$_2$ recognition site can be labeled with high affinity by a number of drugs reported to function as 5-HT antagonists, such as spiperone,[109] mianserin,[110] and ketanserin.[111] Denervation of 5-HT axons fails to increase the binding of ketanserin and spiperone to 5-HT$_2$ recognition sites,[112] although it enhances the binding of mianserin.[112-114] This finding suggests that mianserin does not bind to the same site as that where ketanserin and spiperone bind. Moreover, it raises some doubts about the possibility that the mianserin binding site is a site for a cotransmitter. In fact, the number of cotransmitter recognition sites are not up-regulated by denervation.[139]

The results of studies investigating adaptive changes of 5-HT$_1$ binding sites after long-term daily injections with antidepressants have not contributed a uniform profile. In rats injected repeatedly with desipramine, the B_{max} of 5-HT$_1$ recognition sites is decreased in some studies[64,115] but not in others.[71,73] A lack of modification of the characteristics of 5-HT$_1$ binding sites has also been reported after long-term treatment with daily injections of chlorimipramine,[64,108] amitriptyline,[73] and iprindole.[73] Although the lack of effect of iprindole is not surprising because this drug is a very poor blocker of 5-HT uptake, it is surprising that repeated daily injections of chlorimipramine, a powerful blocker of 5-HT uptake, fail to down-regulate 5-HT$_1$ recognition sites. Also, the sensitivity to 5-HT of the hippocampal adenylate cyclase remained unchanged in rats treated twice daily for 3 weeks with imipramine (Unpublished results from our laboratory). In contrast, repeated daily injections of type A MAO inhibitors such as clorgyline and nialamide[107,108] decrease the specific binding of [^{3}H]5-HT to 5-HT$_1$ recognition sites located in crude synaptic membranes prepared from rat brain.

Of great interest is the recent finding by Peroutka and Snyder[73,117] that the number of 5-HT$_2$ binding sites is decreased by repeated daily injections of typical antidepressants such as amitriptyline, desipramine, and imipramine, atypical antidepressants such as iprindole and mianserin, and type A MAO inhibitors. However, in the cerebral cortex of cats treated daily with imipramine (7.5 mg/kg i.p. twice daily for 20 days), the number of 5-HT$_2$ binding sites is unchanged despite a significant down-regulation of the β-adrenergic receptors.[118] This and other experiments[119] have shown that a down-regulation of 5-HT$_2$ receptors is not associated with every procedure that ameliorates the symptoms of depression. Also, long-term administration of lithium decreases the B_{max} of [^{3}H]spiperone binding to the 5-HT$_2$ recognition sites located in the hippocampus without changing those located in the cortex.[120,121] Moreover, repeated electroconvulsive therapy, one of the most effective treatments for depression, increases rather than decreases the density of 5-HT$_2$ sites located in the rat cortex.[122] Because of these results, it is difficult to conclude that the down-regulation of 5-HT$_2$ recognition sites is operative in mediating the action of antidepressants on the symptoms of depression.

4. ARE NEUROTRANSMITTER RECOGNITION SITES ASSOCIATED WITH THE RECOGNITION SITES FOR ANTIDEPRESSANT DRUGS?

Several lines of investigations discussed above have indicated that many antidepressants can modify the function of a number of receptors for CNS neurotransmitters, such as those for catecholamines, 5-HT, histamine, and acetylcholine. In this section, these interactions are reviewed, and their relationship to the therapeutic effect is discussed.

4.1. Adrenergic

A number of antidepressants have been studied as blockers of central α_1-adrenergic receptors. It has been shown that several antidepressants compete with the binding of [^{3}H]WB-4101, a selective ligand for the recognition sites of the α_1 receptor located in calf brain crude synaptic membranes.[63] In general, tertiary amine derivatives of tricyclic antidepressants, such as doxepine and amitriptyline, are more potent than secondary amine derivatives in competing with [^{3}H]WB 4101.[100] It is likely that sedative and hypotensive side effects known to be associated with the administration of tricyclics containing a tertiary amino group may result from a blockade of α_1 adrenergic receptor function. Typical and atypical antidepressants have a low affinity for [^{3}H]clonidine binding sites[63,64]; clonidine is used to tag α_2 adrenergic receptor recognition sites. Mianserin is an exception because it effectively displaces [^{3}H]clonidine and [^{3}H]WB 4101 from their specific high-affinity binding sites with a K_i of 12 and 56 nM, respectively.[63] Though interactions of mianserin with both α_1 and α_2 adrenergic receptors have also been demonstrated *in vivo*,[56,123-126] the role of the interaction of mianserin with these receptors to explain its antidepressant action requires further investigation. Finally, it is important to note that all antidepressants have very low affinity for β-adrenergic receptor recognition sites as assessed by binding studies using [^{3}H]dihydroalprenolol as a ligand.[63,73]

4.2. Serotonergic

The IC_{50} of a number of typical and atypical antidepressants to displace specific ligands from 5-HT$_1$ and 5-HT$_2$ recognition sites has been investigated using crude synaptic membranes prepared from rat cerebral cortex. Most of the antidepressants are more potent in displacing [^{3}H]spiperone or [^{3}H]D-LSD (*d*-lysergic acid diethylamide) bound to so-called 5-HT$_2$ sites than [^{3}H]5-HT bound to 5-HT$_1$ sites[73,127] This preference must be kept in mind when one discusses the role of serotonergic mechanisms in mediating the antidepressant action elicited by these drugs. Though one does not know what the role of 5-HT$_2$ recognition sites is in serotonergic transmission, it must be mentioned that among the antidepressants tested, mianserin, amitriptyline, and nortriptyline have the highest potency to displace specific ligands of 5-HT$_2$ binding sites (K_i in the nanomolar range).[73,127] However, not all antidepressants display such a high affinity for the 5-HT$_2$ sites; this suggests that the affinity of an antidepressant for the 5-HT$_2$ recognition sites may be one of the mechanisms mediating the action of antidepressants, but this action cannot be generalized as a mechanism for the entire class of compounds. In addition, some neuroleptics, such as chlorpromazine and haloperidol, also have an affinity for 5-HT$_2$ recognition sites comparable to that of the most active antidepressants.[73] This finding suggests that a high-affinity binding to the 5-HT$_2$ recognition sites can even be unrelated to the antidepressant action. Finally, it is our opinion that before a definite decision is made on the role of the so-called 5-HT$_2$ binding sites in the action of antidepressants, the physiological significance of these sites must be studied with greater accuracy.

4.3. Histaminergic

Two types of histamine receptors, H$_1$ and H$_2$, are present in mammalian brain. Both are thought to be operative in mediating biochemical and physiological responses.[128] Some tricyclic antidepressants are extremely potent inhibitors of histamine H$_1$ receptors in brain,[129,130] guinea pig ileum,[131] and neuroblastoma cells.[132] Binding studies show that [^{3}H]doxepin binds to rat brain homogenate with an extremely high affinity (K_d = 0.02 nM) but low capacity (B_{max} = 14 fmol/mg).[130] Various H$_1$ histamine antagonists and some typical antidepressants display nanomolar affinity for this high-affinity doxepin recognition site.[130] However, as pointed out by Richelson and co-workers, there is no correlation between the value of the blood levels of tricyclics that must be reached to relieve the symptoms of depression and their affinity for histamine H$_1$ recognition sites.[130] Hence, it is likely that histamine H$_1$ receptor blockade is unrelated to the antidepressant efficacy of these drugs.

Peroutka and Snyder reported that [^{3}H]mianserin labels the ligand recognition sites of H$_1$ and 5-HT$_2$ receptors.[110] However, we have shown that in the hippocampus, [^{3}H]mianserin binding increases after denervation.[112–114] Since 5-HT$_2$ recognition sites are not innervated by 5-HT-containing axons,[112] we have suggested that 5-HT$_2$ recognition sites may be releated to certain 5-HT$_1$ recognition sites in an allosteric manner.[112,133] Whether the same holds true for the relationship between mianserin binding sites and histamine H$_1$ recognition sites remains to be shown. A possible role for [^{3}H]mianserin binding sites in the expression of the long-term effect if mianserin is discussed in Section 5.4.

Both typical and atypical antidepressants in micromolar concentration are inhibitors of the signal amplification elicited by stimulation of histamine H$_2$ receptors on the cyclic AMP formation by brain homogenates.[134,135] *In vitro* binding assays also demonstrate moderate reactivity of some antidepressants for histamine H$_2$ recognition sites.[135] However, classical antihistamines and neuroleptic phenothiazines are comparable to antidepressants in blocking H$_2$ receptor-linked adenylate cyclase,[134] but they fail to relieve the symptoms of depression. This tends to exclude an involvement of histamine H$_2$ receptors in the mediation of the action of antidepressants.

4.4. Cholinergic

Tricyclic antidepressants have been tested for their ability to act as muscarinic receptor blockers *in vivo* and *in vitro*. Acute amitriptyline, nortriptyline, and imipramine antagonize the hypothermic and tremorogenic activities induced by oxotremorine in mice.[101] The binding constant of these antidepressants to recognition sites of muscarinic receptors located in crude synaptic membranes prepared from mouse brain homogenates can be correlated with their potencies in antagonizing *in vivo* responses[101] to cholinomimetics. The slight increase in the number of muscarinic binding sites elicited by long-term amitriptyline[99] may be the result of a persistent blockade by amitriptyline of muscarinic recognition sites. Because tricyclics differ considerably in their affinity for central muscarinic receptor sites, and other antidepressants such as MAO inhibitors lack the antimuscarinic effect, one might conclude that blockade of central muscarinic receptor function is unrelated to the therapeutic action. Instead, anticholinergic activity of some antidepressants may contribute to drug-induced side effects.[136]

4.5. Opiate and Dopaminergic

The recent suggestion that both endorphins and enkephalins are involved in the etiology of mental illness[137] has stimulated the investigation of possible interactions between receptors for tricyclic antidepressants and endogenous opiates. Various tricyclic antidepressants interact with [^{3}H]naloxone binding sites with low affinity but equal potency (IC_{50} values are about 30 μM).[138] Acute amitriptyline (10 mg/kg i.p.) raises the pain threshold in the rat foot shock test, and this analgesic effect can be antagonized by naloxone.[138] Whether this effect is relevant for the amelioration of depression is not known at this time.

The affinity of typical and atypical antidepressants for dopamine recognition sites is very weak[73,127] and appears to be irrelevant to the action of antidepressants.

5. ARE RECOGNITION SITES FOR ANTIDEPRESSANT DRUGS INVOLVED IN ENDOGENOUS MECHANISMS REGULATING NEUROTRANSMITTER RECEPTOR FUNCTION?

The discovery that benzodiazepines act by increasing the gain at which GABAergic synapses operate[139] and the consequent discovery that benzodiazepines act by binding to a specific recognition site for an endogenous modulator of the GABA recognition site[140–142] prompted the suggestion that the GABAergic synapses are modulated by at least two different chemical signals. Occupancy of the GABA recognition sites by GABA opens the Cl^- channel; occupancy of the benzodiazepine recognition sites determines the gain at which the GABA receptor operates.[139] This site is physiologically occupied by an endogenous effector (cotransmitter), which modulates the action of GABA (primary transmitter).[143] These findings stimulated the search for a high-affinity recognition site for the antidepressants as a starting point to ascertain whether antidepressants also function in a model of synaptic transmissions in which two chemical signals are operative in the exchange of information at synapses.

5.1. High-Affinity Binding Sites for Imipramine

A high-affinity specific binding site for [^{3}H]imipramine was first demonstrated by Langer and co-workers in synaptic membranes prepared from rat brain.[144,145] Their finding was confirmed by us[75] and others.[146–148] Subsequent studies indicate that this imipramine binding site is present in the brain of various mammals[118,146,149] including man.[146] This antidepressant is specifically bound to brain membranes with high affinity (K_d = 5 to 10 nM) and to a limited number of sites.[75,144,146] These binding sites are different from any of the recognition sites for known neurotransmitters.[144,145] The binding of imipramine is effectively displaced by structurally related tricyclic antidepressants and various inhibitors of 5-HT uptake.[145] The efficiency of the displacement by tricyclic antidepressants can be correlated with their clinical potencies,[150] but not all the antidepressants displace [^{3}H]imipramine with high affinity.[144,145] The association of [^{3}H]imipramine to brain membranes has rapid kinetics,[75,145,146] it is Na$^+$ dependent,[151] and the number of binding sites varies in different brain structures[152,153]: the highest is in the hypothalamus, and the lowest is in the cerebellum. In the rat hippocampus and other brain structures, the binding of imipramine to synaptic membranes is rapidly and irreversibly inactivated when the membrane preparation is incubated at 37° for a few minutes.[153,154] Results from experiments using EGTA and leupeptin suggest that this temperature-dependent inactivation is an action of Ca^{2+}-dependent protease.[153,154]

Subcellular fractionation indicates that [^{3}H]imipramine binding sites are concentrated in the synaptosomal fraction, suggesting a neuronal location.[153] Such a location for imipramine binding sites is further supported by their presence in cultured neuroblastoma cells,[153] by their absence in glial fractions prepared from horse striatum,[149] and by the reduction of the B_{max} for [^{3}H]imipramine binding following a lesion of 5-HT nerve terminals (see below). With the exception of the platelets,[155–157] none of the peripheral cell types contain high-affinity [^{3}H]imipramine binding sites.

It has been suggested that imipramine binding sites are associated with the 5-HT uptake system. Their location in 5-HT neurons is supported by the decrease in [^{3}H]imipramine binding sites following the specific lesion of 5-HT neuronal terminals with 5,7-dihydroxytryptamine[113,154]; the decrease in [^{3}H]imipramine binding sites parallels the percent reduction in the content of endogenous 5-HT. Similarly, a selective destruction of 5-HT nerve terminals by electrolytic lesions of the raphe nuclei also cuases a reduction in the B_{max} of [^{3}H]imipramine binding.[147,158,159] In contrast, a selective lesion of catecholaminergic brain neurons fails to change the number of [^{3}H]imipramine binding sites.[148] Moreover, the rank order in the potency of a group of drugs to inhibit [^{3}H]imipramine binding can be correlated with their rank order to inhibit 5-HT uptake.[159,160] Furthermore, in platelets, both the binding of imipramine and the uptake of serotonin can be irreversibly inactivated by 2-nitroimipramine.[161]

However, several lines of recent evidence support the view that imipramine does not label the 5-HT uptake carrier but an allosteric site that interacts with the carrier protein. Such a view is in agreement with the finding that (1) the uptake of 5-HT in the platelets of alcoholic cirrhotic patients is reduced although the binding characteristics of imipramine are unchanged[162]; (2) the affinity of imipramine for the imipramine binding site in brain is about 100-fold stronger than that for the 5-HT recognition sites[145,160]; (3) imipramine binding sites can be down-regulated following daily injections of imipramine repeated for 3 weeks, but the B_{max} for 5-HT uptake is increased[133]; and (4) the B_{max} for imipramine binding can be increased by daily injections of deprenyl repeated for 3 weeks, but the B_{max} for 5-HT uptake is not increased.[119]

The third point above was established by showing that repeated daily injections of imipramine[75,118,133] or desipramine[145,163] decreased by about 40% the number of imipramine binding sites in membranes prepared from brains of rats receiving such a treatment, whereas the V_{max} of 5-HT uptake increased by about 50%.[133] The fourth point was established by showing that the B_{max} for imipramine binding was increased by repeated daily injections of deprenyl without increasing the V_{max} of the 5-HT uptake.[119] The change in imipramine binding without a corresponding change in 5-HT uptake following long-term daily injections of deprenyl is a strong indication that imipramine does not label the 5-HT uptake site; this dissociation is upheld by the increase in 5-HT uptake elicited by repeated daily doses of imipramine, which down-regulate imipramine binding sites.

Perhaps imipramine labels an allosteric site that controls the uptake of 5-HT by a mechanism that remains to be elucidated. The down-regulation of the number of imipramine binding sites elicited by repeated daily imipramine injections brings about a greater efficiency of 5-HT uptake, which functionally may lead to a fast clearance of 5-HT from receptor recognition sites. This, in turn, causes an inhibition of 5-HT transmission, because the uptake of 5-HT is important to terminate serotonergic transmission. This finding implies that in patients receiving long-term treatment with imipramine, the time delay between the onset of treatment and the therapeutic response may be the time required to obtain a significant down-regulation of the number of inhibitory sites for the regulation of 5-HT uptake.

This evidence suggests the novel concept that uptake of 5-HT, like that of Na^+, is modulated by a site at which endogenous effectors act. When this site is activated by an appropraite endogenous agonists, it inhibits the 5-HT transmission. According to this concept, certain endogenous depressions could be characterized by an excess in the 5-HT transmission amplification; this could be regulated by reducing the number of regulatory sites, thereby bringing toward the norm the serotonergic transmission. Formerly, it was assumed that the 5-HT uptake process was not regulated and that the 5-HT transmission regulation depended exclusively on the amount of transmitter release and the number of 5-HT recognition sites at postsynaptic sites; the uptake for 5-HT was not supposed to be a regulatory site but was seen as operating at a maximum. Now we have reason to believe that the uptake of 5-HT is regulated by a regulatory site, which might be occupied by an endogenous effector which,

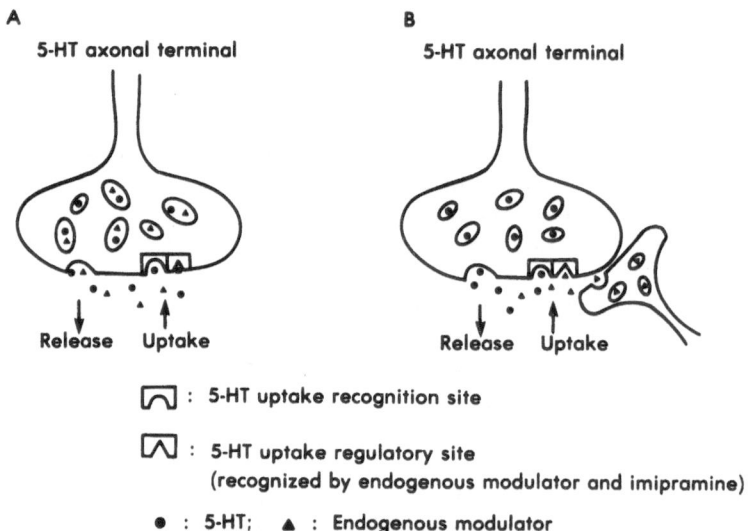

Fig. 2. Two possible mechanisms whereby the uptake of 5-HT is regulated by endogenous modulator. The endogenous modulator, which recognizes the imipramine binding site, may be coreleased with 5-HT (A) or released from an axonal terminal impinging on 5-HT axons (B).

similarly to imipramine, might inhibit 5-HT uptake. This effector either is stored and secreted from the 5-HT axon or is a chemical signal produced by an afferent axon (Fig. 2).

It remains to be demonstrated that, similarly to an ion gate, the 5-HT gate is regulated by a chemical signal probably produced by an axon. This chemical signal (neuromodulator) could regulate 5-HT uptake, and imipramine or desipramine appears to mimic the action of this endogenous ligand, acting with a similar intrinsic activity following its binding on the specific recognition site. The report that other effective antidepressant thereapies, including rapid eye movement sleep deprivation[164] and repeated electroconvulsive shock,[165] also down-regulate impramine binding sites in rat brain further suggests that [³H]imipramine binding sites are relevant to the therapeutic effect of antidepressants.

The question then arises, is there any action that we can ascribe to the reduction of 5-HT synaptic function brought about by the imipramine-elicited down-regulation in the number of regulatory sites for 5-HT uptake that are labeled by imipramine? An answer to this question is given in Section 5.4; here, it is sufficient to say that there are reasons to believe that 5-HT and NE neurons are connected by interneurons and that this connection appears to be regulated transsynaptically by the 5-HT synapses modulated by imipramine recognition sites. It should, however, be mentioned here that the number of imipramine binding sites in the platelets of depressed patients[166,167] and in the cortex of suicides[168] is significantly lower than that found in normals. And this number of binding sites is not reduced further by imipramine. Moreover, the uptake of 5-HT in the platelets of depressed patients is lower than that of normals.[169]

It is difficult to reconcile these data with the data discussed above; indeed, there might be certain limitations in the analogies between platelet and nerve terminal functions.

5.2. High-Affinity Binding Sites for Mianserin

Mianserin is a potent 5-HT and histamine H_1 receptor antagonist in smooth muscles.[170] By using [^{3}H]mianserin as a ligand, it has been shown that it binds with high affinity to synaptic membranes. This binding can be effectively displaced by antagonists of 5-HT$_2$ recognition sites such as spiroperidol[110,147] and *d*-LSD or antagonists of histamine H_1 receptor sites such as triprolidine and *d*-chlorpheniramine.[110] The K_i of 5-HT for displacing [^{3}H]mianserin bound to synaptic membranes varies from 390 nM in calf caudate[171] to 17 μM in rat cortex.[147,172] Based on these results, it has been inferred that [^{3}H]mianserin labels the recognition sites of 5-HT$_2$ receptors and histaminergic H_1 receptors.

However, several pieces of evidence obtained in our laboratory indicate that [^{3}H]mianserin labels a site that may be related to but is not identical to the 5-HT$_2$ recognition site. Moreover, one may question the functional significance of 5-HT$_2$ sites and whether their effector really acts as a transmitter or cotransmitter in 5-HT synapses. In building a model to study the action of mianserin, we have assumed that the 5-HT$_2$ recognition sites are sites for an endogenous effector operative in the modulation of 5-HT recognition sites for their interaction with 5-HT, the primary transmitter in 5-HT synapses. Referring to the previously mentioned model, 5-HT$_1$ is the site for the action of the primary transmitter, and 5-HT$_2$ is the site for the action of the cotransmitter. For example, the number of mianserin binding sites in the hippocampus and other brain areas can be up-regulated when 5-HT axonal terminals are denervated by intracerebral injections of 5,7-dihydroxytryptamine[113,154] or transection of 5-HT pathways afferent to the hippocampus.[173] A persistent blockade of 5-HT synthesis by *p*-chlorophenylalanine[113,154] also increases the density of mianserin binding sites. Under all these conditions, the number of 5-HT$_2$ recognition sites labeled by [^{3}H]spiroperidol or [^{3}H]ketanserin fails to change.[112] This clearly suggests that mianserin acts by occupying a site that modulates the 5-HT transmission.

Nevertheless, though we can differentiate the mianserin binding site from other sites, we cannot establish its identity. Moreover, following protracted treatment with mianserin, imipramine, or desipramine, the number of binding sites for 5-HT$_2$ receptors is decreased,[112,133] but the binding characteristics of [^{3}H]mianserin are unchanged. In view of the lack of modification of benzodiazepine recognition sites in the brain of rats receiving repeated daily injections of diazepam,[174-176] one may suggest that, unlike what is known for the neurotransmitters, repeated administration of cotransmitters fails to change the characteristics of their specific recognition sites. By this extrapolation, one may infer that mianserin binding site is the recognition site for a cotransmitter that participates in 5-HT synaptic function, but we do not understand its function completely. A comparison of the characteristics of [^{3}H]mianserin and

Table I
*Biochemical and Pharmacological Properties of Imipramine and
Mianserin Recognition Sites in Rat Hippocampus*[a]

	Imipramine	Mianserin
K_d (nm)	8.3 ± 1.5	1.45 ± 0.3
B_{max} (fmol/mg prot.)	800 ± 60	160 ± 24
Hill coefficient	1.0	1.0
Na^+ dependence	Yes	No
Temperature sensitivity	Yes	No
Self down-regulation	Yes	No
5,7-DHT lesion	↓	↑
PCPA treatment	—	↑
Fimbria transection	↓	↑
Kainic acid lesion	—	↓

[a] Data and experimental details were described by Kinnier *et al.*[75,153] and Brunello *et al.*[113,114] Abbreviations: DHT, dihydroxytryptamine; PCPA, *p*-chlorophenyla-lanine; ↓, decrease; ↑, increase; —, no change.

[³H]imipramine binding to rat hippocampal synaptic membranes is summarized in Table I.

5.3. High-Affinity Binding Sites for Desipramine

[³H]Desipramine, like [³H]imipramine, binds to a high-affinity site.[148,177–179] This binding is Na^+ dependent,[179] requires low temperatures,[179] and is preferentially located in catecholaminergic axons.[148,179] This binding is preferentially displaced by blockers of NE uptake and not by blockers of 5-HT uptake.[177–179] Moreover, a lesion of 5-HT axons does not reduce high-affinity despiramine binding, which is instead decreased by a destruction of NE axons.[148,179] Although [³H]desipramine in nanomolar concentrations preferentially binds to NE axons, when given in therapeutic doses it also binds to 5-HT axons and displaces imipramine from its specific binding site. These imipramine sites are down-regulated after repeated daily doses[145,163] of desipramine. Presumably, the binding of desipramine to the imipramine binding sites is the one that is operative to explain the antidepressant action of desipramine, which is similar to that of imipramine.[75,133] The binding of desipramine to NE axons could be operative in causing the side effects on the cardiovascular system that are described for desipramine.[136]

5.4. Modes of Action of Imipramine and Mianserin

As we mentioned earlier in this chapter, the delayed effects of imipramine appear to be those that are important in explaining the action of imipramine and other antidepressants on the symptoms of depression. Among the delayed effects elicited by protracted daily treatments with antidepressants are a down-regulation of [³H]imipramine binding, an increase of 5-HT uptake, a reduction in the number of central β-adrenergic receptor recognition sites, and a decline

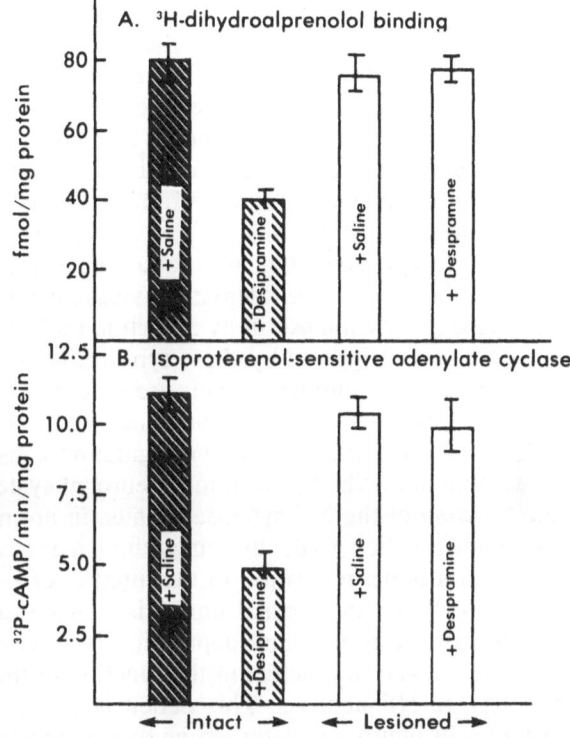

Fig. 3. Effects of lesion of 5-HT axonal terminals on the down-regulation of β-adrenergic receptors induced by desipramine. Rats were injected with 5,7,-dihydroxytryptamine (8 μg) 1 month prior to treatment with desipramine. Following desipramine treatment (10 mg/kg, twice daily i.p.) for 3 weeks, rats were sacrificed, and the binding of [³H]dihydroalprenolol and adenylate cyclase activity sensitive to isoproterenol were measured in the membrane of the cortex. Results were taken from previous publications.[114,133]

of the NE-elicited signal amplification by adenylate cyclase. Although the modification of the number of imipramine binding sites and the change of 5-HT uptake vary in intensity or direction according to the drug, and, in the case of mianserin and others, a postsynaptic action on 5-HT synapses predominates over the presynaptic action of imipramine, in both cases it appears that a reduction of the efficiency of 5-HT transmission is operative.

Such a reduction is also seen after treatment with MAO inhibitors, which reduce the number of 5-HT recognition sites.[107,108] An important question that has been answered only in part is whether the reduction in 5-HT transmission triggers a reduction in the function of central β-adrenergic receptors which is characterized by a decline in the signal amplification elicited by stimulation of β-adrenergic receptors alone or by this action associated with a reduction in the number of recognition sites for the β-adrenergic receptor agonists. One or both of these long-term events has been used frequently as a model system to study the molecular mechanisms that mediate the antidepressant action.

Since imipramine and desipramine act on sites that are used by an endogenous effector to regulate the uptake of 5-HT, we have performed a selective destruction of the 5-HT axons in the rat brain by intracerebral injections of 5,7-dihydroxytryptamine in order to evaluate the role of these sites in the expression of the long-term effects elicited by repeated daily administrations of imipramine and desipramine.[133,180,181] The results in Fig. 3 show that in rats receiving repeated desipramine injections, the down-regulation of the recognition sites for β-adrenergic receptor agonists and the attenuation of the iso-proterenol-stimulated adenylate cyclase are prevented when 5-HT axons are selectively lesioned a few days prior to the onset of repeated daily injections of this antidepressant. Similar effects of lesion were obtained when rats were treated repeatedly with imipramine.[133] In contrast, the attenuation of the NE-sensitive adenylate cyclase in cortical slices of rats treated repeatedly with mianserin is unaffected if a lesion of the 5-HT axons precedes the onset of the mianserin treatment.[133]

These observations indicate that a similar attenuation of the NE-elicited signal amplification on the adenylate cyclase can be elicited by imipramine or desipramine and by mianserin by an action on 5-HT synapses. But the molecular mechanisms triggered by these typical (imipramine and desipramine) and atypical (mianserin) antidepressants are different. Imipramine and desipramine may act by down-regulating the modulatory site of 5-HT uptake located in the 5-HT axonal terminals. This down-regulation causes a decrease in the 5-HT synaptic function, which, by an interneuronal system, triggers the attenuation in the function of the NE synapses. Thus, in normal rats a modification of 5-HT synaptic function gradually brings about a decrease in NE synaptic activity, resulting in a down-regulation in the number of β-adrenergic receptor recognition sites. When this animal model is transferred to brain function modification in man, it implies that depression is associated with an excess of NE activity as a result of a defect in the function of the interneuronal system that links 5-HT to NE neurons. This deficit is caused by a malfunction of a co-transmitter or neuromodulator acting on the circuit that connects 5-HT to NE axons.

To simplify a very complicated issue, one can say that in depression either 5-HT or NE synapses are too active. The precise mechanism whereby imipramine and desipramine modify the function of 5-HT nerve terminals is undefined but may be related to the down-regulation of imipramine binding sites, which causes a reduction of 5-HT synaptic function because they reduce the endogenous inhibition of 5-HT uptake. Since the uptake system of 5-HT has a supramolecular organization that includes imipramine binding sites as the regulatory sites for the 5-HT transport carrier, one may assume that the enhanced uptake of 5-HT is the result of the down-regulation of the number of imipramine binding sites located in 5-HT terminals. This enhanced uptake of 5-HT may diminish the content of 5-HT available in the synaptic cleft, and, as a result, a smaller fraction of the 5-HT released by nerve impulses is available to activate the postsynaptic receptors.

Conversely, mianserin binding sites appear to be located on the postsynaptic 5-HT receptors, because they increase in number following denervation[113,163,173] and are destroyed by kainic acid.[173] Since it is not displaced by 5-HT, mianserin probably occupies a site that interacts with 5-HT recognition sites allosterically. Mianserin probably acts in the opposite direction to the endogenous modulator. Assuming that the endogenous modulator increases the function of 5-HT synapses, mianserin decreases it. Hence, by an antiserotonergic action, mianserin down-regulates 5-HT receptors. A proposed model of the molecular mechanisms whereby imipramine and mianserin bring about down-regulation of β-adrenergic receptors is shown in Fig. 4.

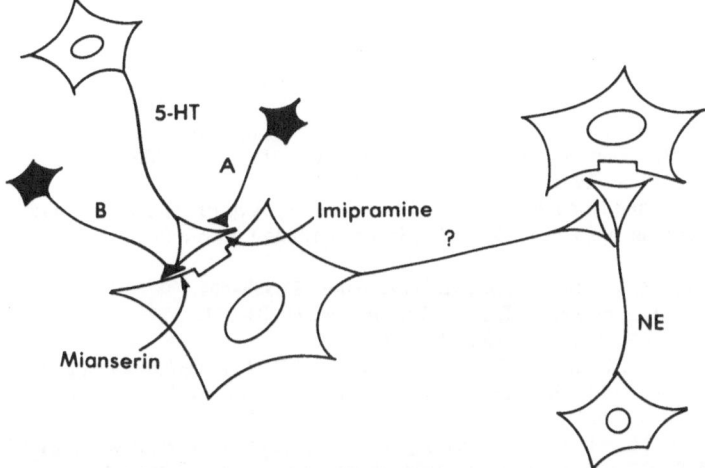

Fig. 4. A proposed model for the mechanisms whereby β-adrenergic receptor is down-regulated by imipramine and mianserin. An interneuronal system integrating the axons of 5-HT and NE neurons is essential for the down-regulation of β-adrenergic receptors. In the serotonin synapses, a persistent interaction between imipramine and presynaptic imipramine binding site or between mianserin and postsynaptic mianserin binding site causes a decrease in 5-HT transmission, leading to a down-regulation of β-adrenergic receptor. Imipramine binding site is normally recognized by an endogenous modulator, which could be released from 5-HT axon or axon A. Mianserin binding site is normally impinged on by another endogenous modulator, which could be released from 5-HT axon or axon B.

6. CONCLUSIONS

Chronic treatment with imipramine or mianserin alters serotonergic transmission. Imipramine acts presynaptically, whereas mianserin acts postsynaptically at the 5-HT synapses to reduce serotonergic transmission. Because of this attenuation of serotonergic transmission, they reduce the activity of a yet hypothetical interneuronal system that links 5-HT to NE axons. The action of antidepressants on 5-HT axons results in a reduction of the stimulus amplification elicited by NE in NE synapses. A current working hypothesis suggests that this modification of the function of the NE synapses reduces the increased NE function that is associated with depression. However, this interaction between NE and 5-HT axons should be seen as one of mutual modulation leading to a balance of the two systems, which is the basis for a balanced mood. Depression might be associated with lack of interaction and balance between 5-HT and NE systems.

REFERENCES

1. Brodie, B. B., Pletsher, A., and Shore, P. A., 1956, *Science* **123**:992–994.
2. Holzbauer, M., and Vogt, M., 1956, *J. Neurochem.* **1**:8–11.
3. Brodie, B. B., and Shore, P. A., 1957, *Ann. N.Y., Acad. Sci.* **66**:631–638.

4. Carlsson, A., Corrodi, H., Fuxe, K., and Hokfelt, T., 1969, *Eur. J. Pharmacol.* **5:**357–366.
5. Carlsson, A., Johanson, J., Lindquist, M., and Fuxe, K., 1959, *Brain Res.* **12:**456–460.
6. Glowinski, J., and Axelord, J., 1964, *Nature* **204:**1318–1319.
7. Ross, S. B., and Renyl, A. L., 1967, *Eur. J. Pharmacol.* **2:**181–186.
8. Thomas, P. C., and Jones, P. B., 1977, *J. Pharm. Pharmacol.* **29:**562–563.
9. Hamberger, B., and Tuck, J. R., 1973, *Eur. J. Clin. Pharmacol.* **5:**229–235.
10. Bunney, W. E., and Davis, J. M., 1965, *Arch. Gen. Psychiatry* **13:**483–494.
11. Schildkraut, J. J., 1965, *Am. J. Psychiatry* **122:**509–522.
12. Post, R. M., and Goodwin, F. K., and 1978, *Handbook of Psychopharmacology*, Volume 13 (L. L. Iversen, Iversen, S. D., and S. H. Snyder, eds.), Raven Press, New York, pp. 1235–1247.
13. Van Praag, H. M., 1978, *Acta Psychiatr. Scand.* **57:**389–404.
14. Garver, D. L., and Davis, J. M., 1979, *Life Sci.* **24:**383–391.
15. Kostowski, W., 1981, *Science* **6:**314–317.
16. Murphy, D. L., Campbell, I., and Costa, J. L., 1978, *Psychopharmacology* (M. A. Lipton, A. DiMascio, and K. K. Killam, eds.), Raven Press, New York, pp. 1235–1247.
17. Leonard, B. E., 1974, *Psychopharmacologia* **36:**221–236.
18. Goodlet, I., Mireylees, S. E., and Sugrue, M. F., 1977, *Br. J. Pharmacol.* **61:**307–313.
19. Rosloff, B. N., and Davis, J. M., 1974, *Psychopharmacologia* **40:**53–64.
20. Ferris, R. M., Maxwell, R. A., Cooper, R. A., and Soroko, F. E., 1982, *Advances in Biochemical Psychopharmacology*, Volume 31 (E. Costa and G. Racagni, eds.), Raven Press, New York, pp. 277–286.
21. Hertting, G., Axelrod, J., and Whitby, L., 1961, *J. Pharmacol. Exp. Ther.* **134:**146–153.
22. Nielsen, M., 1975, *J. Pharmacol.* **27:**206–208.
23. Roffman, M., Kling, M. A., Cassens, G., Orsulak, P. J., Reigle, T. G., and Schildkraut, J. J., 1977, *Psychopharmacol. Commun.* **1:**195–206.
24. Bareggi, S. R., Markey, K., and Genovese, E., 1978, *Eur. J. Pharmacol.* **50:**301–308.
25. Tang, S. W., Helmeste, D. M., and Stancer, H. C., 1978, *Naunyn Schmiedebergs Arch. Pharmacol.* **305:**207–211.
26. Neff, N. H., and Costa, E., 1967, *Antidepressant Drugs* (S. Garattini and M. N. H. Dukes, eds.), Excerpta Medica, Amsterdam, pp. 28–34.
27. Schildkraut, J. J., Roffman, M., Orsulak, P. J., Schatzberg, A. F., Kling, M. A., and Reigle, T. G., 1976, *Pharmakopsychiatrie* **9:**193–202.
28. Sugrue, M. F., 1980, *Life Sci.* **26:**423–429.
29. Schildkraut, J. J., Winokur, A., Draskoczy, P. R., and Hensle, J. H., 1971, *Am. J. Psychiatry* **127:**72–79.
30. Van Wijk, M., Meisch, J. J., and Korf, J., 1977, *Psychopharmacology* **55:**217–223.
31. Corrodi, H., and Fuxe, K., 1969, *Pharmacology* **7:**56–59.
32. Meek, J. L., and Werdinius, B., 1970, *J. Pharm. Pharmacol.* **22:**141–143.
33. Goodlet, I., and Sugrue, M. F., 1974, *Eur. J. Pharmacol.* **29:**241–248.
34. Marcro, E. J., and Meek, J. L., 1979, *Naunyn Schmiedebergs Arch. Pharmacol.* **306:**75–79.
35. Nielsen, M., and Braestrup, C., 1977, *Naunyn Schmiedebergs Arch. Pharmacol.* **300:**87–92.
36. Raiteri, M., Angelini, F., and Bertollini, A., 1976, *J. Pharm. Pharmacol.* **28:**483–488.
37. Kafoe, W. F., DeRidder, J. J., and Leonard, B. E., 1976, *Biochem. Pharmacol.* **25:**2455–2460.
38. Van Riezen, H., 1972, *Arch. Int. Pharmacodyn.* **198:**256–269.
39. Maj, J., Sowinska, H., Baran, L., Gancarczyk, L., and Rawlow, A., 1978, *Psychopharmacology* **59:**79–84.
40. Rosloff, B. N., and Davis, J. M., 1978, *Psychopharmacology* **56:**335–341.
41. Sherman, A., 1979, *Psychopharmacol. Commun.* **3:**1–5.
42. Sanghvi, I., and Gershon, S., 1975, *Biochem. Pharmacol.* **24:**2103–2104.
43. Ross, S. B., Ogren, S. O., and Renyi, A. L., 1976, *Acta Pharmacol. Toxicol.* **39:**152–166.
44. Fuller, R. W., Perry, K. W., and Molloy, B. B., 1975, *J. Pharmacol. Exp. Ther.* **193:**796–803.
45. Wong, D. T., Bymaster, F. P., Horng, J. S., and Malloy, B. B., 1975, *J. Pharmacol. Exp. Ther.* **193:**804–811.

46. Fuxe, K., Ogren, S.-O., and Agnati, L. F., 1979, *Neurosci. Lett.* **13:**307–312.
47. Hwang, E. C., and Van Woert, M. H., 1980, *Psychopharmacol. Commun.* **4:**161–167.
48. Langer, S. Z., 1974, *Biochem. Pharmacol.* **23:**1793–1800.
49. Berthelsen, S., and Pettinger, W. A., 1977, *Life Sci.* **21:**595–606.
50. Starke, K., and Langer, S. Z., 1979, *Presynaptic Receptors* (S. Z. Langer, K. Starke, and M. L. Dubocovich, eds.), Pergamon Press, Oxford, pp. 1–3.
51. U'Prichard, D. C., Bechtel, W. D., Rouot, B. M., and Snyder, S. H., 1979, *Mol. Pharmacol.* **16:**47–60.
52. Langer, S. Z., 1977, *Br. J. Pharmacol.* **60:**481–497.
53. Starke, K., Taube, H. D., and Borowski, E., 1977, *Biochem. Pharmacol.* **26:**259–268.
54. Westfall, T. C., 1977, *Physiol. Rev.* **57:**659–728.
55. Anden, N.-E., Grabowska, M., and Strombom, U., 1976, *Naunyn Schmiedebergs Arch. Pharmacol.* **292:**43–52.
56. Delini-Stula, A., Baumann, P., and Buch, O., 1979, *Naunyn Schmiedebergs Arch. Pharmacol.* **307:**115–122.
57. Crews, F. T., and Smith, C. B., 1978, *Science* **202:**322–324.
58. McMillen, B. A., Warnack, W., German, D. C., and Shore, P. A., 1980, *Eur. J. Pharmacol.* **61:**239–246.
59. Sugrue, M. F., 1981, *Life Sci.* **208:**377–384.
60. Sugrue, M. F., 1981, *Pharmacol. Ther.* **13:**219–249.
61. Johnson, R. W., Reisine, T., Spotnitz, S., Wiech, N., Ursillo, R., and Yamamura, H. I., 1980, *Eur. J. Pharmacol.* **67:**123–127.
62. Smith, C. B., Garcia-Sevilla, J., and Hollingsworth, P. J., 1981, *Brain Res.* **210:**413–418.
63. Tang, S. W., and Seeman, P., 1980, *Naunyn Schmiedebergs Arch. Pharmacol.* **311:**255–261.
64. Maggi, A., U'Prichard, D. C., and Enna, S. J., 1980, *Eur. J. Pharmacol.* **61:**91–98.
65. Fludder, J. M., and Leonard, B. E., 1979, *Psychopharmacology* **64:**329–332.
66. Sugrue, M. F., 1980, *Eur. J. Pharmacol.* **68:**377–380.
67. Tang, S. W., Helmeste, D. M., and Stancer, H. C., 1979, *Can. J. Physiol. Pharmacol.* **57:**435–437.
68. Banerjee, S. P., Kung, L. S., Riggi, S. J., and Chanda, S. K., 1977, *Nature* **268:**455–456.
69. Sarai, K., Frazer, A., Brunswick, D., and Mendels, J., 1978, *Biochem. Pharmacol.* **27:**2179–2181.
70. Wolfe, B. B., Harden, T. K., Sporn, J. R., and Molinoff, P. B., 1978, *J. Pharmacol. Exp. Ther.* **207:**446–457.
71. Bergstrom, D. A., and Kellar, K. J., 1979, *J. Pharmacol. Exp. Ther.* **209:**256–261.
72. Dibner, M. D., and Molinoff, P. B., 1979, *J. Pharmacol. Exp. Ther.* **210:**433–439.
73. Peroutka, S. J., and Snyder, S. H., 1980, *Science* **210:**88–90.
74. Sellinger-Barnette, M. M., Mendels, J., and Frazer, A., 1980, *Neuropharmacology* **19:**447–454.
75. Kinner, W. J., Chuang, D. M., and Costa, E., 1979, *Eur. J. Pharmacol.* **67:**289–294.
76. Rosenblatt, J. E., Pert, C. B., Tallman, J. F., Pert, A., and Bunney, W. E., Jr., 1979, *Brain Res.* **160:**186–191.
77. Clement-Jewery, S., 1978, *Neuropharmacology* **17:**779–781.
78. Minneman, K. P., and Molinoff, P. B., 1980, *Biochem. Pharmacol.* **29:**1317–1323.
79. Minneman, K. P., Dibner, M. D., Wolfe, B. B., and Molinoff, P. B., 1979, *Science* **204:**866–868.
80. Weiss, B., Heydorn, W., and Frazer, A., 1982, *Advances in Biochemical Psychopharmacology*, Volume 31, *Typical and Atypical Antidepressants* (E. Costa and G. Racagni, eds.), Raven Press, New York, pp. 37–54.
81. Chuang, D. M., and Costa, E., 1979, *Proc. Natl. Acad. Sci. U.S.A.* **76:**3024–3028.
82. Chuang, D. M., Kinnier, W. J., Farber, L., and Costa, F., 1980, *Mol. Pharmacol.* **18:**348–355.
83. Chuang, D. M., 1981, *J. Biol. Chem.* **256:**8291–8293.
84. Chuang, D. M., 1982, *Biochem. Biophys. Res. Commun.* **105:**1466–1472.
85. Bergstrom, D. A., and Kellar, K. J., 1979, *Nature* **278:**464–466.
86. Gillespie, D. D., Manier, D. H., and Sulser, F., 1979, *psychopharmacol. Commun.* **3:**191–195.

87. Pandey, G. N., Heinze, W. J., Brown, B. D., and Davis, J. M., 1979, *Nature* **280:**234–235.
88. Schultz, J. E., Siggins, G. R., Shocker, F. W., Turck, M., and Bloom, F. E., 1981, *J. Pharmacol. Exp. Ther.* **216:**28–38.
89. Sulser, F., 1982, *Advances in Biochemical Psychopharmacology*, Volume 31, *Typical and Atypical Antidepressants* (E. Costa and G. Racagni, eds.), Raven Press, New York, pp. 1–20.
90. Sulser, F., 1981, *Towards Understanding Receptors* (J. W. Lamble, ed.), Elsevier/North-Holland, Amsterdam, pp. 99–104.
91. Paul, S. M., and Crews, F. T., 1980, *Eur. J. Pharmacol.* 62:349–350.
92. Sethy, V. H., and Harris, D. W., 1981, *Eur. J. Pharmacol.* **75:**53–56.
93. Vetulani, J., and Sulser, F., 1975, *Nature* **257:**495–496.
94. Vetulani, J., Stawarz, R. J., Dingell, J. V., and Sulser, F., 1976, *Naunyn Schmiedebergs Arch. Pharmacol.* **293:**109–114.
95. Mishra, R., Janowsky, A., and Sulser, F., 1980, *Neuropharmacology* **19:**983–987.
96. Mishra, R., Janowsky, A., and Sulser, F., 1979, *Eur. J. Pharmacol.* **60:**379–382.
97. Schweitzer, J. W., Schwartz, R., and Friedhoff, A. J., 1979, *J. Neurochem.* **33:**377–379.
98. Janowsky, A., Steranka, L. R., Gillespie, D. D., and Sulser, F., 1982, *J. Neurochem.* **39:**290–292.
99. Rehavi, M., Ramot, O., Yavetz, B., and Sokolovsky, M., 1980, *Brain Res.* **294:**443–453.
100. U'Prichard, D. C., Greenberg, D. A., Sheehan, P. P., and Snyder, S. H., 1978, *Science* **199:**197–198.
101. Rehavi, M., Maayani, S., and Sokolovsky, M., 1977, *Biochem. Pharmacol.* **26:**1559–1567.
102. Reisine, T., U'Prichard, D., Wiech, N., Ursillo, R., and Yamamura, H., 1980, *Brain Res.* **188:**587–592.
103. Smith, C. B., Garcia-Sevilla, J. A., and Hollingsworth, P. J., 1981, *Brain Res.* **210:**413–418.
104. Enjalbert, A., Bourgoin, S., Hamon, M., Adrien, J., and Bockaert, J., 1978, *Mol. Pharmacol.* **14:**2–10.
105. Fillion, G., Rousselle, J. C., Beaudoin, D., Pradelles, P., Goiny, M., Dray, F., and Jacob, J., 1979, *Life Sci.* **24:**1815–1822.
106. Barbaccia, M. L., Brunello, N., Chuang, D. M., and Costa, E., 1983, *J. Neurochem.* **40:**1671–1679.
107. Savage, D. D., Frazer, A., and Mendels, J., 1979, *Eur. J. Pharmacol.* **58:**87–88.
108. Savage, D. D., Mendels, J., and Frazer, A., 1980, *J. Pharmacol. Exp. Ther.* **212:**259–263.
109. Peroutka, S. J., and Snyder, S. H., 1979, *Mol. Pharmacol.* **16:**687–689.
110. Peroutka, S. J., and Snyder, S. H., 1981, *J. Pharmacol. Exp. Ther.* **216:**142–148.
111. Leysen, J. E., Niemegeers, C. J. E., Van Neuten, J. M., and Laduron, P. M., 1982, *Mol. Pharmacol.* **21:**301–314.
112. Barbaccia, M. L., Gandolfi, O., Chuang, D. M., and Costa, E., 1983, *Neuropharmacology* **22:**123–126.
113. Brunello, N., Chuang, D. M., and Costa, E., 1982, *Science* **215:**1112–1115.
114. Brunello, N., Chuang, D. M., and Costa, E., 1982, *Advances in the Bioscience: New Vistas in Depression*, Volume 40 (S. Z. Langer, R. Takahashi, and M. Briley eds.), Pergamon Press, Oxford, New York, pp. 141–145.
115. Segawa, T., Mizuta, T., and Nomura, Y., 1979, *Eur. J. Pharmacol.* **58:**75–83.
116. Wirz-Justice, A., Krauchi, K., Lichsteiner, M., and Feer, H., 1978, *Life Sci.* **23:**1249–1254.
117. Peroutka, S. J., and Snyder, S. H., 1980, *J. Pharmacol. Exp. Ther.* **215:**582–587.
118. Briley, M., Raisman, R., Arbilla, S., Casadamont, M., and Langer, S. Z., 1982, *Eur. J. Pharmacol.* **81:**309–314.
119. Zsilla, G., Barbaccia, M. L., Gandolfi, O., Knoll, J., and Costa, E., 1983, *Eur. J. Pharmacol.* **89:**111–117.
120. Maggi, A., and Enna, S. J., 1980, *J. Neurochem.* **34:**888–892.
121. Treiser, S., and Kellar, K. J., 1979, *Eur. J. Pharmacol.* **64:**183–185.
122. Kellar, K. J., Cascio, C. S., Butler, J. A., and Kurtzke, R. H., 1981, *Eur. J. Pharmacol.* **69:**515–518.
123. Borowski, E., Starke, K., Ehrl, H., and Endo, T., 1977, *Neuroscience* **2:**285–296.
124. Doxey, J. C., Everitt, J., and Metcalf, G., 1978, *Eur. J. Pharmacol.* **51:**1–10.

125. Cavero, I., Gomeni, R., LeFevre-Borg, F., and Roach, A. G., 1980, *Br. J. Pharmacol.* **68**:321–332.
126. Clineschmidt, B. V., Flataker, L. M., Faison, E., and Holmes, R., 1979, *Arch. Int. Pharmacodyn. Ther.* **242**:59–76.
127. Hall, H., and Ogren, S. O., 1981, *Eur. J. Pharmacol.* **70**:393–407.
128. Black, J. W., Duncan, W. A. M., Durant, G. J., Ganellin, C. R., and Parson, E. M., 1972, *Nature* **236**:385–390.
129. Tran, V. T., Lebovitz, R., Toll, L., and Snyder, S. H., 1981, *Eur. J. Pharmacol.* **70**:501–509.
130. Taylor, J. E., and Richelson, E., 1982, *Eur. J. Pharmacol.* **78**:279–285.
131. Figge, J., Leonard, P., and Richelson, E., 1979, *Eur. J. Pharmacol.* **58**:479–483.
132. Richelson, E., 1978, *Nature,* **274**:176–177.
133. Barbaccia, M. L., Brunello, N., Chuang, D. M., and Costa, E., 1983, *Neuropharmacology* **22**:373–383.
134. Kanof, P. D., and Greengard, P., 1978, *Nature* **272**:329–333.
135. Green, J. P., and Maayani, S., 1977, *Nature* **269**:165–166.
136. Hollister, L. E., 1978, *N. Engl. J. Med.* **299**:1106–1109, 1168–1172.
137. Bloom, F., Segal, D., Ling, N., and Guillemin, R., 1976, *Science* **194**:630–632.
138. Biegon, A., and Samuel, D., 1980, *Biochem. Pharmacol.* **29**:460–462.
139. Costa, E., 1981, *Towards Understanding Receptors* (J. W. Lamble, ed.), Elsevier/North-Holland, Amsterdam, pp. 176–183.
140. Squires, R. F.,, and Braestrup, C., 1977 *Nature* **266**:732–734.
141. Mohler, H., and Okada, T., 1977, *Life Sci.* **20**:2101–2110.
142. Massoti, M., Guidotti, A., and Costa, E., 1981, *GABA and Benzodiazepine Receptors* (E. Costa, G. DiChiara, and G. L. Gressa, eds.), Raven Press, New York, pp. 19–26.
143. Ebstein, B., Guidotti, A., and Costa, E., 1982, *Problems in GABA Research from Brain to Bacteria* (Y. Okada and E. Roberts, eds.), Excerpta Medica, Amsterdam, pp. 348–354.
144. Raisman, R., Briley, M., and Langer, S. Z., 1979, *Eur. J. Pharmacol.* **54**:307–308.
145. Raisman, R., Briley, M., and Langer, S. Z., 1979, *Nature* **281**:148–150.
146. Rehavi, M., Paul, S. M., Skolnick, P., and Goodwin, F. K., 1980, *Life Sci.* **26**:2273–2279.
147. Dumbrille-Ross, A., Tang, S. W., and Coscina, D. V., 1981, *Life Sci.* **29**:2049–2058.
148. Hirdina, P. D., Elson-Hartman, K., Roberts, D. C. S., and Rappas, B. A., 1981, *Eur. J. Pharmacol.* **73**:375–376.
149. Briley, M. S., Fillion, G., Beaudoin, D., Fillion, M.-P., and Langer, S. Z., 1980, *Eur. J. Pharmacol.* **64**:191–194.
150. Briley, M. S., Raisman, R., Sechter, D., Zarifian, E., and Langer, S. Z., 1980, *Neuropharmacology* **19**:1209–1210.
151. Briley, M. S., and Langer, S. Z., 1981, *Eur. J. Pharmacol.* **72**:377–380.
152. Palkovits, M., Raisman, R., Briley, M., and Langer, S. Z., 1981, *Brain Res.* **210**:493–498.
153. Kinnier, W. J., Chuang, D. M., Gwynn, G., and Costa, E., 1981, *Neuropharmacology* **20**:411–419.
154. Chuang, D. M., Brunello, N., Kinnier, W. J., and Costa, E., 1982, *Advances in the Biosciences: New Vistas in Depression,* Volume 40 (S. Z. Langer, R. Takahashi, and M. Briley, eds.), Pergamon Press, Oxford, New York, pp. 133–139.
155. Briley, M. S., Raisman, R., and Langer, S. Z., 1979, *Eur. J. Pharmacol.* **58**:347–348.
156. Paul, S. M., Rehavi, M., Skolnick, P., and Goodwin, F. K., 1980, *Life Sci.* **26**:953–959.
157. Wennogle, L. P., Beer, B., and Meyerson, L. R., 1981, *Pharmacol. Biochem. Behav.* **15**:975–982.
158. Sette, M., Raisman, R., Briley, M., and Langer, S. Z., 1981, *J. Neurochem.* **37**:40–42.
159. Paul, S. M., Rehavi, M., Rice, K. C., Ittah, Y., and Skolnick, P., 1981, *Life Sci.* **28**:2753–2760.
160. Langer, S. Z., Morat, C., Raisman, R., Dubocovich, M. L., and Briley, M., 1980, *Science* **210**:1133–1135.
161. Rehavi, M., Ittah, Y., Rice, K. C., Skolnick, P., Goodwin, F. K., and Paul, S. M., 1981, *Biochem. Biophys. Res. Commun.* **99**:954–959.
162. Ahtee, L., Briley, M., Raisman, R., Lebrec, D., and Langer, S. Z., 1982, *Life Sci.* **29**:2323–2329.

163. Brunello, N., Chuang, D. M., and Costa, E., 1982, *Advances in Biochemical Psychopharmacology:* Volume 31, *Typical and Atypical Antidepressants* (E. Costa and G. Racagni, eds.), Raven Press, New York, pp. 179–184.
164. Mogilnicka, E., Arbilla, S., Depoortere, H., and Langer, S. Z., 1980, *Eur. J. Pharmacol.* **65:**289–292.
165. Langer, S. Z., Zarifian, E., Briley, M., Raisman, R., and Sechter, D., 1981, *Life Sci.* **29:**211–220.
166. Briley, M. S., Langer, S. Z., Raisman, R., Sechter, D., and Zarifian, E., 1980, *Science* **209:**303–305.
167. Paul, S. M., Rehavi, M., Skolnick, P., Ballenger, J. C., and Goodwin, F. K., 1981, *Arch. Gen. Psychiatry* **38:**1315–1317.
168. Stanley, M., Virgilio, J., and Gershon, S., 1982, *Science* **216:**1337–1339.
169. Tuomoisto, J., Tukiainen, E., and Ahlfors, U. G., 1979, *Psychopharmacology* **65:**141–147.
170. Vargaftig, B. B., Coignet, J. L., deVos, C. J., Grijsen, H., and Bonta, I. L., 1971, *Eur. J. Pharmacol.* **16:**336–346.
171. Whitaker, P. M., and Cross, A. J., 1980, *Biochem. Pharmacol.* **29:**2709–2715.
172. Dumbrille-Ross, A., Tang, S. W., and Seeman, P., 1980, *Eur. J. Pharmacol.* **68:**395–396.
173. Brunello, N., Chuang, D. M., and Costa, E., 1982, *Eur. J. Pharmacol.* **78:**383–384.
174. Mohler, H., Okada, T., and Enna, S. J., 1978, *Brain Res.* **156:**391–395.
175. Braestrup, C., Nielsen, M., and Squires, F. R., 1979, *Life Sci.* **24:**347–350.
176. Massotti, M., Alleva, F. R., Balázs, T., and Guidotti, A., 1980, *Neuropharmacology* **19:**951–956.
177. Rehavi, M., Skolnick, P., Hulihan, B., and Paul, S. M., 1981, *Eur. J. Pharmacol.* **70:**597–599.
178. Langer, S. Z., Raisman, R., and Briley, M., 1981, *Eur. J. Pharmacol.* **72:**423–424.
179. Lee, C.-M., and Snyder, S. H., 1981, *Proc. Natl. Acad. Sci. U.S.A.* **78:**5250–5254.
180. Brunello, N., Barbaccia, M. L., Chuang, D. M., and Costa, E., 1982, *Neuropharmacology* **21:**1145–1149.
181. Chuang, D. M., Barbaccia, M. L., Brunello, N., and Kinnier, W. J., 1982, *Dynamics of Neurotransmission* (I. Hanin, ed.), Raven Press, New York (in press).

12

Opiate Receptors

Eric J. Simon

1. INTRODUCTION AND HISTORICAL OVERVIEW

This review begins with a brief summary of the discovery and earlier studies of opiate receptors. The major portion of this chapter is devoted to the more recent advances in our knowledge of these receptors. Opioid peptides are covered only with respect to their interactions with their receptors, since their physiology and biochemistry are reviewed elsewhere in the *Handbook*.

Since this entire volume is devoted to receptors, it is clearly unnecessary to define the term once more. Let me indicate, however, that the term receptor will be used here somewhat loosely. Instead of distinguishing between the functional receptor unit and the binding site, we shall use the term receptor in situations where the term binding site would be more appropriate. This may upset the purists but should improve the readability of this chapter by permitting us to vary the terminology.

The idea that opiate narcotic analgesics must bind to specific sites in the central nervous system in order to elicit their pharmacological responses dates back to the 1940s. During the large effort mounted by the pharmaceutical industry in the synthesis of morphine analogues, it became apparent that there are important structural and steric constraints for most of the actions of the opiates. Thus, analgesia and addictive liability were found to reside in only one enantiomer of a racemic mixture. Moreover, relatively small changes in certain parts of the morphine molecule can result in drastic alterations in its pharmacology. The most interesting such change is the displacement of the methyl group on the tertiary nitrogen by an allyl or cyclopropylmethyl group, which results in a molecule with potent and specific antagonistic action against morphine and related opiates. These various findings are most easily explained by the existence of binding sites that exhibit complementarity to the important structural and steric features of opiates. Binding to these receptor sites would be the first step that triggers sequential events that result in the observed re-

---

Eric J. Simon • Departments of Psychiatry and Pharmacology, New York University School of Medicine, New York, New York 10016.

sponses. Antagonists are presumed to be analogues with high binding affinity that are incapable of triggering the steps that must follow binding in order to elicit a response.

Though pharmacological evidence for specific receptor sites was abundant for several decades, the biochemical demonstration of specific binding of opiates did not occur until 1973.

Opiates will readily bind to many tissue components, and it proved difficult to demonstrate specific binding that might reflect attachment to receptors in the midst of an abundance of nonspecific binding. Attempts in our laboratory in 1966[1] to demonstrate specificity by measuring nalorphine-displaceable binding were not successful. However, using modifications of a method for measuring stereospecific binding to brain homogenate developed by Goldstein and collaborators,[2] our laboratory[3] and two others[4,5] simultaneously and independently reported stereospecific binding that constituted the major portion of total binding. Stereospecific binding is defined as the fraction of binding that can be displaced by an excess of an unlabeled opiate but not by its inactive enantiomer.

Since that time a great deal has been learned about the characteristics of these opiate binding sites, and considerable evidence suggests that they are the pharmacologically relevant opiate receptors. Highlights of this work carried out in many laboratories will be summarized.

Opiate binding sites have been found in the CNS and in some peripheral tissues of all vertebrates including man[6] and recently also in some invertebrates.[7] They are tightly bound to cell membranes, and cell fractionation experiments suggest that they are located at or near synapses.[8] Stereospecific binding is saturable and of high affinity (10^{-10}–10^{-7} M) for drugs that exhibit strong to moderate opiate activity. The pH optimum for binding occurs in the physiological range (pH 7–8).

Biochemical studies have indicated that opiate binding is highly sensitive to various proteolytic enzymes and to a large number of reagents capable of reacting with amino acids and functional groups present in proteins. Thus, opiate binding is inhibited by various sulfhydryl reagents, such as N-ethylmaleimide and iodoacetate. These results suggest that one or more proteins play an essential role in the specific binding of opiates. The evidence for other structural components is less conclusive. Binding is highly sensitive to some preparations of phospholipase A[9] but not to others[3] or to phospholipases C and D. Phospholipids may have a role in holding the receptor in its proper conformation in the membrane lipid bilayer. This idea is supported by our finding[10,11] that the inhibition of opiate binding by phospholipase A can be reversed by washing the membranes with buffer containing 1% BSA. Evidence for the presence of carbohydrate in the opiate receptor has been obtained only recently and will be discussed later.

Detailed studies of the distribution of opiate binding sites within the CNS have been carried out in human autopsy material,[6] monkeys,[12] and other animals by dissection and homogenization of brain and spinal cord regions and measuring binding of labeled opiates to the homogenates, as well as by autoradiography of brain and spinal cord slices[13–16] after injection of animals with labeled opiate (mainly [³H]etorphine and [³H]diprenorphine). More recently,

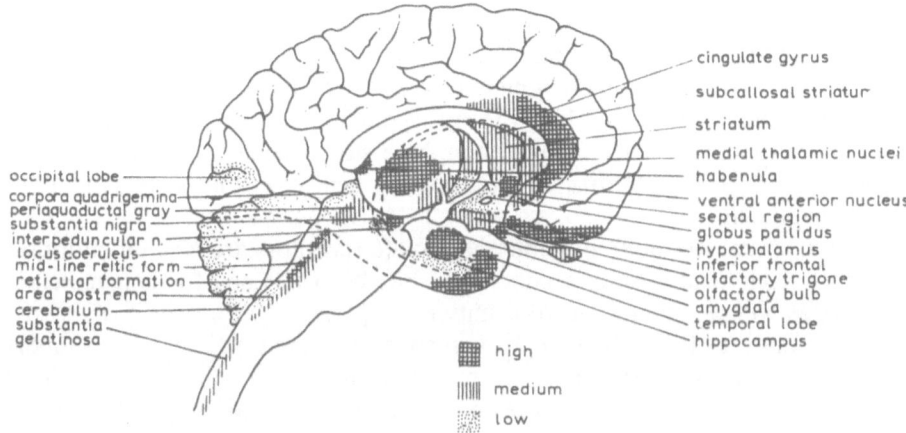

occipital lobe
corpora quadrigemina
periaquaductal gray
substantia nigra
interpeduncular n.
locus coeruleus
mid-line reticformation
reticular formation
area postrema
cerebellum
substantia
gelatinosa

cingulate gyrus
subcallosal striatur
striatum
medial thalamic nuclei
habenula
ventral anterior nucleus
septal region
globus pallidus
hypothalamus
inferior frontal
olfactory trigone
olfactory bulb
amygdala
temporal lobe
hippocampus

high
medium
low

Fig. 1. Distribution of opiate receptors in the central nervous system. This schematic diagram of a human brain represents a synthesis of results from various laboratories. Drawing by Dr. John Pearson.

an effective method has been developed for carrying out the binding to tissue slices *in vitro*, followed by autoradiography.[17] All of these studies can be summarized by saying that large differences in levels of binding sites exist among different regions of the CNS. The regions that are rich in opiate binding sites are in the limbic system and in all of the areas that have been implicated in pathways of pain perception and modulation, including the substantia gelatinosa of the dorsal spinal cord, the nucleus raphe magnus, the medial thalamus, and the periaqueductal and periventricular gray regions. These data are summarized in Fig. 1.

All of the properties of the binding sites as well as their distribution in the CNS are consistent with the notion that they are the pharmacologically relevant opiate receptors. Strong evidence also comes from the excellent correlation that has been observed in several laboratories between the pharmacological potency and receptor binding affinity of a large number of drugs with opiate activity.[18,19]

Opiate receptors have also been found in a number of peripheral tissues. These include endocrine glands such as the pituitary and adrenals as well as the intestinal tract and vasa deferentia of several species.

An important question that is not completely settled is the pre- or postsynaptic location of opiate receptors. There is evidence that seems to support both locations. Thus, dorsal rhizotomy in monkeys resulted in a significant reduction in opiate binding in the dorsal horn of the spinal cord.[20] This suggests a presynaptic location, but, as the authors pointed out, a transsynaptic effect on postsynaptic receptors cannot be ruled out. Perhaps the most direct evidence for presynaptic opiate receptors comes from the finding that high levels of opiate binding are present in the neuritic outgrowth of cultured dorsal root ganglia even when they are grown in the absence of spinal cord.[21] These binding sites are presumed to reside on nerve fibers destined to provide afferent input into the dorsal horn of the spinal cord.

Evidence for postsynaptic opiate receptors comes primarily from electrophysiological experiments. Thus, Zieglgansberger and colleagues[22] have shown that microiontophoretically applied morphine stereospecifically blocks the excitatory response of neurons to glutamate, a neurotransmitter known to act postsynaptically.

It is, of course, quite possible that opiate receptors are located both pre- and postsynaptically. Such a conclusion has been reached for dopamine receptors as well as other neurotransmitter receptors.

The finding of specific receptors in so many animal species gave support to the idea that they must serve an important function for the organism. This in turn sparked the search for endogenous ligands for opiate receptors. Initially, the search was among the already known neurotransmitters and hormones, especially since intracerebral injection of some of them, for instance, acetylcholine, produces analgesia. When it became clear that none of the known neurochemicals were able to bind to opiate receptors with sufficiently high affinity, the tedious search for new substances with opioid activity was begun. The first reports of the existence of such substances came from the laboratories of Terenius and Wahlström[23] and Hughes and Kosterlitz.[24] It was the group of Hughes, Kosterlitz, and collaborators[25] that characterized the first endogenous opioids and found them to be two pentapeptides, Tyr-Gly-Gly-Phe-Met and Tyr-Gly-Gly-Phe-Leu, which they named methionine and leucine enkephalin. Since then, a number of other endogenous opioid peptides have been identified, the most important of which are β-endorphin[26,27] and dynorphin.[28] The full sequence of the 17-amino-acid peptide dynorphin has recently been elucidated by Goldstein *et al.*[29]

This is a rather sketchy review of the background for the studies to be discussed in greater detail. Our focus is on the recent work on the heterogeneity of opiate receptors and on the efforts to isolate and purify the receptor molecules. Finally, we discuss the limited knowledge available concerning the sequelae of opiate binding.

2. MULTIPLE OPIATE RECEPTORS

There were a number of reasons for the suggestions that several types of opiate receptors may exist. Opiates and opioid peptides exert a variety of pharmacological effects that seem unlikely to be mediated via a single receptor. A number of endogenous opioid peptides exist that may each have its own receptor. Finally, multiple receptors are known to exist for the various neurotransmitters. Since it is probable that one or more of the opioid peptides serve as neurotransmitters or modulators, they may, by analogy, also bind to several classes of receptors.

The first evidence for the existence of multiple classes of opiate receptors was obtained by Martin and co-workers in pharmacological experiments performed in chronic spinal dogs.[30,31] Different natural and synthetic opiates were found to exhibit very different pharmacological profiles. Moreover, the analogues were unable to substitute for each other in the prevention of withdrawal symptoms in dogs chronically treated with one of the drugs. Based on these findings, Martin concluded that the existence of three types of opiate receptors had been demonstrated. He named them for the prototype drugs used in the study: μ for morphine, κ for ketocyclazocine, and σ for SKF-10,047 (N-allyl-normetazocine).

2.1. μ and δ Receptors

Kosterlitz and his collaborators[32] were the first to show that opiate receptor heterogeneity could also be recognized in *in vitro* systems. They found that in the isolated guinea pig ileum opiate alkaloids were significantly more effective than enkephalins or peptidase-resistant analogues of enkephalins. In the mouse vas deferens, it was the enkephalins that exhibited greater potency. These studies suggested that different classes of receptors predominated in the two *in vitro* bioassay systems. The major receptor present in the guinea pig ileum was called μ because of its resemblance to Martin's μ receptor, whereas the receptor in the mouse vas deferens was called δ.

The next question, also first investigated in Kosterlitz's laboratory, was whether receptor heterogeneity could be demonstrated through binding studies. Competition binding experiments were performed in homogenates of guinea pig brain. It was observed that enkephalins compete better for receptor binding against a labeled enkephalin than against a labeled opiate (naloxone), whereas opiates compete significantly more effectively against labeled naloxone than against a labeled enkephalin. This result agreed well with those obtained in the bioassay systems, suggesting the existence of an opiate-preferring (μ) and an enkephalin-preferring (δ) receptor. Since that time, considerable additional evidence has been accumulated in support of the existence of μ and δ receptors, which will be summarized briefly.

Early attempts to inactivate selectively either μ or δ binding sites met with failure. Various agents such as lipases, proteolytic enzymes, or sulfhydryl reagents inhibited enkephalin and opiate binding to the same degree. It therefore became necessary to resort to cross-protection experiments for evidence of heterogeneity. In our laboratory,[33] the sulfhydryl reagent N-ethylmaleimide (NEM) was used for the irreversible inactivation of opiate receptors. We had previously shown[34] that the presence of ligand will protect the receptor against inactivation by NEM. When [^{3}H]naltrexone was the ligand, morphine and naloxone were significantly more effective protecting agents than the enkephalins. When the labeled ligand was [^{3}H][D-ala^2,D-leu^5]enkephalin (DADL), unlabeled enkephalins were about 80 times more effective protecting agents than opiate alkaloids. Identical results were obtained by Robson and Kosterlitz[35] using phenoxybenzamine as the irreversible inactivator. These results support the existence of two receptor subtypes.

If receptor subclasses exist, it should be possible to find differences in their regional distribution in the CNS. Evidence for this has also been reported. Chang *et al.*,[36] and our laboratory[37] found differences in ratios of μ to δ receptors in regions of rat brain. The most striking difference was found in the thalamus, where there seems to be a large preponderance of μ receptors. Whereas in most brain areas naloxone was 6–10 times better in competition with [³H]naloxone than with [³H]DADL, this difference is not observed in the thalamus. Frontal cortex was found to be enriched in δ receptors.

The distribution of opiate receptor subclasses was also studied in bovine brain.[38] The ratio of the binding of tritiated morphine (0.5 nM) to that of tritiated DADL (0.5 nM) was found to vary as much as tenfold between brain regions. The highest ratios (high portion of μ sites) were seen in the substantia nigra and the thalamus, whereas the lowest (highest fraction of δ sites) were observed in the frontal cortex and the hippocampus.

We have recently performed similar experiments in human brain[39] obtained at autopsy from the Office of the Chief Medical Examiner of the City of New York. Based on cross-competition studies, human brain shows similar heterogeneity (μ and δ subtypes) to that seen in animal brain. The thalamus shows a striking enrichment in μ receptors; i.e., naloxone competes better (almost twofold) against [³H][Leu]enkephalin than against [³H]naloxone.

Autoradiographic evidence for the existence of at least two receptor subtypes has been obtained.[40,41]

The selective inhibition of one of these two opiate receptor subtypes has recently been achieved in our laboratory.[42] Ethanol and other aliphatic alcohols were shown to inhibit the binding of enkephalin and its analogues at levels at which the binding of naloxone or morphine was essentially unaffected. The inhibitory potency of the alcohols increased exponentially with chain length. The inhibition is completely reversible. Evidence was presented indicating that the alcohols affect the binding site rather than the ligand. Scatchard analysis of binding studies showed that alcohols affect the binding affinity of enkephalin rather than the number of binding sites and that this inhibition is the result of increased rate of dissociation of the ligand–receptor complex. It was concluded from these studies that aliphatic alcohols selectively inhibit δ receptors. When a tissue contains virtually only δ sites, as is the case in neuroblastoma (N4TG1) cells, the binding of the μ-preferring ligands naloxone and morphine is inhibited to a similar extent as that of enkephalins.

An alkylating derivative of naltrexone, β-fumarylnaltrexone (βFNA), has been found to bind covalently to μ receptors and not to δ receptors.[43]

Finally, it has also been possible to distinguish μ and δ subclasses of opiate receptors by the failure of drugs or peptides to produce cross tolerance to each other. For example, Schulz et al.[44] found that the vasa deferentia from mice chronically exposed to DADL were 800-fold less sensitive to this peptide in vitro, although there was no change in sensitivity to normorphine or sufentanyl. Chronic exposure to sufentanyl produced a dramatic decrease (1000-fold) in sensitivity to this drug and to normorphine with little change in sensitivity to DADL.

More recently, the same group[45] has reported similar results when tolerance to behavioral effects was studied in intact rats. Selective tolerance to either DADL or sufentanyl was observed for both catatonia and analgesia, though the differential was far less impressive than with the isolated tissues.

All of the above evidence lends rather impressive support to the existence of μ and δ binding sites. However, proof that they are separate molecular entities is still lacking, and theories about the nature of the μ and δ subtypes differ. Thus, Pert and co-workers[46] have published evidence that they feel is best explained by a single opiate receptor able to exist in different conformations, two of which have the properties of μ and δ sites.

Another hypothesis put forth by Lee *et al.*[47] suggests that μ and δ sites are present on the same receptor complex. The strong chemical resemblance of these sites, already alluded to, and the recent finding that they can occur on the same nerve terminal[48] make this an attractive idea that should be investigated further.

A theory that bears a resemblance to the Lee model was recently presented by Rothman and Westfall.[49,50] These workers found evidence for an allosteric interaction between μ and δ sites. The major evidence in support of allostery was the finding that opiates and enkephalins interact noncompetitively at the receptor; i.e., the Scatchard plots of opiates in the presence and absence of enkephalins are parallel, and the same is true for Scatchard plots of enkephalin binding in the presence and absence of opiates.

Considerable work is still needed to distinguish among all of these hypotheses. One of the difficulties has been the relative nonspecificity of the ligands currently available. They all bind most effectively to their own receptor subtype, but all exhibit more or less overlap of binding to other subclasses. The development of highly specific ligands, agonists as well as antagonists, for each subclass of opiate receptors is essential if real progress in this area is to be achieved. Final proof for the existence of separate receptor molecules may require their isolation and physical separation.

2.2. κ *and* σ *Receptors*

The other two opiate receptor subtypes postulated by Martin were the κ and σ receptors. It now appears that they too can be demonstrated in *in vitro* binding experiments.

Early attempts to demonstrate separate binding sites for ethylketocyclazocine (EK) were unsuccessful.[51-53] This was because of two problems not recognized at the time: EK and similar κ ligands bind almost equally well to μ and δ sites as they do to κ. A simple study of EK binding therefore leads to the conclusion that binding affinities of various ligands and regional distribution of EK binding strongly mimic those of μ ligands. The second problem was the use of rat brain in most of the early studies. The rat proved to have the smallest fraction of κ sites of any species studied to date (about 10–15% of total opiate binding sites).

The use of guinea pig brain, relatively rich in κ receptors, and the use of μ and δ blockers enabled Kosterlitz and his group[54] to demonstrate the existence of separate kappa sites. [D-Ala^2, D-Leu^5]Enkephalin (100 nM) was used to block δ sites, and a new peptide, remarkably specific for mu sites, [D-Ala^2, MePhe^4, gly-ol^5]enkephalin (100 nM), was used as a μ blocker. Sites that prefer to bind benzomorphans were also reported by Chang *et al.*,[55] who used high levels of morphiceptin to block μ sites. Morphiceptin is the amide of a tetra-peptide sequence found in casomorphin, a heptapeptide previously isolated from the milk protein, β-casein.[56] The authors pointed out that these benzo-morphan-preferring sites bound benzomorphans of the κ and σ type and could not be clearly identified as being Martin's κ sites. This is, in fact, still true for all of the so-called κ sites so far investigated.

Evidence for the existence of σ sites was the last to be obtained and is consequently still the most controversial. Binding sites for phencyclidine (PCP) were first reported to exist in animal brain by Vincent *et al.*[57] and by Zukin and Zukin.[58] In a more recent paper, Zukin and Zukin[59] found that opiates of the σ type, such as cyclazocine and SKF 10,047, but not the more classical opiates or the opioid peptides can displace PCP from its binding site. Phen-cyclidine, in turn, is able to displace the σ ligands quite specifically. The authors suggested that the PCP binding site may also be the σ opiate receptor. Support for this intriguing idea came from behavioral experiments by Holtzman.[60] Rats trained to distinguish PCP from saline were able to generalize the drug cue to cyclazocine and SKF 10,047 but not to other opiates. Rats trained to recognize cyclazocine were unable to distinguish this drug from PCP.

Recently, studies by Dr. Itzhak in our laboratory[61] have produced evidence for the existence of high levels of benzomorphan-preferring receptors in regions of human brain. As seen in Table I, the lowest level (23%) was found in the thalamus, and the highest level was observed in the hypothalamus (64%). Al-though a large portion of the binding sites exhibit the characteristics of κ sites, preliminary evidence for the presence of σ sites has been obtained by the use of the PCP analogue, 3-hydroxy-PCP.

2.3. Other Receptor Subclasses

In addition to the four receptor subclasses discussed so far, a number of others have been proposed. They are mentioned only briefly since their ex-istence is even more tenuous than that of the four more thoroughly studied types.

Experiments with the irreversible opiate ligand naloxazone by Pasternak and collaborators[62] have shown that inhibition of the binding of all ligands, regardless of opiate receptor preference, seems to be exerted on their high-affinity sites. This result is based on the assumption that the curved Scatchard plots, almost always observed, represent two binding sites with different bind-ing affinities. Pasternak has suggested that these results are most easily ex-plained by postulating a common binding site to which μ, δ, κ and σ ligands can bind with higher affinity than to their own subclass. He has termed this receptor μ₁. Evidence also obtained in Pasternak's laboratory[63] indicates that analgesia may be mediated via this μ₁ receptor.

A receptor found in the rat vas deferens[64–66] seems to be highly specific for β-endorphin. It has been named the ε receptor.[67] Yet another type of binding site has been reported to be present in dog and rabbit ileum, which Oka has named the ι receptor.[68] Some evidence for heterogeneity within subclasses has also been reported. It appears likely to this reviewer that much of the Greek

Table I
Effect of Blocking μ and δ Sites on Opiate Binding in Human Brain Regions[a]

Brain region		[³H]Naloxone	[³H]DHM	[³H]Bremazocine	[³H]SKF-10047
Hypothalamus	U	23 ± 2	14 ± 2	152 ± 4	35 ± 4
	B	10 ± 1	2 ± 1	97 ± 5	23 ± 2
	B/U	0.41	0.41	0.64	0.66
Parietal	U	20 ± 2	11 ± 0.2	114 ± 4	30 ± 3
	B	7 ± 0.4	0.3	49 ± 3	19 ± 2
	B/U	0.35	0.03	0.43	0.63
Hippocampus	U	29 ± 6	19 ± 1	105 ± 13	54 ± 4
	B	8 ± 2	1.5 ± 0.4	50 ± 5	29 ± 1
	B/U	0.27	0.08	0.48	0.54
Frontal cortex	U	28 ± 1	14 ± 2	129 ± 3	29 ± 8
	B	8 ± 1	2 ± 1	53 ± 3	14 ± 5
	B/U	0.29	0.14	0.41	0.48
Cingulate	U	20 ± 2	11 ± 0.5	91 ± 8	18 ± 4
	B	5 ± 0.5	1 ± 0.4	44 ± 5	8 ± 0.4
	B/U	0.25	0.09	0.48	0.44
Caudate	U	45 ± 6	33 ± 2	120 ± 20	56 ± 1
	B	9 ± 1	1 ± 0.7	45 ± 8	19 ± 2
	B/U	0.20	0.03	0.32	0.34
Thalamus	U	58 ± 6	45 ± 5	140 ± 13	54 ± 5
	B	9 ± 0.4	3 ± 1	32 ± 1	15 ± 2
	B/U	0.15	0.07	0.23	0.28

[a] Specific binding (fmol/mg protein) of [³H]opiates (1.0 nM) in the absence (U, unblocked) and presence (B, blocked) of DAGO (100 nM) and DADL (100 nM) in seven regions of human brain. Results from 3–5 experiments are expressed as mean ± standard error of the mean. From Itzhak *et al.*[61] with permission from Pergamon Press.

alphabet will be used up for opiate receptor subclassification before the inevitable simplification from increased knowledge will come about.

2.4. Putative Endogenous Ligands for Opiate Receptor Subclasses

The relationship between endogenous opioid peptides and receptor subclasses is not yet clarified. The peptides have recently been shown to yield to classification into three major groups based on their genetic origin. The three classes are the gene products of proopiocortin (β-endorphin and related peptides derived from β-lipotropin), of proenkephalin (the enkephalins and related peptides such as the heptapeptide [Met]enkephalin-Phe⁶-Arg⁷), and of the common precursor of dynorphin and α-neoendorphin (dynorphin, α- and β-neoendorphin). It is reasonable to assume that each of these groups will have at least one major opiate receptor class. So far it appears likely that the enkephalins bind mainly to δ receptors, and considerable evidence has recently been reported[69–71] suggesting that the κ receptor may be the preferred binding site of dynorphin and some of its fragments. The major binding site for β-endorphin could be the ε receptor. However, so far it has only been found in a few species and tissues and not yet in the CNS. A very recent paper by Li and co-workers[72] has reported the solubilization of a β-endorphin receptor complex from rat brain homogenate. Whether these receptors are similar to the ϵ receptors of peripheral tissues is not clear. Lee's hypothesis,[47] already mentioned, suggests that β-endorphin may bind to δ sites with its N-terminal (enkephalin) portion and to μ sites with its C-terminal portion on a single μ–δ receptor complex.

The first evidence for opiate receptor heterogeneity came from *in vivo* studies. There are now a large number of *in vivo* systems that support the notion of multiple opiate receptors. These include studies of seizure thresholds, operant behavior, intracerebral injection, discriminative stimuli, analgesia, and others. The reader who wishes more detail about *in vivo* studies of multiple opiate receptors is referred to reviews by Cowan[73] and Adler.[74]

This review of the present state of knowledge of the multiplicity of opiate receptors, although relatively detailed, is not exhaustive. It is intended to impress the reader with the fact that this is still a challenging and promising area for investigation. Much needs to be learned before we can speak with assurance about which receptor types really exist as separate entities, which represent different forms of the same receptor, and which are binding sites on a single receptor complex. The question of the relationship between endogenous opioid ligands and receptor subtypes and the functions of the various opiate receptor subtypes are still important and largely unsettled questions.

3. SOLUBILIZATION AND PARTIAL PURIFICATION OF OPIATE RECEPTORS

The purification to homogeneity is a difficult but very important goal in receptor research. Perhaps a word is in order as to why this is such an essential achievement to be pursued with considerable tenacity.

An idea of what sort of information can be obtained with purified receptor molecules can be gotten by reading the literature about other receptors such as those for the estrogens, insulin, and acetylcholine, where purification has been achieved. It becomes possible to probe the chemical composition and subunit structure, to produce antibodies (both polyclonal and monoclonal) against the receptor, and to determine what effect injection of antibodies has on an organism. This may yield evidence about the function of receptors. As mentioned earlier, purification and separation or inability to separate receptor subtypes will permit us to determine the molecular basis of receptor heterogeneity. Reconstitution of purified receptors into artificial and natural membranes may permit studies of the role of the receptors in steps that follow opiate binding, such as conformational changes in the membrane leading to changes in ionic fluxes or regulation of adenylate cyclase.

Finally, purification of opiate receptors to a point at which monoclonal antibodies can be produced will permit elucidation of the molecular biology of the receptor, including the regulation of its biosynthesis and turnover and characterization of its messenger RNA and gene. It will in all likelihood provide knowledge of the amino acid sequence of receptor subunits before they can be purified and sequenced by classical protein techniques.

3.1. Solubilization

The first step in opiate receptor purification, solubilization of binding sites from cell membranes, to which they are tightly attached, proved to be a difficult one. Efforts to solubilize receptors that retain their ability to bind in the soluble state were unsuccessful for many years. This was at least in part because of the high sensitivity of opiate receptors to even tiny concentrations of detergents, even those of the nonionic type, which have proved useful for solubilization of other receptors.

In 1975, after a great deal of fruitless and frustrating effort, we succeeded in solubilizing with the detergent BRIJ 36T an etorphine–macromolecular complex.[75] This complex had the characteristics consistent with the notion that it was an etorphine–receptor complex. However, our efforts to exchange bound and free opiates or to rebind the receptor after dissociation of the etorphine were unsuccessful. The etorphine–receptor complex had a molecular weight of *ca.* 400,000 determined by gel filtration on a calibrated column of Sepharose 6B. These studies were confirmed by Zukin and Kream,[76] who showed that a similar solubilization can be done with a receptor–enkephalin complex. They also reported that it was possible to link a portion of the bound enkephalin covalently to the receptor by means of the cross-linking agent dimethylsuberimidate. The covalent complex had a molecular weight of about 400,000, in excellent agreement with our results. Zukin *et al.*[77] were also able to find smaller-molecular-weight labeled species that may well be subunits of the receptor.

This was the state of the art for several years. In 1979, several laboratories finally succeeded in solubilizing opiate receptors that can bind opiates in solution.

In collaboration with Urs Ruegg,[78] we reported the solubilization of active opiate receptors from the brain of the toad *Bufo marinus*, using digitonin. The idea of trying to solubilize receptors from a nonmammalian source came from a careful reading of the literature on the β-adrenergic receptor. Active β-adrenergic binding sites have been solubilized from turkey[79,80] and frog erythrocytes,[81] but only drug–receptor complexes could be solubilized from mammalian sources.[82]

Two other laboratories reported solubilization of active opiate receptors. Using a new detergent, 3-[(3-cholamidopropyl)dimethylammonio]-l-propanesulfonate (CHAPS), a zwitterionic derivative of cholic acid synthesized by L. M. Hjelmeland, Simonds *et al.*[83] were able to solubilize opiate receptors from the neuroblastoma × glioma hybrid cell line (NG-108-15). They also reported some success with opiate receptors in rat brain. J. Bidlack and L. Abood[84] reported the solubilization of opiate receptors from rat brain with Triton X-100.

More recently, we have been able to apply digitonin to the solubilization of opiate receptors from mammalian sources.[85] This required a modification that involves carrying out the detergent extraction in 0.5 M NaCl. The presence of salt was found to enhance yields of solubilization by 10- to 20-fold. This effect was highly specific for salts of sodium; no other alkali metals could be

Table II
Recovery of Prebound Rat and Toad Brain Opiate Receptors from Various Lectin Columns[a]

| | | Recovery of specific binding (%) | | | |
| | | Toad fractions | | Rat fractions | |
Lectin	Sugar eluant	1–3	4–5	1–3	4–5
Wheat germ	N-Acetylglucosamine	42 ± 2.8	40 ± 2.6	38 ± 2.7	36 ± 5.4
Concanavalin A	α-Methyl-D-mannoside	80 ± 2.4	2.3 ± 0.7	75	5
Ulex europaeus	α-Fucose	91 ± 12	<1	85 ± 10	<1
Dolichos biflorus	N-Acetylgalactosamine	85 ± 5.7	<1	82	0
Lentil	Sucrose	76 ± 9	2 ± 1	57	7
Peanut	Galactose	90	<1	106	1
Ricin I	Lactose	100	1	97	4
Ricin II	Lactose	90	1	102	3
Horseshoe crab	Sialic acid	105	<1	86	1

[a] The percentage recovered is based on the amount applied as determined by the assay of an aliquot of the sample applied. Data with S.E.M. are the results of at least three experiments. The others are the average of duplicate experiments.

substituted. Yields of active binding sites ranging from 20% to 45% were obtained from cell membranes derived from rat, cow, and human brain.

3.2. Characterization and Partial Purification of Solubilized Opiate Receptors

In general, it has been found that the solubilized binding sites resemble closely their membrane-bound counterparts. From all sources, opiate binding in solution proved to be sensitive to heat, sulfhydryl reagents, and proteolytic enzymes. There is excellent correlation between affinities of binding of a variety of opioids to soluble and membrane-bound receptors.

For opiate receptors solubilized from toad brain, it has also been shown that the sodium effect is retained.[86] As with membrane-bound receptors, the addition of NaCl (10–100 mM) to the incubation mixture increases the binding of antagonists while reducing the binding of agonists. The inhibition of agonist binding by GTP and its analogues was also retained in solution. For receptors extracted from neuroblastoma × glioma hybrid cells, one difference has been observed[87]: the inhibition by GTP affects both agonist and antagonist binding in solution, whereas with membrane-bound receptors only agonist binding is inhibited by GTP.

A result that had not been obtainable with membrane-bound receptors has now been achieved with solubilized binding sites. It has been possible to demonstrate the presence of carbohydrate in the receptor molecule.[88] This was done by determining whether immobilized plant lectins could retain soluble opiate receptors. As shown in Table II, of nine lectins used, only wheat-germ agglutinin (WGA) consistently retained 35–45% of solubilized opiate receptors passed through a column of WGA-agarose. This suggests the presence of N-acetylglucosamine in opiate receptors and provides the first evidence that opiate receptors are glycoproteins. Opiate receptors solubilized from five species were all retained on WGA-agarose. Wheat-germ agglutinin is known to react with sialoproteins also. However, the inability of horseshoe crab lectin, specific for sialic acid, to retain receptors suggests that retention on WGA is likely to be caused by N-acetylglucosamine. It is probable that the carbohydrate moiety contains other sugars, which for some reason are not accessible to the lectins. Further studies using glycosidases should shed light on this question.

The lectin binding site appears to be functionally independent from the opiate binding site. This conclusion is based on two results: (1) binding to WGA-agarose is equally effective whether free or prebound soluble opiate receptors are passed through the column, and (2) the addition of high concentrations of WGA to membrane-bound or soluble receptors has no effect on opiate binding.

Another indication that supports the glycoprotein nature of opiate receptors was recently obtained by Dawson *et al.*[89] in cell culture. In NG108-15 cells as well as in NCB-20 cells, the addition of tunicamycin, an inhibitor of glycoprotein synthesis, greatly reduced the level of opiate binding.

Partial purification has been reported from two laboratories. Bidlack and Abood[90] have reported considerable purification by use of a morphine affinity column. Our laboratory[88] found that purifications up to 50-fold were achieved by a single pass through a WGA-agarose column.

4. BIOCHEMICAL OR BIOPHYSICAL EVENTS FOLLOWING OPIATE BINDING

Relatively little is yet known about the events that are triggered by the binding of exogenous or endogenous opioids to the receptors and ultimately result in the observed pharmacological responses. Some interesting results have been obtained in studies of the CNS. Studies in a cell culture system, tne neuroblastoma × glioma hybrid cells NG108-15, have yielded some fascinating information, but the applicability of these results to the events in the CNS is not yet clearly established.

4.1. Studies in Brain Homogenate and Slices

4.1.1. Biophysical Events

A physical event that may be triggered by opiate binding is a conformational change in the opiate receptor and in the neighboring synaptic membrane. The evidence that opiate receptors can undergo conformational changes, for example, in response to the presence and absence of sodium ions is reasonably well established.[34,91] Evidence for similar conformational changes evoked by acute or chronic treatment of animals (or homogenates) with opiates is more difficult to obtain and is still very scant. The possibility that such a change could translate into an alteration in the neighboring synaptic membrane, which results in altered ion flux, is even more difficult to prove. The well-known fact that opiates inhibit the electrical activity of neurons does, of course, indicate that ion fluxes are altered. However, we do not know whether this is a direct or indirect effect.

The only published evidence for a structural alteration in synaptic membranes of which this reviewer is aware is a rather old report by Kang *et al.*[92] These investigators showed that synaptosomal membranes isolated from rat brain cortex undergo a characteristic structural change that is manifested by a time-dependent increase in ultraviolet absorption (at 265 and 220–230 nm) and by a decrease in intrinsic fluorescence intensity. This transition appears to be blocked by morphine sulfate. However, the effect requires rather high concentrations of opiate (95 μg/ml) and, to the best of my knowledge, was never tested for stereospecificity or naloxone reversibility.

Preliminary results obtained some time ago in our laboratory indicated that quenching of intrinsic fluorescence in partly purified synaptosomal membranes can be produced by the addition of levorphanol but not by a similar concentration of dextrorphan (N. Clendeninn and E. J. Simon, unpublished results).

4.1.2. Biochemical Events

The biochemical studies of sequelae of opiate binding have focused on the enzyme adenylate cyclase. This is the enzyme that synthesizes cyclic AMP from ATP. Cyclic AMP has been known for some time to act as "second messenger" in a number of hormonally regulated phenomena. The finding by Ho *et al.*[93] that cyclic AMP and the phosphodiesterase inhibitor theophylline antagonize the antinociceptive action of morphine suggested to Collier and Roy that cyclic AMP may have a role in the action of opiates. These authors[94] showed that the stimulation of cyclic AMP synthesis by prostaglandins (PG) E_1 or E_2 in rat brain homogenate is inhibited by opiates. There is no inhibition of basal cyclic AMP production. This inhibition was found to be reversible by naloxone. A study of a series of opiates[95] demonstrated that inhibition of PGE-stimulated formation of cyclic AMP was well correlated with antinociceptive potency, opiate receptor binding affinity, and inhibitory potency for electrically stimulated contraction of the isolated guinea pig ileum. Some specificity was suggested by the finding that fluoride-stimulated cyclic AMP production was not inhibited by opiates. Involvement of E prostaglandins was further supported by the finding of Ehrenpreiss *et al.*[96] that they reverse inhibition by morphine of contractions of the guinea pig ileum. Prostaglandins had no effect on opiate binding even in relatively high concentration.[3]

This area of research has advanced relatively slowly, in part because of the difficulty experienced by a number of laboratories in getting reproducible stimulation by PGE and/or inhibition by opiates of cyclic AMP formation. Some related results will be mentioned briefly.

Iwatsubo and Clouet[97] reported that the addition of morphine and other opiates (3–300 nM) to crude synaptosomal membranes from rat caudate nucleus had no effect on the dopamine stimulation of adenylate cyclase. However, when morphine was administered subcutaneously to rats at a dose of 60 mg/kg, significant increases were found in basal and dopamine-stimulated adenylate cyclase in the caudate nucleus between 15 and 120 min following injection.

Bonnet[98] has reported that systemic injection of morphine produced dose-dependent increases in the level of cyclic AMP and adenylate cyclase in the striatum and thalamus of rat brain, whereas in the substantia nigra and the periventricular gray area, the level of cyclic AMP was found to be reduced. Bonnet and Gusik[99] have demonstrated that the increases in thalamic adenylate cyclase were reversible by naloxone, suggesting the involvement of opiate receptors in this effect. They also found that the level of calcium was of great importance. Physiological levels of calcium (up to 1.3 mM) tended to enhance the opiate stimulation of adenylate cyclase in the thalamus, whereas high levels of calcium tended to suppress it.

Minneman[100] reported that morphine at concentrations from 10^{-7} to 10^{-4} M caused a 30–50% decrease in cyclic AMP levels of striatal slices. He also found a highly specific and complete inhibition of dopamine-stimulated cyclic AMP production in the slices. Morphine, up to a concentration of 10^{-3} M, had no effect on the stimulation of cyclic AMP levels elicited by isoproterenol, adenosine, or PGE_1. This selective inhibition of dopamine-sensitive adenylate cyclase, as well as the depression of basal cyclic AMP, was not seen in striatal homogenate and appears, therefore, to require intact cells. Both effects were blocked by naloxone (1 μM), suggesting that they are mediated via specific opiate receptors.

In contrast to the rat striatal system, in which inhibition by opiates was seen with slices but disappeared in homogenates, Makman and co-workers[101] reported that opiates and opioid peptides inhibited the dopamine-stimulated adenylate cyclase activity in homogenates of monkey amygdaloid nucleus. Etorphine was about 10^4 times more active than morphine, and enkephalins and their analogues were about 5–20 times more active than morphine. The inhibition was antagonized by naloxone, and the inactive opiate enantiomer dextrorphan failed to inhibit. These results suggest that inhibition is via classical opiate receptors. Sensitivity of the cyclase to opioids is eliminated by freezing homogenates or preincubating them at 0°C for 90 min. These treatments do not affect the ability of dopamine to stimulate the enzyme. In contrast to the results of Collier's laboratory in rat brain homogenates, no inhibition of prostaglandin-stimulated cyclase was observed.

4.2. Studies in Cell Culture

A great deal of work has been done in a cell culture system that may have potential as an *in vitro* model for opioid action in the CNS. A hybrid neuro-blastoma × glioma cell line (NG108-15) was shown by Klee and Nirenberg[102] to contain opiate receptors. The parental cell lines showed little or no opiate binding. Binding affinities of various opioid ligands to opiate receptors of the cell line correlated well with their pharmacological potencies. More recently, a sympathetic ganglion × neuroblastoma hybrid cell line[103] as well as a neu-roblastoma cell line[104] were reported to possess high levels of opiate receptors. As mentioned earlier, the neuroblastoma and NG108-15 cell lines were reported to have virtually only the δ type binding sites.

More recently, a neuroblastoma–Chinese hamster brain cell hybrid line (NCB20) was shown by McLawhon *et al.*[105] to contain both δ and benzomorphan-preferring opiate receptors.

Following the studies of Collier and Roy in brain homogenates, the effect of opiates on adenylate cyclase in the NG108–15 cells was investigated. It was found that opiates as well as opioid peptides inhibit both basal and PGE_1- and adenosine-stimulated adenylate cyclase.[106,107] This inhibition was stereospecific and reversed by naloxone. Furthermore, affinity of opiate agonists for the receptor correlated well with inhibitory potency for adenylate cyclase. No such inhibition could be observed in cell lines devoid of opiate receptors.

Chronic treatment of these cultures with opioid drugs or peptides has led to results that have been interpreted as being the cell's equivalent of tolerance and physical dependence in animals and man. When the hybrid cells were cultured in the presence of morphine for 4 days, there was a gradual rise in both basal and PGE_1-stimulated adenylate cyclase.[108] After 2 to 3 days, both basal and PGE_1-stimulated enzyme activity measured in the presence of morphine were about the same as in control cells preincubated and assayed in the absence of morphine. Sharma *et al.*[108] suggested that this may be the cells' equivalent of tolerance. Adenylate cyclase was abnormally high in morphine-pretreated cells when assayed in the absence of opiates. This was demonstrated even more dramatically when cyclic AMP levels were measured in cells exposed briefly to the antagonist naloxone after pretreatment with morphine for several days. Such cells showed as much as a fivefold increase in cyclic AMP levels, a phenomenon suggested as analogous to precipitated withdrawal. Adenylate cyclase levels were shown to return to normal within 24 h after withdrawal of morphine from the culture medium. This cell line had also been shown to become "tolerant to" and "dependent on" the endogenously occurring opioid peptide [methionine]enkephalin. A scheme showing the authors' perception of the formation of tolerance and dependence in NG108–15 cultures is depicted in Fig. 2. Results similar to those of Klee and co-workers were reported by Traber *et al.*[109]

A very interesting result has recently been reported by Koski and Klee,[110] which has bearing on the mechanism by which opioids inhibit adenylate cyclase in the neuroblastoma × glioma hybrid cells. It is generally accepted that the regulation of adenylate cyclase by hormones occurs via a guanine-nucleotide-sensitive regulatory protein (G protein). An associated GTPase is believed to mediate inactivation of the enzyme. The enzyme is thought to shuttle between an active GTP-complexed state and an inactive GDP-complexed state. Hormones that stimulate adenylate cyclase may do so by promoting an interaction between their specific receptors and the G protein that favors its conversion to the GTP-complexed form.

On the other hand, the receptor-mediated inhibition of the enzyme could be mediated through stimulation of the GTPase that converts the G protein to the GDP-complexed form. Koski and Klee have been able to demonstrate that opiates and opioid peptides stimulate GTP hydrolysis catalyzed by membranes from NG108–15 cells. The stimulation is rapid, stereospecific, and reversed by naloxone. Potencies of opioids to stimulate GTP hydrolysis are well correlated

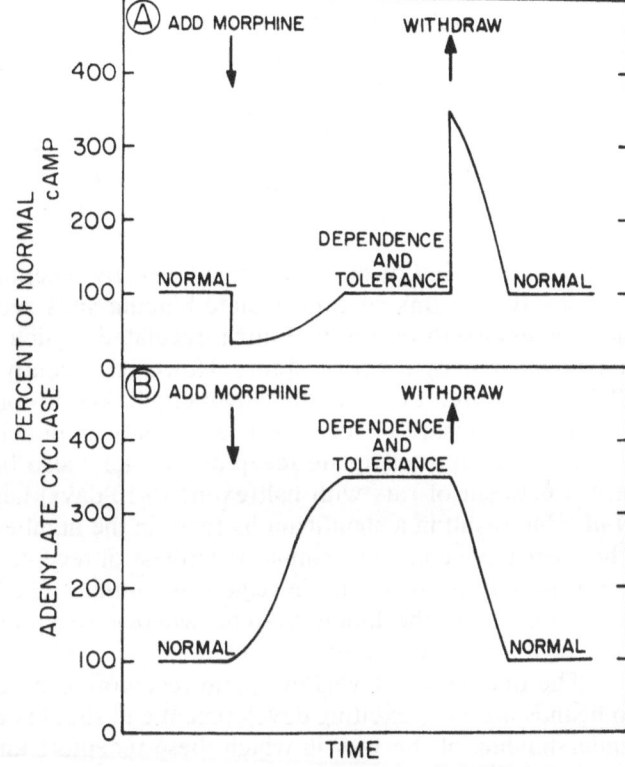

Fig. 2. A model of the role of adenylate cyclase regulation in the development of morphine tolerance and dependence based on results obtained with NG108-15 cells in culture. Part A shows the effect of morphine on cyclic AMP levels, and part B the effect on adenylate cyclase activity as a function of time. (From Sharma *et al.*,[108] with permission.)

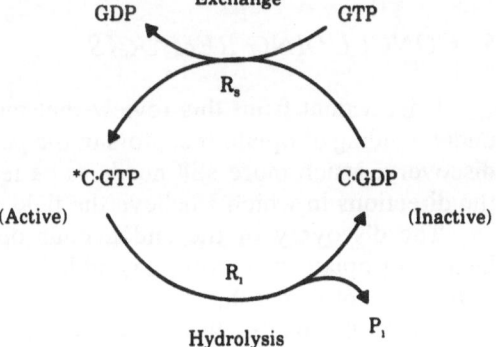

Fig. 3. Scheme for receptor-mediated stimulation and inhibition of adenylate cyclase. R_s are receptors that are coupled to adenylate cyclase so that occupancy produces stimulation. Occupancy of receptor R_i produces inhibition of the enzyme. The latter represent opiate receptors among others. (From Koski and Klee,[110] with permission).

with their ability to inhibit adenylate cyclase. Opioids have no effect on GTPase activity in membranes prepared from the glioma cell line, one of the parents of NG108–15, which lacks opiate receptors. Figure 3 shows a simple scheme taken from the paper of Koski and Klee[110] that summarizes what has just been said about the mechanism for receptor-mediated stimulation and inhibition of adenylate cyclase.

The increase in adenylate cyclase discussed earlier, which was interpreted as being the basis for tolerance and dependence in the cell line, was not observed if the enzyme was assayed in the presence of the nonhydrolyzable GTP analogue pNHppG.[111] This observation implies that activation of adenylate cyclase induced by chronic opiate treatment may occur through a slow, compensatory inhibition of GTPase activity.

4.3. Regulation of Opiate Receptor Levels by the Presence of Ligands

The regulation of opiate binding levels by receptor ligands is a topic that fits loosely into this section. Opiate binding sites seemed for a long time to lack the ability to be up- and down-regulated, which is an important characteristic of so many other receptors. However, recently several laboratories[112–114] have reported evidence for down-regulation of opiate receptor binding by prolonged exposure to receptor ligands, at least in cell culture.

Up-regulation of opiate receptors has now also been reported. The long-term treatment of rats with naltrexone (6-10 days) has been shown by Zukin *et al.*[115] to result in a significant increase in the number of opiate binding sites. This increase seems to be similar regardless of receptor subclass examined and varies greatly from one brain region to another. The highest increase (130%) was observed in the limbic system, whereas receptors in the periaqueductal gray region were virtually unaffected.

The finding that levels of opiate receptors can be regulated by exposure to ligands is a very exciting development and should contribute to our ultimate understanding of the way in which these receptors and their endogenous and exogenous ligands function in the organism.

5. CONCLUDING REMARKS

It is evident from this review that much progress has been made in our understanding of opiate receptors in the period of less than 10 years since their discovery. Much more still needs to be learned, and I shall mention some of the directions in which I believe the field will be moving.

The discovery of the endogenous opioid peptides, the putative natural ligands of opiate receptors, has added greatly to the interest and excitement in this area of research.

These discoveries have already spawned some very promising offspring. They have served as models for the discovery of binding sites for other drugs such as benzodiazepines and phencyclidine. There is even some evidence for specific binding sites for nicotine, presumably separate from nicotinic acetylcholine receptors. There is considerable activity at the present time in many laboratories concerned with efforts to isolate endogenous ligands for these drug receptors.

The most important question in the opioid field is the as yet unanswered one concerning the physiological role of the various opioid peptides and their receptors. There is a general feeling that they represent new neurotransmitter or neuromodulator systems, but there is as yet no real proof for this.

As indicated earlier, there is a surprising lack of information concerning the steps triggered by the binding of opioids to their receptors that result in the pharmacological effects. A knowledge of these steps, be they changes in ion flux across the synapse or the modulation of an enzyme such as adenylate cyclase, would be extremely helpful towards the ultimate elucidation of function.

Another area in which exploration has only begun is the possible changes in opiate receptor number or properties in various physiological and pathological states. Such information could aid in our understanding of endogenous opioid function as well as of disease states of unknown etiology.

An accurate knowledge of the various types of opiate receptors, their ligands, functions, and distribution will have far-ranging theoretical and practical importance. Such information has furthered progress in such areas as cholinergic and catecholaminergic receptors. If subtypes of opiate receptors prove to have distinct functions, it may become possible to synthesize highly specific ligand molecules that have certain desired effects while lacking undesirable ones.

Much of the research that led to the discovery of the endogenous opioid system was carried out in laboratories interested in understanding the biochemical basis of narcotic addiction. It has been a disappointment to many of us that the progress in this research has not yet yielded any new insight into the mechanism of tolerance, physical dependence, and other aspects of drug abuse. Nevertheless, I share the optimistic view that knowledge about an endogenous opioid system cannot fail ultimately to provide information about both acute and chronic actions of exogenous opiates.

REFERENCES

1. Van Praag, D., and Simon, E. J., 1966, *Proc. Soc. Exp. Biol. Med.* **122**:6–11.
2. Goldstein, A., Lowney, L. I., and Pal, B. K., 1971, *Proc. Natl. Acad. Sci. U.S.A.* **68**:1742–1747.
3. Simon, E. J., Hiller, J. M., and Edelman, I., 1973, *Proc. Natl. Acad. Sci. U.S.A.* **70**:1947–1949.
4. Terenius, L., 1973, *Acta Pharmacol. Toxicol. (Kbh.)* **32**:317–320.
5. Pert, C. B., and Snyder, S. H., 1973, *Science* **179**:1011–1014.
6. Hiller, J. M., Pearson, J., and Simon, E. J., 1973, *Res. Commun. Chem. Pathol. Pharmacol.* **6**:1052–1062.
7. Stefano, G. B., Kream, R. M., and Zukin, R. S., 1980, *Brain Res.* **181**:440–445.
8. Pert, C. B., Snowman, A. M., and Snyder, S. H., 1974, *Brain Res.* **70**:184–188.
9. Pasternak, G. W., and Snyder, S. H., 1974, *Mol. Pharmacol.* **10**:183–193.
10. Lin, H.-K., and Simon, E. J., 1978, *Nature* **271**:383–384.
11. Lin, H.-K., Holland, M.-J. C., and Simon, E. J., 1981, *J. Pharmacol. Exp. Ther.* **216**:149–155.
12. Kuhar, M. J., Pert, C. B., and Snyder, S. H., 1973, *Nature* **245**:447–450.
13. Atweh, S. F., and Kuhar, M. J., 1977, *Brain Res.* **124**:53–67.
14. Atweh, S. F., and Kuhar, M. J., 1977, *Brain Res.* **129**:1–12.
15. Atweh, S. F., and Kuhar, M. J., 1977, *Brain Res.* **134**:393–406.
16. Pearson, J., Brandeis, L., Simon, E., and Hiller, J., 1980, *Life Sci.* **26**:1047–1052.
17. Young, W. S., III, and Kuhar, M. J., 1979, *Brain Res.* **179**:255–270.

18. Stahl, K. D., van Bever, W., Janssen, P., and Simon, E. J., 1977, *Eur. J. Pharmacol.* **10**:183–193.
19. Creese, I., and Snyder, S. H., 1975, *J. Pharmacol. Exp. Ther.* **194**:205–219.
20. LaMotte, C., Pert, C. B., and Snyder, S. H., 1976, *Brain Res.* **112**:407–412.
21. Hiller, J. M., Simon, E. J., Crain, S. M., and Peterson, E. R., 1978, *Brain Res.* **145**:396–400.
22. Satoh, M., Zieglgansberger, W., and Herz, A., 1976, *Brain Res.* **115**:99–110.
23. Terenius, L., and Wahlström, A., 1974, *Acta Pharmacol. Toxicol.* (*Kbh.*) **35**(Suppl. 1):55.
24. Hughes, J., 1975, *Brain Res.* **88**:295–308.
25. Hughes, J., Smith, T. W., Kosterlitz, H. W., Fothergill, L. A., Morgan, B. A., and Morris, H. R., 1975, *Nature* **258**:577–579.
26. Cox, B. M., Goldstein, A., and Li, C. H., 1976, *Proc. Natl. Acad. Sci. U.S.A.* **73**:1821–1823.
27. Bradbury, A., Smyth, D., Snell, C., Deakin, J., and Wendlant, S., 1977, *Biochem. Biophys. Res. Commun.* **74**:748–754.
28. Goldstein, A., Tachibana, S., Lowney, L. I., Hunkapiller, M., and Hood, L., 1979, *Proc. Natl. Acad. Sci. U.S.A.* **76**:6666–6670.
29. Goldstein, A., Fischli, W., Lowney, L. I., Hunkapiller, M., and Hood, L., 1981, *Proc. Natl. Acad. Sci. U.S.A.* **78**:7219–7223.
30. Martin, W. R., Eades, C. G., Thompson, J. A., Huppler, R. E., and Gilbert, P. E., 1976, *J. Pharmacol. Exp. Ther.* **197**:517–532.
31. Gilbert, P. E., and Martin, W. R., 1976, *J. Pharmacol. Exp. Ther.* **198**:66–82.
32. Lord, J. A. H., Waterfield, A. A., Hughes, J., and Kosterlitz, H. W., 1977, *Nature* **267**:495–500.
33. Smith, J. R., and Simon, E. J., 1980, *Proc. Natl. Acad. Sci. U.S.A.* **77**:281–284.
34. Simon, E. J., and Groth, J., 1975, *Proc. Natl. Acad. Sci. U.S.A.* **72**:2404–2407.
35. Robson, L. E., and Kosterlitz, H. W., 1979, *Proc. R. Soc. Lond.* [*Biol.*] **205**:425–432.
36. Chang, K.-J., Cooper, B. R., Hazum, E. R., and Cuatrecasas, P., 1979, *Mol. Pharmacol.* **16**:91–104.
37. Simon, E. J., Bonnet, K. A., Crain, S. M., Groth, J., Hiller, J. M., and Smith, J. R., 1981, *Neural Peptides and Neuronal Communication* (E. Costa and M. Trabucchi, eds.), Raven Press, New York, pp. 335–346.
38. Ninkovic, M., Hunt, S. P., Emson, P. C., and Iversen, L. L., 1981, *Brain Res.* **214**:163–167.
39. Bonnet, K. A., Groth, J., Gioannini, T., Cortes, M., and Simon, E. J., 1981, *Brain Res.* **221**:437–440.
40. Goodman, R. R., Snyder, S. H., Kuhar, M. J., and Young, W. S., 1980, *Proc. Natl. Acad. Sci. U.S.A.* **77**:6239–6243.
41. Herkenham, M., and Pert, C. B., 1980, *Proc. Natl. Acad. Sci. U.S.A.* **77**:5532–5536.
42. Hiller, J. M., Angel, L., and Simon, E. J., 1981, *Science* **214**:468–469.
43. Ward, S. J., Portoghese, P. S., and Takemori, A. E., 1980, *Soc. Neurosci. Abstr.* **6**:525.
44. Schulz, R., Wuster, K., Dreuss, H., and Herz, A., 1980, *Nature* **285**:242–243.
45. Schulz, R., Wuster, M., and Herz, A., 1981, *Pharmacol. Biochem. Behav.* **14**:75–79.
46. Bowen, W. D., Gentleman, S., Herkenham, M., and Pert, C. B., 1981, *Proc. Natl. Acad. Sci. U.S.A.* **78**:4818–4822.
47. Lee, N. M., and Smith, A. P., 1980, *Life Sci.* **26**:1459–1464.
48. Egan, T. M., and North, R. A., 1981, *Science* **214**:923–924.
49. Rothman, R. B., and Westfall, T. C., 1982, *Mol. Pharmacol.* **21**:538–547.
50. Rothman, R. B., and Westfall, T. C., 1982, *Mol. Pharmacol.* **21**:548–557.
51. Hiller, J. M., and Simon, E. J., 1980, *J. Pharmacol. Exp. Ther.* **214**:516–519.
52. Harris, D. W., and Sethy, V. H., 1980, *Eur. J. Pharmacol.* **66**:121–123.
53. Pasternak, G. W., 1980, *Proc. Natl. Acad. Sci. U.S.A.* **77**:3691–3694.
54. Kosterlitz, H. W., Paterson, S. J., and Robson, L. E., 1981, *Br. J. Pharmacol.* **73**:939–949.
55. Chang, K.-J., Hazum, E., and Cuatrecasas, P., 1981, *Proc. Natl. Acad. Sci. U.S.A.* **78**:4141–4145.
56. Brantl, V., Teschemacher, H., Henschen, A., and Lottspeich, F., 1979, *Hoppe Seylers Z. Physiol. Chem.* **360**:1211–1216.
57. Vincent, J. P., Kartalovski, B., Geneste, P., Kamenka, J. M., and Lazdunski, M., 1979, *Proc. Natl. Acad. Sci. U.S.A.* **76**:4578–4582.

58. Zukin, S. R., and Zukin, R. S., 1979, *Proc. Natl. Acad. Sci. U.S.A.* **76:**5372–5376.
59. Zukin, R. S., and Zukin, S. R., 1981, *Mol. Pharmacol.* **20:**246–254.
60. Holtzman, S. F., 1980, *J. Pharmacol. Exp. Ther.* **214:**614–619.
61. Itzhak, Y., Bonnet, K. A., Groth, J., Hiller, J. M., and Simon, E. J., 1982, *Life Sci.* **31:**1363–1366.
62. Pasternak, G. W., 1981, *Neurology (N.Y.)* **31:**1311–1315.
63. Zhang, Z., and Pasternak, G. W., 1981, *Life Sci.* **29:**843–851.
64. Lemaire, S., Magnan, J., and Regnoli, D., 1978, *Br. J. Pharmacol.* **64:**327–329.
65. Miranda, H., Huidobro, F., and Huidobro-Toro, J. P., 1979, *Life Sci.* **24:**1511–1518.
66. Schulz, R., Faase, E., Wüster, M., and Herz, A., 1979, *Life Sci.* **24:**843–850.
67. Wüster, M., Schultz, R., and Herz, A., 1980, *Neurosci. Lett.* **15:**1511–1518.
68. Oka, T., Negishi, K., and Suda, M., 1981, *Advances in Endogenous and Exogenous Opioids* (H. Takagi and E. J., Simon, eds.), Kodansha, Tokyo, pp. 21–23.
69. Chavkin, C., James, I. F., and Goldstein, A., 1982, *Science* **215:**413–415.
70. Yoshimura, K., Huidobro-Toro, J. P., Lee, N. M., Loh, H. H., and Way, E. L., 1982, *J. Pharmacol. Exp. Ther.* **222:**71–79.
71. Corbett, A. D., Paterson, S. J., McKnight, A. T., Magnan, J., and Kosterlitz, H. W., 1982, *Nature* **299:**79–81.
72. Hammonds, R. G., Jr., Nicolas, P., and Li, C. H., 1982, *Proc. Natl. Acad. Sci. U.S.A.* **79:**6494–6496.
73. Cowan, A., 1981, *Life Sci.* **28:**1559–1570.
74. Adler, M. W., 1981, *Life Sci.* **28:**1543–1545.
75. Simon, E. J., Hiller, J. M., and Edelman, I., 1975, *Science* **190:**389–390.
76. Zukin, R. S., and Kream, R. M., 1979, *Proc. Natl. Acad. Sci. U.S.A.* **76:**1593–1597.
77. Zukin, R. S., Federoff, G. D., and Kream, R. M., 1980, *Endogenous and Exogenous Opiate Agonists and Antagonists* (E. L. Way, ed.), Pergamon Press, New York, pp. 221–224.
78. Ruegg, U. T., Hiller, J. M., and Simon, E. J., 1980, *Eur. J. Pharmacol.* **64:**367–368.
79. Pike, L. J., and Lefkowitz, R. J., 1976, *Mol. Pharmacol.* **14:**370–375.
80. Vauquelin, G., Geynet, P., Hanoune, J., and Strosberg, A. D., 1977, *Proc. Natl. Acad. Sci. U.S.A.* **74:**3710–3714.
81. Caron, M. G., and Lefkowitz, R. J., 1976, *J. Biol. Chem.* **251:**2374–2384.
82. Strauss, W. L., Ghai, G., Fraser, C. M., and Venter, J. C., 1979, *Arch. Biochem. Biophys.* **196:**566–573.
83. Simonds, W. F., Koski, G., Streaty, R. A., Hjelmeland, L. M., and Klee, W. A., 1980, *Proc. Natl. Acad. Sci. U.S.A.* **77:**4623–4627.
84. Bidlack, J. M., and Abood, L. G., 1980, *Life Sci.* **27:**331–340.
85. Howells, R. D., Gioannini, T. L., Hiller, J. M., and Simon, E. J., 1982, *J. Pharmacol. Exp. Ther.* **222:**629–634.
86. Ruegg, U. T., Cuenoud, S., Hiller, J. M., Gioannini, T. L., Howells, R. D., and Simon, E. J., 1981, *Proc. Natl. Acad. Sci. U.S.A.* **78:**4635–4638.
87. Koski, G., Simonds, W. F., and Klee, W. A., 1981, *J. Biol. Chem.* **256:**1536–1538.
88. Gioannini, T., Foucaud, B., Hiller, J. M., Hatten, M. E., and Simon, E. J., 1982, *Biochem. Biophys. Res. Commun.* **105:**1128–1134.
89. Dawson, G., McLawhon, R. W., and Ellory, C. J., 1982, *Fed. Proc.* **41:**1160.
90. Bidlack, J. M., Abood, L. G., Osei-Gyimah, P., and Archer, S., 1981, *Proc. Natl. Acad. Sci. U.S.A.* **78:**636–639.
91. Pert, C. B., and Snyder, S. H., 1974, *Mol. Pharmacol.* **10:**868–879.
92. Kang, S., Sessa, G., and Green, J. P., 1973, *Res. Commun. Chem. Pathol. Pharmacol.* **5:**359–388.
93. Ho, I. K., Loh, H. H., and Way, E. L., 1973, *J. Pharmacol. Exp. Ther.* **185:**336–346.
94. Collier, H. O. J., and Roy, A. C., 1974, *Nature* **248:**24–27.
95. Collier, H. O. J., and Roy, A. C., 1974, *Prostaglandins* **7:**361–376.
96. Ehrenpreiss, S., Greenberg, J., and Belman, S., 1973, *Nature* **245:**280–282.
97. Iwatsubo, K., and Clouet, D. H., 1975, *Biochem. Pharmacol.* **24:**1499–1503.
98. Bonnet, K. A., 1975, *Life Sci.* **16:**1877–1882.
99. Bonnet, K. A., and Gusik, S., 1976, *Neurosci. Abstr.* **2:**849.

100. Minneman, K. P., 1977, *Br. J. Pharmacol.* **59:**480p–481p.
101. Walczak, S. A., Wilkening, D., and Makman, M. H., 1979, *Brain Res.* **160:**105–116.
102. Klee, W. A., and Nirenberg, M., 1974, *Proc. Natl. Acad. Sci. U.S.A.* **71:**3474–3477.
103. Blosser, J., Abbot, J., and Shain, W., 1976, *Biochem. Pharmacol.* **25:**2395–2399.
104. Chang, K.-J., Miller, R. J., and Cuatrecasas, P., 1978, *Mol. Pharmacol.* **14:**961–970.
105. McLawhon, R. W., West, R. E., Miller, R. J., and Dawson, G., 1981, *Proc. Natl. Acad. Sci. U.S.A.* **78:**4309–4313.
106. Sharma, S. K., Nirenberg, M., and Klee, W. A., 1975, *Proc. Natl. Acad. Sci. U.S.A.* **72:**590–594.
107. Traber, J., Fischer, K., Latzin, S., and Hamprecht, B., 1974, *FEBS Lett.* **49:**260–263.
108. Sharma, S. K., Klee, W. A., and Nirenberg, M., 1975, *Proc. Natl. Acad. Sci. U.S.A.* **72:**3092–3096.
109. Traber, J., Gullis, R., and Hamprecht, B., 1975, *Life Sci.* **16:**1863–1868.
110. Koski, G., and Klee, W. A., 1981, *Proc. Natl. Acad. Sci. U.S.A.* **78:**4185–4189.
111. Sharma, S. K., Klee, W. A., and Nirenberg, M., 1977, *Proc. Natl. Acad. Sci. U.S.A.* **74:**3365–3369.
112. Chang, K.-J., Eckel, R. W., and Blanchard, S. G., 1982, *Nature* **296:**446–448.
113. Simantov, R., Baram, D., and Dornay, M., 1982, *Regulatory Peptides: From Molecular Biology to Function* (E. Costa and M. Trabucchi, eds.), Raven Press, New York, pp. 291–300.
114. Law, P. Y., Horn, D. S., and Loh, H. H., 1982, *Mol. Pharmacol.* **22:**1–4.

Peptide Receptors

David R. Burt and Najam A. Sharif

1. INTRODUCTION

1.1. Peptides as Transmitter Candidates

The general area of peptide neurotransmission/neuromodulation has been reviewed extensively in recent years[1-5] and is covered elsewhere in the present series. Neuropeptide receptors have also been reviewed previously.[6,7] This chapter concentrates on the modest progress that has been made towards identifying receptors for nonopioid peptides in the vertebrate central nervous system (CNS) by binding and other biochemical measurements. Only limited reference is made to results of anatomic, electrophysiological, or behavioral experiments in the CNS or to results in various peripheral systems. Peptides are included because they are present in the CNS and have effects there. In most cases, additional criteria for neurotransmitter identification have been met as well. Many of the peptides discussed here have prominent hormonal or other peripheral roles so that their receptors have been best studied outside the CNS. These receptors may or may not be the same as CNS receptors. In some cases, there are questions about the exact identities of the peptides exciting CNS receptors, since the peptides' presence has been established by radioimmunoassay or immunohistochemistry—methods that can be detecting families of structurally related peptides. In other cases, there are questions about the sources of peptides exciting these receptors, i.e., whether they are of central or peripheral origin.

Some aspects of peptides as transmitter candidates predict properties of their receptors. Available evidence indicates that most neuronal peptides are synthesized via larger precursor peptides on ribosomes in the perikaryon and transported by axoplasmic flow to nerve terminals, where they are only used once; i.e., there is no local reuptake or resynthesis.[8] Since there is so much cellular energy invested in each peptide molecule, economy of function suggests that peptides should be more potent than simpler transmitter candidates, i.e., that their receptors should be of higher affinity. In general, this appears to be the case, such that peptide binding and receptor activation occur in the picomolar or low nanomolar range.

David R. Burt and Najam A. Sharif • Department of Pharmacology and Experimental Therapeutics, University of Maryland School of Medicine, Baltimore, Maryland 21201.

1.2. Criteria for Receptors

The identification of a peptide binding site as a receptor requires that it fulfill the usual criteria of saturability, high affinity, appropriate kinetics, appropriate pharmacology, and localization to appropriate brain or tissue regions or subcellular fractions.[9,10] Exact correlation of peptide levels in brain regions with receptor levels cannot be expected when peptides are destined for export rather than local action. As a prime example, the hypothalamus is relatively much higher in levels of many peptides than in their receptors (see individual peptides below). The correlation of the pharmacology of peptide binding to that of responses presents similar problems in that peptide analogues may show enhanced activity in behavioral tests because of enhanced ability to reach receptors (through peptidase resistance or lipid solubility) rather than greater affinity. The life-span of exogenous peptides in the brain is typically very short, and it may be very difficult to generate conventional dose–response curves. The most reliable guide to pharmacology of neuronal peptide responses is likely to be obtained *in vitro*, but relatively little work of this nature has been done with CNS preparations.

1.3. Methodological Features

Neuropeptide receptor binding assays share most of their methodological features with assays of receptors for other neurotransmitters or for peripheral peptide hormones, both reviewed widely elsewhere.[11–15] Three important features are considered briefly below (see also Burt *et al.*[16]).

1.3.1. Peptidases

The brain and many other tissues are rich in peptidases.[17,18] This represents an efficient mechanism for inactivating peptide neurotransmitters *in vivo* but creates problems *in vitro* in peptide receptor binding assays. Many of the relevant enzymes have not been characterized in detail, and specific inhibitors are lacking. Peptide ligand degradation is considerably slowed by running incubations at 0–4°C, at the expense of slowing the approach to binding equilibrium. In other cases, incubations at 37°C have been possible through addition of various nonspecific peptidase inhibitors, including bacitracin, aprotinin, benzamidine, phenylmethylsulfonyl fluoride, and EDTA. Use of washed membranes will reduce many soluble peptidases.

1.3.2. Stickiness

Peptides tend to stick to glass and other surfaces. It is usual to reduce losses on glassware by addition of bovine serum albumin (BSA, 0.1% or greater), serum (e.g., horse), gelatin, or other carrier protein/peptide. Some of these carriers, especially BSA or serum, may have peptidase activity associated with them. Substitution of selected plasticware or siliconized glassware should further reduce adsorptive losses. The large surface area of filters often leads to so much binding (generally nonsaturable) of peptide ligands as to preclude use of filtration to separate bound radioactivity. Membrane filters may work when glass fiber filters do not, or presoaking filters in BSA or polylysine may reduce the problem. Otherwise, various forms of centrifugation assays may be necessary.

1.3.3. Ligands

An almost unique advantage of peptides is the ease with which those containing noncritical Tyr or His residues may be reacted with ^{125}I to give a high-specific-activity radioactive ligand. Other peptides with noncritical free amino groups may be similarly iodinated with the Bolton–Hunter reagent. These and other methods of labeling peptide ligands are more extensively discussed elsewhere.[11,16] Besides high affinity, a feature to seek in a peptide ligand is built-in resistance to degradation, sometimes achievable by selected D-amino acid substitutions or other modifications that do not adversely affect binding affinity.

1.4. Peptide Response Mechanisms

The response mechanisms linked to most peptide receptors are still little explored and poorly understood, especially in the brain. As discussed further under the individual peptides, many peptides, including VIP, seem to produce responses through an increase in cyclic AMP. Others, including AII and TRH, seem more clearly linked to changes in Ca^{2+} fluxes. These and other changes ultimately affect the firing of neurons, either directly or through modulation of the release of or response to other neurotransmitters. There is thus far no evidence that the smaller neuropeptides discussed in this chapter have any response mechanisms uniquely associated with them as opposed to other classes of neurotransmitters.

2. ANGIOTENSIN

Angiotensin II (AII, Asp-Arg-Val-Tyr-Ile-His-Pro-Phe) as a peripheral hormone plays a major role in maintenance of extracellular fluid volume and blood pressure. It also produces a variety of central effects on intraventricular administration, many of which appear linked to its peripheral effects.[19-22] These include increased drinking, vasopressin release, and sympathetic outflow. The interpretation of these effects has been problematical, since many seem to be exerted at least in part through receptors in the circumventricular organs, outside the blood–brain barrier and accessible to circulating AII. These include the subfornical organ, organum vasculosum of the lamina terminalis, and area postrema.

If circulating hormone were producing all of AII's central effects, detailed discussion of its receptors in this chapter would be unwarranted. There is limited and suggestive evidence for considering AII or a related peptide to be a neurotransmitter candidate as well as a hormone, however. Although the subject has generated considerable controversy,[23,24] and endogenous AII in the brain appears to be very low or absent in the hands of some workers,[25,26] biochemical results suggest that the brain contains a complete renin–angiotensin system[19,27] including angiotensinogen, renin, angiotensin I, converting enzyme, and AII. Furthermore, immunohistochemistry has revealed widely distributed AII-like activity in the CNS,[28,29] with high density in areas of the spinal cord, the median eminence, and the central amygdaloid nucleus. Finally, the iontophoretic application of the AII antagonist seralasin ([Sar^1, Ala^8]AII) slows background firing of AII-sensitive neurons as well as inhibiting the stimulation of firing by exogenous AII.[30]

Receptors for angiotensin have been extensively studied in the periphery.[31–33] Analysis of dose–response curves for AII agonist and antagonist analogues has suggested only a single class of peripheral AII receptors for many responses,[34] but both high- and low-affinity binding components for [^{125}I]AII or other ligands are evident in corresponding tissues, and low-affinity sites appear coupled to responses in at least a few.[33] Properties of the peripheral receptors include localization to plasma membrane fractions of target tissues; pharmacology matching responses; sensitivity of the high-affinity component to guanine nucleotide reduction of agonist binding affinity and increase in dissociation rate; apparent linkage of receptor occupation to increased Ca^{2+} fluxes and phosphatidylinositol hydrolysis, with possible secondary changes in protein phosphorylation, prostaglandin synthesis, and cyclic nucleotide levels; apparent internalization of occupied receptors; modulation of binding by ions, particularly Na^+ and Ca^{2+}, that is tissue specific; and an apparent lability during attempts at solubilization (see cited reviews for details and references). High-affinity binding sites have apparent affinities for AII corresponding to half-maximal occupation at about 0.1–1 nM; this is about tenfold greater than circulating levels of AII.

Several laboratories have reported apparent receptor binding of [^{125}I]AII or related ligands to various regions of mammalian brain.[35–44] The affinities of these sites have been generally similar to those measured in peripheral tissues, but where detailed comparisons were performed, clear differences in properties emerged. Thus, calf cerebellar sites were much more sensitive to stimulation of [^{125}I]AII binding by Na^+ (150 mM) than those in calf adrenals or rabbit uterus[39] and showed marked differences in pharmacology for AII analogues from sites in calf adrenals.[40]

There appear to be great species differences in the brain regional distribution of AII receptor binding. These were apparent even in the first report,[35] in which rat brain gave highest binding in the thalamus–hypothalamus, midbrain, and brainstem, whereas calf brain binding was localized almost exclusively to the cerebellum. More detailed studies in rat[36,37] found highest binding in the lateral septum and caudal region of the superior colliculi, regions possibly connected with AII's known pressor and dipsogenic effects. However, many other sites of binding, notably including calf cerebellum, have no known functional correlates. The picture is further clouded by the recent demonstration that, even among six rodent species, there are marked differences in relative and absolute levels of AII receptor binding in different brain regions.[43] One species (dogs) showed no detectable CNS binding at all.

A possible explanation for some aspects of these findings arises from an ontogenetic study in the rat that demonstrated a rapid postnatal increase in [^{125}I]AII binding in several brain regions to a maximum at 1–2 weeks of age, followed by a gradual decline over the next 4–6 weeks to much lower adult levels.[41] The authors suggest that circulating AII may have physiological roles in the newborn brain, when an immature blood–brain barrier allows access, that are lost in the adult. Thus, binding in many adult brain regions could be a "developmental remnant" not associated with the presence of neuronal AII and much more subject to species variation than binding associated with current (i.e., retained) function. The presence of the same phenomenon in the newborn of other species needs to be explored.

Considerable electrophysiological and immunohistochemical data do suggest that some (if not all) biochemically identified receptors are coupled to a response and that neuronally derived AII-like peptide(s) are available to occupy these receptors in the adult. However, the most interesting connection between AII binding and function, an increased level of binding in the organum vasculosum of the lamina terminalis of spontaneously hypertensive rats, has appeared only in abstract form[45] or review.[20]

Technically, these studies illustrate many features mentioned in the introduction. By adding several peptidase inhibitors, all workers were able to incubate at elevated temperatures (22–37°C), typically for 30–60 min. Bound radioactivity was separated by filtration (glass fiber filters) or centrifugation. The usual ligand was [^{125}I]AII, which seemed to retain most of the activity of the parent compound, as estimated by comparing its affinity as determined in saturation curves with that of AII in competition curves. Additionally, one study[39] used the antagonist ligand [^{125}I] [Sar^1,Leu^8]AII, with generally similar results.

In conclusion, there remain many problems with the identification of CNS angiotensin receptors. It is not clear that they are the same as peripheral AII receptors, whose properties have been much more thoroughly studied. The CNS regions where responses are best characterized, the circumventricular organs, contain too little tissue for convenient binding studies. The regional distribution is clouded by major species differences and generally poor correlations with other markers of the renin–angiotensin system. Further attention to regions where an AII-like peptide is likely to be acting as a neurotransmitter, including the dorsal horn of the spinal cord, should prove rewarding, as should studies in a well-chosen *in vitro* preparation.

3. NEUROTENSIN

Neurotensin (NT, pGlu-Leu-Tyr-Glu-Asn-Lys-Pro-Arg-Arg-Pro-Tyr-Ile-Leu-OH) was discovered serendipitously during the purification of substances P.[46] Although originally isolated from bovine hypothalami, it is widely distributed in the CNS and gut. Highest levels are in the N cells of ileal mucosa. Neurotensin has a variety of central and peripheral effects.[47–49] After peripheral administration, it produces vasodilatation (used as a bioassay in its purifica-

tion), hyperglycemia, and various endocrine effects. Following central administration, it produces hypothermia, analgesia, reduced spontaneous locomotor
activity, muscle relaxation, prolongation of barbiturate and ethanol narcosis
(in the former case in part because of effects on drug metabolism), naloxone-
insensitive analgesia, increased turnover of brain monoamines and acetylcholine, excitatory or inhibitory effects on firing of certain neurons, etc. Many of
its central effects are opposite to those of thyrotropin-releasing hormone (TRH)
and/or appear to reflect possible antagonism of certain dopaminergic systems.[49–51] Neurotensin has no apparent central effects on peripheral administration, indicating that it poorly penetrates the blood–brain barrier and has
no relevant receptors on circumventricular organs. Its heterogeneous distribution in the CNS is discussed below with that of its receptors. With its apparent
localization to neurons and nerve terminals[52,53] and Ca^{2+}-sensitive release from
depolarized brain slices,[54] NT fulfills many of the criteria for being a neurotransmitter or neuromodulator.

Three reports of apparent NT receptor binding in the brain appeared within
a few months of each other and within 2 years of NT's sequence determination.[55–57] Although these studies differed considerably in methodology, all three
found dissociation constants (K_ds) in the low nanomolar range (2–8 nM) for
binding of [^{3}H]NT or [^{125}I]NT, and all three demonstrated impressive pharmacological specificity. Unfortunately, for overlapping NT analogues (e.g.,
$NT_{8–13}$), there were some major discrepancies in described relative potencies.
Similarly, between the two papers describing the regional distribution of binding,[56,57] there were again a few surprising discrepancies. Both agreed that the
rat thalamus, hypothalamus, midbrain, and cerebral cortex had higher binding
than the medulla–pons and cerebellum, in general accord with NT's regional
distribution in rat brain.[58,59] A more detailed comparison of binding with NT
levels was possible in calf brain, for which the same group reported binding
data and levels.[57,60] There was a clear excess of levels over binding in hypothalamus and basal ganglia and a relative excess of binding over levels in cerebral cortical regions. Many other regions showed generally parallel values for
the two types of measurement. Low values of NT binding and levels in the
cerebellum parallel findings for many other peptides. In spite of the discrepancies mentioned above and others, it appears that all three groups were looking
at NT receptors.

This is less clear for reports of [^{125}I]NT binding to rat mast cells,[61,62] for
which a low binding affinity (K_d = 154 nM) and a high potency of bradykinin
as a competing agent are potential problems. However, binding sites for [^{3}H]NT
in longitudinal muscle of guinea pig ileum[63] and to a cell line (HT 29) derived
from a human colon carcinoma[64] appear remarkably similar to those in brain.[55]
A detailed pharmacological comparison between NT analogues' abilities to
compete for [^{3}H]NT binding to HT 29 cells and rat brain and their abilities to
stimulate contraction of the longitudinal muscle of the guinea pig ileum yielded
excellent correlations[64] The latter *in vitro* response is thought to reflect release
of acetylcholine.[65] These detailed pharmacological data are probably the best
evidence to date for NT receptor identification. Note that several NT analogues, including [D-Tyr^{11}]NT,[D-Phe^{11}]NT, and [D-Leu^{11}]NT, appear relatively

much more potent *in vivo* than they did in the various *in vitro* tests. This is thought to reflect their resistance to inactivation and emphasizes the hazards of attempting to correlate any kind of *in vivo* pharmacology with binding pharmacology for peptide receptors. Similar modifications of the Tyr-11 residue have been reported to produce NT antagonists for certain peripheral responses.[66]

Neurotensin receptors are one of the few types of peptide receptors that have been visualized to date by light microscopic autoradiographic techniques following ligand binding to slide-mounted sections.[67] Besides a detailed description of NT receptors, which generally but not completely paralleled the distribution of NT-like immunoreactivity localized by immunohistochemistry,[52,53]these methods have already yielded useful results on the cell types involved in the binding. Thus, local 6-hydroxydopamine lesions of the zona compacta of the substantia nigra greatly reduced NT receptor binding there, suggesting a receptor localization to dopaminergic cell bodies,[68] whereas dorsal root section failed to decrease NT receptor binding in the dorsal spinal cord, although opioid receptor binding was decreased 40% in layers I and II.[69] The latter finding suggests that even though both NT and enkephalins are distributed similarly in the dorsal spinal cord and both produce analgesic responses,[70] the sites of the relevant receptors, and, consequently, the mechanisms of the responses, are quite different. The combination of *in vitro* receptor autoradiography with specific lesions promises to be a powerful technique in unraveling sites and mechanisms of peptide responses in the brain.

4. BOMBESIN

Bombesin (BN, pGlu-Gln-Arg-Leu-Gly-Asn-Gln-Trp-Ala-Val-Gly-His-Leu-Met-NH_2) is a tetradecapeptide isolated from the skin of frogs.[71] Bombesin itself is evidently not found in mammalian brain. Rather, there is a larger related peptide that cross reacts with antisera to BN[72,73] and shares its biological effects. Bombesinlike immunoreactivity is heterogeneously distributed in the brain, with highest levels in the hypothalamus and lowest in the cerebellum. It is also found in the gut and lungs. Bombesin has a variety of central actions including somatostatin-reversible hypothermia and hyperglycemia,[74] increased locomotor activity, and naloxone-insensitive analgesia.[75]

Apparent receptor binding of [^{125}I-Tyr^4]BN has been described in brain[75,76] and pancreas.[77] Although all these studies had common authors, methodology varied somewhat, with incubations of 5–24 min at 25°C or 37°C in the presence of bacitracin and separation of bound radioactivity by filtration (Whatman GF/B filters presoaked in 1% bovine serum albumin, initial study) or by brief centrifugation. Results in brain and pancreas were similar, with apparent K_ds of 2–4 nM in both tissues and a similar order of potencies for at least three related peptides ([Tyr^4]BN > BN > litorin). In brain, these and a variety of other BN-related peptides were shown to compete for [^{125}I-Tyr^4]BN binding with relative potencies generally resembling their potencies for inducing hypothermia on intracisternal injection. In the pancreas, the three listed peptides were shown to compete for binding with absolute potencies about 10% of those observed in stimulating $^{45}Ca^{2+}$ efflux, amylase release, and cyclic GMP accumulation, leading the authors to suggest that 25% BN receptor occupation is sufficient to elicit a maximal biological response. The distribution of binding among gross regions of rat brain[76] did not closely parallel the earlier-reported distribution of BN-like immunoreactivity,[73] with the usual relative excess of peptide levels in the hypothalamus and a relative excess of binding in the hippocampus. These discrepancies probably do not bear on receptor identification, as discussed earlier. A more recent detailed study of the distribution of [^{125}I-Tyr^4]BN binding in rat brain yielded highest binding in amygdala, hypothalamus, frontal pole, hippocampus, and the mesencephalic periaqueductal gray.[75] Overall, the available evidence, particularly the pharmacological data on related peptides, strongly suggests that the three cited studies were looking at receptors for a BN-like peptide.

5. CHOLECYSTOKININ

Cholecystokinin (CCK, Lys-Ala-Pro-Ser-Gly-Arg-Val-Ser-Met-Ile-Lys-Asn-Leu-Gln-Ser-Leu-Asp-Pro-Ser-His-Arg-Ile-Ser-Asp-Arg-Asp-Tyr-Met-Gly-Trp-Met-Asp-Phe-NH_2), a 33-amino-acid peptide originally isolated from porcine intestine,[78] has been localized in the periphery and brain by radioimmunoassays[79] and immunohistochemistry.[80,81] Numerous molecular forms of CCK (e.g., CCK_{39}, CCK_{12}, CCK_8, CCK_4) have been detected and separated by gel chromatography following reaction with specific antibodies.[82] Demonstrated CNS properties of CCK strongly suggest a possible neurotransmitter function for this peptide.[83] Among the CNS effects of CCK are its ability to suppress appetite on central (or peripheral) administration, its regulation of pain perception, and modulation of oxytocin and/or vasopressin release from the posterior pituitary gland.[83]

The first few biochemical studies of CCK receptors involved binding of [^{125}I-Bolton–Hunter]CCK$_{33}$ ([^{125}I-BH]CCK) to pancreatic acinar cells.[84,85] Saturation isotherms of binding were compatible with labeling of two sites in rat pancreas (K_{d1} = 64 pM, and K_{d2} = 21 nM), whereas in the guinea pig only a single class (K_d = 0.5 nM) was found,[85] from which the label dissociated in a biphasic manner.

In contrast, a homogeneous population of high-affinity ($K_d \approx 0.3$–1.7 nM) receptor binding sites for [^{125}I-BH]CCK has been detected in brains of rats,[86,87] guinea pigs,[85] and mice.[88] In the latter study, binding was extremely pH sensitive, being optimum at pH 6.5 (K_d = 0.44 nM), with both affinity and capacity reduced at higher pH (7.4). It also displayed an absolute requirement for Mg^{2+} and EGTA. The regional distribution of specific [^{125}I-BH]CCK binding in rat and mouse brain[86,88] (cortex > olfactory bulbs > caudate) correlated well with that reported for CCK-like immunoreactivity in rat.[89] The involvement of receptors in binding to the pancreas[84] and rat cortical membranes[86] was supported by the pharmacology. Thus, CCK_8 (sulfated) was three times more active than CCK_{33} (CCK) at displacing the binding, whereas the desulfated CCK_8 and

CCK_4 possessed a quarter of the potency of CCK_{33}. These findings are in accord with the fact that the biological activity of the hormone resides in the carboxyl terminal octapeptide (CCK_8). Furthermore, the relative potency ratios for displacing pancreatic and brain [^{125}I-BH]CCK binding compare well with those for stimulation of cyclic GMP levels, $^{45}Ca^{2+}$ release, and other functional responses (cited in ref. 84). However, the absolute potencies of CCK analogues for inducing amylase secretion were tenfold greater than those for inhibiting pancreatic binding.

Both pancreatic and cerebral [^{125}I-BH]CCK binding were inhibited by monovalent cations and guanine nucleotides. Divalent cations at low concentrations were stimulatory in both tissues. However, although dibutyrl cyclic GMP was a potent competitive inhibitor of pancreatic binding and amylase release,[85,90] this nucleotide was only a weak inhibitor of brain [^{125}I-BH]CCK binding,[88] suggesting possible dissimilarity of peripheral and central receptors (or a species difference).

An elevated cortical CCK receptor density in genetically obese rodents has been reported.[91,92] A possible mechanism of CCK-induced satiety is suggested by the observations that abdominal and gastric vagotomy abolish it in rodents,[83] with a concomitant decrease in fast axoplasmic transport of CCK receptors to the periphery.[93]

Additional observations include the age-related development of CCK receptors[94] and their reduction following intrastriatal kainate injections.[87] A similar loss of [^{125}I-BH]CCK binding sites in Huntington's disease[95] has suggested CCK-regulated extrapyramidal functions.

6. SUBSTANCE P

Although its existence has long been known,[96] the sequence of the undecapeptide, substance P (SP, Arg-Pro-Lys-Pro-Gln-Gln-Phe-Phe-Gly-Leu-Met-NH₂), was only determined relatively recently.[97] Substance P displays a heterogeneous regional distribution in the brain and enrichment within synaptosomal fractions, from which it can be released[98] and subsequently inactivated by "specific" peptidases.[99] In electrophysiological experiments, SP has potent, slowly initiated, but persistent actions, generally excitatory, on mammalian neurons. Its modulatory actions are best exemplified by augmentation of dopamine release in the mesolimbic regions to enhance locomotor activity, although the apparent mediation of nociception in the spinal cord is the result of direct actions on dorsal horn neurons. Other behavioral responses involving SP include an antidipsogenic action in rats. These properties, discussed in more detail elsewhere,[100-102] and the demonstration of apparent receptor binding, described below, suggest a transmitter role for SP.

Using paradoxical incubation conditions (0°C for 1 min), Nakata *et al.*[103] first reported high-affinity (K_d = 2.7 nM) [^{3}H]SP binding sites in the rabbit CNS. However, a more detailed study in rat,[104] also employing previously frozen crude membranes, demonstrated a relatively slow [^{3}H]SP–receptor association ($k_1 = 5 \ \mu M^{-1} s^{-1}$), requiring incubation at 4°C for 20 min to approach equilibrium. The observed K_d (0.38 nM) differed markedly from the previous study.[103] However, the regional distribution and the pharmacological specificity of the binding sites in the two studies agreed remarkably well. Thus, the highest density of [³H]SP binding sites was in the hypothalamus and other midbrain regions, whereas the cerebellum and cortex possessed fewer sites. This profile of receptor density resembled the distribution of SP-like immunoreactivity.[105] Furthermore, [³H]SP binding to rabbit[103] and rat[104] brain membranes was inhibited by C-terminal fragments of SP in a rank order of potency consistent with results of iontophoretic application in the CNS[106] and by other bioassays.[107] Thus, the hexapeptide (SP$_{6-11}$) had twice the potency of native SP and was some 3–5 times more active than the heptapeptide (SP$_{5-11}$) and decapeptide (SP$_{2-11}$), respectively, at displacing [³H]SP binding.

Structurally related SP-like peptides (tachykinins), physalaemin (PSM) and eledoisin, also possessed substantial inhibitory activity. Subsequently, a labeled derivative of PSM, [¹²⁵I]PSM, has been utilized to assess SP-type binding sites on dispersed pancreatic and parotid acinar cells.[108,109] A very low number (200/cell) of [¹²⁵I]PSM sites were detected, whose pharmacological selectivity bore little resemblance to [³H]SP binding specificity in the CNS. Studies of this type, and the differential pharmacological efficacy of SP fragments and tachykinins at causing tissue contractions[107] and at competing for CNS [³H]SP binding,[104] have suggested that there exist distinct peripheral and central receptors for SP.

Another approach to label "SP receptors" has been to use [¹²⁵I]SP. However, although this radiopeptide exhibited high-affinity binding ($K_d = 0.32$ nM) to rat synaptic vesicle fractions, they showed little affinity for SP analogues, binding was inhibited by Ca^{2+} and Mg^{2+} ions and trypsin, and PSM paradoxically enhanced [¹²⁵I]SP binding.[110] The fact that binding was mostly to lipids rather than to protein moieties[110] was confirmed by organic extractions and has been supported somewhat by the finding that [³H]SP binding to similar preparations[111] was resistant to proteolysis but was reduced by delipidation agents and lipolytic enzymes. The results of these latter studies are difficult to interpret and may reflect interaction of the radioligand with a proteolipid associated with SP storage[110–112] rather than to binding sites coupled to effector mechanisms. No attempts were made to correlate binding properties with any responses.

The lack of specific antagonists has hampered progress in this area in the past, and, therefore, the recent synthesis of a purported competitive SP blocker, [D-Pro²,D-Phe⁷,D-Trp⁹]SP,[113,114] provide a new impetus for further characterization of both peripheral[107] and central SP receptors.[104] Unfortunately, the excitement generated by this antagonist of SP-induced guinea pig ileum contractions[113] and the SP-evoked excitation of locus coeruleus neurons[115] has been rather short-lived because of its possible neurotoxic effects.[116] Evidently, the discovery of better SP-related analgesics must await further research.

7. VASOACTIVE INTESTINAL PEPTIDE

Vasoactive intestinal peptide (VIP, porcine = His-Ser-Asp-Ala-Val-Phe-Thr-Asp-Asn-Tyr-Thr-Arg-Leu-Arg-Lys-Glu-Met-Ala-Val-Lys-Lys-Tyr-Leu-

Asn-Ser-Ile-Leu-Asn-NH$_2$) is a basic octosapeptide isolated from hog intestine by Said and Mutt.[117] It is related in structure to secretin, glucagon, and other gut peptides. In the gut and elsewhere in the periphery, it seems to occur primarily in neurons and is also widely distributed in neurons of the CNS.[118] Studies of its distribution have been complicated by the existence of VIP molecular variants and cross reactions of some antisera with other secretinlike peptides. In the periphery, VIP neurons occur in sensory ganglia, sympathetic and parasympathetic autonomic ganglia (where it may coexist with acetylcholine in neurons innervating exocrine glands[119]), and the submucous (Meissner's) plexus of the gut wall. Besides exocrine glands and other neurons, structures prominently innervated by VIP nerves include blood vessels in a variety of locations, including the brain, and smooth muscle, e.g., of gastrointestinal sphincters. In the CNS, VIP-like immunoreactivity is high in cerebral cortical areas, limbic areas including hippocampus and amygdala, suprachiasmatic nucleus and elsewhere in the anterior hypothalamus, and amacrine cells of the retina. The VIP terminals are rather sparsely distributed in the brainstem and spinal cord compared to many other peptides.

Vasoactive intestinal peptide has a variety of peripheral effects, including vasodilatation of most vascular beds, relaxation of smooth muscle in many other sites, and stimulation of secretion from many exocrine and some endocrine glands. The CNS actions of VIP include depolarization and excitation of neurons in several regions, possible regulation of pituitary prolactin secretion as a hypothalamic hormone, and regulation of the release of other hypothalamic releasing hormones. Little in the way of behavioral pharmacology of VIP has been described, perhaps because of its expense. Vasoactive intestinal peptide has fulfilled many of the criteria for being a neurotransmitter or neuromodulator in the CNS and peripheral nervous system.[120-122]

Apparent receptor binding for VIP has been described in a variety of peripheral tissues, including pituitary cells,[123] pancreas,[124] liver,[125] intestinal epithelium,[126,127] adrenal cells,[128] uterus,[129] and fat cells.[130] The VIP receptors may be distinguished from receptors for secretin, to which [^{125}I]VIP may also bind, by their higher affinity for VIP and for [Val^5]secretin, compared to secretin and secretin$_{7-17}$.[131] Some tissues, e.g., pancreatic acinar cells,[124] appear to possess both types of receptors. Even though receptors in many of these tissues are likely responding to VIP released from nerves, space does not permit detailed review of the cited studies. Results of most were generally similar to each other and to studies in the CNS described below. In most tissues, binding results correlated well with observations on the stimulation of cyclic AMP formation by VIP and its analogues, even though the two measurements were made under dissimilar conditions. High-affinity K_ds were typically near 1 nM for VIP, with secretin being at least 100-fold less potent. There appeared to be receptor heterogeneity within some tissues and between some tissues and species. Methodological problems included relatively rapid degradation of [^{125}I]VIP and VIP receptors at higher temperatures.

Three groups have studied binding of [^{125}I]VIP to various CNS preparations.[132-135] These studies revealed a strong apparent species difference in binding affinity between rat (K_d ca. 1 nM)[134,135] and guinea pig (K_d ca. 36 nM).[132] The latter species also appeared to have a second, even lower-affinity site (K_d

ca. 285 nM). Only one group reported such a second site in rat brain (K_d *ca.* 125 nM).[135] The results in the guinea pig must be interpreted with caution because of evidence that guinea pig VIP may differ in structure from VIP of many other mammals[136] and because the authors' curved Scatchard plot[132] appears to have been analyzed inappropriately. The K_d reported in the rat brain,[134,135] which varied from 1 to 6 nM depending on the method used to measure it, was very similar to that reported in many peripheral tissues. As far as could be told in the absence of a side-by-side comparison, the pharmacology of binding in both rat brain and guinea pig brain[133] was also generally similar to that in other tissues, with secretin about 100 times less potent than VIP.

As is the case in the periphery, VIP stimulates adenylate cyclase activity in the brain.[137–140] In either slices or membrane particulate preparations, and in both rat and guinea pig, the concentration of VIP needed to observe this effect is higher by one or two orders of magnitude than that needed to observe inhibition of [^{125}I]VIP binding (under different incubation conditions). Many aspects of the pharmacology of the two types of measurement are similar, however. Brain VIP-sensitive cyclase appears to differ from that in many peripheral regions in being relatively insensitive to stimulation by secretin and to potentiation by GTP. The regional distributions of cyclase stimulation by VIP in rat brain found in three studies[138–140] differed markedly from each other and from that expected on the basis of the distribution of VIP-like immunoreactivity[141–144] and [^{125}I]VIP binding.[134] In particular, all three studies reported appreciable cyclase stimulation in the cerebellum, a region relatively devoid of VIP-like immunoreactivity and receptor binding. Overall, the relationship between cyclase stimulation by VIP and CNS receptors for VIP remains unclear.

The regional distribution of [^{125}I]VIP binding in rat brain[134] was in reasonable agreement with the distribution of VIP-like immunoreactivity.[141–144] Highest binding was in the striatum, hippocampus, cerebral cortex, and thalamus. The major discrepancy was in the hypothalamus, which was fairly low in binding but possessed considerable immunoreactive VIP. As discussed previously for other peptides, this discrepancy may be resolved by assuming that much hypothalamic VIP is destined for export.

Methodological features of the VIP binding experiments were generally similar to those for other peptides. The [^{125}I]VIP was prepared by the chloramine T method and appeared to retain full activity. With addition of peptidase inhibitors bacitracin and aprotinin, incubations could be run at 20°C or 37°C for 10 or 20 min. Separation of bound radioactivity was by filtration (0.45-μm cellulose acetate Millipore® filters) or centrifugation through buffered 0.32 M sucrose.

An intriguing effect of VIP in the cat submandibular salivary gland is the recently reported enhancement of agonist binding to muscarinic receptors.[145] The concentration dependence of this effect (half-maximal near 1 nM VIP) is appropriate for mediation via typical VIP receptors. It will be interesting to see whether this effect is mediated via cyclic AMP or other mechanisms. The possible involvement of additional mechanisms of action in the pituitary gland is suggested by the apparent presence of intact VIP inside prolactin-secreting cells[146] a finding that parallels observations for other releasing hormone candidates.[147-150]

8. THYROTROPIN-RELEASING HORMONE

Thyrotropin-releasing hormone (TRH, thyroliberin, pGlu-His-Pro-NH$_2$) is a tripeptide that was identified and named on the basis of its ability to stimulate the release of thyrotropin (TSH) from the anterior pituitary gland.[151,152] There it also stimulates the release of prolactin. Thyrotropin-releasing hormone was later found to be widely distributed outside the hypothalamus and to have a variety of central effects apparently unrelated to its endocrine role(s).[153-156] This and other evidence have suggested an additional neurotransmitter or neuromodulator role for TRH in the CNS.

The discovery of TRH was followed shortly by the demonstration that [^{3}H]TRH binds to apparent receptors in pituitary plasma membranes[157] or pituitary-derived cell lines.[158-160] The strongest evidence that these binding sites indeed represented TRH receptors was the close correlation between the potencies of a variety of TRH analogues in competing for binding and their potencies in stimulating release of thyrotropin[161] and prolactin.[162] This evidence, and the whole subject of TRH receptors, is reviewed in greater detail elsewhere.[163,164]

The initial demonstration of the presence of pituitarylike high-affinity binding sites for [^{3}H]TRH in rat brain by Burt and Snyder[165] was hampered by very high blank values, in part because of a large excess of lower-affinity but saturable sites (K_d ca. 5 μM). Better characterization of the apparent CNS receptors for TRH was achieved in later work in sheep retina[166] and nucleus accumbens.[167] These tissues proved to be relatively enriched in receptors, raising the proportion of specific binding from the 15 or 20% seen in rat brain in earlier experiments to 50% or more. The high-affinity binding sites (K_d ca. 20–40 nM) were found to closely resemble sheep pituitary receptors in affinity and pharmacology for TRH analogues. This resemblance was the major evidence for identifying the CNS binding sites as TRH receptors. Discrepancies between behavioral and endocrine potencies of certain TRH analogues[168-171] were attributed to differences in their ability to reach CNS and pituitary receptors or to the possible existence of additional, undetected classes of CNS TRH receptors. The presence of these sites in retina was compatible with evidence for the existence of high levels of immunoreactive TRH in rat retina,[172-175] although this is not without controversy.[176] Their presence in the nucleus accumbens was consistent with immunohistochemical evidence for high levels of TRH there in rats[177] and with behavioral evidence for dopamine-mediated stimulation of locomotor activity by TRH in this region.[178,179] However, later work (see below) has suggested that the amygdala is in fact the brain region highest in TRH receptors in most mammals.

In spite of the relative success in sheep retina and nucleus accumbens, the use of [^{3}H]TRH as ligand was severely limiting in CNS regions because of high blanks. Most modifications in all three amino acid residues reduce receptor affinity, but the analogue with a methyl group on the 3-nitrogen of the histidine ring ([3-Me-His^2]TRH, MeTRH) has long been known to be more potent than TRH in the pituitary gland,[180] and this enhanced potency was later shown to extend to the CNS as well.[165,181,182] Taylor and Burt[183,184] prepared this analogue in radioactive form and showed that it binds to the same sites in the pituitary gland and CNS as [^{3}H]TRH, only with approximately eightfold higher affinity (K_d *ca.* 3 nM), giving lower blanks. This improved ligand ([^{3}H]MeTRH) has recently become commercially available (New England Nuclear, Boston, MA).

Use of [^{3}H]MeTRH made practical for the first time the screening of a large number of brain regions in a variety of species.[185,186] These studies were undertaken when the distribution of TRH receptor binding in the rat CNS was found to differ from that in sheep. Binding in the amygdala was higher than in the nucleus accumbens, a finding that extended to most other tested species, and binding in the retina was highest of all, a situation apparently unique to the rat. Retinal TRH receptors exhibited remarkable species variation, with density in the rat about 100 times that in the dog. Lesser but still extensive species differences were detected in other regions, with the amygdala and hypothalamus being particularly high in the guinea pig, the spinal cord and septal area high in the rabbit, the anterior and posterior pituitary gland high in the sheep, etc. In the rat, the distribution of receptor binding appeared to be in reasonable agreement with that reported earlier for TRH-like immunoreactivity with some exceptions: a typical excess of levels over binding in the hypothalamus, a similar excess in the olfactory bulb, and an excess of binding over levels in the amygdala. The density of TRH receptor binding sites in the three highest regions tested (0.2–0.4 pmol/mg crude membrane protein or *ca.* 20–30 pmol/g wet weight in sheep pituitary, rat retina, and guinea pig amygdala) was still only about 10–20% of that reported in pituitary cell lines,[187] which typically have about 100,000 sites per cell.[160,188]

The species differences in distribution of TRH receptors do not appear to be accompanied by any major differences in other binding properties of these receptors. The high-affinity binding sites identified as receptors look basically the same even in birds[189] and fish.[189a] No binding identifiable as representing TRH receptors has been reported outside the pituitary or CNS.

The methodology of studies of TRH receptor binding has varied considerably, ranging from use of intact pituitary-derived cells in physiological media at 37°C[159,160,188] to homogenates or membrane fractions in hypotonic buffer at 0°C.[157,160,165] The latter conditions appear to be the most favorable in the CNS, not only because they increase the apparent number of binding sites and affinity in broken-cell preparations.[187] These effects remain surprisingly prominent even for temperatures quite near 0°C.[190] At 0°C, the attainment of equilibrium with [^{3}H]MeTRH may take several hours. Mono- and divalent cations inhibit binding; and, at least in the CNS, receptor binding is optimal in 20 mM sodium phosphate or 50 mM HEPES (pH 7.4), whereas Tris citrate and Krebs bicarbonate buffers are clearly less favorable (N. A. Sharif and D. R. Burt, unpublished data). Separation of bound radioactivity by filtration has usually employed Whatman GF/B glass fiber filters and high tissue concentrations (equivalent to 40–50 mg wet weight/ml) to minimize the contribution to blanks of binding to the filter. Methodological features of TRH receptor binding assays are discussed in much greater detail elsewhere.[164]

An unresolved question in binding studies to date is that of heterogeneity of TRH receptors. Two early binding studies using [³H]TRH in the pituitary reported two classes of high-affinity site,[159,191] but this has not been confirmed by most others for [³H]TRH[157,160,166,187,188,192] or [³H]MeTRH.[183] The situation is similar in the CNS, with a single, very limited two-site report for [³H]TRH,[193] which contradicts more extensive reports of linear (one-site) Scatchard plots of slopes in the nanomolar range for both ligands.[166,167,184,190] There is functional evidence for possible receptor heterogeneity in the pituitary.[194] Several considerations make receptor heterogeneity seem likely in the CNS (e.g., both excitatory and inhibitory electrophysiological responses, probable presynaptic and postsynaptic localization, variety of behavioral responses, some with distinct pharmacology). Recent studies with sulfhydryl reagents and heavy metal cations have indicated the involvement of reactive thiol residues in [³H]MeTRH binding and provided further indirect evidence for the resemblance between TRH receptors in rat pituitary and CNS (N. A. Sharif and D. R. Burt, unpublished data). Other unresolved issues in the CNS include receptor structure, regulation, response mechanisms, and detailed localizations and roles.

9. SOMATOSTATIN

Somatostatin (SS, Ala-Gly-Cys-Lys-Asn-Phe-Phe-Trp-Lys-Thr-Phe-Thr-Ser-Cys-OH) is a cyclic tetradecapeptide isolated from sheep hypothalamus on the basis of its ability to inhibit the release of pituitary growth hormone.[195] Recently, a larger related peptide elongated at the N-terminus, somatostatin 28 (SS-28, Ser-Ala-Asn-Ser-Asn-Pro-Ala-Met-Ala-Pro-Arg-Glu-Arg-Lys-SS), has been identified in extracts of gut[196] and brain.[197,198] Somatostatin 28 appears to be more potent than SS in some test systems[199,200] but represents only a small portion of total SS-like activity in brain,[201] which includes at least one component that is larger still. In the periphery, SS inhibits the release of an astonishing variety of hormones besides growth hormone, including insulin, glucagon, gastrin, secretin, VIP, CCK, motilin, pepsin, parathyroid hormone, renin, aldosterone, calcitonin, and thyrotropin.[202] Considerable evidence has suggested a possible neurotransmitter role for SS besides its multiple endocrine roles.[203] A variety of central effects of SS have been described.[204,205]

The initial demonstration of binding of [¹²⁵I-Tyr¹]SS to apparent SS receptors was performed in GH_4C_1 clonal pituitary tumor cells,[206] although there had been earlier mention of an observation of a degree of saturable binding in the brain.[56,207] The K_d in intact pituitary tumor cells at 37°C was 0.6 nM. The evidence for receptor identification included the match between the concentration dependence of binding and of biological response (growth hormone inhibition) for SS and between the presence of a response and the presence of binding sites in three of five related clones. Success appeared to have largely depended on the presence of a high concentration of receptors on a homogeneous population of cells. A later study in bovine anterior pituitary membranes[208] observed sites of such low apparent affinity (30 nM and 8 μM) that it is not clear that they were receptors. This study and many other early attempts to look at SS receptors were hampered by very high blanks and ligand breakdown.

In the last 2 years, several groups appear to have successfully identified receptors for SS and/or SS-28 in brain,[209-213] anterior pituitary gland,[214-217] pancreatic tumors (insulinomas),[218] and adrenal cortex.[219] Besides [^{125}I-Tyr[1]]SS, ligands used in these studies included [^{125}I-Tyr[11]]SS, [Leu[8], D-Trp[22], ^{125}I-Tyr[25]]SS-28, [^{125}I-N-Tyr,D-Trp[8]]SS, and, most recently, [des-Ala[1], Gly[2]-desamino-Cys[3]-^{125}I-Tyr[11]-dicarba[3,14]]SS.[220] The reasons for recent success, where earlier attempts to study SS receptors in heterogeneous preparations had largely failed, appear to include the fact that, in some tissues at least, the newer ligands are of higher affinity and/or are more stable than [^{125}I-Tyr[1]]SS, the use in some studies[210] of extensive preliminary subcellular fractionation to obtain enriched synaptic or plasma membrane preparations, and presumably, the use of more favorable incubation conditions or separation techniques. The extent of specific binding has ranged from only 15–30% in crude membrane preparations of rat cerebral cortex using [^{125}I-Tyr[1]]SS[213] to over 70% in a similar membrane preparation using [Leu[8],D-Trp[22],[125]I-Tyr[25]]SS-28.[212] All of these recent studies using different tissues and ligands have reported similar high binding affinities, with K_ds ranging from less than 0.1 nM for [^{125}I-Tyr[11]]SS in rat brain[210] to 2.4 nM for [^{125}I-Tyr[11]]SS in rat pituitary.[216].

A major point of interest in many of these studies was whether there exist pharmacologically distinct types of SS receptors in different tissues. Somatostatin analogues have potential application in diabetes through inhibition of glucagon release, and some progress had already been made in developing selective SS analogues.[221,222] Most available data from binding studies are consistent with the idea of several receptor types for SS,[209-211] although detailed results have not been fully consistent among groups, and not all agree on receptor multiplicity.[213] The most potent SS analogue on brain receptors reported thus far is [D-5-F-Trp[8]]SS, which is about 16[218] or 32[209] times as potent as SS in inhibiting binding. Although suitable brain response measurements are not yet available, it is worth noting that in all other tissues, excellent matches have been reported between receptor binding affinities and response potencies of SS analogues.

Evidence for receptor identification of binding sites for SS analogues in brain is still weak in the absence of response data but includes a general resemblance (aside from some pharmacological differences) to sites in other tissues where receptor identification is stronger and a regional distribution consistent with that of SS-like immunoreactivity. Highest binding in rat was observed in hippocampus, amygdala, and olfactory tubercle, and lowest in cerebellum.[212] There was the usual discrepancy with SS levels for the hypothalamus, which is highest in SS levels[223] but only modest in binding.

Technical features of these studies have resembled those for other peptides. All incubations were conducted at elevated temperatures (20–37°C), usually in the presence of various peptidase inhibitors. Bound radioactivity was separated by centrifugation or filtration (Whatman GH/C filters presoaked in bovine serum albumin). In the pituitary, one study[216] reported reduced binding in frozen tissues, although frozen membranes prepared from fresh tissue seemed comparable to fresh tissue. Some studies found the iodinated ligands stored frozen at $-20°C$ to be stable for up to a month.

Response mechanisms for SS are poorly understood in all tissues, so that this will be an important area for future investigation. The ability of SS to inhibit secretion of such a wide variety of substances suggests a mechanism involving calcium, but evidence is limited.[163] Numbers of SS receptors on pituitary cells appear to be modulated by TRH,[224] but other aspects of SS receptor regulation await discovery.

10. MORE PEPTIDES

This section considers briefly a number of peptides for which relatively little information about CNS receptors yet exists.

10.1. Carnosine

A neurotransmitter role for carnosine (β-Ala-His) in the mammalian CNS, especially in the olfactory bulb, has been suggested by substantial neurochemical evidence,[225] including the presence of high concentrations of carnosine and its metabolizing enzymes; the reduction of both following olfactory bulb deafferentation; and the axonal transport and synaptosomal release of the dipeptide.[226] However, recent iontophoretic investigations have provided conflicting conclusions about the neuroactivity of carnosine.[227,228]

The first binding study for this peptide[229] described a low-affinity ($K_d = 0.77$ μM), saturable [³H]carnosine interaction with mouse olfactory bulb membranes. Binding was sensitive to pH and ions. At best, the stereospecific binding component represented 30% of the total. Although [³H]carnosine binding exhibited some features suggestive of receptor interaction,[229,230] its low affinity and anamolous pharmacology[229] presented unresolved problems.

10.2. Bradykinin

The potent vasodilator peptide bradykinin (Arg-Pro-Pro-Gly-Phe-Ser-Pro-Phe-Arg) possesses central as well as peripheral actions.[231,232] To date, however, receptor binding for this nonapeptide has been reported only in peripheral tissues. With [¹²⁵I-Tyr¹]kallidin, a bradykinin analogue, dual high-affinity (K_ds 0.1 and 20 nM) binding sites in bovine myometrium particulate fractions have been detected,[233] but other peripheral organs possessed negligible specific binding. Unfortunately, brain regions were not tested.

[¹²⁵I-Tyr¹]Kallidin binding was inhibited by cations and was displaced most efficaciously by structural analogues in a rank order of potency similar to their oxytocic, physiological activities; thus, [^{127}I-Tyr^1]kallidin > bradykinin > [^{127}I-Tyr^8]bradykinin > [^{127}I-Tyr^5]bradykinin.

The affinity, pharmacology, and other binding parameters suggest that kinin receptors were being studied in the periphery. The recent demonstration of bradykininlike immunofluorescence in central areas[234] and the central hypertensive properties of bradykinin[232] suggest that similar sites await demonstration in the CNS.

10.3. Vasopressin

Vasopressin (VP, antidiuretic hormone, Cys-Tyr-Ile-Gln-Asn-Cys-Pro-Leu-Gly-NH$_2$) is well known as the neurosecretory hormone that controls the body's state of hydration.[235] Recently, some hypothalamic VP neurons have been shown to project to many extrahypothalamic loci,[236] and there is an extensive literature documenting VP's effects on learning behavior.[237,238] These findings and others predict the presence of CNS VP receptors, but these have yet to be demonstrated by binding measurements. Vasopressin receptors and their mechanisms have been well studied in the kidney.[239]

10.4. Oxytocin

Oxytocin (OT, Cys-Tyr-Phe-Gln-Asn-Cys-Pro-Arg-Gly-NH$_2$) is the other major posterior pituitary peptide of hypothalamic origin, with prominent actions on smooth muscle of the breast and uterus. Oxytocin neurons also project widely in the CNS,[240] and OT has effects on maternal behavior.[241] Although OT receptor binding has been studied in the periphery,[242] it has not yet been seen in the CNS.

10.5. Luteinizing Hormone-Releasing Hormone

Luteinizing hormone-releasing hormone (LHRH, luliberin, pGlu-His-Trp-Ser-Tyr-Gly-Leu-Arg-Pro-Gly-NH$_2$) is the second of the identified releasing hormones.[151,152] Although it does affect the firing of hypothalamic and other CNS neurons and has some behavioral effects related to its endocrine role,[207,243] LHRH has a fairly limited CNS distribution[207,243] and did not arouse great interest as a peptide neurotransmitter candidate until the recent demonstration that a similar peptide underlies late slow excitatory potentials in sympathetic ganglia.[244–246] Interestingly, this finding was anticipated histochemically by the observation of apparent LHRH receptors in the adrenal medulla.[247] The pharmacology of LHRH responses in ganglia resembles that of the anterior pituitary gland.[245] There have been extensive binding studies of LHRH receptors using a variety of ligands in the pituitary[163] and gonads,[248] but little success has been reported yet in nervous tissue.

10.6. Prolactin

Prolactin is one of several large peripheral peptide hormones of uncertain origin and function in the brain. Prolactinlike immunoreactivity appears localized to hypothalamic nerve terminals,[249] and prolactin has electrophysiological[250,251] and behavioral[252] activity in the CNS. Much of this activity may be related to feedback effects on dopamine neurons. Limited biochemical studies have demonstrated apparent receptor binding sites for prolactin in the hypothalamus.[253,254] There have been more extensive studies in peripheral tissues.[255]

10.7. Insulin

Insulin and its receptors are widely and heterogeneously distributed in the brain, as in most other tissues.[256] The region highest in receptor binding is the olfactory bulb. The insulin appears to be of local origin and to be localized to a specific population of neurons. The nature of the responses coupled to biochemically identified CNS insulin receptors is as yet unclear.

10.8. Still More Peptides

The list of neuropeptides whose receptors are just being demonstrated or await demonstration seems never ending. A few more in this category include secretin,[257,258] pancreatic polypeptides,[259] MIF-1 (Pro-Leu-Gly-NH$_2$),[260] and proctolin and a variety of other invertebrate neuropeptides.[261].

11. CONCLUSIONS

Receptor binding methodology has much to contribute to the study of neuropeptide function. The most obvious contribution is detailed, relatively unambiguous knowledge of pharmacology. Moreover, binding measurements provide a relatively simple biochemical marker of presumed peptide response that may often precede the discovery of the nature of the response(s). Radioreceptor assays can provide a less sensitive but more relevant alternative to radioimmunoassays for detecting the presence of related peptides. Changes in receptors may reveal important regulatory mechanisms or pathological processes. Careful comparison of binding sites in different tissues may permit information about peptide response mechanisms in nonnervous tissue (e.g., smooth muscle) to be extrapolated, at least tentatively, to neurons.

Study of most peptide receptors, or at least the correlation with response, has been hindered by the absence of suitable receptor antagonists. Exceptions include AII, NT, SP, and LHRH. In some cases,[262,263] use of specific peptide antisera may provide an alternative.

The preceding review has indicated that this area of research is still in its infancy. We hope that the chapter has conveyed some of its excitement and promise.

ACKNOWLEDGMENTS. During preparation of this review, the authors were supported in part by USPHS grant MH 29671, NSF grant BNS 8025469, and USAMRDC contract DAMD-17-81-C1279.

REFERENCES

1. Gainer, H. (ed.), 1977, *Peptides in Neurobiology*, Plenum Press, New York.
2. Emson, P. C., 1979, *Prog. Neurobiol.* **13:**61–76.
3. Hökfelt, T., Johansson, O., Ljungdahl, A., Lundberg, J. M., and Schultzberg, M., 1980, *Nature* **284:**515–521.
4. Snyder, S. H., 1980, *Science* **209:**976–983.
5. Krieger, D. T., and Martin, J. B., 1981, *N. Engl. J. Med.* **304:**876–885, 944–951.
6. Burt, D. R., 1980, *Neurotransmitter Receptors*, Part 1, *Receptors and Recognition, Series B*, Volume 9 (S. J. Enna and H. I. Yamamura, eds.), Chapman Hall, London, pp. 149–205.
7. Frederickson, R. C. A., 1980, *The Endocrine Functions of the Brain* (M. Motta, ed.), Raven Press, New York, pp. 233–270.
8. McKelvy, J. F., Charli, J.-L., Joseph-Bravo, P., Sherman, T., and Loudes, C., 1980, *The Endocrine Functions of the Brain* (M. Motta, ed.), Raven Press, New York, pp. 171–193.
9. Burt, D. R., 1978, *Neurotransmitter Receptor Binding* (H. I. Yamamura, S. J. Enna, and M. J. Kuhar, eds.), Raven Press, New York, pp. 41–55.
10. Hollenberg, M. D., and Cuatrecasas, P., 1979, *The Receptors*, Volume 1 (R. D. O'Brien, ed.), Plenum Press, New York, pp. 193–214.
11. Cuatrecasas, P., and Hollenberg, M. D., 1976, *Adv. Protein Chem.* **30:**251–451.
12. Catt, K. J., and Dufau, M. L., 1977, *Annu. Rev. Physiol.* **39:**529–557.
13. Baxter, J. D., and Funder, J. W., 1979, *N. Engl. J. Med.* **301:**1149–1161.
14. Gardner, J. D., 1979, *Gastroenterology* **76:**202–214.
15. Schulster, D., and Levitzki, A. (eds.), 1980, *Cellular Receptors for Hormones and Neurotransmitters*, John Wiley & Sons, New York.
16. Burt, D. R., Rossie, S. S., and Miller, R. J., 1980, *Receptor Binding Techniques, Syllabus, Society for Neuroscience Short Course, Cincinnati, Nov. 8–9, 1980*, Society for Neuroscience, Bethesda, pp. 150–167.
17. Witter, A., 1975, *Biochem. Pharmacol.* **24:**2025–2030.
18. Marks, N., 1978, *Frontiers in Neuroendocrinology* Volume 5 (W. F. Ganong and L. Martini, eds.), Raven Press, New York, pp. 329–377.
19. Ganten, P., and Speck, G., 1978, *Biochem. Pharmacol.* **27:**2379–2389.
20. Phillips, M. I., 1978, *Neuroendocrinology* **25:**354–377.
21. Simpson, J. B., 1981, *Neuroendocrinology* **32:**248–256.
22. Felix, D., 1982, *Trends Pharmacol. Sci.* **3:**208–210.
23. Reid, I. A., 1977, *Circ. Res.* **41:**147–153.
24. Ramsay, D. J., 1979, *Neuroscience* **4:**313–321.
25. Horvath, J. S., Baxter, G., Furby, F., and Tiller, D. J., 1977, *Prog. Brain Res.* **47:**161–165.
26. Meyer, D. K., Phillips, M. I., and Eiden, L., 1982, *J. Neurochem.* **38:**816–820.
27. Felix, D., and Schelling, P., 1982, *Trends Pharmacol. Sci.* **3:**230.
28. Ganten, P., Fuxe, K., Phillips, M. E., Mann, J. F. E., and Ganten, U., 1978, *Frontiers in Neuroendocrinology*, Volume 5 (W. F. Ganong and L. Martini, eds.), Raven Press, New York, pp. 61–99.
29. Quinlan, J. T., and Phillips, M. I., 1981, *Brain Res.* **205:**212–218.
30. Phillips, M. I., and Felix, D., 1976, *Brain Res.* **109:**531–540.
31. Devynck, M. A., and Meyer, P., 1978, *Biochem. Pharmacol* **27:**1–5.
32. Catt, K. J., and Aguilera, G., 1980, *Cellular Receptors for Hormones and Neurotransmitters* (D. Schulster and A. Levitzki, eds.), John Wiley & Sons, New York, pp. 233–251.
33. Peach, M. J., 1981, *Biochem. Pharmacol.* **30:**2745–2751.
34. Regoli, D., 1979, *Can. J. Physiol. Pharmacol.* **57:**129–139.
35. Bennett, J. P., Jr., and Snyder, S. H., 1976, *J. Biol. Chem.* **251:**7423–7430.
36. Sirett, N. E., McLean, A. S., Bray, J. J., and Hubbard, J. I., 1977, *Brain Res.* **122:**299–312.
37. Sirett, N. E., Thornton, S. N., and Habbard, J. I., 1979, *Brain Res.* **166:**139–148.
38. Cole, F. E., Frohlich, E. D., and Macphee, A. A., 1978, *Brain Res.* **154:**178–181.
39. Bennett, J. P., Jr., and Snyder, S. H., 1980, *Eur. J. Pharmacol.* **67:**1–10.
40. Bennett, J. P., Jr., and Snyder, S. H., 1980, *Eur. J. Pharmacol.* **67:**11–25.

41. Baxter, C. R., Horvath, J. S., Duggin, G. G., and Tiller, D. J., 1980, *Endocrinology* **106**:995–999.
42. Van Houten, M., Schiffrin, E. L., Mann, J. F. E., Posner, B. I., and Boucher, R., 1980, *Brain Res.* **186**:480–485.
43. Harding, J. W., Stone, L. P., and Wright, J. W., 1981, *Brain Res.* **205**:265–274.
44. Cole, F. E., Blakesley, H. L., Graci, K. A., Frohlich, E. D., and Macphee, A. A., 1981, *Peptides* **2**:441–444.
45. Stamler, J. F., Raizada, M. K., Phillips, M. I., and Fellows, R. E., 1978, *Physiologist* **21**:115.
46. Carraway, R., and Leeman, S. E., 1973, *J. Biol. Chem.* **248**:6854–6861.
47. Bissette, G., Manberg, P., Nemeroff, C. B., and Prange, A. J., Jr., 1978, *Life Sci.* **23**:2173–2182.
48. Fernstrom, M. H., Carraway, R. E., and Leeman, S. E., 1980, *Frontiers in Neuroendocrinology*, Volume 6 (L. Martini and W. F. Ganong, eds.), Raven Press, New York, pp. 103–127.
49. Nemeroff, C. B., Luttinger, D., and Prange, A. J., Jr., 1980, *Trends Neurosci.* **3**:212–215.
50. Nemeroff, C. B., 1980, *Biol. Psychiatry* **15**:283–302.
51. Haubrich, D. R., Martin, G. E., Pflueger, A. B., and Williams, M., 1982, *Brain Res.* **231**:216–221.
52. Uhl, G. R., Goodman, R. R., and Snyder, S. H., 1979, *Brain Res.* **167**:72–91.
53. Uhl, G. R., and Snyder, S. H., 1981, *Neurosecretion and Brain Peptides* (J. B. Martin, S. Reichlin, and K. L. Bick, eds.), Raven Press, New York, pp. 87–106.
54. Iversen, L. L., Iversen, S. D., Bloom, F. E., Douglas, C., Brown, M., and Vale, W., 1978, *Nature* **273**:161–163.
55. Kitabgi, P., Carraway, R., Van Rietschoten, J., Granier, C., Morgat, J. L., Menez, A., Leeman, S., and Freychet, P., 1977, *Proc. Natl. Acad. Sci. U.S.A.* **74**:1846–1850.
56. Lazarus, L. H., Brown, M. R., and Perrin, M. H., 1977, *Neuropharmacology* **16**:625–629.
57. Uhl, G. R., Bennett, J. P., Jr., and Snyder, S. H., 1977, *Brain Res.* **130**:299–313.
58. Carraway, R., and Leeman, S. E., 1976, *J. Biol. Chem.* **251**:7045–7052.
59. Kobayashi, K. M., Brown, M. R., and Vale, W., 1977, *Brain Res.* **126**:584–588.
60. Uhl, G. R., and Snyder, S. H., 1976, *Life Sci.* **19**:1827–1832.
61. Lazarus, L. H., Perrin, M. H., and Brown, M. R., 1977, *J. Biol. Chem.* **252**:7174–7179.
62. Lazarus, L. H., Perrin, M. H., Brown, M. R., and Rivier, J. E., 1977, *J. Biol. Chem.* **252**:7180–7183.
63. Kitabgi, P., and Freychet, P., 1979, *Eur. J. Pharmacol.* **55**:35–42.
64. Kitabgi, P., Poustis, C., Granier, C., Van Rietschoten, J., Rivier, J., Morgat, J.-L., and Freychet, P., 1980, *Mol. Pharmacol.* **18**:11–19.
65. Kitabgi, P., and Freychet, P., 1979, *Eur. J. Pharmacol.* **56**:403–406.
66. Rioux, F., Quirion, R., Regoli, D., Leblanc, M. A., and St. Pierre, S., 1980, *Eur. J. Pharmacol.* **66**:273–279.
67. Young, W. S. III, and Kuhar, M. J., 1981, *Brain Res.* **206**:273–285.
68. Palacios, J. M., and Kuhar, M. J., 1981, *Nature* **294**:587–589.
69. Ninkovic, M., Hunt, S. P., and Kelly, J. S., 1981, *Brain Res.* **230**:111–119.
70. Clineschmidt, B. V., McGuffin, J. C., and Bunting, P. B., 1979, *Eur. J. Pharmacol.* **54**:129–139.
71. Anastasi, A., Erspamer, V., and Bucci, M., 1971, *Experientia* **27**:166–167.
72. Brown, M., Allen, R., Villarreal, J., Rivier, J., and Vale, W., 1978, *Life Sci.* **23**:2721–2728.
73. Villarreal, J. A., and Brown, M. R., 1978, *Life Sci.* **23**:2729–2734.
74. Brown, M., and Vale, W., 1979, *Trends Neurosci.* **2**:95–97.
75. Pert, A., Moody, T. W., Pert, C. B., DeWald, L. A., and Rivier, J., 1980, *Brain Res.* **193**:209–220.
76. Moody, T. W., Pert, C. B., Rivier, J., and Brown, M. R., 1978, *Proc. Natl. Acad. Sci. U.S.A.* **75**:5372–5376.
77. Jensen, R. T., Moody, T., Pert, C., Rivier, J., and Gardner, J. D., 1978, *Proc. Natl. Acad. Sci. U.S.A.* **75**:6139–6143.
78. Jorpes, J. E., and Mutt, V., 1973, *Secretin, Cholecystokinin, Pancreozymin and Gastrin* (J. E., Jorpes and V. Mutt, eds.), Springer-Verlag, New York, pp. 1–179.

79. Rehfeld, J. F., 1978, *J. Biol. Chem.* **253**:4022–4030.
80. Vanderhaeghan, J. J., Signeau, J. C., and Gepts, W., 1975, *Nature* **257**:604–605.
81. Strauss, E., Miller, J. E., Choi, H. S., Paronetto, F., and Yalow, R. S., 1977, *Proc. Natl. Acad. Sci. U.S.A.* **74**:3033–3034.
82. Larsson, L. I., and Rehfeld, J. F., 1979, *Brain Res.* **165**:201–218.
83. Morley, J. E., 1982, *Life Sci.* **30**:474–493.
84. Sankarant, H., Goldfine, I. D., Deveney, C. W., Wong, K. Y., and Williams, J. A., 1980, *J. Biol. Chem.* **255**:1849–1853.
85. Innis, R. B., and Snyder, S. H., 1980, *Proc. Natl. Acad. Sci. U.S.A.* **77**:6917–6921.
86. Saito, A., Sankarant, H., Goldfine, I. D., and Williams, J. A., 1980, *Science* **208**:1155–1156.
87. Hays, S. E., Meyer, D. K., and Paul, S. M., 1981, *Brain Res.* **219**:208–213.
88. Saito, A., Goldfine, I. D., and Williams, J. A., 1981, *J. Neurochem.* **37**:483–490.
89. Dockray, G. J., 1976, *Nature* **264**:568–570.
90. Pieken, S. R., Costenbader, C. L., and Gardner, J. D., 1979, *J. Biol. Chem.* **254**:5321–5327.
91. Hays, S. E., and Paul, S. E., 1981, *Eur. J. Pharmacol.* **70**:591–592.
92. Saito, A., Williams, J. A., and Goldfine, I. D., 1981, *Endocrinology* **109**:984–986.
93. Zarbin, M. A., Wamsley, J. K., Innis, R. B., and Kuhar, M. J., 1982, *Life Sci.* **29**:697–705.
94. Hays, S. E., Goodwin, F. K., and Paul, S. M., 1981, *Peptides (Suppl.)* **2**:21–26.
95. Hays, S. E., Goodwin, F. K., and Paul, S. M., 1981, *Brain Res.* **225**:452–456.
96. Von Euler, U. S., and Gaddum, J. H., 1931, *J. Physiol. (Lond.)* **72**:74–87.
97. Chang, M. M., and Leeman, S. E., 1970, *J. Biol. Chem.* **245**:4784–4790.
98. Schenker, C., Mroz, E. A., and Leeman, S. E., 1976, *Nature* **264**:790–792.
99. Lee, C. M., Sandberg, B. E., Hanley, M. R., and Iversen, L. L., 1981, *Eur. J. Biochem.* **114**:315–327.
100. Bury, R. W., and Mashford, M. L., 1977. *Aust. J. Exp. Biol. Med. Sci.* **55**:671–735.
101. Nicoll, R. A., Schenker, C., and Leeman, S. E., 1980, *Annu. Rev. Neurosci.* **3**:227–268.
102. Hanley, M. R., and Iversen, L. L., 1980, *Neurotransmitter Receptors*, Part 1, *Receptors and Recognition, Series B*, Volume 9 (S. J. Enna and H. I. Yamamura, eds.), Chapman Hall, London, pp. 73–103.
103. Nakata, Y., Kusaka, Y., Segawa, T., Yajima, H., and Kitagewa, K., 1978, *Life Sci.* **22**:259–268.
104. Hanley, M. R., Sandberg, B. E. B., Lee, C. M., Iversen, L. L., Brundish, D. E., and Wade, R., 1980, *Nature* **286**:810–812.
105. Kanazawa, I., and Jessell, T. M., 1976, *Brain Res.* **117**:362–367.
106. Otsuka, M., and Kanishi, S., 1976, *Cold Spring Harbor Symp. Quant. Biol.* **40**:135–143.
107. Lee, C. M., Iversen, L. L., Hanley, M. R., and Sandberg, B. E. B., 1982, *Naunyn Schmiedebergs Arch. Pharmacol.* **318**:281–287.
108. Jensen, R. T., and Gardner, J. D., 1979, *Proc. Natl. Acad. Sci. U.S.A.* **76**:5679–5683.
109. Putney, J. W., Jr., Van De Walle, C. M., and Wheeler, C. S., 1980, *J. Physiol.* **301**:205–212.
110. Mayer, N., Lembeck, F., Saria, A., and Gamse, R., 1979, *Naunyn Schmiedebergs Arch. Pharmacol.* **306**:45–51.
111. Nakata, Y., Kusaka, Y., Yajima, H., Kitagawa, K., and Segawa, T., 1980, *Naunyn Schmiedebergs Arch. Pharmacol.* **314**:211–214.
112. Lembeck, F., Mayer, N., and Schindler, G., 1978, *Naunyn Schmiedebergs Arch. Pharmacol.* **303**:79–86.
113. Folkers, K., Horig, J., Rosell, S., and Bjorkroth, U., 1981, *Acta Physiol. Scand.* **111**:505–506.
114. Rosell, S., and Folkers, K., 1982, *Trends Pharmacol. Sci.* **3**:211–212.
115. Engberg, G., Svensson, T. H., Rossell, S., and Folkers, K., 1981, *Nature* **293**:222–223.
116. Piercy, M. F., Schroeder, L. A., Folkers, K., Xu, J.-C., and Horig, J., 1981, *Science* **214**:1361–1363.
117. Said, S. I., and Mutt, V., 1970, *Science* **169**:1217–1218.
118. Larsson, L.-I., Fahrenkrug, J., Schaffalitzky de Muckadell, O., Sundler, F., Hakanson, R., and Rehfeld, J., 1976, *Proc. Natl. Acad. Sci. U.S.A.* **73**:3197–3200.
119. Lundberg, J. M., Änggärd, A., Fahrenkrug, J., Hökfelt, T., and Mutt, V., 1980, *Proc. Natl. Acad. Sci. U.S.A.* **77**:1651–1655.

120. Fahrenkrug, J., 1982, *Vasoactive Intestinal Peptide* (S. I. Said, ed.), Raven Press, New York, pp. 361–372.
121. Fahrenkrug, J., 1980, *Trends Neurosci.* 3:1–2.
122. Marley, P., and Emson, P., 1982, *Vasoactive Intestinal Peptide* (S. I. Said, ed.), Raven Press, New York, pp. 341–360.
123. Bataille, D., Peillon, F., Besson, J., and Rosselin, G., 1979, *C. R. Acad. Sci.* [D] (*Paris*) 288:1315–1317.
124. Christopher, J. P., Conlon, T. P., and Gardner, J. D., 1976, *J. Biol. Chem.* 251:4629–4634.
125. Desbuquois, B., 1974, *Eur. J. Biochem.* 46:439–450.
126. Amiranoff, B., Laburthe, M., and Rosselin, G., 1980, *Biochim. Biophys. Acta* 627:215–224.
127. Binder, H. J., Lemp, G. F., and Gardner, J. D., 1980, *Am. J. Physiol.* 238:G190–G196.
128. Morera, A. M., Cathiard, A. M., Laburthe, M., and Saez, J. M., 1979, *Biochem. Biophys. Res. Commun.* 90:78–85.
129. Ottesen, B., Staun-Olsen, P., Gammeltoft, S., and Fahrenkrug, J., 1982, *Endocrinology* 110:2037–2043.
130. Bataille, D., Freychet, P., and Rosselin, G., 1974, *Endocrinology* 95:713–721.
131. Robberecht, P., Chatelain, P., Waelbroeck, M., and Christophe, J., 1982, *Vasoactive Intestinal Peptide* (S. I. Said, ed.), Raven Press, New York, pp. 323–332.
132. Robberecht, P., DeNeef, P., Lammens, M., Deschodt-Lanckman, M., and Christophe, J. P., 1978, *Eur. J. Biochem.* 90:147–154.
133. Robberecht, P., König, W., Deschodt-Lanckman, M., DeNeef, P., and Christophe, J., 1979, *Life Sci.* 25:879–884.
134. Taylor, D. P., and Pert, C. B., 1979, *Proc. Natl. Acad. Sci. U.S.A.* 76:660–664.
135. Staun-Olsen, P., Ottesen, B., Bartels, P. D., Nielsen, M. H., Gammeltoft, S., and Fahrenkrug, J., 1982, *J. Neurochem.* 39:1242–1251.
136. Hutchinson, J. B., Dimaline, R., and Dockray, G. J., 1981, *Peptides* 2:23–30.
137. Deschodt-Lanckman, M., Robberecht, P., and Christophe, J., 1977, *FEBS Lett.* 83:76–80.
138. Quik, M., Iversen, L. L., and Bloom, S. R., 1978, *Biochem. Pharmacol.* 27:2209–2213.
139. Borghi, C., Nisosia, S., Giachetti, A., and Said, S. I., 1979, *Life Sci.* 24:65–70.
140. Kerwin, R. W., Pay, S., Bhoola, K. D., and Pycock, C. J., 1980, *J. Pharm. Pharmacol.* 32:561–566.
141. Besson, J., Rotsztehn, W., Laburthe, M., Epelbaum, J., Beaudet, A., Kordon, C., and Rosselin, G., 1979, *Brain Res.* 165:79–85.
142. Emson, P. C., Gilbert, R. F. T., Loren, I., Fahrenkrug, J., Sundler, F., and Schaffalitzky de Muckadell, O. B., 1979, *Brain Res.* 177:437–444.
143. Loren, I., Emson, P. C., Fahrenkrug, J., Björklund, A., Alumets, J., Hakanson, R., and Sundler, F., 1979, *Neuroscience* 4:1953–1976.
144. Roberts, G. W., Woodhams, P. L., Bryant, M. G. Crow, T. J., Bloom, S. R., and Polak, J. M., 1980, *Histochemistry* 65:103–119.
145. Lundberg, J. M., Hedlund, B., and Bartfai, T., 1982, *Nature* 295:147–149.
146. Morel, G., Besson, J., Rosselin, G., and Dubois, P. M., 1982, *Neuroendocrinology* 34:85–89.
147. Sternberger, L. A., and Petrali, J. P., 1975, *Cell. Tissue Res.* 162:141–176.
148. Hopkins, C. R., and Gregory, H., 1977, *J. Cell Biol.* 75:528–540.
149. Hazum, E., Cuatrecasas, P., Marian, J., and Conn, P. M., 1980, *Proc. Natl. Acad. Sci. U.S.A.* 77:6692–6695.
150. Childs (Moriarty), G. V., Cole, D. E., Kubek, M., Tobin, R. B., and Wilber, J. F., 1978, *J. Histochem. Cytochem.* 26:901–908.
151. Blackwell, R. E., and Guillemin, R., 1973, *Annu. Rev. Physiol.* 35:357–390.
152. Schally, A. V., Arimura, A., and Kastin, A., 1973, *Science* 179:341–350.
153. Morley, J. E., 1979, *Life Sci.* 25:1539–1550.
154. Yarbrough, G. G., 1979, *Prog. Neurobiol.* 12:291–312.
155. Breese, G. R., Mueller, R. A., Mailman, R. B., and Frye, G. D., 1981, *The Role of Peptides and Amino Acids as Neurotransmitters* (J. B. Lombardini and A. D. Kenney, eds.), Alan R. Liss, New York, pp. 99–116.
156. Jackson, I. M. D., 1982, *N. Engl. J. Med.* 306:145–154.

157. Labrie, F., Barden, N., Poirier, G., and De Lean, A., 1972, *Proc. Natl. Acad. Sci. U.S.A.* **69**:283–287.
158. Grant, G., Vale, W., and Guillemin, R., 1972, *Biochem. Biophys. Res. Commun.* **46**:28–34.
159. Gourdji, D., Tixier-Vidal, A., Morin, A., Pradelles, P., Morgat, J. L., Fromageot, P., and Kerdelhué, B., 1973, *Exp. Cell Res.* **82**:39–46.
160. Hinkle, P. M., and Tashjian, A. H., Jr., 1973, *J. Biol. Chem.* **248**:6180–6186.
161. Vale, W., Grant, G., and Guillemin, R., 1973, *Frontiers in Neuroendocrinology* (W. F. Ganong and L. Martini, eds.), Oxford University Press, London, pp. 375–413.
162. Hinkle, P. M. Woroch, E. L., and Tashjian, A. H., Jr., 1974, *J. Biol. Chem.* **249**:3085–3090.
163. Tixier-Vidal, A., and Gourdji, D., 1981, *Physiol. Rev.* **61**:974–1011.
164. Burt, D. R., 1983, *Methods in Neurobiology* (P. J. Marangos, I. Campbell, and R. M. Cohen, eds.), Academic Press, New York (in press).
165. Burt, D. R., and Snyder, S. H., 1975, *Brain Res.* **93**:309–328.
166. Burt, D. R., 1979, *Exp. Eye Res.* **29**:353–365.
167. Burt, D. R., and Taylor, R. L., 1980, *Endocrinology* **106**:1416–1423.
168. Cott, J. M., Breese, G. R., Cooper, B. R., Barlow, T. S., and Prange, A. J., Jr., 1976, *J. Pharmacol. Exp. Ther.* **196**:594–604.
169. Veber, D. F., Holly, F. W., Varga, S. L., Hirschmann, R., Nutt, R. F., Lotti, V. S., and Porter, C. C., 1977, *Peptides 1976, Proceedings of the 14th European Peptide Symposium* (A. Loffett, ed.), University of Brussels Press, Brussels, (pp. 453–461).
170. Bissette, G., Nemeroff, C. B., Loosen, P. T., Breese, G. R., Burnett, G. B., Lipton, M. A., and Prange, A. J., Jr., 1978, *Neuropharmacology* **17**:229–237.
171. Nutt, R. F., Holly, F. W., Homnick, C., Hirschmann, R., Veber, D. F., and Arison, B. H., 1981, *J. Med. Chem.* **24**:692–698.
172. Schaeffer, J. M., Brownstein, M. J., and Axelrod, J., 1977, *Proc. Natl. Acad. Sci. U.S.A.* **74**:3579–3581.
173. Brammer, G. L., Morley, J. E., Geller, E., Yuwiler, A., and Hershman, J. M., 1979, *Am. J. Physiol.* **236**:E146–E420.
174. Kellokumpu, S., Vuolteenaho, O., and Leppäluoto, J., 1980, *Life Sci.* **26**:475–480.
175. Martino, E., Seo, H., Lernmark, A., and Refetoff, S., 1980, *Proc. Natl. Acad. Sci. U.S.A.* **77**:4345–4348.
176. Eskay, R. L., Long, R. T., and Iuvone, P. M., 1980, *Brain Res.* **196**:554–559.
177. Hökfelt, T., Fuxe, K., Johansson, O., Jeffcoate, S., and White, N., 1975, *Eur. J. Pharmacol.* **34**:389–92.
178. Miyamoto, M., and Nagawa, Y., 1977, *Eur. J. Pharmacol.* **44**:143–152.
179. Heal, D. J., and Green, A. R., 1979, *Neuropharmacology* **18**:23–31.
180. Vale, W., Rivier, J., and Burgus, R., 1971, *Endocrinology* **89**:1485–1488.
181. Wei, E., Loh, H., and Way, E. L., 1976, *Eur. J. Pharmacol.* **36**:227–229.
182. Nicoll, R. A., 1977, *Nature* **165**:242–243.
183. Taylor, R. L., and Burt, D. R., 1981, *Neuroendocrinology* **32**:310–316.
184. Taylor, R. L., and Burt, D. R., 1981, *Brain Res.* **218**:207–217.
185. Burt, D. R., and Taylor, R. L., 1982, *Exp. Eye Res.* **35**:173–182.
186. Taylor, R. L., and Burt, D. R., 1982, *J. Neurochem.* **38**:1649–1656.
187. Hinkle, P. M., Lewis, D. G., and Greer, T. L., 1980, *Endocrinology* **106**:1000–1005.
188. Gershengorn, M. C., 1978, *J. Clin. Invest.* **62**:937–943.
189. Thompson, D. F., Taylor, R. L., and Burt, D. R., 1981, *Gen. Comp. Endocrinol.* **44**:77–81.
189a. Burt, D. R., and Ajah, M. A., 1983, *Gen. Comp. Endocrinol.*, (in press).
190. Simasko, S. M., and Horita, A., 1982, *Life Sci.* **30**:1793–1799.
191. Grant, G., Vale, W., and Guillemin, R., 1973, *Endocrinology* **92**:1629–1633.
192. Gershengorn, M. C., Marcus-Samuels, B. E., and Geras, E., 1979, *Endocrinology* **105**:171–176.
193. Ogawa, N., Yamawaki, Y., Kuroda, H., Ofuji, T., Itoga, E., and Kito, S., 1981, *Brain Res.* **205**:169–174.
194. Dannies, P. S., and Markell, M. S., 1980, *Endocrinology* **106**:107–112.
195. Brazeau, P., Vale, W., Burgus, R., Ling, N., Butcher, M., Rivier, J., and Guillemin, R., 1973, *Science* **179**:77–79.

196. Pradayrol, L., Jörnvall, H., Mutt, V., and Ribet, A., 1980, *FEBS Lett.* **109**:55–58.
197. Böhlen, P., Brazeau, P., Benoit, R., Ling, N., Esch, F., and Guillemin, R., 1980, *Biochem. Biophys. Res. Commun.* **96**:725–734.
198. Schally, A. V., Huang, W.-Y., Chang, R. C. C., Arimura, A., Redding, T. W., Millar, R. P., Hunkapiller, M. W., and Hood, L. E., 1980, *Proc. Natl. Acad. Sci. U.S.A.* **77**:4489–4493.
199. Spiess, J., Villarreal, J., and Vale, W., 1981, *Biochemistry* **20**:1982–1988.
200. Brown, M., Rivier, J., and Vale, W., 1981, *Endocrinology* **108**:2391–2393.
201. Spiess, J., and Vale, W., 1980, *Biochemistry* **19**:2861–2866.
202. Efendic, S., Hökfelt, T., and Luft, R., 1978, *Adv. Metab. Dis.* **9**:367–424.
203. Luft, R., Efendic, S., and Hökfelt, T., 1978, *Diabetologia* **14**:1–13.
204. Kastin, A. J., Coy, D. H., Jacquet, Y., Schally, A. V., and Plotnikoff, N. P., 1978, *Metabolism* **27**(Suppl. 1):1247–1252.
205. Havlicek, V., and Friesen, H. G., 1979, *Central Nervous System Effects of Hypothalamic Hormones and Other Peptides* (R. Collu, A. Barbeau, J. G. Rochefort, and J. R. Ducharme, eds.), Raven Press, New York, pp. 381–402.
206. Schonbrunn, A., and Tashjian, A. H., Jr., 1978, *J. Biol. Chem.* **253**:6473–6483.
207. Vale, W., Rivier, C., and Brown, M., 1977, *Annu. Rev. Physiol.* **39**:473–527.
208. Leitner, J. W., Rifkin, R. M., Maman, A., and Sussman, K. E., 1979, *Biochem. Biophys. Res. Commun.* **87**:919–927.
209. Srikant, C. B., and Patel, Y. C., 1981, *Endocrinology* **108**:341–343.
210. Srikant, C. B., and Patel, Y. C., 1981, *Proc. Natl. Acad. Sci. U.S.A.* **78**:3930–3934.
211. Srikant, C. B., and Patel, Y. C., 1981, *Nature* **294**:259–260.
212. Reubi, J. C., Perrin, M. H., Rivier, J. E., and Vale, W., 1982, *Life Sci.* **28**:2191–2198.
213. Epelbaum, J., Tapia-Arancibia, L., Kordon, C., and Enjalbert, A., 1982, *J. Neurochem.* **38**:1515–1523.
214. Aguilera, G., and Parker, D. S., 1982, *J. Biol. Chem.* **257**:1134–1137.
215. Enjalbert, A., Tapia-Arancibia, L., Rieutort, M., Brazeau, P., Kordon, C., and Epelbaum, J., 1982, *Endocrinology* **110**:1634–1640.
216. Reubi, J. C., Perrin, M., Rivier, J., and Vale, W., 1982, *Biochem. Biophys. Res. Commun.* **105**:1538–1545.
217. Srikant, C. B., and Patel, Y. C., 1982, *Endocrinology* **110**:2138–2144.
218. Reubi, J. C., Rivier, J., Perrin, M., Brown, M., and Vale, W., 1982, *Endocrinology* **110**:1049–1051.
219. Aguilera, G., Parker, D. S., and Catt, K. J., 1982, *Endocrinology* **111**:1376–1384.
220. Czernik, A. J., and Petrack, B., 1982, *Soc. Neurosci. Abstr.* **8**:980.
221. Vale, W., Rivier, J., Ling, N., and Brown, M., 1978, *Metabolism* **27**(Suppl. 1):1391–1401.
222. Veber, D. F., Holly, F. W., Nutt, R. F., Bergstrand, S. J., Brady, S. F., Hirschmann, R., Glitzer, M. S., and Saperstein, R., 1979, *Nature* **280**:512–514.
223. Brownstein, M., Arimura, A., Sato, H., Schally, A. V., and Kizer, J. S., 1975, *Endocrinology* **96**:1456–1461.
224. Schonbrunn, A., and Tashjian, A. H., Jr., 1980 *J. Biol. Chem.* **255**:190–198.
225. Margolis, F. L., 1980, *The Role of Peptides in Neuronal Function* (J. L. Barker and T. Smith, eds.), Marcel Dekker, New York, pp. 545–572.
226. Rochel, S., and Margolis, F. L., 1982, *J. Neurochem.* **38**:1505–1514.
227. MacLeod, N. K., and Straughan, D. W., 1979, *Exp. Brain Res.* **34**:183–188.
228. Nicoll, R. A., Elgar, B. E., and Jahr, C. E., 1980, *Proc. R. Soc. Lond.* [*Biol.*] **210**:133–149.
229. Hirsch, J. D., Grillo, M., and Margolis, F. L., 1978, *Brain Res.* **158**:407–422.
230. Hirsch, J. D., and Margolis, F. L., 1979, *Brain Res.* **174**:81–94.
231. Regoli, D., and Barabé, J., 1980, *Pharmacol. Rev.* **32**:1–46.
232. Correa, F. M. A., and Graeff, F. G., 1975, *J. Pharmacol. Exp. Ther.* **192**:670–676.
233. Odya, C., Goodfriend, T. L., and Pena, C., 1980, *Biochem. Pharmacol.* **29**:175–185.
234. Correa, F. M. A., Innis, R. B., Uhl, G. R., and Snyder, S. H., 1979, *Proc. Natl. Acad. Sci. U.S.A.* **76**:1489–1493.
235. Handler, J. S., and Orloff, J., 1981, *Annu. Rev. Physiol.* **43**:611–624.
236. Swanson, L. W., Sawchenko, P. E., Wiegand, S. J., and Price, J. L., 1980, *Brain Res.* **198**:190–195.

237. DeWied, D., and Versteeg, D. H. G., 1979, *Fed. Proc.* **38:**2348–2354.
238. DeKloet, R., and DeWied, D., 1980, *Frontiers in Neuroendocrinology*, Volume 6 (L. Martini and W. F. Ganong, eds.), Raven Press, New York, pp. 157–201.
239. Bockaert, J., Roy, C., Rajerison, R., and Jard, S., 1973, *J. Biol. Chem.* **248:**5922–5931.
240. Nilaver, G., Zimmerman, E. A., Wilkins, J., Michaels, J., Hoffman, D., and Silverman, A. J., 1980, *Neuroendocrinology* **30:**150–158.
241. Pedersen, C. A., Ascher, J. A., Monroe, Y. L., and Prange, A. J., Jr., 1982, *Science* **216:**648–650.
242. Soloff, M., Swartz, T., Morrison, M., and Saffran, M., 1973, *Endocrinology* **92:**104–107.
243. Moss, R. L., 1979, *Annu. Rev. Physiol.* **41:**617–631.
244. Jan, L. Y., Jan, Y. N., and Kuffler, S. W., 1979, *Nature* **288:**380–382.
245. Jan, L. Y., and Jan, Y. N., 1981, *Fed. Proc.* **40:**2560–2564.
246. Adams, P. R., and Brown, D. A., 1980, *Br. J. Pharmacol.* **68:**353–355.
247. Bernardo, L. A., Petrali, J. P., Weiss, L. P., and Sternberger, L. A., 1978, *J. Histochem. Cytochem.* **26:**613–617.
248. Reeves, J. J., Sëguin, C., Lefebvre, F.-A., Kelly, P. A., and Labrie, F., 1980, *Proc. Natl. Acad. Sci. U.S.A.* **77:**5567–5571.
249. Fuxe, K., Hökfelt, T., Eneroth, P., Gustafsson, J. A., and Skett, P., 1977, *Science* **196:**899–900.
250. Clemens, J. A., Gallo, R. V., Whitmoyer, D. I., and Sawyer, C. H., 1971, *Brain Res.* **25:**371–379.
251. Yamada, K., 1975, *Neuroendocrinology* **18:**263–271.
252. Scapagnini, U., Rizza, V., Drago, F., Canonico, P. L., Pellegrini-Quarantotti, B., Ragusa, N., Clementi, C., Prato, A., Marchetti, B., and Gessa, G. L., 1980, *Central and Peripheral Regulation of Prolactin Function* (R. M. MacLeod and U. Scapagnini, eds.), Raven Press, New York, pp. 293–309.
253. Walsh, R. J., Posher, B. I., Kopriwa, B. M., and Brawer, J. R., 1978, *Science* **201:**1041–1043.
254. DiCarlo, R., and Muccioli, G., 1981, *Brain Res.* **230:**445–450.
255. Shiu, R. P. C., and Friesen, H. G., 1974, *Biochem. J.* **140:**301–311.
256. Underhill, L. H., Rosenzweig, J. L., Roth, J., Brownstein, M. J., Young, W. S., III, and Havrankova, J., 1982, *Front Horm. Res.* **10:**96–110.
257. VanCalker, D., Muller, M., and Hamprecht, B., 1980, *Proc. Natl. Acad. Sci. U.S.A.* **77:**6907–6911.
258. Fremeau, R. T., Jensen, R. T., O'Donohue, T. L., and Moody, T. W., 1982, *Soc. Neurosci. Abstr.* **8:**980.
259. Floyd, J. C., Jr., Fajans, S. S., Pek, S., and Chance, R. E., 1977, *Recent Prog. Hormone Res.* **33:**519–570.
260. Chiu, S., Paulose, C. S., and Mishra, R. K., 1981, *Science* **214:**1261–1262.
261. O'Shea, M., 1982, *Trends Neurosci.* **5:**69–73.
262. Prasad, C., Jacobs, J. J., and Wilber, J. F., 1980, *Brain Res.* **193:**580–583.
263. Kovacs, G. L., Vecsei, L., Medue, L., and Telegdy, G., 1980, *Exp. Brain Res.* **38:**357–361.

14

Cholinergic Systems and Cholinergic Pathology

Patrick L. McGeer and Edith G. McGeer

1. INTRODUCTION

The history and chemistry of acetylcholine as a neurotransmitter have been discussed in other chapters in this book. In this chapter we are concerned with the neuroanatomy of cholinergic systems, including the location and nature of the receptors, and with the apparent involvement of such systems in various neurological disorders.

2. ANATOMY OF CHOLINERGIC NEURONS

It is now part of classical neuroanatomy that in the spinal cord all anterior horn cells supplying nerves to voluntary muscles and all lateral horn cells supplying preganglionic nerves to autonomic ganglia are cholinergic. It is also part of classic neuroanatomy that ganglion cells giving rise to postganglionic parasympathetic fibers are nearly all cholinergic. In addition, those cranial nerves containing voluntary motor fibers (III, IV, VI, IX, XI, and XII) and those containing preganglionic parasympathetic fibers (III, VII, IX, and X) also have cholinergic cell bodies. The reader is referred to standard textbooks of neuroanatomy for the details of the tracts and to standard pharmacology texts for the classical evidence on which the cholinergic assignment has been made. The sympathetic postganglionic fibers to sweat glands are also cholinergic, and there is suggestive evidence, which needs to be pursued, of an even more widespread distribution of such postganglionic cholinergic fibers in the sympathetic system.

Patrick L. McGeer and Edith G. McGeer • Kinsmen Laboratory of Neurological Research, Department of Psychiatry, University of British Columbia, Vancouver, British Columbia V6T 2A1, Canada.

2.1. Techniques

The main challenge at the present time is to understand the distribution of cholinergic pathways in the central nervous system (CNS). Considerable progress has been made in recent years in this regard. This has come about as a result of substantial improvements in the basic techniques necessary for gathering data with respect to these systems. The techniques are the following: (1) immunohistochemistry of choline acetyltransferase (ChAT); (2) histochemistry of acetylcholinesterase (AChE) following administration of the irreversible cholinesterase inhibitor diisopropyl fluorophosphate (DFP); (3) lesioning of suspected cholinergic pathways in the CNS, particularly with kainic acid and ibotenic acid which destroy cells but spare terminals and axons of passage. Ancillary techniques include autoradiographic localization of receptors using labeled ligands presumed to be specific for cholinergic receptors, high-affinity choline uptake, and ACh release.

Since each of these principal techniques has its limitations, extensive confirmation of all data is necessary for there to be firm confidence in the results. Although in theory the localization of the specific synthetic enzyme (ChAT) at the cellular and subcellular level by immunohistochemistry should be the definitive way of establishing the nature and presence of cholinergic systems, in practice the technique has imperfections. Immunohistochemical methods are fraught with difficulties, and these seem to apply particularly to ChAT. Good immunohistochemical localization must be performed on fixed tissue. The fixation process can obviously alter the nature of the recognition sites on the enzyme, and these sites form the basis of the antigen–antibody reaction, thus reducing its intensity. For specific staining to be detected, there must be a relatively strong reaction to overcome the nonspecific staining introduced by the complicated sandwich technique that marks the reaction. That this is a particular problem with ChAT is evidenced by several reports from laboratories that have apparently been successful in raising monospecific antibodies to ChAT only to have weak or negative immunohistochemical results.

Some highly satisfactory results have, however, been obtained using Fab fragments of high-titer rabbit antibodies against human ChAT; these allowed a mapping of the cholinergic structures in cat brain.[1] This mapping followed a number of earlier and incomplete studies from several laboratories that demonstrated the potential of the immunohistochemical technique for ChAT.[2–5] Examples of the staining achieved in the study of Kimura *et al.*[1] are shown in Figs. 1–4. Figure 1 is a photomicrograph of a cholinergic cell taken from the interstitial nucleus of the anterior commissure. Figure 4 is a terminal field in the superficial molecular layer of the cerebral cortex. Figures 2 and 3 represent the kinds of cells that have produced some problems in mapping. Figure 2 is an example of a purely cholinoceptive cell with heavy terminal densities around the cell soma observed in the presubicular cortex. No diffuse staining is seen in this cell. Figure 3 is a cholinergic–cholinoceptive cell from a magnocellular field. It shows both smooth staining of the cytoplasm and the intense dots of terminal boutons. Detecting differences between the cell types illustrated in Figs. 2 and 3 obviously requires excellent fixation. Diffusion in the terminal

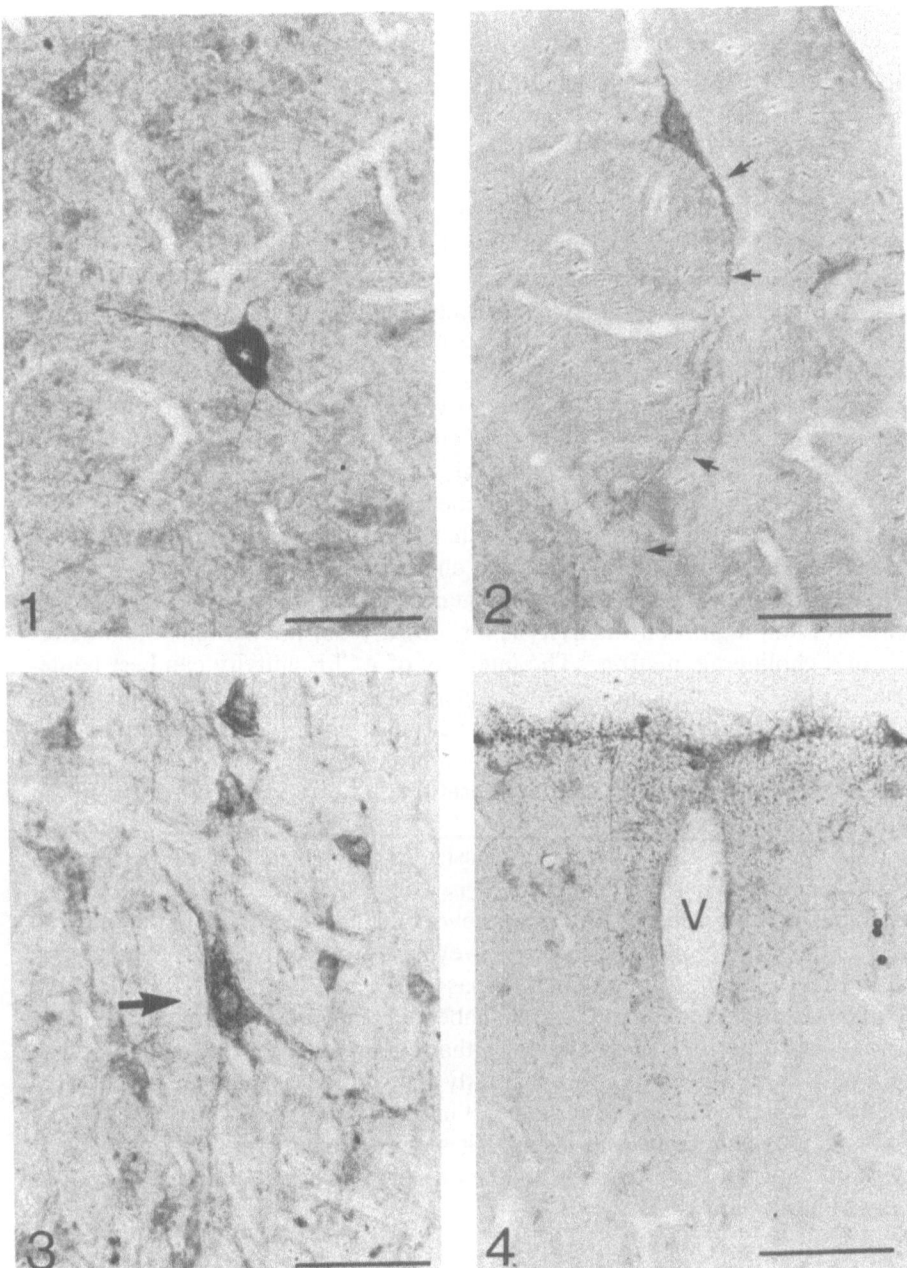

Figs. 1–4. Examples of cholinergic structures. Bars indicate 100 μm. (1) A pure cholinergic cell observed in the interstitial nucleus of the anterior commissure. Note the dense and even stain throughout the cytoplasm and extending into the fine processes. A few short stained fibers from other cholinergic neurons can also be seen in the field. (2) A pure cholinoceptive cell observed in the presubicular cortex. Note that the surface of the long process (arrows) is contacted by tiny dots (probably nerve terminals) but that no diffuse staining produce is seen. (3) A cholinergic cholinoceptive cell (arrow) seen in the magnocellular terminal field. See text for further explanation. (4) An example of a terminal field observed in the superficial molecular layer of the cerebral cortex and characterized by fine dots (probably nerve terminals). Such fine dots can also be seen around the blood vessel (v).

fields or poor fixation of tissues could easily cause a cholinoceptive cell to appear cholinergic. This is a major difficulty with the method and one of the main reasons for remaining uncertainties with respect to certain cholinergic cell types in the brain.

Next to ChAT, the most valuable marker for establishing cholinergic systems is AChE. Acetylcholinesterase is an enzyme with activity generally two orders of magnitude higher than ChAT. It is stable, permitting reliable assay of its presence at the cellular level using histochemical, as opposed to immunohistochemical, methods. It cannot, however, be regarded as a definitive marker for cholinergic neurons. In most regions, there is an excellent correlation between AChE and ChAT activity (Table 1), but in some areas, for example, the substantia nigra and the cerebellum, the ratio between the two is radically different. Some noncholinergic neurons, such as dopaminergic neurons of the substantia nigra and noradrenergic neurons of the locus coeruleus, contain considerable AChE. On the other hand, all neurons that have been proven beyond reasonable doubt to be cholinergic have high levels of AChE and an excellent capacity for rapid regeneration of the enzyme. It may be, therefore, that high AChE activity is a necessary but not sufficient characteristic of a cholinergic neuron.[6] The intensity of AChE activity can best be determined by the technique of AChE regeneration after irreversible inhibition with DFP. Used in this fashion, AChE activity can be valuable in confirming the existence of cholinergic neurons when other evidence suggests their presence, and, in addition, it can provide useful clues for systems that are still in doubt.

The technique of lesioning is a classical one for establishing specific biochemical neuronal pathways in brain. It is the use of this technique that helped to establish the presence of the first known cholinergic pathway in brain from the septum to the hippocampus.[7–9] However, a problem with this technique is the possibility of interrupting fibers of passage. Thus, the early suggestion that a cholinergic tract extended from the habenula to the interpeduncular nucleus has needed to be revised on the basis that cholinergic fibers emanating from the medial basal forebrain are apparently interrupted by habenular lesions.[10] The use of kainic acid as a lesioning tool has helped to alleviate some potential problems with this technique because it spares axons of passage and has allowed resolution of some doubtful data on pathways in cholinergic as well as many other systems.[11,12]

2.2. Cholinergic Structures

The detailed map of cholinergic structures in cat brain prepared by the immunohistochemical procedure for ChAT indicates the existence of at least five major cholinergic cellular systems in brain.[1] In all cases, the cholinergic cells are large to giant cells. The major systems so far identified are the following:

1. The medial basal forebrain complex, which serves the cerebral cortex and limbic system. This includes adjoining cell groups in the olfactory

Table I

Levels of ChAT, AChE, ACh, [³H]QNB (Muscarinic) Binding, and [¹²⁵I]α-BTX (Nicotinic) Binding in Some Human and Rat Brain Regions[a]

Area	Human				Rat				
	ChAT[46]	AChE[46]	QNB[264]	α-BTX[267]	ChAT[84]	AChE[85]	ACh[268]	QNB[266]	α-BTX[145]
Putamen	114	16	472	—	51[b]	30[b]	17[b]	625[b]	5[b]
Caudate	107	15	480	1.6	—	—	—	503	—
N. accumbens	83	14	—	—	32	33	6	484	—
Olfactory tubercle	59	8.6	—	7.0	61[b]	41	11	280	—
Amygdala	21	3.9	297	5.7	28	6.7	6	216	—
Preoptic area	16	5.5	—	—	7.4	3.4	5	205	—
Septal area	14	2.5	—	—	12	3.5	5	—	10
Int. globus pallidus	7.4	5.1	60[c]	—	16[c]	8[c]	4[c]	11[c]	—
Ext. globus pallidus	19	5.0	—	—	—	—	—	—	—
Hippocampus	8.7	2.2	196	3.8	20	4	6	508	24
Cortex	5.7	1.5	294	4.7	9	6	2	518	15
Thalamus	7.1	1.8	182	—	29	—	2.5	269	8
Hypothalamus	7.8	3.5	49	—	8.9	3.6	7	170	26
Substantia nigra	4.3	3.9	18	—	3.0	—	4.5	116	—
Cerebellar cortex	3.6	3.9	—	0.7	—	—	2	55	1
Medial habenula	—	—	—	3.2[d]	69	6.1	11[d]	87	—
Lateral habenula	—	—	—	—	15	11	—	160	—
Interpeduncular n.	—	—	—	3.6	240	—	24	160	—

[a] ChAT is in nmol/mg protein per hr; AChE in μmol/mg protein per hr; ACh in pmol/mg tissue; and [³H]QNB and [¹²⁵I]α-BTX binding in fmol/mg protein. More detailed data are available in the references from which these data were taken.
[b] Caudate plus putamen.
[c] Internal plus external globus pallidus.
[d] Medial plus lateral habenula.

tubercle, medial septum, the nucleus of the diagonal band of Broca, and the substantia innominata complex.

2. The large cells of the caudate, putamen, and nucleus accumbens, which serve the extrapyramidal system.
3. The large motor neurons of the cranial nerve nuclei and spinal cord, which serve the voluntary skeletomuscular and preganglionic autonomic systems.
4. The parabrachial nuclear complex.
5. The giganto- and magnocellular elements of the tegmental fields of the pons and medulla.

The systems served by the last two groups are unknown. Other cholinergic cell clusters are found in the red nucleus, lateral reticular formation, superior olivary complex, some vestibular nuclei, and a few other areas, where they probably serve a more restricted function.

As might be judged from the ubiquitous nature of acetylcholine (ACh), ChAT, and AChE in the nervous system (Table I), coupled with the limited sources of cholinergic cell bodies, the occurrence of cholinoceptive cells and terminal fields is much broader than the cell bodies themselves. This means that many cholinergic cells must have axons that diverge widely, thereby serving large populations of neurons. No area of the brain has been found that is totally devoid of cholinergic innervation. However, some areas are more richly served than others. The immunohistochemical findings of Kimura *et al.*[1] on the apparent density of cholinergic innervation of various brain regions are in accord with biochemical data on the distribution of ChAT (Table I). Thus, the relatively high activities of ChAT found in the interpeduncular nucleus, habenula, olfactory tubercle, and amygdala bear witness to the extensive cholinergic innervation of these nuclei.

It is noteworthy that even areas having relatively low values of ChAT nevertheless possess a synthesizing capacity for ACh comparable to the highest synthesizing capacities for catecholamines or serotonin that exist in brain.

Distribution data on ACh are more limited and of lower reliability. Acetylcholine is a labile material that can be rapidly destroyed unless it is sequestered from AChE. As a result, values for ACh often vary considerably from laboratory to laboratory depending on the methods of preparing tissues and extracts for analysis. The most reliable values for brain are now thought to be those obtained when animals are sacrificed by microwave irradiation. This technique permits very rapid heating of brain tissue, so that within a second the enzymes become inactivated. Since ACh is stable at the temperatures generated, it is unaffected by the treatment. Some idea of the variability in ACh and choline values with the method of sacrifice has been given by Jenden.[13] For whole rat brain a value of 24.8 nmol/g of ACh was obtained by microwave irradiation, whereas only 14.9 was obtained after standard cervical dislocation followed by the normal dissection, which takes about 5 min. Choline, on the other hand, increased from 26.3 to 148.4 nmol/g during this period.

Some values for ACh in rat brain obtained by using the microwave irradiation technique are shown in Table I. These values are higher than those

Table II
Probable (and Suggested) Cholinergic Pathways

Substantia innominata complex to cortex
(Cortical interneurons)
Olfactory tubercle to olfactory bulb
Diagonal band and substantia innominata to olfactory bulb
Diagonal band and substantia innominata to amygdala
(Lateral preoptic area to amygdala)
Medial septum and diagonal band to hippocampus
Medial septum and diagonal band to cingulate gyrus
Diagonal band to interpeduncular nucleus
(Habenular–interpeduncular tract)
Neostriatal interneurons
Nucleus accumbens interneurons
(Thalamostriatal tract)
(Lateral tegmental nucleus to anteroventral thalamus)
(Hypothalamic interneurons)
(Olivocochlear efferents)
Some mossy fiber efferents to the cerebellum
Motor nuclei of cranial nerves III–VII, IX–XII
Anterior horn cells to all voluntary muscles
Lateral horn cells to all autonomic ganglia
Postganglionic parasympathetic fibers
Some postganglionic sympathetic fibers
Some retinal amacrine cells

usually quoted using techniques that would only measure bound ACh. Nevertheless, they follow the same general pattern and parallel to some degree the relative ChAT values.

2.3. Cholinergic Pathways

Establishment of cholinergic pathways requires synthesis of information obtained from immunohistochemical staining for ChAT, histochemical staining for AChE, biochemical measurements of ChAT and AChE following electrolytic or kainic acid lesioning, plus other techniques, such as the effects of lesions on uptake or ACh release. A list of probable and suggested cholinergic pathways is given in Table II. A resume of the evidence on which this table is based, insofar as it concerns CNS pathways, is given in following paragraphs.

2.3.1. Innervation of the Cortex[14]

The presence of cholinoceptive cells in the cerebral cortex and the release of acetylcholine from the cortex have long been established,[15] but there is still considerable doubt as to the nature of the afferent cholinergic systems.[14] Reductions of about 65–80% in ChAT and AChE staining have been reported in isolated cortical slabs[9,16,17] and in the anterior and middle portion of the cortex in rats with large lesions of the globus pallidus[18,19] or substantia innominata.[20]

However, Ulmar et al.[21] found only a 45% decrease in AChE and a 10% decrease in ChAT after undercutting the rat cerebral cortex. Shute and Lewis[22] originally proposed a projection to the neocortex from the globus pallidus on the basis that AChE-rich axons projecting to the cortex appear to originate from this area. Others[23-26] demonstrated that horseradish peroxidase (HRP) was transported from the cortex to AChE-rich neurons of the nucleus basalis of the substantia innominata. Lesions of the basalis neurons sparing the globus pallidus brought about drops in cortical ChAT, indicating that this area, and not the globus pallidus, was the source of a major afferent connection to the cortex.[20,25] However, it appears that this is not the sole source of cholinergic afferents to the cortex. Following extensive lesions to the basal forebrain, ChAT activity still persists in the cortex in an amount approximating at least 30% of the total activity and possibly much higher, depending somewhat on the time between lesioning and sacrifice.[21,27] This means that either cortical cholinergic interneurons exist or that some unidentified subcortical system also supplies afferents to the neocortex.

Cortical interneurons thought to be ChAT containing have been observed,[2] but, as mentioned previously, this may merely represent intensely cholinoceptive neurons. Johnston and Coyle[28] argued for the existence of some neocortical cholinergic interneurons on the basis of changes in the activities of AChE and ChAT in cortical layers following either lesions to the nucleus basalis cortical cholinergic pathway or induced cortical hypoplasia through fetal administration of methylazoxymethanol acetate. Lesions of the nucleus basalis produced marked reductions in the activities of cholinergic markers in all cortical layers plus an elimination of the uneven distribution of these markers within the cortex. On the other hand, fetally induced hypoplasia of the cortex produced enrichment in all layers. These data were interpreted as being consistent with the existence of cholinergic interneurons. Further, more exact experiments will be required to determine the precise nature of cholinergic cortical innervation not associated with the magnocellular basal forebrain region.

2.3.2. Innervation of the Olfactory Bulb

Lesions caudal to the olfactory bulb lead to almost complete loss of ChAT activity in the bulb according to Ross et al.,[29] and we have found in unpublished work decreases of 71% in ChAT and 86% in AChE with an insignificant change in GAD. Youngs et al.[30] showed decreases of from 22 to 65% in ChAT in olfactory bulbs contralateral to lesions placed in the magnocellular preoptic area (nucleus of the horizontal limb of the diagonal band) in the hamster, suggesting this is one source of cholinergic innervation. De Olmos et al.[31] reported, following HRP studies, that the afferent connections to the olfactory bulb originated in the nucleus of the horizontal limb of the diagonal band, the vertical limb, and the substantia innominata, amongst other areas. These areas all contain cholinergic cell bodies.[1,32] Thus, it seems likely that these areas innervate the olfactory bulb via the olfactory tract.

2.3.3. Innervation of the Extrapyramidal System

The caudate and putamen are extremely rich in all cholinergic indices, and the evidence is convincing that all, or almost all, of the activity is in intrinsic neurons. Initial evidence for such intrinsic neurons came from experiments showing that lesioning of the known afferents and efferents did not cause any significant decrease in ACh or in ChAT or AChE activities in the caudate–putamen.[33,34] A cholinergic input from the thalamus to the head of the striatum has been suggested on the basis of local changes in AChE, ChAT, and ACh following lesions of the parafascicular nucleus of the thalamus,[35-37] but this cannot be confirmed in our laboratory. Simke and Saelens[36] have reported a decrease in only the anterolateral tip of the caudate nucleus involving less than 15% of the tissue. Since the thalamostriatal tract is a massive projection, it would certainly appear that the bulk of the fibers must be served by some other transmitter(s). Moreover, no evidence of neurons staining for ChAT[1] or intensely staining for AChE[38] has been seen in the parafascicular nucleus or other nuclei of the thalamus from which striatal afferents might originate. The report that scopolamine blocks the response of striatal cells to thalamic stimulation[39] may reflect measurement of a polysynaptic action involving ACh interneurons.

Immunohistochemical studies of ChAT in the cat neostriatum have, however, provided morphological evidence of large-sized cholinergic neurons in the striatum, which is consistent with previous indications from AChE studies.[40,41] This type of neuron is thought to comprise less than 1% of the total neuronal population of the neostriatum,[42] and whether this is sufficient to account for the extremely high levels of ChAT and AChE cannot yet be regarded as a certainty. However, although this large aspiny neuron was once considered to be the sole source of descending projections from the striatum, this view has been revised since it has been shown through HRP studies that at least 50% of the medium spiny neurons have descending projections.[43] It seems probable now that the large aspiny neuron does not project significantly beyond the neostriatum, but the extent to which it ramifies within the neostriatum is not yet established. Lesions to the substantia innominata region that cause sharp decreases in ChAT in the cortex do not affect levels in the neostriatum, indicating that the projection field passes through but does not terminate in this region.

The nucleus accumbens resembles the caudate–putamen in its high concentration of ChAT, AChE, ACh, and muscarinic binding sites. As in the caudate–putamen, ChAT activity in the accumbens is seriously depleted following local injections of kainic acid but is not affected by hemitransections or other lesions that would influence extrinsic afferents.[44,45] Since the accumbens possesses large-sized cholinergic neurons similar to those of the caudate and putamen, it is provisionally assumed that the cholinergic innervation is the same as in the neostriatum.

The globus pallidus and entopeduncular nucleus contain only about 20–30% of the levels of ChAT and AChE found in the striatum (see Table I). Nevertheless, they contain high levels of cholinoceptive neurons and terminal fields. They also contain some intrinsic cholinergic neurons in their medial aspects and have large-sized cholinergic neurons around their margins.[1] These adjacent neurons are obvious candidates for innervation of the globus pallidus, as are the large neurons of the neostriatum. Unfortunately, the exact source is not yet known.

The substantia nigra (SN) has relatively high AChE compared to ChAT activity.[46] Histochemical studies on AChE in normal and lesioned cats and monkeys led to the suggestion of the possible existence of striatopallidal and striatonigral cholinergic systems, with the latter constituting a massive indirect feedback to the dopaminergic system.[47] Studies in rats with hemitransections between the caudate and globus pallidus (anterior lesions) or SN (posterior lesions) indicated, however, no significant decreases of ChAT in either the pallidum or the SN.[48] Confirmatory reports have come from several laboratories working with rats, cats, and baboons.[33,49,50] These data make it seem unlikely that the striatopallidal and the striatonigral tracts contain significant cholinergic components. Other evidence indicates that a major problem in using AChE data in this system is that there is considerable AChE activity in dopaminergic neurons.[6,34,41] Nigral injections of kainic acid that caused extensive destruction of neuronal cell bodies in the SN, as evidenced by both histological examination and by losses of up to 50% in SN GAD and up to 95% in striatal tyrosine hydroxylase activity, had no significant effect on ChAT activity in the nigra but did decrease AChE activity by about 50%.[51] These data argue in favor of much of the ChAT activity in the substantia nigra being in afferents, but hemitransection data suggest that they could not come from a rostrolateral direction. As yet the source is unknown.

2.3.4. Innervation of the Limbic System

The septohippocampal pathway was the first cholinergic path to be reasonably well established and has been the most widely investigated by electrophysiological and chemical techniques.[52] Lesions of the septal area cause large decreases of ACh, AChE, ACh turnover, ChAT, and high-affinity choline uptake in the hippocampus on the operated side.[7,8,53–57] It has also been shown that the turnover and release of ACh in the hippocampus are increased by septal stimulation,[58–60] and the excitatory action of such stimulation on dentate granule cells is blocked by atropine.[61] Results obtained using kainic acid are consistent with the localization of all, or almost all, of the ChAT activity in the hippocampus in afferent neurons. Injections of kainic acid into the hippocampus lead to decreases of at least 70% in hippocampal glutamate decarboxylase (GAD) but cause no significant change in hippocampal ChAT activity.[62,63]

The probability that a substantial portion of the AChE activity in the hippocampus is located within processes of cholinergic neurons is indicated by the parallel decreases in the two enzymes in lesion experiments[64] and by the remarkably constant ratio between AChE and ChAT activities in various regions of the hippocampus.[8] Thus, in this instance, AChE histochemistry is probably a reliable tool and has given considerable information about the precise localization of the cholinergic neuronal elements in the septohippocampal system. Lewis and Shute[54] found that AChE-containing afferents to the hip-pocampus appeared to rise from the medial septal nucleus and the nucleus of the diagonal band and travel via the medial supracallosal stria of Lancisi, the dorsal fornix, the alveus, and the fimbria.

Within the hippocampus itself, AChE staining was shown by Storm-Mathisen[65] to be localized to very discrete bands. The staining seen agrees with the finding of Fonnum[8] that the regions of the hippocampus with the highest AChE and ChAT activities are the narrow infrapyramidal zone of the striatum oriens, which contains the basal dendrites of the pyramidal cells, and the supragranular and hilus fasciae dentatae of the area dentata. In light microscopic studies of the autoradiographic localization of tentative cholinergic muscarinic receptors in the hippocampus, very high grain densities have been observed in regions that contain the dendrites of pyramidal and granule cells.[66] Electrophysiological studies have been consistent in that they have indicated excitation of both pyramidal and (more strongly) granule cells by iontophoretically applied ACh.[67] Studies of the regional localization of grains in the hippocampus following injections of radioactive leucine into the medial septal nucleus have also been generally consistent with this distribution of septal efferents.[68]

The source of cholinergic efferents to the amygdala was suggested from lesion experiments to be the lateral preoptic area,[69] although Nagai *et al.*[70] have used double labeling with a retrograde tracer and AChE histochemistry to show that cells that send projections to the amygdala and are intensely reactive for AChE are located mainly in the nucleus of the substantia innominata. Some are also found in the ventral globus pallidus, the nucleus of the horizontal limb of the diagonal band of Broca, and the interstitial nucleus of the ansa lenticularis. A correspondence in both distribution and morphology was established between these cells and cells staining for ChAT by immunohistochemistry. Noncholinergic neurons that send their axons to the amygdala were also found in the substantia innominata complex.[70]

The interpeduncular nucleus (IPN) has the highest concentration of ChAT thus far found in brain. It is known to receive a massive input from the habenula via the fasciculus retroflexus. The cholinergic nature of many neurons in the habenulointerpeduncular tract was initially indicated by the sharp fall of ChAT activity in the IPN following lesions of this tract[71] and was subsequently confirmed by further measurements of ChAT, choline uptake, and other cholinergic markers in the IPN in lesioned animals.[72-75] One group of workers[76,77] suggested that the cholinergic innervation of the IPN originated in the lateral habenula and that this also sent cholinergic fibers to the medial habenula. Others have argued in favor of the medial habenula being the principal source of the IPN cholinergic afferents on the basis of the higher ChAT activity and much more intense immunohistochemical staining in the medial, as compared to the lateral, habenula.[78]

More recent lesion and histochemical evidence suggests that much if not all of the cholinergic input to the IPN arises in areas such as the stria terminalis or the nucleus of the diagonal band of Broca (NDB), with the axons passing through and innervating the habenula before descending through the fasciculus retroflexus to the IPN. Electrolytic or surgical lesions of the stria medullaris or NDB cause drops of 45–52% in ChAT activity in the IPN.[10,79] Kainic acid injections into the habenula cause much smaller drops in ChAT activity in the IPN than do electrolytic lesions of the same nucleus.[80,81] An early report of positive immunohistochemical localization of ChAT to the medial habenula may reflect the intensely cholinoceptive nature of those neurons.[78] At this stage it must be concluded that some, but not necessarily all, cholinergic innervation of the interpeduncular nucleus comes from forebrain cholinergic neurons in the diagonal band area.

On the basis of AChE staining, Wilson and Watson[82] suggested that the interpedunculotegmental tract might be cholinergic. Kimura *et al.*[1] saw no cholinergic cells in the interpeduncular nucleus, but the strong terminal field staining in that area may have made detection of positive cells difficult.

2.3.5. Innervation of the Diencephalon

At present, there are no known cholinergic cell bodies in the thalamus, but there are extensive and widely distributed terminal fields and cholinoceptive neurons. The source of these might be complex, and, at this stage, little information exists with respect to the afferent cholinergic input to any part of the thalamus. Hoover and Baisden[83] have reported an AChE-rich tract from the lateral tegmental nucleus of the rat to the anteroventral thalamic nucleus.

Within the hypothalamus of rats, Walaas and Fonnum[45] have reported that AChE staining and ChAT activity are concentrated in the median eminence and the arcuate nucleus with relatively little in the ventromedial nucleus. Intermediate amounts are found in the dorsomedial nucleus. Other authors,[84,85] using different dissection techniques, have not found such high activities in the arcuate nucleus, but all agree on the relatively high concentration of ChAT in the median eminence. When neurons of the arcuate nucleus are destroyed by neonatal administration of sodium glutamate, significant decreases in ChAT activity and AChE staining in both the median eminence and the arcuate nucleus are found.[86–89] There were no significant changes in other hypothalamic regions examined, in the amygdala, or in the habenula. These results are interpreted as indicating the existence of cholinergic fibers in the tuberoinfundibular tract. Meyer and Brownstein[90] measured the concentration of ChAT in the supraoptic nucleus after a variety of lesions. Only lesions that separated the nucleus from the posterior part of the lateral hypothalamus slightly decreased its concentration of ChAT. They concluded that the bulk of cholinergic neurons innervating that structure are either in the nucleus itself or in its immediate vicinity.

All of these data suggest the possibility of intrinsic cholinergic neurons in the hypothalamus despite the fact that they have not been identified by immunohistochemistry.[1] The alternative seems to be innervation of these hypothalamic regions by the areas immediately adjacent to the lateral hypothalamus.

2.3.6. Innervation of the Cerebellum

Kan *et al.*[5,91] have reported the specific immunohistochemical localization of ChAT to the mossy fibers and the glomeruli of rabbit and human cerebellar folia, but not all mossy fibers are cholinergic. This immunohistochemical finding is consistent with previous reports that ChAT occurs in the white matter and in the granular cell layer but not in the molecular layer of the rabbit cerebellum,[92] that AChE staining is dense in glomeruli of the granular layer,[93-96] that mossy fiber- and glomeruli-enriched fractions of the cerebellum have relatively high concentrations of ACh[97] and ChAT,[98] and that the level of ChAT in the cerebellum of various mutant mouse strains is correlated with the density of mossy fibers.[99] The cholinergic mossy fibers presumably originate outside of the cerebellum, since ChAT levels drop sharply in the cerebellar folia after transection of the cerebral peduncles[100] or surgical isolation of the vermis.[101] Consistent with an extracerebellar origin is the finding that mouse cerebellar cultures only show significant ChAT activity if they include dorsal pontine tissue.[102] As yet, the source of these mossy fiber cholinergic inputs is unknown.

It has been suggested, on the basis of AChE staining, that Golgi cells of the cerebellum may be cholinergic.[96,101] However, these cells do not stain for ChAT, and present evidence would suggest that they are cholinoceptive, GABAergic cells that receive an input from the mossy fibers.[103] The presence of AChE in cholinoceptive as well as cholinergic cells is not unexpected, particularly in view of the unusually high ratio of AChE to ChAT activity in the cerebellum.[46]

2.3.7. Innervation of the Retina

Physiological evidence suggests the presence of cholinergic neurons in the retina.[104,105] Within the retina both ChAT and AChE are believed to be in a population of amacrine cells[104,106-109] that symmetrically line both margins of this layer.[110] Autoradiographic studies on [³H]choline uptake support this localization,[105] as do studies in rats treated with sodium glutamate in the neonatal period[111] or in chicks injected intraocularly with kainic acid.[112,113.] Almost no ChAT activity was found in the lesioned retinas, and there is considerable morphological evidence that these treatments preferentially destroy retinal ganglion cells and interneurons while leaving the photoreceptor and Muller cells intact. Uptake and lesion studies argue against cholinergic ganglion cells; the scattered cells in the ganglionic cell layer that take up [³H]choline are believed to be displaced amacrine cells.[114]

2.3.8. Auditory Pathways

Iontophoretic application of ACh has the same effect as stimulation of the olivocochlear bundle,[115,116] and the inhibitory effect elicited by stimulation is blocked by α-bungarotoxin (α-BTX).[117] This suggested that the olivocochlear bundle may be largely cholinergic, and the distribution of ChAT in various auditory nuclei in the mouse is said to support this possibility.[118] On the other hand, lesion data are said to indicate that ACh is not a transmitter between the auditory nerve and the cochlear nucleus[119] or between the cochlear or cortex and the inferior colliculus.[120]

2.3.9. Summary on Cholinergic Pathways

The foregoing discussion indicates that much has recently been learned regarding the nature of cholinergic systems in the CNS. Although doubt still exists regarding the presence or absence of cholinergic cell bodies in a few areas of brain, it seems probable that most have been located and assigned in cat brain. Immunohistochemical studies in progress on human, monkey, baboon, and rat brain and those reported recently on rabbit forebrain[121] suggest that there are some species differences but that the general patterns are similar. Considerable information is available about the intensity and distribution of terminal fields throughout the brain. However, much information is needed about the precise connections between the cell body groups and the terminal fields. A few of the major pathways have been well defined, but much work obviously remains to be done. So far, also, only a little information is available regarding the ultrastructural details of the terminals[78,122] the presumed cholinergic cells,[123,124] or the interconnections with other transmitter systems.[122]

An interesting possibility with regard to cholinergic neurons is that some may contain a second neurotransmitter, although whether both transmitters are physiologically active in such cases remains to be demonstrated. Acetylcholine has been said to occur along with vasoactive intestinal peptide (VIP) in autonomic ganglia of the cat[125,126] and along with dopamine in neurons of the bladder.[127] In connection with the possible cooccurrence of ACh with a catecholamine, it is interesting that most fetal neurons, at least from peripheral systems, appear to have the potential for either cholinergic or adrenergic function.[128] Glucocorticoids may play a major role in modulating development of such cells since, in culture, they specifically stimulate development of catecholamine synthetic enzymes while inhibiting development of ChAT as well as causing alterations in cell morphology.[129–132.] Medium conditioned on heart cell cultures, on the other hand, contain a factor that appears to influence development towards the cholinergic side.[130] Nerve growth factor (NGF) has been said by some to enhance activity of glucocorticoids on tyrosine hydroxylase (TH) synthesis[131] and, on the other hand, to promote ChAT while suppressing TH.[132]

3. THE CHOLINERGIC RECEPTOR

3.1. Types of Cholinergic Receptors

Although the locus of action of a neurotransmitter is determined by the anatomic distribution of the nerve endings of its parent neuron, the nature of that action is determined by the type of receptor on the postsynaptic neuron. In the case of ACh, two types of cholinergic receptors have been distinguished for years. In this, as in so many other aspects of neurotransmitter function, the property of multiple receptors was first discovered in the cholinergic system and subsequently was found to have a much wider application. Multiple receptors are now presumed to exist for almost every known neurotransmitter.

The classical descriptive terms for the two types of cholinergic receptors, "muscarinic" and "nicotinic," derive from the fact that the physiological effects of ACh at certain sites can be mimicked by nicotine, an alkaloid from *Nicotiana tobacum*, whereas others can be mimicked by muscarine, a drug obtained from the fungus *Amanita muscaria*.

The nicotinic receptor occurs in autonomic ganglia of both the sympathetic and parasympathetic systems in cells that are innervated by preganglionic cholinergic fibers. The neuromuscular junction, innervated by cholinergic fibers from anterior horn cells, also has a nicotinic-type receptor. α-Bungarotoxin (α-BTX) is the specific blocking agent. The muscarinic receptor occurs in smooth muscle innervated by postganglionic fibers of the parasympathetic system. This also includes cardiac muscle and certain exocrine glands. Atropine is the classic blocking agent.

The nicotinic receptor is not stereospecific, whereas the muscarinic receptor is highly stereospecific.[133] The nicotinic receptor is linked to ionotropic action, whereas the muscarinic receptor is linked to metabotropic action.[134]

The classic source of nicotinic receptors is the electroplax organ of the eel *Electrophorus electricus*, which has 10^{-9}–10^{-10} receptors per organ.[133] This concentration of receptors, combined with the strong binding characteristics of cobratoxin, paved the way for isolation and purification of the nicotinic receptor.

The nicotinic receptor is a glycoprotein of molecular weight about 250,000.[135] The receptor appears to be a hydrophilic protein of elongated shape, ideal for traversing the thickness of the membrane.[136] It is made up of subunits (most likely four), each of which has a molecular weight of about 40,000 and binds one α-BTX molecule.

In the past, it has often been suggested that AChE corresponds to the ACh receptor. This is clearly incorrect. Acetylcholinesterase can be solubilized from synaptic membranes by high salt concentrations, whereas the receptor is tightly bound. The binding agent α-BTX interferes with cholinergic transmission but not with AChE *in vivo*. Its action is confined to binding with and blocking the nicotinic cholinergic receptor, which it does in both mammalian and submammalian species. Furthermore, it has been shown by autoradiography that labeled α-BTX is attached to the surface of the folded myoneural junction, whereas AChE seems to be distributed throughout the postsynaptic membrane, including the depths of the folds.[137]

Work on isolation of the muscarinic cholinergic receptor is less advanced, although the specific binding sites for radioactive ligands such as QNB and propylbenzilylcholine mustard have been solubilized; binding seems to be to a single protein with a molecular weight of about 80,000.[138]

Central synapses have muscarinic- and nicotinic-like receptors, which respond somewhat differently from their peripheral counterparts. Considerable information is now rapidly accumulating on the distribution and pharmacology

Table III
Some Representative Binding Constants Obtained with Various Ligands and Tissues[a]

Radioactive ligand	Tissue	K_D (nM)	B_{max} (fmol/mg)	Ref.
QNB	Human hippocampus	0.07	20	264
QNB	Rat hippocampus	0.4	105	183
QNB	Rat hippocampus	0.6	1280	269
CD	Rat brain	1.8 and 123	8.5 and 132	270
α-BTX	Rat hippocampus	1.2	43	269
α-Naja toxin	Rat cortex	—	0.7	271

[a] As illustrated, there is considerable variation from laboratory to laboratory, although relative data from region to region are generally more consistent. Data given in terms of fmol/mg protein have been converted on the assumption of 10% protein. CD, [^{3}H]cis-methyldioxolane, a muscarinic agonist.

of such "receptors" by studies of the selective high-affinity binding of radioactively labeled agonists or antagonists; whether or not all such binding is to physiologically active receptors in vivo remains open to question. Typical affinity constants in such binding studies with various ligands are indicated in Table III. There has been considerable controversy as to whether α-BTX binding is really to true nicotinic sites in brain. These, like central muscarinic sites, appear to have some pharmacological differences from peripheral receptors. Such subtypes of receptors seem, however, to be relatively common in various neurotransmitter systems, and the balance of evidence indicates that the binding is to classical nicotinic sites in brain.[139-141] Some of the strongest evidence is that autoradiographic localization of α-BTX binding sites indicates good correlation with ACh pathways,[142-144] particularly in the hippocampus, where such sites appear primarily in layers innervated by septal afferents.[145,146] As exemplified in the section on innervation of the limbic system, the autoradiographic localization of muscarinic binding sites in brain also gives results consistent with known ACh systems.[66,147-149]

Such binding studies suggest that the hypothalamus, hippocampus, and tectum may be particularly rich in nicotinic receptors, whereas the striatum, hippocampus, and cortex are particularly rich in muscarinic receptors (Table I). It is estimated that brain, overall, has a 100-fold predominance of muscarinic receptors. It is known that some cholinoceptive cells, such as the noradrenergic cells of the superior cervical ganglion, possess both nicotinic and muscarinic receptors, so that a mixed action is possible on a given cholinoceptive cell. It is not known how commonly this may occur.

The basic question that needs to be asked is why should there be two different responses to ACh? What different physiological purposes are being achieved? For answers to such questions we have to look at the iontophoretic effects of ACh at various types of receptors and consider the mechanisms that may underlie the effects observed. Possible differences in the locations of different types of receptors, e.g., pre- as opposed to postsynaptic, and development of supersensitivity or subsensitivity also need to be considered.

3.2. Iontophoretic Effects of ACh on Central Neurons

• The iontophoretic action of ACh on brain cells varies widely. It may cause an excitation with rapid onset (nicotinic), an excitation with slow onset (muscarinic), or an inhibition with slow onset (muscarinic).

The classical example of nicotinic action of ACh on CNS cells is that of motor axon collaterals on Renshaw cells. The Renshaw cells respond with an extremely short latency to stimulation by ventral roots. This can be duplicated by iontophoretic application of ACh or nicotine. The activity is blocked by the nicotinic agonists dihydro-β-erythroidine or hexamethonium. Other areas where somewhat similar effects can be observed are the medulla, areas of the thalamus and hypothalamus, and the cerebellum. Occasional cortical cells also respond similarly.[15,150,151] The depolarizing effects at such receptors are ionotropic; i.e., they are associated with a decrease in membrane resistance and the mechanism, like that at the neuromuscular junction, involves opening of Na^+ and K^+ channels by the neurotransmitter.

A much more commonly observed response to iontophoretically applied ACh on CNS cells is a slow and prolonged excitatory action that is blocked by atropine. With extracellular application of ACh and intracellular recording of the response, it can be quite clearly demonstrated that this depolarizing effect is not associated with the decrease in membrane resistance that would accompany the opening of typical sodium and other ionic channels. Instead, the membrane resistance increases, with the depolarizing action having a reversal level close to -100 mV. This result can be explained by a decrease in the conductance of either Cl^- or K^+, which have equilibrium potentials in this vicinity. Since intracellular injections of Cl^- cause large positive shifts in the Cl^- equilibrium potential but do not change the character of response to ACh, it must be concluded that a reduction in K^+ conductance (G_K) is associated with the effect. Such slow excitatory effects, blocked by atropine, are seen with cells in the caudate, hippocampus, pyriform cortex, lateral and medial geniculate bodies, ventrobasal thalamus, and other areas. Even Renshaw cells have some of these muscarinic receptors.[151]

Acetylcholine can also depress the action of a wide variety of CNS cells in the cortex, hypothalamus, pons, medulla, and other areas.[15,150,151] This inhibitory action is also clearly muscarinic because it is slow in onset and can easily be blocked by atropine. In this case, however, the membrane changes are thought to be associated with an increase rather than a decrease in K^+ conductance. And for both the excitatory and inhibitory responses, the change in K^+ conductance is believed to be secondary to chemical reactions in the membrane induced by the neurotransmitter. Hence, such responses have been termed "metabotropic" in nature.[134]

What are the chemical reactions in the membrane initiated by ACh at muscarinic receptors? Since the pioneering work of Sutherland, great attention has been focused on the role of cyclic nucleotides as the intracellular second messengers for the action of many hormones on their target cells.[152,153] This concept has been extended by Greengard and his colleagues to include their role as intracellular second messengers for certain neurotransmitters. For example, cyclic adenosine-3',5'-monophosphate (cyclic AMP) has been suggested as the second messenger for dopamine, whereas cyclic guanosine-3',5'-monophosphate (cyclic GMP) has been suggested as the second messenger for the muscarinic actions of ACh.[154] Cyclic GMP is not exclusive to such muscarinic receptors; some histamine and β-adrenergic receptors also seem to be coupled to a guanylate cyclase.

The hypothesis is that the neurotransmitter (in this case, ACh) does not directly open ionic channels. Instead, it induces a conformational change on the external surface of the membrane, which activates a generator of the second messenger (in this case, guanylate cyclase) in the membrane. The second messenger, cyclic GMP, then initiates a series of chemical reactions. These commence with the phosphorylation of protein kinases that are dependent on the cyclic nucleotide. It is this family of protein kinases that is actually responsible for the variety of actions that might take place in the postsynaptic cell. The specificity of action would then depend on the kinases and their substrate proteins.

Among these substrate proteins could be ones controlling membrane permeability and altering either the ion conductance or the electrogenic pump. Thus, a metabotropic neurotransmitter could indirectly produce ionotropic effects of either an excitatory or an inhibitory nature, although the effects would have to be slower in onset because of the chemical method of initiation. Termination of the action of cyclic GMP is brought about by phosphodiesterase hydrolysis, and that of protein kinases by phosphoprotein phosphatases.

Most of the chemical components of the proposed system have been identified, but it still remains speculation whether ACh or any other transmitter actually operates through such a metabotropic scheme.

There is considerable evidence from *in vitro* work that ACh does stimulate cyclic GMP formation in brain slices and that the effect can be blocked by muscarinic but not by nicotinic antagonists. Cholinomimetic agents with a predominantly muscarinic action such as methacholine, pilocarpine, or bethanechol also cause an increase in cyclic GMP, whereas nicotinic cholinomimetics do not. In these experiments, little or no effect is seen on the content of cyclic AMP, which is believed to be associated with dopamine and some other types of receptors.[155-157] These results are all compatible with the hypothesis that the nicotinic actions of ACh are ionotropic and the muscarinic ones are metabotropic, with cyclic GMP as the postsynaptic mediator. A direct correlation between the slowly developing membrane changes seen in physiological experiments and the increased cyclic GMP formation found in biochemical studies has never been made; however, it does not seem unreasonable to hypothesize that they are causally related.

3.3. Location and Plasticity of Receptors

There is considerable evidence for autoreceptors on some, but not all, ACh nerve endings through which there is feedback control of synaptosomal activity.[158] Both nicotinic and muscarinic receptors may apparently occur preas well as postsynaptically, although there is probably considerable regional variation, and the presynaptic autoreceptors seem to be predominantly muscarinic. Pharmacological evidence, for example, suggests that both nicotinic and muscarinic presynaptic sites may regulate ACh release in the hippocampus,[159] but almost all of these autoreceptors seem to be muscarinic.[160,161] In the cortex, nicotinic sites are said to be postsynaptic, whereas muscarinic sites are both pre- and postsynaptic.[162,163] Autoreceptors presumably form only a small part of the total receptor population, since lesion experiments indicate that the great majority of binding sites for both QNB and α-BTX in areas such as the hippocampus are on postsynaptic membranes.[164–166]

Some receptors may also be on other types of nerve endings, so that action of ACh at these sites may regulate the release of other neurotransmitters. Thus, for example, some QNB binding sites in the hippocampus have been reported to be on nerve endings of the entorhinal projection.[167]

Cholinergic receptors may also exist on nerve endings presynaptic to cholinergic systems so that a synaptic dialogue can be used as a feedback control mechanism. There is good evidence, for example, that cholinergic agents modulate the release of dopamine in the striatum.[168–175] Although some have suggested that such regulation is through axoaxonal contacts, electron microscopic work suggests that an apparatus exists for the release of ACh from the postjunctional dendrite.[176] Binding data, however, are controversial, with some, but not all, laboratories reporting decreases in [^{125}I]α-BTX or [^{3}H]QNB binding in the striatum of rats following 6-hydroxydopamine-induced degeneration of the nigrostriatal dopamine tract.[177–180]

Acetylcholine receptors appear to have considerable plasticity. It has been demonstrated that the number of [^{3}H]QNB binding sites in areas such as the cortex, striatum, and hippocampus decreases in rats chronically exposed to excess cholinergic stimulation by treatment with DFP[181] or oxotremorine[182]; the numbers increase after chronic blockade with atropine[183,184] or scopolamine.[182] Denervation also induces increased QNB binding in the hippocampus[184] and cortex.[185] Some behavioral evidence for supersensitivity to ACh or oxotremorine has also been reported following chronic atropine administration.[186] Chronic treatment with drugs also appears to affect the sensitivity of autoreceptors. In hippocampal synaptosomes from rats chronically stimulated with paraoxon (an AChE inhibitor) or blocked with scopolamine, extracellular ACh is, respectively, less or more efficient in inhibiting release of ACh elicited by depolarization.[187]

Although denervation supersensitivity has been reported in skeletal muscle, there is as yet little evidence for plasticity of central nicotinic receptors.

Although much remains to be learned about cholinergic receptors, it is already apparent that there are probably multiple types, occurring in a variety of locations, and serving such diverse functions as modulation of the postsynaptic membrane, feedback control of acetylcholine release, and modulation of the release of other neurotransmitters. There is also already ample evidence that abnormalities in cholinergic receptors may play as important a role in human pathology as abnormalities in the cholinergic neurons.

4. CHOLINERGIC SYSTEMS IN HUMAN PATHOLOGICAL CONDITIONS

The most significant development relating cholinergic systems to human pathology is the discovery that myasthenia gravis is related to the production of antibodies to the nicotinic cholinergic receptors at the neuromuscular junction.[188] Myasthenia gravis is a neuromuscular disease characterized by weakness and fatigability of muscle. As already mentioned, purification of the ACh receptor protein from electric eels was made possible using α-cobra toxin attached to Sepharose beads to bind the solubilized ACh receptor, which could be subsequently eluted by means of a specific competing ligand. Injection of the purified ACh receptor glycoprotein into rabbits in an attempt to raise antibodies led to the development in the animals of marked muscular weakness and respiratory insufficiency. Further studies in various species indicated many other similarities between the animal model of experimental allergic myasthenia gravis (EAMS) produced by the injection of purified receptor protein and the human disease. Measurements of [^{125}I]α-BTX binding sites indicate a 70–90% reduction in the number of such sites in biopsies from myasthenic muscles as compared to controls, and antireceptor antibodies have now been identified in the majority of myasthenic gravis patients. It is now accepted that myasthenia gravis is an autoimmune disease in which the circulating antibody both blocks the receptor and causes accelerated degradation.[189,190] The autoimmune attack is directed specifically against the ACh nicotinic receptors. The use of thymectomy and adrenal corticosteroids in the treatment of myasthenia gravis is predicated on their apparent interference with the autoimmune reaction.[188,191,192]

An even more exciting approach to the therapy of autoimmune diseases in general and myasthenia gravis in particular has been suggested by Fuchs.[193] She found that injection of a methylated derivative of the ACh receptor protein into rabbits did not induce EAMS but, in fact, had both preventive and therapeutic effects on EAMS induced by the injection of the purified parent protein. The chemically modified ACh receptor preparation did elicit the formation of antibodies that cross reacted with the intact receptor, as shown by studies using [^{125}I]α-BTX. One hypothesis is that effective blockade of the receptor and promotion of its degradation requires that the antibody react with the receptor at more than one site; if the antibody to the modified receptor protein reacts with the receptor only at a single site, it may be ineffective in blocking the receptor and inducing its degradation and, in addition, may prevent the antibodies to the unmodified receptor from establishing the normal antigen–antibody linkage. Whether or not such modified antigens can be of real therapeutic value remains to be seen. In any case, myasthenia gravis is a clear example of a peripheral disorder involving pathology of a cholinergic system.

Work in the CNS is much more difficult because of the greater complexity and relative inaccessibility of the systems involved. Nevertheless, there is clear evidence that cholinergic systems in the brain are involved in the pathology of Huntington's disease and senile dementia of the Alzheimer type (SDAT). They may also be important in the neurological symptomatology produced by pro-

Table IV
Choline Acetyltransferase Activity in SDAT as Percent of That in Controls

Region	Data of Rossor et al.[202]	Unpublished data
Frontal cortex	72%[a]	57%[a]
Parietal cortex	47%[a]	37%[a]
Motor cortex	69%	70%
Mid-temp. gyrus	67%	63%
Anterior hippocampus	87%	—
Posterior hippocampus	46%[a]	—
Caudate nucleus	83%	78%
Globus pallidus	107%	—

[a] Significant difference between SDAT and control data.

longed drug treatment (tardive and *levo*-DOPA-induced dyskinesias), certain viruses (scrapie), thiamine deficiency, or such toxins as lead and DDT. Involvement of cholinergic systems in the etiology of epilepsy, abnormal aggressivity, and psychotic diseases has also frequently been hypothesized, but the evidence varies from slim to inconclusive. In every instance involving the CNS, the available evidence suggests pathological involvement of other transmitter systems as well. We shall discuss here senile dementia and Huntington's disease as examples of conditions in which cholinergic pathology seems to be important. Possible cholinergic involvement in a variety of other pathological conditions has been reviewed elsewhere.[194]

4.1. Senile Dementia of the Alzheimer Type (SDAT)

This condition is characterized by the appearance of numerous senile plaques and neurofibrillary tangles in the brain, particularly in hippocampal and cortical areas. Clinically, the symptoms include loss of memory and dementia and are frequently of such severity that the victims require institutional care. It is estimated that dementia is now the most prominent illness in 60% of the residents in nursing homes, and the problem is growing with the age of the population. Although SDAT is a frequent cause of such symptoms, it must be recognized that clinical dementia can occur in patients from other causes such as vascular pathology. Such cases do not show the classical tangles and plaques, and there may be no significant involvement of central cholinergic systems.[195]

Considerable excitement has been generated by reports of specific defects in the cholinergic systems of the hippocampus and cortex in SDAT. There are indications that losses similar in kind but lesser in extent to those of SDAT may also occur in normal aging[196] and that central cholinergic pathways may be involved in human age-related memory degeneration. As a result, it has been speculated that choline agonists may help to restore some aspects of memory in aged normal individuals as well as in persons with SDAT. Actual clinical data are so far not highly encouraging.[197-199]

The data in Table IV are qualitatively typical of those from a large number of laboratories, indicating that some cortical and hippocampal tissue from cases of SDAT show markedly less ChAT activity than do similar samples from age-matched normal controls. There are some quantitative variations from laboratory to laboratory, with Davies,[200,201] for example, finding losses of ChAT in the hippocampus in SDAT of more than 90%, whereas others[202–212] find losses of only 50–70%. There is also some controversy as to whether there are significant losses in ChAT in other areas of brain such as the caudate, or in muscarinic binding sites in the hippocampus or cortex.[200,211–214] These discrepancies may reflect differences in dissection, diagnosis, or severity of the disease. The losses, according to Rossor *et al.* (Table IV), may be limited to certain small regions within affected nuclei, and Perry *et al.*[209] found that persons dying with multiinfarct dementia showed considerably higher ChAT activities than those with SDAT.

The losses of ChAT in SDAT might reflect a dying back of nerve endings in the hippocampus and cortex, a decreased axonal transport of enzyme protein from the cell bodies to the nerve endings, and/or a loss of some of the cholinergic cells. As mentioned in Section 2, the cholinergic innervation of the hippocampus comes from the septal area, whereas much of that of the cortex comes from the substantia innominata (SI) (which also projects to the amygdala). Rossor *et al.*[215] recently reported that cases of SDAT showed decreased ChAT activity in the SI and amygdala but not in the adjacent lentiform nucleus or hypothalamus. Using Nissl-stained histological sections, Whitehouse *et al.*[216,217] showed very dramatic (80–90%) losses of neurons in the SI in six cases of SDAT. Using the immunohistochemical method of ChAT, we have also seen evidence of cell shrinkage and loss in the SI in SDAT,[218] although the decreases we have seen are not as dramatic as those reported by Whitehouse *et al.*

The apparent losses in cholinergic function in the cortex and hippocampus in SDAT are of particular interest with regard to the defects of memory in this condition, since there is considerable evidence linking ACh and these areas to memory. Administration of anticholinergics such as scopolamine to normal subjects produces amnesic effects that do not correlate with the drowsiness produced by the drug.[219–223] Similarly, administration of cholinomimetics such as arecoline or physostigmine has been found to improve some aspects of learning in normal subjects.[223–225] An association between hippocampal lesions and memory dysfunction in humans has been recognized for many years.[226] There is, moreover, considerable evidence from pharmacological and lesion studies in animals that cholinergic systems are involved in learning and memory, although there is often difficulty in studying an effect on memory *per se* as opposed to effects on movement and motivation in such animal work.[227–231]

The findings of this decreased cholinergic activity in the hippocampus and cortex have led to extensive trials of possible acetylcholine precursors, such as 2-dimethylaminoethanol (Deanol), lecithin, or choline itself, for the treatment of senile dementia and, in a few research studies, for the possible improvement of memory in "normal" aged individuals. Significant clinical improvement has been noted in only a small percentage of the cases of senile dementia that have been treated with these substances.[197–199,232]

Several possible explanations illustrate the difficulties in such pharmacological trials. First, there is no definitive evidence that these materials actually increase ACh levels in human brain. Even in rats given these materials as diet supplements, where ACh measurements in brain can be made, the data are controversial.[233-242] Second, even if such supplements were to produce elevated ACh levels in brains with normally functioning cholinergic neurons, it is not certain that they would do so in brains where these neurons were defective. This is because these materials are not direct cholinergic agonists; they are rather presumed precursors that depend on enzymatic reactions in the appropriate neurons to convert them to the neurotransmitter.

Two essential differences exist between the successful use of the precursor DOPA as a supplement to raise dopamine levels in parkinsonism and the use of a choline derivative as a supplement to raise ACh levels in memory deficiency. The first is that the enzyme ChAT is specific to cholinergic neurons; if these are defective, there is no other source of the enzyme to make the ACh. By contrast, the decarboxylase that converts DOPA to dopamine is present in excess in brain and is in serotoninergic as well as in dopaminergic and noradrenergic neurons. The second and perhaps more important difference is that decarboxylation to dopamine is the only important metabolic path through which brain DOPA may pass. Choline, on the other hand, can be converted to a variety of phospholipids and intermediates of phospholipid metabolism, and only a minor amount is converted to ACh. The free choline in brain accounts for less than a tenth of a percent of total choline derivatives.[243] Moreover, even if these supplements did result in enhanced ACh levels in diseased human brain, the greatest enhancement might occur in areas where the cholinergic neurons are normal or near normal rather than in those where they are grossly defective; hence, undesirable side effects may outweigh the desired action.

Some of these difficulties might be circumvented by the use of direct cholinergic agonists or anticholinesterases safer than physostigmine. Much effort is being directed along such lines. A serious problem that remains, however, is the uncertainty as to whether cholinoceptive and other neuronal types in the affected areas are also damaged. As already mentioned, available data about the binding sites are conflicting. The appearance of tangles and plaques in senile dementia would suggest that neurons of the hippocampus and cortex degenerate, and the chemical identity of these neurons has not yet been defined. They are probably not cholinergic neurons, since available evidence suggests that there are no cholinergic perikarya in the hippocampus and few if any in the cortex. Types of neurons known to exist in the cortex and hippocampus include those using GABA, glutamate/aspartate, vasoactive intestinal peptide (VIP), choleocystokinin (CCK), and probably somatostatin. There is reasonably good evidence that GABA and VIP neurons are not adversely affected in Alzheimer's disease. No data are yet available on glutamate/aspartate systems[244] except that levels of ornithine δ-transaminase, an enzyme possibly involved in glutamate formation, is not decreased in the frontal cortex.[205] Somatostatin and CCK, however, are decreased in some cortical areas.[245-247] There are a few reports suggesting losses of catecholamines in SDAT, but these are unconvincing, and it is generally agreed that neither the catecholamines nor serotonin are specifically affected in most cases[244]; in any case, losses in aromatic amines, like those in ChAT, could not be attributed to neuronal death in the cortex or hippocampus.

It appears that the pathology of SDAT is more complicated than a simple loss of cholinergic function, but it is still hoped that cholinergic agonists may be useful in maintaining function during early stages of the disease.

4.2. Extrapyramidal Disorders

Huntington's disease (HD) is a well-known degenerative disease of the extrapyramidal system. It is inherited in an autosomal dominant pattern with complete penetrance. Its symptoms include psychosis, dementia, and involuntary choreoathetoid movements of the trunk and extremities; these movements decrease markedly during sleep. It has long been known that the most striking pathological finding in the brains of patients with HD is a marked atrophy in the basal ganglia, especially in the neostriatum. There is also some atrophy in the cerebral cortex.[249]

The possible involvement of cholinergic systems of the striatum in HD was postulated as early as 1963 on clinical pharmacological grounds.[250] It was known then that dopamine and acetylcholine were both heavily concentrated in the striatum, that agents that depleted or blocked dopamine could produce a parkinsonianlike syndrome that could be alleviated by treatment with anticholinergics, and that the occasional patient receiving overdoses of anticholinergics would develop choreic symptoms. On this basis, it was suggested that there was a balance between dopaminergic and cholinergic systems in extrapyramidal function such that a relative loss of dopaminergic function would lead to parkinsonian symptoms and a relative loss of cholinergic function to choreic symptoms. Experimental evidence indicating a loss in dopaminergic activity in the extrapyramidal system in Parkinson's disease was rapidly accumulated, but it was not until 1973 that data began to appear indicating a specific loss of ChAT in the neostriatum of patients dying with HD.[251] Rather similar percentage losses of this enzyme in the putamen and caudate nucleus have been reported from a number of laboratories.[205,252–254]

Two groups have reported that the loss of cholinergic systems in the putamen and caudate is patchy in HD, with some areas showing normal values and others extremely low values[251,252]; in similar dissections of the brains from neurologically normal humans, no such great diversity of ChAT values have been found within the caudate and putamen,[46] although there is evidence from one group that the rostromedial caudate may contain somewhat less ChAT than other regions of this nucleus.[252]

Cholinergic systems in most other areas of brain do not seem to be affected in HD. Aquilonius et al.[252] looked at ChAT activity in 17 other areas of brain and found no significant abnormality, whereas McGeer and McGeer[46] assayed some 55 brain regions and found significantly low ChAT activity only in the caudate and locus coeruleus; the latter region was only studied in three patients dying with HD and five controls, but the mean for the HD patients was only

40% of the control mean. Reisine *et al.*[255] recently reported data on the amygdala, hippocampus, globus pallidus, and anterior cerebellar vermis that again indicate no significant abnormalities, although the means are all less than the means in control cases.

The data indicating a cholinergic dysfunction in HD have led to numerous clinical trials of agents such as choline[256-258] or Deanol in HD (W. R. Brown and P. L. McGeer, unpublished observations).[259] Lecithin has also been used as a presumptive precursor of ACh.[198] Most reports indicate that these treatments have marginally significant or no beneficial effect on HD. The direct choline agonist arecoline has even been reported to exacerbate choreic movements.[260] Some general reasons have already been given as to why supplements such as choline and lecithin may be clinically inactive in conditions involving a cholinergic deficiency. In HD, particularly, it is becoming evident that the cholinergic deficiency in the caudate–putamen is only a small part of the chemical pathology of this disease; insofar as the biochemical neuroanatomy of the extrapyramidal system is known, it appears that all neurons with somata in the caudate and putamen may be affected in HD, whereas neurons with their soma lying outside that region are generally spared. Thus, striatal neurons using as neurotransmitters GABA, substance P, enkephalin, ACh, and probably other unidentified substances are all affected.[261]

In view of this generalized loss of neurons in the caudate and putamen, it is not surprising that the density of [^{3}H]QNB binding sites ("muscarinic cholinergic receptors") in the striatum is decreased in HD,[262-265] and cholinergic agonists would certainly not be expected to help greatly in patients with such a loss in the cholinoceptive neurons.

5. CONCLUSIONS

Modern histochemical, biochemical, and pharmacological techniques are rapidly providing long sought after information regarding cholinergic systems and their functions in the CNS. Many of the major cholinergic cell groups in brain are now known, and data regarding anatomic pathways are currently being accumulated. The location and types of cholinergic receptors in brain are also being assessed now that powerful and specific ligands are available. New interpretations of old pharmacological data are now possible, and more meaningful approaches to the roles of cholinergic systems in human disease can be undertaken.

Purification of the nicotinic receptor from electric organs and preparation of antibodies to it were critical to the development of an understanding of myasthenia gravis as an autoimmune disease that destroyed nicotinic receptors at the myoneural junction.

Centrally, the involvement of cholinergic systems in senile dementia of the Alzheimer type (SDAT) and in Huntington's disease has been established. The possible involvement of cholinergic systems in epilepsy, mental disease, tardive dyskinesia, and other disorders is under investigation.

REFERENCES

1. Kimura, H., McGeer, P. L., Peng, J. H., and McGeer, E. G., 1981, *J. Comp. Neurol.* **200:**151–201.
2. McGeer, P. L., McGeer, E. G., Singh, V. K., and Chase, W. H., 1974, *Brain Res.* **81:**373–379.
3. Eng, L. F., Uyeda, C. T.. Chao, L. P., and Wolfgram, F., 1974, *Nature* **250:**243–245.
4. Cozzari, C., and Hartman, B. K., 1977, *Proc. Int. Soc. Neurochem.* **6:**140.
5. Kan, K. S. K., Chao, L. P., and Eng, L. F., 1978, *Brain Res.* **146:**221–230.
6. Lehmann, J., and Fibiger, H. C., 1979, *Life Sci.* **25:**1939–1947.
7. McGeer, E. G., Wada, J. A., Terao, A., and Jung, E., 1969, *Exp. Neurol.* **24:**277–284.
8. Fonnum, F., 1970, *J. Neurochem.* **17:**1029–1037.
9. Shute, C. C. D., and Lewis, P. R., 1963, *Nature* **199:**1160–1164.
10. Gottesfeld, Z., and Jacobowitz, D. M., 1979, *Brain Res.* **176:**391–394.
11. Coyle, J. T., Molliver, M. E., and Kuhar, M. J., *J. Comp. Neurol.* **180:**301–324.
12. McGeer, P. L., and McGeer, E. G., 1982, *Critical Reviews in Toxicology*, Volume 10 (L. Goldberg, ed.), CRC Press, Boca Raton, Florida, pp. 1–26.
13. Jenden, D. J., 1975, *Cholinergic Mechanisms* (P. G. Waser, ed.), Raven Press, New York, pp. 87–98.
14. Wenk, H., Bigl, V., and Meyer, U., 1981, *Brain Res. Rev.* **2:**295–316.
15. Phillis, J. W., 1975, *Neurohumoral Coding of Brain Function* (R. D. Myers and R. R. Drucker-Colin, eds.), Plenum Press, New York, pp. 55–77.
16. Green, J. R., Halpern, L. M., and Van Niel, S., 1970, *Brain* **93:**57–64.
17. Hebb, C. O., Krnjevic, K., and Silver, A., 1963, *Nature* **198:**692.
18. Kelly, P. H., and Moore, K. E., 1978, *Exp. Neurol.* **61:**479–484.
19. Hartgraves, S. L., Mensah, P. L., and Kelly, P. H., 1979, *Soc. Neurosci.* **5:**590.
20. Johnston, M. V., and Coyle, J. T., 1979, *Soc. Neurosci. Abstr.* **5:**116.
21. Ulmar, G., Ljungdahl, A., and Hokfelt, T., 1975, *Exp. Neurol.* **46:**199–208.
22. Shute, C. C. D., and Lewis, P. R., 1967, *Brain* **90:**497–522.
23. Divac, I., 1975, *Brain Res.* **9:**385–398.
24. Jones, E. G., Burton, H., Saper, C. B., and Swanson, L. W., 1976, *J. Comp. Neurol.* **167:**385–420.
25. Lehmann, J. C., Nagy, J. I., Atmadja, S., and Fibiger, H. C., 1980, *Neuroscience* **5:**1161–1174.
26. Mesulam, M. M., and Van Hoesen, G. W., 1976, *Brain Res.* **109:**152–157.
27. Pedata, F., Conte, G. L., Sorbi, S., Marconcini-Pepeu, I., and Pepeu, G., 1982, *Brain Res.* **233:**359–367.
28. Johnston, M. V., and Coyle, J. T., 1979, *Brain Res.* **170:**135–155.
29. Ross, C. D., Godfrey, D. A., Williams, A. D., and Matschinsky, F. M., 1978, *Soc. Neurosci.* **4:**264.
30. Youngs, W. M., Nadi, N. S., Davis, B. J., Margolis, F. L., and Macrides, F., 1979, *Soc. Neurosci.* **5:**36.
31. De Olmos, J., Hardy, H., and Heimer, L., 1978, *J. Comp. Neurol.* **181:**213–244.
32. Kimura, H., McGeer, P. L., Peng, F., and McGeer, E. G., 1980, *Science* **208:**1057–1059.
33. McGeer, P. L., McGeer, E. G., Fibiger, H. C., and Wickson, V., 1971, *Brain Res.* **35:**308–314.
34. Butcher, S. G., and Butcher, L. L., 1974, *Brain Res.* **71:**167–171.
35. Wagner, A. Hassler, R., and Kim, J. S., 1975, *Proc. Int. Soc. Neurochem.* **5:**116.
36. Simke, J. P., and Saelens, J. K., 1977, *Brain Res.* **126:**487–495.
37. Saelens, J. K., Edwards-Neale, S., and Simke, J. P., 1976, *J. Neurochem.* **32:**1093–1094.
38. Jacobowitz, D. M., and Palkovits, M., 1974, *J. Comp. Neurol.* **157:**13–28.
39. Spehlmann, R., Norcross, K., and Grimmer, E. J., 1978, *Brain* **101:**649–660.
40. Butcher, L. L., Talbot, K., and Bilezikjian, L., 1975, *J. Neural. Transm.* **37:**127–153.
41. Lehmann, J., and Fibiger, H. C., 1978, *J. Neurochem.* **30:**615–624.
42. Kemp, J. M., and Powell, T. P. S., 1971, *Phil. Trans.* [*Biol.*] **262:**383–401.

43. Graybiel, A. M., Ragsdale, C. W., and Edley, S. M., 1979, *Exp. Brain Res.* **34**:189–195.
44. Fonnum, F., Walaas, I., and Iversen, E., 1977, *J. Neurochem.* **29**:221–230.
45. Walaas, I., and Fonnum, F., 1979, *Brain Res.* **177**:325–336.
46. McGeer, P. L., and McGeer, E. G., 1976, *J. Neurochem.* **26**:65–76.
47. Olivier, A., Parent, A., Simard, H., and Poirier, L. J., 1970, *Brain Res.* **18**:273–282.
48. McGeer, E. G., Fibiger, H. C., McGeer, P. L., and Brooke, S., 1973, *Brain Res.* **52**:289–300.
49. Fonnum, F., Gottesfeld, Z., and Grofova, I., 1978, *Brain Res.* **143**:125–128.
50. Kataoka, K., Bak, I. J., Hassler, R., Kim, J. S., and Wagner, A., 1974, *Exp. Brain Res.* **19**:217–227.
51. Nagy, N. I., Vincent, S. R., Lehmann, J., Fibiger, H. C., and McGeer, E. G., 1978, *Brain Res.* **149**:431–441.
52. Kuhar, M. J., 1976, *The Hippocampus*, Volume 1 (R. L. Isaacson and K. H. Pribram, eds.), Plenum Press, New York, pp. 269–284.
53. Kuhar, M. J., Sethy, V. H., Roth, R. H., and Aghajanian, G. K., 1973, *J. Neurochem.* **20**:581–593.
54. Lewis, P. R., and Shute, C. C. D., 1967, *Brain* **90**:521–542.
55. Marshall, K. C., Flumerfelt, B. A., and Gwyn, D. G., 1980, *Brain Res.* **190**:493–504.
56. Pepeu, G., Mulas, A., Ruffi, A., and Sotgiu, P., 1971, *Life Sci.* **10**:181–184.
57. Sethy, V. H., Roth, R. H., Kuhar, M. J., and Van Woert, M. H., 1973, *Neuropharmacology* **12**:819–823.
58. Atwah, S. T., and Kuhar, M. J., 1976, *Eur. J. Pharmacol.* **37**:311–319.
59. Dudar, J. D., 1975, *Brain Res.* **83**:123–133.
60. Smith, C. M., 1974, *Life Sci.* **14**:2159–2166.
61. Wheal, H. V., and Miller, J. J., 1980, *Brain Res.* **182**:145–155.
62. Fonnum, F., and Walaas, I., 1978, *J. Neurochem.* **31**:1173–1181.
63. Schwarcz, R., Zaczek, R., and Coyle, J. T., 1978, *Eur. J. Pharmacol.* **50**:209–220.
64. Oderfeld-Nowak, B., Narkiewicz, O., Bialowas, J., Dabrowska, J., Wieraszko, A., and Gradkowska, M., 1974, *Acta Neurobiol. Exp.* **34**:583–586.
65. Storm-Mathisen, J., 1970, *J. Neurochem.* **17**:739–750.
66. Kuhar, M. J., and Yamamura, H. I., 1976, *Brain Res.* **110**:229–243.
67. Bland, B. H., Kostopoulos, G. K., and Phillis, J. W., 1974, *Can. J. Physiol. Pharmacol.* **52**:966–971.
68. Rose, A. M., Hattori, T., and Fibiger, H. C., 1976, *Brain Res.* **108**:170–174.
69. Emson, P. C., Paxinos, G., Le Gal La Salle, G., Ben-Ari, Y., and Silver, A., 1979, *Brain Res.* **165**:271–282.
70. Nagai, T., McGeer, P. L., and McGeer, E. G., 1982, *J. Neurosci.* **2**:513–520.
71. Kataoka, K., Nakamura, Y., and Hassler, R., 1973, *Brain Res.* **62**:264–267.
72. Kuhar, M. J., DeHaven, R. N., Yamamura, H. I., Rommelspacher, H., and Simon, J. R., 1975, *Brain Res.* **97**:265–275.
73. Leranth, C. S., Brownstein, M., Zaborsky, L., Jaranyi, Z. S., and Palkovits, M., 1975, *Brain Res.* **99**:124–128.
74. Mata, M. M., Schrier, B. K., and Moore, R. Y., 1977, *Exp. Neurol.* **57**:913–921.
75. Sorimachi, M., and Katoaka, K., 1974, *Brain Res.* **72**:350–353.
76. Cuello, A. C., Emson, P. C., Paxinos, T., and Jessell, T., 1978, *Brain Res.* **149**:413–423.
77. Emson, P. C., Cuello, A. C., Paxinos, G., Jessel, T., and Iversen, L. L., 1977, *Acta Physiol. Scand. [Suppl.]* **452**:43–46.
78. Hattori, T., McGeer, E. G., Singh, V. K., and McGeer, P. L., 1977, *Exp. Neurol.* **55**:666–679.
79. Gottesfeld, Z., and Jacobowitz, D. M., 1978, *Brain Res.* **156**:329–332.
80. McGeer, E. G., Scherer-Singler, U., and Singh, E. A., 1979, *Brain Res.* **168**:375–376.
81. Vincent, S. R., Staines, W. A., McGeer, E. G., and Fibiger, H. C., 1980, *Brain Res.* **195**:479–484.
82. Wilson, P. W., and Watson, C., 1980, *Brain Res.* **201**:418–422.
83. Hoover, D. B., and Baisden, R. H., 1980, *Brain Res. Bull.* **5**:519–524.
84. Brownstein, M., Kobayashi, R., Palkovira, M., and Saavedra, J. M., 1975, *J. Neurochem.* **24**:35–38.

85. Hoover, D. B., Muth, E. A., and Jacobowitz, D. M., 1978, *Brain Res.* **153**:295–306.
86. Carson, K. A., Nemeroff, C. B., Rone, M. S., Youngblood, W. W., Prange, A. J., Jr., Hanker, J. S., and Kizer, J. S., 1977, *Brain Res.* **129**:169–173.
87. Nemeroff, C. B., Lipton, M. A., and Kizer, J. S., 1978, *Dev. Neurosci.* **1**:102–109.
88. Walaas, I., and Fonnum, F., 1978, *Brain Res.* **153**:549–562.
89. Karscú, S., Tóth, L., Király, E., and Jancsó, G., 1981, *Brain Res.* **206**:203–207.
90. Meyer, D. K., and Brownstein, M. J., 1980, *Brain Res.* **193**:566–569.
91. Kan, K.-S. K., Chao, L. P., and Forno, L. S., 1978, *Brain Res.* **193**:165–171.
92. McCaman, R. E., and Hunt, J. M., 1965, *J. Neurochem.* **12**:253–259.
93. Kasa, P., Joo, F., and Csillik, B., 1965, *J. Neurochem.* **12**:31–35.
94. Csillik, B., Joo, F., and Kasa, P., 1963, *J. Histochem. Cytochem.* **11**:113–114.
95. Friede, R. L., and Fleming, L. M., 1964, *J. Neurochem.* **11**:1–17.
96. Shute, C. C. D., and Lewis, P. R., 1965, *Nature* **205**:242–246.
97. Israel, M., and Whittaker, V. P., 1965, *Experientia* **21**:325–326.
98. Balázs, R., Hajos, F., Johnson, A. L., Reynierf, G. L., Tapia, R., and Wilkin, G. P., 1975, *Brain Res.* **86**:17–30.
99. Mikoshiba, K., and Changeux, J. P., 1978, *Neurochem. Res.* **3**:680.
100. Fonnum, F., 1972, *Adv. Biochem. Psychopharmacol.* **6**:75–88.
101. Kasa, P., and Silver, A., 1969, *J. Neurochem.* **16**:389–396.
102. Woodward, W. R., Blank, N. K., and Seil, F. J., 1982, *Brain Res.* **241**:323–327.
103. Eccles, J. C., Ito, M., and Szentagothai, J., 1967, *The Cerebellum as a Neuronal Machine*, Springer-Verlag, Berlin, Heidelberg.
104. Ross, C. G., and McDougall, D. B., 1976, J. Neurochem. **26**:521–526.
105. Braughman, R. W., and Bader, C. R., 1977, *Brain Res.* **138**:469–485.
106. Braughman, R. W., 1979, *Developmental Neurobiology of Vision* (R. D. Freeman, ed.), Plenum Press, New York, pp. 421–432.
107. Graham, L. T., 1974, *The Eye*, Volume 6 (H. Davson and L. T. Graham, eds.), Academic Press, New York, pp. 283–342.
108. Hayashi, T., 1980, *Acta Histochem. Cytochem.* **13**:330–342.
109. Nichols, C. W., and Koelle, G. B., 1968, *J. Comp. Neurol.* **133**:1–15.
110. Masland, R. H., 1980, *Neurochem. Int.* **1**:501–518.
111. Karlsen, R. L., and Fonnum, F., 1976, *J. Neurochem.* **27**:1437–1441.
112. Morgan, I. G., and Ingham, C. A., 1981, *Neurosci. Lett.* **21**:275–280.
113. Schwarcz, R., and Coyle, J. T., 1977, *Invest. Ophthalmol. Vis. Sci.* **16**:141–148.
114. Hayden, J., Mills, J. W., and Masland, R. M., 1980, *Science* **210**:435–437.
115. Robertson, D., and Johnstone, B. M., 1978, *Hearing Res.* **1**:31–34.
116. Comis, S. D., and Leng, G., 1979, *Exp. Brain Res.* **36**:119–128.
117. Fex, J., and Adams, J. C., 1979, *Brain Res.* **159**:440–444.
118. Contreras, N. E. I. R., and Bachelard, H. S., 1979, *Exp. Brain Res.* **36**:573–584.
119. Wenthold, R. J., and Morest, D. K., 1976, *Soc. Neurosci. Abstr* **2**:28.
120. Adams, J. C., and Wenthold, R. J., 1980, *Neuroscience* **4**:1947–1951.
121. Chao, L. P., Kan, K.-S. K., and Hung, F.-M., 1982, *Brain Res.* **235**:65–82.
122. Hattori, T., Singh, V. K., McGeer, E. G., and McGeer, P. L., 1976, *Brain Res.* **102**:164–173.
123. Kaiya, H., Kreutzberg, G. W., and Namba, M., 1980, *Brain Res.* **187**:369–382.
124. Kaiya, H., Iwata, T., Namba, M., Ago, Y., Mayahara, H., and Ogawa, K., 1979, *Exp. Brain Res.* **34**:395–400.
125. Lundberg, J. M., Hokfelt, T., Schultzberg, M., Uvnas-Wallensten, K., Kohler, C. P., and Said, S. I., 1979, *Neuroscience* **4**:1539–1559.
126. Lundberg, J. M., Anggard, A., Fahronksrug, J., Hokfelt, T., and Mutt, V., 1980, *Proc. Natl. Acad. Sci. USA* **77**:1651–1655.
127. Hoyes, A. D., and Barber, P., 1978, *J. Anat.* **127**:533–542.
128. Higgins, D., Iacovitti, L., Joh, T. H., and Burton, H., 1981, *J. Neurosci.* **1**:126–131.
129. McLennan, I. S., Hill, C. E., and Hendry, I. A., 1980, *Nature* **283**:206–207.
130. Fukada, K., 1980, *Nature* **287**:553–555.
131. Schubert, D., LaCorbiere, M., Klier, F. G., and Steinbach, J. H., 1980, *Brain Res.* **190**:67–79.

132. Edgar, D. H., and Thoenen, H., 1978, *Brain Res.* **154**:186–190.
133. Heilbronn, E., 1975, *Cholinergic Mechanisms* (P. G. Waser, ed.), Raven Press, New York, pp. 343–364.
134. McGeer, P. L., Eccles, Sir J. C., and McGeer, E. G., 1978, *Molecular Neurobiology of the Mammalian Brain*, Plenum Press, New York, pp. 141–142.
135. Briley, M. S., and Changeux, J.-P., 1977, *Int. Rev. Neurobiol.* **20**:31–63.
136. Potter, L. T., and Smith, D. S., 1977, *Tissue Cell* **9**:585–594.
137. Fertuk, H. C., and Salpeter, M. M., 1974, *Proc. Natl. Acad. Sci. U.S.A.* **71**:1376–1378.
138. Aronstam, R. S., Triggle, D. J., and Eldefrawi, M. E., *Mol. Pharmacol.* **15**:227–234.
139. Morley, B. J., and Kemp, G. E., 1981, *Brain Res. Rev.* **3**:81–104.
140. Morley, B. J., Kemp, G. E., and Salvaterra, P., 1979, *Life Sci.* **24**:559–672.
141. Oswald, R. E., and Freeman, J. A., 1981, *J. Neurochem.* **37**:1586–1593.
142. Hunt, S., and Schmidt, J., 1978, *Brain Res.* **157**:213–232.
143. Arimatsu, Y., Seto, A., and Amano, T., 1981, *J. Comp. Neurol.* **198**:603–631.
144. Arimatsu, Y., Seto, A., and Amano, T., 1978, *Brain Res.* **147**:165–169.
145. Segal, M., Dudai, Y., and Amsterdam, A., 1978, *Brain Res.* **148**:105–119.
146. Hunt, S. P., and Schmidt, J., 1978, *Brain Res.* **142**:152–159.
147. Kuhar, M. J., Taylor, N., Wamsley, J. K., Hulme, E. C., and Birdsall, N. J. M., 1981, *Brain Res.* **216**:1–10.
148. Wamsley, J. K., Zarbin, M., Birdsall, N., and Kuhar, M. J., 1980, *Brain Res.* **200**:1–12.
149. Wamsley, J. K., Lewis, M. S., Scott Young, W. III, and Kuhar, M. J., 1981, *J. Neurosci.* **1**:176–191.
150. Karczmar, A. G., 1979, *Brain Acetylcholine and Animal Electrophysiology* (K. L. Davis and P. A. Berger, eds.), Plenum Press, New York, pp. 265–308.
151. Krnjevic, K., 1974, *Physiol. Rev.* **54**:418–540.
152. Rall, T. W., 1972, *Pharmacol. Rev.* **24**:399–410.
153. Drummond, G. I., and Ma. Y., 1973, *Progr. Neurobiol.* **2**:119–176.
154. Greengard, P., 1976, *Nature* **260**:101–108.
155. Lee, T. P., Kuo, J. F., and Greengard, P., 1973, *Proc. Natl. Acad. Sci. U.S.A.* **69**:3287–3291.
156. Bartfai, J., Study, R. E., and Greengard, P., 1977, *Cholinergic Mechanisms and Psychopharmacology* (D. S. Jenden, ed.), Plenum Press, New York, pp. 285–295.
157. Black, A. C., Jr., Sandquist, D., West, J. R., Wamsley, J. K., and Williams, T. H., 1979, *J. Neurochem.* **33**:1165–1168.
158. Marchi, M., Paudice, P., and Raiteri, M., 1981, *Eur. J. Pharmacol.* **73**:75–79.
159. Nordstrom, O., and Bartfai, T., 1979, *Int. Soc. Neurochem. Abstr.* **6**:93.
160. Nordstrom, O., and Bartfai, T., 1980, *Acta Physiol. Scand.* **108**:347–353.
161. Nordstrom O., and Bartfai, T., 1980, *Brain Res.* **213**:467–471.
162. Aguilar, J. S., Criado, M., and De Roberts, E., 1979, *Eur. J. Pharmacol.* **57**:227–230.
163. Nordstrom, O., Westlind, A., Unden, A., Meyerson, B., Sachs, C., and Bartfai, T., 1982, *Brain Res.* **234**:287–297.
164. Dudai, Y., and Segal, M., 1978, *Brain Res.* **154**:167–171.
165. Kamiya, H. O., Rotter, A., and Jacobowitz, D. M., 1981, *Brain Res.* **209**:432–439.
166. Overstreet, D. H., Speth, R. C., Hruska, R. E., Ehlert, F., Dumont, Y., and Yamamura, H. I., 1980, *Brain Res.* **195**:203–207.
167. Monaghan, D. T., Mena, E. E., and Cotman, C. W., 1982, *Brain Res.* **234**:480–485.
168. Lehmann, J., and Langer, S. Z., 1982, *Brain Res.* **248**:61–69.
169. De Belleroche, J., and Bradford, H. F., 1980, *J. Neurochem.* **35**:1227–1234.
170. De Belleroche, J., and Bradford, H. F., 1978, *Brain Res.* **142**:53–68.
171. De Bellaroche, J., Gardiner, I., Herberg, L. J., Winn, P., Murzi, E., and Williams, S., 1981, *Int. Soc. Neurochem. Abstr.* **7**:294.
172. Giorguieff, M. F., Le Floch, M. L., Glowinski, J., and Besson, M. J., 1977, *J. Pharmacol. Exp. Ther.* **200**:535–544.
173. Aqueros, L., Naguira, D., and Zunino, E., 1978, *Biochem. Pharmacol.* **27**:2667–2674.
174. Giorguieff-Chesselet, M. F., Kemel, M. L., Wandscheer, D., and Glowinski, D., 1979, *Life Sci.* **25**:1257–1262.

175. Westfall, T. C., 1974, *Neuropharmacology* **13**:693–700.
176. Hattori, T. McGeer, P. L., and McGeer, E. G., 1979, *Brain Res.* **170**:71–81.
177. McGeer, P. L., McGeer, E. G., and Innanen, V. T., 1979, *Brain Res.* **169**:433–441.
178. Nomura, Y., Kajiyama, H., Nakata, Y., and Segawa, T., 1979, *Eur. J. Pharmacol.* **58**:125–131.
179. De Belleroche, J., Luqmani, Y., and Bradford, H. F., 1979, *Neurosci. Lett.* **11**:209–213.
180. Reisine, T. D., Nagy, J. I., Fibiger, H. C., and Yamamura, H. I., 1979, *Brain Res.* **169**:209–214.
181. Schiller, G. G., 1979, *Life Sci.* **24**:1159–1163.
182. Ben-Barak, J., Gazit, H., Silman, I., and Duda, Y., 1981, *Eur. J. Pharmacol.* **74**:73–81.
183. Takeyasu, K., Uchida, S., Noguchi, Y., Fujiota, N., Saito, K., Hafa, F., and Yoshida, H., 1979, *Life Sci.* **25**:585–592.
184. Westlind, A., Grynfarb, M., Hedlund, B., Bartfai, T., and Fuxe, K., 1981, *Brain Res.***225**:131–141.
185. McKinney, M., and Coyle, J. T., 1982, *J. Neurosci.* **2**:97–105.
186. Herman, Z. S., and Slominska-Zurek, J., 1979, *Psychopharmacology* **64**:337–340.
187. Raiteri, M., Marchi, M., and Paudice, P., 1981, *Eur. J. Pharmacol.* **74**:109–110.
188. Drachman, D. B., 1978, *N. Engl. J. Med.* **298**:136–142,186–298.
189. Drachman, D. B., Angus, C. W., Adams, R. N., Michelson, J. O., and Hoffman, G. J. 1978, *N. Engl. J. Med.* **298**:1116–1121.
190. Reiness, C. G., Weinberg, C. B., and Hall, Z. W., 1978, *Nature* **274**:68–70.
191. Scadding, G. K., Wester, A. D. B., Ross, M., Thomas, H. C., and Howard, C. W. H., 1979, *Neurobiology* **29**:502–505.
192. Vincent, A., Thomas, H. C., Scadding, C. K., and Newsome-Davis, J., 1978, *Lancet* **1**:305–307.
193. Fuchs, S., 1979, *Int. Soc. Neurochem. Abstr.* **7**:18.
194. McGeer, E. G., and McGeer, P. L., 1981, *Neuropharmacology of Central Nervous System and Behavioral Disorders*, Academic Press, New York, pp. 479–505.
195. Spillane, J. A., Goodhart, M. J., White, P., Bowen, D. M., and Davison, A. N., 1977, *Lancet* **2**:826–827.
196. McGeer, E. G., and McGeer, P. L., 1982, *Geriatrics 1* (D. Platt, ed.), Springer-Verlag, Berlin, Heidelberg, New York, pp. 263–282.
197. Smith, C. M., Swash, M., Exton-Smith, G. A., Phillips, M. J., Overstall, P. W., Piper, M. E., and Bailey, M. R., 1978, *Lancet* **2**:318.
198. Boyd, W. D., Graham-White, J., Blackwood, G., Glen, I., and McQueen, J., 1977, *Lancet* **2**:711.
199. Barbeau, A., Growdon, J. H., and Wurtman, R. J., eds., 1979, *Choline and Lecithin in Brain Disorders*, Raven Press, New York.
200. Davies, P., 1979, *Brain Res.* **171**:319–327.
201. Davies, P., 1979, *Res. Publ. Assoc. Res. Nerv. Ment. Dis.* **57**:153–160.
202. Rossor, M., Fahrenkrug, J., Emson, P., Mountjoy, C., Iversen, L., and Roth, M., 1980, *Brain Res.* **201**:249–253.
203. Pepeu, G., Gori, G., and Antuono, P., 1979, *Cerebrovascular Disorders* (G. Tognoni and S. Garattini, eds.), Elsevier, Amsterdam, pp. 3–12.
204. McGeer, E. G., 1981, *Prog. Neuropsychopharmacol.* **5**:435–443.
205. Wong, P. T.-H., McGeer, P. L., Rossor, M., and McGeer, E. G., 1982, *Brain Res.* **231**:466–471.
206. Yates, C. M., Allison, Y., Simpson, J., Maloney, A. F. J., and Gordon, A., 1979, *Lancet* **2**:851.
207. Perry, E. K., Perry, R. H., Blessed, G., and Tomlinson, B. E., 1977, *Lancet* **1**:189.
208. Perry, E. K., Gibson, P. H., Blessed, G., Perry, R. H., and Tomlinson, B. E., 1977, *J. Neurol. Sci.* **34**:247–265.
209. Perry, E. K., Perry, R. H., Gibson, P. H., Blessed, G., and Tomlinson, B. E., 1977, *Neurosci. Lett.* **6**:85–89.
210. Perry, E. K., 1980, *Age Aging* **9**:1–8.
211. White, P., Goodhardt, M. J., Keet, J. P., Hiley, C. R., Carrasco, L. H., Williams, I. E. I., and Bowen, D. M., 1977, *Lancet* **1**:668–671.

212. Reisine, T. D., Yamamura, H. I., Bird, E. D., Spokes, E., and Enna, S. J., 1978, *Brain Res.* **159**:477–481.
213. Davies, P., and Verth, A. H., 1977, *Brain Res.* **138**:385–392.
214. Perry, E. K., Perry, R. H., Blessed, G., and Tomlinson, B. E., 1978, *Neuropathol. Appl. Neurobiol.* **4**:273–278.
215. Rossor, M. N., Svendsen, C., Hunt, S. P., Mountjoy, C. Q., Roth, M., and Iversen, L. L., 1982 *Neurosci. Lett.* **28**:217–222.
216. Whitehouse, P. J., Price, D. L., Struble, R. G., Clark, A. W., Coyle, J. T., and DeLong, M. R., 1982, *Science* **215**:1237–1239.
217. Whitehouse, P. J., Price, D. L., Clark, A. W., Coyle, J. T., and DeLong, M. R., 1981, *Ann. Neurol.* **10**:122.
218. Nagai, T., McGeer, P. L., Peng, J. H., McGeer, E. G., and Dolman, C. E., 1983, *Neurosci. Lett.* **36**:195–199.
219. Drachman, D. A., and Leavitt, J., 1974, *Arch. Neurol.* **30**:113–121.
220. Drachman, D. A., 1979, *Fed. Proc.* **38**:2613–2615.
221. Crow, T. J., and Grove-White, I. G., 1973, *Br. J. Pharmacol.* **49**:322–327.
222. Petersen, R. C., 1977, *Psychopharmacologia* **52**:283–289.
223. Sitaram, N., Weingartner, H., and Gillin, J. C., 1978, *Science* **201**:274–276.
224. Sitaram, N., Weingartner, H., Caine, E. D., and Gillin, J. C., 1978, *Life Sci.* **22**:1555–1560.
225. Davis, K. L., Mohs, R. C., Tinklenberg, J. R., Pfefferbaum, A., Hollister, L. E., and Kopell, B. S., 1978, *Science* **201**:272–274.
226. Scoville, W. B., and Milner, W., 1957, *J. Neurol. Neurosurg. Psychiatry* **20**:11–21.
227. Bartus, R. L., 1978, *Pharmacol. Biochem. Behav.* **9**:833–836.
228. Davis, K. L., and Yamamura, H. I., 1978, *Life Sci.* **23**:1729–1734.
229. Smith, C. M., and Swash, M., 1978, *Ann. Neurol.* **3**:471–475.
230. Jaffard, R., Destrade, C., Soumireu-Mourat, B., Durkin, T., and Ebel, A., 1980, *Neurosci. Lett.* **19**:349–352.
231. Durkin, T., Ayad, G., Ebel, A., and Mandel, P., 1977, *Brain Res.* **136**:475–486.
232. Peters, B. H., and Levin, H. S., 1979, *Ann. Neurol.* **6**:219–221.
233. Eckernas, S. A., 1977, *Acta. Physiol. Scand. [Suppl.]* **449**:1–62.
234. Flentge, F., and Van den Berg, C. J., 1979, *J. Neurochem.* **32**:1331–1333.
235. Fernstrom, M. H., and Wurtman, R. J., 1979, *Brain Res.* **165**:358–361.
236. Brunello, N., Cheney, D. L., and Costa, E., 1982, *J. Neurochem.* **38**:1160–1163.
237. Haubrich, D. R., Wang, P. F. L., Clody, P. E., and Wedeking, P. W., 1975, *Life Sci.* **17**:975–980.
238. Wecker, L., and Schmidt, D. E., 1979, *Life Sci.* **25**:75–84.
239. Wecker, L., and Schmidt, D. E., 1980, *Brain Res.* **184**:234–238.
240. Hirsch, M. J., and Wurtman, R. J., 1978, *Science* **202**:223–225.
241. Wecker, L., and Dettbarn, W. D., 1979, *J. Neurochem.* **32**:961–967.
242. Hirsch, M. J., Growdon, J. H., and Wurtman, R. J., 1977, *Brain Res.* **125**:383–385.
243. Kewitz, H., Pleul, O., Dross, K., and Schwartzkopff, T., 1975, *Cholinergic Mechanisms* (P. G. Waser, ed.), Raven Press, New York, pp. 131–135.
244. Anon., 1981, *Science* **211**:1032–1033.
245. Rossor, M. N., Emson, P. C., Mountjoy, C. Q., Roth, M., and Iversen, L. L., 1980, *Neuroscience Lett.* **20**:373–377.
246. Davies, P., Katzman, R., and Terry, R. D., 1980, *Nature* **288**:279–280.
247. Perry, E. K., Tomlinson, B. E., Blessed, G., Perry, R. H., Cross, A. J., and Crow, T. J., 1981, *J. Neurol. Sci.* **51**:279–287.
248. Davies, P., and Terry, R. D., *Neurobiol. Aging* **2**:9–14.
249. Mattisson, B., Gottfries, C. G., Ross, B. E., and Winblad, B., 1974, *Acta Psychiatr. Scand. [Suppl.]* **255**:269–277.
250. McGeer, P. L., 1963, *J. Neuropsychiatry* **4**:247–250.
251. McGeer, P. L., McGeer, E. G., and Fibiger, H. C., 1973, *Neurology (Minneap.)* **23**:912–917.
252. Aquilonius, S. M., Eckernas, S. A., and Sundwall, A., 1975, *J. Neurol. Neurosurg. Psychiatry* **38**:669–677.
253. Bird, E. D., and Iversen, L. L., 1974, *Brain* **97**:457–472.

254. Stahl, W. L., and Swanson, P. D., 1974, *Neurobiology (Minneap.)* **24:**813–819.
255. Reisine, T. D., Fields, J. Z., Yamamura, H. I., Bird, E. D., Spokes, E., Schreiner, P. S., and Enna, S. J., 1977, *Life Sci.* **21:**1123–1128.
256. Aquilonius, S. M., and Eckernas, S. A., 1977, *Neurology (Minneap.)* **27:**887–889.
257. Growdon, J. H., Cohen, E. L., and Wurtman, R. J., 1977, *Ann. Neurol.* **1:**418–422.
258. Davis, K. L., Hollister, L. E., Barchas, J. D., and Berger, P. A., 1976, *Life Sci.* **22:**1865–1872.
259. Caraceni, T. A., Girotti, F., Celano, I., Parati, E., and Balboni, L., 1978, *J. Neurol. Neurosurg. Psychiatry* **41:**1114–1118.
260. Nutt, J. G., Rosin, A., and Chase, T. N., 1978, *Neurology (Minneap.)* **28:**1061–1063.
261. McGeer, E. G., McGeer, P. L., Hattori, T., and Vincent, S. R., 1979, *Advances in Neurology* (T. N. Chase, N. Wexler, and A. Barbeau, eds.), Raven Press, New York, pp. 577–591.
262. Enna, S. J., Bennett, J. P., Bylund, D. B., Snyder, S. H., Bird, E. D., and Iversen, L. L., 1976, *Brain Res.* **116:**531–537.
263. Enna, S. J., Bird, E. D., Bennett, J. P., Bylund, D. B., Yamamura, H. I., Iversen, L. L., and Snyder, S., 1976, *N. Engl. J. Med.* **294:**1305–1309.
264. Wastek, G. J., Stern, L. Z., Johnson, P. C., and Yamamura, H. I., 1976, *Life Sci.* **19:**1033–1040.
265. Fields, J. Z., Reisine, T. D., and Yamamura, H. I., 1978, *Life Sci.* **23:**569–574.
266. Kobayashi, R. M., Palkovits, M., Hruska, R. E., Rothchild, R., and Yamamura, H. I., 1978, *Brain Res.* **154:**13–24.
267. Volpe, B. T., Francis, A., Gazzaniga, M. S., and Schechter, N., 1979, *Exp. Neurol.* **66:**737–744.
268. Vizi, S. E., and Palkovits, M., 1978, *Brain Res. Bull.* **3:**93–96.
269. Salvaterra, P., Matthews, D. A., and Foders, R., 1980, *J. Neurochem.* **35:**1253–1257.
270. Ehlert, F. J., Dumont, Y., Roeske, W.R., and Yamamura, H. I., 1980, *Life Sci.* **26:**961–967.
271. Tindall, R. S., Kent, M., Baskin, F., and Rosenberg, R. N., 1978, *J. Neurochem.* **30:**859–863.

Neurotransmitter Uptake

Edith D. Hendley

1. INTRODUCTION

Neurotransmitter uptake is here defined as the process of translocation of the released neurotransmitter from the extracellular to an intracellular compartment without chemical modification. Most of our knowledge concerning the disposition of the released transmitter was originally elucidated in the peripheral nervous system, where acetylcholine (ACh) and norepinephrine (NE) were the only known neurochemical mediators. For many decades it has been known that ACh released from nerve terminals was hydrolyzed and inactivated rapidly and efficiently by the enzyme acetylcholinesterase. A similar enzymatic degradative process was invoked for the inactivation of the catecholamines, both NE released from sympathetic nerves and epinephrine released as a hormone by the adrenal medulla. The catecholamine-degrading enzymes monoamine oxidase (MAO) and catechol-O-methyltransferase (COMT) were assigned the role of rapid enzymatic removal of the sympathetic nerve transmitter, analogous to the role of acetylcholinesterase in cholinergic transmission. However, several lines of experimental evidence, beginning with Burn in 1932,[1] led to the rejection of the enzymatic hypothesis and to the discovery in the 1950s of uptake as the major mode of inactivation of catecholamines (see Iversen[2] for historical review). A key finding was that the pressor actions of the catecholamines injected intravenously were neither greatly prolonged nor markedly potentiated when both MAO and COMT were simultaneously inhibited by pharmacological blockade[3] (a paradigm analogous to the known marked potentiation and prolongation of ACh's actions following blockade of acetylcholinesterase by physostigmine).

Further, more direct evidence for uptake emerged from the introduction in the 1950s of high-specific-activity radiolabeled catecholamines as research tools for following the disposition of the catecholamines after their administration in experimental animals in low, physiological concentrations. Using

Edith D. Hendley • Department of Physiology and Biophysics, University of Vermont College of Medicine, Burlington, Vermont 05405.

these tools, Axelrod and his colleagues were able to demonstrate that a significant portion of intravenously injected catecholamines in mice and cats was rapidly and selectively removed from the circulation in unchanged form by tissues densely innervated by sympathetic nerves[4,5] and, interestingly, that a number of psychoactive drugs inhibited that removal.[6] The blood–brain barrier to the catecholamines was circumvented by intraventricular administration of [^{3}H]NE in the rat brain, and these experiments likewise demonstrated that NE is rapidly taken up, stored, and then slowly metabolized by MAO and COMT in the brain as in the periphery.[7]

During this same fruitful era of neurochemical discoveries in the 1960s, the other biogenic amine and amino acid neurotransmitters were discovered in the brain, and it soon became evident that uptake and not enzymatic hydrolysis was the major mode of inactivation of all of the other known and suspected neurotransmitters in the brain, including serotonin (5-HT), dopamine (DA), γ-aminobutyric acid (GABA), glutamate, aspartate, glycine and taurine. Indeed, in cholinergic nerve transmission itself, the choline generated from enzymatic hydrolysis of ACh was found likewise to be recaptured from the extracellular space by a high-affinity, sodium-dependent uptake process,[8,9] thus illustrating that uptake is the rule rather than the exception as the rapid inactivating mechanism in neurochemical transmission.

Since the discovery in the 1970s that the neurally active peptides, such as enkephalin, substance P, vasoactive intestinal polypeptide, are probably neurotransmitters or cotransmitters in the central and autonomic nervous systems,[10] our thinking has come full circle regarding transmitter inactivation processes. For example, there is no evidence of concentrative (uphill) transport, i.e., uptake, for substance P transmission,[11,12] whereas there is evidence for substance P-degrading peptidase activity localized in specific neuronal elements involved in substance P transmission.[13] Also, several lines of evidence indicate that enkephalinase, a specific, high-affinity enkephalin-degrading dipeptidyl carboxypeptidase, subserves an inactivating role in enkephalinergic transmission. For example, enkephalinase is associated with nerve membranes, and its distribution follows the heterogeneous distribution of the opiate receptors in the brain.[14] When enkephalinase is blocked by a specific inhibitor, the analgesic actions of enkephalins are markedly potentiated, and these effects are blocked by the opiate receptor antagonist naloxone.[15]

Despite these findings, it is premature to conclude that peptidergic transmission in general involves enzymatic inactivation of the released transmitter. To do so requires that a specific, high-affinity peptidase for each peptide in question be localized on the external surface of the plasma membranes at peptidergic synapses, and such evidence has not yet been obtained in any of the peptidergic systems.

General reviews of the uptake processes for amine and amino acid neurotransmitters can be found in the following sources: Iversen,[2,16–18] Snyder *et al.*,[19] Krnjevic,[20] Levi and Raiteri,[21] Fonnum *et al.*,[136] and selected chapters in the book edited by Paton.[22] These reviews outline the progress of the research in this area that led to our current understanding of each transmitter uptake process. So far, all of the known high-affinity transmitter uptake processes have in common the characteristics of other carrier-mediated active transports that utilize Na^+,K^+-ATPase, the "sodium pump," as energy source. These characteristics include:

1. Dependence on a sodium gradient across the cell membrane and inhibition of transport by ouabain.
2. Ability to transport substrate "uphill," against a concentration gradient.
3. Saturability; i.e., initial uptake rates follow Michaelis–Menten kinetics.
4. Dependence on metabolic energy, e.g., inhibition by dinitrophenol, cyanide, and other metabolic poisons.
5. Stereoselectivity.
6. Temperature sensitivity with Q_{10} greater than 2.
7. Specificity for the inhibition of uptake by structural analogues of the transmitter and by other specific inhibitors.

Current research activity in this field continues to be centered on the search for more and more specific uptake inhibitors that are highly selective for one neuronal uptake process with virtually no affinity for any other neuronal uptake mechanism. Such efforts have met with only limited success so far, and more work in this area is clearly warranted.

Another line of current research activity involves the study of the regulation of uptake as a biological process and not simply as a biological substrate for pharmacological manipulation. For example, uptake was previously regarded as a fixed property of the plasma membrane of the nerve terminal, subject to alteration only by drugs. It is now known that uptake can also be altered rapidly and reversibly by behavioral state[23] and by rate of impulse flow in neurons.[24]

A third area of current research interest is a search for the molecular identity of uptake carriers in specific neuronal systems. The use of rigid analogues or conformationally restricted forms of the transmitter or of its competitors has been an important tool in estimating the preferred conformation of these agents at the uptake site, receptor sites, or enzymatic sites with which they interact.[25,26] Such studies have also provided insight into the topology of the carrier on its membrane site.[27]

Finally, radioactive ligand binding techniques are currently in use to identify the 5-HT uptake carrier with [³H]imipramine[28] and the NE uptake carrier with [³H]desmethylimipramine.[29,30] Irreversible ligand binding using photoaffinity labeling is also being used to isolate the DA uptake carrier[31] and the NE uptake carrier (studies in progress in the author's laboratory). In light of this background or overview of the uptake field, it is the purpose of this chapter to summarize the salient features of the individual amine and amino acid neuronal uptake processes and to assess the current status of our knowledge, or lack of knowledge, concerning each of these neuronal uptake mechanisms.

2. NOREPINEPHRINE

More is known about the uptake process in noradrenergic neurons that in any other neuronal system. The early studies in this field were carried out in the isolated perfused rat heart,[32,33] isolated rat heart slices,[34] and isolated slices of the cat brain,[35] mouse brain,[36] rat brain,[37,38] and monkey brain.[39] With the introduction of subcellular fractionation techniques,[40,41] synaptosomes or pinched-off nerve ending particles have been extremely useful in elucidating the catecholamine uptake process in brain.[42–45]

Whether studied in sympathetic nerves or in brain noradrenergic neuronal systems, NE uptake has all of the characteristics of a saturable, sodium-dependent, carrier-mediated, concentrative transport mechanism, as summarized adequately in other reviews.[18,22] Stereoselectivity for the *levo* isomer of NE is four- to fivefold greater than for the *dextro* isomer. The apparent K_m for ($-$)-NE uptake is 0.1–0.2 μM, and noradrenergic neuronal uptake can be blocked potently and selectively by desmethylimipramine (IC_{50}, 5.6 nM)[46] or nisoxetine (IC_{50}, 1 nM).[47]

The structural requirements for uptake of phenethylamines by the noradrenergic neuronal uptake mechanism include the absence of bulky substituents on the amine group in the side chain, the absence of methoxy groups on the catechol ring, and the presence of at least one phenolic hydroxy group in the ring.[18]

Many psychotropic agents are potent inhibitors of NE (and of other biogenic amine) uptake mechanisms, and these actions have usually been invoked in ascribing their psychotropic actions. Among these are many central stimulant drugs,[6,48,49] including cocaine and amphetamine, and agents that are effective in combating depression, including tricyclic and bicyclic antidepressant drugs[46,47,50–52] and certain monoamine oxidase inhibitors that have antidepressant efficacy.[2,53] In that regard, considering the clinical treatments in current use for the affective disorders, lithium, an antimania agent that also moderates bipolar depressive symptoms, has the unusual pharmacological action of increasing NE uptake.[54] Of further interest, electroconvulsive shock, the single most effective treatment for severe depression, decreases the affinity for NE uptake in mouse and rat brain cerebral cortical synaptosomes.[55–57]

The characteristics of the NE uptake mechanism so far described pertain to the sodium-dependent, high-affinity, intraneuronal transport process, designated as uptake$_1$ by Iversen.[2] Iversen[58] also described a low-affinity, high-capacity transport of NE and epinephrine in rat heart, designated uptake$_2$, that appears to be localized extraneuronally in various peripheral organs innervated by sympathetic nerves (see Paton[22] for extensive coverage of extraneuronal accumulation of NE in peripheral organs). Uptake$_2$ has the characteristics of a facilitated diffusion, and the structural requirements for transport of phenethylamines by the process of uptake$_2$ in rat heart are in distinct contrast with, and frequently opposite to, the requirements for transport by uptake$_1$. Uptake$_2$ in rat heart is saturable with an apparent K_m of 252 μM for ($\pm$)-norepinephrine.[58] The methoxy derivatives of catecholamines, which are inactive as inhibitors of uptake$_1$, are the best inhibitors of uptake$_2$. Of further interest is the finding that the steroid hormones corticosterone, estradiol, and progestrone are potent inhibitors of uptake$_2$, as are also phenoxybenzamine and clonidine.[59–61]

Hendley *et al.*[62] described a process in rat brain slices with many of the characteristics of uptake$_2$ in the heart. This study used the uptake of

[^{3}H]normetanephrine, the methoxylated derivative of NE, to describe this process. Burrows *et al.*[63] recently described a novel NE accumulation process in rat brain that was revealed when uptake$_1$ was blocked by cocaine, desmethylimipramine, or sodium replacement. This accumulation of NE had a rapid time course and a high capacity for accumulation of NE, and it appeared to have a synaptosomal localization in rat cerebral cortex. A key feature of this cocaine-insensitive accumulation process is that it was highly dependent on the presence of calcium in the incubation medium; thus, we have named this process calcium-sensitive accumulation (CSA). The existence of a high-capacity, calcium-sensitive uptake process in brain, in addition to the well-known sodium-sensitive, low-capacity, uptake$_1$ process, raises the intriguing possibility that changing concentrations of sodium and calcium in the extracellular space during nerve terminal depolarization may dictate which of these two processes is operative at any given moment for removal of the transmitter efficiently, rapidly, and more or less completely during nerve activity. Another proposed biological function for CSA is that it may modify the interactions of noradrenergic systems with drugs that alter calcium availability.

In addition to alteration by various pharmacological agents, NE uptake has been shown to be altered by a variety of behavioral and environmental interventions and by genetic inbreeding of certain behavioral or disease states in rodents. In a series of experiments by Hendley and colleagues, we reported that fighting behavior and certain forms of stress lead to rapid and reversible changes in apparent K_m and/or V_{max} for NE uptake in brain of rats and mice.[23,55,64] These findings were confirmed by Hadfield and Weber using whole-brain synaptosomes from fighting mice.[65]

Moisset *et al.*[66] reported genetic differences in NE uptake kinetic constants between two inbred strains of mice of opposite emotionality or reactivity to a mild stress. By recombinant inbreeding of two such mouse strains of opposite behavioral reactivity, Moisset[67] was further able to show that the high-reactivity trait was highly correlated with increased K_m for NE uptake, and low reactivity with low apparent K_m. She further observed that the recombinant strains showed a bimodal distribution of K_m values across all strains and concluded that the NE uptake carrier in mouse brain probably exists in one of two conformational states, one with K_m around 0.2 and the other around 0.4 μM NE.

Work in our laboratory indicated th-.. genetically inbred spontaneously hypertensive rats (SHR) are also highly reactive and hyperactive in behavior, and these animals likewise exhibit elevated uptake rates of NE when compared with appropriate control rats.[68] Changes in uptake were observed in all regions of the brain tested and were present as early as 1 to 2 weeks postnatally. Increased uptake (elevated V_{max}) was associated with down-regulation of β-adrenergic receptor number in the SHR, as observed using [^{3}H]dihydroalprenolol binding in brain membranes during development,[68] and as a decrease in the responsiveness of the NE-stimulated adenylate cyclase associated with the β-adrenergic receptor in the cerebral cortex of the SHR.[69]

As a final consideration of the regulation of the NE uptake process, Brenneman and Rutledge[70] recently reported that the NE uptake kinetic constants can also be influenced by lipid composition of the diet. Neonatal rat pups subjected *in utero* to a maternal diet high in saturated fat throughout gestation exhibited a marked decrease in affinity (elevated K_m) and a less marked increase in V_{max} for NE uptake during early postnatal development. Lipid analysis of neuronal membranes indicated altered phospholipids induced by exposure to saturated fats during gestation.

3. DOPAMINE

The DA neurons, which are more circumscribed in their distribution within the CNS than either the noradrenergic or serotonergic systems, are comprised of three major neuronal projection systems. These include the nigrostriatal tract, the mesolimbic and mesocortical projections from the A10 region of the mesencephalon, and the tuberoinfundibular system originating in the arcuate nucleus and terminating in the median eminence within the hypothalamus. In addition to these central systems, dopaminergic neurons are also localized within a population of amacrine cells in the retina and as interneurons [small intensely fluorescent (SIF) cells] within sympathetic autonomic ganglia.

It has been known from the early histofluorescence studies of Hamberger[71] that dopaminergic neurons of the rat brain exhibit a catecholamine uptake process that is similar to that in noradrenergic neurons; i.e., it is reserpine resistant, energy requiring, sodium dependent, and inhibited by cocaine and amphetamine. It was also apparent in this early study that dopaminergic uptake of catecholamines was different from the noradrenergic neuronal uptake process in that desmethylimipramine (DMI) was a very potent inhibitor of the latter uptake process and ineffectual in inhibiting uptake in the dopaminergic neurons. This important early pharmacological distinction is still the best means available today for distinguishing noradrenergic from dopaminergic neuronal uptakes, which in many other respects are very similar.

Most of our information concerning dopaminergic neuronal uptake has been elucidated in the rat neostriatum, where the population of dopaminergic nerve terminals is the most highly enriched of any area in the CNS (approximately 15% of the synaptosomal content is dopaminergic), and where noradrenergic terminals exist in insignificant numbers. The key studies that described the characteristics of catecholamine uptake in neostriatal homogenates are those of Snyder and Coyle,[43,44,72] Harris and Baldessarini,[73] and Horn.[74] Holz and Coyle[75] extended the characterization using purified synaptosomes of rat neostriatum.

The results of these studies indicated that dopaminergic neuronal uptake is a rapid, saturable, temperature-dependent process that utilizes the Na^+,K^+-ATPase sodium pump as energy source. Another characteristic similar to the noradrenergic neuronal uptake system is that it is inhibited by cocaine and amphetamine. However, the antiparkinson agent benztropine had 20 times the potency in inhibiting dopaminergic uptake than noradrenergic uptake,[72] a trait that has led to the popular use of benztropine as a more selective blocker of dopaminergic uptake than of noradrenergic neuronal uptake.

Because of the fine structure and high density of the nerve terminals of striatal dopaminergic neurons, the uptake of DA is very high, and tissue : medium ratios of 200:1 can be observed within 2 min of incubation of purified synaptosomes with [^{3}H]DA at 37°C.[75] This high capacity for uptake is also reflected in the high V_{max} one obtains in the striatum as compared with the other regions of the rat brain.[38,43]

In contrast with noradrenergic neuronal uptake, dopaminergic neuronal uptake failed to demonstrate stereoselectivity towards the isomers of NE and amphetamine,[44,76] and this was also observed in uptake studies using noradrenergic and dopaminergic systems in peripheral organs.[77] However, numerous other studies failed to confirm the lack of stereoselectivity of striatal uptake with respect to the stereoisomers of amphetamine,[78–81] and no obvious differences in methodologies can be invoked to explain the discrepancies between these opposing findings.

Horn[74] conducted a detailed study comparing the structural requirements for catecholamine uptake in dopaminergic (striatal) versus noradrenergic (hypothalamic) synaptosomes and noted a striking similarity in the structure–activity relationships of phenethylamine analogues for these two uptake systems, as also observed by Raiteri *et al.*[82] Thus, both noradrenergic and dopaminergic synaptosomes show increased affinity on the addition of phenolic hydroxyl groups and with α-methylation of the side chain of phenethylamines. Similarly, methoxylation of the phenolic ring, β-hydroxylation of the side chain, and the addition of bulky substituents on the amine group reduced affinity for uptake in both uptake systems. Interestingly, noradrenergic uptake sites were quantitatively more sensitive to structural changes in phenethylamines than were the striatal uptake sites, and Horn[74] rationalized that this might be expected considering that NE and not DA possesses an asymmetric center at the β-carbon, thereby increasing the stringency of steric requirements for noradrenergic uptake affinity.

In contrast with the detailed studies of DA uptake in the neostriatum, relatively little attention has been given to uptake of DA in nonstriatal dopaminergic neurons. Horn *et al.*[83] prepared homogenates of the mesolimbic structures, the olfactory tubercles, and nucleus accumbens in the rat brain and compared the characteristics of DA uptake in these structures with those of the neostriatum. They were able to conclude that the DA uptake process is very similar in both systems, including their sensitivities to the uptake-blocking actions of benztropine and amphetamine.

Following the demonstration by Thierry *et al.*[84,85] of the mesocortical dopaminergic neuronal projections from cell bodies in the ventral tegmental (A10) region to the terminations in the frontal, cingulate, and entorhinal cortices of the rat brain, Tassin *et al.*[86] were able to demonstrate a dopaminergic neuronal uptake process in the rat cerebral cortex that was blockable by benztropine and that could be revealed in the presence of blockade of noradrenergic neuronal uptake by DMI or by destruction of the noradrenergic neuronal input to the cerebral cortex.

In the tuberoinfundibular dopaminergic neuronal system, Cuello *et al.*[87] provided preliminary data on the uptake of dopamine in homogenates of the rat median eminence. By using DMI to block the entry of dopamine into noradrenergic uptake sites, they revealed an uptake of [^{3}H]DA that was inhibited by benztropine with an IC_{50} similar to that in neostriatal homogenates, suggesting a similarity of DA uptake mechanisms regardless of regional origin of the dopaminergic system. However, when more detailed studies were carried out in the tuberoinfundibular system, it was clearly evident that this was in fact not the case. Kinetic studies of initial uptake rates of [^{3}H]DA in homogenates of rat median eminence indicated a very-low-affinity uptake process in the median eminence (apparent K_m 1.6 μM) as compared with the high-affinity process in striatal homogenates (apparent K_m 0.58 μM).[88]

In a more detailed analysis by Annunziato *et al.*[89] using synaptosomes (resuspended P_2) of rat median eminence, olfactory tubercle, and neostriatum, researchers confirmed that a high-affinity uptake of [^{3}H]DA was demonstrable in the mesolimbic and striatal systems (apparent K_m was 0.06 μM in the olfactory tubercle and 0.04 μM in the striatum). However, synaptosomes of the median eminence exhibited an apparent K_m of 1.8 μM, and similar low-affinity kinetics were observed whether the rats were male or female, whether they had been previously reserpinized or not, and whether P_2 had been preincubated with DMI or not. These interesting results strongly suggest that the dopaminergic neurons of the tuberoinfundibular system are distinctly different from their mesolimbic and neostriatal counterparts. Functionally, they are more suitably adapted to their neurosecretory endocrine role whereby DA released from these neurons enters the surrounding portal vessels and is transported intact to the anterior pituitary where it serves as a prolactin-inhibitory factor.

In view of the major role of the DA neurons in behavioral regulation and in neurological and psychiatric disorders, it is not surprising that most major psychotropic drugs have been screened for potency in inhibiting dopaminergic neuronal uptake, usually in striatal synaptosomes, as compared with noradrenergic and/or serotonergic uptake systems (see refs. 46,51,52,90,91). Agents that inhibit DA uptake selectively are eagerly sought as potentially useful agents in treating Parkinson's disease,[72,76] and it is also suggested that agents that facilitate DA uptake, such as low concentrations of certain tricyclic antidepressants, may be of some relevance to the treatment of depression[52] and indeed act as antipsychotic agents.

As mentioned previously, tricyclic antidepressants are generally more potent in inhibiting uptake in noradrenergic and/or serotonergic neurons than they are as inhibitors of uptake in dopaminergic neurons.[46,51,71] On the other hand, central stimulants and/or anorectic drugs such as amphetamine, methylphenidate, pipradrol, cocaine, and mazindol are all potent inhibitors of dopaminergic neuronal uptake, although their potencies as inhibitors of noradrenergic neuronal uptake are frequently not much less than their potencies at dopaminergic uptake sites.[46,52,78,90–94] For this reason, it is exceedingly difficult to identify an inhibitor of dopaminergic neuronal uptake that is not also a potent inhibitor of other monoaminergic uptake systems. Benztropine has been the most useful agent in this regard so far, although the studies of Koe[46] indicate that it is not more potent in inhibiting dopaminergic than noradrenergic neuronal uptake. Bupropion has recently been introduced as a selective DA uptake inhibitor that possesses behaviorally stimulating and antidepressant actions; however, its potency in inhibiting dopaminergic neuronal uptake is only twice that of noradrenergic uptake inhibition.[95]

The DA uptake carrier has been shown to be regulated by behavioral state, as previously noted in the NE uptake system. Hadfield[96] made the interesting observation that fighting in mice (isolation-induced aggression) produced a marked increase in both apparent K_m and V_{max} for [³H]DA uptake in synaptosomes from the mesolimbic–mesocortical brain areas, whereas no significant changes were observed in neostriatal DA uptake. In our own studies,[68] SHR rats, which are not only hypertensive but markedly hyperactive in locomotor behaviors, likewise exhibited altered DA uptake rates in the mesocortical brain region and not in the neostriatum during their development.

4. SEROTONIN

Serotonergic neuronal systems are diffusely distributed throughout the CNS, originating from cell groups of the raphe system within the brainstem. In the periphery, the gastrointestinal tract contains serotonergic neurons within the myenteric plexuses, and these neurons take up, synthesize, store, and release 5-HT and exist in addition to the enterochromaffin cells that synthesize and release 5-HT within the gut.[97,98] Another peripheral source of 5-HT is the blood platelet cell, whose serotonergic mechanisms are physiologically and pharmacologically remarkably similar to those of serotonergic nervous mechanisms.[99]

A carrier-mediated, high-affinity transport process for 5-HT was demonstrated in rat, mouse, and rabbit brain when low concentrations of radiolabeled 5-HT were used in *in vitro* studies.[100–104] These studies revealed that 5-HT uptake is localized in the synaptosomal fraction of brain tissues and that it is sodium dependent, saturable, reserpine resistant, energy dependent, and subject to inhibition by cocaine, amphetamine, and tricyclic antidepressants. Although these are very similar characteristics to the catecholaminergic uptake processes described above, it became evident early on that the potency of drugs inhibiting 5-HT uptake differed markedly from their potencies at catecholaminergic uptake sites. The most notable distinction was that the tertiary amine tricyclics were more potent at 5-HT uptake sites than were their corresponding secondary amine tricyclics.[105,106]

Initial rates of uptake of [³H]5-HT in rat brain slices[102] or homogenates[107] revealed two uptakes, one of high affinity (apparent K_m 0.1 μM) and a second of low affinity and high capacity (apparent K_m 8 μM). The high-affinity uptake represents the entry of 5-HT into the serotonergic neuronal uptake sites, and the low-affinity uptake was attributed to the entry of 5-HT into catecholaminergic nerve terminals.[102,107] These conclusions are also supported by the studies of Iversen[16] who noted that the uptake of high concentrations of 5-HT (5 μM) could be diminished by the selective destruction of catecholamine nerve terminals in the brain, whereas the uptake of a low concentration (0.05 μM) of 5-HT was unaffected by destruction of the catecholamine nerve terminals.

This ability of 5-HT to enter catecholaminergic uptake sites, often with affinities that exceed those of the catecholamines themselves, was further demonstrated in the sympathetic nerves innervating the pineal gland[108–110] and the vas deferens[111,112] and in the chromaffin granules of the adrenal medulla.[113]

The structural requirements for uptake of tryptamines in serotonergic neurons have been described by Horn and his colleagues.[114,115] Their studies indicated that the 5-hydroxy group was required for uptake. As in the case of the catecholamine systems, methoxylation of the ring hydroxyl group diminished uptake affinity, and α-alkylation of the side chain increased affinity. Increasing the length of the side chain to three carbons instead of two altered affinity only slightly, whereas reduction of the length to one carbon drastically reduced uptake affinity. Among the hydroxylated tryptamines, the neurocytotoxic compound 5,6-dihydroxytryptamine (5,6-DHT) was the most potent and 5,7-DHT, which is also cytotoxic, was the least potent.[115] In this series, cytotoxicity was not correlated with 5-HT uptake inhibitory potency but rather with the chemical instability imparted to these compounds by the positioning of the hydroxyl groups, as reflected also in their tendency to autooxidize.

It is generally recognized from many studies in this field that the tricyclic compound chlorimipramine is a potent and selective inhibitor of serotonergic neuronal (and platelet) uptake of 5-HT,[46,102,103,105,116,117] thus providing the researcher with a useful tool for selectively targeting the serotonergic uptake sites, as DMI does the noradrenergic uptake sites. More recently, two bicyclic compounds, fluoxetine[118] and nisoxetine,[47] were introduced as potent and selective inhibitors of serotonergic and noradrenergic uptake mechanisms, respectively. They have been suggested as more appropriate tools than the tricyclics, as their potencies and selectivity in distinguishing serotonergic from noradrenergic uptake systems in functional, *in vivo* assays are even more striking than their selectivity in *in vitro* assays.[119]

A number of other selective and potent 5-HT uptake inhibitors have recently been developed as potential research tools and/or as possible therapeutic agents by virtue of their antidepressant actions. Among these are zimelidine and its more potent desmethylated derivative norzimelidine, which is also formed *in vivo* following the administration of zimelidine.[120] The potency of norzimelidine was 25-fold greater at 5-HT uptake sites than at noradrenergic uptake sites and 170-fold greater than at dopaminergic uptake sites. Citalopram was found to have a K_i of 1 nM as an inhibitor of serotonergic neuronal uptake in brain, and NE was ineffectual in preventing this action of citalopram.[121] Organon's compound ORG 6582 is 16 times more potent in inhibiting serotonergic than noradrenergic neuronal uptake and 72 times more potent in inhibiting serotonergic than dopaminergic neuronal uptake.[122] Another antidepressant compound, trazodone, is 60-fold more potent at 5-HT than at NE uptake sites and 100-fold more potent at 5-HT than at dopaminergic uptake sites.[117]

Several approaches have been used to gain some insight into the topography of the 5-HT uptake carrier and in estimating the preferred conformation of the substrates that use this carrier. Some of these studies include the structure–activity relationships of tryptamine analogues and of tricyclic compounds by Horn and his colleagues[51,114,115,123] and the use of conformationally restricted analogues of 5-HT[124] or of other selective compounds with more or less rigid structures as studied by Svante Ross and his colleagues (see Ross[125] for a review).

5. γ-AMINOBUTYRIC ACID

In 1958, Elliott and van Gelder[126] used a bioassay method for detecting GABA and demonstrated that slices of rat cerebral cortex accumulate exogenous GABA in the occluded (nonmetabolized) form by means of an energy-requiring process that could concentrate GABA by as much as 40-fold over its concentration in the medium. This first and many subsequent studies by others have firmly established that GABA's transmitter actions are terminated by an uptake process and not by degradation via the mitochondrial enzyme GABA : glutamate transaminase (GABA-T), which is the only GABA-degrading enzyme in brain tissue.

Iversen and Neal[127] used more finely sliced rat cerebral cortex and radiolabeled GABA to measure uptake and metabolism of GABA and confirmed that uptake, using very low concentrations of [³H]GABA, reached tissue : medium concentrations of nearly 100 in 60 min. This uptake process was sodium dependent, energy dependent, and saturable, with an apparent K_m of 22 μM. Competition studies showed that the neurally active amino acids glycine, glutamate, and aspartate did not inhibit GABA uptake, whereas two GABA analogues, β-hydroxy-GABA and β-guanidinopropionic acid, did inhibit uptake.[127]

Regional distribution of GABA uptake in the rat brain revealed that [³H]GABA uptake rate generally followed the uneven regional distribution of endogenous GABA levels in the rat brain.[128] Subcellular localization studies demonstrated that the synaptosomal fraction was the site of the carrier-mediated transport of low concentrations of GABA in rat[128,129] and mouse[130,131] brain.

In addition to neuronal uptake of GABA, there is an important uptake of GABA into glial elements that Henn and Hamberger[132] reported to be of even higher affinity (apparent K_m 0.42 μM) than they observed in their neuronally enriched fractions of the rabbit brain. Henn[133] has pointed out that approximately half of the particles formed in the preparation of synaptosomes using either sucrose or Ficoll gradient techniques are of glial origin rather than totally of neuronal origin, thus imparting considerable functional significance to the proposed removal of GABA by the surrounding glia during GABAergic neurotransmission.

Further differences between neuronal and glial GABA uptake mechanisms were seen in their differential sensitivities to specific uptake-blocking agents, and these studies are summarized in several excellent reviews to which the reader is referred.[134-136]

The structural requirements for GABA uptake appear to be very stringent, as many structural analogues of GABA are not inhibitory, and those that are do not compete with GABA for the uptake site.[137] However, when conformationally restricted analogues of GABA were tested as uptake inhibitors, two potent compounds were identified, *trans*-4-aminocrotonic acid and 4-amino-tetrolic acid, and their inhibition of GABA uptake was of the competitive type.[138] In an extended study screening further analogues and drugs that interact with GABA systems, Beart and Johnston[139] identified 2-hydroxy-GABA as an even more potent competitive inhibitor (IC_{50} 1 μM) than the conformationally restricted analogues. The analogue L-2,4-diaminobutyric acid (DABA), a potent competitive antagonist of GABA uptake specifically in neurons, has also been shown to be taken up, localized, and released in a manner very similar to that of GABA itself, suggesting that DABA may serve as a false transmitter in GABAergic neurons.[140] Other recently identified competitive inhibitors of GABA uptake include nipecotic acid[141] and guvacine.[142]

The GABA molecule possesses great conformational flexibility because of the ease with which each of its bonds can rotate in aqueous solutions.[143] Studies using GABA analogues of more or less restricted conformation have already described at least three different conformational states assumed by GABA in its interactions with the carrier site, the receptor–binding site, and its GABA-T, enzymatic substrate site (see Fonnum *et al.*[136] for a review of these findings). The studies of Beart *et al.*[138] suggest that the preferred conformation of GABA at the uptake site is the extended form.

6. GLYCINE

By virtue of the uneven endogenous distribution of glycine (GLY) within the cat spinal cord, Aprison and Werman[144] postulated that GLY may serve as an inhibitory neurotransmitter in the spinal cord. Neal and Pickles[145,146] described the uptake of low concentrations of [^{14}C]GLY in slices of rat and rabbit spinal cord. This uptake achieved a 30-fold concentration in the tissue over that in the medium, and the uptake rates paralleled the uneven endogenous distribution of GLY in the CNS. Glycine uptake in the spinal cord exhibited all of the characteristics of a saturable, carrier-mediated active transport, and it had an apparent K_m of 31 μM in rat spinal cord. None of the other neurally active amino acids, including GABA, competed for the uptake of GLY. Interestingly, one of the inhibitors of GLY uptake in these studies,[146] *p*-hydroxymercuribenzoate, had also been demonstrated previously to potentiate the inhibitory actions of GLY on spinal neurons,[147] thus providing functional evidence for the role of uptake as an inactivating mechanism for the candidate transmitter, GLY.

Logan and Snyder[148,149] used kinetic studies in homogenates of rat spinal cord and cerebral cortex and reported that high-affinity uptake of GLY (apparent K_m 27 μM) could be observed in the spinal cord and not in the cerebral cortex, whereas a second, low-affinity uptake was present in both areas. Bennett *et al.*[150] further revealed that the high-affinity uptake process for GLY, as well as for the other neurally active amino acids, was highly dependent on sodium in the medium, and the criterion of a sodium-dependent, high-affinity (low K_m) uptake process is now a consistently useful means of identifying candidate neurotransmitters, including amino acids,[151] choline,[24] and taurine.[152]

Using autoradiography for localizing radiolabeled GLY uptake in slices[153] or homogenates[130] of rat spinal cord, these studies revealed a prominent localization of GLY in the nerve terminals of a subpopulation of spinal cord neurons, as also confirmed by Arregui *et al.*[154] in subcellular fractions of spinal cord separated on sucrose density gradients.

In addition to a nerve ending localization of GLY uptake, Hokfelt and Ljungdahl[153] also observed the uptake of GLY into glial elements within the rat spinal cord. Henn[133] extended these observations by demonstrating that glial fractions of rabbit spinal cord preparations exhibited a high-affinity uptake of GLY (K_T 13 μM) and that similarly prepared glial fractions of cerebral cortex exhibited approximately half the uptake rates of the spinal cord glial preparations. These findings lent credence to the hypothesis that glia serve a major role in the inactivation of amino acid transmitters in the central nervous system, as mentioned in the previous section.

7. GLUTAMIC ACID AND ASPARTIC ACID

Glutamic acid (GLU) and aspartic acid (ASP) are neurally excitatory amino acids whose transmitter status has been very difficult to establish because of their ubiquitous presence in all neural tissues as metabolic intermediates of carbohydrate and protein metabolism. The identification of a sodium-dependent, high-affinity uptake of both GLU and ASP provided strong support for a transmitter function for both of these amino acids.[149,150,155,156] The uptakes of GLU and ASP are very similar, and both appear to compete for the same uptake sites, as do other similar endogenous amino acids.[155,156] Balcar and Johnston[155-157] carried out extensive studies on the structural specificity for inhibition of the high-affinity uptake of GLU and ASP and noted that the requirements are fairly stringent. Those analogues with the highest inhibitory potency towards the high-affinity uptake of L-GLU, including L-cysteate and D-aspartate, were even more potent than L-GLU itself. Interestingly, the compound *p*-chloromercuriphenylsulfonate, a mercurial agent that potentiates the excitatory actions of L-GLU in electrophysiological experiments,[158] is one of the drugs that inhibits the high-affinity uptake of L-GLU,[155] and this finding strengthened the proposed transmitter status for GLU.

The localization of GLU uptake in nerve terminals of selected neuronal populations is suggested by the studies of Iversen and Storm-Mathisen[159] and of Hokfelt and Ljungdahl.[160] Henn *et al.*[161] confirmed that glia, either prepared by bulk isolation from rabbit brain or as obtained in C-6 glioma cultures of astrocytelike cells, exhibited both high- and low-affinity uptake processes for GLU, whereas HeLa cells in culture exhibited only a single low-affinity process.

8. CHOLINE

Sodium-dependent, high-affinity choline uptake has been described by several groups in the early 1970s.[8,9,162-164] Its apparent K_m is 1.4 μM in the rat striatum and 3.1 μM in cerebral cortex.[8] The regional localization of the high-affinity uptake followed the distribution of identified cholinergic neurons in the CNS as well as the activity of choline acetyltransferase, emphasizing the role of choline uptake as a regulator of the synthesis of ACh in these neurons.[8,24,162–164]

Simon and Kuhar[165] found that the sodium-dependent, high-affinity uptake of choline is well correlated with the rate of impulse flow in cholinergic neurons in the CNS, whether the impulse flow is experimentally increased or decreased.

The structural requirements for choline uptake have been described, and the most potent inhibitor of the high-affinity process is hemicholinium-3.[8,9,162] Interestingly, ACh itself is also a potent inhibitor of choline uptake.

9. CONCLUDING REMARKS

Neurotransmitter uptake plays a major role in all of the known and suspected monoaminergic and amino-acid-mediated neuronal systems. These uptake mechanisms are substrates for the pharmacological actions of many psychotropic drugs, and they can also be modulated by neuronal impulse flow and by behavioral state. All of the uptake processes studied so far possess not only a high-affinity mechanism but, in addition, a low-affinity, high-capacity uptake mechanism that operates at high concentrations of substrate and probably insures that all extracellularly placed transmitter molecules are rapidly and completely internalized within cells and thereby neutralized as neuroactive agents.

There are now indications that in spite of the selectivity of each transmitter uptake process for its own natural substrate, other transmitters can also make use of the same uptake process. For example, the catecholamines NE and DA readily compete with each other for the same uptake sites, and there is ample opportunity for them to do so where their terminals coexist in the same brain regions. Furthermore, one might postulate a similar participation of catecholamine uptake systems for inactivating the transmitter epinephrine now thought to exist in certain neuronal systems in the brain.

The uptake of 5-HT into noradrenergic nerve endings in the pineal gland, and probably in other brain areas as well, is another example of this phenomenon and is probably of major significance in limiting the availability of 5-HT for the synthesis of melatonin in the pineal gland.

Among the amino acid-type transmitters, GLU and ASP and probably other similar neuroexcitatory endogenous amino acids readily compete for the same uptake sites. The high-affinity uptake of GABA, GLY, and GLU into glial cells as well as neuronal uptake sites is particularly important in limiting the neuroactivity of these potent, ubiquitous neuroactive agents.

High-affinity uptake measurements are very useful in localizing transmitter-specific nerve terminals during development and in following the appearance or disappearance of nerve terminals during nerve regeneration or degeneration studies, respectively. The selectivity of neuronal uptake mechanisms is also useful in targeting specific neuronal populations, while sparing others, during treatment with neurocytotoxic agents in experimental animal models.

As a final consideration, Svante Ross and his colleagues[166] pointed out yet another aspect of the importance of neuronal uptake mechanisms. They demonstrated that drugs that inhibit monoamine uptake down-regulate the release of the monoamine when such drugs are administered over several weeks of time. Thus, considering that peptides often coexist with the monoamines in certain neuronal systems, this down-regulation by uptake-inhibiting drugs will also decrease the corelease of the neuropeptide in these neurons. They suggest that this modulation of peptide release may be an important mechanism by which drugs that inhibit uptake can alter behavior.

REFERENCES

1. Burn, J. H., 1932, *J. Pharmacol. Exp. Ther.* **46**:75–95.
2. Iversen, L. L., 1967, *The Uptake and Storage of Noradrenaline in Sympathetic Nerves*, Cambridge University Press, Cambridge.
3. Crout, J. R., 1961, *Proc. Soc. Exp. Biol. Med.* **108**:482–484.
4. Axelrod, J., Weil-Malherbe, H., and Tomchick, R., 1959, *J. Pharmacol. Exp. Ther.* **127**:251–256.
5. Whitby, L. G., Axelrod, J., and Weil-Malherbe, H., 1961, *J. Pharmacol. Exp. Ther.* **132**:193–201.
6. Axelrod, J., Whitby, L. G., and Hertting, G., 1961, *Science* **133**:383–384.
7. Glowinski, J., Kopin, I. J., and Axelrod, J., 1965, *J. Neurochem.* **12**:25–30.
8. Yamamura, H. I., and Snyder, S. H., 1973, *J. Neurochem.* **21**:1355–1374.
9. Haga, T., and Noda, H., 1973, *Biochim. Biophys. Acta* **291**:564–575.
10. Hokfelt, T., Johansson, O., Ljungdahl, A., Lundberg, J. M., and Schultzberg, M., 1980, *Nature* **284**:515–521.
11. Segawa, T., Nakata, Y., Nakamura, K., Yajima, H., and Kitagawa, K., 1976, *Jpn. J. Pharmacol.* **26**:757–760.
12. Iversen, L. L., Jessell, T., and Kanazawa, I., 1976, *Nature* **264**:81–83.
13. Quik, M., and Emson, P. C., 1979, *Neurosci. Lett.* **15**:217–222.
14. Malfroy, B., Swerts, J. P., Llorens, C., and Schwartz, J. C., 1979, *Neurosci. Lett.* **11**:329–334.
15. Roques, B. P., Fournie-Zaluski, M. C., Soroca, E., Lecomte, J. M., Malfroy, B., Llorens, C., and Schwartz, J. C., 1980, *Nature* **288**:286–288.
16. Iversen, L. L., 1970, *Advances in Biochemical Psychopharmacology*, Volume 2, *Biochemistry of Simple Neuronal Models* (E. Costa and E. Giacobini, eds.), Raven Press, New York, pp. 109–131.
17. Iversen, L. L., 1974, *Biochem. Pharmacol.* **23**:1927–1935.
18. Iversen, L. L., 1975, *Handbook of Psychopharmacology* Volume 3, *Biochemistry of Biogenic Amines* (L. L. Iversen, S. D. Iversen, and S. H. Snyder, eds.), Plenum Press, New York, pp. 381–442.
19. Snyder, S. H., Kuhar, M. J., Green, A. I., Coyle, J. T., and Shaskan, E. G., 1970, *Int. Rev. Neurobiol.* **13**:127–158.
20. Krnjevic, K., 1974, *Physiol. Rev.* **54**:418–540.
21. Levi, G., and Raiteri, M., 1976, *Int. Rev. Neurobiol.* **19**:51–74.
22. Paton, D. M., 1976, *The Mechanism of Neuronal and Extraneuronal Transport of Catecholamines*, Raven Press, New York.
23. Hendley, E. D., Welch, B. L., and Moisset, B., 1973, *Science* **180**:220–221.
24. Kuhar, M., and Murrin, L. C., 1978, *J. Neurochem.* **30**:15–21.
25. Horn, A. S., and Snyder, S. H., 1972, *J. Pharmacol. Exp. Ther.* **180**:523–530.
26. Tuomisto, J., and Tuomisto, L., 1974, *Med. Biol.* **52**:176–180.
27. DePaulis, T., Kelder, D., and Ross, S. B., 1978, *Mol. Pharmacol.* **14**:596–606.

28. Langer, S. Z., Zarifian, E., Briley, M., Raisman, R., and Sechter, D., 1981, *Life Sci.* **29**:211–220.
29. Lee, C.-M., and Snyder, S. H., 1981, *Proc. Natl. Acad. Sci. U.S.A.* **78**:5250–5254.
30. Langer, S. Z., Raisman, R., and Briley, M., 1981, *Eur. J. Pharmacol.* **72**:423–424.
31. Davies, B., Abood, L., and Tometsko, A. M., 1980, *Life Sci.* **26**:85–88.
32. Iversen, L. L., 1963, *Br. J. Pharmacol.* **21**:523–537.
33. Burgen, A. S. V., and Iversen, L. L., 1965, *Br. J. Pharmacol.* **25**:34–49.
34. Bogdanski, D. F., and Brodie, B. B., 1966, *Life Sci.* **5**:1563–1569.
35. Dengler, J. J., Michaelson, I. A., Spiegel, H. E., and Titus, E., 1962, *Int. J. Neuropharmacol.* **1**:23–38.
36. Ross, S. B., and Renyi, A. L., 1964, *Acta Pharmacol. Toxicol. (Kbh.)* **21**:226–239.
37. Hamberger, B., and Masuoka, D., 1965, *Acta Pharmacol. Toxicol. (Kbh.)* **22**:363–368.
38. Snyder, S. H., Green, A. I., and Hendley, E. D., 1968, *J. Pharmacol. Exp. Ther.* **164**:90–102.
39. Snyder, S. H., Hendley, E. D., and Gfeller, E., 1969, *Brain Res.* **16**:469–477.
40. De Robertis, E., DeIraldi, A. P., Rodriguez, G., and Gomez, J., 1961, *J. Biophys. Biochem. Cytol.* **9**:229–235.
41. Gray, E. G., and Whittaker, V. P., 1962, *J. Anat.* **96**:79–88.
42. Colburn, R. W., Goodwin, F. K., Murphy, D. L., Bunney, W. E., and Davis, J. M., 1968, *Biochem. Pharmacol.* **17**:957–964.
43. Snyder, S. H., and Coyle, J. T., 1969, *J. Pharmacol. Exp. Ther.* **165**:78–86.
44. Coyle, J. T., and Snyder, S. H., 1969, *J. Pharmacol. Exp. Ther.* **170**:221–231.
45. Baldessarini, R. J., and Vogt, M., 1971, *J. Neurochem.* **18**:951–962.
46. Koe, B. K., 1976, *J. Pharmacol. Exp. Ther.* **199**:649–661.
47. Wong, D. T., Horng, J. S., and Bymaster, F. P., 1975, *Life Sci.* **17**:755–760.
48. Carlsson, A., Fuxe, K., Hamberger, B., and Lindqvist, M., 1966, *Acta Physiol. Scand.* **67**:481–497.
49. Berti, F., and Shore, P., 1967, *Biochem. Pharmacol.* **16**:2091–2094.
50. Maxwell, R. A., Keenan, P. D., Chaplin, E., Roth, B., and Eckhardt, S. B., 1969, *J. Pharmacol. Exp. Ther.* **166**:320–329.
51. Horn, A. S., 1976, *Postgrad. Med. J.* **52**(Suppl. 3):25–30.
52. Randrup, A., and Braestrup, C., 1977, *Psychopharmacology* **53**:309–314.
53. Hendley, E. D., and Snyder, S. H., 1968, *Nature* **220**:1330–1331.
54. Colburn, R. W., Goodwin, F. K., Bunney, W. E., and Davis, J. M., 1967, *Nature,* **215**:1395–1397.
55. Welch, B. L., Hendley, E. D., and Turek, I., 1974, *Science* **183**:220–221.
56. Hendley, E. D., and Welch, B. L., 1975, *Life Sci.* **16**:45–54.
57. Hendley, E. D., 1976, *Psychopharmacol. Commun.* **2**:17–25.
58. Iversen, L. L., 1965, *Br. J. Pharmacol.* **25**:18–33.
59. Iversen, L. L., and Salt, P. J., 1970, *Br. J. Pharmacol.* **40**:528–530.
60. Salt, P. J., and Iversen, L. L., 1973, *Naunyn Schmiedebergs Arch. Pharmacol.* **279**:381–386.
61. Iversen, L. L., and Langer, S. Z., 1969, *Br. J. Pharmacol.* **37**:627–637.
62. Hendley, E. D., Taylor, K. M., and Snyder, S. H., 1970, *Eur. J. Pharmacol.* **12**:167–179.
63. Burrows, G. H., Myers, M. M., Whittemore, S. R., and Hendley, E. D., 1981, *Eur. J. Pharmacol.* **69**:301–312.
64. Hendley, E. D., Burrows, G. H., Robinson, E. S., Heidenreich, K. A., and Bulman, C. A., 1977, *Pharmacol. Biochem. Behav.* **6**:197–202.
65. Hadfield, M. G., and Weber, N. E., 1975, *Biochem. Pharmacol.* **24**:1538–1540.
66. Moisset, B., Hendley, E. D., and Welch, B. L., 1975, *Brain Res.* **92**:157–164.
67. Moisset, B., 1977, *Brain Res.* **121**:113–120.
68. Myers, M. M., Whittemore, S. R., and Hendley, E. D., 1981, *Brain Res.* **220**:325–338.
69. Palmer, G. C., 1979, *Biochem. Pharmacol.* **28**:2847–2849.
70. Brenneman, D. E., and Rutledge, C. O., 1979, *Brain Res.* **179**:295–304.
71. Hamberger, B., 1967, *Acta Physiol. Scand. [Suppl.]* **295**:1–56.
72. Coyle, J. T., and Snyder, S. H., 1969, *Science* **166**:899–901.
73. Harris, J. E., and Baldessarini, R. J., 1973, *Life Sci.* **13**:303–312.

74. Horn, A. S., 1973, *Br. J. Pharmacol.* **47**:332–338.
75. Holz, R. W., and Coyle, J. T., 1974, *Mol. Pharmacol.* **10**:746–758.
76. Horn, A. S., Coyle, J. T., and Snyder, S. H., 1971, *Mol. Pharmacol.* **7**:66–80.
77. Hendley, E. D., and Snyder, S. H., 1972, *Eur. J. Pharmacol.* **19**:56–66.
78. Ferris, R. M., Tang, F. L. M., and Maxwell, R. A., 1972, *J. Pharmacol. Exp. Ther.* **181**:407–416.
79. Thornburg, J. E., and Moore, K. E., 1973, *Res. Commun. Chem. Pathol. Pharmacol.* **5**:81–89.
80. Harris, J. E., and Baldessarini, E. J., 1973, *Neuropharmacology* **12**:669–679.
81. Ross, S. B., and Renyi, A. L., 1975, *Acta Pharmacol. Toxicol. (Kbh.)* **36**:56–66.
82. Raiteri, M., Cerrito, F., Cervoni, A. M., del Carmine, R., Ribera, M. T., and Levi, G., 1978, *Advances in Biochemical Psychopharmacology*, Volume 19 (P. J. Roberts, G. N. Woodruff, and L. L. Iversen, eds.), Raven Press, New York, pp. 35–56.
83. Horn, A. S., Cuello, A. C., and Miller, R. J., 1974, *J. Neurochem.* **22**:265–270.
84. Thierry, A. M., Blanc, G., Sobel, A., Stinus, L., and Glowinski, J., 1973, *Science* **182**:499–501.
85. Thierry, A. M., Stinus, L., Blanc, G., and Glowinski, J., 1973, *Brain Res.* **50**:230–234.
86. Tassin, J. P., Thierry, A. M., Blanc, G., and Glowinski, J., 1974, *Naunyn Schmiedebergs Arch. Pharmacol.* **282**:239–244.
87. Cuello, A. C., Horn, A. S., Mackay, A. V. P., and Iversen, L. L., 1973, *Nature* **243**:465–467.
88. Demarest, K. T., and Moore, K. E. 1979, *Brain Res.* **171**:545–551.
89. Annunziato, L., Leblanc, P., Kordon, C., and Weiner, R. I., 1980, *Neuroendocrinology* **31**:316–320.
90. Schacht, U., Leven, M., and Backer, G., 1977, *Br. J. Clin. Pharmacol.* **4**:77S–87S.
91. Van der Zee, P., and Hespe, W., 1978, *Neuropharmacology* **17**:483–490.
92. Heikkila, R. E., Cabbat, F. S., and Mytilineou, C., 1977, *Eur. J. Pharmacol.* **45**:329–333.
93. Hendley, E. D., Snyder, S. H., Fauley, J. J., and LaPidus, J. B., 1972, *J. Pharmacol. Exp. Ther.* **183**:103–116.
94. Ross, S. B., 1979, *Life Sci.* **24**:159–168.
95. Cooper, B. R., Hester, T. J., and Maxwell, R. A., 1980, *J. Pharmacol. Exp. Ther.* **215**:127–134.
96. Hadfield, M. G., 1981, *Brain Res.* **222**:172–176.
97. Gershon, M. D., Robinson, R. G., and Ross, L. L., 1976, *J. Pharmacol. Exp. Ther.* **198**:548–561.
98. Gershon, M. D., 1981, *J. Physiol. (Paris)* **77**:257–265.
99. Paasonen, M. K., 1968, *Ann. Med. Exp. Fenn.* **46**:416–422.
100. Ross, S. B., and Renyi, A. L., 1967, *Life Sci.* **6**:1407–1415.
101. Blackburn, K. J., French, P. C., and Merrills, R. J., 1967, *Life Sci.* **6**:1653–1663.
102. Shaskan, E. G., and Snyder, S. H., 1970, *J. Pharmacol. Exp. Ther.* **175**:404–418.
103. Carlsson, A., 1970, *J. Pharm. Pharmacol.* **22**:729–732.
104. Takatsuka, K., Segawa, T., and Takagi, H., 1971, *Jpn. J. Pharmacol.* **21**:57–67.
105. Carlsson, A., Corrodi, H., Fuxe, K., and Hokfelt, T., 1969, *Eur. J. Pharmacol.* **5**:357–366.
106. Ross, S. B., and Renyi, A. L., 1969, *Eur. J. Pharmacol.* **7**:270–277.
107. Wong, D. T., Horng, J.-S., and Fuller, R. W., 1973, *Biochem. Pharmacol.* **22**:311–322.
108. Jaim-Etcheverry, G., and Zieher, L. M., 1968, *Z. Zellforsch. Mikrosk. Anat.* **86**:393–400.
109. Jaim-Etcheverry, G., and Zieher, L. M., 1971, *J. Pharmacol. Exp. Ther.* **178**:42–48.
110. Tilders, F. J. H., Ploem, J. S., and Smelik, P. G., 1974, *J. Histochem. Cytochem.* **22**:967–975.
111. Snipes, R. L., Thoenen, H., and Tranzer, J. P., 1968, *Experientia* **24**:1026–1027.
112. Zieher, L. M., and Jaim-Etcheverry, G., 1971, *J. Pharmacol. Exp. Ther.* **178**:30–41.
113. Carlsson, A., Hillarp, N.-A., and Waldeck, B., 1963, *Acta Physiol. Scand. [Suppl.]* **215**:1–38.
114. Horn, A. S., 1973, *J. Neurochem.* **21**:883–888.
115. Horn, A. S., Baumgarten, H. G., and Schlossberger, H. G., 1973, *J. Neurochem.* **21**:233–236.

116. Heikkila, R. E., Goldfinger, S. S., and Orlansky, H., 1976, *Res. Commun. Chem. Pathol. Pharmacol.* **13**:237–250.
117. Riblet, L. A., Gatewood, C. F., and Mayol, R. F., 1979, *Psychopharmacology* **63**:99–101.
118. Wong, D. T., Horng, J. S., Bymaster, F. P., Hauser, K. L., and Molloy, B. B., 1974, *Life Sci.* **15**:471–479.
119. Fuller, R. W., and Wong, D. T., 1977, *Fed. Proc.* **36**:2154–2158.
120. Ross, S. B., and Renyi, A. L., 1977, *Neuropharmacology* **16**:57–63.
121. Hyttel, J., 1978, *Psychopharmacology* **60**:13–18.
122. Mireylees, S. E., Goodlet, I., and Sugrue, M. F., 1978, *Biochem. Pharmacol.* **27**:1023–1027.
123. Horn, A. S., and Trace, R. C. A. M., 1974, *Br. J. Pharmacol.* **51**:399–403.
124. Friedman, E., Meller, E., and Hallock, M., 1981, *J. Neurochem.* **36**:931–937.
125. Ross, S. B., 1980, *Pharmacology* **21**:123–131.
126. Elliott, K. A. C., and van Gelder, N. M., 1958, *J. Neurochem.* **3**:28–41.
127. Iversen L. L., and Neal, M. J., 1968, *J. Neurochem.* **15**:1141–1149.
128. Hokfelt, T., Jonsson, G., and Ljungdahl, A., 1970, *Life Sci.* **9**:203–212.
129. Iversen, L. L., and Bloom, F. E., 1972, *Brain Res.* **41**:131–143.
130. Varon, S., Weinstein, H., Kakefuda, T., and Roberts, E., 1965, *Biochem. Pharmacol.* **14**:1213–1224.
131. Kuriyama, K., Weinstein, H., and Roberts, E., 1969, *Brain Res.* **16**:479–492.
132. Henn, F. A., and Hamberger, A., 1971, *Proc. Natl. Acad. Sci. U.S.A.* **68**:2686–2690.
133. Henn, F. A., 1976, *J. Neurosci. Res.* **2**:271–282.
134. Hertz, L., 1979, *Prog. Neurobiol.* **13**:277–323.
135. Stewart, R. M., and Rosenberg, R. N., 1979, *Int. Rev. Neurobiol.* **21**:275–309.
136. Fonnum, F., Karlsen, R. L., Malthe-Sorenssen, D., Sterri, S., and Walaas, I., 1980, *The Cell Surface and Neuronal Function* (C. W. Cotman, G. Poste, and G. L. Nicholson, eds.), Elsevier/North Holland Biomedical Press, Amsterdam, pp. 455–504.
137. Iversen, L. L., and Johnston, G. A. R., 1971, *J. Neurochem.* **18**:1939–1950.
138. Beart, P. M., Johnston, G. A. R., and Uhr, M. L., 1972, *J. Neurochem.* **19**:1855–1861.
139. Beart, P. M., and Johnston, G. A. R., 1973, *J. Neurochem.* **20**:319–324.
140. Weitsch-Dick, F., Jessell, T. M., and Kelly, J. S., 1978, *J. Neurochem.* **30**:799–806.
141. Johnston, G. A. R., Stephanson, A. L., and Twitchin, B., 1976, *J. Neurochem.* **26**:1029–1032.
142. Johnston, G. A. R., Krogsgaard-Larsen, P., and Stephanson, A., 1975, *Nature* **258**:627–628.
143. Pullman, B., and Berthod, H., 1975, *Theor. Chim. Acta (Berl.)* **36**:317–328.
144. Aprison, M. H., and Werman, R., 1965, *Life Sci.* **4**:2075–2083.
145. Neal, M. J., and Pickles, H. G., 1969, *Nature* **223**:679–680.
146. Neal, M. J., 1971, *J. Physiol. (Lond.)* **215**:103–117.
147. Curtis, D. R., Hosli, L., and Johnston, G. A. R., 1968, *Exp. Brain Res.* **6**:1–18.
148. Logan, W. J., and Snyder, S. H., 1971, *Nature* **234**:297–299.
149. Logan, W. J., and Snyder, S. H., 1972, *Brain Res.* **42**:413–431.
150. Bennett,, J. P., Logan, W. J., and Snyder, S. H., 1973, *J. Neurochem.* **21**:1533–1550.
151. Bennett, J. P., Mulder, A. H., and Snyder, S. H., 1974, *Life Sci.* **15**:1045–1056.
152. Hruska, R. E., Padjen, A., Bressler, R., and Yamamura, H. I., 1978, *Mol. Pharmacol.* **14**:77–85.
153. Hokfelt, T., and Ljungdahl, A., 1971, *Brain Res.* **32**:189–194.
154. Arregui, A., Logan, W. J., Bennett, J. P., and Snyder, S. H., 1972, *Proc. Natl. Acad. Sci. U.S.A.* **69**:3485–3489.
155. Balcar, V. J., and Johnston, G. A. R., 1972, *J. Neurochem.* **19**:2657–2666.
156. Balcar, V. J., and Johnston, G. A. R., 1973, *J. Neurochem.* **20**:529–539.
157. Balcar, V. J., Johnston, G. A. R., and Twitchin, B., 1977, *J. Neurochem.* **28**:1145–1146.
158. Curtis, D. R., Duggan, A. W., and Johnston, G. A. R., 1970, *Exp. Brain Res.* **10**:447–462.
159. Iversen, L. L., and Storm-Mathisen, J., 1976, *Acta Physiol. Scand.* **96**:22A–23A.
160. Hokfelt, T., and Ljungdahl, A., 1972, *Adv. Biochem. Psychopharmacol.* **6**:1–37.
161. Henn. F. A., Goldstein, M. N., and Hamberger, A., 1974, *Nature* **249**:663–664.
162. Guyenet, P., Lefresne, P., Rossier, J., Beaujouan, J. C., and Glowinski, J., 1973, *Mol. Pharmacol.* **9**:630–639.

163. Guyenet, P., Lefresne, P., Rossier, J., Beaujouan, J. C., and Glowinski, J., 1973, *Brain Res.* **62:**523–529.
164. Kuhar, M. J., Sethy, N. H., Roth, R. H., and Aghajanian, G. K., 1973, *J. Neurochem.* **20:**581–593.
165. Simon, J. R., and Kuhar, M. J., 1975, *Nature* **255:**162–163.
166. Ross, S. B., Hall, H., Renyi, A. L., and Westerlund, D., 1981, *Psychopharmacology* **72:**219–225.

Release of Catecholamines, Serotonin, and Acetylcholine from Isolated Brain Tissue

Maurizio Raiteri, Mario Marchi, and Guido Maura

1. INTRODUCTION

Release of transmitter substances is a fundamental step of chemical neurotransmission that occurs in the living brain, under physiological conditions, when presynaptic nerve endings are invaded by an action potential.[5] The available evidence suggests that *in vivo* the release process is intimately connected with many other steps of neurotransmission including supply of transmitter precursors, transmitter synthesis, compartmentation, reuptake, and catabolism. The release of several transmitters appears to be under the control of specific feedback mechanisms mediated by receptors localized on presynaptic nerve terminals. Moreover, the release of one transmitter can be regulated by other transmitters or modulators through receptor-mediated interactions varying among different brain regions. Neuro- and psychoactive drugs can interfere, at various levels, with neurotransmitter release.

The complexity of the organization of the living brain makes it very difficult to study *in vivo* the mechanisms of neurotransmitter release at a cellular, subcellular, and molecular level. This can be better accomplished through an *in vitro* approach by using isolated central nervous tissue preparations such as brain slices or synaptosomes.

The present chapter deals with the release *in vitro* of four central transmitters: acetylcholine (ACh) and the three major biogenic amines, norepinephrine (NE), dopamine (DA), and 5-hydroxytryptamine (5-HT). Attention is focused on some important aspects of the problem whose recent evolution has been most significant. For information about other aspects of neurotransmitter

Maurizio Raiteri, Mario Marchi, and Guido Maura • Istituto di Farmacologia e Farmacognosia, Università di Genova, 16146 Genoa, Italy.

release, the reader may refer to previous reviews[1-4] and to other chapters of this *Handbook*.

2. GENERAL CONSIDERATIONS AND TECHNICAL ASPECTS RELATED TO STUDIES OF BIOGENIC AMINE AND ACETYLCHOLINE RELEASE FROM CENTRAL NERVE ENDINGS

Brain slices[5] represent the *in vitro* preparation most widely used in experiments of transmitter release in the central nervous system. However, an increasing number of reports indicate that superfused synaptosomes[2,6] may also be very useful, particularly when investigating the mechanisms of transmitter release. Both systems have strengths and weaknesses, which will be discussed in Section 7.4. and in other parts of the chapter.

Brain tissue takes up exogenous NE, DA, and 5-HT as well as their biosynthetic precursors. By using the radioactive amines, it is therefore possible to label specific families of nerve endings (see below) and then study how the labeled transmitters are released when the tissue is exposed to various agents. It is not known whether the exogenous amines taken up distribute homogeneously within the intraterminal pools. Therefore, the results of these studies may reflect release only from the transmitter pool just acquired by the nerve endings through the reuptake process, which is considered to be the major physiological mechanism for the synaptic inactivation of the amines.[7,8] In order to obtain information on the pool of recent synthesized transmitter, brain tissue can be exposed to labeled amine precursors in order to analyze the release of the transmitters just synthesized. A comparison between the effects of various agents on the release of the transmitter contained in the "reuptake pool" or in the "new synthesis pool" can thus be made by studying in parallel brain tissue preparations prelabeled with exogenous radioactive transmitters or with their radioactive precursors.

Since ACh is not taken up into cholinergic nerve endings, release of labeled ACh can be studied following exposure to a labeled precursor. The most widely used ACh precursor is choline, which is taken up into cholinergic terminals by a high-affinity system and rapidly converted into ACh.[9]

The release of endogenous biogenic amines[10-13] or ACh[14,15] present in nerve terminals has been measured only rarely with respect to that of radioactive transmitters because of technical difficulties. Although the determination of endogenous transmitters does not allow one to distinguish among the functional pools from which the transmitter is released, it offers the possibility of studying release without any previous manipulation of the transmitter pools. With some recently introduced methods, for instance, high-performance liquid chromatography with electrochemical detection of biogenic amines[16] and chemoluminescent determination of ACh,[17] the analysis of endogenous NE, DA, 5-HT, and ACh in the effluent from superfused slices or synaptosomes should become more feasible.

When release studies are performed with radioactive compounds, it may be risky to rely, as it is often done, on measurements of the total radioactivity released, even if inhibitors of transmitter-metabolizing enzymes are present. As pointed out by Trendelenburg,[18] in several cases, a high percentage of the total radioactivity released from tissues previously loaded with [³H]NE was accounted for by deaminated metabolites, even when inhibitors of monoamine oxidase (MAO) were employed. If one considers that the use of enzyme inhibitors in general should be avoided because of several unwanted intra- and extraterminal actions, the importance of the separation between transmitter and metabolites in the effluent fluids becomes evident; this is particularly true when different releasing stimuli are used because of the possible involvement of different mechanisms of transmitter efflux from nerve endings (see also Section 6).

The choice of the brain region appears to be of critical importance in studying release of catecholamines, 5-HT, or ACh. For example, one area may be more suitable than others because it is relatively rich in terminals of a given population (see below). Moreover, when transmitter interactions are investigated, it should be considered that a given transmitter is likely to regulate the release of another transmitter only in some areas of the brain but not in others. Large regional differences were observed when the rates of release of [³H]NE, [³H]DA, and [³H]ACh were compared in 12 regions of the rat brain.[19]

As a rule, slices or synaptosomes from the mammalian brain contain many different families of nerve endings. The possibility that a given amine is taken up into and subsequently released from a family of nerve endings normally not utilizing that amine as a transmitter is well documented[7,8,20] and may represent a drawback in studying release from brain tissue prelabeled with radioactive transmitters.

The problem of the coexistence of different nerve ending populations can often be satisfactorily circumvented, so that it is possible to study the properties of a given type of terminals with little or no interference by the other populations.

In the case of the cholinergic system, the high-affinity uptake of choline appears to be sufficiently specific for cholinergic nerve endings[9,21] that the exogenous radioactive choline generally used in these studies is likely to label exclusively nerve terminals of the cholinergic type. Moreover, the choline high-affinity uptake is so intimately linked to the enzyme choline acetyltransferase[9,22] that the choline taken up is quickly and almost completely converted into ACh. Thus, the various aspects of ACh release can be conveniently studied in slices or synaptosomes from the mammalian brain prelabeled with choline.

When brain tissue is labeled with exogenous catecholamines or serotonin, the possibility exists that each of the three transmitters enters the others' nerve endings. This certainly occurs if brain slices or synaptosomes obtained from a region in which different aminergic nerve terminals coexist (for instance, the cerebral cortex), are incubated with relatively high concentrations of the labeled transmitter. The use of low concentrations (10 nM for instance) minimizes the labeling of catecholaminergic nerve terminals by 5-HT and *vice versa*. As far as catecholamines are concerned, the fact that NE and DA have similar affinities for each other's membrane carrier[8] makes it impossible to obtain specific labeling only by decreasing the concentration of the labeled amines. The problem has been circumvented in some cases by taking into account, for example, that the corpus striatum (rich in DA terminals) contains relatively few NE nerve endings or that the cerebellum contains NE but not DA nerve endings or that DA terminals in the hypothalamus (an area rich in NE nerve endings) are few and possess a low-affinity uptake system.[23] Furthermore, some new specific uptake inhibitors[24–28] make it possible to obtain selective labeling practically in every brain area. For example, one can label with radioactive DA the dopaminergic terminals present in a cortical preparation following inactivation of the NE and 5-HT transport systems with selective NE and 5-HT uptake blockers.

There may be some difficulty in studying the release of catecholamines newly synthesized from labeled tyrosine, since NE and DA share this biosynthetic precursor. The problem can be partly circumvented by choosing brain areas with a very high ratio of NE to DA nerve endings or *vice versa*.

Our work with *d*-amphetamine[20] provides a good example of the importance of selecting the appropriate brain region when studying the effect of drugs on catecholamine release. Amphetamine was unable to augment in superfusion the release of [³H]NE previously taken up in noradrenergic synaptosomes from hypothalamus or cerebellum, whereas the release of [³H]DA from dopaminergic striatal synaptosomes was greatly potentiated by the drug. However, after [³H]NE was taken up as a "false transmitter" into striatal nerve endings, it became sensitive to the releasing action of amphetamine.[20]

3. SPONTANEOUS RELEASE AND STIMULATED RELEASE

3.1. Spontaneous Release

The release of transmitters that is observed *in vitro* when nerve terminals are exposed to a physiological solution at physiological termperature is referred to as spontaneous or basal release. The rate of the spontaneous release can be reduced by various agents or, more frequently, increased (stimulated or evoked release). Some considerations indicate that the stimulated release may not be interpreted in all cases as a potentiation of the spontaneous release. First of all, some releasing stimuli require the presence of Ca^{2+}, whereas others are Ca^{2+} independent. Moreover, the same transmitter can exit from nerve endings by different mechanisms, depending on the releasing stimulus (see Section 6).

The spontaneous release of [³H]DA newly synthesized from [³H]tyrosine was Ca^{2+} dependent and inhibited by tetrodotoxin (TTX)[29] in analogy with the electrically stimulated release. Both basal and electrically stimulated release of endogenous ACh were Ca^{2+} dependent and TTX sensitive.[14] Equally Ca^{2+} dependent were the basal and the K^+-evoked release of [³H]ACh from hippocampal synaptosomes.[30] In hypothalamic synaptosomes, extracellular 5-HT inhibited both K^+-evoked and spontaneous release of [³H]5-HT through the activation of presynaptic autoreceptors[31] (see also Section 7.1.2). Moreover, both the depolarization-induced and the spontaneous release of NE, DA, and 5-HT occurred independently of the membrane carriers[32–34] (see also Section 6). All together, these results suggest that the spontaneous release of biogenic amines and ACh occurs, at least in part, through a mechanism similar to that of the Ca^{2+}-dependent depolarization-evoked release.

3.2. Stimulated Release

3.2.1. Depolarizing Stimuli

Depolarizing stimuli induce a release of NE, DA, 5-HT, and ACh (endogenous, previously taken up, or newly synthesized; from brain slices or from synaptosomes) that is strictly Ca^{2+} dependent[1–4].

Electrical stimulation and exposure to solutions containing a high concentration of K^+ or containing the alkaloid veratridine have been used to induce Ca^{2+}-dependent release of various transmitters. The properties of these depolarizing stimuli, including their mechanism of action, have been discussed in recent reviews[35,36]; we therefore limit ourselves to a few considerations originating from results of recent acquisition.

There is increasing evidence that the concentration of K^+ (or the frequency of the electrical stimulus) and the concentration of extracellular Ca^{2+} are of critical importance in studying depolarization-evoked release. For example, inhibition of [3H]NE release by α-agonist drugs through presynaptic autoreceptors (see Section 7.1.1) could be observed at low K^+ concentrations (13–26 mM) but not when 56 mM KCl was used unless extracellular Ca^{2+} was lowered from 1.2 to 0.2 mM. Similarly, α agonists decreased the release of [3H]NE evoked from cortical slices by electrical stimulation at 5 Hz but not at 10 Hz.[38] Other relevant examples are reported throughout the text.

It is generally believed that high K^+ induces neurotransmitter release by depolarizing the axon terminals directly. For this reason, the K^+-evoked release has been found to be insensitive to TTX,[39] which acts by blocking the traffic of action potentials. However, it was recently reported[40] that the release of [3H]NE evoked by relatively low K^+ concentrations (13 mM, but not 25 mM) was as sensitive to TTX as the release produced by electrical field stimulation, indicating the involvement of Na^+ channel activation in the release of [3H]NE evoked by both stimuli.

Tetrodotoxin has been often used in a test (that will be termed the "high-K^+–TTX text") to exclude the involvement of interneuronal pathways in the regulation by a given transmitter either of its own release or of the release of another transmitter. If the regulation of the K^+-evoked release also persists in the presence of TTX, it is concluded that the regulating transmitter does not act by way of action potentials in interneurons[41] (see also Section 7.4).

Whereas the use of high K^+ is being accepted by an increasing number of fans of electrical stimulation, the fortune of veratridine as a depolarizing agent appears to be in decline. Although the cationic fluxes elicited by the alkaloid (concomitant influx of Na^+ and Ca^{2+} and efflux of K^+, all specifically blocked by TTX)[39] resemble more the fluxes occurring during physiological depolarization than those taking place during depolarization with high extracellular K^+ concentrations,[42] some recent results indicate that the use of veratridine has some limitations. Strikingly, whereas the veratridine-induced release of [^{3}H]NE and [^{3}H]DA was largely Ca^{2+} dependent, the release of GABA and glutamate (previously accumulated or endogenous) evoked by the alkaloid from superfused synaptosomes was inhibited by Ca^{2+} ions.[43] The K^+-stimulated release of [^{3}H]GABA was, however, Ca^{2+} dependent.[44] Cholinergic autoreceptors (see Section 7.1.4) could be activated (and [^{3}H]ACh release decreased) by ACh or muscarinic agonists in synaptosomes exposed to high K^{+}[45,46] but not to veratridine.[46] It was found that the alkaloid, even at concentrations lower than those used as a depolarizing agent, produced a dramatic decrease in muscarinic receptor sites,[46,47] possibly through damage to the synaptosomal membrane.[46]

3.2.2. Voltage-Sensitive Calcium Channels and Calcium Ionophores

Depolarization of nerve endings induces an increase in Ca^{2+} conductance; Ca^{2+} ions penetrate into the terminals, moving down their electrochemical gradient. This passive entry of Ca^{2+} apparently occurs through Ca^{2+}-selective channels whose opening is triggered by depolarizing stimuli (for reviews see Blaustein[48]; Hagiwara and Byerly[49]).

When the kinetics of the entry of $^{45}Ca^{2+}$ was analyzed in K^+-depolarized synaptosomes, an initial "fast" phase and a subsequent "slow" phase of K^+-stimulated Ca^{2+} influx were observed.[50] These results are consistent with the presence of two distinct populations of voltage-regulated channels for Ca^{2+} entry into presynaptic brain nerve endings.

A number of organic compounds that have been named Ca^{2+} antagonists (verapamil, nifedipine, diltiazem, felodipine, nimodipine, nisoldipine, D-600) have recently been developed.[49] Some of them have been introduced in clinical practice to treat hypertension, angina, and other vasoconstrictive conditions.[51] Calcium antagonists block the penetration of Ca^{2+} into cells by acting at the voltage-dependent Ca^{2+} channels. The finding that Ca^{2+} antagonists block Ca^{2+} entry differentially gives further support to the idea of the existence of subtypes of Ca^{2+} channels (for review see ref. 49).

Pentobarbital was found to inhibit the depolarization-induced Ca^{2+} influx across the synaptosomal membrane.[52] Tolerance seems to develop to this inhibition with the same time course as behavioral tolerance. In fact, in synaptosomes from animals showing decreased hypnotic response to barbiturate administration, the K^+-induced $^{45}Ca^{2+}$ influx was not depressed by pentobarbital. Whether barbiturates act at the voltage-sensitive Ca^{2+} channels is not known.

The relationship between transmitter release and extracellular Ca^{2+} concentration was studied using rat neocortex slices under various experimental conditions.[40] The electrically evoked release of [^{3}H]NE increased sigmoidally with the external Ca^{2+} concentration up to 1.2 mM. When high K^+ (13 or 20 mM) was used as a depolarizing agent, [^{3}H]NE release reached a maximal value at 0.6 or at 0.9 mM Ca^{2+}, respectively, and then decreased at higher Ca^{2+} concentrations. As previously suggested by Brehm and Eckert,[53] the increase of Ca^{2+} entry through the voltage-dependent channels may be counteracted by an autoinhibitory process that becomes apparent at lower Ca^{2+} concentration (0.6 mM) when the Ca^{2+} channels are less activated (13 mM KCl).

According to the "calcium hypothesis," an increased availability of "free" intracellular Ca^{2+} in nerve endings is sufficient to trigger neurotransmitter release.[54] This hypothesis has received further support by studies in which Ca^{2+} ionophores (see Pressmann[55] and Triggle[56] for reviews) elicited, in the absence of nerve depolarization, the release of transmitters from peripheral and central nerve endings.

The Ca^{2+} ionophore most widely used in neurotransmitter release studies is A23187. This compound permits the entry into nerve endings of Ca^{2+} ions down their gradient by a process that bypasses the physiological route of Ca^{2+} penetration. The different transmitter systems are not equally sensitive to the releasing action of A23187. In rat brain synaptosomes, the release of DA[33,57] was consistently increased by concentrations of ionophore two orders of magnitude lower than those required to elicit release of NE,[32,58] 5-HT,[34] ACh,[59] or GABA.[60] The reason for the greater sensitivity of DA terminals is not clear. The finding that the K^+-evoked [^{3}H]DA release[33] was depressed less than that of [^{3}H]NE[61] when the Ca^{2+} concentration in the medium was reduced may suggest that the critical concentration of Ca^{2+} necessary to trigger DA release is lower than that required for the other transmitters. The release of catecholamines induced by the ionophore A23187 may mimic the physiological release and occur, at least in part, through an exocytosis-like process.[2,56]

It has been recently reported[62] that phosphatidate, a neuronal phospholipid, when incorporated into the membrane of striatal synaptosomes, increased $^{45}Ca^{2+}$ uptake and stimulated Ca^{2+}-dependent [^{3}H]DA release. Several other phospho- and glycolipids were inactive. The authors[62] suggest that phosphatidate may be a natural Ca^{2+} ionophore serving as a link between depolarization and release.

3.2.3. Coupling between Calcium Influx and Transmitter Release

It was proposed[63] that Ca^{2+} ions entering the nerve endings during depolarization inhibit Na^+,K^+-ATPase. Inhibition of the enzyme would be the trigger of transmitter release under physiological conditions. This idea is based in part on the findings that a variety of experimental conditions known to inhibit Na^+,K^+-ATPase (Na^+ deprivation, K^+ deprivation, exposure to ouabain) increase transmitter release. It should be noted, however, that in some cases ouabain, at concentrations (20–100 μM) that certainly inhibited the enzyme, could not stimulate release of NE,[60] 5-HT,[64] or ACh.[64] Since high K^+ did evoke release of the three transmitters, one should assume that the Ca^{2+} entering during K^+ depolarization could inhibit $NA,^+ K^+$-ATPase more effectively than exposure to ouabain. Furthermore, if Ca^{2+}-dependent stimuli and ouabain (even without Ca^{2+}) triggered release by inhibiting Na^+,K^+-ATPase, it would be difficult to explain how the release of [^{3}H]DA evoked by high K^+ occurred by a mechanism (carrier-independent) different from that (carrier-mediated) of the ouabain-stimulated release[33] (see also Section 6).

Evidence is being provided that calmodulin, a major Ca^{2+}-binding protein in brain[65] which is contained in synaptic vesicles as well as in the synaptoplasm, mediates many of the actions of Ca^{2+} at the synaptic level. Trifluoperazine, a phenothiazine that inactivates calmodulin, and two inhibitors of Ca^{2+}–calmodulin protein kinase, phenytoin and diazepam, blocked the release of NE and ACh evoked in synaptosomes by high K^+ or by the Ca^{2+} ionophore A23187.[66] The Ca^{2+} ions entering into synaptosomes during depolarization simultaneously stimulated Ca^{2+}–calmodulin-dependent protein phosphorylation, vesicle–presynaptic membrane interactions, and transmitter release. All three processes were inhibited by trifluoperazine. It was proposed by De Lorenzo[66] that the binding of the newly available Ca^{2+} to calmodulin activates tubulin kinases associated with presynaptic membranes and vesicles, leading to phosphorylation of tubulin in these two structures. Tubulin phosphorylation would then result in physicochemical changes in the surface of vesicles and presynaptic membrane favoring their fusion, which culminates in neurotransmitter release.

An involvement of transmethylation reactions in stimulation–secretion coupling was suggested by the finding that the depolarization-dependent release of [^{3}H]NE and of ACh from the pheochromocytoma clone PC12 was enhanced by 3-deazaadenosine, an inhibitor of S-adenosylmethionine-dependent transmethylations.[67]

3.2.4. Other Releasing Stimuli

Release of biogenic amines and ACh can also be elicited by stimuli that are either Ca^{2+} independent or possess only a minor Ca^{2+}-dependent component. A number of reviews are available containing detailed discussions about release induced by lack of Na^+, by lack of K^+, or by ouabain,[2,35,68] release induced by neurotoxins,[2,4,69,70] and release induced by phenylethylamines.[2,18]

4. DISCRIMINATION BETWEEN TRUE RELEASERS AND APPARENT RELEASERS (REUPTAKE INHIBITORS)

The large majority of studies on the release of neurotransmitters, both previously taken up and newly synthesized, have been performed using brain slices in incubation or in superfusion and synaptosomes in incubation conditions.

In release studies performed with brain slices, or even with synaptosomes in incubation, the efficient reuptake process that continuously recaptures a large fraction of the transmitter just released represents a complicating factor. For example, it is difficult, in these experimental conditions, to distinguish between agents that stimulate release directly and agents that, by preventing reuptake, indirectly increase the efflux of the transmitter.

The discrimination between action on release and on reuptake is important for the characterization of a drug. Although both actions may lead to an increase of neurotransmitter in the synaptic cleft, some presynaptic homeostatic mechanisms may be differently affected by a drug that activates release and by a reuptake inhibitor. For example, neurotransmitter turnover is reduced by amine reuptake blockers such as tricylic antidepressants,[71,72] whereas it is generally increased by "true" amine-releasing agents such as amphetamine.[73]

The problem of studying the release in conditions preventing the reuptake of the released transmitters can be overcome if a thin layer of synaptosomes is superfused under appropriate conditions[6,74] (for technical details see Raiteri and Levi[2]). The technique of synaptosome superfusion has been shown to be very efficient in preventing the reuptake of several neurotransmitters[2] including NE, DA, 5-HT, GABA, and glutamate. Since the basal release of all of these substances was not increased when reuptake blockers were added to the superfusion medium, one can conclude that reuptake does not occur in this experimental condition because the transmitters are removed by the superfusion fluid as soon as they are released.

When studied in superfusion with hypothalamic synaptosomes prelabeled with [^{3}H]NE, *d*-amphetamine, known for decades as a stimulant of NE release in noradrenergic nerve endings, appeared to be devoid of releasing activity. The drug, which is a strong NE reuptake inhibitor, may therefore act only as an indirect NE releaser.[20,75]

The central stimulant effects of amphetamine and methylphenidate have been attributed to the ability of the drugs to release DA in the central nervous system.[76,77] However, in superfused striatal synaptosomes, only amphetamine appeared to be a strong releaser of DA, either previously accumulated or newly synthesized from labeled tyrosine,[20,78,79] whereas methylphenidate was almost inactive[80] (also G. Maura and M. Raiteri, unpublished observation).

The antidepressant nomifensine is a strong inhibitor of both NE and DA uptake.[81–84] Because of the increase of striatal dihydroxyphenylacetic acid and homovanillic acid[85,86] and of the decrease of prolactin secretion[87] seen following *in vivo* administration, nomifensine has also been considered to be a DA releaser. However, in superfusion conditions, the drug did not alter the release (spontaneous or evoked by 56 mM KCl) of [^{3}H]DA previously taken up or newly synthesized from [^{3}H]tyrosine in rat striatal synaptosomes[33,78,79] or the release of [^{3}H]DA previously accumulated in rat median eminence synaptosomes.[88] Nomifensine only increased [^{3}H]DA release from synaptosomes in incubation, most likely by preventing reuptake.[80]

In conclusion, the central stimulants of the amphetaminelike group (which includes amphetamine and other phenylethylamines) appear to be true DA-releasing agents, whereas methylphenidatelike drugs (including methylphenidate, nomifensine, benztropine, and cocaine) may act mainly as uptake inhibitors, as proposed on the basis of indirect evidence by Ross and Kelder.[89]

The discrepancies between the results obtained with nomifensine *in vivo* and *in vitro* could be explained by the recent observation that in striatal synaptosomes the synthesis of DA was consistently activated by DA uptake blockers including nomifensine, cocaine, and benztropine.[90–93] If this also occurred in the living brain, preventing reuptake (for example, with nomifensine or methylphenidate) would lead to an increase in DA release, which is in keeping with the *in vivo* observations,[83,85,86,94] but it is in fact consequent to the increase in synthesis.

5. RELATIONSHIPS BETWEEN NEUROTRANSMITTER COMPARTMENTATION AND RELEASE

Intraterminal NE, DA, and 5-HT are considered to be largely stored in synaptic vesicles; only a small fraction of these transmitters appears to exist in a free form in the cytoplasm in normal conditions. *In vitro* studies indicate that the NE taken up by nerve endings is rapidly bound. In fact, only a minimal fraction of the [^{3}H]NE accumulated by synaptosomes was lost during procedures such as osmotic shock,[95] cold shock,[96] and exposure to antisynaptosome antibodies plus complement,[97] all causing large losses of cytoplasmic constituents. In experiments in which endogenous NE was determined in hypothalamic synaptosomes osmotically shocked on Millipore® filters, it was found that the "free" amine did not exceed 15%.[98] Moreover, the radioactive NE, DA, or 5-HT taken up are largely protected from intraterminal monoamine oxidase and can be subsequently released as unmetabolized amines by depolarizing stimuli in a Ca^{2+}-dependent way,[1–4] in keeping with an exocytotic type of release.[99,100]

In central cholinergic nerve endings, the localization of ACh in both cytoplasm and synaptic vesicles is generally accepted (see MacIntosh[101]).

At least in the case of NE (reviewed in ref. 18) and 5-HT,[102,103] extravesicular binding sites seem also to exist in nerve terminals. A high-affinity 5-HT binding protein associated with serotonergic tracts was isolated and proposed to represent an extravesicular soluble storage form for 5-HT.[102] Thus, NE and 5-HT would be distributed into three distinct compartments: a vesicular pool of bound amine and two cytoplasmic pools, one of bound and the other of free amine.

Understanding the relationships among these compartments and the transmitter pools functionally involved in the release processes is not easy. Evidence has been provided that newly synthesized NE,[104] DA,[105,106] and ACh[107,108] are preferentially released with respect to the less recently synthesized transmitters. Since the depolarization-evoked release of biogenic amines is thought to directly involve the synaptic vesicles, it has been proposed that the preferentially released newly synthesized amine originates from vesicles in close proximity to the external membrane containing the newly synthesized amine (see Zimmermann[109]).

This idea is not unanimously accepted in the case of ACh release, however. There is considerable controversy concerning the origin of the ACh released by depolarizing stimuli (for an extensive list of reviews on this topic see MacIntosh[101]). According to the classical view, ACh is synthesized in the cytoplasm and stored in synaptic vesicles from which it is released through an exocytotic process[110,111]; however, evidence has also been provided suggesting that Ca^{2+}-dependent stimuli elicit release of ACh from the cytoplasm and that vesicular ACh represents a reserve pool.[112–114] As an alternative, the possibility that ACh originates from both the vesicular and the cytoplasmic pool has been recently proposed.[114,115] It should be noted that ACh compartmentation can be affected by various experimental conditions such as methods of animal sacrifice, ionic composition of the medium used during subcellular fractionation, and so on (reviewed in ref. 109).

When the release of biogenic amines is evoked by agents other than the Ca^{2+}-dependent releasing stimuli, the transmitter may not be released directly from vesicles but rather from the cytoplasm through a process mediated by the neuronal membrane carrier. This implies that a transient shifting of the amine from the bound vesicular form into a free cytoplasmic form takes place before the exit from the nerve ending.

A stable cytoplasmic pool can be artificially established by excluding the vesicular pool with reserpine and protecting the amine from catabolism with a MAO inhibitor.[116] Following this treatment, *d*-fenfluramine largely lost its [³H]5-HT-releasing potency in superfused synaptosomes, whereas its active metabolite *d*-norfenfluramine became more potent than in the absence of reserpine treatment.[117] The K^+-evoked release of [³H]NE or [³H]DA was strongly diminished by reserpine pretreatment, supporting the view that depolarization mobilizes vesicular catecholamines.[118,119]

The fact that the amines present in nerve endings are derived not only from synthesis but also from reuptake of the released amine raises the problem of the intraterminal compartmentation and of the functional significance of the amine taken up. When the sensitivity of the ''new synthesis pool'' to various releasing stimuli is compared with that of the ''reuptake pool,'' synaptosomes in superfusion present some advantages with respect to brain slices or incubated synaptosomes. In fact, in these latter systems, the amine newly synthesized, which is in part released (spontaneously) during exposure to the labeled precursor, can be taken up into the ''reuptake pool.'' In contrast, in synaptosomes superfused with the radioactive precursor, the newly formed transmitter that is spontaneously released is removed by the superfusion fluid before reuptake can occur. When the release pattern of [³H]DA originated from [³H]tyrosine was compared in striatal superfused synaptosomes with that of [³H]DA originated by uptake, it was found[79] that the two species of [³H]DA not only were equally responsive to a number of releasing stimuli (high K^+, lack of Na^+, *d*-amphetamine), but their efflux from the nerve ending occurred through an identical mechanism (carrier-independent in the case of the high-K^+-evoked release, carrier-mediated when the release was promoted by the absence of Na^+ or by *d*-amphetamine; see also Section 6). The data suggest that newly synthesized and recaptured DA have similar compartmentation. It is difficult, on the basis of the present evidence, to determine whether the amine taken up also distributes in other intraterminal pools.

6. MODE OF EXIT FROM NERVE ENDINGS: CARRIER-MEDIATED AND CARRIER-INDEPENDENT RELEASE

As already mentioned, it is generally believed that NE, DA, and 5-HT are largely bound in storage granules and that only a small fraction is free in the cytoplasm. A dynamic equilibrium seems, however, to exist between bound and free amine. Neurotransmitter amines could be released in a biologically active form either directly from vesicles by exocytosis, a mechanism amply discussed in several reviews,[4,99,100,109] or following transfer into the cytoplasm from which, because of the relatively high polarity of their ionized form, they could be transported out by a membrane carrier.

Assuming that this carrier is similar to that involved in the Na^+-dependent amine uptake,[7,8] one would expect to see stimulation of release when the cytoplasmic concentration of the amine is increased and/or the inward Na^+ gradient across the plasma membrane is inverted. Moreover, this stimulated release should be sensitive to specific uptake inhibitors.

Three relatively selective uptake inhibitors, desipramine, nomifensine, and chlorimipramine, were used to investigate, respectively, the release of NE, DA, and 5-HT elicited by various stimuli from superfused synaptosomes, with the aim of discriminating between carrier-mediated and carrier-independent release.[32–34,79]

The Ca^{2+}-dependent release evoked by depolarizing stimuli was not affected when the carriers of the three amines were immobilized by the specific inhibitors. Similar results were obtained when release was stimulated by the Ca^{2+} ionophore A23187. Also, the basal release of NE, DA, or 5-HT was almost unaffected by the presence of the superfusion fluid of the appropriate carrier blocker. These results, indicating that Ca^{2+}-dependent evoked release and spontaneous release occurred independently of a carrier, are compatible with an exocytosis-like mechanism.

The stimulation of NE, DA, or 5-HT release observed when synaptosomes, either "normal" or treated with reserpine and a MAO inhibitor (in order to establish a cytoplasmic compartmentation of the amine), were superfused with a Na^+-free medium was largely prevented by desipramine, nomifensine, or chlorimipramine, respectively, indicating an involvement of the membrane carriers. Also, the release of DA evoked by ouabain from striatal synaptosomes was nomifensine sensitive.[33]

Several drugs stimulated amine release through a mechanism involving the membrane transport systems. In particular, the release of [³H]NE promoted by phenylethylamines from hypothalamic synaptosomes was prevented by desipramine[32]; the release of [³H]DA evoked by *d*-amphetamine or other phenylethylamines from striatal synaptosomes was blocked by nomifensine[33,79]; the release of [³H]5-HT elicited by *p*-chloroamphetamine, fenfluramine, tryptamine, or mianserin from whole-brain synaptosomes was inhibited by chlorimipramine.[34]

The release stimulated by phenylethylamines and tryptamine deserves some comments. It is generally believed that carrier inhibitors block the releasing action of phenylethylamines, including the amphetamines, by preventing their carrier-mediated entry into the nerve terminals and the consequent displacement of the transmitter amines.[120,121] However, the high lipophilicity of several phenylethylamines, together with the findings[32,34] that phenylethylamines and tryptamine largely retained their releasing activity in Na^+-poor media, i.e., in conditions in which their uptake through the amine carriers should be largely impaired, militate in favor of a penetration into the nerve terminals by passive diffusion and not through a carrier. Desipramine, nomifensine, chlorimipramine, and other amine carrier inhibitors would therefore block the release of NE, DA, or 5-HT elicited by phenylethylamine and tryptamine derivatives largely, although not exclusively, by preventing the carrier-mediated efflux of the transmitters. This view is corroborated by some *in vivo* data. Although the stereotyped behavior induced in the rat by amphetamine through a release of DA was blocked by DA uptake inhibitors, the lowering of dihydroxyphenylacetic acid (an intraneuronal metabolite of DA) caused by amphetamine through intraneuronal MAO inhibition[122] was not abolished.[123] Thus, DA uptake inhibitors appear to prevent the amphetamine-induced carrier-mediated exit of DA into the synaptic cleft but not the entry of the phenylethylamine into the neuron.

Acetylcholine is not actively taken up by a carrier present in cholinergic nerve terminals.[7,8] Therefore, even if the transmitter is present both in synaptic vesicles and in the cytoplasm, and, according to some authors,[112–114] it can be released from the latter compartment, the possibility that it is transported out of nerve endings by a carrier working in the inside–outside direction does not seem very likely. In contrast, the precursor of ACh, choline, is actively taken up into cholinergic nerve terminals.[9] A hemicholinium-sensitive carrier-mediated choline release was observed in cortex synaptosomes exposed to ouabain.[124] A process of homoexchange between extra- and intrasynaptosomal choline was also found to occur in synaptosomes.[125] This carrier-mediated process seems to be more marked in newborn than in adult rats (M. Marchi, A. Caviglia, P. Paudice, and M. Raiteri, unpublished data).

7. REGULATION OF RELEASE MEDIATED BY PRESYNAPTIC RECEPTORS

In the present chapter the term "presynaptic receptor" will be limited to those release-regulating receptors that are localized on the external surface of the nerve endings. At least three physiological functions can be considered for presynaptic receptors. (1) Autoregulation of release occurs by transmitters activating a type of presynaptic receptors that are termed autoreceptors (for reviews see refs. 41,126–128). (2) Heteroregulation of the release of a given transmitter by another transmitter or modulating substance secreted from neighboring cells, in particular from neighboring neurons, can occur through the activation of receptors that we call presynaptic heteroreceptors (for reviews see refs. 41,127–129). It has to be noted that although these receptors are located on a presynaptic terminal, they may also be considered in some cases to be postsynaptic receptors with respect to the structure (a nerve ending in the case of an axoaxonic synapse) that releases the modulating compound. One important aspect of the question is that presynaptic heteroreceptors may differ pharmacologically from the corresponding presynaptic autoreceptors. On the other hand, it is known that autoreceptors differ in general from the classic postsynaptic receptors on neuronal cell bodies and dendrites (see below). Such pharmacological differences open the possibility of selective activation or blockade by drugs of the different classes of receptors. (3) Modulation of neurotransmitter release can occur by agents originating from a remote part of the body (see Osborne[130]).

A complicating factor in the above scheme may be represented by the existence of cotransmitters (see Hökfelt[127]). A cotransmitter might activate presynaptic receptors on the nerve terminal from which it is released, leading to changes in the rate of release of the primary transmitter. For example, DA neurons in the mesolimbic system contain and release cholecystokininlike peptides, which can inhibit DA release through the activation of specific cholecystokinin sites.[131]

The precise localization of presynaptic receptors on the nerve ending membrane is not known. In an axoaxonic synapse between nerve endings A and B, it would seem reasonable to assume that presynaptic heteroreceptors on B are concentrated in the area corresponding to the synaptic contact, possibly separated from the active zone[132] from where transmitter release from B occurs. As far as autoreceptors are concerned, a likely possibility would be that they are concentrated in the region of the active zone of release, where they would become activated when the transmitter concentration acting at postsynaptic receptors is raised above a threshold level[126–128] (but see Section 7.1.3).

7.1. Presynaptic Autoreceptors

7.1.1. Noradrenergic Autoreceptors

Presynaptic inhibitory autoreceptors of the α-$_2$-adrenergic type have been reported to exist both in the peripheral and in the central noradrenergic system (for reviews see Langer[126,127]; Starke[41,128]).

Release studies with cortical[133] and hypothalamic[92] superfused synaptosomes have yielded results that are similar to those obtained using brain slices, thus supporting the existence of presynaptic autoreceptors on the external membrane of central noradrenergic terminals.

Starting from the findings that some α-adrenoceptor blocking drugs are also effective purine antagonists and that activation of α receptors can evoke purine release, Stone[134] proposed in a recent review that, as an alternative to the α-autoreceptor idea, NE (synaptically released or exogenous) may induce a release of purines, which then become the active principles in effecting the feedback inhibition of NE release. However, the fact that the release of [^{3}H]NE can be inhibited by exogenous NE in a superfused thin layer of synaptosomes,[92] a condition in which indirect effects are unlikely,[2] militates against the above hypothesis.

It is important to note that the pharmacological characteristics of the presynaptic autoreceptors clearly differ from those of the classical adrenoceptors mediating the response of the postsynaptic neuron in the CNS.[41,127]

7.1.2. Serotonergic Autoreceptors

The presence of serotonergic autoreceptors in central nerve endings was first proposed on the basis of results obtained in experiments with brain slices, utilizing LSD or ergocornine as 5-HT receptor agonists.[135,136]

The natural agonist 5-HT directly inhibited the K^+-induced release of [^{3}H]5-HT in superfused hypothalamic synaptosomes. The action of 5-HT was antagonized by methiothepin but not by other known serotonin receptor blockers, e.g., cyproheptadine, methysergide, or mianserin, indicating that, in analogy with the noradrenergic system, 5-HT presynaptic autoreceptors differ from 5-HT postsynaptic receptors.[31] Extracellular 5-HT also reduced the spontaneous release of [^{3}H]5-HT; the action was counteracted by methiothepin, suggesting that spontaneous release occurs, at least in part, through a mechanism similar to the exocytosis-like release elicited by depolarization, possibly following random collison of the storage vesicles with the plasma membrane.

Our data with superfused hypothalamic synaptosomes[31] were confirmed in electrically stimulated rat brain cortical slices.[137] The only 5-HT antagonist examined was methiothepin, which increased the release of [^{3}H]5-HT evoked by field stimulation, probably by removing the inhibition by the amine just released (see also Section 7.4).

7.1.3. Dopaminergic Autoreceptors

There has been controversy about the existence of autoreceptors directly modulating DA release from central dopaminergic nerve endings. The first suggestion for the existence of DA autoreceptors was based on the inhibition by apomorphine and the enhancement by chlorpromazine of the depolarization-evoked release of [^{3}H]DA from rat striatal slices.[138] However, subsequent studies with striatal slices failed to reproduce these results.[139,140] It was found that the butyrophenones haloperidol and spiroperidol paradoxically inhibited the electrically stimulated overflow of [^{3}H]DA, whereas apomorphine was without effect.

In superfused rat striatal synaptosomes, we have been unable to demonstrate the existence of DA autoreceptors. Under a large variety of experimental conditions, DA agonists failed to decrease the depolarization-evoked release of [^{3}H]DA, either previously taken up or endogenously synthesized from labeled tyrosine.[141,142]

More recently, results supporting the presence of a mechanism of autoregulation of DA release were obtained using slices of rabbit caudate nucleus.[143,144] Although the high-K^+–TTX test tends to exclude the involvement of interneurons,[145] the exact location of the DA release-modulating receptors needs further investigation.

The classical idea[146] that autoreceptors are localized on the presynaptic membrane facing the postsynaptic receptor area and become activated when the transmitter concentration in the cleft is raised above a threshold level would not be easily compatible with the findings that postsynaptic DA receptors of the D_2 type are sensitive to DA concentrations ($\sim$5000 nM) that are much higher than those ($\sim$3 nM) required to activate D_3 receptors, a category in which presynaptic DA autoreceptors are included.[147]

7.1.4. Cholinergic Autoreceptors

In superfused brain slices, the release of [^{3}H]ACh evoked by depolarizing stimuli was decreased by acetylcholinesterase inhibitors or by cholinomimetic agents; muscarinic antagonists restored or enhanced the release of the transmitter. These actions were interpreted, also on the basis of the TTX test, as being caused by the presence of a negative feedback mechanism mediated by presynaptic autoreceptors.[107,148–152]

In a cell-free system of incubated nerve terminals, the muscarinic agonist carbachol reduced the K^+-evoked [^{3}H]ACh release. The action was antagonized by atropine, which by itself enhanced the release of the transmitter.[30]

Direct evidence that the natural transmitter ACh inhibits its own release in cholinergic nerve endings has recently been provided.[45,153] The Ca^{2+}-dependent release of [^{3}H]ACh evoked by 15 mM KCl from rat hippocampal synaptosomes prelabeled with [^{3}H]choline was inhibited by exogenous ACh. This action was antagonized by atropine or scopolamine. The muscarinic agonists oxotremorine and carbachol also inhibited ACh release.[45] Interestingly, another well known muscarinic agonist, bethanechol, was inactive on ACh release, suggesting that in the CNS, in analogy with the noradrenergic and serotonergic systems, presynaptic muscarinic autoreceptors may differ from classical postsynaptic muscarinic receptors.[45]

7.1.5. Supersensitivity or Subsensitivity of Autoreceptors Regulating Transmitter Release

It is now widely accepted that modulation of receptor sensitivity is one of the major mechanisms by which synaptic chemical transmission is regulated (for reviews see Schwartz *et al.*[154]; Gnegy and Costa[155]; Reisine[156]). In general, desensitization of transmitter receptors has been reported to occur following long-lasting receptor activation with agonists; conversely, chronic treatments with receptor antagonist drugs or reduced supply of the transmitter to its receptor appear to lead to supersensitivity phenomena.

In most of the reports available in the literature, modulations of receptor sensitivity have been inferred from changes of receptor site density as measured in binding experiments with radioactive ligands. However, these changes in receptor site density may not reflect proportional changes in the functional effect mediated by the agonist–receptor interaction. In fact, according to the "spare receptors" concept, it may not be necessary to activate all of the receptors present on the effector organ in order to elicit a maximal response. Therefore, it is essential to ascertain whether, following chronic drug treatment, the functional response elicited by a given concentration of agonist is indeed lower (subsensitivity) or higher (supersensitivity) than that observed in the absence of chronic drug treatment or following acute administration.

In addition, with binding experiments, it may sometimes be difficult to establish the localization (whether pre- or postsynaptic) of the receptors undergoing sensitivity changes, even if binding measurements are performed with membranes from animals carrying specific nerve fiber lesions. In fact, degeneration of presynaptic terminals following lesions may induce a compensatory increase in postsynaptic binding sites, which can obscure the primary decrease of presynaptic sites.

The development of sub- and supersensitivity is a well-established phenomenon for the classical postsynaptic receptors that mediate responses in the peripheral and in the central nervous system. Presynaptic receptors involved in the auto- or heteroregulation of neurotransmitter release may also undergo sensitivity changes. For instance, the α_2 autoreceptors could become super- or subsensitive towards extraterminal NE following long-term treatment with α_2 antagonists or α_2 agonists, respectively. To examine this possibility in the brain, one could use more or less indirect tests, such as the determination of the cerebral levels of 3-methoxy-4-hydroxyphenylethyleneglycol sulfate ($MHPG \cdot SO_4$) or of normetanephrine, taken as an index of NE release. Alternatively, one could determine the kinetic parameters of the binding of [^{3}H]clonidine, a ligand considered to be selective for α_2 adrenoceptors. However, since α_2 autoreceptors control NE release, the most direct approach is to ascertain whether the potency of extraterminal NE as an inhibitor of NE release has been modified following chronic treatment with α_2 antagonists or α_2 agonists.

We recently demonstrated directly, in release studies, that the autoreceptors modulating NE release can develop supersensitivity towards the natural agonist NE.[157] Rats were administered the antidepressant mianserin,[158] a drug possessing α_2-blocking activity in the CNS,[159] either acutely or chronically. Hypothalamic synaptosomes were then prepared from the animals killed 4 days after the last injection, prelabeled with [^{3}H]NE, and depolarized in superfusion with 15 mM KCl. Exogenous NE decreased the K^+-evoked [^{3}H]NE release more effectively in nerve endings chronically exposed to mianserin.[157]

Also, the muscarinic autoreceptors in cholinergic nerve endings can undergo adaptive changes during long-term treatment with agonist or antagonist drugs. Synaptosomes were prepared from the hippocampus of rats chronically treated with the acetylcholinesterase inhibitor paraoxon (in order to cause long-term receptor activation by endogenous ACh) or with the muscarinic antagonist scopolamine. When the inhibitory potency of exogenous ACh on [^{3}H]ACh release was analyzed, it was found that in rats chronically treated with paraoxon the release was significantly less inhibited than in synaptosomes from acutely treated animals. The opposite was true in the case of rats chronically treated with scopolamine.[153]

Chronic methiothepin administration has recently been found to induce supersensitivity of 5-HT autoreceptors (G. Maura, P. Versace, A. Gemignani, and M. Raiteri, unpublished data).

The above results represent the first demonstration, obtained directly with release experiments, that receptors specifically involved in the autocontrol of neurotransmitter release (presynaptic NE, ACh, and 5-HT autoreceptors) can undergo sensitivity changes.

7.2. Regulation of Release through Presynaptic Heteroreceptors

7.2.1. Heteroregulation of Norepinephrine Release

The K^+-evoked release of [^{3}H]NE was facilitated by GABA in rat occipital cortex slices.[160] However, GABA inhibited the release of [^{3}H]NE induced by high K^+ in cerebellar cortex slices.[161] The possibility that the two brain regions contained different GABA receptors was ruled out when it was found that in cerebellar cortex slices GABA could either inhibit or facilitate [^{3}H]NE release depending on the concentration of the KCl used to depolarize the nerve endings. Both GABA and its β-chlorophenyl derivative baclofen reduced the release evoked by 35 mM or 25 mM KCl. The inhibition was bicuculline insensitive. When 15 mM KCl was used, GABA enhanced the release of [^{3}H]NE, whereas baclofen was still inhibitory. Bicuculline converted the increase produced by GABA into a reduction comparable to that produced by baclofen. These results were interpreted[161] as indicating the presence in the cerebellum of two GABA receptors, mediating, respectively, facilitation and inhibition of NE release depending on the state of depolarization of noradrenergic terminals.

7.2.2. Heteroregulation of Serotonin Release

The actions of GABA on the spontaneous and K^+-evoked release of [^{3}H]5-HT were investigated in rat hippocampal synaptosomes.[162] GABA increased the spontaneous release but inhibited the Ca^{2+}-dependent release promoted by high K^+. Diazepam enhanced the inhibitory action of GABA on the depolarization-induced release. The two actions of GABA may be mediated by different receptors (see Section 7.2.1), since the action on spontaneous release was antagonized by bicuculline, whereas that on the K^+-evoked release was not. Picrotoxin antagonized both GABA actions on [^{3}H]5-HT release.

Also, in cortical slices GABA inhibited the K^+-evoked [^{3}H]5-HT release through bicuculline-insensitive receptors.[161] These receptors were selectively activated by the GABA derivative (−)-baclofen, a drug inactive at the classical bicuculline-sensitive GABA receptors.

Experiments *in vivo* using push–pull cannulae showed that the application of GABA in the cat substantia nigra reduced the release of newly synthesized [^{3}H]5-HT, whereas picrotoxin induced the opposite effect.[163]

The depolarization-induced release of [^{3}H]5-HT from rat hippocampal slices was inhibited by NE. The action was antagonized by the α-adrenoceptor blocker phentolamine.[164] We found that NE also inhibited the release of [^{3}H]5-HT evoked by 15 mM KCl in hippocampal and cortical superfused synaptosomes. Moreover, the action of NE was antagonized by the $α_2$ blocker yohimbine but not by the $α_1$ antagonist prazosin. These results indicate that 5-HT release can be modulated by adrenoceptors of the $α_2$ type located on presynaptic serotonergic nerve terminals (G. Maura and M. Raiteri, unpublished data).

On the basis of a study carried out on the spontaneous release of [^{3}H]5-HT newly synthesized from [^{3}H]tryptophan in rat hypothalamus slices, it was reported that cholinergic receptors of the muscarinic and nicotinic type are involved in the control of 5-HT release.[165] Activation of muscarinic or nicotinic receptors resulted in inhibition or stimulation of [^{3}H]5-HT release, respectively. The localization of the receptors, whether on presynaptic 5-HT terminals or on interneurons, was not investigated.

7.2.3. Heteroregulation of Dopamine Release

The regulation of DA release by other transmitters has been the object of several *in vitro* studies carried out almost exclusively using striatal slices.

The spontaneous release of [³H]DA from striatal slices continuously superfused with [³H]tyrosine was stimulated by nicotine; this action, which was TTX insensitive, was counteracted by nicotinic antagonists.[29,166] Also, ACh stimulated [³H]DA spontaneous release; however, this action was only in part prevented by nicotinic blockers, the remaining being counteracted by atropine. Finally, the potentiation of [³H]DA release seen with the muscarinic agonist oxotremorine was antagonized by atropine but not by nicotinic blockers. It was concluded that in the corpus striatum ACh can potentiate the spontaneous release of newly synthesized DA through the activation of both nicotinic and muscarinic receptors.[29] It should be noted that the two types of receptors need not coexist on the same DA terminal. The presence of inhibitory muscarinic receptors on DA striatal terminals was proposed on the basis of experiments with striatal synaptosomal beds.[167]

In *in vivo* experiments,[168] it was found that the release of [³H]DA newly formed from [³H]tyrosine was stimulated when ACh was delivered into the striatum with a push–pull cannula. However, the stimulation was no longer seen during activation of the DA neurons in the substantia nigra. Recently, we were able to observe that the K^+-evoked release of [³H]DA from superfused striatal synaptosomes was potentiated by ACh when the K^+ concentration was 15 mM, whereas no effect was detected in the presence of 56 mM KCl (M. Marchi, P. Paudice, and M. Raiteri, unpublished data). Thus, the regulation of DA release by presynaptic striatal cholinergic receptors appears to depend on the level of polarization of the terminals.

The K^+-evoked release of [³H]DA previously accumulated in striatal slices was depressed by 5-HT and other indoleamines.[169] The action was unaffected by TTX, suggesting that an interneuron was not involved. This interaction between 5-HT and DA seems to be mediated by "prepostsynaptic" 5-HT receptors different from 5-HT presynaptic autoreceptors. In fact, methysergide, which is ineffective at 5-HT autoreceptors,[31] was the most potent among the antagonists tested at the 5-HT receptors inhibiting DA release.[169] Agonists at these receptors might represent useful agents in the treatment of tardive dyskinesias caused by overactivity in the nigrostriatal DA system.

In striatal slices GABA was reported to be ineffective towards the spontaneous release of [³H]DA previously taken up.[170,171] In contrast, the spontaneous release of [³H]DA from slices continuously superfused with [³H]tyrosine was markedly increased by GABA, muscimol, or baclofen. The potentiation was counteracted by picrotoxin and disappeared in the absence of Ca^{2+} or in the presence of TTX.[172] The last finding suggested to the authors that the GABA receptors are not located presynaptically on DA terminals but on neurons or neuronal afferents within the striatum.

In regard to the depolarization-evoked DA release, GABA was found to potentiate the K^+-induced release of [³H]DA previously accumulated in striatal slices.[171] The potentiation was not counteracted by bicuculline or picrotoxin.

The GABA agonists muscimol or baclofen were inactive. The potentiation of [³H]DA release by GABA disappeared in the high-K⁺–TTX test or when the size of the slices was reduced, suggesting the involvement of an interneuron.[173]

In contrast with the above results, an inhibition by GABA of the K⁺-induced [³H]DA release was reported.[161] The effect was mimicked by baclofen but was not antagonized by bicuculline.

The reasons for these discrepancies are not known. Differences in the K⁺ concentrations used by the various investigators, as in the case of the already discussed regulation by GABA of the release of NE, do not seem to be responsible.

The spontaneous release of [³H]DA from striatal slices superfused with [³H]tyrosine was stimulated not only by GABA but also by L-glutamic acid[174] and by glycine.[175] The action of glycine was abolished by TTX, suggesting that the glycinergic receptors are located on neurons in contact with DA terminals, although the authors "do not exclude that they are on DA terminals, due to the possibility that the toxin, by changing the polarization state of membranes, influences directly presynaptic receptors."[175] The stimulation of [³H]DA release induced by L-glutamic acid persisted in the presence of TTX, suggesting that the amino acid may exert a direct effect on the DA nerve terminals.

In contrast to what has been observed with GABA (see above), L-glutamic acid potentiated not only the spontaneous release of newly synthesized [³H]DA[174] but also the spontaneous release of the previously accumulated exogenous catecholamine.[176]

7.2.4. Heteroregulation of Acetylcholine Release

The release of [³H]ACh evoked by 20–26 mM KCl was decreased by GABA in neostriatal slices.[177,178] The findings that GABA concomitantly increased the release of [³H]DA[177] (but see ref. 161) and that the action of GABA on [³H]ACh release was abolished when DA synthesis was blocked[177] support the hypothesis[179] that GABA inhibits ACh release through a facilitation of DA release. The inhibition of striatal ACh release by DA is well established (see below). The above effects of GABA were insensitive to bicuculline and picrotoxin. The inhibition of ACh release by GABA could not be seen in the cortex.

Quite different results were recently obtained.[180,181] Concentrations of GABA similar to those used previously increased the release of ACh evoked by high K⁺ (12.5–25 mM). The action of GABA was sensitive to bicuculline and picrotoxin and very pronounced not only in the caudate nucleus but also in the cerebral cortex. The high-K⁺–TTX test suggested to the authors that the GABA receptors may be located on ACh terminals. The only evident difference between the two groups of experiments is that in one case[177,178] release of [³H]ACh was measured, whereas release of endogenous ACh was analyzed in the other.[180,181]

In striatal slices DA and dopaminergic agonists reduced both the basal release of ACh[182] and the release evoked by high K⁺,[177,183] by electrical stimulation,[184] and by ouabain.[182] The inhibition of ACh release appeared to be mediated by neuroleptic-sensitive receptors.[184] On the basis of the results of the high-K^+–TTX test, the receptors may be located on cholinergic axon terminals.[185]

In guinea pig cortical slices, NE reduced the release of ACh evoked by electrical stimulation but not that elicited by 25 mM KCl.[186] However, NE did reduce the release of ACh promoted by 25 mM KCl or by ouabain in rat cortical slices.[187,188] Both spontaneous and evoked ACh release were higher in slices where the NE input was impaired (rats pretreated with 6-hydroxydopamine or lesioned in the locus coeruleus), suggesting that in the cortex ACh release is continuously controlled by NE released from nerves arising from the locus coeruleus.[188]

It has been suggested that 5-HT can also modulate ACh release in the striatum. The increase of ACh levels induced by 5-HT receptor stimulants was related to a decrease of ACh release.[189,190] Recently it was found[15] that L-(m-chlorophenyl)-piperazine, a 5-HT receptor stimulant, and d-fenfluramine, a 5-HT releaser, reduced the ouabain-evoked release of endogenous ACh in rat striatal slices. Moreover, both the spontaneous and the ouabain-evoked release of ACh were higher in slices where serotonergic input was impaired (rats pretreated with 5,7-dihydroxytryptamine or lesioned in the raphe nuclei). The authors propose that 5-HT released from raphe–striatal neurons is able to inhibit the striatal release of ACh. The location of the 5-HT receptors involved was not investigated.

In conclusion, there seems to be good evidence that NE, DA, and 5-HT can inhibit ACh release in the CNS by activating specific receptors. To our knowledge, the presynaptic location of these receptors has not been investigated with isolated nerve endings.

7.2.5. Adenosine as a Modulator of Neurotransmitter Release

An increasing number of reports seem to indicate that adenosine and its analogues act as modulators of several central neurotransmitter systems (for a recent review see Stone[134]).

A release of adenosine has been shown to occur from cerebral cortex preparations.[191,192] Some of the characteristics of labeled adenosine release (for instance, the fact that the K^+-evoked release is weak and Ca^{2+} independent) are unlike those of compounds generally accepted as being true transmitters and may better fit with the definition of a neuromodulator.[36,130]

Adenosine was reported to inhibit the K^+-evoked release of ACh, DA, and 5-HT from slices of rat corpus striatum[193] and of NE and GABA from slices of rat cerebral cortex.[194,195]

Adenosine decreased the $^{45}Ca^{2+}$ uptake in rat cerebral cortex synaptosomes depolarized with high K^+, suggesting that the compound may inhibit neurotransmitter release by preventing the entry of Ca^{2+} into the nerve endings through the voltage-dependent channels.[196]

Adenosine, or its analogue 2-chloroadenosine, decreased the K^+-evoked release of [^{3}H]DA from striatal synaptosomes.[197] Dipyridamole, a blocker of the neuronal adenosine uptake,[198] did not prevent the action of adenosine.

These results are suggestive of the presence of specific inhibitory adenosine receptors on the external surface of striatal DA nerve terminals.

A direct modulatory action of adenosine on ACh release could not be demonstrated in cortical synaptosomes depolarized with 33 mM KCl.[199]

7.3. Coupling between Activation of Presynaptic Receptors and Regulation of Release

Little is known about the pathways of intermediate steps linking prejunctional receptor activation to transmitter release regulation. Evidence has been provided that the availability of Ca^{2+} for stimulus–secretion coupling may be modified during activation of presynaptic receptors. In particular, it was proposed that the site of action of presynaptic regulators in noradrenergic[200] and serotonergic[201] neurons is the voltage-sensitive permeability channel for Ca^{2+}, whose affinity for Ca^{2+} would be decreased when presynaptic receptors are activated. According to Starke,[41] many presynaptic auto- and heteroreceptors appear to be coupled with Ca^{2+} channels as one effector system.

Results of experiments carried out in hippocampal slices with penetrating cyclic GMP analogues suggest that cyclic GMP may act as a second messenger when the muscarinic autoreceptor on cholinergic nerve terminals is activated.[202] The same derivatives of cyclic GMP specifically inhibited the initial rate of K^+-stimulated $^{45}Ca^{2+}$ uptake in rat brain slices and synaptosomes.[203] It was therefore postulated that activation of muscarinic autoreceptors leads to cyclic GMP production, which, in turn, inhibits Ca^{2+} penetration into nerve terminals and, consequently, ACh release. Of course, this could not be a generalized mechanism, since not all transmitters trigger cyclic GMP formation as does ACh acting at muscarinic receptors.

On the basis of the observation that the depolarization-induced release of NE from neocortical slices was reduced by phosphodiesterase inhibitors, it was proposed that an increase of cyclic AMP formation could be involved in the autoreceptor-mediated inhibition of NE release.[204]

A different hypothesis is based on the possibility that α-adrenoceptors act via modulation of membrane Na^+,K^+-ATPase activity.[63,205] According to this idea, NE acting on α_2 autoreceptors stimulates the enzyme, normally inhibited by the Ca^{2+} entering into nerve endings during depolarization, leading to depression of the evoked NE release.

7.4. Complementarity between Slices and Synaptosomes in Studies of Receptors Regulating Transmitter Release

Although the drawbacks of synaptosomes as compared to more intact brain tissue preparations are self-evident and unanimously accepted, we think that the statement by Kelly and co-workers in their recent review[4] that "despite the vast amount of work on synaptosomes it is different to ascertain the novel contributions these studies have made to our biochemical understanding of transmitter release at the nerve terminal" can only reflect a lack of information about the work on synaptosomes. Instead, we believe that in studies of neurotransmitter release, slices and synaptosomes should possibly be used concomitantly in the same laboratory because they complement each other. Since the interconnections between neurons as well as between nerve and glial cells are often maintained in brain slices, this preparation permits the study of intercellular interactions. On the other hand, synaptosomes may be the preparation of choice if attention is focused on presynaptic events.

Superfused synaptosomes offer two distinct advantages over slices in studies on presynaptic receptors involved in the auto- or in the heterocontrol of transmitter release. First, in brain slices, presynaptic autoreceptors can be activated by the transmitter being released during depolarization, particularly when a reuptake inhibitor (in the case of the amine transmitters) or an anticholinesterase agent (when studying ACh release) increases its extraterminal concentration. Therefore, the different release systems are already modulated (in general, inhibited) before the addition of the exogenous agonists or antagonists under study (see also Section 8). One of the consequences is that, in slices, antagonists at inhibitory autoreceptors potentiate the depolarization-evoked release by disinhibiting the receptors. In superfused synaptosomes, the transmitters released by depolarization are rapidly removed, and presynaptic autoreceptors are completely available to interact with agonists or antagonists added to the superfusion medium. In this condition, antagonists are inactive *per se*; they only prevent the activation of autoreceptors by agonists added to the superfusion medium, in keeping with the classic pharmacological definition of an antagonist.

Secondary, brain slices consist of an intricate network of fibers, often with many interneurons still intact. Therefore, when a compound A (transmitter or drug) modifies the release of a transmitter B, a primary action on the cell bodies or dendrites of interneurons must always be considered; the direct involvement of receptors located on B terminals can only be hypothesized on the basis of indirect experiments, for example, by determining the sensitivity to TTX of the effect of A on B release.

In our opinion, the results of the high-potassium–TTX test should be interpreted with caution. First of all, the fact that the modulation by transmitter A of the K^+-evoked release of B through specific receptors is insensitive to TTX does not necessarily imply that the receptors are located on B terminals. The test possibly indicates that A does not act by way of action potentials in interneurons. Secondly, one may wonder whether in a slice exposed to high K^+ concentrations a traffic of action potentials with activation of TTX-sensitive Na^+ channels is likely to take place. In fact, high K^+, at concentrations generally used, will depolarize directly not only nerve terminals but every cell in the slice by a mechanism that is insensitive to TTX. Tetrodotoxin might therefore be a valuable discriminative tool only in conditions of spontaneous release (see for example ref. 175) or when the slices are exposed to K^+ concentrations not higher than 15 mM, concentrations that apparently trigger the opening of TTX-sensitive Na^+ channels in the neuronal membrane.[40]

Coming back to the complementarity between slices and synaptosomes, once it is observed in slices from a particular brain area that compound A regulates the release of the transmitter B, the hypothesis of a direct interaction of A on B nerve terminals should be verified by analyzing the release of B from synaptosomes prepared from the same brain region and exposed to A.

In general, one should expect to see a larger number of neurotransmitter interactions in slices than in isolated nerve endings. For instance, if the compound A (transmitter or modulator) produced a change in the release of B through an interneuron X, no action of A on B release would be seen in synaptosomes, whereas three interactions (A → X, X → B, and A → B) could be detected in brain slices. Needless to say, one should never exclude the existence of a release regulatory system in the brain on the basis of a negative result obtained with synaptosomes.

A seemingly trivial consideration concerns the fact that in a slice, the Ca^{2+}-dependent release of a given transmitter (endogenous, newly synthesized, or taken up) evoked by electrical field or high-K^+ stimulation may be the resultant of several interactions. In fact, the stimulus evokes the concomitant release of all the transmitters present in the slice, some of which can modulate (directly or indirectly) the release of the transmitter under study. It may be quite surprising to consider that the release of DA promoted from a striatal slice by what is often called a "quasiphysiological stimulus" is the resultant of the following (so far known) modulations: stimulation by GABA, glutamic acid, ACh (nicotinic), glycine, angiotensin II, substance P; inhibition by GABA (bicuculline insensitive), 5-HT, DA (autoreceptors), ACh (muscarinic), opiates, prostaglandins, adenosine. Of course, not every DA terminal in the striatum will receive all of the above modulatory messages; however, this is irrelevant, since the signal measured is the release of DA from a striatal slice. In conclusion, in brain slices, the action of an exogenous compound on the release of the transmitter under study may originate through complex pathways. Interactions among endogenous transmitters may also occur in beds of purified synaptosomes,[206] whereas this is unlikely in a thin layer of superfused crude brain synaptosomes.[2]

8. DEVELOPMENTAL ASPECTS OF BIOGENIC AMINE AND ACETYLCHOLINE RELEASE

The appearance during development of a Ca^{2+}-dependent transmitter release represents a valuable index of neuronal and, in particular, of synaptic maturation. An even more indicative parameter is the appearance of presynaptic receptors modulating transmitter release. Few and fragmentary are the data available in the literature concerning the ontogenetic aspects of the release of biogenic amines and of ACh.

The high-K^+-induced release of [^{3}H]NE and [^{3}H]5-HT from synaptosomes of 8-day-old rats was completely Ca^{2+} dependent. In contrast, high K^+ promoted a release of [^{3}H]GABA that did not show any Ca^{2+} dependence.[207] A Ca^{2+}-dependent K^+-evoked [^{3}H]GABA release could be detected only after the 14th day of age.[208]

A Ca^{2+}-dependent high-K^+-evoked release of [^{3}H]DA was reported to exist in brain on day 18 of gestation and to increase during postnatal development.[209] The spontaneous release of [³H]DA was augmented by ACh in striatal slices from 3-day-old rats but not from 1-day-old animals. This potentiation was dependent on Ca^{2+} and antagonized by both muscarinic and nicotinic receptor blockers[209] (see Section 7.2.3 for results in adult rats).

On the basis of results from experiments of [³H]5-HT release in brain slices from 7-day-old rats, it was proposed that the negative feedback mechanism controlling 5-HT release through presynaptic autoreceptors may appear early after birth.[210] In keeping with this idea, exogenous 5-HT decreased the K^+-evoked release of [³H]5-HT in superfused synaptosomes from 4-day-old rats (G. Maura, P. Versace, A. Gemignani, and M. Raiteri, unpublished data).

The development of the α_2-autoreceptor-mediated presynaptic regulation of NE release during the first postnatal month was studied in superfused slices or synaptosomes from rat cerebral cortex.[211] The results indicate that the regulatory mechanism already exists on the day of birth. Interestingly, this study illustrates one of the disadvantages of using slices instead of synaptosomes (see Section 7.4 for discussion): in slices, but not in synaptosomes, exogenous NE caused inhibition of [³H]NE release that was stronger in neonatal than in adult cortical tissue. This decrease of efficiency with age of the release autoregulatory mechanism was only apparent, however, since it reflected the fact that autoinhibition of release by the amine (endogenous and radioactive) being released and accumulating in the extracellular spaces was higher in slices from adult than from neonatal rats.

The maturation of some processes related to ACh release has been the object of a recent comparative study carried out using rat cortex synaptosomes (M. Marchi, A. Caviglia, P. Paudice, and M. Raiteri, unpublished data). In analogy with the peripheral nervous system,[212] choline uptake also developed early in the CNS. At 10 days of age, the synaptosomal accumulation of [³H]choline was 65% of the adult value. In contrast to choline uptake, ACh synthesis and release developed later. At 10 days, only 25–30% of the choline taken up was converted into ACh, compared to the 65–70% in the adult. A Ca^{2+}-dependent K^+-evoked release of [³H]ACh was detectable in cortex synaptosomes from 10-day-old rats, but the percentage of [³H]ACh released was much lower than in synaptosomes from adult animals. Autoreceptors do not seem to be present at 10 days; autoregulation of ACh release was shown to exist in synaptosomes from 14-day-old rats.

The Ca^{2+}-dependent K^+-evoked release of [¹⁴C]ACh from mouse brain slices declined with age from 100% (3 months) to 58% (10 months) and 25% (30 months). According to the authors,[213] this decrease of release reflected a reduction with aging of ACh synthesis.

9. CONCLUDING REMARKS

Brain slices and superfused synaptosomes appear to be equally useful in studying neurotransmitter release. In many cases, the two preparations complement each other. Brain slices, in which most of the nerve endings maintain the physiological spatial relation with other tissue elements, permit the study of intercellular interactions. Synaptosomes may be more suitable when investigating some aspects of the release process at the presynaptic level. For example, synaptosomes in superfusion represent the best approach in studies aimed at discriminating between true releasers and apparent releasers (reuptake inhibitors) or at ascertaining the presence on axon terminals of receptors regulating neurotransmitter release.

The release of NE, DA, 5-HT, and ACh can be stimulated *in vitro* by several agents. Depending on the releasing stimulus, NE, DA, or 5-HT can exit from their nerve endings by a mechanism independent of the membrane carrier or by a carrier-mediated mechanism. Stimuli requiring the presence of Ca^{2+} (electrical stimulation, high K^+, Ca^{2+} ionophores) promote a carrier-independent release that is compatible with an exocytosis-like process involving the synaptic storage vesicles. In contrast, a carrier-mediated release of NE, DA, or 5-HT from a cytoplasmic compartment is evoked by treatments leading to changes in the transmembrane Na^+ gradient and/or causing an increase in the amine cytoplasmic concentrations. Whereas the carrier-independent release is likely to reflect the release of biogenic amines occurring in the living brain, the possible physiological and pathological significance of the carrier-mediated efflux remains to be established. Some drugs, including phenylethylamines, act indirectly by inducing a carrier-mediated release of biogenic amines. In cholinergic nerve endings, only a carrier-mediated release of choline can apparently occur because of the absence of a carrier for ACh transport. Whether the ACh released by Ca^{2+}-dependent stimuli originates from the vesicular or from the cytoplasmic pool is still a matter of controversy.

In the last few years an increasing number of processes of auto- and heteroregulation of neurotransmitter release have been reported to exist in the central nervous system. Experiments with slices and synaptosomes have provided compelling evidence that autoreceptors controlling the release of NE, 5-HT, and ACh are located on the respective axon terminals. Such evidence is not yet available for the autoregulatory process of DA release reported to be present in slices of the caudate nucleus. Noradrenergic, serotonergic, and cholinergic presynaptic autoreceptors appear to differ from the corresponding postsynaptic receptors. Long-term activation or blockade of these autoreceptors with specific drugs induce, respectively, subsensitivity or supersensitivity of the mechanism of release autoregulation. Thus, not only postsynaptic receptors but also presynaptic autoreceptors can undergo sensitivity changes.

As far as the receptors involved in the heteroregulation of release are concerned, in only a few cases has a location on presynaptic terminals been demonstrated unequivocally. It should be borne in mind that the often used high-potassium–tetrodotoxin test possibly suggests that an interneurone is not involved in the transmitter interaction under study, but it can not be taken as a proof that the receptors are located on presynaptic axon terminals. This proof can only come from experiments with isolated presynaptic nerve endings.

There is increasing evidence that the intensity of the depolarizing stimulus applied to the preparation represents a factor of major importance in studies on the modulation of neurotransmitter release. Depending on the characteristics of the depolarizing stimulus used to evoke release, a given transmitter may in fact modulate up or down the release of another transmitter or be ineffective. It would be important to ascertain whether these findings have physiological implications.

ACKNOWLEDGMENTS. The research reviewed here that was carried out in the authors' laboratory was supported by grants from the Italian National Research Council (C.N.R.) and from the Italian Ministry of Education. The authors wish to thank Mrs. Maura Agate for typing the manuscript.

REFERENCES

1. Baldessarini, R. J., 1975, *Handbook of Psychopharmacology*, Volume 3, *Biochemistry of Biogenic Amines* (L. L. Iversen, S. D. Iversen, and S. H. Snyder, eds.), Plenum Press, New York, pp. 37–137.
2. Raiteri, M., and Levi, G., 1978, *Reviews of Neurosciences*, Volume 3 (S. Ehrenpreis and I. Kopin, eds.), Raven Press, New York, pp. 77–130.
3. Paton, D. M. (ed.), 1979, *The Release of Catecholamines from Adrenergic Neurons*, Pergamon Press, Oxford.
4. Kelly, R. B., Deutsch, J. W., Carlson, S. S., and Wagner, J. A., 1979, *Annu. Rev. Neurosci.* 2:399–446.
5. Yamamoto, C., and McIlwain, H., 1966, *J. Neurochem.* 13:1333–1343.
6. Raiteri, M., Angelini, F., and Levi, G., 1974, *Eur. J. Pharmacol.* 25:411–414.
7. Iversen, L. L., 1970, *Advances in Biochemical Psychopharmacology*, Volume 2, *Biochemistry of Simple Neuronal Models* (E. Costa and E. Giacobini, eds.), Raven Press, New York, pp. 109–132.
8. Iversen, L. L., 1974, *Biochem. Pharmacol.* 23:1927–1935.
9. Kuhar, M. J., and Murrin, L. C., 1978, *J. Neurochem.* 30:15–21.
10. Lane, J. D., and Aprison, M. H., 1977, *Life Sci.* 20:665–672.
11. Reubi, J. C., and Emson, P. C., 1978, *Brain Res.* 139:164–168.
12. Hamlet, M. A., Rorie, D. K., and Tyce, G. M., 1981, *Brain Res.* 217:315–325.
13. Wightman, R. M., Bright, C. E., and Caviness, J. N., 1981, *Life Sci.* 28:1279–1286.
14. Moroni, F., Bianchi, C., Tanganelli, S., Moneti, G., and Beani, L., 1981, *J. Neurochem.* 36:1691–1697.
15. Vizi, E. S., Harsing, L. G., and Zsilla, G., 1981, *Brain Res.* 212:89–99.
16. Mefford, I. N., 1981, *J. Neurosci. Methods* 3:207–224.
17. Israel, M., and Lesbats, B., 1981, *Neurochem. Int.* 3:81–90.
18. Trendelenburg, U., 1979, *The Release of Catecholamines from Adrenergic Neurons* (D. M. Paton ed.), Pergamon Press, Oxford, pp. 333–354.
19. Haycock, J. W., and Meligeni, J. A., 1977, *Life Sci.* 21:1837–1844.
20. Raiteri, M., Bertollini, A., Angelini, F., and Levi, G., 1975, *Eur. J. Pharmacol.* 34:189–195.
21. Kuhar, M. J., Sethy, V. H., Roth, R. H., and Aghajanian, G. K., 1973, *J. Neurochem.* 20:581–593.
22. Guyenet, P., Lefresne, P., Rossier, J., Beaujouan, J. C., and Glowinski, J., 1973, *Mol. Pharmacol.* 9:630–639.
23. Annunziato, L., Leblanc, P., Kordon, C., and Weiner, R. I., 1980, *Neuroendocrinology* 31:316–320.
24. Ross, S. B., Ogren, S. O., and Renyi, A. L., 1976, *Acta Pharmacol. Toxicol.* (*Kbh.*) 39:152–166.
25. Waldmeier, P. C., Baumann, P. A., Wilhelm, M., Bernasconi, R., and Maître, L., 1977, *Eur. J. Pharmacol.* 46:387–391.
26. Wong, D. T., and Bymaster, F. P., 1978, *Life Sci.* 23:1041–1048.
27. Van der Zee, P., Koger, H. S., Gootjes, J., and Hespe, W., 1980, *Eur. J. Med. Chem.* 15:363–370.

28. Fuller, R. W., 1980, *Annu. Rev. Pharmacol. Toxicol.* **20**:111–127.
29. Giorguieff, M. F., Le Floc'h, M. L., Glowinski, J., and Besson, M. J., 1977, *J. Pharmacol. Exp. Ther.* **200**:535–544.
30. Nordström, O., and Bartfai, T., 1980, *Acta Physiol. Scand.* **108**:347–353.
31. Cerrito, F., and Raiteri, M., 1979, *Eur. J. Pharmacol.* **57**:427–430.
32. Raiteri, M., del Carmine, R., Bertollini, A., and Levi, G., 1977, *Mol. Pharmacol.* **13**:746–758.
33. Raiteri, M., Cerrito, F., Cervoni, A. M., and Levi, G., 1979, *J. Pharmacol. Exp. Ther.* **208**:195–202.
34. Maura, G., Gemignani, A., Versace, P., Martire, M., and Raiteri, M., 1982, *Neurochem. Int.* **4**:219–224.
35. Paton, D. M., 1979, *The Release of Catecholamines from Adrenergic Neurons* (D. M. Paton, ed.), Pergamon Press, Oxford, pp. 323–332.
36. Orrego, F., 1979, *Neuroscience* **4**:1037–1057.
37. Dismukes, K., de Boer, A. A., and Mulder, A. H., 1977, *Naunyn-Schmiedeberg's Arch. Pharmacol.* **299**:115–122.
38. Starke, K., and Montel, H., 1973, *Neuropharmacologia* **12**:1073–1080.
39. Blaustein, M. P., 1975, *J. Physiol. (Lond.)* **247**:617–655.
40. Schoffelmeer, A. N. M., Wemer, J., and Mulder, A. H., 1981, *Neurochem. Int.* **3**:129–136.
41. Starke, K., 1981, *Annu. Rev. Pharmacol. Toxicol.* **21**:7–30.
42. Goddard, G. A., and Robinson, J. D., 1976, *Brain Res.* **110**:331–350.
43. Levi, G., Gallo, V., and Raiteri, M., 1980, *Neurochem. Res.* **5**:281–295.
44. Raiteri, M., Federico, R., Coletti, A., and Levi, G., 1975, *J. Neurochem.* **24**:1243–1250.
45. Marchi, M., Paudice, P., and Raiteri, M., 1981, *Eur. J. Pharmacol.* **73**:75–79.
46. Nordström, O., Westlind, A., Hedlund, B., Unden, A., and Bartfai, T., 1981, *Cholinergic Mechanisms: Phylogenetic Aspects, Central and Peripheral Synapses and Clinical Significance* (G. Pepeu and H. Ladinsky, eds.), Plenum Press, New York, pp. 579–586.
47. Luqmani, Y. A., Bradford, H. F., Birdsall, N. J. M., and Hulme, E. C., 1979, *Nature* **277**:481–483.
48. Blaustein, M. P., 1979, *The Release of Catecholamines from Adrenergic Neurons* (D. M. Paton ed.), Pergamon Press, Oxford, pp. 39–58.
49. Hagiwara, S., and Byerly, L., 1981, *Annu. Rev. Neurosci.* **4**:69–125.
50. Nachshen, D. A., and Blaustein, M. P., 1980, *J. Gen. Physiol.* **76**:709–728.
51. Henry, P. D., 1980, *Am. J. Cardiol.* **46**:1047–1057.
52. Blaustein, M. P., and Hector, C. A., 1975, *Mol. Pharmacol.* **11**:369–378.
53. Brehm, P., and Eckert, R., 1978, *Science* **202**:1203–1206.
54. Katz, B., and Miledi, R., 1970, *J. Physiol. (Lond.)* **207**:789–801.
55. Pressman, B. C., 1976, *Annu. Rev. Biochem.* **45**:501–530.
56. Triggle, D. J., 1979, *The Release of Catecholamines from Adrenergic Neurons* (D. M. Paton ed.), Pergamon Press, Oxford, pp. 303–322.
57. Holz, R. W., 1975, *Biochim. Biophys. Acta* **375**:138–152.
58. Colburn, R. W., Thoa, N. B., and Kopin, I. J., 1976, *Life Sci.* **17**:1395–1400.
59. Wonnacott, S., Marchbanks, R. M., and Fiol, C., 1978, *J. Neurochem.* **30**:1127–1134.
60. Levi, G., Roberts, P. J., and Raiteri, M., 1976, *Neurochem. Res.* **1**:409–416.
61. Vargas, O., and Orrego, F., 1976, *J. Neurochem.* **26**:31–34.
62. Harris, R. A., Schmidt, J., Hitzemann, B. A., and Hitzemann, R. J., 1981, *Science* **212**:1290–1291.
63. Vizi, E. S., 1978, *Neuroscience* **3**:367–384.
64. O'Fallon, J. V., Brosemer, R. W., and Harding, J. W., 1981, *J. Neurochem.* **36**:369–378.
65. Cheung, W. Y., 1980, *Science* **207**:19–27.
66. De Lorenzo, R., 1982, *Neurotransmitter Interaction and Compartmentation* (H. F. Bradford, ed.), Plenum Press, New York **48**:101–120.
67. Rabe, C. S., Williams, T. P., and McGee, R., 1980, *Life Sci.* **27**:1753–1759.
68. Paton, D. M., 1976, *The Mechanisms of Neuronal and Extraneuronal Transport of Catecholamines* (D. M. Paton, ed.), Raven Press, New York, pp. 155–174.
69. Howard, B. D., and Gundersen, C. B., 1980, *Annu. Rev. Pharmacol. Toxicol.* **20**:307–336.

70. Wonnacott, S., 1980, *Synaptic Constituents in Health and Disease* (M. Brzin, D. Sket, and H. Bachelard, eds.), Pergamon Press, Oxford, pp. 690–724.
71. Corrodi, H., and Fuxe, K., 1969, *Eur. J. Pharmacol.* 7:56–59.
72. Nielsen, M., and Braestrup, C., 1977, *Naunyn-Schmiedeberg's Arch. Pharmacol.* 300:93–99.
73. Costa, E., Groppetti, A., and Naimzada, M. K., 1972, *Br. J. Pharmacol.* 44:742–750.
74. Mulder, A. H., Van der Berg, W. B., and Stoof, J. C., 1975, *Brain Res.* 99:419–424.
75. Raiteri, M., Levi, G., and Federico, R., 1974, *Eur. J. Pharmacol.* 28:237–240.
76. Thornburg, J. E., and Moore, K. E., 1973, *Neuropharmacology* 12:853–860.
77. Chiueh, C. C., and Moore, K. E., 1975, *J. Pharmacol. Exp. Ther.* 193:559–565.
78. Raiteri, M., Cerrito, F., Cervoni, A. M., del Carmine, R., Ribera, M. T., and Levi, G., 1978, *Advances in Biochemical Psychopharmacology*, Volume 19, *Dopamine* (P. J. Roberts, G. N. Woodruff, and L. L. Iversen, eds.), Raven Press, New York, pp. 35–56.
79. Cerrito, F., Casazza, G., Levi, G., and Raiteri, M., 1980, *Neurochem. Res.* 5:115–121.
80. Hunt, P., Raynaud, J. P., Leven, M., and Schacht, U., 1979, *Biochem. Pharmacol.* 28:2011–2016.
81. Schacht, U., and Heptner, W., 1974, *Biochem. Pharmacol.* 23:3413–3422.
82. Hunt, P. P., Kannengiesser, M. H., and Raynaud, J. P., 1975, *J. Pharm. Pharmacol.* 26:370–371.
83. Gerhards, H. J., Carenzi, A., and Costa, E., 1974, *Naunyn-Schmiedeberg's Arch. Pharmacol.* 286:49–63.
84. Samanin, R., Bernasconi, S., and Garattini, S., 1975, *Eur. J. Pharmacol.* 34:377–380.
85. Ponzio, F., Achilli, G., Perego, C., and Algeri, S., 1981, *Neurosci. Lett.* 27:61–67.
86. Algeri, S., Ponzio, F., Achilli, G., and Perego, C., 1982, *Typical and Atypical Antidepressants* (E. Costa and G. Racagni, eds.), Raven Press, New York pp. 219–228.
87. Camanni, F., Genazzani, A. R., Massara, F., De Leo, V., Molinatti, G. M. and Muller, E. E., 1980, *J. Clin. Endocrinol. Metab.* 51:650–653.
88. Annunziato, L., Cerrito, F., and Raiteri, M., 1981, *Neuropharmacology* 20:727–731.
89. Ross, S. B., and Kelder, D., 1979, *Acta Pharmacol. Toxicol. (Kbh.)* 44:329–335.
90. Cerrito, F., and Raiteri, M., 1980, *Eur. J. Pharmacol.* 68:465–470.
91. Cerrito, F., Maura, G., and Raiteri, M., 1981, *Apomorphine and Other Dopaminemimetics* (G. L. Gessa and G. U. Corsini, eds.), Raven Press, New York, pp. 123–132.
92. Raiteri, M., Maura, G., and Cerrito, F., 1982, *Typical and Atypical Antidepressants* (E. Costa and G. Racagni, eds.), Raven Press, New York, pp. 199–209.
93. Maura, G., and Raiteri, M., 1982, *Neurochem. Int.* 4:225–231.
94. Braestrup, C., and Scheel-Kruger, J., 1976, *Eur. J. Pharmacol.* 38:305–312.
95. White, T. D., and Archibald, J. T., 1974, *Brain Res.* 82:360–364.
96. Raiteri, M., and Levi, G., 1973, *Nature* 243:180–183.
97. Raiteri, M., and Levi, G., 1973, *Nature* 245:89–91.
98. West, D. P., and Fillenz, M., 1980, *J. Neurochem.* 35:1323–1328.
99. Douglas, W. W., 1968, *Br. J. Pharmacol.* 34:451–474.
100. Smith, A. D., 1979, *The Release of Catecholamines from Adrenergic Neurons* (D. M. Paton ed.), Pergamon Press, Oxford, pp. 1–16.
101. MacIntosh, F. C., 1980, *Synaptic Constituents in Health and Disease* (M. Brzin, D. Sket, and H. Bachelard, eds.), Pergamon Press, Oxford, pp. 11–52.
102. Tamir, H., and Rapport, M. M., 1976, *Res. Commun. Chem. Pathol. Pharmacol.* 13:225–235.
103. Halaris, A. E., and Freedman, D. X., 1977, *J. Pharmacol. Exp. Ther.* 203:575–586.
104. Kopin, I. J., Breeze, G. R., Krauss, R., and Weiss, V. K., 1968, *J. Pharmacol. Exp. Ther.* 161:271–278.
105. Besson, M. J., Cheramy, A., Feltz, P., and Glowinski, J., 1969, *Proc. Natl. Acad. Sci. U.S.A.* 62:741–748.
106. De Belleroche, J. S., Bradford, H. F., and Jones, D. G., 1976, *J. Neurochem.* 26:561–571.
107. Molenaar, P. C., Nickolson, V. J., and Polak, R. L., 1973, *Br. J. Pharmacol.* 47:97–108.
108. Weiler, M. H., Gundersen, C. B., and Jenden, D. J., 1981, *J. Neurochem.* 36:1802–1812.
109. Zimmermann, H., 1979, *Neuroscience* 4:1773–1804.
110. Whittaker, V. P., and Zimmermann, H., 1974, *Synaptic Transmission and Neuronal Interaction* (M. V. L. Bennett, ed.), Raven Press, New York, pp. 217–238.

111. Carroll, P. T., and Aspry, J. A. M., 1980, *Science* **210**:641–642.
112. Dunant, Y., Gautron, J., Israel, M., Lesbats, B., and Manaranche, R., 1974, *J. Neurochem.* **23**:635–643.
113. Marchbanks, R. M., 1975, *Int. J. Biochem.* **6**:303–312.
114. Israel, M., Lesbats, B., and Manaranche, R., 1981, *Nature* **294**:474–475.
115. Jope, R. S., 1981, *J. Neurochem.* **36**:1712–1721.
116. Axelrod, J., 1973, *Harvey Lect.* **67**:175–197.
117. Mennini, T., Borroni, E., Samanin, R., and Garattini, S., 1981, *Neurochem. Int.* **3**:289–294.
118. Vargas, O., de Lorenzo, M. D., Saldate, M. C., and Orrego, F., 1977, *J. Neurochem.* **28**:165–170.
119. de Langen, C. D. J., and Mulder, A. H., 1980, *Naunyn Schmiedebergs Arch. Pharmacol.* **311**:169–177.
120. Sanders-Bush, E., and Massari, V. J., 1977, *Fed. Proc.* **36**:2149–2153.
121. Fuller, R. W., 1980, *Neurochem. Res.* **5**:241–245.
122. Green, A. L., and El Hait, M. A. S., 1978, *J. Pharm. Pharmacol.* **30**:262–263.
123. Fuller, R. W., and Snoddy, H. D., 1979, *J. Pharm. Pharmacol.* **31**:183–184.
124. Vyas, S., and Marchbanks, R. M., 1981, *J. Neurochem.* **37**:1467–1474.
125. Marchbanks, R. M., Wonnacott, S., and Rubio, M. A., 1981, *J. Neurochem.* **36**:379–393.
126. Langer, S. Z., 1977, *Br. J. Pharmacol.* **60**:481–497.
127. Langer, S. Z., 1981, *Pharmacol. Rev.* **32**:337–362.
128. Starke, K., 1977, *Rev. Physiol. Biochem. Pharmacol.* **77**:1–124.
129. Vizi, E. S., 1979, *Prog. Neurobiol.* **12**:181–290.
130. Osborne, N. N., 1981, *Neurochem. Int.* **3**:3–16.
131. Hökfelt, T., Johansson, O., Ljungdahl, A., Lundberg, J. M., and Schultzberg, M., 1980, *Nature* **284**:515–521.
132. Couteaux, R., and Pecot-Dechavassine, M., 1974, *C. R. Acad. Sci.* [*D*] (*Paris*) **278**:291–293.
133. Mulder, A. H., de Langen, C. D. J., de Regt, V., and Hogenboom, F., 1978, *Naunyn Schmiedebergs Arch. Pharmacol.* **303**:193–196.
134. Stone, T. W., 1981, *Neuroscience* **6**:523–555.
135. Farnebo, L. O., and Hamberger, B., 1974, *J. Pharm. Pharmacol.* **26**:642–643.
136. Hamon, M., Bourgoin, S., Jagger, J., and Glowinski, J., 1974, *Brain Res.* **69**:265–280.
137. Göthert, M., and Weinheimer, G., 1979, *Naunyn Schmiedebergs Arch. Pharmacol.* **310**:93–96.
138. Farnebo, L. O., and Hamberger, B., 1971, *Acta Physiol. Scand.* [*Suppl.*] **371**:35–44.
139. Seeman, P., and Lee, T., 1975, *Science* **188**:1217–1219.
140. Dismukes, K., and Mulder, A. H., 1977, *Naunyn Schmiedebergs Arch. Pharmacol.* **297**:23–29.
141. Raiteri, M., Cervoni, A. M., del Carmine, R., and Levi, G., 1978, *Nature* **274**:706–708.
142. Raiteri, M., Cerrito, F., Casazza, G., and Levi, G., 1980, *Advances in Biochemical Psychopharmacology*, Volume 24, *Long-Term Effects of Neuroleptics* (F. Cattabeni, G. Racagni, P. F. Spano, and E. Costa, eds.), Raven Press, New York, pp. 37–43.
143. Starke, K., Reimann, W., Zumstein, A., and Hertting, G., 1978, *Naunyn Schmiedebergs Arch. Pharmacol.* **305**:27–36.
144. Kamal, L. A., Arbilla, S., and Langer, S. Z., 1981, *J. Pharmacol. Exp. Ther.* **216**:592–598.
145. Jackisch, R., Zumstein, A., Hertting, G., and Starke, K., 1980, *Naunyn Schmiedebergs Arch. Pharmacol.* **314**:129–133.
146. Langer, S. Z., 1974, *Biochem. Pharmacol.* **23**:1793–1800.
147. Seeman, P., 1980, *Pharmacol. Rev.* **32**:229–313.
148. Molenaar, P. C., and Polak, R. L., 1970, *Br. J. Pharmacol.* **40**:406–417.
149. Polak, R. L., 1971, *Br. J. Pharmacol.* **14**:660–666.
150. Szerb, J. C., 1977, *Cholinergic Mechanisms and Psychopharmacology* (D. J. Jenden, ed.), Plenum Press, New York, pp. 49–60.
151. Rospars, J. P., Lefresne, P., Beaujouan, J. C., and Glowinski, J., 1977, *Naunyn Schmiedebergs Arch. Pharmacol.* **300**:153–161.
152. Szerb, J. C., 1980, *Presynaptic Receptors* (S. Z. Langer, K. Starke, and M. L. Dubocovich, eds.), Pergamon Press, Oxford, pp. 293–298.

153. Raiteri, M., Marchi, M., and Paudice, P., 1981, *Eur. J. Pharmacol.* **74:**109–110.
154. Schwartz, J. C., Costentin, J., Martres, M. P., Protais, P., and Baudry, M., 1978, *Neuropharmacology* **17:**665–685.
155. Gnegy, M. E., and Costa, E., 1980, *Essays in Neurochemistry and Neuropharmacology,* Volume 4 (M. B. H. Youdim, W. Lovenberg, D. F. Sharman, and J. R. Lagnado, eds.), John Wiley & Sons, New York, pp. 249–282.
156. Reisine, T., 1981, *Neuroscience* **6:**1471–1502.
157. Cerrito, F., and Raiteri, M., 1981, *Eur. J. Pharmacol.* **70:**425–426.
158. Brogden, R. N., Heel, R. C., Speight, T. M., and Avery, G. S., 1978, *Drugs* **16:**273–301.
159. Baumann, P. A., and Maître, L., 1977, *Naunyn Schmiedebergs Arch. Pharmacol.* **300:**31–37.
160. Arbilla, S., and Langer, S. Z., 1979, *Naunyn Schmiedebergs Arch. Pharmacol.* **306:**161–168.
161. Bowery, N. G., Hill, D. R., Hudson, A. L., Doble, A., Middlemiss, D. N., Shaw, J., and Turnball, M., 1980, *Nature* **283:**92–94.
162. Balfour, D. J. K., 1980, *Eur. J. Pharmacol.* **68:**11–16.
163. Soubrié, P., Montastruc, J. L., Bourgoin, S., Reisine, T., Artaud, F., and Glowinski, J., 1981, *Eur. J. Pharmacol.* **69:**483–488.
164. Frankhuyzen, A. L., and Mulder, A. H., 1980, *Eur. J. Pharmacol.* **63:**179–182.
165. Héry, F., Bourgoin, S., Hamon, M., Ternaux, J. P., and Glowinski, J., 1977, *Naunyn Schmiedebergs Arch. Pharmacol.* **296:**91–97.
166. Giorguieff-Chesselet, M. F., Kemel, M. L., Wandscheer, D., and Glowinski, J., 1979, *Life Sci.* **25:**1257–1262.
167. De Belleroche, J. S., and Bradford, H. F., 1978, *Brain Res.* **142:**53–68.
168. Giorguieff-Chesselet, M. F., Kemel, M. L., and Glowinski, J., 1979, *Neurosci. Lett.* **14:**177–182.
169. Ennis, C., Kemp, J. D., and Cox, B., 1981, *J. Neurochem.* **36:**1515–1520.
170. Stoof, J. C., and Mulder, A. H., 1977, *Eur. J. Pharmacol.* **46:**177–180.
171. Starr, M. S., 1979, *Eur. J. Pharmacol.* **53:**215–226.
172. Giorguieff, M. F., Kemel, M. L., Glowinski, J., and Besson, M. J., 1978, *Brain Res.* **139:**115–130.
173. Ennis, C., and Cox, B., 1981, *Eur. J. Pharmacol.* **70:**417–420.
174. Giorguieff-Chesselet, M. F., Besson, M. J., Cheramy, A., and Glowinski, J., 1979, *J. Physiol. (Paris)* **75:**611–613.
175. Giorguieff-Chesselet, M. F., Kemel, M. L., Wandsheer, D., and Glowsinki, J., 1979, *Eur. J. Pharmacol.* **60:**101–104.
176. Roberts, P. J., and Anderson, S. D., 1979, *J. Neurochem.* **32:**1539–1545.
177. Stoof, J. C., Den Breejen, E. J. S., and Mulder, A. H., 1979, *Eur. J. Pharmacol.* **57:**35–42.
178. Scatton, B., and Bartholini, G., 1981, *Cholinergic Mechanisms: Phylogenetic Aspects, Central and Peripheral Synapses, and Clinical Significance* (G. Pepeu and H. Ladinsky, eds.), Plenum Press, New York, pp. 771–780.
179. Ladinsky, H., Consolo, S., Bianchi, S., and Jori, H., 1976, *Brain Res.* **108:**351–361.
180. Beani, L., Bianchi, C., and Tanganelli, S., 1981, *Cholinergic Mechanisms: Phylogenetic Aspects, Central and Peripheral Synapses, and Clinical Significance* (G. Pepeu and H. Ladinsky, eds.), Plenum Press, New York, pp. 763–770.
181. Bianchi, C., Tanganelli, S., Marzola, G., and Beani, L., 1982, *Naunyn Schmiedebergs Arch. Pharmacol.* **318:**253–258.
182. Vizi, E. S., Hársing, L. G., and Knoll, J., 1977, *Neuroscience* **2:**953–961.
183. Miller, J. C., and Friedhoss, A. J., 1979, *Life Sci.* **25:**1249–1256.
184. Bianchi, C., Tanganelli, S., and Beani, L., 1979, *Eur. J. Pharmacol.* **58:**235–246.
185. Hertting, G., Zumstein, A., Jackisch, R., Hoffmann, I., and Starke, K., 1980, *Naunyn Schmiedebergs Arch. Pharmacol.* **315:**111–117.
186. Beani, L., Bianchi, C., Giacomelli, A., and Tamberi, F., 1978, *Eur. J. Pharmacol.* **48:**179–193.
187. Vizi, E. S., Ronai, A., Harsing, L. G., Jr., and Knoll, J., 1977, *Cholinergic Mechanisms and Psychopharmacology* (D. J. Jenden, ed.), Plenum Press, New York, pp. 587–603.
188. Vizi, E. S., 1980, *Neuroscience* **5:**2139–2144.

189. Euvrard, C., Javoy, F., Herbet, A., and Glowinski, J., 1977, *Eur. J. Pharmacol.* **41**:281–289.
190. Ladinsky, H., Consolo, S., Peri, G., Crunelli, V., and Samanin, R., 1977, *Cholinergic Mechanisms and Psychopharmacology* (D. J. Jenden, ed.), Plenum Press, New York, pp. 615–627.
191. Pull, I., and McIlwain, H., 1972, *Biochem. J.* **126**:965–970.
192. Daval, J. L., Barberis, C., and Gayet, J., 1980, *Brain Res.* **181**:161–168.
193. Harms, H. H., Wardeh, G., and Mulder, A. H., 1979, *Neuropharmacology* **18**:577–582.
194. Harms, H. H., Wardeh, G., and Mulder, A. H., 1978, *Eur. J. Pharmacol.* **49**:305–400.
195. Hollins, C., and Stone, T. W., 1980, *J. Physiol. (Lond.)* **303**:73–78.
196. Ribeiro, J. A., Sa-Almeida, A. M., and Namorado, J. M., 1979, *Biochem. Pharmacol.* **28**:1297–1300.
197. Michaelis, M. L., Michaelis, E. K., and Myers, S. L., 1979, *Life Sci.* **24**:2083–2092.
198. Barberis, C., Minn, A., and Gayet, J., 1981, *J. Neurochem.* **36**:347–354.
199. Corrieri, A. G., Barberis, C., and Gayet, J., 1981, *Biochem. Pharmacol.* **30**:2732–2734.
200. Göthert, M., 1979, *Naunyn Schmiedebergs Arch. Pharmacol.* **307**:29–37.
201. Göthert, M., 1980, *Naunyn Schmiedebergs Arch. Pharmacol.* **314**:223–230.
202. Nordström, O., and Bartfai, T., 1981, *Brain Res.* **213**:467–471.
203. Ichida, S., Osugi, T., Noguchi, Y., and Yoshida, H., 1981, *Brain Res.* **213**:472–475.
204. Dismukes, R. K., and Mulder, A. H., 1976, *Eur. J. Pharmacol.* **39**:383–388.
205. Powis, D. A., 1981, *Biochem. Pharmacol.* **30**:2389–2397.
206. De Belleroche, J. S., and Bradford, H. F., 1973, *Prog. Neurobiol.* **1**:275–298.
207. Levi, G., Gallo, V., Ciotti, T., and Raiteri, M., 1979, *J. Neurochem.* **33**:1043–1053.
208. Redburn, D. A., Broome, D., Ferkany, J., and Enna, S. J., 1978, *Brain Res.* **152**:511–519.
209. Nomura, Y., Yotsumoto, I., and Segawa, T., 1981, *Dev. Brain Res.* **1**:171–177.
210. Bourgoin, S., Artaud, F., Enjalbert, A., Héry, F., Glowinski, J., and Hamon, M., 1977, *J. Pharmacol. Exp. Ther.* **202**:519–531.
211. Wemer, J., and Mulder, A. H., 1981, *Brain Res.* **208**:299–309.
212. Marchi, M., Hoffman, D. W., Giacobini, E., and Volle, R., 1981, *Dev. Neurosci.* **4**:442–450.
213. Gibson, G. E., and Peterson, C., 1981, *J. Neurochem.* **37**:978–984.

Release of Putative Transmitter Amino Acids

Giulio Levi

1. INTRODUCTION

In order to exert its biological action, any neurotransmitter must be released from the competent storage sites in nerve endings and must interact with specific receptors localized in the vicinity of the release sites.

In the last 10 years or so, our knowledge of the possible neurotransmitter role of amino acids has greatly increased, and it is not surprising that this brought about a multitude of studies on amino acid release and a great refinement of the techniques adopted for this purpose.

Since the time of the earlier studies, it has been clear that, in contrast to other neurotransmitters, the transmitter amino acids are present at high concentrations in the central nervous system (CNS) and take part in intermediary (and, at times, protein) metabolism in addition to subserving a neurotransmitter role. Moreover, unlike other transmitters, which are mainly stored in synaptic vesicles, the amino acids are largely "free" in the cytoplasm, even in the nerve terminals. For these reasons, it was hypothesized from the beginning that only a distinct neurotransmitter pool of the amino acids could be related to their neurotransmitter role.

The studies on amino acid release can be subdivided into several groups: (1) studies on the phenomenological characteristics of the release process; (2) studies on the molecular mechanisms of release; (3) studies aimed at determining the neurotransmitter role of a given amino acid on the basis of the characteristics of the release process; (4) studies on the origin of the amino acids released, both in terms of the releasing structures and of the functional pools available for release; (5) studies on alterations of release following lesions applied to specific pathways or cells; (6) studies on drugs whose mechanism of action may consist in the alteration of the release of a certain transmitter;

Giulio Levi • Istituto di Biologia Cellulare, Consiglio Nazionale delle Ricerche, 00196 Rome, Italy.

(7) studies on the mechanisms of control of the release process. As we shall see, the majority of these studies concern the release of amino acids induced by depolarizing stimuli.

Most of the work was performed *in vitro* for the obvious reasons of greater simplicity, better control of experimental conditions, and greater flexibility, and in the present chapter the analysis is limited to *in vitro* studies on slices, synaptosomes, ganglia, and cultured cells. The aim of this chapter is to discuss, with the help of a large number of examples, the problems that have been the object of the most intense investigation in recent years in the field of amino acid release rather than to give complete coverage of the topic. Some recent publications dealing totally or in part with neurotransmitter release may be consulted in order to have a more systematic picture.[1-8] Several methodological aspects have been treated previously.[4,6] Three recent books also cover topics (such as putative transmitter amino acid release from retina or *in vivo* release) that could not be accommodated in the present chapter.[9,10]

In order to avoid confusion, it may be appropriate to define the word "release" as used in the present article. "Release" is used to mean "exit from an intracellular to an extracellular compartment" irrespective of the agent or conditions causing the exit and of the mechanism whereby exit occurs. Other definitions (such as: release = exocytosis; efflux = carrier-mediated exit; leakage = outward diffusion) may be more precise but also too restrictive in a field in which the exact mechanism of exit is, in many cases, still uncertain.

2. DEPOLARIZATION-INDUCED RELEASE

2.1. General Overview

In the living brain, depolarization is the physiological stimulus eliciting the release of neurotransmitters from nerve terminals. The abrupt influx of Ca^{2+} into the nerve ending caused by depolarization is believed to be the triggering event of neurotransmitter release (see, for example, refs. 5,8). Thus, the existence of a depolarization-induced, Ca^{2+}-dependent release of a given substance is generally regarded as one of the most valid biochemical criteria for considering that substance to be a neurotransmitter candidate. Conversely, a compound not released by depolarizing stimuli or released similarly in the presence and in the absence of Ca^{2+} is not believed to qualify for a neurotransmitter role. The above theory, which was initially developed on the basis of experiments on acetylcholine and then on catecholamines, now appears to apply to all neurotransmitters, including amino acids, although in several instances, differences have been noted between the behavior of amino acids and that of other transmitters (see following sections and refs. 4,6,11,12).

The importance attributed to the Ca^{2+} dependence of the depolarization-evoked release explains the enormous increase in the number of studies on this topic that occurred in the last decade or so. Over this same time period, the refinement of the techniques and the improvement of the analytical methods have made it possible to perform experiments that were unthinkable at the time of the pioneering studies in this field.

In order to provide a general view of the results on the depolarization-induced release of the most widely studied amino acids (GABA, Glu, Asp, Gly, Tau*) and on the extent to which the evoked release is dependent on extracellular Ca^{2+}, a great deal of published data has been grouped in tabular form in Tables I–VI. The figures presented in these tables, giving a quantification of the evoked release and of its Ca^{2+}-dependent component, were often recalculated from graphs or histograms and are at times unavoidably approximate. By even a superficial glance at the data summarized in the tables, it will be appreciated that there are large variations in the results obtained by the various authors. These variations can at times be attributed to differences in the experimental conditions that cannot be discussed in detail here (for example, differences in the type of electrical stimulation, in the rates of superfusion, in the frequency of collection of samples, in the composition of media, etc.). The differences in the Ca^{2+}-dependency values may have resulted, in some cases, from the addition of Ca^{2+} chelators to the Ca^{2+}-free media or from the presence of Mg^{2+} or Mn^{2+} in these media. Differences in the magnitude of the calculated evoked release may at times be caused by differences in the basal rates of "spontaneous" release, which, for reasons difficult to explain, are highly variable. In a number of cases, there are large variations in the results even when the experimental conditions are apparently very similar.

It should be noted that the method of calculation of the evoked release as percent increase (at the peak of stimulated release) over base line does not give any information on the absolute amounts released, for which the reader is referred to the original publications.

The results in the tables and, more generally, the data on depolarization-induced release of amino acids can be analyzed according to different criteria (such as type of stimulus, type of amino acid, type of pool, type of tissue, etc.). We shall attempt such an analysis in the following paragraphs.

2.2. Depolarizing Stimuli

2.2.1. High K^+ Concentrations

This is the most frequently used depolarizing method. Whether or not depolarization with high $[K^+]$ represents a more "physiological" stimulus than other methods (see below) has been and still is an object of discussion. In fact, the depolarization produced does not involve the activation of the voltage-dependent Na^+ channels and, consequently, unlike physiological depolarization, is not prevented by tetrodotoxin. It has been claimed that high $[K^+]$ causes an aspecific increase in the membrane permeability, leading to release of transmitter as well as nontransmitter substances.[57,72] Moreover, high $[K^+]$ does not act specifically on neuronal structures and is capable of releasing preaccumulated neurotransmitters from glia as well (see Section 2.7.1). In spite of these objections, this method of depolarization has given the most consistent results.

---

* Abbreviations used: GABA, γ-aminobutyric acid; Glu, L-glutamic acid; Asp, L-aspartic acid; Gly, glycine; Tau, taurine.

Table I

Depolarization-Induced Release of GABA from CNS Slices

Type of tissue	Type of stimulus	Method and temperature	Pool studied or precursor type	Release (% over base line)	Ca^{2+}-dependent release (%)	References
Rat cortex	K$^+$ 68 mM	Superfusion 37°C	Exogenous	686	95	Lopez-Colomé et al.[13]
Rat cortex	K$^+$ 60 mM	Incubation 37°C	Endogenous	380	71	Cutler and Young[14]
Rat cortex	K$^+$ 50 mM	Superfusion 37°C	Exogenous	2200	87	Szerb[15]
Rat cortex	K$^+$ 47 mM	Superfusion 37°C	[^{14}C]Pyruvate	280	95	Gauchy et al.[16]
Rat cortex	K$^+$ 45 mM	Superfusion 37°C	Exogenous	325	90	Johnston[17]
Rat cortex	K$^+$ 40 mM	Superfusion 37°C	Exogenous	209	81	Mulder and Snyder[18]
Rat cortex	K$^+$ 40 mM	Superfusion 37°C	Exogenous	150	42	Srinivasan et al.[19]
Rat cortex	K$^+$ 40 mM	Superfusion 37°C	Exogenous	90	95	Vargas et al.[20]
Rat cortex	K$^+$ 25 mM	Superfusion 25°C	Exogenous	620	52–82	Neal and Bowery[21]
Rat olfact. cortex	K$^+$ 25 mM	Cup, room temp.	Endogenous	430	100	Collins[22]
Rat occip. cortex	K$^+$ 30 mM	Superfusion 25°C	Exogenous	1200	10	Arbilla et al.[23]
Frog hemisphere	K$^+$ 40 mM	Superfusion 15°C	Exogenous	741	97	Davidoff and Adair[24]
Rat hippocampus	K$^+$ 25 mM	Superfusion 37°C	Exogenous	600	70	Collingridge et al.[25]
Rat striatum	K$^+$ 25 mM	Superf. room temp.	Exogenous	600	80	Collingridge et al.[26]
Rat striatum	K$^+$ 47 mM	Superfusion 37°C	[^{3}H]Gln	1700	91	Reubi and Cuénod[27]
Pigeon opt. tectum	K$^+$ 47 mM	Superfusion 37°C	[^{3}H]Gln	810	75	Reubi et al.[28]
Rat midbrain	K$^+$ 50 mM	Superfusion 37°C	Endogenous	1500	69	Waller and Richter[29]
Rat s. nigra	K$^+$ 50 mM	Superf. room temp.	Exogenous	>3500	>90	Collingridge et al.[26]
Rat s. nigra	K$^+$ 25 mM	Superf. room temp.	Exogenous	>1000	>70	Collingridge et al.[26]

Rat s. nigra	K$^+$ 30 mM	Superfusion 25°C	Exogenous	2600	85	Arbilla et al.[23]
Cat cochlear n.	K$^+$ 50 mM	Superfusion 37°C	[^{3}H]Gln	620	85	Čanžek and Reubi[30]
Cat cochlear n.	K$^+$ 50 mM	Superfusion 37°C	[^{14}C]Glu	180	68	Čanžek and Reubi[30]
Rat cerebellum	K$^+$ 55 mM	Superfusion 37°C	Endogenous	>5000	>90	Flint et al.[31]
Rat cerebellum	K$^+$ 55 mM	Superfusion 37°C	Endogenous	150	90	Foster and Roberts[32]
Guinea pig cereb.	K$^+$ 50 mM	Superfusion 37°C	Exogenous	900	74	Okamoto and Namina[33]
Rat sp. cord	K$^+$ 68 mM	Superfusion 37°C	Exogenous	40	100	Lopez-Colomé et al.[13]
Frog sp. cord	K$^+$ 40 mM	Superfusion 15°C	Exogenous	150	95	Adair and Davidoff[34]
Rat cortex	protoverat. 200 μM	Superfusion 37°C	Exogenous	270	58	Hammerstad et al.[35]
Rat cortex	veratridine 50 μM	Superfusion 37°C	Exogenous	1800	10	Szerb[15]
Rat cortex	protoverat. 10 μM	Superfusion 37°C	Exogenous	420	None	Minchin[36]
Rat cortex	veratridine 5 μM	Incubation 37°C	Endogenous	570	None	Cutler and Young[14]
Rat cortex	veratridine 5 μM	Superfusion 25°C	Exogenous	470	None	Neal and Bowery[21]
Rat cortex	electrical	Superfusion 37°C	Exogenous	237	96	Hammerstad and Cutler[37]
Rat cortex	electrical	Superfusion 37°C	Endognous	80	95	Valdès and Orrego[38]
Rat cortex	electrical	Superfusion 37°C	Exogenous	230	None^a	Srinivasan et al.[19]
Rat cortex	electrical	Superfusion 36°C^b	Exogenous	140	40	Potashner[39]
Rat cortex	electrical	Superfusion 36°C^b	[^{14}C]Glucose	600	80	Potashner[39]
Rat cortex	electrical	Incubation 37°C	Exogenous	0–2000	None	Orrego and Miranda[40]
Rat olfact. cortex	electrical	Cup, room temp.	Endogenous	1900	85	Collins[22]
Rat striatum	electrical	Superfusion 37°C	Exogenous	115	None	Katz et al.[41]
Rat s. nigra	electrical	Incubation 37°C	Exogenous	200	None	Okada and Hassler[42]
Rat sp. cord	electrical	Superfusion 37°C	Exogenous	117	68	Hammerstad et al.[43]

^{a} 55% Ca^{2+}-dependent using a Ca^{2+}-free medium with 10 mM Mg^{2+}.
^{b} Superfusion in glucose-free medium.

Table II

Depolarization-Induced Release of GABA from Synaptosomal Preparations

Type of preparation	Type of stimulus	Method and temperature	Pool studied	Release (% over base line)	Ca²⁺-dependent release (%)	References
Rat brain P$_2$[c]	K$^+$ 56 mM	Superfusion 37°C	Endogenous	458	72	Levi et al.[44]
Rat brain syn.[e]	K$^+$ 56 mM	Superfusion 37°C	Exogenous	290	52	Raiteri et al.[45]
Rat brain syn.[e]	K$^+$ 56 mM	Superf. room temp.	Exogenous	350	60	Redburn et al.[46]
Mouse brain P$_2$	K$^+$ 15 mM (Ca^{2+} pulse)	Superf. room temp.	Exogenous	—[a]	51	Cotman et al.[47]
Mouse brain P$_2$	K$^+$ 15 mM (Ca^{2+} pulse)	Superf. room temp.	Endogenous	—[a]	87	Cotman et al.[47]
Rat cortex syn. beds	K$^+$ 56 mM	Incubation 37°C	Endogenous	183[b]	95	De Belleroche et al.[48]
Rat cortex P$_2$	K$^+$ 56 mM	Superfusion 37°C	Exogenous[c]	500	77	Placheta et al.[49]
Rat cortex P$_2$	K$^+$ 56 mM	Superfusion 37°C	Exogenous[c]	204	48	Sieghart et al.[50]
Rat cortex P$_2$	K$^+$ 40 mM	Superf. room temp.	Exogenous[c]	776	65	Haycock et al.[51]
Rabbit cortex syn.	K$^+$ 15 mM	Superfusion	Exogenous[c]	76	54	Sellström and Hamberger[52]
Rat sp. cord–medulla syn. beds	K$^+$ 56 mM	Incubation 37°C	Endogenous	58[b]	55	Osborne et al.[53]
Rat brain syn.	Veratridine 100 µM	Superf. room temp.	Exogenous	410	75	Redburn et al.[46]
Rat brain syn.	Veratridine 5–75 µM	Superfusion 37°C	Exogenous	200–900	None[d]	Levi et al.[11]
Rat brain syn.	Veratridine 10 µM	Superfusion 37°C	Endogenous	323	None[d]	Levi et al.[11]
Sheep cortex syn.	Electrical	Incubation 37°C	Endogenous	>1000	≃90	Bradford et al.[54]
Rat sp. cord–medulla syn. beds	Electrical	Incubation 37°C	Endogenous	86[b]	73	Osborne et al.[53]
Rat brain syn.	Electrical	Superf. room temp.	Exogenous	0–900	None	Redburn et al.[46]

[a] No base-line efflux given.
[b] Calculated as ratio between GABA released during the whole period of incubation (which included a period of stimulation) and GABA released during a corresponding period of incubation without stimulation.
[c] Loaded into incubated slices from which the P$_2$ fractions were prepared.
[d] Release was inhibited by Ca^{2+} > 1 mM
[e] Abbreviations used: syn., purified synaptosomes; P$_2$, crude synaptosomal fractions

Table III

Depolarization-Induced Release of Acidic Amino Acids from CNS Slices

Type of tissue	Type of stimulus	Method and temperature	Amino acid pool and precursor	Release (% over base line)	Ca²⁺-dependent release (%)	References
Rat cortex	K⁺ 60 mM	Incubation 37°C	Glu, endogenous	130	34	Cutler and Young[14]
Rat cortex	K⁺ 60 mM	Incubation 37°C	Asp. endogenous	39	None	Cutler and Young[14]
Rat cortex	K⁺ 55 mM	Incubation 37°C.	Glu, exogenous	180	80	Arnfred and Hertz[55]
Rat cortex	K⁺ 45 mM	Superfusion 37°C	D-Asp, exogenous	217	77	Davies et al.[56]
Rat cortex	K⁺ 45 mM	Superfusion 37°C	Glu, exogenous	60	100	Vargas et al.[57]
Rat cortex	K⁺ 40 mM	Superfusion 37°C	Glu, exogenous	128	82	Mulder and Snyder[18]
Rat cortex	K⁺ 40 mM	Superfusion 37°C	Asp, exogenous	196	69	Mulder and Snyder[18]
Rat olfact. cortex	K⁺ 25 mM	Cup, room temp.	Glu, endogenous	45	None	Collins[22]
Rat olfact. cortex	K⁺ 25 mM	Cup, room temp.	Glu, endogenous	346	78	Collins[22]
Rat striatum	K⁺ 47 mM	Superfusion 37°C	Asp[³H]Asn	1300	95	Reubi et al.[58]
Rat striatum	K⁺ 47 mM	Superfusion 37°C	Glu[³H]Gln	2900	95	Reubi and Cuénod[27]
Rat dentate gyrus	K⁺ 56 mM	Superfusion 37°C	Glu, endogenous[a]	545	68	Hamberger et al.[59]
Rat dentate gyrus	K⁺ 56 mM	Superfusion 37°C	Glu, endogenous[b]	294	75	Hamberger et al.[59]
Rat dentate gyrus	K⁺ 56 mM	Superfusion 37°C	Glu,[¹⁴C]glucose[a]	525	85	Hamberger et al.[59]
Rat dentate gyrus	K⁺ 56 mM	Superfusion 37°C	Glu,[¹⁴C]Gln[b]	141	68	Hamberger et al.[59]
Rat dentate gyrus	K⁺ 56 mM	Superfusion 37°C	Glu,[¹⁴C]glucose[c]	500	61	Hamberger et al.[59]
Pigeon optic tectum	K⁺ 47 mM	Superfusion 37°C	Glu,[³H]Gln	250	85	Reubi et al.[28]
Rat midbrain	K⁺ 50 mM	Superfusion 37°C	Glu, endogenous	150	70	Waller and Richter[29]
Rat midbrain	K⁺ 50 mM	Superfusion 37°C	Asp, endogenous	150	70	Waller and Richter[29]
Cat cochlear n.	K⁺ 50 mM	Superfusion 37°C	Glu, exogenous	610	75	Čanžek and Reubi[30]
Cat cochlear n.	K⁺ 50 mM	Superfusion 37°C	Glu,[³H]Gln	310	68	Čanžek and Reubi[30]
Rat cerebellum	K⁺ 55 mM	Superfusion 37°C	Glu, endogenous	3500	95	Flint et al.[31]
Rat cerebellum	K⁺ 55 mM	Superfusion 37°C	Asp, endogenous	700	85	Flint et al.[31]
Rat cerebellum	K⁺ 55 mM	Superfusion 37°C	Glu, endogenous	228	100	Foster and Robert[32]
Rat cerebellum	K⁺ 55 mM	Superfusion 37°C	Asp, endogenous	12	None	Foster and Roberts[32]
Rat cortex	Veratridine 5 µM	Incubation 37°C	Glu, endogenous	218	None	Cutler and Young[14]
Rat cortex	Veratridine 5 µM	Incubation 37°C	Asp, endogenous	75	None	Cutler and Young[14]
Rat cortex	Electrical	Superfusion 37°C	Glu, exogenous	658	89	Hammerstad and Cutler[37]

(Continued)

Table III. (Continued)

Type of tissue	Type of stimulus	Method and temperature	Amino acid pool and precursor	Release (% over base line)	Ca²⁺-dependent release (%)	References
Rat cortex	Electrical	Superfusion 36°C[d]	Glu,[¹⁴C]glucose	1000	94	Potashner[39]
Rat cortex	Electrical	Superfusion 36°C[d]	Glu, exogenous	300	35	Potashner[39]
Rat cortex	Electrical	Superfusion 36°C[b]	Asp,[¹⁴C]glucose	700	90	Potashner[39]
Rat cortex	Electrical	Superfusion 36°C[b]	Asp, exogenous	100	45	Potashner[39]
Rat, olfact. cortex	Electrical	Cup, room temp.	Glu, endogenous	73	None	Collins[22]
Rat, olfact. cortex	Electrical	Cup, room temp.	Asp, endogenous	151	77	Collins[22]
Rat dentate gyrus	Electrical	Superfusion 37°C	Glu, endogenous	700	90	Cotman and Hamberger[60]

[a] Medium with 3 mM glucose and [¹⁴C]glucose throughout the experimental period.
[b] Medium with 10 mM glucose, 0.4 mM Gln, and [¹⁴C]Gln throughout the experimental period.
[c] Medium with 0.4 mM Gln, 3 mM glucose, and [¹⁴C]glucose throughout the experimental period.
[d] Superfusion with glucose-free medium.

Table IV

Depolarization–Induced Release of Acidic Amino Acids from Synaptosomal Preparation

Type of preparation	Type of stimulus	Method and temperature	Amino acid pool and precursor	Release (% over base line)	Ca^{2+}-dependent release (%)	References
Rat cortex syn. beds	K$^+$ 56 mM	Incubation 37°C	Glu, endogenous	155ᵃ	99	De Belleroche and Bradford[48]
Rat cortex syn. beds	K$^+$ 56 mM	Incubation 37°C	Asp, endogenous	128ᵃ	100	De Belleroche and Bradford[48]
Rat cortex P₂	K$^+$ 30 mM	Superfusion, 37°C	Glu, exogenous	150	59	Levi et al.[61]
Rat cortex P₂	K$^+$ 30 mM	Superfusion, 37°C	Glu, endogenous	2960	84	Levi (unpublished data)
Rat hippocampus syn.	K$^+$ 56 mM	Superf. room temp.	Glu, endogenous	—ᵇ	75	Sandoval et al.[62]
Rat hippocampus syn.	K$^+$ 56 mM	Superf. room temp.	Glu, exogenous	—ᵇ	75	Sandoval et al.[62]
Rat cerebellum P₂	K$^+$ 53 mM	Superf. room temp.	Glu, exogenous	200	None	Sandoval and Cotman et al.[63]
Rat cerebellum P₂	K$^+$ 53 mM	Superf. room temp.	Glu, endogenous	200	65	Sandoval and Cotman[63]
Rat cerebellum P₂	K$^+$ 30 mM	Superfusion, 37°C	D-Asp, exogenous	300	0–35	Levi et al.[64] (unpublished data)
Rat cerebellum P₂	K$^+$ 30 mM	Superfusion, 37°C	Glu, endogenous	675	74	Levi et al.[64] (unpublished data)
Rat cerebellum P₂	K$^+$ 30 mM	Superfusion, 37°C	Glu, [¹⁴C]Gln	414	71	Levi (unpublished data)
Rat cerebellum P₂	K$^+$ 30 mM	Superfusion, 37°C	Asp, endogenous	132	None	Levi (unpublished data)
Rat cerebellum Pli2	K$^+$ 30 mM	Superfusion, 37°C	Asp,[¹⁴C]Gln	288	26	Levi (unpublished data)
Rat sp. cord–medulla syn. beds	K$^+$ 56 mM	Incubation 37°C	Glu, endogenous	55ᵃ	82	Osborne et al.[53]
Rat sp. cord–medulla syn. beds	K$^+$ 56 mM	Incubation 37°C	Asp, endogenous	32ᵃ	93	Osborne et al.[53]
Rat brain syn.	Veratridine 25 μM	Superfusion 37°C	Glu, exogenous	216	Noneᶜ	Levi et al.[11]
Sheep cortex syn.	Electrical	Incubation 37°C	Glu, endogenous	300	73	Bradford et al.[54]
Sheep cortex syn.	Electrical	Incubation 37°C	Asp, endogenous	300	85	Bradford et al.[54]

(Continued)

Table IV. (Continued)

Type of preparation	Type of stimulus	Method and temperature	Amino acid pool and precursor	Release (% over base line)	Ca^{2+}-dependent release (%)	References
Rat sp. cord–medulla syn. beds	Electrical	Incubation 37°C	Glu, endogenous	55[a]	90	Osborne et al.[53]
Rat sp. cord–medulla syn. beds	Electrical	Incubation 37°C	Asp, endogenous	54[a]	100	Osborne et al.[53]

[a] Calcuated as ratio between Glu or Asp released during the whole period of incubation (which included a period of stimulation) and Glu or Asp released during a corresponding period of incubation without stimulation.
[b] No base-line efflux given.
[c] Release was inhibited by 2.7 mM Ca^{2+}.

Table V
Depolarization-Induced Release of Glycine from CNS Slices and Synaptosomal Preparations

Type of preparation	Type of stimulus	Method and temperature	Pool studied or precursor type	Release (% over base line)	Ca²⁺-dependent release (%)	References
Slices						
Rat cortex	K⁺ 68 mM	Superfusion 37°C	Exogenous	50	10	Lopez-Colomé et al.[13]
Rat cortex	K⁺ 60 mM	Incubation 37°C	Endogenous	41	None	Cutler and Young[14]
Rat cortex	K⁺ 45 mM	Superfusion 37°C	Exogenous	40	100	Vargas et al.[57]
Rat cortex	K⁺ 40 mM	Superfusion 37°C	Exogenous	26	None	Mulder and Snyder[18]
Mouse cortex	K⁺ 60 mM	Superfusion 37°C	Exogenous	None	None	O'Fallon et al.[65]
Frog hemisphere	K⁺ 40 mM	Superfusion 15°C	Exogenous	25	63	Davidoff and Adair[24]
Frog optic tectum	K⁺ 40 mM	Superfusion 15°C	Exogenous	65	76	Davidoff and Adair[24]
Rat midbrain	K⁺ 50 mM	Superfusion 37°C	Endogenous	60	(100)[a]	Waller and Richter[24]
Rat s. nigra	K⁺ 40 mM	Superfusion 37°C	Exogenous	100	75	James and Starr[66]
Rat cerebellum	K⁺ 68 mM	Superfusion 37°C	Exogenous	13	—	Lopez-Colomé et al.[13]
Rat cerebellum	K⁺ 55 mM	Superfusion 37°C	Endogenous	None	None	Flint et al.[31]
Rat cerebellum	K⁺ 55 mM	Superfusion 37°C	Endogenous	15	None	Foster and Roberts[32]
G. pig cerebellum	K⁺ 50 mM	Superfusion 37°C	Exogenous	100	30	Okamoto and Namina[33]
Rat sp. cord	K⁺ 68 mM	Superfusion 37°C	Exogenous	61	100	Lopez-Colomé et al.[13]
Rat sp. cord	K⁺ 40 mM	Superfusion 37°C	Exogenous	100–300	None	Hopkin and Neal[67]
Rat sp. cord	K⁺ 40 mM	Superfusion 37°C	Exogenous	201	75	Mulder and Snyder[18]
Rat cortex	Veratridine 5 μM	Incubation 37°C	Endogenous	95	None	Cutler and Young[55]

(*Continued*)

Table V. (*Continued*)

Type of preparation	Type of stimulus	Method and temperature	Pool studied or precursor type	Release (% over base line)	Ca^{2+}-dependent release (%)	References
G. pig cortex	Electrical	Superfusion 37°C	Exogenous	None	None	McIlwain and Snyder[68]
Rat sp. cord	Electrical	Superfusion 37°C	Exogenous	119	94[b]	Hammerstad et al.[72]
Rat sp. cord	Electrical	Superfusion 37°C	Exogenous	110	None	Hopkin and Neal[69]
Synaptosomes						
Rat cortex syn. beds	K$^+$ 56 mM	Incubation 37°C	Endogenous	16[c]	None	De Belleroche and Bradford[48]
Rat cortex P$_2$	K$^+$ 30 mM	Superfusion 37°C	Exogenous	104	78	Levi et al.[69]
Rat cortex P$_2$	K$^+$ 30 mM	Superfusion 37°C	Endogenous	83	84	Levi et al.[69]
Rat cortex P$_2$	K$^+$ 30 mM	Superfusion 37°C	[^{14}C]Serine	220	80	Levi et al.[69]
Rat sp. cord–medulla syn. beds	K$^+$ 56 mM	Incubation 37°C	Endogenous	45[c]	51	Osborne et al.[53]
Rat sp. cord–medulla P$_2$	K$^+$ 30 mM	Superfusion 37°C	Exogenous	183	95	Levi et al.[69]
Rat sp. cord–medulla P$_2$	K$^+$ 30 mM	Superfusion 37°C	Endogenous	230	92	Levi et al.[69]
Sheep cortex syn.	Electrical	Incubation 37°C	Endogenous	<25[c]	None	Bradford et al.[54]
Rat sp. cord–medulla syn. beds	Electrical	Incubation 37°C	Endogenous	51[c]	76	Osborne et al.[53]

[a] Not statistically significant.

[b] Long preincubation in Ca^{2+}-free medium containing Ca^{2+} chelator.

[c] Calculated as ratio between Gly released during the whole period of incubation (which included a period of stimulation) and Gly released during a corresponding period of incubation without stimulation.

Table VI

Depolarization-Induced Release of Taurine from CNS Slices and Synaptosomal Preparations

Type of preparation	Type of stimulus	Method and temperature	Pool studied or precursor type	Release (% over base line)	Ca^{2+}-dependent release (%)	References
Slices						
Rat cortex	K$^+$ 68 mM	Superfusion 37°C	Exogenous	None	None	Lopez-Colomé et al.[13]
Rat olfact. cortex	K$^+$ 25 mM	Cup, room temp.	Endogenous	97	None	Collins[22]
Rat cerebellum	K$^+$ 55 mM	Superfusion 37°C	Endogenous	None	None	Flint et al.[31]
G. pig cerebellum	K$^+$ 50 mM	Superfusion 37°C	Exogenous	260	30	Okamoto and Namina[33]
Rat cortex	Electrical	Superfusion 37°C	Exogenous	320	35	Kaczmarek and Davison[70]
Rat olfact. cortex	Electrical	Cup, room temp.	Endogenous	497	None	Collins[22]
Synaptosomes						
Rat cortex P$_2$	K$^+$ 56 mM	Superfusion 37°C	Exogenous	100	None	Placheta et al.[49]
						Sieghart and Heckl[71]

In general, all neurotransmitter amino acids are released by high $[K^+]$, and the evoked release is 50–100% dependent on extracellular Ca^{2+} (Tables I–V), whereas substances that are not believed to act as transmitters are generally not released or are released to a small extent in a Ca^{2+}-independent manner (see Section 2.4 for more details). Even aside from the problem of Ca^{2+} dependence, the lack of correlation between regional amino acid levels and K^+-evoked release (see Section 2.5) and the small effect of high $[K^+]$ on the release of taurine, whose concentration in nervous tissues is high (Table VI), seem to exclude the possibility[72] that the proportionately larger release of Glu, Asp, and GABA compared to that of other amino acids can be accounted for by their greater intracellular chemical (or electrochemical) potentials.

2.2.2. Veratridine

Veratridine is known to depolarize excitable cells by blocking the inactivation of the Na^+ conductance.[73] Isolated nerve endings are depolarized by micromolar concentrations of the alkaloid,[74] which cause a large influx of Na^+ and a concomitant influx of Ca^{2+} and efflux of K^+, all specifically abolished by tetrodotoxin.[75,76] Thus, the cationic fluxes elicited by veratridine more closely resemble the fluxes occurring during physiological depolarization than do those taking place during depolarization with high $[K^+]$.[77] Other apparent advantages of the alkaloid are (1) that it can be used without altering the ionic composition of the incubation or superfusion media, and (2) that it can be utilized to discriminate between release from neuronal structures and release from glia. In fact, glial cells are not known to possess tetrodotoxin-sensitive Na^+ channels, and amino acid release from glia is generally scarcely if at all affected by veratridine[21,34,78,79] (however, see ref. 80). In spite of these advantages, veratridine has proven to be an inadequate tool for studying the Ca^{2+} dependence of the depolarization-induced release of amino acids. In fact, in contrast with the behavior of transmitter amines, which are released by veratridine in a Ca^{2+}-dependent manner (see refs. 11, 21, and 36 for bibliography), the evoked release of putative transmitter amino acids has been generally described as non-Ca^{2+}-dependent in slices[14,15,21,36,78] and in synaptosomes[11] and is even inhibited by relatively low concentrations of Ca^{2+}.[7,11,21,36] In some cases, however, the release of GABA induced by the alkaloid was enhanced in the presence of extracellular Ca^{2+}.[35,46,80,81] In a study[11] undertaken to try to resolve this discrepancy, it was observed that in only one of the 11 different experimental conditions tested (temperature of incubation and superfusion of 22–23°C instead of 37°C) did Ca^{2+} potentiate somewhat the veratridine-induced release of GABA from rat brain synaptosomes. This observation appeared to be consistent with the fact that the previous reports describing a Ca^{2+}-dependent, veratridine-induced release of GABA from synaptosomes referred to experiments run at room temperature.[46,81] The problem, however, does not appear to be completely solved in the case of slice experiments, since there is a report of Ca^{2+} dependence at 37°C[35] and one of Ca^{2+} independence at 25°C.[21]

The inhibitory effect of Ca^{2+} on GABA and Glu release evoked by veratridine has been attributed to the antagonism by Ca^{2+} of the veratridine-induced activation of the action potential Na^+ channels.[11,21,36] The observation that in brain slices, Ca^{2+} inhibited the uptake of $^{22}Na^+$ caused by protoveratrine[36] supports this interpretation.

It can be concluded that the veratridine-evoked release of GABA and Glu is caused more by the alkaloid-induced depolarization (and consequent release from the cytoplasmic pool) than by the accompanying influx of Ca^{2+} (which might cause release from a small vesicular pool). When the concentration of extracellular Ca^{2+} is high enough to counteract in part the action of the alkaloid at the level of the Na^{2+} channels, the release of the amino acids may be depressed because the increase in release caused by the entry of Ca^{2+} into the nerve terminals is smaller than the decrease in release caused by the lower depolarization.[11] It has been suggested that the GABA released by veratridine in the presence of Ca^{2+} originates from a pool different from that affected by the alkaloid in the absence of Ca^{2+}.[15]

The difference between the behavior of amino acids and that of biogenic amines (which are also released by veratridine in a Ca^{2+}-dependent way in those conditions in which the release of amino acids is inhibited by Ca^{2+}[11,21]) most probably relates to the fact that, unlike amino acids, the transmitter amines are largely stored in synaptic vesicles. In the case of amines, the main component of release would be exocytotic and triggered by the influx of Ca^{2+} elicited by veratridine, whereas the depolarization would have, by itself, little or no releasing effect.[11]

2.2.3. Electrical Stimuli

Electrical stimulation proved to be more difficult to use than other depolarizing stimuli, at least judged from the variability of the results obtained by different groups of investigators (Tables I–VI). Amino acid transmitters appear to be released by electrical stimuli less readily than other transmitters,[12,40,46,72,82] and in many stimulation conditions, the evoked release was not Ca^{2+} dependent[19,40–42,46,67] or was accompanied by an aspecific release of other compounds unrelated to neurotransmission.[40,72,82] With synaptosomal preparations, electrical stimulation was used very sparingly (Tables II, IV, and V) and, because of the experimental conditions adopted (fairly long incubation with single sample collection[48,53,54]) the results are difficult to evaluate in quantitative terms. At any rate, a preferential Ca^{2+}-dependent release of putative transmitter amino acids was clearly apparent in these studies.

In slices, in a number of cases, appropriate electrical stimuli did cause a substantial release of putative neurotransmitter amino acids not accompanied by aspecific release of other compounds[22,43,67] and largely Ca^{2+} dependent (Tables I, III, and V).

Judged from experiments of the type reported in Tables I–VI, field depolarization with electrical stimuli does not seem to provide appreciable advantages compared to depolarization with high $[K^+]$ or veratridine. However, as we shall see later (Section 2.9.1), the electrical stimulation of specific systems of fibers provides, even *in vitro*, a powerful tool for studying selectively the release of transmitters from specific populations of nerve terminals.

2.3. Tissue Preparations

2.3.1. Slices

Slices are the oldest and still the most widely used preparations. They have the advantage over other preparations of being quickly prepared and of suffering the least manipulation. Moreover, they can provide a tissue maintaining intact neuronal circuits and connections, which can be exploited for combined electrophysiological and biochemical studies. The existence of intact neuronal circuits in slices may at times render more complex the interpretation of biochemical data. For example, in view of the influence that many neurotransmitters have on the release of others (see Section 3.2), one should be aware that the release of the compound under study may be influenced by other neurotransmitters or neuromodulators being released by the applied depolarizing stimulus.

Slices also have some disadvantages: the complexity of their structure and the fact that the existence of several types of neurons and of glial cells may complicate the interpretation of the results. It is too often assumed that whatever is released from slices by depolarizing stimuli in a Ca^{2+}-dependent fashion derives exclusively from nerve endings. Moreover, slices have an extracellular space and diffusion barriers, and even in superfusion studies, reuptake of neurotransmitters cannot be completely prevented. This may be, together with the better preservation of the tissue, the reason why the "spontaneous" release of transmitter amino acids is generally lower in slices than in synaptosomal preparations, whereas the stimulated release, taken as a percentage of the basal release, is often higher. A description of the various types of slices that have been used is beyond the scope of this chapter, but it is worth mentioning that the method of slice preparation may affect the results obtained.[84,85]

2.3.2. Synaptosomal Preparations

The advantages and disadvantages of these preparations are opposite to those described for slices. They are less rapidly prepared, they suffer more manipulations, but, on the other hand, they make it possible to study more specifically the phenomena occurring at the level of nerve endings, and the interpretation of the results is often simpler, since diffusion barriers, reuptake, and neuronal interactions can be eliminated[6] (see Chapter 16). They are particularly useful for discriminating between reuptake inhibitors and true releasers[6] (Chapter 16) and for analyzing the modulation of neurotransmitter release mediated by presynaptic receptors[6] (see also Sections 3.1 and 3.2 and Chapter 16).

It is often said that synaptosomal preparations may be heavily contaminated by glial fragments (gliosomes) and that this could profoundly affect the results obtained, particularly when the release of exogenous radioactive amino acids, which are also accumulated by glial cells, is studied. This possibility may be difficult to disprove when specific glial uptake inhibitors are lacking and is indeed likely when glial uptake is more active than neuronal uptake, as seems to be the case for acidic amino acids in the cerebellum.[61] On the other hand, on the basis of the available evidence,[11,17,34,50,52,85] the release of GABA from "gliosomes" does not appear to appreciably affect the data obtained with synaptosomal preparations. Other subcellular particles that may be present even in large amounts in synaptosomal preparations (mitochondria, microsomes, myelin) have generally been shown not to release amino acids in a Ca^{2+}-dependent way when subjected to depolarizing stimuli.

2.3.3. Other Preparations

Release of neurotransmitter amino acids has also been studied in other preparations, including peripheral ganglia, retina, and cultured nerve or glial cells. The release from ganglia is analyzed in a section dealing with glial release (Section 2.7.1); that from cultured cells is treated in Section 2.9.3.

2.4. Calcium-Dependent Release: Neurotransmitter versus Nonneurotransmitter Amino Acids

It is apparent from Tables I and II that the depolarization-induced release of GABA (whether exogenous, endogenous, or previously synthesized from precursors) has a substantial Ca^{2+}-dependent component (50 to 100%) both in slices and in synaptosomal preparations. In only a few cases did the evoked release of GABA not appear to be Ca^{2+} dependent, namely, in some of the experiments in which the release was stimulated electrically or with veratridine alkaloids, as discussed in Sections 2.2.3 and 2.2.2, respectively.

The behavior of Glu (Tables III and IV) seems to follow essentially the same pattern. The lack of a clear Ca^{2+}-dependence in the release of exogenous Glu (or of its analogue D-aspartate) in P_2 cerebellar fractions[61,63] (Table IV) must probably be attributed to the high accumulation of the acidic amino acids into glial fragments contaminating the preparation.[61] The lack of Ca^{2+}-dependent release of Glu in the olfactory cortex (Table III) may be because Glu does not have a neurotransmitter role in this area of the brain (however, see also other data reported in Section 2.9.1).

It is interesting that Asp, which has been generally less studied than Glu, behaves at times differently from Glu.[22,31,32,61] This is particularly relevant when the release of the two endogenous amino acids was studied in a comparative way (see the data on rat olfactory cortex and cerebellum in Tables III and IV) and suggests that the two amino acids are not released from the same sites. It will be noted, however, that the data on endogenous Asp release from cerebellar preparations are somewhat contradictory and do not help much in elucidating whether Asp can be considered a neurotransmitter candidate in this area of the brain (see also Section 2.9.2). The recent observation that, unlike Glu, Asp is only minimally synthesized and released in cultured cerebellar granule cells seems to confirm that Glu, and not Asp, is the excitatory transmitter of these cells.[86]

Altogether, the data on GABA and Glu release are consistent with the idea that these amino acids have a neurotransmitter role in several areas of the CNS.

Fewer studies are available on Gly release (Table V). In the spinal cord, where the transmitter role of Gly is well established (see bibliography in refs. 1, 9), in only one case was the depolarization-evoked release of this amino acid reported to be Ca^{2+} independent,[67] whereas in other reports Gly release was shown to depend on extracellular Ca^{2+}.[13,18,34,43,53] The results obtained in supraspinal centers, where the neurotransmitter role of Gly is more uncertain, generally show little evoked Gly release and dubious Ca^{2+} dependence (Table V). However, in a recent study[69] performed with superfused rat cortical synaptosomes, a relatively low concentration of K^+ (30 mM) was shown to stimulate the release of exogenous and endogenous Gly and, to a greater extent, of Gly previously synthesized from the precursor [^{14}C]serine. In all cases, the release was roughly 80% Ca^{2+} dependent (Table V). These data, together with the electrophysiological evidence provided[69] and the recent discovery of high-affinity binding sites for [^{3}H]Gly in the rat cerebral cortex,[87] are compatible with a neurotransmitter role of Gly in higher brain centers of some animal species as well.

Taurine release, on the other hand, does not have the characteristics of the release of a neurotransmitter (Table VI), being difficult to be elicited and not Ca^{2+} dependent. It has to be noted, however, that, in contrast with the results summarized in Table VI, 40 mM K^+ was reported to elicit a totally Ca^{2+}-dependent release of preaccumulated [^{3}H]Tau from the perfused cerebellar cortex.[88] This discrepancy may have several explanations: (1) in the *in vivo* perfusion experiments, Tau may be accumulated into and be released from a limited number of cells of the molecular layer; the release from these cells may be overshadowed by nonspecific release from other structures in *in vitro* studies; (2) the Tau-releasing cells or terminals may be labile and not survive *in vitro*; (3) Tau may behave as a false transmitter; (4) *in vivo*, the depolarizing stimulus may have released some other transmitter, which affects the release of taurine.

In a number of reports, the release of amino acids that are not believed to subserve a neurotransmitter role has been studied in parallel with that of putative neurotransmitters. In the large majority of cases, depolarizing stimuli did not enhance the release of nonneurotransmitters or caused a small, Ca^{2+}-independent, increase in release. Such studies have been carried out for glutamine,[22,32,48,54,55,60,88-90] alanine[18,22,32,48,54,68,88-91] α-aminoisobutyric acid,[18,19] serine,[14,18,48,54,88-90] valine and threonine,[91] proline,[43,81] leucine,[13,18,24,42,55,60,90] lysine,[18,43,55,90] and arginine, histidine, tyrosine, and phenylalanine.[18,90]

A Ca^{2+}-dependent release of exogenous proline[18] and endogenous alanine[31] has been reported in cerebral cortex slices and in cerebellar slices, respectively. Exogenous glutamine was released by high [K^+] in a 50% Ca^{2+}-dependent way from cat cochlear nucleus slices.[30] Calcium-independent, electrically induced release of nonneurotransmitter amino acids has occasionally been described.[41,72] High [K^+] and veratridine have been shown to stimulate the release of the taurine precursor [³⁵S]hypotaurine from cerebral cortex slices, but the Ca^{2+} dependence of the evoked release was not studied.[92]

Two lysine metabolites that have been suggested to have a neuronal function, L-α-aminoadipic acid[93] and pipecolic acid,[94] can be released by high [K^+] in a largely Ca^{2+}-dependent way.

Two analogues of GABA, sharing its neuronal transport system, [³H]diaminobutyric acid[95] and [³H]*cis*-3-aminocyclohexanecarboxylic acid,[96] showed a Ca^{2+}-dependent release in depolarized cortical slices. In contrast, the evoked release of [³H]β-alanine, an inhibitor of glial GABA uptake, does not appear to be enhanced by extracellular Ca^{2+}[17,34,50-80] (see also Section 2.7.1).

To conclude this section, it can be said that the existence of a depolarization-induced, Ca^{2+}-dependent release of a given amino acid from CNS preparations can be considered as an important biochemical criterion supporting its candidacy for a neurotransmitter role, but it should be added that the validity of the criterion is increased if the ratio between evoked and "spontaneous" release is high and if the component of the evoked release that depends on extracellular Ca^{2+} is at least 50–60%.

2.5. Regional Differences

The existence of regional differences in the depolarization-evoked release of neurotransmitter amino acids and of its Ca^{2+} dependence can be deduced from the data reported in Tables I–VI. However, because of the sometimes large quantitative differences in signal/noise ratio obtained by the various authors even in similar experimental conditions, regional comparisons are valid only when the results were obtained by the same author in the same experimental conditions. Regional studies on depolarization-induced release are rarely systematic and often concern the release of exogenously loaded amino acids.[13,18,23,24,41,43,91,97-99] These studies generally show a poor correlation between the percent evoked release of a given amino acid and its endogenous concentration or uptake capacity (see, for example, ref. 97). In general, however, amino acids that are known to play a particularly important neurotransmitter role in a given area of the CNS are released from that area to a larger extent than in other areas, where they may even not be released at all. Comparatively high release of GABA and glutamate from the cerebral cortex, of GABA from substantia nigra, of Glu from striatum and hippocampus, and of Gly from spinal cord has been reported in several papers (see Tables I–V).

2.6. Developmental Aspects

The problem of the ontogenesis of amino acid depolarization-induced release has been analyzed in some detail only in the case of GABA.[44,100]

Although biogenic amines are already released in a Ca^{2+}-dependent way in synaptosomal preparations of newborn rats,[44,100] the evoked release of exogenous[44,100] and of endogenous GABA[44] is still not Ca^{2+} dependent in 8-day-old animals and starts to show some dependence on extracellular Ca^{2+} in animals at 14 days.[100] This is in sharp contrast with the full development of GABA high-affinity transport in 1-week-old animals.[44,100] The release of exogenous and endogenous GABA evoked by 56 mM KCl in synaptosomes from 8-day-old animals (in the presence or in the absence of Ca^{2+}) was similar to the release evoked in adult animals in the absence of Ca^{2+}[44] (about 130% over base line). Thus, it appears that it is the Ca^{2+}-dependent component of the evoked release that is missing in the preparations from immature animals. Interestingly, even in 8-day-old animals, a Ca^{2+}-dependent release of [³H]GABA became apparent if the synaptosomal preparations had been depleted of Na^+ prior to depolarization with high [K^+]; moreover, the calcium ionophore A23187 was able to elicit GABA release.[44] In conclusion, the data suggest that, although a Ca^{2+}-sensitive pool of GABA may also be present in immature synaptosomes, this pool is not released in normal conditions, probably because the high intrasynaptosomal Na^+ level prevents a sufficient membrane depolarization in the immature GABA-releasing terminals.[44]

Judged from a study in the developing rat cerebellum, the evoked release of Glu also acquires a Ca^{2+}-dependent component fairly late in development (after the eighth day of postnatal life in the rat).[61] At partial variance with the above results, it has been reported that the K^+-evoked release of exogenous GABA and Gly from newborn rat cortical "minislices" is small but Ca^{2+} dependent.[103]

That the establishment of a Ca^{2+}-dependent depolarization-induced release is related to neuronal development and differentiation is also demonstrated by studies with tissue cultures. Glutamate (endogenous and synthesized from [³H]glutamine) release induced by high [K^+] was modest and not Ca^{2+} dependent in cerebellar granule cell cultures at 2 days *in vitro*, but at later stages it became large and about 80% Ca^{2+} dependent.[86] Similarly, the Ca^{2+}-dependent release of exogenous GABA was shown to increase in mixed primary cultures from rat cerebral cortex from 7 to 28 days *in vitro* in parallel with the development of other parameters of GABAergic function.[102]

We are aware of only one study concerning neurotransmitter release in aged animals.[103] This study shows a modest increase in the Ca^{2+}-dependent component of the release of [¹⁴C]GABA elicited by 55 mM K^+ in 12-month-old, as compared to 2-month-old mice.

2.7. Origin of the Amino Acids Released by Depolarization

2.7.1. Neuronal versus Glial Release

When release studies are conducted using preparations containing intact glial cells (such as slices) or resealed glial fragments (such as synaptosomes), the question may arise whether amino acids are released from glial rather than from neuronal structures, particularly when the analysis is restricted to preloaded exogenous amino acids, which in some cases may be taken up more effectively by glial than by neuronal cells.[61] Therefore, there has been a search for criteria that would make it possible to discriminate between neuronal and glial release. In general (but not unanimously[80,104]), the Ca^{2+} dependence of the K^+- or electrically evoked release and the ability of veratridine to induce a release antagonized by tetrodotoxin are taken as evidence for a neuronal origin of the released amino acids.[17,21,34,50,52,61,78,79,85,105,106] In the case of Glu and GABA, the evidence may become even stronger if the above criteria are applied to the release of the amino acids previously synthesized from radioactive glutamine, which is considered as a specific precursor of the neuronal neurotransmitter pool of these amino acids (see Section 2.7.2). Moreover, the signal/noise ratio of neuronal evoked release appears to be much higher than that of glial release (see, for example, refs. 17, 21, and 80).

The following observations support the validity of the criteria of Ca^{2+} dependence and veratridine and tetrodotoxin sensitivity mentioned above:

1. Depolarization with high $[K^+]$ evoked a Ca^{2+}-independent release of radioactive GABA preaccumulated by glial cells in sympathetic[21,105] and sensory[21] ganglia, cerebellar glial cultures,[85] brain bulk-isolated glia,[52] and rat retina.[21]

2. Electrical stimulation evoked a Ca^{2+}-independent and tetrodotoxin-insensitive release of exogenous GABA preaccumulated by glial cells of sympathetic ganglia[105] and posterior pituitary gland.[107]

3. Veratridine did not stimulate the release of GABA preloaded into sensory ganglia,[21,79] trigeminal ganglia,[80] sympathetic ganglia, and rat retina.[21]

4. Radioactive β-alanine preaccumulated into glial cells of brain slices through the GABA transport system[108] was released by high $[K^+]$ in a Ca^{2+}-independent manner,[17,34] even when release was studied from synaptosomes prepared from the prelabeled slices.[50]

5. Depolarization with high $[K^+]$ in the presence of Ca^{2+} did not evoke release of exogenous acidic amino acids preaccumulated by cerebellar glial cultures[61,85] or glial cells in peripheral or central (optic) nerve trunks.[106] Electrical stimulation induced a Ca^{2+}-independent release of Glu from the latter preparations.[106]

Some data in the literature are in partial conflict with these observations: the K^+-induced release of radioactive GABA from gliocytes of the posterior pituitary,[107] from sensory ganglia,[79,104] and from a preparation of central glia[80] (deafferented olfactory nerve layer) was depressed by omission of Ca^{2+}, particularly if a high concentration of Mg^{2+} was added to the Ca^{2+}-free medium. Moreover, veratridine stimulated the release of [³H]GABA (but not of [³H]β-alanine) from the olfactory nerve layer preparation, and this release was prevented by tetrodotoxin.[80] Finally, regional differences in the K^+-evoked release of [³H]β-alanine from slices were observed, which could not be explained on the basis of differences in the density of glial cells.[65]

The Ca^{2+}-dependent release of GABA from glia, however, was shown to have atypical features. For example, the Ca^{2+} antagonist D-600 (a methoxy derivative of verapamil) did not block, as expected, the Ca^{2+}-evoked release of [³H]GABA.[79,80] Diphenylhydantoin and La^{3+} were also ineffective in blocking the evoked release,[79] and high $[K^+]$ promoted GABA release only in the presence of aminooxyacetic acid,[79] which, in certain cases, has been shown to increase cell excitability.[40] Finally, high concentrations of Mg^{2+} in Ca^{2+}-

free solutions were necessary to see an 80% reduction of GABA release,[80] but high Mg^{2+} levels have been shown also to inhibit the Ca^{2+}-independent K^+-evoked release of GABA and glutamate from CNS slices.[81]

In conclusion, when taken with a certain caution, the criteria mentioned at the beginning of this section for discriminating between neuronal and glial release appear to be valid. The suggestion that CNS glia may behave differently from peripheral glia[80] cannot be indiscriminately accepted, since amino acid release from rat retina,[21] optic nerve trunks,[106] and cultured cerebellar astrocytes[61,86] does not seem to have the characteristics described in the case of the glial olfactory nerve layer preparation.[80] However, the results of the latter study point to the possibility that not all glial cells behave in the same way. Moreover, even the same cells may release or not release preaccumulated acidic amino acids depending on whether they are in the form of disrupted fragments in a homogenate or in the form of intact cells in a slice or in a culture dish.[61]

2.7.2. Cytoplasmic versus Vesicular Release

By analogy with the behavior of other neurotransmitters[5,6] (Chapter 16), it is often implied that the amino acids released by depolarization in a Ca^{2+}-dependent way originate from synaptic vesicles. However, experimental evidence in support of this concept is lacking, and the investigators who have directly approached this problem have reached the conclusion that amino acids are released mainly from the cytoplasmic pool.[89,109,110] This is in keeping with the fact that amino acids are largely "free" in the cytoplasm[109,111,112] and that no stable form of association of these compounds with synaptic vesicles has been described. On the other hand, some recent data obtained with GABA suggest that the amino acid released in a Ca^{2+}-dependent way by high $[K^+]$ originates preferentially from a distinct intraterminal pool, which is less heavily labeled than the main cytoplasmic pool by exogenous radioactive GABA.[51,113] Moreover, both veratridine,[113] which stimulated GABA release independently of Ca^{2+} (see Section 2.2.2), and a high $[K^+]$, Ca^{2+}-free medium[51] evoked GABA release from a pool (probably the main cytoplasmic pool) with higher than average specific radioactivity.

Although it is tempting to suggest that the less-labeled pool is localized in synaptic vesicles, the possibility that such a suggestion represents only a wishful thought remains open. In fact, it has to be remembered that the neurotransmitters that are mainly associated with synaptic vesicles and released by a mechanism of exocytosis[5,8] (Chapter 16) behave in several respects differently from putative transmitter amino acids. For example, they are released more easily than the amino acids by the Ca^{2+} ionophore A23187[4,47,114] and by electrical stimuli,[40,72,82] and, unlike amino acids, they are consistently released by veratridine in a Ca^{2+}-dependent way[11] (see Section 2.2.2).

From a pharamcological point of view, the differences become even more striking. Vesicle depleters or releasers analogous to reserpine or amphetamine, respectively, do not exist in the field of neurotransmitter amino acids. It is likely that these differences derive from the predominant cytoplasmic localization of amino acids (in this respect, it is interesting that, in a comparative study of norepinephrine and GABA release from brain slices,[20] norepinephrine release became similar to GABA release after reserpine treatment), and it is difficult to determine whether the Ca^{2+}-dependent release of these compounds involves a vesicular pool in rapid equilibrium with the main pool of the cytoplasm or a distinct cytoplasmic pool displaced by the intraterminal increase of Ca^{2+} through a mechanism different from exocytosis. The existence of a distinct pool (if not a vesicular pool) is also suggested by the fact that the amino acids not related to neurotransmission, which are also "free" in the cytoplasm, are not released in a Ca^{2+}-dependent manner by depolarizing stimuli (see Section 2.4).

2.7.3. Releasable Pools

Even if anatomic localization of the releasable amino acid pools within the nerve endings is difficult (see preceding paragraph), from a biochemical point of view different pools can be distinguished and analyzed in terms of their availability to be released. These studies have generally concentrated on three pools: the "reuptake pool" (the pool labeled by exogenous radioactive amino acids via the high-affinity transport system), the "new synthesis pool" (the pool labeled by a radioactive neurotransmitter precursor), and the total "endogenous pool." Before these studies are analyzed, it must be mentioned that the results obtained may be greatly influenced by the experimental conditions adopted. Important variables that at times have been considered include (1) the concentration of the exogenous amino acid,[33,115] (2) the concentration and type of precursor,[30,59,116–119] (3) the presence of the precursor only before or also during the application of the releasing stimulus,[59] (4) the presence of an oxidizable substrate in the bathing medium,[120] (5) the type of superfusion protocol, including the duration of the washing phase before the depolarizing stimulus is applied or, more generally, the sequence of medium changes,[51,121] (6) the type of preparation used[62] (slices, crude or purified synaptosomes), and (7) the type of depolarizing stimulus.[11,35,113,116,122] Since the data reported in the literature were rarely obtained under comparable conditions, their evaluation is often difficult.

There is generally good agreement in considering exogenous glutamine as the best precursor for labeling[28,30,59,60,117–119,123,124] and for replenishing[59,117,118,123,124] the neurotransmitter pools of Glu and GABA releasable by depolarizing stimuli. It is interesting that in brain slices exogenous glutamine is an even better precursor than exogenous Glu for labeling the releasable GABA pool,[30,119] probably because exogenous Glu is also accumulated into pools not directly related to GABA synthesis, whereas glutamine would selectively feed an intraterminal glutamate pool immediately available for the synthesis of GABA.

In appropriate conditions (presence of high concentration of radioactive glutamine during incubation and during stimulation), the labeled Glu derived from glutamine was released preferentially from slices[59] and synaptosomes[123] with respect to the unlabeled amino acid. However, whereas with slices the specific radioactivity of the Glu released was higher than that found in the tissue only if [^{14}C]glutamine was also present in the medium during the stimulation phase,[59] with synaptosomes the specific radioactivity of the Glu released was also higher when the radioactive glutamine was removed from the medium prior to stimulation.[123] With cerebellar synaptosomal preparations prelabeled with a low concentration of [^{14}C]glutamine and then superfused in the absence of glutamine, the specific radioactivity of the glutamine released by high [K$^+$] was moderately higher than that present in synaptosomes at the beginning and substantially higher than that present at the end of the superfusion period, again suggesting a preferential release of the newly synthesized [^{14}C]glutamate.[61]

When [^{14}C]glucose was used as a Glu or GABA precursor, no preferential stimulated release of newly synthesized radioactive amino acids was found either in slices[59] or in synaptosomes,[48] although in rather extreme conditions (glucose-free media), electrical stimulation promoted a greater release of newly synthesized than of exogenous Glu and GABA from cortical slices.[120] In synaptosomes preincubated with [^{14}C]glucose and then subfractionated, the specific radioactivity of GABA found in synaptic vesicles was lower than that of the free cytoplasmic pool.[110] Thus, if the Ca^{2+}-dependent release of amino acids does indeed involve synaptic vesicles, no preferential evoked release of amino acids synthesized from [^{14}C]glucose should be expected.

Other precursors have been used sparingly. [^{14}C]Pyruvic acid was shown to label a small, rapidly released GABA pool.[16] [^{3}H]Asparagine was successfully used as a precursor of Asp,[58] but it is not known whether Asp synthesis from asparagine is physiologically important.

The problem of whether or not exogenous amino acids previously taken up are released by depolarizing stimuli more readily than the bulk of the endogenous amino acid pool has not been solved unequivocally. Although it is frequently assumed that the behavior of exogenous and endogenous amino acids is similar, this assumption was proven not to be valid in all cases, and the conclusions drawn from experiments run in any particular condition should be restricted to that condition.

A few examples will clarify this point. (1) In a study with synaptosomes, it was shown that the fractional [^{14}C]GABA release dependent on Ca^{2+} was decreased by a relatively small increase in the concentration of [^{14}C]GABA (from 0.4 to 10 μM) during the labeling phase.[115] With the higher [^{14}C]GABA concentration, the net amount of [^{14}C]GABA released was obviously higher but was less than expected from the greater accumulation. Thus, the specific radioactivity of the GABA released did not increase in proportion to the synaptosomal GABA specific radioactivity, and a greater percentage of the accumulated [^{14}C]GABA was stored in less readily releasable pools.[115] (2) Other studies with prelabeled synaptosomes[46,51,113] show that the specific radioactivity of the GABA released by high [K$^+$] in the presence of Ca^{2+} was similar to that found in synaptosomes, suggesting no preferential release of exogenous GABA (however, see ref. 125). On the other hand, a second period of stimulation resulted in a much depressed release of radioactive GABA unless a second labeling phase was interposed between the two stimulations.[121,126] So, in the latter conditions, the most recently accumulated amino acid was released more readily than that taken up during the first exposure.(3) In experiments in which dentate gyrus slices were prelabeled with 0.4 μM [^{14}C]Glu and then superfused in the presence of 0.4 mM glutamine, the Ca^{2+}-dependent release of [^{14}C]Glu elicited by high [K^+] dropped rapidly on repetitive stimulation, whereas that of endogenous Glu remained constant,[60] probably because of the continuous supply, by glutamine, of readily releasable unlabeled Glu.[59,118] Although the latter results supported the conclusion that most of the accumulated [^{14}C]Glu is inaccessible to releasable pools,[60] other experiments performed with prelabeled synaptosomes superfused with a glucose-containing medium devoid of glutamine suggested that the exogenous Glu is released more promptly than the amino acid synthesized from glucose.[113] An opposite conclusion (greater release of [^{14}C]glucose-derived than of exogenous Glu) was reached when cortex slices were stimulated in a glucose-free medium.[120]

2.8. Calcium-Independent Release

It has almost invariably been observed that the depolarization-induced release of putative transmitter amino acids has a component that is not dependent on the presence of extracellular Ca^{2+} (see Tables I–V).

Although the magnitude of the Ca^{2+}-independent release may vary depending on the experimental conditions (for example, media with high Mg^{2+} concentrations have been shown to block not only the Ca^{2+}-dependent release but also the Ca^{2+}-independent component of the evoked release[81]), several observations suggest that depolarization in Ca^{2+}-free conditions may elicit amino acid release from the same structures (even if not from the same pool) competent for Ca^{2+}-dependent release. (1) The Ca^{2+}-independent fraction of release observed with slices is only at times[60] larger than that observed with synaptosomal preparations (Tables I–IV) in spite of the greater heterogeneity of structures present in slices. (2) The Ca^{2+}-independent release could be reduced, but not eliminated, by using purer synaptosomal preparations.[46,51,60,62] (3) Calcium-independent release was observed not only in studies with preaccumulated exogenous amino acids but also in studies on endogenous amino acids and on amino acids previously synthesized from radioactive precursors thought to label quite specifically the neurotransmitter releasable pools.[59,61,117] (4) The large release of GABA and Glu elicited by depolarization with veratridine is tetrodotoxin sensitive but generally not Ca^{2+} dependent (see Section 2.2.2) and is unlikely to originate from structures other than nerve endings.[11] (5) In most cases, the release of nonneurotransmitter amino acids was not stimulated by depolarization (see Section 2.4).

Two hypotheses have been advanced in order to explain the mechanism of the Ca^{2+}-independent release of amino acids elicited by depolarization. According to the first, GABA release could occur via the amino acid carrier working in an outward direction.[51] However, this hypothesis is unlikely, since it contrasts with several experimental observations: (1) carrier-mediated release of GABA from synaptosomes could not be evoked by removing extracellular Na^+ and thus reversing the normal Na^+ gradient[4,45]; (2) the Ca^{2+}-independent release of GABA induced by veratridine and the high-[K$^+$]-induced GABA release (which in the experimental conditions adopted was 40–50% Ca^{2+} independent[45]) were both unaffected by a block of the GABA carrier with the quasiirreversible inhibitor diaminobutyric acid.[35,127,129]

According to the second hypothesis, the depolarization-induced release of amino acids independent of extracellular Ca^{2+} could occur through mobilization of intracellular Ca,$^{2+}$[129] which, according to the "calcium hypothesis,"[132] (for other references, see refs. 5 and 8), should be sufficient to trigger transmitter release provided that an increase in the intraterminal concentration of free Ca^{2+} is achieved. However, this hypothesis is also difficult to reconcile with several observations: (1) in synaptosomes depleted of intracellular Ca^{2+} by superfusion with a Ca^{2+}-free medium containing a Ca^{2+} chelator and the Ca^{2+} ionophore A23187, the K$^+$-induced Ca^{2+}-independent release of GABA was unaffected[51]; (2) in the absence of extracellular Ca^{2+}, the release of catecholamines and acetylcholine was scarcely, if at all, increased by high [K$^+$][51,131] (Chapter 16) or veratridine,[11] in contrast with the behavior of amino acids. Mobilization of intramitochondrial Ca^{2+} would, if anything, be expected to release more easily the transmitters associated with vesicles than the amino acids. (3) As already mentioned, high-[K$^+$], Ca^{2+}-free solutions and veratridine appear to release GABA from an intrasynaptosomal pool different from the Ca^{2+}-sensitive pool,[51,113] a finding difficult to reconcile with the hypothesis that release occurs through mobilization of intraterminal Ca^{2+}.

In conclusion, it seems likely that the depolarization of the nerve ending membrane can, by itself, cause an increased release of transmitter amino acids that is totally independent of Ca^{2+} and is not mediated by the membrane carriers. It has been suggested that, in the living brain, transmitter amino acids may be released from depolarized neurons of immature animals even if a Ca^{2+}-sensitive pool has not yet developed.[44]

2.9. Release from Specific Pathways or Cells

The large majority of the data on the depolarization-evoked release of amino acids was obtained by applying stimuli causing a generalized depolarization of complex and heterogeneous preparations such as slices or synaptosomes. The degree of specificity of the information obtained with such studies is limited by the widespread distribution of the putative transmitter amino acids in the CNS, in particular if one wants to determine which is "the" transmitter utilized in a particular pathway or by a particular cell type. In certain cases, however, the specificity of the information can be greatly increased by studying amino acid release during the stimulation of specific fiber tracts or after producing specific lesions or by analyzing the behavior of selected cell types cultured *in vitro*. In the following paragraphs, some of the results obtained by these methods are briefly summarized.

2.9.1. Stimulation of Specific Pathways

Calcium-dependent release of exogenous GABA was observed on rostral electrical stimulation of the spinal tracts in the hemisected frog spinal cord.[132]

Specific release of acidic amino acids was obtained in the olfactory cortex by stimulation of the lateral olfactory tract,[22,133] in the stratum radiatum of the hippocampus by stimulating the Schaffer's collaterals,[134–136] and in the dorsolateral septum after stimulation of the fimbria.[137] The latter study shows that only the selective stimulation of the afferent fibers released [3H]D-aspartate in a specific way: direct stimulation of the dorsolateral septum caused an aspecific (Ca^{2+}-independent) release of D-asparate and GABA similar to that evoked by stimulating the fimbria with higher than just supramaximal stimuli. These observations may explain some of the inconsistencies in the results obtained with electrical field stimulation (see Section 2.2.3).

In spite of the elegance of the studies mentioned above, some disagreement persists as to whether Glu or Asp is specifically released from the olfactory cortex[22,133] or from the dentate gyrus.[135,136]

2.9.2. Lesion Studies

The destruction of a specific population of nerve endings or cells should lead to a specific decrease in the evoked release of the transmitter utilized by the lesioned structures. Selective lesions have been obtained by surgical methods, X-irradiation, and neurotoxins. It often happens that lesioned tissues show a decreased level, synthesis, uptake, and release of the putative transmitter amino acid under study. Thus, it may at times be difficult to evaluate whether the depression of release is a primary phenomenon or a consequence of the concomitant decrease in pool size or uptake, and the method of expression of the results may become critical for their interpretation. In general, it can be thought that the release process is specifically affected if the evoked release is depressed substantially more than the level, the synthesis, or the uptake of the compound under study.

Surgical lesion of the afferent fibers brought about a depression of the evoked release of endogenous and newly synthesized Glu in the rat hippocampal dentate gyrus,[59,117] or [3H]D-aspartate in the stratum radiatum of the rat hippocampus,[134] of newly synthesized Glu and Asp in the cat cochlear nucleus,[30] and of Asp in the rat olfactory cortex.[22] Ablation of the frontoparietal cortex in the rat brought about a large decrease in the striatal release of Glu (both exogenous and synthesized from [3H]glutamine[27]) and of Asp synthesized from [3H]asparagine.[58] Interestingly, in the latter experiments, the striatal level of Glu, but not that of Asp, was significantly decreased. Destruction by X-irradiation of cerebellar granule cells and of the parallel fibers originating from them led to a decreased release of Glu in studies utilizing slices[31] or synaptosomal preparations.[63] Irradiation with higher doses, which also destroy the stellate cells in the molecular layer, also caused a depression in the release of GABA.[31] Destruction of the cerebellar climbing fibers with 3-acetylpyridine had no effect on endogenous Glu or Asp Ca^{2+}-dependent release in one study[31] but caused a significant reduction of Asp release in another study.[90] The latter report, which also shows retrograde labeling of the inferior olive neurons after intracerebellar injection of [3H]D-aspartate, keeps open the possibility that Asp is the climbing fiber transmitter.

Destruction of striatal neurons by local injection of kainic acid led to a reduction in the release of exogenous GABA and taurine from superfused P_2 fractions.[138]

In conclusion, lesion studies, particularly if combined with studies on the effect of stimulation of specific pathways,[22,60,134] have become a powerful tool for the identification of neurotransmitters.

2.9.3. Release from Cultured Cells

This type of study is potentially very useful for the identification of putative neurotransmitters and for the correlation between development of neurotransmitter release processes and cell differentiation.

Depolarization-induced, Ca^{2+}-dependent release of exogenous [^{3}H]GABA was shown to be present in differentiated cultures of cortical,[102] spinal,[139] and cerebellar[85] neurons. Autoradiographic examination showed that in the spinal cultures, about 50% of the neurons were labeled by [^{3}H]GABA.[139] The percentage was much lower (7%) in cerebellar cultures, where about 8% of the labeled cells was represented by astrocytes; these, however, were unable to release [^{3}H]GABA on depolarization.[85]

In cortical cultures, [^{14}C]GABA synthesized from [^{14}C]glucose was released by high [K^+] more than the [^{3}H]GABA derived from [^{3}H]acetate, suggesting a preferential release from neurons. [^{14}C]Glutamate was also synthesized from [^{14}C]glucose and released by these cultures.[102]

In differentiated cerebellar neuronal cultures consisting of about 80% granule cells,[140] depolarization with high [K^+] caused a selective Ca^{2+}-dependent release of endogenous Glu and of [^{3}H]glutamate previously synthesized from [^{3}H]glutamine.[86] Release of endogenous GABA was not measurable, and [^{3}H]GABA synthesis from [^{3}H]glutamine was negligible, showing that the presence of GABA neurons in the culture was minimal. [^{3}H]Aspartate was synthesized at least ten times less efficiently than [^{3}H]glutamate, and the evoked release of Asp (endogenous and newly synthesized) was small and Ca^{2+} independent.[86]

2.10. Mechanism of Ca^{2+}-Dependent Release

2.10.1 Artifact or Stimulus-Coupled Secretion?

Arguments in favor of the idea that the depolarization-induced, Ca^{2+}-dependent release of putative transmitter amino acids observed in *in vitro* studies does not represent an experimental artifact but is a phenomenon related to the physiological release process can be found in the preceding pages. However, in view of the doubts that have occasionally been raised on this issue,[8,40,72,82] a list of points substantiating the contention that the Ca^{2+}-dependent release observed is not an experimental artifact follows. (1) The depolarizing stimuli used do indeed depolarize brain slices and synaptosomes,[47,65,74,141] and depolarization is accompanied by increased Ca^{2+} influx.[142–144] (2) Calcium-dependent, depolarization-induced release can also be observed in the absence of extracellular Na^+,[47,81,145] whereas agents such as ouabain, which can increase intracellular Na^+ without altering the membrane potential, elicit amino acid release independently of Ca^{2+}.[65] These observations exclude the possibility that the increased influx of Ca^{2+} triggering neurotransmitter release is merely related to an increased concentration of intracellular Na^+. (3) Ba^{2+} and $Sr2^+$ can effectively replace Ca^{2+}, whereas the effect of Ca^{2+} can be blocked by Mg^{2+} and Mn^{2+}, a behavior similar to that of adrenal and neuromuscular preparations.[47,145] (4) The release of putative transmitter amino acids from structures not related to neurotransmission is not Ca^{2+} dependent (see Sections 2.3.2 and 2.7.1). (5) With only few exceptions, the release of nonneurotransmitter amino acids is not stimulated by depolarizing stimuli and, if stimulated, is not Ca^{2+} dependent (see Section 2.4). (6) The development of Ca^{2+}-dependent release is related to neuronal maturation and differentiation (see Section 2.6). (7) There are quantitative and qualitative regional differences in the Ca^{2+}-dependent release of putative transmitter amino acids. Certain amino acids may be released only in certain areas of the CNS (see Sections 2.5 and 2.9 and Tables I–V). (8) Differential release of neurotransmitter amino acids belonging to different functional pools has been often observed (see Section 2.7.3).

2.10.2. Nature of the Ca^{2+}-Dependent Release Process

The role of Ca^{2+} in neurotransmitter release and the mechanisms linking Ca^{2+} influx to the release process have always been discussed in terms of exocytotic release (see refs. 5, 8, and Chapter 16 for a more extensive bibliography; for recent hypotheses on the possible role of phosphatidic acid and of calmodulin, see refs. 146 and 147, respectively). The fact that certain amino acids have a neurotransmitter function and that they are released in a Ca^{2+}-dependent manner may suggest, but does not necessarily indicate, that they are released by the same exocytotic mechanism that appears to be responsible for the release of biogenic amines and acetylcholine. As already mentioned (see Section 2.7.2), the possibility that transmitter amino acids are contained in, and released from, synaptic vesicles has neither been totally disproved nor convincingly proved. Even if they were not released from vesicles, they appear to be released from a Ca^{2+}-sensitive cytoplasmic pool distinct from the main amino acid pool (see Sections 2.7.2 and 2.8).

It is interesting to note that although an increase in the concentration of intraterminal free Ca^{2+} is considered to be a condition necessary and sufficient to trigger catecholamine or acetylcholine release, the depolarization in the absence of Ca^{2+} is ineffective[5,6,8≈130]; amino acids can also be effectively released by depolarization in the absence of Ca^{2+} (see Sections 2.2.2, 2.2.3, 2.8, and Tables I–V), particularly when the depolarization is achieved through the activation of the Na^+ channels. Since amino acid transport is highly Na^+ dependent, and carriers can work in an inward and outward direction (see Section 4), it could be hypothesized that the release of amino acid transmitters evoked by depolarization is mediated by the operation of their specific carriers in an inside–outside direction (see, for example, ref. 57). However, the observation that Ca^{2+}-dependent release is also present in the absence of extracellular

Na^+[47,81,145] and is absent at a development stage at which high-affinity transport is fully developed[44,100] and the fact that the GABA carrier blocker diaminobutyric acid does not reduce the high-$[K^+]$- or veratridine-induced release of GABA from synaptosomes[35,122,129] argue against this possibility. As is discussed in Section 4.2, however, it has been proposed[128] that carrier-mediated outward transport of transmitter amino acids may function as an additional release mechanism superimposed on the carrier-independent, Ca^{2+}-dependent release process.

3. MODULATION OF AMINO ACID RELEASE

3.1. Presynaptic Autoreceptors

The possibility that the depolarization-induced release of neurotransmitters is subjected to a negative feedback control mechanism mediated by presynaptic autoreceptors has been the object of intense investigation in the case of biogenic amines and acetylcholine. The topic is amply treated in Chapter 16 of this book, to which the reader is referred. According to this mechanism, when the concentration of a neurotransmitter in the synaptic cleft reaches a critical level, the neurotransmitter could inhibit its own release by activating specific autoreceptors localized in the presynaptic terminal membrane. The molecular events leading from the activation of receptors to the depression of neurotransmitter release are still largely unknown but do not appear to involve variations in the membrane potential of the presynaptic membrane.

In the case of putative transmitter amino acids, few reports have appeared on the subject, and all of them except one[148] concern GABA release. The data on GABA are summarized in Table VII. In the various preparations and experimental conditions used, low concentrations of GABA agonists were able to inhibit the depolarization-induced release of GABA (see, however, ref. 153). The effective concentrations appeared to be in the low nanomolar range in one of the studies with synaptosomes[151] (Table VII), which is intriguing because, if GABA were effective *in vivo* at concentrations comparable to those effective *in vitro* (namely, around 10^6 times lower than the presumed concentration of GABA in nerve terminals[154]), a constant autoinhibition of GABA release might be expected. The reason that bicuculline was not able to antagonize the effect of muscimol in the experiments with nigral slices (the drug unexpectedly inhibited the evoked release of GABA in the absence of GABA agonists[23]) is not clear.

The experiments with synaptosomes[151,152] are the only ones providing convincing evidence for a presynaptic localization of the GABA receptor responsible for the effects observed (with slices, indirect effects are difficult to rule out completely; see Section 3.2).

The possibility that Glu release is also subjected to a negative feedback control is suggested by a study in which the high-$[K^+]$-induced release of the Glu analogue [^{3}H]D-aspartate from hippocampal slices was shown to be reduced 20–40% by 100 μM L- or D-glutamate, L-cysteate, and L- or DL-homocysteate.[148]

Table VII

Autoreceptor-Mediated Depression of Depolarization-Induced Release of GABA

Type of preparation and experimental procedure	Agonist and percent inhibition of the evoked release		Effectiveness of antagonists	References
Rat frontal cortex slices Superfusion at 37°C Stimulation with K$^+$ 15 mM Exogenous GABA	GABA, 0.3 µM Muscimol, 0.1 µM APS^a, 1 µM	38 22 30	Picrotoxin, 1 µM Bicuculline, 1 µM	Mitchell and Martin[149]
Rat s. nigra slices Superfusion at 25°C Stimulation with K$^+$ 30 mM Exogenous GABA	GABA, 1µM Muscimol, 1 µM Muscimol, 0.1 µM	37 34 24	Picrotoxin, 10 µM (Bicuculline, 1–10 µM, not effective)	Arbilla et al.[23]
Rat olfactory cortex slices Cup, room temperature Electrical stimulation (direct or lateral olfactory tract) Endogenous GABA	Muscimol, 2 µM Muscimol, 10 µM Muscimol, 10 µM (direct stimulation)	17 60 70	Picrotoxin 15 µM (Bicuculline not tested)	Collins[150]
Rat cortex synaptosomes Superfusion at 37°C Stimulation with K$^+$ 55 mM Exogenous GABA	Muscimol 16 nM P4S^a 36 nM Isoguvacine 49 nM THIP^a 110 nM δ-Aminolevulinic acid 10 µM	IC$_{50}$ IC$_{50}$ IC$_{50}$ IC$_{50}$ IC$_{50}$	Picrotoxin 1 µM Bicuculline 1 µM	Brennan et al.[151]
Rat brain synaptosomes (P$_2$) Superfusion at room temp. Stimulation with high [K$^+$] Exogenous GABA	Muscimol 1 µM Muscimol 25 µM	66 88	Bicuculline 100 µM^b	Snodgrass[152]

^{a} Abbreviations used: APS, 3-aminopropanesulfonic acid; P4S, piperidine-4-sulfonic acid; THIP, 4,5,6,7-tetrahydroisoxazolo-(5,4-C)-pyridin-3-ol.
^{b} At this concentration bicuculline had a potentiating effect on the evoked release of GABA. In the presence of 100 µM bicuculline and 25 µM muscimol, GABA release was 64% inhibited.

The effect of these agonists was prevented by a group of antagonists known to interact with the glutamate-preferring receptors, whereas agonists and antagonists of the aspartate-preferring receptor were without effect.[148]

It is not known whether mechanisms comparable to those described here are operating in the living brain under physiological conditions. One should also consider the possibility that such mechanisms may represent a sort of safety device which becomes operative only in particular pathological conditions. The answer to these questions is of extreme importance from a pharmacological and clinical point of view. In fact, the autoreceptors described in this section resemble, in terms of structural requirements and sensitivity to antagonists, the classical postsynaptic receptors. As a consequence, any drug directly or indirectly causing an increased release of transmitter or behaving as a transmitter agonist would be expected to produce two opposite effects, one of potentiation (through the postsynaptic receptors) and another of depression (through the presynaptic autoreceptors) of neurotransmission.

3.2. Neurotransmitters or Neuromodulators Affecting Amino Acid Release

Several recent reports suggest that the release (in general, the depolarization-induced release) of a number of neurotransmitters, including the amino acids, can be enhanced or depressed by other neurotransmitters or, more generally, by neuromodulatory substances. (For a recent review on modulatory actions of neurotransmitters, see ref. 155.) The mechanism by which the release of a given transmitter is affected by a neuromodulator does not necessarily involve conventional synaptic interactions and consequent changes in the potential of the postsynaptic membrane. However, it does appear to involve the activation of presynaptic receptors localized in the membrane of the nerve ending containing the neurotransmitter whose release is modulated. Although it is implied that the alteration of release is a consequence of intraterminal biochemical events following the activation of the competent receptors,[156,157] the nature of these events has not been determined, and studies on the characterization of the receptors are just starting to appear.[157] In the present chapter, only compounds affecting amino acid release are analyzed. Neuromodulatory control by amino acids of the release of biogenic amines and acetylcholine is treated in Chapter 16 of this book.

Most of the studies on this topic (summarized in Table VIII) were performed with brain slices, a preparation that maintains "normal" anatomic connections among cells and local neuronal circuits. This makes it at times very difficult to ascribe the effect of a given neurotransmitter on the release of another transmitter to a neuromodulatory action mediated by presynaptic receptors, since other conventional mechanisms are possible. For example, the inhibitory effect of muscimol on the release of Asp from olfactory cortex slices (Table VIII) can be interpreted in terms of classical presynaptic inhibition.[150] Moreover, a neurotransmitter (A) added to the fluid bathing the slices may influence the evoked release of another transmitter (B) indirectly, e.g., by affecting the activity of interneurons controlling the release of (B). Even when

Table VIII

Effect of Various Neurotransmitters or Neuromodulators on Amino Acid Release

Active substance	Amino acid affected	Type of preparation	Type of stimulus	Effect on release and antagonists preventing it		References
				(+), potentiation.	(−), depression	
Dopamine, 5–500 μM[a]	Exogenous GABA	Rat s. nigra, slices	None	(+)22%–47%	Fluphenazine Haloperidol	Reubi et al.[156]
Met-enkephalin 1 μM–1 pM	Exogenous GABA	Rat brain, synaptosomes	K+ 55 mM	(−)15%–75%	Naloxone	Brennan et al.[158]
Adenosine or AMP, 10–1000 μM	Exogenous GABA	Rat cortex, slices	K+ 36 mM	(−) 9%–35%	Theophylline	Hollins and Stone[159]
Taurine, 10–100 mM[b]	Exogenous GABA	Rat cortex, slices	K+ 30 mM	(+)30%–170%	—	Leach[160]
Dopamine, 100 μM[c]	Endogenous Glu	Rat striatum, slices	K+ 50 mM	(−)48%	Haloperidol	Rowlands and Roberts[161]
GABA or muscimol, 100 μM	Exogenous Glu	Rat striatum, slices	K+ 30 mM	(+)55%	Bicuculline, picrotoxin	Mitchell[162]
Isoguvacine or APS. 100 μM	Exogenous Glu	Rat striatum. slices	K+ 30 mM	(+)90%–100%	Bicuculline, picrotoxin	
GABA or muscimol, 2–20 μM[d]	Exogenous D-asp	Rat cerebellum, P2	K+ 15–56 mM	(+)25%–180%	Bicuculline, picrotoxon	Levi and Gallo[64]
GABA, 20 μM	Glu synth. from [3H]Gln	Rat cerebellum, P2	K+ 30 mM	(+)80%	Bicuculline, picrotoxin	
Muscimol, 10 μM	Endogenous Asp	Rat olfact. cortex, slices	electrical[e]	(−)50%–75%	Picrotoxin	Collins[150]
Muscimol, 10 μM	Endogenous Tau	Rat olfact. cortex, slices	electrical[e]	(−)46–86%	Picrotoxin	
GABA or muscimol, 100 μM	Exogenous Gly	10-Day-old rat s. nigra, slices	K+ 50 mM	(−)100%	Picrotoxin	Kerwin and Pycock[163]

[a] Stimulation also with ADTN (4-amino-6,7-dihydroxy-1,2,3,4-tetrahydronaphtalene), dibutyryl cyclic AMP, and amphetamine (the latter effect is probably indirect, through dopamine release[156].

[b] Active at somewhat lower concentrations when release was stimulated by veratridine or ouabain.

[c] Inhibition also with ADTN, apomorphine, and 3,4-dihydroxinomifensine (100 μM). Benztropine potentiated the dopamine effect.

[d] At 20 μM, the following order of potency was observed: GABA ≈ muscimol ≈ APS > P4S > isoguvacine > THIP (G. Levi and V. Gallo. unpublished observations). For abbreviations, see legend for Table VII).

[e] The first values refer to "indirect" electrical stimulation; the second values, which are higher, refer to stimulation of the lateral olfactory tract.

the effect of (A) is maintained in the presence of tetrodotoxin, this and other possibilities are difficult to discard completely, particularly if the depolarizing stimulus causes an indiscriminate depolarization of all the cells in the tissue.

Another complicating factor in the experiments with slices is that the addition of exogenous substances affecting neurotransmitter release may interfere with a complex situation determined by the influence that the same substances, present endogenously in the tissue, have on the same release process being studied. These difficulties can be partly circumvented by appropriate controls (for example, destruction of the interneurons by kainic acid[161]); however, the use of superfused synaptosomes (which lack intercellular connections and presumably have presynaptic receptors free of endogenous ligands) represents at present the method of choice for determining whether the neuromodulatory effect of a given substance on neurotransmitter release occurs by the activation of presynaptic receptors[6] (Chapter 16).

The studies reported in Table VIII are largely self explanatory, but some of them are worth a brief comment. The potentiation by GABA agonists of evoked release of Glu observed in cerebellar synaptosomal preparations[64,157] appears to be mediated by a receptor whose characteristics (in terms of order of potency of different GABA agaonists, see note *e* in Table VIII) are similar to those of GABA autoreceptors (see Table VII) and of the GABA receptors revealed by binding studies.[158] On the other hand, the GABA-induced potentiation of Glu release observed in striatal slices[162] may occur through the activation of a different type of GABA receptor, as suggested by the different order of potency of various GABA agonists (Table VIII) and by the observation that muscimol and GABA were unable to potentiate acidic amino acid release from striatal synaptosomal preparations in conditions in which a maximal effect was observed in the cerebellum.[64]

The cerebellar receptor has been characterized in terms of its ionic requirements. Chloride was apparently the only ion necessary for activity, and the threshold concentration of Cl^- required became lower when the depolarizing stimulus (KGl concentration) was increased.[157]

The Cl^- requirement and the fact that the potentiation by GABA and muscimol of Glu depolarization-induced release could not be attributed to an effect on the nerve ending membrane potential[64] led to the suggestion that an increased Cl^- influx represents the first of a series of otherwise unknown biochemical events leading to increased Glu release.[157]

It is premature to say whether and to what extent the mentioned effects of neurotransmitters (or neuromodulators) on amino acid release observed *in vitro* also take place in the living brain in normal or in pathological conditions. The effective agonist concentrations used in the experiments reported in Table VIII were at times very high, but in other cases they were within the "physiological" range. As noted in the discussion of the data on presynaptic autoreceptors controlling amino acid release (see Section 3.1), the existence *in vivo* of neuromodulatory mechanisms controlling neurotransmitter release would have enormous pharmacoclinical implications.

4. CARRIER-MEDIATED RELEASE

A direct demonstration that release can occur through the membrane carriers working in an outward direction can be obtained only by studying release in conditions in which the carrier is blocked. Experimentally, the problem can be rather easily approached in the case of biogenic amines, since carrier-blocking drugs not transported by the carrier and not themselves causing release are available[6] (Chapter 16). In the case of transmitter amino acids, L-diaminobutyric acid is, so far, the only tool for studying directly whether release is carrier mediated. L-Diaminobutyric acid is a GABA analogue transported by the GABA carrier and presumably binding irreversibly to the carrier, because synaptosomal GABA transport remains greatly depressed when the inhibitor is removed from the medium.[11] As already discussed (Sections 2.8, 2102), on the basis of the evidence obtained with GABA, the depolarization-induced release of amino acids is unlikely to be mediated by the membrane transport systems.

4.1. Spontaneous Release

The term spontaneous release should be restricted to the release occurring in basal conditions of incubation or superfusion and should not be applied to cases in which the basal release rate is increased by modifications in the composition of the bathing fluid (for example, addition of ouabain, amino acids, drugs, etc.). In fact, an increase in the basal rate of release is often caused by the activation of a release mechanism different from the "spontaneous" release.

The rate of "spontaneous" release of amino acids (exogenous or endogenous) observed with brain slices and synaptosomes is extremely variable (being, as a rule, lower with the better preparations) and probably reflects passive leakage from damaged tissue or particles more than an active, carrier-mediated process. This idea is corroborated by the observation that the spontaneous release of GABA from synaptosomes was not inhibited by prior incubation of the suspension with diaminobutyric acid[11] and was scarcely affected by a decrease in temperature.[164] We are aware of only one case in which evidence was provided for a carrier-mediated "spontaneous" efflux of amino acids from brain slices.[165] In these experiments, the slices were preloaded with millimolar concentrations of radioactive amino acids (D-glutamate, α-aminoisobutyrate, lysine, or leucine), and it was shown that the release of these compounds into an amino-acid-free medium was decreased if the slices had also previously accumulated another amino acid sharing the transport system of the radioactive compound under study. It is possible that the carrier-mediated component of release from slices preloaded with high concentrations of amino acids becomes high compared to the passive leakage.

It is generally thought that making the intracellular/extracellular Na^+ gradient more favorable to outward transport should facilitate release through the carrier; for example, the increased amino acid release evoked by ouabain (see, for example, refs. 37,48,65,132,166,167) has been often interpreted as resulting from accelerated outward transport. However, experiments with superfused synaptosomes showed that the ouabain-induced release of GABA was not blocked by the transport inhibitor diaminobutyric acid[11] unless, as we shall see later, extracellular GABA was also present.[7,128] (Incidentally, it has to be noted that small concentrations of GABA may be present extracellularly in experiments with slices or incubated synaptosomes.) Moreover, the sudden establishment of a favorable Na^+ gradient by removal of extracellular Na^+ did not stimulate GABA release from synaptosomes.[45,164] In contrast, catecholamine release was enhanced in Na^+-free media, and this release as well as the release induced by ouabain was inhibited by transport blockers[6] (Chapter 16).

4.2. Homoexchange and Heteroexchange

Amino acid carrier-mediated release, however, can be activated by a process of exchange when an amino acid sharing the same transport system is present extracellularly. In synaptosomal preparations under appropriate experimental conditions, homo- and heteroexchange of putative transmitter amino acids can be shown with concentrations of extracellular amino acids in the range of high-affinity transport systems[45,71,128,129,164,168] (see also refs. 4, 7, and 169 for more complete treatment). A notable exception is provided by taurine, whose release from slices [170] or synaptosomes[71] was slightly or not at all affected even by high extracellular taurine concentrations. Countertransport has also been demonstrated with brain slices; however, the concentrations of extracellular amino acids necessary to detect it were generally higher than with synaptosomal preparations.[35,132,167,170–172]

The specific radioactivity of the GABA released by heteroexchange from superfused synaptosomes prelabeled with [^{3}H]GABA was moderately higher than that found in the synaptosomes,[113] suggesting that the amino acid released originated from the main cytoplasmic pool and not from the small pool with lower specific radioactivity that may be related to the Ca^{2+}-dependent, depolarization-induced release (see Section 2.7.2).

In steady state and standard ionic conditions, it was shown that the amount of GABA entering into and exiting from synaptosomes through the high-affinity transport system approximately equaled each other.[45,164,168,173] (Release via exchange is easily underestimated in experimental conditions not adequate to measure it; this has resulted at times in different conclusions.[174,175])

The stoichiometry of the exchange process could be altered in experimental conditions producing changes in the fluxes of ions into and out of the synaptosomes. In particular, in the presence of ionic fluxes resembling those of physiological depolarization (Na^+ and/or Ca^{2+} influx and K^+ efflux), the amount of GABA released by exchange exceeded by far the amount entering the synaptosomes.[7,128] This extra release, which was superimposed on the carrier-independent release induced by the alteration of the ionic fluxes[128] (as observed in the absence of extracellular GABA), was largely inhibited by pretreatment of the synaptosomes with the carrier blocker diaminobutyric acid and therefore appeared to be mediated by the membrane transport system.[7,128]

Experiments on glutamate exchange, although not as complete as those on GABA, suggested the existence of comparable phenomena.[169]

The observations reported above on the flexibility of GABA homo- or heteroexchange in relation to cationic fluxes led to the hypothesis that carrier-mediated release through the exchange process may represent a means for potentiating the carrier-independent release of neurotransmitter amino acids elicited by physiological depolarization.[128,169] According to this hypothesis, the GABA initially released by the depolarizing stimulus would activate the exchange process and thus cause a net extra release of the amino acid (see above). Alternatively, it is also conceivable that minute amounts of the amino acid are always present extracellularly in view of the huge concentration of GABA existing within the GABAergic nerve terminals.[154] In this case, the basal 1:1 stoichiometry of GABA homoexchange present in resting conditions would switch to a stoichiometry producing net outward transport (efflux > influx) during depolarization.[7] Inactivation of the released GABA would occur through the same exchange system working with an opposite stoichiometry when the ionic fluxes are reversed during repolarization.[7,128]

The possibility of increasing the release of neuroactive amino acids by heteroexchange may have interesting pharmacological possibilities. In fact, the administration of compounds capable of exchanging with neuroactive amino acids should potentiate their action by a dual mechanism: increased release and inhibition of reuptake. In accord with this expectation, it has recently been shown that the increased release of GABA elicited by a group of analogues sharing the GABA transport system (including nipecotic acid and 3-aminocyclohexane carboxylic acid) was probably responsible for the enhanced depolarization produced by the exposure of olfactory cortex slices to these compounds.[176] On the other hand, it can be predicted that the pharmacological action of amino acid releasers of this type may be of short duration. In fact, the compounds capable of releasing by exchange will themselves be accumulated and will eventually replace the endogenous transmitter stores, thus inhibiting competitively the carrier-mediated release of the transmitter and behaving as false transmitters.

5. EFFECTS OF DRUGS ON AMINO ACID RELEASE

The search for drugs specifically affecting amino acid release is justified by the success obtained in the field of biogenic amines and by the clinical usefulness of biogenic amine releasers or depleters. The fact that the achievements in the amino acid field have been very limited most likely reflects the predominant localization of amino acids in a "free" cytoplasmic pool within the nerve terminals. Thus, releasers of the type of amphetamine or reserpine[6] (Chapter 16) are difficult to conceive in the field of amino acid transmitters. Other difficulties arise from the ubiquitous distribution of neurotransmitter amino acids within the CNS and from their participation in normal brain metabolism. A brief account of the experimental observations on this topic is provided below.

5.1. Barbiturates

This classical group of drugs has been the most studied because for a long time it has been thought that their central depressant effects might be related to a potentiation of the GABAergic inhibitory tonus. This could obviously be achieved in various ways, and we are considering here only the effects that barbiturates have an amino acid release.

The most studied of these drugs, pentobarbital, was shown to inhibit spontaneous GABA release[177] and GABA release elicited by high $[K^+]$, both in slices[29,177-179] and in synaptosomes,[180] at concentrations ranging from 0.1 to 3 mM in slices and 0.02 to 0.2 mM in synaptosomes. In one case, the inhibitory effect of 0.1 mM pentobarbital on Ca^{2+}-dependent, K^+-evoked GABA release was reported to be negligible.[181] In a couple of studies, pentobarbital was shown to stimulate the evoked release of GABA and not that of acid amino acids (in olfactory cortex slices stimulated electrically through the lateral olfactory tract[150] and in cortex slices stimulated electrically[182]). The effect of the drug on the depolarization-evoked release was not specific for GABA: Asp and Tau release from olfactory cortex slices on electrical stimulation of the lateral olfactory tract,[150] Glu and/or Asp release induced by high $[K^+]$ from cortical[178,179,183] or midbrain slices,[29] and K^+-evoked norepinephrine release from synaptosomes[180] were also found to be reduced. Amobarbital [177] and thiopental[179] appeared to behave similarly to pentobarbital, but phenobarbital did not affect the spontaneous or the stimulated release of GABA or Glu either in synaptosomes[184] or in slices[185] at 0.5–1 mM concentrations.

Other barbiturates, including some convulsants, were studied in some cases,[179,183] but no convincing correlation between pharmacological activity and effects on amino acid release could be demonstrated. The ability of some barbiturates to inhibit the depolarization-induced release of transmitters is likely to be a function of the ability of these compounds to inhibit the influx of Ca^{2+} induced by depolarizing stimuli. In fact, pentobarbital, but not phenobarbital, inhibited the Ca^{2+} accumulation induced by high $[K^+]$ in synaptosomal preparations.[142,186] A regional specificity of the effect of pentobarbital on Ca^{2+} uptake has been reported.[144] Interestingly, the inhibitory effect of pentobarbital on evoked GABA release was not present when the release of the amino acid was stimulated with the Ca^{2+} ionophore A23187, confirming that the drug does not affect the viability of the release process subsequent to the entry of Ca^{2+} but rather decreases the ability of depolarization to activate the Ca^{2+} channels.[180]

5.2. Baclofen [β-(p-Chlorophenyl) GABA]

Baclofen is a neuronal depressant that has been recently shown to interact with a distinct type of GABA receptors (called $GABA_B$ receptors) present in the peripheral[187] and central nervous systems.[188] The interaction of GABA and of the (−) form of baclofen with these receptors causes an inhibition of the release of several biogenic amines (norepinephrine, dopamine, serotonin) elicited by moderately high K^+ concentrations (25–35 mM). The $GABA_B$ receptor differs from the classical GABA receptor in that muscimol is less effective on it than GABA and in that the action of GABA and baclofen is not antagonized by picrotoxin or bicuculline.[187,188] Its localization has been suggested to be presynaptic, but definite proof is still lacking.

Since the GABA$_B$ receptor might be a general regulator of neurotransmitter release, it is worth briefly analyzing the studies (generally preceding the characterization of the baclofen-preferring receptor) dealing with the effect of the drug on the release of transmitter amino acids. The picture emerging from these studies is rather confusing. Baclofen appears to have a depressing effect on the depolarization-induced release of acidic amino acids. In a study with cortex slices stimulated electrically, the drug (4 μM) had a selective effect on the release of Glu and Asp previously synthesized from [^{14}C]glucose, although it indiscriminately depressed the release of exogenously loaded amino acids, including GABA.[189] In another study, however, 4 μM baclofen inhibited the release of [^{3}H]D-aspartate from cortex slices depolarized with 40 mM K$^+$ or 100 μM protoveratrine but did not affect the release of exogenous GABA.[190] The effect was specific for the ($-$) isomer but was not shared by GABA (10–100 μM), and the author concluded that a GABA receptor was unlikely to be involved. With striatal slices, baclofen was reported not to influence [^{3}H]glutamate release evoked by 30 mM K$^+$.[162] The release of GABA induced by depolarizing stimuli does not seem to be affected by the drug.[153,190] At relatively high concentrations (50 μM or 0.3–1 mM), baclofen stimulated the "spontaneous" release of GABA from synaptosomes[191] and pallidal slices.[192] In the latter experiments, the ($+$) rather than the ($-$) form was active, and since the drug was shown also to inhibit GABA transport, the releasing effect is likely to reflect heteroexchange and not a receptor-mediated process.

The existence of different GABA receptors (see Section 3.2) with different agonist sensitivities involved in the control of neurotransmitter release may explain some of the apparent contradictions found in the literature on the modulatory effect of GABA on the release of other transmitters (see, for example, the discussion in ref. 193). Experiments on the effect of GABA on norepinephrine release evoked by different K$^+$ concentrations[188] suggest the interesting possibility that the activation of GABA$_B$ receptors prevails over that of other (presynaptic?) receptors with relatively intense depolarizing stimuli, whereas the opposite could be true with weaker depolarizing stimuli.

5.3. Other Drugs

A number of anticonvulsant, convulsant, or CNS depressant drugs were tested under various experimental conditions for their ability to interfere with amino acid (mainly GABA) release processes. For the convenience of the reader, a list of the effects observed is provided below. A discussion of all of these effects would be too speculative and would take too much space.

5.3.1. Anticonvulsants

Dipropylacetate had no effect (concentrations up to 1 mM) on the spontaneous release of exogenous GABA in cortical synaptosomes.[184]

Phenytoin (up to 1 mM) had no effect on the spontaneous release of exogenous GABA in synaptosomes[153,184] and in cortical slices.[185] The high-[K$^+$]-evoked release of GABA from slices was increased by 0.2 mM phenytoin,[185] but no effect on Ca^{2+}-dependent GABA release could be detected in synaptosomes.[153]

Trimethadione, phenacetamide, and ethosuccimide (0.5 mM) had no effect on the spontaneous release of exogenous GABA in cortical slices but potentiated the release evoked by 30 mM K$^+$.[185]

The GABA-transaminase inhibitors *γ-acetylenic and γ-vinyl-GABA* (administered *in vivo* and tested before the brain concentrations of GABA started to increase or added to the superfusion media *in vitro*) stimulated the "spontaneous" release of endogenous and exogenous GABA from superfused synaptosomes. The effect was partly counteracted by tetrodotoxin, was blocked by the Ca^{2+} antagonist verapamil, was additive to that of veratridine, and was tentatively interpreted as occurring through a depolarizing action of the drugs.[194]

5.3.2. Convulsants

Picrotoxin (0.1 mM or higher) was found to inhibit the high-[K$^+$]-induced Ca^{2+}-dependent release of exogenous GABA in synaptosomes[153] and the electrically stimulated release of exogenous GABA in rat cortex slices[195] and in the frog hemisected spinal cord.[132]

Bicuculline (0.05 mM or higher) enhanced the electrically induced release of exogenous GABA in cortex slices[195] and in hemisected frog spinal cord.[132] The Ca^{2+}-dependent K$^+$-evoked synaptosomal release of exogenous GABA was reported to be enhanced*152* or unaffected.[15] If the effect of bicuculline in slices and spinal cord preparations is interpreted as an antagonism with GABA at the level of presynaptic autoreceptors (see Section 3.1), the reason picrotoxin has an opposite effect remains obscure.

Strychnine was reported to inhibit the high-[K$^+$]-evoked[185] and the electrically induced[195] release of exogenous GABA from cortex slices at concentrations of 200 μM or 50–100 μM, respectively. However, 1 mM strychnine did not affect GABA release from the rostrally stimulated hemisected frog spinal cord.[132] A depression of "spontaneous" GABA release has been reported.[132,185]

Pentylenetetrazole (1–100 μM) depressed the spontaneous release of exogenous GABA in cortical slices but had no effect on the release induced by electrical stimulation.[195]

Penicillin (34 mM) abolished the Ca^{2+}-dependent, K$^+$-evoked release of endogenous GABA in cortex slices without affecting that of other amino acids.[14] In another report, benzylpenicillin (100 μM) was shown to enhance the Ca^{2+}-independent, K$^+$-evoked release of exogenous GABA from synaptosomes.[153]

The Ca^{2+} antagonist *ruthenium red*, injected intracisternally, inhibited the Ca^{2+}-dependent release of exogenous GABA from superfused synaptosomes[196] prepared from treated animals.

5.3.3. CNS Depressants

Diazepam (1 μM) was shown to stimulate the release of exogenous GABA induced by 15 mM K^+ in small cortex slices and to inhibit it at a higher concentration (100 μM).[197] At this concentration, [³H]glutamate release from striatal slices was also inhibited.[198] With superfused synaptosomal preparations, the drug was shown to stimulate the Ca^{2+}-independent, K^+-evoked GABA release and to inhibit the Ca^{2+}-dependent release with an IC_{50} of 30 μM.[153] The spontaneous release of [³H]GABA from superfused synaptosomes was not affected by another benzodiazepine, carbamazepine.[184]

Diazepam at a concentration of 0.7 μM enhanced the potentiating effect of muscimol (but not that of 3-aminopropanesulfonic acid) on depolarization-induced [³H]glutamate release from striatal slices[198] (see Section 3.2). This effect was interpreted as a facilitation by diazepam of the muscimol effect at the level of the GABA receptor regulating Glu release but might be also attributed to the stimulation of GABA release [effected by the drug at a similar concentration (0.1 μM[197])].

Chlorpromazine, imipramine, and *haloperidol* were shown to inhibit the Ca^{2+}-dependent, K^+-evoked release of [³H]GABA from synaptosomes with IC_{50} values of 4, 8, and 80 μM, respectively.[153]

6. EFFECTS OF NEUROTOXINS ON AMINO ACID RELEASE

Several neurotoxins have recently been shown to have a presynaptic site of action and to affect neurotransmitter release. Their mechanisms of action, often complex and not fully understood, are treated in detail in two recent reviews[8,199] and in the proceedings of a symposium held in 1981.[200] For this reason, and also because it would be impossible to deal with this subject without analyzing the action of neurotoxins on the release of transmitters other than the amino acids (which in some cases has been studied in much more detail), we shall limit ourselves to providing a list of the effects observed on amino acid release and advise the reader interested in more details to consult the abovementioned monographs.

Tetanus toxin, administered *in vivo* and added to the incubation media, caused a large decrease (70%) in the electrically evoked release of endogenous Gly from rat spinal synaptosomes and smaller reductions in GABA and Asp release.[201] After *in vivo* administration, the high-[K^+]-induced release of exogenous GABA from nigral and striatal slices was 30–35% inhibited.[26] The evoked release of exogenous GABA was also reduced in hippocampal slices pretreated for 3 h with a very high dose of toxin.[25]

The snake neurotoxin β-*bungarotoxin* was shown to release exogenous GABA[202,203] and Glu[203] and endogenous Glu, GABA, and Asp[204] from synaptosomes. The releasing activity was Ca^{2+} dependent.

The scorpion toxin *tityustoxin* was also capable of releasing endogenous GABA and Glu (to a lesser extent, also Asp), and the effect was partially Ca^{2+} dependent.[205]

α-Latrotoxin, isolated from the black widow spider venom, enhanced the release of exogenous GABA from synaptosomes[206,207] and from cortex slices.[208] The effect was decreased in Ca^{2+}-free media and abolished in Na^+-free, Ca^{2+}-free media.[206]

The tremorgenic mycotoxins *penitrem A* and *verruculogen* administered *in vivo* caused an increased release of endogenous amino acids from cortical synaptosomes.[209] GABA was affected more than Glu or Asp by penitrem A, whereas verruculogen greatly increased Asp and Glu release without influencing the release of GABA. Neither toxin had any effect on amino acid release from spinal synaptosomes.

Avermectin B_1a, a macrocyclic lactone antiparasitic agent, enhanced [³H]GABA but not [³H]glutamate released from superfused synaptosomes. Glycine and Asp were also unaffected, and the effect on [³H]GABA was maintained in the absence of Ca^{2+}.[210]

ACKNOWLEDGMENTS. I thank Mrs. M. T. Ciotti for helping with the bibliography and Mrs. A. Sebastiano for typing the manuscript. The original work reported was supported in part by Research Grant 058.80 DI of the North Atlantic Treaty Organization and by a Grant of the Italian National Rsearch Council in the framework of the Italy–United Kingdom bilateral projects.

REFERENCES

1. Curtis, D. R., and Johnston, G. A. R., 1974, *Ergeb. Physiol.* **69**:97–188.
2. Tapia, R., 1975, *Handbook of Psychopharmacology,* Volume 4 (L. L. Iversen, S. D. Iversen, and S. H. Snyder, eds.), Plenum Press, New York, PP. 1–58.
3. Levi, G., Battistin, L., and Lajtha, A. (eds.), 1976, *Transport Phenomena in the Nervous System,* Plenum Press, New York.
4. Levi, G., and Raiteri, M., 1976, *Int. Rev. Neurobiol.* **19**:51–74.
5. Llinas, R . R., and Heuser, J. E., 1977, *Neurosci. Res. Prog. Bull.* **15**:557–687.
6. Raiteri, M., and Levi, G., 1978, *Reviews of Neuroscience,* Volume 3 (S. Ehrenpreis and I. Kopin, eds.), Raven Press, New York, pp. 77–130.
7. Levi, G., Banay-Schwartz, M., and Raiteri, M., 1978, *Amino Acids as Chemical Transmitters* (F. Fonnum, ed.), Plenum Press, New York, pp. 327–350.
8. Kelly, R. B., Deutsch, J. W., Carlson, S. S., and Wagner, J. A., 1979, *Annu. Rev. Neurosci.* **2**:399–446.
9. Fonnum, F. (ed.), 1978, *Amino Acids as Chemical Transmitters,* Plenum Press, New York.
10. Bradford, H. F. (ed.), 1982, *Neurotransmitter Interaction and Compartmentation,* Plenum Press, New York.
11. Levi, G., Gallo, V., and Raiteri, M., 1980, *Neurochem Res.* **5**:281–295.
12. Orrego, F., Jankelevich, J., Ceruti, L., and Ferrera, E., 1974, *Nature* **251**:55–57.
13. Lopez-Colomé, A. M., Tapia, R., Salceda, R., and Pasantes-Morales, H., 1978, *Neuroscience* **3**:1069–1074.
14. Cutler, R. W. P., and Young, J., 1979, *Brain Res.* **170**:157–163.
15. Szerb, J. C., 1979, *J. Neurochem.* **32**:1565–1573.
16. Gauchy, C. M., Iversen, L. L., and Jessell, T. M., 1977, *Brain Res.* **138**:374–379.
17. Johnston, G. A. R., 1977, *Brain Res.* **121**:179–181.
18. Mulder, A. H., and Snyder, S. H., 1974, *Brain Res.* **76**:297–308.
19. Srinivasan, V., Neal, M. J., and Mitchell, J. F., 1969, *J. Neurochem.* **16**:1235–1244.
20. Vargas, O., Doria De Lorenzo, M. C., Saldate, M. C., and Orrego, F., 1977, *J. Neurochem.* **28**:165–170.

21. Neal, M. J., and Bowery, N. G., 1979, *Brain Res.* **167**:337–343.
22. Collins, G. G. S., 1979, *J. Physiol. (Lond.)* **291**:51–60.
23. Arbilla, S., Kamal, L., and Langer, S. Z., 1979, *Eur. J. Pharmacol.* **57**:211–217.
24. Davidoff, R. A., and Adair, R., 1976, *Brain Res.* **118**:403–415.
25. Collingridge, G. L., Thompson, P. A., Davis, J., and Mellanby, J., 1981, *J. Neurochem.* **37**:1039–1041.
26. Collingridge, G. L., Collins, G. G. S., Davis, J., James T. A., Neal, M. J., and Tongroach, P., 1980, *J. Neurochem.* **34**:540–547.
27. Reubi, J. C., and Cuénod, M., 1979, *Brain Res.* **176**:185–188.
28. Reubi, J. C., van den Berg, C., and Cuénod, M., 1978, *Neurosci. Lett.* **10**:171–174.
29. Waller, M. B., and Richter, J. A., 1980, *Biochem. Pharmacol.* **29**:2189–2198.
30. Čanžek, V., and Reubi, J. C., 1980, *Exp. Brain Res.* **38**:437–441.
31. Flint, R. S., Rea, M. A., and McBride, W. J., 1981, *J. Neurochem.* **37**:1425–1430.
32. Foster, A. C., and Roberts, P. J., 1980, *J. Neurochem.* **35**:517–519.
33. Okamoto, K., and Namina, M., 1978, *J. Neurochem.* **31**:1393–1402.
34. Adair, R., and Davidoff, R. A., 1977, *J. Neurochem.* **29**:213–220.
35. Hammerstad, J. P., Cawthon, M. L., and Lytle, C. R., 1979, *J. Neurochem.* **32**:195–202.
36. Minchin, M. C. W., 1980, *Biochem. J.* **190**:333–339.
37. Hammerstad, J. P., and Cutler, R. W. P., 1972, *Brain Res.* **47**:401–413.
38. Valdés, F., and Orrego, F., 1978, *Brain Res.* **141**:357–363.
39. Potashner, S. J., 1978, *J. Neurochem.* **31**:187–195.
40. Orrego, F., and Miranda, R., 1976, *J. Neurochem.* **26**:1033–1038.
41. Katz, R. I., Chase, T. N., and Kopin, I. J., 1969, *J. Neurochem.* **16**:961–967.
42. Okada, Y., and Hassler, H., 1973, *Brain Res.* **49**:214–217.
43. Hammerstad, J. P., Murray, J. E., and Cutler, R. W. P., 1971, *Brain Res.* **35**:357–367.
44. Levi, G., Gallo, V., Ciotti, T., and Raiteri, M., 1979, *J. Neurochem.* **33**:1043–1053.
45. Raiteri, M., Federico, R., Coletti, A., and Levi, G., 1975, *J. Neurochem.* **24**:1243–1250.
46. Redburn, D. A., Shelton, D., and Cotman, C. W., 1976, *J. Neurochem.* **26**:297–303.
47. Cotman, C. W., Haycock, J. W., and White, W. F., 1976, *J. Physiol. (Lond.)* **254**:475–505.
48. De Belleroche, J. S., and Bradford, H. F., 1972, *J. Neurochem.* **19**:585–602.
49. Placheta, P., Singer, E., Sieghart, W., and Karobath, M., 1979, *Neurochem. Res.* **4**:703–712.
50. Sieghart, W., and Singer, E., 1979, *Brain Res.* **170**:203–208.
51. Haycock, J. W., Levy, W. B., Denner, L. A., and Cotman, C. W., 1978, *J. Neurochem.* **30**:1113–1125.
52. Sellström, Å., and Hamberger, A., 1977, *Brain Res.* **119**:189–198.
53. Osborne, R. H., Bradford, H. F., and Jones, D. G., 1973, *J. Neurochem.* **21**:407–419.
54. Bradford, H. F., Bennett, G. W., and Thomas, A. J., 1973, *J. Neurochem.* **21**:495–505.
55. Arnfred, T., and Hertz, L., 1971, *J. Neurochem.* **18**:259–265.
56. Davies, L. P., Johnston, G. A. R., and Stephanson, A. L., 1975, *J. Neurochem.* **25**:387–392.
57. Vargas, O., Doria De Lorzeno, M. C., and Orrego, F., 1977, *Neuroscience* **2**:383–390.
58. Reubi, J. C., Toggenburger, G., and Cuénod, M., 1980, *J. Neurochem.* **35**:1015–1017.
59. Hamberger, A., Chiang, G. H., Nylén, E. S., Scheff, S. W., and Cotman, C. W., 1979, *Brain Res.* **168**:513–530.
60. Cotman, C. W., and Hamberger, A., 1978, *Amino Acids as Chemical Transmitters* (F. Fonnum, ed.), Plenum Press, New York, pp. 379–412.
61. Levi, G., Gordon, R. D., Gallo, V., Wilkin, G. P., and Balázs, R., 1982, *Brain Res.* (in press).
62. Sandoval, M. E., Horch, P., and Cotman, C. W., *Brain Res.* **142**:285–299.
63. Sandoval, M. E., and Cotman, C. W., 1978, *Neuroscience* **3**:199–206.
64. Levi, G., and Gallo, V., 1981, *J. Neurochem.* **37**:22–31.
65. O'Fallon, J. W., Brosemer, R. W., and Harding, J. W., 1981, *J. Neurochem.* **36**:369–378.
66. James T. A., and Starr, M. S., 1979, *Eur. J. Pharmacol.* **57**:115–125.
67. Hopkin, J., and Neal, M. J., 1971, *Br. J. Pharmacol.* **42**:215–223.
68. McIlwain, H., and Snyder, S. H., 1970, *J. Neurochem.* **17**:521–530.
69. Levi, G., Bernardi, G., Cherubini, E., Gallo, V., Marciani, M. G., and Stanzione, P., 1982, *Brain Res.* (in press).
70. Kaczmarek, L. K., and Davison, A. N., 1972, *J. Neurochem.* **19**:2355–2362.

71. Sieghart, W., and Heckl, K., 1976, *Brain Res.* **116:**538–543.
72. Orrego, F., and Doria De Lorenzo, M. C., 1980, *Neurochem. Res.* **5:**523–536.
73. Ohta, M., Narahashi, T., and Keeler, R. F., 1973, *J. Pharmacol. Exp. Ther.* **184:**143–154.
74. Blaustein, M. P., and Goldring, J. M., 1975, *J. Physiol. (Lond.)* **247:**589–615.
75. Li, P. P., and White, T. D., 1977, *J. Neurochem.* **28:**967–975.
76. Blaustein, M. P., 1975, *J. Physiol. (Lond.)* **247:**617–655.
77. Goddard, G. A., and Robinson, J. D., 1976, *Brain Res.* **110:**331–350.
78. Benjamin, A. M., and Quastel, J. H., 1972, *Biochem. J.* **128:**631–646.
79. Minchin, M. C. W., 1975, *J. Neurochem.* **24:**571–577.
80. Jaffé, E. H., and Cuello, A. C., 1981, *J. Neurochem.* **37:**1457–1466.
81. Nadler, J. V., White, W. F., Vaca, K. W., Redburn, D. A., and Cotman, C. W., 1977, *J. Neurochem.* **29:**279–290.
82. Orrego, F., Miranda, R., and Saldate, C., 1976, *Neuroscience* **1:**325–332.
83. Levi, G., and Raiteri, M., 1973, *Life Sci.* **12**(1):81–88.
84. Garthwaite, J., Woodhams, P. L., Collins, M. J., and Balázs, R., 1979, *Brain Res.* **173:**373–377.
85. Pearce, B. R., Currie, D. N., Beale, R., and Dutton, G. R., 1981, *Brain Res.* **206:**485–489.
86. Levi, G., Gallo, V., Ciotti, M. T., Coletti, A., and Aloisi, F., 1982, Transactions Amer. Soc. Neurochem. **13:**250.
87. Kishimoto, H., Simon, J. R., and Aprison, M. H., 1981, *J. Neurochem.* **37:**1015–1024.
88. Edwardson, J. A., Bennett, G. W., and Bradford, H. F., 1972, *Nature* **240:**554–556.
89. Osborne, R. H., and Bradford, H. F., 1975, *J. Neurochem.* **25:**31–41.
90. Wiklund, L., Toggenburger, G., and Cuénod, M., 1982, *Science* (in press).
91. Hardy, J. A., De Belleroche, J. S., Border, D., and Bradford, H. F., 1980, *J. Neurochem.* **34:**1130–1139.
92. Korpi, E. R., Kontro, P., Nieminen, K., Marnela, K. M., and Oja, S. S., 1981, *Life Sci.* **29:**811–816.
93. Charles, A. K., and Chang, Y. F., 1981, *J. Neurochem.* **36:**1127–1136.
94. Nomura, Y., Okuma, Y., Segawa, T., Schmidt-Glenewinkel, T., and Giacobini, E., 1979, *J. Neurochem.* **33:**803–805.
95. Weitsch-Dick, F., Jessel, T. M., and Kelly, J. S., 1978, *J. Neurochem.* **30:**799–806.
96. Bowery, N. G., and Neal, M. J., 1978, *Br. J. Pharmacol.* **62:**431P.
97. Haycock, J. W., and Meligeni, J. A., 1978, *Life Sci.* **21:**1837–1844.
98. Nadler, J. V., White, W. F., Vaca, K. W., and Cotman, C. W., 1977, *Brain Res.* **131:**241–258.
99. Bondy, S. C., Burks, J. S., and Harrington, M. E., 1979, *Arch. Neurol.* **36:**540–543.
100. Redburn, D. A., Broome, D., Ferkany, J., and Enna, S. J., 1978, *Brain Res.* **152:**511–519.
101. Davies, L. P., and Johnston, G. A. R., 1976, *J. Neurochem.* **26:**1007–1014.
102. Snodgrass, S. R., White, W. F., Biales, B., and Dichter, M., 1980, *Brain Res.* **190:**123–138.
103. Haycock, J. W., White, W. F., McGaugh, J. L., and Cotman, C. W., 1977, *Exp. Neurol.* **57:**873–882.
104. Roberts, P. J., 1974, *Brain Res.* **74:**327–332.
105. Bowery, N. G., Brown, D. A., and Marsh, S., 1979, *J. Physiol. (Lond.)* **293:**75–101.
106. Weinreich, D., and Hammerschlag, R., 1975, *Brain Res.* **84:**137–142.
107. Minchin, M. C. W., and Nordmann, J. J., 1975, *Brain Res.* **90:**75–84.
108. Schon, F., and Kelly, J. S., 1975, *Brain Res.* **86:**243–257.
109. De Belleroche, J. S., and Bradford, H. F., 1973, *J. Neurochem.* **21:**441–451.
110. De Belleroche, J. S., and Bradford, H. F., 1977, *J. Neurochem.* **29:**335–343.
111. Raiteri, M., and Levi, G., 1973, *Nature (New Biol.)* **243:**180–183.
112. Raiteri, M., and Levi, G., 1973, *Nature (New Biol.)* **245:**89–91.
113. Levi, G., Banay-Schwartz, M., and Raiteri, M., 1981, *Neurochem. Res.* **6:**275–285.
114. Raiteri, M., Cerrito, F., Cervoni, A. M., and Levi, G., 1979, *J. Pharmacol. Exp. Ther.* **208:**195–201.
115. Cotman, C. W., and Haycock, J. W., 1976, *Br. J Pharmacol.* **58:**569–572.
116. Minchin, M. C. W., 1977, *Exp. Brain Res.* **29:**215–226.
117. Hamberger, A., Chiang, G. H., Sandoval, E., and Cotman, C. W., 1979, *Brain Res.* **168:**531–541.

118. Hamberger, A., Chiang, G., Nylén, E. S., Scheff, S. W., and Cotman, C. W., 1978, *Brain Res.* **143**:549–555.
119. Tapia, R., and Gonzales, R. M., 1978, *Neurosci. Lett.* **10**:165–169.
120. Potashner, S. J., 1978, *J. Neurochem.* **31**:177–186.
121. Levy, W. B., Haycock, J. W., and Cotman, C. W., 1976, *Brain Res.* **115**:243–256.
122. Szerb, J. C., Ross, T. E., and Gurevich, E., 1981, *J. Neurochem.* **37**:1186–1192.
123. Bradford, H. F., Ward, H. K., and Thomas, A. J., 1978, *J. Neurochem.* **30**:1453–1459.
124. Shank, R. P., and Aprison, M. H., 1977, *J. Neurochem.* **28**:1189–1196.
125. Ryan, L. D., and Roskoski, R., Jr., 1975, *Nature* **258**:254–256.
126. Haycock, J. W., Levy, W. B., and Cotman, C. W., 1978, *Brain Res.* **155**:192–195.
127. Levi, G., Rusca, G., and Raiteri, M., 1976, *Neurochem. Res.* **1**:581–590.
128. Levi, G., and Raiteri, M., 1978, *Proc. Natl. Acad. Sci. U.S.A.* **75**:2981–2985.
129. Sandoval, M. E., 1980, *J. Neurochem.* **35**:915–921.
130. Miledi, R., 1973, *Proc. R. Soc. Lond. [Biol.]* **183**:421–425.
131. Raiteri, M., Levi, G., and Federico, R., 1975, *Pharmacol. Res. Commun.* **7**:181–187.
132. Collins, G. G. S., 1974, *Brain Res.* **66**:121–137.
133. Bradford, H. F., and Richards, C. D., 1976, *Brain Res.* **105**:168–172.
134. Malthe-Sörenssen, D., Skrede, K. K., and Fonnum, F., 1979, *Neuroscience* **4**:1255–1263.
135. Spencer, H. J., Tominez, G., and Halpern, B., 1981, *Brain Res.* **212**:194–197.
136. Wieraszko, A., and Lynch, G., 1979, *Brain Res.* **160**:372–376.
137. Malthe-Sörenssen, D., Skrede, K. K., and Fonnum, F., 1980, *Neuroscience* **5**:127–133.
138. Placheta, P., Singer, E., Schönbeck, G., Heckl, K., and Karobath, M., 1979, *Neuropharmacology* **18**:399–402.
139. Farb, T. H., Berg, D. K., and Fischbach, G. D., 1979, *J. Cell. Biol.* **80**:651–661.
140. Balázs, R., Regan, C. H., Gordon, R. D., Annunziata, P., Kingsbuty, A. E., and Meier, E., 1982, *Neurotransmitter Interaction and Compartmentation* (H. F. Bradford, ed.), Plenum Press, New York **48**:515–534.
141. Creveling, C. R., McNeal, E. T., McCulloch, D. H., and Daly, J. W., 1980, *J. Neurochem.* **35**:922–932.
142. Blaustein, M. P., and Ector, A. C., 1975, *Mol. Pharmacol.* **11**:369–378.
143. Chandler, D. E., and Williams, J. A., 1977, *Nature* **268**:659–660.
144. Elrod, S. V., and Leslie, S. W., 1980, *J. Pharmacol. Exp. Ther.* **212**:131–136.
145. Levy, W. B., Haycock, J. W., and Cotman, C. W., 1974, *Mol. Pharmacol.* **10**:438–449.
146. Harris, R. A., Schmidt, J., Hitzemann, B. A., and Hitzemann, R. J., 1981, *Science* **212**:1290–1291.
147. De Lorenzo, R. J., 1982, *Neutrotransmitter Interaction and Compartmentation* (F. Fonnum, ed.), Plenum Press, New York **48**:101–120.
148. McBean, G. J., and Roberts, P. J., 1981, *Nature* **291**:593–594.
149. Mitchell, P. R., and Martin, I. L., 1978, *Nature* **274**:904–905.
150. Collins, G. G. S., 1980, *Brain Res.* **190**:517–528.
151. Brennan, M. J. W., Cantrill, R. C., Oldfield, M., and Krogsgaard-Larsen, P., 1981, *Mol. Pharmacol.* **19**:27–30.
152. Snodgrass, S. R., 1978, *Nature* **273**:392–394.
153. Olsen, R. W., Ticku, M. K., Van Ness, P. C., and Greenlee, D., 1978, *Brain Res.* **139**:277–294.
154. Fonnum, F., Grofovà, I., Rinvik, E., Storm-Mathisen, J., and Walberg, F., 1974, *Brain Res.* **71**:77–92.
155. Kupferman, I., 1979, *Annu. Rev. Neurosci.* **2**:447–465.
156. Reubi, J. C., Iversen, L. L., and Jessell, T. M., 1977, *Nature* **268**:652–654.
157. Levi, G., and Gallo, V., 1982, *Neurotransmitter Interaction and Compartmentation* (H. F. Bradford, ed.), Plenum Press, New York **48**:53–78.
158. Brennan, M. J. W., Cantrill, R. C., and Wylie, B. A., 1980, *Life Sci.* **27**:1097–1101.
159. Hollins, C., and Stone, T. W., 1980, *Br. J. Pharmacol.* **69**:107–112.
160. Leach, M. J., 1979, *J. Pharm. Pharmacol.* **31**:533–535.
161. Rowlands, G. J., and Roberts, P. J., 1980, *Eur. J. Pharmacol.* **62**:241–242.
162. Mitchell, R., 1980, *Eur. J. Pharmacol.* **67**:119–122.

163. Kerwin, R., and Pycock, C., 1980, *Eur. J. Pharmacol.* **64:**169–172.
164. Simon, J. R., Martin, D. L., and Kroll, M., 1974, *J. Neurochem.* **23:**981–991.
165. Levi, G., Blasberg, R., and Lajtha, A., 1966, *Arch. Biochem. Biophys.* **114:**339–351.
166. Cutler, R. W. P., Murray, J. E., and Hammerstad, J. P., 1972, *J. Neurochem.* **19:**539–542.
167. Cutler, R. W. P., 1976, *Transport Phenomena in the Nervous System* (G. Levi, L. Battistin, and A. Lajtha, eds.), Plenum Press, New York, pp. 435–446.
168. Sellström, Å., Venema, R., and Henn, F., 1976, *Nature* **264:**652–653.
169. Levi, G., and Raiteri, M., 1980, *Nerve Cells, Transmitters and Behaviour* (R. Levi-Montalcini, ed.), Elsevier, Amsterdam, pp. 217–232.
170. Crnic, D. M., Hammerstad, J. P., and Cutler, R. W. P., 1973, *J. Neurochem.* **20:**203–209.
171. Cutler, R. W. P., Hammerstad, J. P., Cornick, L. R., and Murray, J. E., 1971, *Brain Res.* **35:**337–355.
172. Benjamin, A. M., and Quastel, J. H., 1976, *J. Neurochem.* **26:**431–441.
173. Storm-Mathisen, J., Fonnum, F., and Malthe-Sörenssen, D., 1976, *GABA in Nervous System Function* (E. Roberts, T. N. Chase, and D. B. Tower, eds.), Raven Press, New York, pp. 387–394.
174. Levi, G., Bertollini, A., Chen, J., and Raiteri, M., 1974, *J. Pharmacol. Exp. Ther.* **188:**429–438.
175. Pastuszko, A., Wilson, D. F., and Ericinska, M., 1981, *Proc. Natl. Acad. Sci. U.S.A.* **78:**1242–1244.
176. Brown, D. A., Collins, G. G. S., and Galvan, M., 1980, *Br. J. Pharmacol.* **68:**251–262.
177. Cutler, R. W. P., Markowitz, D., and Dudzinski, D. S., 1974, *Brain. Res.* **81:**189–197.
178. Cutler, R. W. P., and Dudzinski, D. S., 1975, *Brain Res.* **88:**415–423.
179. Cutler, R. W. P., and Young, J., 1979, *Neurochem. Res.* **4:**319–329.
180. Haycock, J. W., Levy, W. B., and Cotman, C. W., 1977, *Biochem. Pharmacol.* **26:**159–161.
181. Olsen, R. W., Lamar, E. E., and Bayless, J. D., 1977, *J. Neurochem.* **28:**299–305.
182. Cutler, R. W. P., and Dudzinski, D. S., 1974, *Brain Res.* **67:**546–548.
183. Willow, M., Bornstein, J. C., and Johnston, G. A. R., 1980, *Neurosci. Lett.* **18:**185–190.
184. Abdul-Ghani, A. S., Coutinho-Netto, J., Druse, D., and Bradford, H. F., 1981, *Biochem. Pharmacol.* **30:**363–368.
185. Tappaz, M., and Pacheco, H., 1973, *J. Pharmacol. (Paris)* **4:**433–452.
186. Ondrusek, M. G., Belknap, J. K., and Leslie, S. W., 1979, *Mol. Pharmacol.* **15:**386–395.
187. Bowery, N. G., Doble, A., Hill, D. R., Hudson, A. L., Shaw, J. S., Turnbull, M. J., and Warrington, R., 1981, *Eur. J. Pharmacol.* **71:**53–70.
188. Bowery, N. G., Hill, D. R., Hudson, A. L., Doble, A., Middlemiss, D. N., Shaw, J., and Turnbull, M., 1980, *Nature* **283:**92–94.
189. Potashner, S. J., 1979, *J. Neurochem.* **32:**103–109.
190. Johnston, G. A. R., Hailstone, M. H., and Freeman, C. G., 1980, *J. Pharm. Pharmacol.* **32:**230–231.
191. Roberts, P. J., Gupta, H. K., and Shargill, N. S., 1978, *Brain Res.* **155:**209–212.
192. Kerwin, R., and Pycock, C., 1978, *J. Pharm. Pharmacol.* **30:**622–627.
193. Sawynok, J., and Labella, F. S., 1981, *Eur. J. Pharmacol.* **70:**103–110.
194. Abdul-Ghani, A. S., Norris, P. J., Smith, C. C. T., and Bradford, H. F., 1981, *Biochem. Pharmacol.* **30:**1203–1209.
195. Johnston, G. A. R., and Mitchell, J. F., 1971, *J. Neurochem.* **18:**2441–2446.
196. Meza-Ruiz, G., and Tapia, R., 1978, *Brain Res.* **154:**163–166.
197. Mitchell, P. R., and Martin, I. L., 1978, *Neuropharmacology* **17:**317–320.
198. Mitchell, R., 1980, *Eur. J. Pharmacol.* **68:**369–372.
199. Howard, B. D., and Gundersen, C. B., Jr., 1980, *Annu. Rev. Pharmacol. Toxicol.* **20:**307–336.
200. Rochat, H. (ed.), 1982, *Proceedings of the 4th European Symposium on Animal, Plant and Microbiology Toxins*, Pergamon Press, Oxford (in press).
201. Osborne, R. H., and Bradford, H. F., 1973, *Nature (New Biol.)* **244:**157–158.
202. Wernicke, J. F., Vanker, A. D., and Howard, B. D., 1975, *J. Neurochem.* **25:**483–496.
203. Tse, C. K., Dolly, J. O., and Diniz, C. R., 1980, *Neuroscience* **5:**135–143.
204. Smith, C. C. T., Bradford, H. F., Thompson, E. J., MacDermot, J., 1980, *J. Neurochem.* **34:**487–494.

205. Coutinho-Netto, J., Abdul-Ghani, A. S., Norris, P. J., Thomas, A. J., and Bradford, H. F., 1980, *J. Neurochem.* **35:**558–565.
206. Grasso, A., and Senni, M. I., 1979, *Eur. J. Biochem.* **102:**337–344.
207. Grasso, A., Pelliccia, M., and Alemà, S., 1982, *Proceedings of the 4th European Symposium on Animal, Plant and Microbiol. Toxins*, Pergamon Press, Oxford (in press).
208. Tzeng, M. C., Cohen, R. S., and Siekevitz, P., 1978, *Proc. Natl. Acad. Sci. U.S.A.* **75:**4016–4020.
209. Norris, P. J., Smith, C. C. T., De Belleroche, J., Bradford, H. F., Mantle, P. J., Thomas, A. J., and Penny, R. H. C., 1980, *J. Neurochem.* **34:**33–42.
210. Pong, S. S., Wang, C. C., and Fritz, L. C., 1980, *J. Neurochem.* **34:**351–358.

Transmitter Specificity in Neurons

Neville N. Osborne

1. INTRODUCTION

For the past 10 years considerable thought has centered on the question of whether neurons can utilize more than one transmitter substance.[1-6] This recent interest originated from biochemical studies on isolated invertebrate neurons,[3,7,8] and also from the development of specific immunofluorescent procedures to visualize transmitter-type molecules.[6,9,10] Before this era, it was widely accepted that each neuron had the ability to synthesize, store, and release only one transmitter substance. This concept, referred to as Dale's principle by Eccles,[11] was based on a vast quantity of experimental data. Although neither Dale nor Eccles stated categorically that neurons have the capacity to produce and release only one transmitter substance, this was generally believed to be the case, as embodied in Dale's principle. However, even before the present era of sophisticated methodology, this belief was constantly questioned. For example, in 1959, Burn and Rand[12] suggested that acetylcholine was involved in certain forms of adrenergic transmission in order to explain several pharmacological inconsistencies from the traditional point of view. It was proposed that acetylcholine is present together with norepinephrine in adrenergic axons and forms an intermediate link between nerve impulses and the release of norepinephrine from the nerve terminal.[13,14] Stimulation of sympathetic postganglionic nerves produces cholinergic contraction of newborn rabbit intestine but a predominantly adrenergic relaxation several days later.[15] This caused Burn to suggest[16] that certain cholinergic fibers may become adrenergic as a result of development. The purpose of this chapter is to examine the evidence that favors the opinion that neurons may employ more than one neurotransmitter and the functional implications it may have.

2. PHYLOGENIC CONSIDERATIONS

Two essentially opposing theories have been proposed for the mechanism whereby neurotransmitters evolve.[17] One view is that neurons have multiple

Neville N. Osborne • Nuffield Laboratory of Ophthalmology, University of Oxford, Oxford OX2 6AW, England.

origins (polygeny); the other opinion is that they have a common phylohisto-genetic root. One of the main proponents of the first hypothesis is Sakharov,[18–20] who bases his tenet on the following assumptions: (1) independent origin of neurons, each with a particular chemistry related to secretion, (2) chemical specificity of neurons from different origins, (3) divergence of neurotransmitter molecules within each inherited chemical type; i.e., dopamine cells diverge to be additionally noradrenergic or octopaminergic, (4) reduction in the number of transmitters as a result of selection within each chemical type; i.e., dopaminergic cells either remain dopaminergic or become noradrenergic or octopaminergic. Support for Sakharov's polygeny theory comes from the finding that certain secretory cells (tanning cells) in the ectoderm layer of some living hydroids do contain dopamine.[21] Such cells, of an ancient ancestral metazoan, could have lost their original function and joined a diffuse nerve net. These cells would eventually develop into nervous cell types specialized to synthesize and secrete dopamine. Eventually, some of the same cells would evolve to utilize norepinephrine. In this respect, it is of interest to note that dopamine is the only, or predominant, neuronal catecholamine in primitive nervous systems.

The theory of a common phylohistogenetic origin of neurons is based on the assumption that all neurons have a common origin and differentiate into various chemical types in the course of evolution.[22,23] This implies that all neurons have the genes available for producing all transmitter-type molecules and that they have evolved to produce one or more transmitters through suppression of specific genes. Support for this theory comes from work on cultured neurons, as will be described later.

It may well be that during the course of evolution both mechanisms operated; either way, it seems unlikely that a new nerve type with a new biochemical machinery suddenly appeared at some point in evolutionary development. It is more probable that there was a gradual evolutionary transition from one neurotransmitter to another. If this is the case, one might expect to find certain neurons utilizing more than one transmitter. Nor would one be surprised to find that neurons in certain lower forms of invertebrates have trace amounts of a second neurotransmitter coexisting with the true transmitter simply because suppression of certain genes had not proceeded to its end point. It may well be that two transmitters can therefore coexist in a neuron, one in high concentrations and having a functional role, and the other in low concentrations without a definite function. Dale's principle would still then be valid, even though two transmitter molecules actually exist in the same neuron.

3. NEURONAL CULTURE STUDIES

It has been suggested that neurons grown in culture can, at their initial stage of growth, be induced to produce different transmitter substances. There is now compelling evidence that under certain conditions *in vitro*, a single sympathetic neuron may at different times utilize norepinephrine, acetylcholine, or a mixture of these two transmitter substances.[24–31] It was shown by

Table I

Transmitter Synthesis by Single Neurons Grown in Various Conditions[a]

Growth condition	ACh neurons	NE neurons	ACh and NE neurons	ACh:NE ratio
Control	0	18	0	<0.02
20% CM	2	21	0	0.15
50% CM	15	15	2	2.4

[a] Neurons were grown from primary sympathetic neurons of the rat in control medium and in 20% and 50% conditioned medium (CM). Neurons grown with 20% and 50% conditioned medium were labeled 25–28 days after plating. The values are the numbers of single neurons making only acetylcholine (ACh), the numbers making only norepinephrine (NE), the number making both ACh and NE, and the average ratio of ACh to NE synthesis in the pooled results. The number of neurons that made neither NE nor ACh was the same as the number of cells producing ACh plus NE. (Modified from Reichardt and Patterson.[31]

Patterson and colleagues, for example,[26,28,31] that sympathetic neurons grown in the virtual absence of nonneuronal cells developed the ability to synthesize and accumulate radioactive norepinephrine from tyrosine and [^{3}H]acetylcholine from labeled choline (Table I). In the presence of nonneuronal cells or a medium conditioned by them, the neurons produced vast amounts of acetylcholine from choline, and their ability to form norepinephrine was inversely related to the percentage of conditioning medium present. Thus, the conditioning medium acts by altering the differentiated fate of individual neurons, showing that the neurons have the genetic machinery to produce more than one transmitter molecule.

Hill and Hendry[32] and Ross *et al.*[33] showed independently that the development of cholinergic characteristics in adrenergic neurons is age dependent. They revealed a diminution in the ability of superior cervical ganglia explants from rats older than 2–5 days to produce choline acetyltransferase in culture. It has also been shown that sympathetic neurons in culture eventually lose their ability to respond to conditioning medium.[34]

The elegant electrophysiological studies by Furshpan and his colleagues[30] on single isolated sympathetic neurons grown in previously dissociated heart cells from newborn rats provided evidence that they can release more than one transmitter. It was shown that some neurons inhibited, some excited, and others first inhibited and then excited the cardiac myocytes. Neurons that inhibited are thought to release acetylcholine, neurons that excited, norepinephrine, and neurons that both inhibited and excited, to release first acetylcholine and then norepinephrine. This opinion is supported by vast amounts of pharmacological and electron microscopic data.

The experiments described so far show that sympathetic neurons in culture exhibit plasticity at an early stage and may develop to utilize acetylcholine, norepinephrine, or both substances as transmitters, depending on the environmental stimulus. Le Douarin and her colleagues[35,36] have shown that transmitter functions of autonomic neurons can also be influenced by environmental factors during development *in vivo*. In heterospecific transplantation experiments between chick and quail, they found that whether an autonomic neuron precursor population produces adrenergic or cholinergic neurons depends on the site of transplantation. Thus, autonomic neurons developing both *in vivo* and in culture are at least transiently plastic with respect to transmitter function.

Recent experiments by Mudge[37] show that sensory neurons, like sympathetic neurons, also respond to a factor from nonneuronal cells that dramatically influences neuropeptide (transmitter?) synthesis. Sensory neurons from embryonic chick dorsal root ganglia that were grown together with ganglionic nonneuronal cells or with medium "conditioned" by incubation with such cells produced increased amounts of somatostatin. This increase was accompanied by neither an increase in substance P nor a detectable change in neuronal survival and was therefore different from what occurs when nerve growth factor is added. Nerve growth factor inclusion increased survival of sensory neurons without affecting the levels of somatostatin or substance P. Even though cultured neurons can contain more than one transmitter-type molecule, it is difficult to know whether they can be released and function postsynaptically as transmitters. Recent experiments by Livett *et al.*[38] have thrown more light on this subject. These authors showed that enkephalin and catecholamines are released together from primary cultures of bovine adrenal medullary chromaffin cells by nicotine in a calcium-dependent manner.

4. BIOCHEMICAL EVIDENCE FOR THE COEXISTENCE OF TRANSMITTERS

4.1. Cell Bodies

Biochemical analysis of single perikarya has been restricted to the invertebrate animals, in particular the gastropod (e.g., common snail *Helix*) and opisthobranch (e.g., *Aplysia*) molluscs. These animals have very large perikarya, which can be specifically localized, dissected, and then biochemically analyzed (see refs. 39,40). Two papers appeared in 1974[7,8] that concluded that specific neurons in *Helix* and *Aplysia* contain more than one transmitter-type molecule.

The results of these experiments have, however, been challenged.[3,4] For example, a single neuron in *Aplysia* known as R-14 was shown by Brownstein *et al.*[7] to contain 10^{-5} M serotonin, 10^{-5} M histamine, 10^{-4} M octopamine, and 10^{-3} M glutamate. Yet neither McCaman and McCaman,[41] who used a radioactive method, nor Farnham *et al.*,[42] who employed gas chromatography and mass spectrometry, could find octopamine in the R-14 cell body. In the case of the study by Hanley *et al.*,[8] it was reported that a well-known serotonergic neuron situated in each metacerebral ganglion[39,40] also contained choline acetyltransferase sufficient to produce 20 pmol of authentic acetylcholine per cell per hour. However, Osborne[3] could not confirm this finding, which was also inconsistent with an earlier study by Emson and Fonnum.[43] A number of other studies can be quoted that have produced inconsistencies in the biochemical analysis of single cells. Table II summarizes some studies of *Aplysia* neurons that claim the coexistence of transmitter-type molecules. Whether data derived from hand-dissected neurons are meaningful or not is often difficult to

Table II
Putative Neurotransmitters in Identified Neurons of Aplysia

Transmitter[a]	R-2	R-14	R-11	C-1
Serotonin[7]	1.8×10^{-5}	3.44×10^{-5}	1.1×10^{-5}	9.4×10^{-4}
Histamine[7]	3.0×10^{-6}	7.0×10^{-6}	4.5×10^{-6}	14.0×10^{-6}
Octopamine[81]	2.5×10^{-6}	1.5×10^{-5}	0.9×10^{-5}	N.D.
Acetylcholine[82]	3.9×10^{-4}	N.D.	3.3×10^{-4}	N.D.
Glutamate[83]	7.3×10^{-3}	3.2×10^{-3}	6.0×10^{-3}	3.5×10^{-3}
Glutamine[83]	3.5×10^{-3}	2.0×10^{-3}	2.1×10^{-3}	2.0×10^{-3}

[a] Dopamine, norepinephrine, and phenylethanolamine were also assayed but were not detected in any of the single cells analyzed. Results are reported as molarity ± S.E.M. Molarity was calculated from volume estimations obtained by measurement of greatest and smallest cell diameter at time of dissection. N.D. indicates not detected.

judge, simply because of the problems inherent in the dissection process.[39] The dissection of a "clean" neuron is impossible, and the use of highly sensitive assay procedures would amplify a "background noise" resulting from any impurities. During the dissection, a neuron will undoubtedly come into contact with exogenous substances released from other damaged neurons and incur considerable contamination. Any case for the coexistence of transmitter candidates in perikarya derived from hand dissection has therefore to be viewed with extreme caution.

4.2. Vesicles

There is now information available on the occurrence of the coexistence of transmitter-type molecules in certain vesicles. Since it is thought that vesicles are involved in the mechanism of release of transmitters, these observations are of major significance. The initial studies by Lewis et al.[44] and Viveros et al.[45] provided biochemical evidence for the coexistence of opioid peptides and catecholamines in granules (vesicular contents) of the adrenal medullary gland cells. If ATP is to be thought of as a neurotransmitter, then the finding by Dowdall et al.[46] can also be claimed as providing proof for the coexistence of two transmitter-type molecules, ATP and acetylcholine, in synaptic vesicles of the electric organ of the *Torpedo*. A recent biochemical study by Wilson et al.[47] also shows that opiatelike peptides and norepinephrine can probably coexist in large dense-cored noradrenergic vesicles of the splenic nerve. All of the studies briefly described are based on the biochemical analysis of a highly purified population of vesicles, which are assumed to be homogeneous. If this assumption is false, it could explain the occurrence of more than one type of transmitter molecule in such vesicle homogenates. The latest immunohistological studies at electron microscopic level by Pelletier et al.[48] are therefore of significance. These authors localized substance P and serotonin on consecutive ultrathin sections of raphe nuclei and dorsal horn by immunoelectron microscopy and came to the conclusion that about 20% of serotonin-containing nerve terminals were also positive for substance P. In these terminals, both substances were found in the same large dense-core vesicles (60–90 nm in diameter).

5. HISTOLOGICAL EVIDENCE FOR THE COEXISTENCE OF TRANSMITTERS IN NEURONS

Evidence for the histochemical demonstration of more than one transmitter-type molecule in a neuron is substantial. All studies have been based on the use of immunohistochemistry originally as introduced by Coons and collaborators.[49] Several problems are inherent in the technique, particularly that of specificity. The appropriate controls can easily be carried out to reveal any unspecific adsorption of antibodies to sections. However, an antiserum (or monoclonal antibody) may react with antigens structurally related to the antigen against which the antiserum was raised, and it is therefore impossible to identify a substance with absolute certainty by immunohistochemistry. Because of this, expressions such as "somatostatinlike immunoreactivity" or "somatostatin immunoreactivity" are often used. It is therefore possible for the immuno-reaction in two cells situated in the same section to be caused by different structurally related antigens, and the possibility of overinterpretation of data from immunohistochemistry must always be taken into consideration.

5.1. Serotonin with Dopamine

Kerkut and co-workers,[50] using fluorescence histochemical procedures, noticed that in the snail brain there exist not only yellow (serotonin-containing) or green (dopamine-containing) fluorescing neurons but also some cells showing both colors simultaneously. On the basis of these results and from the observations following injection of snails with precursors of serotonin and dopamine, it was concluded that specific neurons contain both dopamine and serotonin. However, it can be argued that this conclusion is incorrect, for it is known that when the concentration of dopamine is very high, the fluorescent reactive product does not manifest the normal green coloration but appears yellow. It is therefore possible that certain snail neurons have different concentrations of dopamine in their cytoplasm. In addition, since it is generally accepted that DOPA decarboxylase and 5-HTP decarboxylase are the same enzyme, it is not surprising to find that either dopamine or 5-hydroxytryptamine was formed in the neurons, depending on the nature of the precursor. For similar reasons, the results of Welsh and Williams[51] are questionable; these authors also suggest that certain neurons situated in planarians contain dopamine and 5-hydroxy-tryptamine, a conclusion based purely on fluorescence histochemical observations.

5.2. Serotonin with Substance P

Chan-Palay and colleagues[9] combined autoradiography, fluorescence microscopy, microspectrophotometry, and immunofluorimetry to produce the first convincing evidence for the coexistence of substance P and serotonin in the one raphe neuron. They found that certain neurons that display both an uptake–storage capacity for tritiated serotonin and a formaldehyde-induced fluorescence with spectral characteristics identical to those of the serotonin fluorophor also exhibited detectable substance-P-like immunoreactivity. Thus, some neurons exist that contain either serotonin, substance P, or serotonin and substance P. Hökfelt *et al.*[10] using immunohistochemical procedures, also produced evidence for the presence of substance-P-like and serotonin-like immunoreactivity in the same neuron of the CNS. Furthermore, after intraventricular injections of neurotoxins 5,6- or 5,7-dihydroxytryptamine, certain serotonin fibers and fibers exhibiting substance-P-like immunoreactivity disappeared. The neurotoxins are specific for the destruction of serotonin neurons, so these studies were interpreted as showing that substance P and serotonin coexist not only in the perikarya of the neuron but also in its endings.[10]

Serotonin and substance-P-like immunoreactivity have also been shown in the same neuron in the snail CNS.[52] Snail neurons are very large, measuring about 120 μm in diameter, which makes it possible to cut serial sections through them and process the sections for a number of putative transmitter molecules. As illustrated in Fig. 1, the neuron that contains both substance-P- and serotoninlike immunoreactivity does not have antigens to CCK (cholecystokinin), VIP (vasointestinal polypeptide), or bombesin.

5.3. Dopamine with CCK

Using indirect immunofluorescence histochemistry combined with retrograde tracing, Hökfelt and collaborators[6,53] have shown the coexistence of a CCK-like peptide in a subpopulation of dopamine-containing mesencephalic neuron. These CCK–dopamine cells are present predominantly in the ventral tegmental area (A10) but are also found in the substantia nigra pars lateralis and parts of the rostral substantia nigra zona compacta.

5.4. Serotonin with CCK

The only example in literature of the existence of serotonin and CCK-like immunoreactivity in the same neuron comes from studies on the snail CNS.[54] Analysis of a specific neuron in the cerebral ganglia of the snail that is known to contain serotonin[39,40] and is designated GSC (giant serotonin cell) was shown to possess CCK-like material as well. No other neuron in the ganglia contained both CCK- and serotoninlike immunoreactivity.

5.5. Norepinephrine with Enkephalin

Evidence exists for the occurrence of enkephalin- and norepinephrinelike immunoreactivity in the same cells of the adrenal medulla.[55] Immunohistochemical data also suggest that cetecholamines (dopamine and/or norepinephrine) and enkephalins coexist in cells in the rat superior cervical ganglia and in the carotid bodies of cat, dog, and monkey.[6,56]

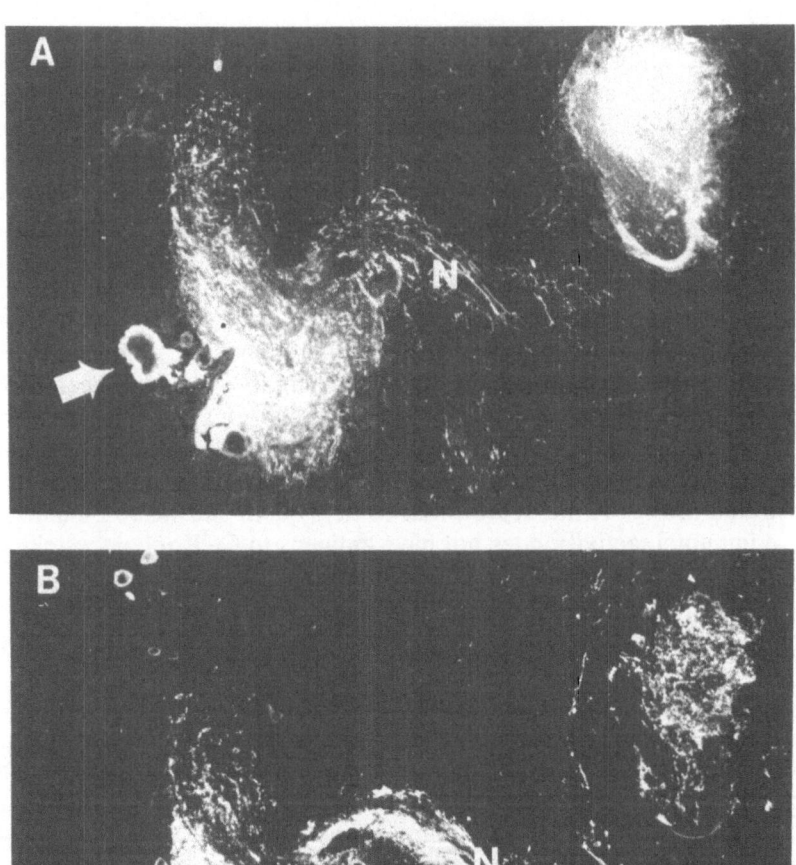

Fig. 1.   A series of consecutive sections taken through the brain of the garden snail (*Helix aspersa*) and processed for visualizing serotonin (A), substance P (B), cholecystokinin (C), bombesin (C), and vasointestinal peptide (D). It can be seen that the cytoplasm of one very large neuron (arrows) reacts positively for serotonin and substance P but not for the other neuropeptides. Note that fibers in the neuropile (N) and other cell bodies react positively for each of the substances. The large neuron that reacts positively for substance P and serotonin has a diameter of about 120 μm.

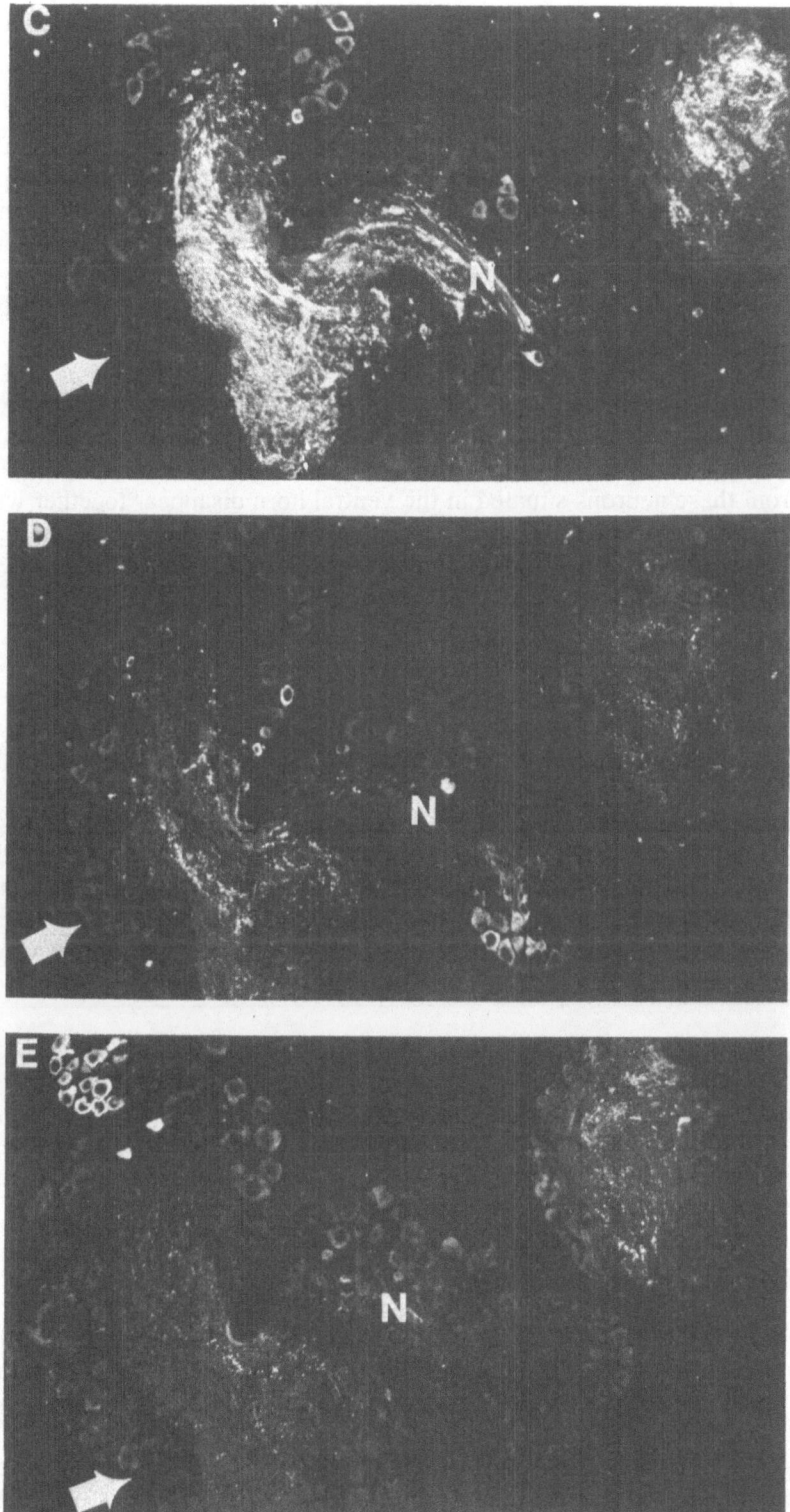

Fig. 1. (Continued)

5.6. Norepinephrine with Somatostatin

It has been shown that in certain sympathetic ganglia, particularly in the prevertebral ganglia of the guinea pig, the majority of noradrenergic sympathetic ganglion cells contain a somatostatinlike peptide.[57] From the immunohistochemical data, it was estimated that about 65% of all principal ganglion cells contained both somatostatin and norepinephrine.

5.7. Serotonin with Thyrotropin-Releasing Hormone

Immunohistochemical studies have shown that a thyrotropin-releasing hormone (TRH)-like peptide is present in serotonin neurons situated in the lower medulla oblongata of rat. The TRH-immunoreactive nerve terminals of perikarya from these neurons situated in the ventral horn disappear together with serotonin-immunoreactive terminals after treatment with the neurotoxin 5,6-dihydroxytryptamine or 5,7-dihydroxytryptamine. There is thus persuasive evidence for the coexistence of TRH and serotonin.[6]

5.8. Acetylcholine with Vasointestinal Polypeptide

Histochemical evidence has recently been obtained for the presence of VIP-like peptide in certain cholinergic neurons.[58] By staining sections from the cat stellate, L7 and S1 sympathetic ganglia for both acetylcholinesterase and VIP-like immunoreactivity, it could be established that a VIP-like peptide was present in almost all acetylcholinesterase-rich ganglion cells. Furthermore, the distribution of nerve terminals and axons was similar. After ganglionectomy, a loss of both acetylcholinesterase-positive and immunoreactive nerve terminals was observed around the sweat glands known to be inervated by ganglion cells in the stellate, L7, and S1 sympathetic ganglia.

5.9. Serotonin with Both Substance P and TRH

Johansson et al.[59] have found that some cell bodies in the medullary raphe nuclei and adjacent areas in rats contain three transmitter-type substances: serotonin, substance P, and TRH. This is the first histochemical study showing the coexistence of more than two transmitter-type molecules in a neuron.

5.10. Coexistence of Neuropeptides

Relatively few examples of the coexistence of neuropeptides in one neuron are reported in literature. More general is the occurrence of a neuropeptide with a classical neurotransmitter-type molecule, e.g., serotonin, norepinephrine, or acetylcholine. Certain hypothalamic neurons have been described as containing β-endorphin and ACTH.[60] Schultzberg et al.[61] have also shown that some gastrointestinal neurons contain both somatostatin and a gastrin–CCK-like peptide. From the same laboratory has come a report concerned with certain medullary raphe neurons containing both substance P and TRH.[6,62]

There has also been a recent publication on the occurrence of substance P and VIP in enteric *p*-type neurons of the guinea pig colon.[63]

6. SPECIFICITY OF RELEASE BY NEURONS

In order to establish that two or more transmitter-type molecules in a neuron function as neurotransmitters, it is important to show that they can be released following a physiological stimulation of the neuron. However, the demonstration that a substance can be released from a neuron does not necessarily mean that it has a transmitter role. The release of a nontransmitter substance from a neuron has been quite well established.[64] The problem lies in deciding what qualifies as a nontransmitter but is nevertheless released from nerve endings. By definition, a transmitter is released into the synaptic cleft and has an effect on postsynaptic receptors; nontransmitters, in contrast, would not be able to influence any known receptors in the synaptic region. It is known that chromagranins and dopamine-β-hydroxylase are released from adrenergic endings,[65] but no postsynaptic effects of these substances have been found or postulated. It also appears that certain lysosomal hydrolases are secreted from the adrenal medulla,[64] and nigral dopamine neurons that contain acetylcholinesterase[66] also "release" the enzyme on stimulation.[69]

Examples of the release of two transmitter-type molecules from the same neuron are rare. The best and most persuasive example comes from work on the GSC (giant serotonin cell) in the snail *Helix*.[68] The following data establish unequivocally that the GSCs use serotonin as a transmitter:

1. Endogenous serotonin and its synthetic enzymes are present in the neuron.
2. Iontophoretically applied serotonin mimics one of the postsynaptic potential changes produced by stimulation of the nerve cell.
3. Inhibition or potentiation of efferent transmission under pharmacological conditions was shown to impede or facilitate, respectively, the action of serotonin.

Although not as extensive, the same type of study by Cottrell and collaborators[69] argues that the GSC also utilizes acetylcholine as a transmitter. It is suggested that acetylcholine is used as a transmitter, evoking the stimulation-induced postsynaptic potentials preceding those attributed to the action of serotonin. As discussed above, the biochemical evidence that the GSCs contain acetylcholine or the enzyme involved in its synthesis, choline acetyltransferase, is equivocal.

Two main arguments have been countered against the pharmacological–electrophysiological findings from Cottrell and collaborators.[69] First, it is not certain whether the neuronal response to application of serotonin and acetylcholine in the vicinity of the serotonin follower cells was, in fact, recorded from the same neuron. Second, the observation that the fast postsynaptic response of the follower neuron to repetitive stimulation of the serotonin cell is blocked by hexamethonium is difficult to interpret. Apart from the possibility that the effect is an example of the known nonspecific influences of the relatively high drug concentration used (1 mM), it could be possible that the block was only minor, given (1) the problems inherent in reproducibly quantifying graded responses to repeated stimulation and (2) the variability of neuronal reactions to acetylcholine in gastropod molluscs.

A number of other studies have suggested that more than one transmitter may be released from the same neuron. Singh, in a series of papers,[70,71] claimed that no fewer than three putative transmitters (acetylcholine, serotonin, and histamine) were released during stimulation of the vagus nerves supplying the stomach of the frog but that with the change of season one or more became the dominant transmitter. The main argument against the suggestion is that the vagus nerve in the frog contains a mixture of separate axons and that these fibers may be either histaminergic, serotonergic, or cholinergic in function.

There also appears to be good evidence from the simultaneous release of ATP and catecholamines from the adrenal medullary vesicles in perfused adrenal glands and in certain adrenergic nerves where norepinephrine is a transmitter.[1,64] A case has been made for ATP being a transmitter in the gastrointestinal tract and in certain other organs too,[72] and if this is accepted, then the data suggest that two transmitters, norepinephrine and ATP, are released from the same neuron. It still, however, remains to be shown that they have postsynaptic effects characteristic of transmitter molecules. In cholinergic cells, e.g., those associated with the electric organs of certain fish, the storage vesicles, like those from the adrenal medulla, contain a fixed amount of ATP in relation to their acetylcholine content.[46] Indirect evidence suggests that acetylcholine and ATP are secreted from cholinergic neurons.

There is also evidence that adenosine can be released from central neurons and can stimulate cyclic AMP formation through specific receptors and alter postsynaptic electrical activity.[73] A transmitter role has been attributed to adenosine.[74] Adenosine is known to coexist with classical transmitter-type molecules in the same neurons.

The corelease of enkephalins and catecholamines from cultured adrenal chromaffin cells has recently been reported.[38] These authors used cultures of bovine adrenal medullary chromaffin cells and showed that nicotine stimulated the release of norepinephrine, epinephrine, and leu-enkephalin from these cells. The release of all of these substances was calcium dependent.

7. SPECIFICITY OF POSTSYNAPTIC RECEPTORS

Should a single axonal process of a neuron release more than one neurotransmitter, it is to be expected that more than one type of receptor would be present on the postsynaptic membrane with which the axonal ending makes a synapse. Unfortunately, there are few data to support this assumption. However, some recent electrophysiological and histological experiments show that the possibility does exist. For example, Silinski[75] has produced convincing evidence for specific adenosine receptors at cholinergic nerve endings and suggests that these receptors are located presynaptically. It could equally well be argued, however, that some of these receptors are located postsynaptically. Another example of the presence of two types of receptors at the same synapse comes from the recent finding by Möhler *et al.*[76] These authors used an autoradiographic method for localizing benzodiazepine receptors in γ-aminobutynergic synapses identified by immunocytochemical procedures. Again, one cannot be certain whether both receptor sites are located pstsynaptically or whether benzodiazepines actually exist in neurons and are actually able to be released from them.

8. FUNCTIONAL ASPECTS

The functional significance of the coexistence of two transmitter type molecules in the same neuron is at present unclear. Evidence of the corelease of substances from the one neuron and the occurrence of specific receptors for each substance on the immediate postsynaptic area are unfortunately still inconclusive. Thus, the paucity of overall relevant data means that the validity of Dale's principle still stands but at the same time indicates that it may not be as convincing at it might have appeared in the 1950s.

Should two or more transmitterlike molecules be released from the same neuron, there are several possible ways in which they can act. They may have completely different functions; one may be a transmitter, and the other substance may have a trophic or modulatory role. If they both were, however, to have transmitter functions, they could act in a cooperative manner, i.e., work together to achieve a certain physiological response. Both substances may practice their roles on the same postsynaptic cell; one substance may activate some postsynaptic receptors, whereas the second substance may act on autoreceptors; one substance may act on the postsynaptic neuron, whereas the second may act on presynaptic receptors located in different neurons; one substance may act on presynaptic receptors of one cell and the other substance on postsynaptic receptors from another cell; a particular stimulus may release one transmitter from the cell, whereas another stimulus parameter has the effect of releasing both transmitter substances. Clearly, if neurons employ more than one transmitter, it is possible to envisage a number of ways in which they can function. The functional possibilities for neurons employing more than two transmitters would be even greater in number.

9. WHAT ARE TRANSMITTER-TYPE MOLECULES?

What substances always qualify as transmitter-type molecules? The answer to this is *none*. For example, acetylcholine exists in the corneal epithelium in large amounts, where it definitely has a nonneurotransmitter function.[77] It is most important to recognize that one particular substance should not be thought of as functioning in one particular way. Therefore, the presence of two transmitter-type molecules (i.e., substances that normally have transmitter functions or are thought to) in a neuron does not reflect any functional significance. One of the transmitters may have a vestigial presence because the suppression of certain genes has not proceeded to its end point. This could well explain why it often appears that when two transmitter-type molecules occur in a neuron (especially in vertebrates), one of the substances exists in very low amounts. It is also possible for one of the substances to have a modulatory function. For example, norepinephrine, a classical transmitter molecule, is also thought to modulate the release of acetylcholine from cholinergic nerves in the gut.[78,79] It might also be that one of the substances functions as an hormone. One of the well known "hormone" actions of tuberoinfundibular dopamine neurons is the inhibition of prolactin secretion,[80] yet it has a neurotransmitter function in a variety of the neurons in the CNS. Thus, dopamine can act as a hormone or a neurotransmitter.

10. CONCLUSION

A number of examples have been reviewed showing that certain neurons contain more than one substance possibly involved in the process of chemical transmission at synapses. There is, however, no conclusive evidence that neurons can utilize more than one transmitter. This is chiefly because of a lack of unequivocal proof of a neuron's ability to release more than one substance that has postsynaptic effects. This lack of proof may result from inadequate methodology, for there is no scientific reason for assuming that neurons can utilize only a single transmitter, a dogma known as Dale's principle. The problem is complicated by the fact that substances may have various functions in different situations.

ACKNOWLEDGMENT. The author is grateful to the Stiftung Volkswagenwerk for financing part of his research.

REFERENCES

1. Burnstock, G., 1976, *Neuroscience* **1**:239–248.
2. Burnstock, G., 1978, *Prog. Neurobiol.* **11**:205–222.
3. Osborne, N. N., 1977, *Nature* **270**:622–623.
4. Osborne, N. N., 1979, *Trends Neurosci.* **2**:73–75.
5. Osborne, N. N., 1981, *Neurochem. Int.* **3**:3–16.
6. Hökfelt, T., Lundberg, J. M., Schultzberg, M., Johansson, O., Ljungdahl, Å., and Rehfeld, J., 1980, *Neural Peptides and Neuronal Communication* (E. Costa and M. Trabuchhi, eds.), Raven Press, New York, pp. 1–23.
7. Brownstein, M. J., Snowedra, J. M., Axelrod, J., Zemen, G. H., and Carpenter, D. O., 1974, *Proc. Natl. Acad. Sci. U.S.A.* **75**:5732–5736.
8. Hanley, M. R., Cottrell, G. A., Emson, P. C., and Fonnum, F., 1974, *Nature (New Biol.)* **251**:631–633.
9. Chan-Palay, V., Jonsson, G., and Palay, S. L., 1978, *Proc. Natl. Acad. Sci. U.S.A.* **75**:1582–1586.
10. Hökfelt, T., Ljungdahl, H., Steinbusch, H., Verhofstad, A., Nilsson, G., Brodin, E., Pernow, B., and Goldstein, M., 1978, *Neuroscience* **3**:517–538.

11. Eccles, J. D., 1957, *The Physiology of Nerve Cells*, The Johns Hopkins University Press, Baltimore.
12. Burn, J. H., and Rand, M. J., 1959, *Nature* **184**:163–165.
13. Koelle, G. B., 1962, *J. Pharm. Pharmacol.* **14**:65–90.
14. Burn, J. H., and Rand, M. J., 1965, *Annu. Rev. Pharmacol.* **5**:163–182.
15. Day, M. D., and Rand, M. J., 1961, *Br. J. Pharm. Chem.* **17**:245–260.
16. Burn, J. H., 1968, *Br. J. Pharm. Chem.* **32**:575–582.
17. Osborne, N. N., 1982, *Trends*. Volume 2 (S. Kalsner, ed.), Urban Schwarzenberg, Baltimore, Munich (in press).
18. Sakharov, D. A., 1974, *J. Neural Transm. [Suppl.]* **11**:43–59.
19. Sakharov, D. A., 1976, *Gastropoda Brain* (J. Salanki, ed.), Akademiai Kiado, Budapest, pp. 27–40.
20. Sakharov, D. A., 1978, *Advances in Pharmacology and Therapeutics*, Volume 8 (G. Olive, ed.), Pergamon Press, Oxford, pp. 275–283.
21. Knight, D. P., 1970, *Tissue Cell* **2**:467–477.
22. Lentz, T. L., 1968, *Primitive Nervous Systems*, Yale University Press, New Haven.
23. Scharrer, B., 1976, *Prog. Brain Res.* **45**:125–137.
24. O'Lague, P. H., Obata, K., Claude, P., Furshpan, E. J., and Potter, D. D., 1974, *Proc. Natl. Acad. Sci. U.S.A.* **71**:3602–3606.
25. Purves, R. D., Hill, C. E., Chamley, J., Mark, G. E., Fry, D. M., and Burnstock, G., 1974, *Pfluegers Arch.* **350**:1–7.
26. Patterson, P. H., and Chun, L. L. Y., 1977, *Dev. Biol.* **56**:263–280.
27. Burton, H., and Bunge, R. O., 1975, *Brain Res.* **97**:157–162.
28. Patterson, P. H., Reichardt, L. F., and Chun, L. L. Y., 1976, *Cold Spring Harbor Symp. Quant. Biol.* **40**:389–397.
29. Hill, C., Purves, R. D., Watanake, H., and Burnstock, G., 1976, *Pfluegers Arch.* **361**:127–134.
30. Furshpan, E. J., Macleish, P. R., O'Lague, P. H., and Potter, D. D., 1976, *Proc. Natl. Acad. Sci. U.S.A.* **73**:4225–4229.
31. Reichardt, L. F., and Patterson, P. H., 1977, *Nature* **270**:147–151.
32. Hill, C. E., and Hendry, I. A., 1977, *Neuroscience* **2**:741–749.
33. Ross, D., Johnson, M., and Bunge, R., 1977, *Nature* **267**:536–539.
34. Chun, L. L. Y., and Patterson, P. H., 1977, *J. Cell Biol.* **75**:694–704.
35. Le Douarin, N. M., Renaud, D., Teillet, M. A., and Le Douarin, G. H., 1975, *Proc. Natl. Acad. Sci. U.S.A.* **72**:728–732.
36. Le Douarin, N. M., Smith, J., Teillet, M.-A., Le Lievre, C. S., and Ziller, C., 1980, *Trends Neurosci.* **3**:39–42.
37. Mudge, A. W., 1981, *Nature* **292**:764–766.
38. Livett, B. G., Dean, D. M., Whelan, L. G., Udenfriend, S., and Rossier, J., 1981, *Nature* **289**:317–319.
39. Osborne, N. N., 1974, *Microchemical Analysis of Nervous Tissue*, Pergamon Press, Oxford.
40. Osborne, N. N., 1980, *Trends Pharmacol.* **1**:290–292.
41. McCaman, R. E., and McCaman, M. W., 1978, *Brain Res.* **141**:347–352.
42. Farnham, P. J., Navak, R. A., and McAdoo, D. J., 1978, *J. Neurochem.* **20**:1173–1176.
43. Emson, P. C., and Fonnum, F., 1972, *J. Neurochem.* **22**:1079–1088.
44. Lewis, R. V., Stern, A. S., Rossier, J., Stein, S., and Udenfriend, S., 1979, *Biochem. Biophys. Res. Commun.* **89**:822–829.
45. Viveros, O. H., Diliberto, E. J., Jr., Hazun, E., and Chang, K.-J., 1979, *Mol. Pharmacol.* **16**:1101–1108.
46. Dowdall, M. J., Boyne, A. F., and Whittaker, V. P., 1974, *Biochem. J.* **140**:1–12.
47. Wilson, S. P., Klein, R. L., Chang, K.-J., Gasparis, M. S., Viveros, H., and Yang, W.-H., 1980, *Nature* **288**:707–709.
48. Pelletier, G., Steinbusch, H. W. M., and Verhofstad, A. A. J., 1981, *Nature* **293**:71–72.
49. Coons, A. H., 1958, *Gen ral Cytochemical Methods* (J. F. Danielli, ed.), Academic Press, New York, pp. 399–422.
50. Kerkut, G. A., Sedden, C. L., and Walker, R. J., 1967, *Comp. Biochem. Physiol.* **23**:159–162.

51. Welsh, J. H., and Williams, L. D., 1970, *J. Comp. Neurol.* **138**:103–116.
52. Osborne, N. N., and Dockray, G. J., 1982, *Neurochem. Int.* **4**:175–180.
53. Hökfelt, T., Rehfeld, J. F., Skirboll, L., Ivemark, B., Goldstein, M., and Markey, K., 1980, *Nature* **285**:476–478-
54. Osborne, N. N., Cuello, A. C., and Dockray, G. J., 1982, *Science* **216**:409–411.
55. Schultzberg, M., Hökfelt, T., Terenius, L., Elfvin, L. G., Lundberg, J. M., Brandt, J., Elde, R., and Goldstein, M., 1978, *Neuroscience* **3**:1169–1186.
56. Schultzberg, M., Hökfelt, T., Terenius, L., Elfvin, L. G., Lundberg, J. M., Brandt, J., Elde, R., and Goldstein, M., 1979, *Neuroscience* **4**:249–270.
57. Hökfelt, T., Elfvin, L. G., Elde, R., Schultzberg, M., Goldstein, M., and Lufe, R., 1977, *Proc. Natl. Acad. Sci. U.S.A.* **74**:3587–3591.
58. Lundberg, J. M., Hökfelt, T., Schultzberg, M., Uvnas-Wallenstein, K., Kohler, L., and Said, S., 1979, *Neuroscience* **4**:1539–1559.
59. Johansson, O., Hökfelt, T., Pernow, B., Jeffcoate, S. L., White, N., Steinbusch, H. W. M., Verhofstad, A. A. J., Emson, P. C., and Spindel, E., 1981, *Neuroscience* **6**:1857–1882.
60. Nilaver, G., Zimmerman, E. A., Defendini, R., Liotta, A. S., Krieger, D. T., and Brownstein, M. J., 1979, *J. Cell. Biol.* **81**:50–58.
61. Schultzberg, M., Hökfelt, T., Nilsson, G., Terenius, L., Rehfeld, J. F., Brown, M., Elde, R., Goldstein, M., and Said, S., 1980, *Neuroscience* **5**:689–744.
62. Hökfelt, T., Lundberg, J. M., Schultzberg, M., Johansson, O., Skirboll, L., Ånggård, A., Fredholm, B., Hamberger, B., Pernow, B., Rehfeld, J., and Goldstein, M., 1980, *Proc. R. Soc. Lond.* [*Biol.*] **210**:63–77.
63. Probert, L., De Mey, J., and Polak, J. M., 1981, *Nature* **294**:470–471.
64. Chubb, I. A., 1977, *Synapses* (G. A. Cottrell and P. N. E. Usherwood, eds.), Blackie, Glasgow, pp. 264–290.
65. Smith, A. D., de Potter, V. P., Moerman, E. J., and Schaepdryver, A. F., 1970, *Tissue Cell* **2**:547–568.
66. Dray, A., 1979, *Neuroscience* **4**:1407–1439.
67. Greenfield, S., Cheramy, A., Leviel, V., and Glovinski, J., 1980, *Nature* **284**:355–357.
68. Osborne, N. N., 1982, *Co-Transmission* (A. C. Cuello, ed.) Macmillan Press, London pp. 207–222.
69. Cottrell, G. A., 1977, *Neuroscience* **2**:1–18.
70. Singh, I., 1964, *Arch. Int. Physiol. Biochem.* **72**:843–851.
71. Singh, I., and Singh, S. I., 1966, *Arch. Int. Physiol. Biochem* **74**:365–373.
72. Burnstock, G., 1975, *J. Exp. Zool.* **194**:103–133.
73. Fox, J. H., and Kenney, W. N., 1978, *Annu. Rev. Biochem.* **47**:655–686.
74. McIlwain, H., 1973, *Cerebral Nervous System* (E. Genazzani and H. Herken, ed.), Springer-Verlag, Berlin, pp. 1–11.
75. Silinski, E. M., 1980, *Br. J. Pharma. Chem.* **71**:191–194.
76. Möhler, H., Richards, J. G., and Wu, J.-Y., 1981, *Proc. Natl. Acad. Sci.* **78**:1935–1938.
77. Fogle, J. A., and Neufeld, A. M., 1979, *Invest. Ophthalmol. Vis. Sci.* **18**:1212–1215.
78. Paton, W. D. M., and Vizi, E. S., 1969, *Br. J. Pharm. Chem.* **35**:10–28.
79. Vizi, E. A., 1973, *Br. J. Pharm. Chem.* **47**:765–777.
80. Neill, J. D., Plotsky, P. M., and de Greep, W. J., 1979, *Trends Neurosci.* **2**:60–63.
81. Saavedra, J. M., Brownstein, M. J., Carpenter, D. O., and Axelrod, J., 1974, *Science* **185**:364–365.
82. McCaman, R. T., Weinreich, D., and Borys, H., 1973, *J. Neurochem.* **21**:473–476.
83. Borys, H. K., Weinreich, D., and McCaman, R. E., 1973, *J. Neurochem.* **21**:1349–1351.

Phospholipid Methylation

Fusao Hirata

1. INTRODUCTION

Phospholipid methylation is a reaction in which phosphatidylethanolamine (PtdEth) is converted to phosphatidylcholine (PtdCho). This reaction was first described by Bremer *et al.*[1] However, the enzyme activity of phospholipid methylation is quite low compared with CDP-choline transferese activity in PdtCho synthesis. For long time, this enzyme activity had not been detected in tissues other than lung and liver. Recently, many laboratories have reported the existence of PtdEth methyltransferase activity in a variety of organs and tissues of mammals. This enzyme activity in many tissues if not all, has now been connected to certain biological functions such as receptor-mediated signal transduction.[2] In this chapter, recent studies on phospholipid methylation and its role in biological functions are described.

2. PROPERTIES OF PHOSPHATIDYLETHANOLAMINE METHYLTRANSFERASE

2.1. Assay of the Methyltransferase

Phosphatidylethanolamine methyltransferase (PMTase) catalyzes three successive incorporations of methyl groups into the amino residue of PtdEth from S-adenosylmethionine (S-AdoMet). When [methyl-^{3}H]S-AdoMet is used as substrate, the enzyme activity can be measured by the racioactivity of the [^{3}H] methyl groups incorporated into the phospholipids after extraction with organic solvents such as a chloroform–methanol (2 : 1, v/v) mixture.[1-3] This enzyme can also methylate the plasmalogen type of PtdEth, which has an alkyl group in the 1 position of glycerol moiety.[4] This compound is quite rich in the brain but is labile in acidic solutions. Therefore, to stop the reaction, neutralized

Fusao Hirata • Laboratory of Cell Biology, National Institute of Mental Health, Bethesda, Maryland 20205.

trichloracetic acid (TCA) is recommended.[5] Methylated phospholipids extracted with organic solvents contain monomethyl, dimethyl, and trimethyl derivatives of PtdEth and their corresponding plasmalogens. These compounds can be separated by thin-layer chromatography or high-pressure liquid chromatography.[5,6] However, the plasmalogen-type lipids generally comigrate with the corresponding PtdEth derivatives. When they are hydrolyzed in alkaline solutions, PtdEth derivatives are converted to glycerophosphorylethanolamine derivatives, but the plasmalogen-type lipids are converted to their lysoforms. Thus, the two related phospholipids can be separately quantified by such chemical treatment.

2.2. Possible Involvement of Two Methyltransferases in the Reaction

When the methyltransferase activity is assayed at varying concentrations of S-AdoMet, the substrate–velocity curve is a typical Michaelis–Menten type with an inflection point around 5 to 20 μM. The Lineweaver–Burk plot of the data distinguish the two components; one has an apparent K_m of approximately 1 μM, and the other has an apparent K_m of 100 μM. The main product formed with 4 μM S-AdoMet is monomethyl PtdEth, whereas PtdCho (trimethyl PtdEth) is mainly formed with 100 μM S-AdoMet. Furthermore, the optimal pH for the reaction at 4 μM S-AdoMet is 7.5, whereas that at 100 μM S-AdoMet is 10.5.[5] Therefore, the involvement of two enzymes in the conversion of PtdEth to PtdCho has been suspected; the first enzyme (PMTase I) catalyzes the methylation of PtdEth to form monomethyl PtdEth, and the second enzyme (PMTaseII) adds two more methyl groups to monomethyl PtdEth. Attempts have been made to purify and separate these enzymes. The two enzymes are copurified when they are solubilized by detergents.[7,8] However, when the membrane fraction is sonicated, PMTase II is partially solubilized, whereas PMTase I remains in the membranes.[6] However, the solubilized enzymes are unstable and easily inactivated regardless of the method used. Attempts have so far failed to purify these enzymes further.

The involvement of the two PMTases in the conversion of PtdEth to PtdCho was first suggested by experiments with variants of *Neurospora crassa*.[9] Recently, McGivney *et al.* reported that mutant cell lines of rat basophilic leukemia cells are deficient in either of PMTase I or PMTase II.[10] Since the resistance of these cell lines to drugs such as chloramphenicol and thioguanine is different, hybridization of the two cell lines can be performed. The hybrid cells have the capacity to synthesize PtdCho by transmethylation *in vivo* and *in vitro*, although the parental cells can not synthesize this lipid by transmethylation. These results confirm the hypothesis that PMTase I and II are necessary to convert PtdEth to PtdCho. However, the methyl incorporation into phospholipids by PMTases is greatly enhanced by addition of exogenous monomethyl and dimethl PtdEth but not by addition of PtdEth. The main products formed under these conditions are dimethyl PtdEth and PtdCho, respectively.[5,6] These results do not rule out the possibility that three enzymes are involved in the reaction.

2.3. Asymmetric Distribution of PMTases and Flip–Flop of Methylated Phospholipids

Studies on the topographical arrangement of membrane components have indicated that both lipids and proteins, major components of membranes, are asymmetrically distributed.[11,12] The erythrocyte membrane represents a simple model membrane and has been studied extensively; PMTases are present in rat erythrocytes.[13] The digestion of the right-side-out ghosts with trypsin results in the inactivation of PMTase II but not PMTase I, whereas the same treatment of the inside-out ghosts inactivates PMTase I but not PMTase II. These results suggest that PMTase I faces on the cytoplasmic side, whereas PMTase II is located on the cell surface. When the resealed right-side-out and inside-out ghosts are incubated with [methyl-^{3}H]S-AdoMet, only the inside-out ghosts can incorporate the [^{3}H]methyl group into the membrane lipids, suggesting that the initial reaction of PMTases starts on the cytoplasmic side. These results are consistent with the observation that PMTases are asymmetrically distributed across membranes. Such spatial segregation of the two PMTases makes it possible to translocate methylated phospholipids across membranes. When the topography of methylated phospholipids is examined by the treatment with phospholipase C or phospholipase A_2, PtdCho synthesized by transmethylation is found mainly on the cell surface. Since PtdEth is largely located on the cytoplasmic side, where the transmethylation reaction is initiated, these results suggest that the methylated lipids are flip–flopped across biomembranes, although spontaneous flip–flop of lipids is a very slow process. These techniques using proteases and phospholipases have successfully been applied to rat synaptosomes[14] and liver microsomes[15] and have demonstrated the asymmetrical distribution of PMTases and flip–flop of methylated phospholipids in these tissues. The PtdCho molecule synthesized by CDP-choline transferase, a major synthetic pathway, can not be translocated but stays on the cytoplasmic side, where this molecule is formed.

2.4. Molecular Species of Phosphatidylcholine Synthesized by Transmethylation

Phosphatidylcholine is synthesized mainly by two pathways, the CDP-choline pathway and the transmethylation pathway.[16] The specific activity of PMTases is only a few percent of the CDP-choline transferase activity in the rat liver. Nevertheless, the transmethylation pathways accounts for one-tenth of the total PtdCho formed in intact hepatocytes and basophilic leukemia cells.[10,17] The principal species of PtdCho formed by the transmethylation is one containing arachidonic acid (20:4) and eicosatrienoic acid (20:3), although that synthesized by the CDP-choline pathway contains more linoleic acid (18:1) and oleic acid (18:2) in its side chain. Since the *in vivo* activities of PMTases and CDP-choline transferase are usually much lower than *in vitro* activities of these enzymes, limited availability of certain species of phospholipids to these enzymes has been suggested. The lipids are reported to be distributed in specific domains based on the degree of unsaturation of their side-chain fatty acids.[18]

The level of PMTase activity is reported to be dependent on the amount of PtdEth available.[19] Rats fed with choline-deficient diets also have higher activities of PMTases,[20] suggesting the coordinate regulation of these two pathways of PtdCho synthesis, the CDP-choline pathway and the transmethylation pathway. In fact, inhibition of transmethylation by inhibitors such as 3-deazaadenosine (see below) causes increased activity of the CDP-choline pathway in rat hepatocytes.[21] Since arachidonic acid and eicosatrienoic acid in the PtdCho synthesized by transmethylation are released by stimulation of various receptors,[2] the turnover of phospholipids appears to reflect the higher capacity of transmethylation synthesis of PtdCho *in vivo* compared with its activity *in vitro*.

Plasmalogens present in the tissues are mostly of the ethanolamine type. A small but significant amount of choline plasmalogen has been detected (22). This choline plasmalogen is converted to platelet-activating factor (1-alkyl-2-acetylglycerophosphorylcholine) by the action of phospholipase A_2 and acetyl-CoA transferase.[23] Platelet-activating factor (PAF) has a variety of biological actions such as platelet aggregation and blockade of α-adrenergic receptors.[24] Ethanolamine plasmalogen is very rich in the brain, but no reports on the formation of PAF by transmethylation in the brain have been available.

2.5. Distribution of PMTases

Phosphatidylethanolamine methyltransferase activity can be detected in a variety of tissues and organs of mammals including brain, lung, liver, lymphocytes, and red blood cells.[2] When subcellular distribution of rat brain PMTases is examined, the highest specific activities of PMTase are associated with plasma membranes, followed by microsomal and mitochondrial fractions.[5]

3. PHOSPHATIDYLETHANOLAMINE METHYLTRANSFERASES IN INTACT CELLS

3.1. Pharmacological Perturbation of Phospholipid Methylation

S-Adenosylhomocysteine (S-AdoHcy), a product of S-AdoMet, is a potent inhibitor of various transmethylation reactions.[23] In intact cells, S-AdoHcy is rapidly degraded by S-AdoHcy hydrolase and adenosine deaminase. Adenosine analogues such as N^6-methyladenosine and 3-deazaadenosine can form corresponding derivatives of AdoHcy intracellularly. In addition to these agents, analogues of S-AdoHcy such as 3-deazaadenosyl-S-isobutylmercaptan (3-deaza-SIBA) or adenosyl-S-isobutylmercaptan (SIBA) can inhibit AdoHcy hydrolase directly. When these agents are administered *in vivo*, the cellular level of S-AdoHcy increases, and subsequently a variety of transmethylation reactions are inhibited.

Under specific conditions, however, treatment with these agents brings about the selective inhibition of PMTases.[2] The PMTase activities in intact cells can be measured by the incorporation of [^{3}H]methyl groups into the phospholipid fraction from [methyl-^{3}H]methionine, a precursor of S-AdoMet. Since intact cells transport S-AdoMet slowly, [methly-^{3}H]methionine is recommended for measurements of transmethylation in intact cells. The incubation period required to equilibrate newly formed [methly-^{3}H]S-AdoMet with nanlabeled S-AdoMet is dependent on cell types and culture conditions. In spite of these facts, these drugs have actions other than inhibition of transmethylations. Adenosine analogues can increase cyclic AMP levels of various cells by inhibiting phosphodiesterase or by stimulating adenosine receptors coupled to adenylate cyclase.[26] Such an increase in intracellular cyclic AMP exerts many biological effects.

3.2. Membrane Fluidity Changes by Transmethylation

Since PMTases are asymmetrically distributed across biomembranes and translocate PtdEth from the cytoplasmic side to the outer surface with its simultaneous conversion to PtdCho, the charges and structures of membranes will be altered by phospholipid methylation. When red blood cell ghosts are incubated with S-AdoMet to produce the methylation of phospholipids, a marked increase of membrane fluidity can be observed as measured by anisotropy of fluorescent probes such as dinitrophenylhexatriene (DPH). The degree of increase in membrane fluidity appears to be proportional to the amounts of the intermediates accumulated during the reaction.[27] These changes can be abolished by adding S-AdoHcy, an inhibitor of transmethylation.

Similar membrane fluidity changes induced by phospholipid methylation can be detected in chicken erythrocytes stimulated by concanavaline A (Con A), a mitogen.[28] When chicken erythrocytes labeled with a stearic acid derivative spin label are treated with Con A, electron spin resonance (esr) spectra show marked changes immediately after the start of the treatment. These changes, reflecting altered membrane fluidity, can be divided into three phases. The first phase, whose peak is attained 10 min after stimulation, is attributable to phospholipid methylation; SIBA, an inhibitor of transmethylation, can block the increase in both phospholipid methylation and membrane fluidity. The second and third phases, which reach maximum at 30 min and 60 min, respectively, might correspond to changes in the cytoskeletons and metabolism of phospholipids occurring after Ca^{2+} influx. This increase in membrane fluidity can be reduced by cytochalasins (cytoskeleton-blocking agent) and EDTA (Ca^{2+}-chelating agent), respectively. Interestingly, alterations of membrane structures as inspected by freeze-fracture replicas are reversible up to 30 min after the treatment and thereafter are irreversible.

Since many membrane enzymes such as Ca^{2+}-ATPase and Na^+,K^+-ATPases are embedded in the lipid bilayers, and their enzyme activities are affected by their microenvironments, the increased membrane fluidity induced by phospholipid methylation would bring about changes in the activities of these enzymes. In fact, an increased activity of Ca^{2+}-ATPase by phospholipid methylation has been reported.[29]

4. BIOLOGICAL ROLES OF PHOSPHATIDYLETHANOLAMINE METHYLTRANSFERASES

4.1. Interaction with Receptors

When rat reticulocyte ghosts are incubated with S-AdoMet, there is an increase in the number of β-adrenergic receptors. The increase in the number of β-adrenergic receptors as measured by [³H]dihydroalprenolol binding parallels the amount of PtdCho synthesized.[30] Since the ghost preparation has no ability to synthesize new receptors, it is suggested that phospholipid methylation unmasks a latent or cryptic form of β-adrenergic receptors. An increased number of receptors following phospholipid methylation is also reported in binding of human growth hormone to the lactogenic receptors.[31] However, in the case of chemotactic receptors in guinea pig macrophages, the affinity of receptors for ligands is affected by phospholipid methylation.[32] This is explained by changes in membrane fluidity, which modify the binding characteristics of receptors. Human mature red blood cells have no β-adrenergic receptors on the cell surface. When β-adrenergic receptors are partially purified from turkey red blood cells and implanted into human red blood cell ghosts, suppression of PMTases in human red blood cell ghosts can be observed. This suppression of PMTases is a function of the number of β-adrenergic receptors implanted. Addition of ligands such as isoproterenol results in the restoration of full PMTase activity, suggesting that the suppressive effects of receptors can be reversed by the binding of ligands.[33]

These results indicate a close association between certain types of receptors and PMTases. The regulation of receptor number by PMTases occurs even in intact cells. The induction of β-adrenergic receptors in HeLa cells by sodium butyrate is blocked by the treatment with N^6-methyladenosine or 3-deazaadenosine.[34] When these agents are removed from culture media, the number of β-adrenergic receptors is restored to control level within hours. This recovery can not be blocked by protein synthesis inhibitors, actinomycin D, and cycloheximide. The injection of 3-deazaadenosine (10 mg/kg) in rats for 1 week causes a decrease in the number of β-adrenergic receptors in the hypothalamus and brainstem without changing the affinity for dihydroalprenolol (F. Hirata, unpublished data). Phospholipid methylation in the brain and liver is inhibited by this treatment.

4.2. Stimulation and Inhibition of PMTases by Receptor Stimulation

Stimulation of many receptors, if not all, in a variety of cells may increase or decrease phospholipid methylation.[2] A decrease in phospholipid methylation is generally caused by rapid turnover of methylated lipids rather than by inhibition of phospholipid methylation.[35] Such changes in phospholipid methylation depend on the types of receptors and cells. Stimulation of neuronal cells with nerve growth factors achieves the maximal increase in phospholipid methylation in approximately several seconds.[36] On the other hand, stimulation of β-adrenergic receptors or benzodiazepine receptors in C_6 astrocytoma cells requires 30 min to obtain the maximal enhancement of phospholipid methylaction.[37] Since PtdCho synthesized by the transmethylation pathway contains more arachidonic acid in its side chain,[38] this molecule is rapidly degraded by the action of phospholipase A_2 to release free arachidonic acid. Since lyso-PtdCho can not be extracted by cholorform–methanol, an apparent decrease can be detected.

The coupling between synthesis and degradation of methylated lipids is dependent on pool sizes of lipid domains in the vicinity of receptors. When a pool is larger, one may obtain an apparent dissociation between receptor stimulation and phospholipid methylation. For example, the treatment of plateletes with 3-deazaadenosine causes instant inhibition of phospholipid methylation, but there is a considerable lag phase before the inhibition of thrombin-induced aggregation.[39] Thrombin-induced aggregation of platelets requires arachidonic acid, which might come from phosphatidylinositol (PtdIno) and PtdCho synthesized by the transmethylation.[40,41] On the other hand, in rabbit neutrophils, stimulation with synthetic chemoattractants causes immediate degradation of PtdCho formed by transmethylation.[35] The inhibition of phospholipid methylation results in the inhibition of chemotaxis in these cells. Thus, an increase in phospholipid methylation appears to take place in the vicinity of the receptors. Indeed, stimulation of phospholipid methylation in neurons from rat superior cervical ganglia with nerve growth factor can be observed in growing neutrites but not in perikarya.[36] Since the receptors for nerve growth factors are mainly localized in growing neurites, these results suggest that PMTases in the vicinity of the nerve growth factor receptors are stimulated.

Stimulation of IgE receptors on rat mast cells with antireceptor antibody causes a transient increase of phospholipid methylation.[42] When homogenates of rat mast cells are fractionated subcellularly, IgE receptors are primarily located in the plasma membrane, and PMTases are present in both plasma membrane and mitochondrial fractions. Stimulation of the plasma membrane fraction but not of the mitochondrial fraction with anti-IgE receptor antibody results in the stimulation of phospholipid methylation. These results also support the hypothesis that the activities of PMTases in the domains of individual receptors increase after binding of ligands. However, it should be noted that some receptors have no connection with PMTases. Neuroblastoma cells have muscarinic acetylcholine receptors, which lead to cyclic GMP formation. Stimulation of these cells with carbamylcholine or batrachotoxin, a sodium channel activator, brings about a decrease in phospholipid methylation. Inhibition of phospholipid methylation by 3-deaza-SIBA or 3-deazaadenosine has no effect on cyclic GMP level (F. Hirata, unpublished data).

4.3. Interaction of PMTases with the Adenylate Cyclase System

Some receptors such as β-adrenergic receptors are well known to be coupled to adenylate cyclase.[43] The coupling of receptors to adenylate cyclase is greatly influenced by membrane fluidity.[44,45] Increased membrane fluidity caused by vaccenic acid enhances cyclic AMP formation stimulated by β-adrenergic receptors, whereas decreased membrane fluidity caused by cholesterol reduces it. Since phospholipid methylation can alter membrane fluidity, it may affect cyclic AMP formation. When rat reticulocyte ghosts, which have β-adrenergic receptors coupled to adenylate cyclase, are incubated with varying concentration of S-AdoMet, cyclic AMP formation stimulated by β-adrenergic ligands is enhanced, but its basal level is not affected.[46] Under these conditions, the number of β-adrenergic receptors is not altered, and cyclic AMP itself does not increase phospholipid methylation. Since the direct interaction between β-adrenergic receptors and PMTases is suggested, these results indicate that methylated phospholipids formed in the domains of β-adrenergic receptors induce changes in membrane fluidity, which, in turn, increase adenylate cyclase activity.

Plasma membranes form mast cells that bear IgE receptors can synthesize cyclic AMP when they are stimulated by IgE plus anti-IgE antibody or by anti-IgE-receptor antibody. This stimulation of cyclic AMP production can be blocked by 3-deaza-SIBA or S-AdoHcy, and such inhibition can be overcome by adding S-AdoMet.[42] Recently, Ross reported that the interaction of the catalytic protein of adenylate cyclase with its regulatory protein is promoted by PtdCho but not by other phospholipids.[47] These results support the observation that phospholipid methylation alters the composition and structure of lipid bilayers, which in turn modify the coupling of the β-adrenergic receptor and/or regulatory proteins to adenylate cyclase.

However, in intact cells, inhibition of cyclic AMP formation is hardly obtained, even when phospholipid methylation in intact cells is blocked by 3-deazaadenosine or N^6-methyladenosine (F. Hirata, unpublished data). In some cases, these treatments enhance cellular cyclic AMP formation. This might be attributable to direct stimulation by adenosine analogues of adenosine receptors, which are also coupled to adenylate cyclase,[43] or to inhibition of phosphodiesterase by S-AdoHcy analogues that are formed intracellularly from these adenosine analogues.[26] On the other hand, C_6 rat glioma cells infected by paramyxovirus have impaired capacity to stimulate cyclic AMP formation mediated through β-adrenergic receptors as well as altered activity of PMTases.[48] To study the role of PMTases in biological functions, mutants of cells that have impaired PMTases or a deficiency provide better insights.

4.4. Interaction between Cytoskeletons and PMTases

Lateral diffusion of membrane components such as proteins is regulated by membrane fluidity, which is affected by many factors including cholesterol content, degree of unsaturation of side-chain fatty acids of phospholipids, and the ratio of PtdCho to PtdEth. However, much evidence has suggested that most protein components, if not all, have the possibility of interacting with underlying cytoskeletons, leading to the idea of cytoskeletal control of diffusion.[49]

The mechanism of cytoskeletal control might involve the attachment of proteins to cytoskeletal elements or indirect effects through the matrix of the cytoskeletal network. Tubulin, a basic structural unit of microtubules, has been implicated in a number of membrane-linked processes such as secretion, transport, and cell shape.[50] This purified protein can be spontaneously incorporated into PtdCho vesicles at the lipid phase transition temperature. The stable vesicle–tubulin complexes thus formed can be fused to form larger structures by adding Ca^{2+} and other metal ions.[51] An erythrocyte membrane protein, ankyrin, has been suggested to interact with spectrin, one of the cytoskeletal elements.[52] However, purified ankyrin cannot provide binding sites to spectrin. Spectrin has the potential for binding ankyrin only in the native membrane or in reconstituted liposomes.[53] Thus, the composition and structure of lipid bilayers, a matrix of membranes, appear to affect the functions and activities of membrane proteins associated with cytoskeletal elements. The inhibitors of tubulins such as colchicine, vincristine, and vinblastine can inhibit PMTases in intact macrophages and neutrophils, whereas the antimicrofilament drug cytochalasin B can not.[54] Conversely, the alterations in membrane lipids by PMTases should have some effects on the function of microtubules. As described above, phospholipid methylation can enhance cyclic AMP formation. Since adenylate cyclase is reported to interact with cytoskeletons,[55] the structural and compositional alteration of membrane lipids induced by phospholipid methylation should have some effects on the activity of adenylate cyclase attached to the cytoskeletons. Addition of exogenous monomethyl and dimethyl PtdEth enhances such adenylate cyclase activity.

4.5. Regulation of Ca^{2+} Influx by PMTases

Many biological functions of various cells, such as chemotaxis, biogenic amine release, and tumor-killing activity, can be blocked by inhibitors of transmethylation.[2,56] In these biological events, Ca^{2+} plays an important role.[57] The regulation of Ca^{2+} influx by PMTases was first suggested by inhibitors of transmethylation. Treatment of rat mast cell and murine lymphocytes with 3-deazaadenosine or 3-deaza-SIBA can block Ca^{2+} influx induced by Con A or anti-IgE-receptor antibody.[58,59] The degree of inhibition of Ca^{2+} influx as measued by ^{45}Ca uptake parallels that of phospholipid methylation. On the other hand, the treatment of neutrophils with 3-deazaadenosine can inhibit Ca^{2+} influx by 40%, which leads to approximately 50% inhibition of chemotaxis of these cells,[60] whereas threatment of neuroblastoma cells with 3-deazaadenosine and 3-deaza-SIBA has no effect on cyclic GMP formation induced by carbamylcholine although it results in approximately 40% inhibition of Ca^{2+} influx under the same conditions. These differences in the sensitivity of Ca^{2+} channels to treatment with transmethylation inhibitors might be related to compartmentation or sequestration and/or the structure of Ca^{2+} channels in the cells and/or receptors.

Nevertheless, one type of Ca^{2+} channel is more closely associated with PMTases. Rat basophilic leukemia cells require Ca^{2+} influx to release histamine when they are stimulated by IgE plus antigen. Variants of these cells that are defective in either PMTase I or PMTase II can not produce Ca^{2+} influx and histamine release.[10] Hybrids of these two cell lines retain the capacity to synthesize PtdCho by transmethylation and to carry out Ca^{2+} influx and histamine release. These results more definitely support the hypothesis that phospholipid methylation can regulate the opening of Ca^{2+} channels, although the mechanism remains to be explored.

The turnover of methylated phospholipids is faster than that of PtdCho synthesized by the CDP-choline pathway because PtdCho formed by transmethylation contains more arachidonic acid.[6] One product, lysoPtdCho, also has a variety of biological effects. It increases glucose uptake and Na^+ permeability.[62] Since these effects are not the primary effects of phospholipid methylation, the details will be described elsewhere.

4.6. Development and PMTases

Cellular functions generally alter as cells develop or differentiate. Transformed cells generally have higher PMTase activities, whereas differentiated cells have lower PMTase activities.[63,64] Interestingly, one type of cells, a hybridoma line of mouse plasmacytoma and mouse spleen cells, requires ethanolamine for its growth.[65] Since adminstration of ethanolamine causes more production of PtdCho followed by phospholipid methylation, phospholipid methylation appears to play an important role in cell growth, namely, mitogenesis. On the other hand, treatment of 3T3 fibroblasts with 3-deazaadenosine results in differentiation of the cells into adipocytes.[65]

In animals, there is a significant decrease in the amount of PtdCho with a concomitant increase in the amount of cholesterol during development. Consequently, membrane fluidity decreases with age. The activities of PMTases in the rat brain are highest 14 days after birth and then gradually decrease with age.[66] Since the synaptic PtdCho synthesized by transmethylation may be a primary source for free choline, which might be involved in the regulation of acetylcholine synthesis,[67] the alteration of PMTase activities in the brain with age might reflect brain function. The number of β-adrenergic receptors increases in rat hearts hypertrophied by thyroid hormones or abdominal aortic constriction.[68] Under these conditions, the activities of PMTases increase. Thus, enhanced phospholipid methylation may have functional implications *in vivo*.

5. CONCLUSIONS

In this chapter, I have described the biological roles of phospholipid methylation in a variety of tissues and organs. Most of the evidence is based on pharmacological manipulation of phospholipid methylation by adenosine and its analogues. Recently, Schanche *et al.* extensively studied the inhibitory effects of these compounds on phospholipid methylation in isolated rat hepatocytes.[69] Generally, the degree of inhibition of phospholipid methylation correlates with the amount of S-AdoHcy accumulated inside cells as a result of inhibition of S-AdoHcy hydrolase. However, adenosine and its analogues can be degraded by adenosine deaminase, and, consequently, inhibition of phospholipid methylation by some analogues is transient. Furthermore, the increase and decrease of phospholipid methylation after stimulation of receptors depend on the population of receptors and the sizes of lipid domains. When PMTase activity not connected with receptors is higher, such changes in phospholipid methylation can hardly be detected.

Methylated lipids turn over rapidly in many cells. Therefore, in experiments with mixed populations of cells, mixed effects on PMTase activities would be observed. Thus, caution is necessary in drawing any conclusion from the apparent experimental results.

Recent experiments with mutants of *Saccharomyces* demonstrated that PMTase I and II are controlled by different genes and that PMTase I adds one methyl group to PtdEth, whereas PMTase II can catalyze the conversions of monomethyl PtdEth to dimethly PtdEth and of dimethyl PtdEth to PtdCho.[70] Interestingly, PMTase activites can be repressed by both inositol and choline. These results suggest that phospholipid methylation is regulated by the turnover of other phospholipids, especially PtdCho synthesized by the CDP-choline pathway and PtdIno. The turnover of PtdIno is also stimulated in a receptor-mediated fashion.[71] Thus, synthesis and degradation of one species of phospholipids are affected by those of others, and they in turn regulate the biological functions of hormone and neurotransmitter receptors and other cellular activities.

REFERENCES

1. Bremer, K., and Greenberg, D. M., 1961, *Biochem. Biophys. Acta* **46**:205–216.
2. Hirata, F., and Axelrod, J., 1980, *Science* **209**:1082–1090.
3. Hirata, F., Viveros, H. O., Deliverto, E. J., Jr., and Axelrod, J., 1979, *Proc. Natl. Acad. Sci. U.S.A.* **75**:1718–1721.
4. Mozzi, R., Andreoli, V., and Porcellati, G., 1980, *FEBS Lett.* **131**:115–118.
5. Crews, F. T., Hirata, F., and Axelrod, J., 1980, *J. Neurochem.* **34**:1491–1498.
6. Sastry, R. B. V., Satham, C. M., Axelrod, J., and Hirata, F., 1981, *Arch. Biochem. Biophys.* **211**:762–773.
7. Schneider, W. J., and Vance, D. E., 1979, *J. Biol. Chem.* **254**:3886–3891.
8. Tanaka, Y., Doi, O., and Akamatsu, Y., 1979, *Biochem. Biophys. Res. Commun.* **87**:1109–1115.
9. Scarborough, G. A., and Nyc, J. F., 1967, *J. Biol. Chem.* **242**:238–342.
10. McGivney, A., Crews, F. T., Hirata, F., Axelrod, J., and Siraganian, R. P., 1981, *Proc. Natl. Acad. Sci. U.S.A.* **78**:6176–6180.
11. Op Den Kamp, A. F., 1979, *Annu. Rev. Biochem.* **48**:47–71.
12. Bergelson, L. D., and Barsucov, L. I., 1977, *Science* **197**:224–230.
13. Hirata, F., and Axelrod, J., 1978, *Proc. Natl. Acad. Sci. U.S.A.* **75**:2348–2352.
14. Crews, F. T., Hirata, F., and Axelrod, J., 1980, *Neurochem. Res.* **5**:983–991.
15. Higgins, J. A., 1981, *Biochim. Biophys. Acta* **640**:1–15.
16. Wakil, S. J., 1970, *Lipid Metabolism*, Academic Press, New York.
17. Trewhella, M. A., and Collins, F. D., 1973, *Biochim. Biophys. Acta* **296**:51–61.
18. Klauser, R. D., Kleinfeld, A. M., Hoover, R. J., and Karnovsky, M. J., 1980, *J. Biol. Chem.* **255**:1286–1295.
19. Akesson, S., 1978, *FEBS Lett.* **92**:177–180.
20. Schneider, W. J., and Vance, D. E., 1979, *Eur. J. Biochem.* **85**:181–187.
21. Pritchard, P. H., Chiang, P. K., Cantoni, G. L., and Vance, D. K., 1982, *J. Biol. Chem.* **237**:6362–6367.
22. Sugiura, T., Onuma, Y., Sekiguchi, N., and Waku, K., 1982, *Biochim. Biophys. Acta* **712**:515–522.

23. Albert, D. H., and Snyder, F., 1983, *J. Biol. Chem.* **258**:97–102.
24. Vargatfig, B. B., Chignard, M., Benvenite, J., Lefort, J., and Wal, F., 1981, *Ann. N.Y. Acad. Sci.* **77**:119–136.
25. Cantoni, G. L., Richard, H. H., and Chiang, P. K., 1979, *Transmethylation* E. Usdin, R. T. Borchardt, and C. R. Creveling, Elsevier, Amsterdam, pp. 155–165. .
26. Zimmerman, T. P., Schmitges, C. J., Wolberg, G., Deeprose, R. D., Duncan, G. S., Cuatrecasas, P., and Elion, G. B., 1980, *Proc. Natl. Acad. Sci. U.S.A.* **77**:5639–5643.
27. Hirata, F., and Axelrod, J., 1980, *Nature* **275**:219–220.
28. Nakajima, M., Tamura, E., Irimura, T., Toyoshima, S., Hirano, H., and Osawa, T., 1981, *J. Biochem.* **89**:665–675.
29. Strittmatter, W. J., Hirata, F., and Axelrod, J., 1979, *Biochem. Biophys. Res. Commun.* **88**:147–153.
30. Strittmatter, W. J., Hirata, F., and Axelrod, J., 1979, *Science* **204**:1207–1209.
31. Bhattacharya, A., and Vonderhaar, B. K., 1979, *Proc. Natl. Acad. Sci. U.S.A.* **76**:4487–4492.
32. Pike, M. C., and Snyderman, R., 1982, *Cell* **28**:107–114.
33. Hirata, F., 1982, *Biochemistry of S-Adenosylmethionine and Related Compounds.* (E. Usdin, R. T. Borchardt, and C. R. Creveling, eds.), Macmillan, London, pp. 109–118.
34. Tallman, J. F., Henneberry, R. C., Hirata, F., and Axelrod, J., 1979, *Catecholamines: Basic and Clinical Frontiers*, Volume 1. (E. Usdin, I. J. Kopin, and J. Barchas, eds.), Pergamon Press, New York, pp. 489–491.
35. Hirata, F., Corcoran, B., Venkatasubramanian, K., Schiffmann, E., and Axelrod, J., 1979, *Proc. Natl. Acad. Sci. U.S.A.* **76**:2640–2643.
36. Pfenniger, K. H., and Johnson, M. P., 1981, *Proc. Natl. Acad. Sci. U.S.A.* **78**:7797–7800.
37. Strittmatter, W. J., Hirata, F., Axelrod, J., Mallorga, P., Tallman, J. F., and Henneberry, R. C., 1979, *Nature* **282**:857–859.
38. Crews, F. T., Morita, Y., McGivney, A., Hirata, F., Siraganian, R. P., and Axelrod, J., 1981, *Arch. Biochem. Biophys.* **212**:561–571.
39. Randon, J., Lecompte, T., Chignard, M., Siess, W., Marlas, G., Dray, F., and Vargraftig, B. B., 1981, *Nature* **293**:660–662.
40. Kannagi, R., Koizumi, K., and Masuda, T., 1981, *J. Biol. Chem.* **256**:1177–1184.
41. Rittenhouse-Simmons, S., 1979, *J. Clin. Invest.* **63**:580–587.
42. Ishizaka, T., Hirata, F., Sterk, A. R., Ishizaka, K., and Axelrod, J., 1981, *Proc. Natl. Acad. Sci. U.S.A.* **78**:6167–6180.
43. Ross, E. M., and Gillman, A. G., 1980, *Annu. Rev. Biochem.* **49**:533–564.
44. Hirata, F., Strittmatter, W. J., and Axelrod, J., 1979, *Proc. Natl. Acad. Sci. U.S.A.* **76**:368–372.
45. Rimmon, G., Hanski, E., Braum, S., and Levitzke, A., 1978, *Nature* **276**:394–397.
46. Klein, I., Moore, L., and Pastan, I., 1978, *Biochem. Biophys. Acta* **506**:42–53.
47. Ross, E. M., 1982, *J. Biol. Chem.* **257**:10751–10758.
48. Münzel, P., and Kosehel, K., 1982, *Proc. Natl. Acad. Sci. U.S.A.* **79**:3692–3696.
49. Edelman, G., 1976, *Science* **192**:218–226.
50. Soifer, D., and Czosnek, H., 1980, *Microtubules and Microtuble Inhibitors*, (M. DeBrabander, and J. DeMey, eds.), Elsevier/North-Holland, Biomedical Press, Amsterdam, pp. 429–447.
51. Kumar, N., Blumenthal, R., Henkart, M., Weinstein, J. N., and Klausner, R. D., 1982, *J. Biol. Chem.* **257**:15137–15144.
52. Marchese, S. L., Steers, E., Marchesi, V. T., and Tillalk, T. W., 1969, *Biochemistry* **9**:50–57.
53. Hargraves, W. R., Giedd, K. N., Verkleji, A., and Branton, D., 1980, *J. Biol. Chem.* **256**:11965–11972.
54. Pike, M., Kredich, N. M., and Snyderman, R., 1980, *Cell* **20**:373–379.
55. Sahyoun, N. E., Levine, H. III, Devis, J., Hobdon, G. E., and Cuatrecasas, P., 1981, *Proc. Natl. Acad. Sci. U.S.A* **78**:6158–6162.
56. Hoffman, T., Hirata, F., Bongnonx, P., Fraser, B. A., Goldfarb, R. H., Herberman, R. B., and Axelrod, J., 1981, *Proc. Natl. Acad. Sci. U.S.A.* **78**:3839–3843.
57. Rasmussen, H., and Goodman, D. B. P., 1977, *Physiol. Rev.* **57**:421–509.
58. Ishizaka, T., Hirata, F., Ishizaka, K., and Axelrod, J., 1980, *Proc. Natl. Acad. Sci. U.S.A.* **77**:1903–1909.

59. Toyoshima, S., Hirata, F., Axelrod, J., Beppu, M., Osawa, T., and Waxdal, M. J., 1982, *Mol. Immunol.* **19:**229–234.
60. Bareis, D. L., Hirata, F., Schiffmann, E., and Axelrod, J., 1982, *J. Cell Biol.* **93:**690–697.
61. Mallorga, P., Tallman, J. F., Henneberry, R. C., Hirata, F., Strittmatter, W. J., and Axelrod, J., 1980, *Proc. Natl. Acad. Sci. U.S.A.* **77:**1341–1345.
62. Shitara, N., McKeever, P. E., Cummins, C., Smith, B. H., Kornblith, P. L., and Hirata, F., 1982, *Biochem. Biophys. Res. Commun.* **109:**753–761.
63. Cimazierre, C., Maziere, J. C., Mora, L., and Polonovski, J., 1981, *FEBS Lett.* **129:**67–69.
64. Honma, Y., Kasukabe, T., and Hozumi, M., 1981, *Biochim. Biophys. Acta* **664:**441–444.
65. Chiang, P. K., 1980, *Science* **211:**1164–1166.
66. Hitzemann, R., 1982, *Life Sci.* **30:**1297–1303.
67. Blusztajn, J. K., Ziesel, S. H., and Wurtman, R. J., 1982, *Biochemistry of S-Adenosylmethionine and Related Compounds.* (E. Usdin, R. T. Borchardt, and C. R. Creveling, eds.), Macmillan, London, pp. 155–164.
68. Limas, C. J., 1980, *Circ. Res.* **47:**536–541.
69. Schanche, J. S., Sanche, T., and Ueland, P. M., 1982, *Biochim. Biophys. Acta* **721:**399–407.
70. Yamashita, S., Oshima, A., Nikawa, J., and Hosaka, K., 1982, *Eur. J. Biochem.* **128:**589–595.
71. Michel, R. H., 1975, *Biochim. Biophys. Acta* **415:**81–147.

Protein Phosphorylation
Role in the Function, Regulation, and Adaptation of Neural Receptors

Yigal H. Ehrlich

1. INTRODUCTION

The phosphorylation and dephosphorylation of proteins is a cyclic process providing the means to carry out reversible conformational changes that have been implicated in the control and regulation of numerous cellular functions.[1-4] In this process, the enzyme protein kinase transfers a phosphate group from ATP (or GTP) onto hydroxyl groups in amino acid residues (serine, threonine, or tyrosine) of protein substrates. The phosphoester linkage formed is hydrolyzed, in turn, by phosphoprotein phosphatases, and the protein substrate resumes its original structure. A detailed review on protein kinases and phosphatases in neural tissues has been included in Volume 4 of this *Handbook*.[5]

Cells in the nervous system contain exceptionally high levels of protein kinase, phosphatase, and their substrates.[5-9] During the past decade, a large body of evidence has accumulated demonstrating the ubiquitous role of protein phosphorylation in neuronal function.[5-9] This evidence is based on two main lines of investigation. The first is the demonstration that various neurohormones, neurotransmitters, and neurotropic factors, acting either directly or indirectly (via second messenger systems), regulate the phosphorylation of many different proteins in neural tissue. The second line of investigation has implicated phosphorylative activity in the regulation of various neuronal processes. These include the metabolism and release of neurotransmitters, the induction and biosynthesis of neural-specific enzymes, axonal transport, chemically and electrically induced depolarization, receptor sensitivity, neuronal maturation, and differentiation. Moreover, reports from several laboratories have demonstrated that modifications in certain protein phosphorylation systems are involved in adaptive processes whereby neurons respond with long-

Yigal H. Ehrlich • Neuroscience Research Unit, Departments of Psychiatry and Biochemistry, University of Vermont College of Medicine, Burlington, Vermont 05405.

lasting alterations to persistent environmental, hormonal, or pharmacological stimulations. In addition to these neuronal-specific functions, protein phosphorylation in the nervous system is most likely involved, as has been shown in other tissues, in the regulation of several key cellular mechanisms. Thus, processes such as DNA, RNA, and protein synthesis, intermediary energy metabolism, mitochondrial activity, cell division, and transitions in cell cycles have all been shown to involve protein phosphorylation/dephosphorylation cycles (for recent reviews, see chapters in the volume edited by Rosen and Krebs[3]).

The present chapter is focused on studies concerning the role of protein phosphorylation in the function of receptors in the nervous system. These studies can be divided into four main aspects. These aspects are presented here in an order that also provides some historical perspective or the research development:

1. Protein phosphorylation as an intracellular messenger in the action of neuronal receptors (see Section 2).
2. Regulation of neurotransmitter synthesis and release by protein phosphorylation (see Section 3).
3. Studies on the role of phosphoproteins in receptor-mediated signal transduction across plasma membranes (Section 4).
4. Protein phosphorylation in receptor modification and neuronal adaptation (Section 5).

1.1. On the Specificity in the Regulation Exerted by Protein Phosphorylation Systems

In spite of the great diversity of metabolic and physiological events regulated by protein phosphorylation, only a few different protein kinases have been identified in brain tissue (see Rodnight[5] for a most recent review). The first clue to the question of the source of specificity in this mode of regulation was provided by studies in the laboratory of Paul Greengard and his colleagues,[10,11] who reported that preparations of synaptic membranes from rat cerebrum contain two different protein substrates whose phosphorylation is regulated by cyclic AMP.

Subsequent studies in other laboratories[12–14] have confirmed this finding and identified six additional proteins in synaptic membranes whose phosphorylation and dephosphorylation are carried out by protein kinase(s) and phosphatase(s) that constitute part of the structure of these membranes. Following the discovery of calcium-dependent protein kinase activity in synaptic membranes,[15,16] it was found that the specific phosphoproteins not studied by Ueda et al.[11] are the endogenous substrates for this activity and that the saturation of calcium binding sites in the membrane, which occurs during the lysis of synaptosomes in a buffer containing 50 μM $CaCl_2$,[12] is sufficient to fully support the endogenous phosphorylation of these proteins.[17] Subcellular fractionation studies[18,19] have revealed over 20 different endogenously phosphorylated proteins in membranes and over 40 in the cytoplasmic fraction from rat brain.

The membrane-bound endogenous phosphorylation systems were found to be unequally distributed among various sybsynaptic organelles: synaptic plasma membranes, synaptic vesicles, postsynaptic densities, etc.[19] Each such endogenous phosphorylation system showed differential sensitivity and responsivity to various factors that affect phosphorylative activity, such as ATP concentration,[11] reaction time,[12] presence of the divalent cations Mg^{2+},[11-13] Ca^{2+},[16] and Mn^{2+},[20] cyclic nucleotides,[10-14] and others. Thus, brain tissue and, in particular, synaptic membranes contain a multiplicity of endogenous phosphorylation systems, each phosphorylating a different protein substrate. The differences in subcellular location and in phosphorylative properties among various phosphoproteins strongly suggest that each phosphorylation system may subserve a different role in the regulation of synaptic function. The specificity in this mode of regulation is provided, in all likelihood, by the nature of the phosphoprotein substrate.[9]

The conclusion that the simultaneous action of a multiplicity of endogenous phosphorylation systems with differential sensitivity is indicative of differences in function implies that such multiple systems should operate within the same cell. The studies cited above, however, investigated heterogeneous preparations from brain tissue containing subcellular organelles from many different cells. It is, therefore, of importance to note here that membranous and cytoplasmic fractions prepared from a homogeneous population of a cloned neuroblastoma cell line were found to have multiple endogenous phosphorylation systems that are as complex as those observed in brain tissue.[21,22] Furthermore, selective alterations in the phosphorylation of specific proteins have been detected when such cells were induced to differentiate in culture and acquired many properties characteristic of mature neurons.[21-23] Thus, different phosphoproteins of an individual nerve cell may subserve different physiological functions, and as transitions in functions occur at different phases of the cell's life cycle, alterations take place in the relative activity of various phosphorylation systems. This conclusion is also supported by the demonstrated differences in ontogeny[24,25] between various endogenous phosphorylation systems of rat brain.

Only a few of the numerous proteins that serve as endogenous substrates for protein phosphorylation activity in brain have been identified and assigned a known function. These include the microtubular associated protein MAP_2[26] (mol. wt. >300,000) and the α subunit of the sodium channel[27] (mol. wt. approximately 270,000), both of which are substrates for a cyclic-AMP-dependent protein kinase; proteins Ia and Ib, discovered by Greengard's laboratory (mol. wts. 80,000 and 76,000), which serve as substrates for both cyclic-AMP- and Ca^{2+}-dependent protein kinases[28] and are discussed in greater detail later in this chapter; the autophosphorylated regulatory subunit RII of cyclic-AMP-dependent protein kinase[29] (mol. wt. 54,000); the α subunit of pyruvate dehydrogenase in brain mitochondria[30,31] (mol. wt. 40–42,000); the α and β subunits of tubulin as substrates for a Ca^{2+}-dependent activity[32]; brain guanylate cyclase, recently shown to be a phosphorylated protein[33]; and finally, a group of proteins we have designated H^{12} in the mol. wt. range of 15–25,000, which undoubtedly contain myelin basic proteins[34] but in addition may include the neural equivalent of myosin light chain[35] and similar proteins with potential for involvement in the transient response of neurons to electroconvulsive stimulation.[17] In a review of the early literature on protein phosphorylation (covering studies until 1974), Rubin and Rosen[36] have indicated that the establishment of direct causal relationships between protein phosphorylation and defined cellular functions must await the identification, isolation, and purification of proteins that serve as specific substrates for various protein kinases. Although much progress has been made in this research during the past 8 years, it can be stated without reservation that in studies of protein phosphorylation in the nervous system, the majority of the required effort still lies ahead.

1.2. A Comment on the Methodology

Many of the suggestions on the possible role of certain phosphoproteins in various aspects of neuronal function have been based on the assumption that the properties that an individual endogenous phosphorylation system reveals *in vitro*, in cell-free assays, may be indicative of the function that this system plays *in vivo*. Such conclusions, however, should be taken with great caution, particularly in relation to phosphorylation systems operating in the central nervous system (CNS). A case in point was made by studies on the effects of the method used to sacrifice animals in experiments on endogenous phosphorylative activity in brain membranes.[9,17,37-39] In all studies of rapidly metabolized substances within the brain, the experimental animals are killed by methods enabling rapid enzyme inactivation, such as microwave irradiation or cryogenic techniques (for a recent review, see Lenox *et al.*[40] in Volume 2 of this *Handbook*). Sacrifice methods that produce rapid postmortem changes, such as decapitation, are avoided in these studies. The phosphorylation/dephosphorylation cycles of membrane-bound proteins from brain are very rapid, particularly under conditions of limited ATP availability.[11-13] Decapitation is known to produce rapid ATP depletion in brain tissue[41] and, therefore, may be expected to produce changes in protein phosphate metabolism.

This issue has been investigated by examining the phosphorylation of specific proteins in cortical membranes prepared from rats sacrificed by decapitation as compared to corresponding preparations from rats sacrificed by head immersion in liquid nitrogen.[17,39] As compared to membranes from quick-frozen rats, decapitation caused selective and bidirectional changes in the *in vitro* phosphorylation of specific proteins. Phosphorylation of proteins with apparent mol. wts. of 56,000 and 52,000 increased twofold after decapitation, whereas phosphorylation of a protein with a mol. wt. of 47,000 showed a 50% decrease. In spite of the variety of manipulation and length of time required to prepare the membrane fragments used in these experiments, the consequences of decapitation were sufficiently stable to permit the detection of changes in protein phosphorylation by *post-hoc* analysis, utilizing an *in vitro* assay.

Since changes in different phosphoprotein bands occurred in opposing directions, the results cannot be explained merely on the basis of ATP depletion. Our interpretation of these findings has been that decreases and increases in the state of phosphorylation of specific proteins induced by a stimulus *in vivo* result in respective increases and decreases of [^{33}P]phosphate incorporation into these protein bands in the *post-hoc* assay carried out *in vitro*. These findings perhaps best illustrate the great importance of considering the mode of sacrifice when interpreting data on protein phosphorylation, even when results are obtained by assays of enzymatic activity carried out *in vitro* and not only when labeling is performed by injecting [^{33}P]orthophosphate *in vivo*.

Another important methodological issue related to the interpretation of data obtained by *in vitro* assays of endogenous phosphorylation concerns the concentration of ATP used in these assays.[13,42,43] Not only is the time course of phosphorylation/dephosphorylation determined by this concentration, as would be expected, but also the relative phosphorylation state achieved by different protein substrates is very sensitive to this variable. This finding indicates that different endogenous phosphorylation systems of brain membranes may have different K_ms for ATP, as shown more directly in recent studies.[44] The final answer to the question of the correct interpretation of *in vitro* data in relation to the physiological function of any given protein phosphorylation system[42,43] must await, therefore, more accurate knowledge of the ATP concentration existing within the cell at the microenvironment where such an endogenous system is located.

A comment on the methodology cannot be concluded without a word on the comparison of results obtained by labeling phosphoproteins with [^{32}P]orthophosphate *in vivo* (or in intact cells grown in culture) to *in vitro* labeling by incubation with [γ-^{32}P]ATP. General aspects of this issue have been discussed in detail.[45] In brain, only one such detailed study has been published to date, and it showed striking differences in the pattern of phosphate incorporation into synaptic membrane proteins of rat brain labeled *in vivo* and *in vitro*.[46] It may not be concluded, however, that the pattern obtained by labeling brain tissue *in vivo* provides a better indication of processes that take place in intact cells than data obtained from *in vitro* assays, since both methods of analysis would be influenced by the mode of sacrifice and by conditions of handling and preincubation[12,47] of the membranes prior to the analysis of phosphorylation patterns and/or phosphorylative activity. As pointed out by Berman *et al.*[46] protein phosphorylation should be investigated at various orders of tissue organization in order to enable accurate assessment of the physiological significance of the findings. It was shown recently[48] that the cytoplasmit cyclic-AMP-dependent protein kinase phosphorylates different membranous proteins from those that are phosphorylated by the endogenous membrane-bound enzyme. Thus, the question of tissue and cellular organization is very crucial in interpreting data on protein phosphorylation in general and in relation to neuronal function in particular. Subsequent sections of this chapter include, therefore, pertinent information on the methodology utilized in each study reviewed here, and this information should be considered by the reader when evaluating the physiological relevance of the data presented.

1.3. Phosphorylative Responses to Neuronal Stimulation

Over a quarter of a century ago, Heald[49] reported that a brief depolarization of respiring slices from cerebral cortex by electrical pulses caused a significant increase in the amount of protein-bound phosphate in the tissue. These studies have been continued and expanded in the laboratory of R. Rodnight. The electrical stimulation-induced increase in protein-bound phosphate has been localized to membranes[50] and within them in neuron-enriched but not in glial-enriched fractions.[51] Subsequent studies from the same laboratory have demonstrated that not only electrical but also hormonal stimulation can induce an increase in protein-bound phosphate of cortical slices. Thus, the neurotransmitters norepinephrine, dopamine, GABA, histamine, and serotonin are capable of increasing [^{33}P]phosphate incorporation into proteins of respiring slices (reviewed by Williams and Rodnight[52]). The specific protein substrates whose phosphorylation is increased by either electrical or neurohormonal stimulation of brain slices have not been identified in these studies. Nonetheless, these reports were the first to indicate that activation of receptors by neurotransmitters results in altered membrane-bound protein phosphorylation activity.

The above studies on the effects of stimulation on protein phosphorylation in cortical slices used [^{32}P]orthophosphate to label intracellular ATP pools and then determined phosphoprotein labeling by measuring [^{32}P] incorporation into alkali-labile phosphate of the slices. It is likely that one of the reasons for the difficulty in identifying specific phosphorylated protein components in these studies was the low specific activity obtained by this method of labeling. On the other hand, *post-hoc* assays that analyzed endogenous phosphorylation activity *in vitro* after stimulation of brain slices *in situ* have proven useful in the investigation of specific phosphoproteins involved in the response of hippocampal slices to electrical[53,54] and neurohormonal[55,56] stimulation. The same experimental paradigm, namely, *post-hoc* analysis, has enabled the identification of specific protein components in cortical membranes whose phosphorylation responds to neuronal stimulation in the intact CNS.

Massive stimulation of central neurons is induced by electroconvulsive shock (ECS), and we have used this input to examine the rate and reversibility of phosphorylative changes induced by neuronal stimulation *in vivo*.[17] Rats were administered ECS and then killed by rapid head freezing during tonic seizures, during clonic seizures, and after behavioral recovery. Cortical membranes incubated with [γ-^{32}P]ATP *in vitro* revealed that phosphorylation of the bands designated group H (mol. wt. 15 to 25,000) increased 100% over sham-shocked controls during the tonic phase, reached maximum (400% increase) during the clonic phase, declined to 200% above control levels 2 min after shock, and returned to control level on behavioral recovery of the animals from the effects of the electroshock. Thus, neuronal stimulation can produce rapid and reversible changes in the phosphorylation of specific proteins of brain tissue. The mechanism underlying this phenomenon may involve activation of receptors by neurotransmitters. Subsequent sections of this chapter review studies in which this possibility has been investigated directly.

2. PROTEIN PHOSPHORYLATION AS AN INTRACELLULAR MESSENGER IN THE ACTION OF NEURAL RECEPTORS

The mechanisms underlying the process whereby stimulation of neural receptors results in altered protein phosphorylation may involve a reaction in

Table I
Alternate Modes for Involvement of Protein Phosphorylation Systems in Receptor Function

Receptor type	Associated protein kinases
1. Receptors coupled to adenylate cyclase (e.g., catecholamines)	Cyclic-AMP-dependent protein kinases
2. Receptors that activate gaunylate cyclase (e.g., muscarinic cholinergic)	Cyclic-GMP-dependent protein kinase(s)
3. Receptors that regulate uptake and/or sequestration of calcium ions (e.g., opiate receptors)	Ca^{2+}/calmodulin-dependent protein kinase
4. Receptors that regulate phospholipid metabolism (e.g., α_1-adrenergic, thrombin)	Diacylglycerol/phosphatidylserine-sensitive protein kinase (C-kinase)
5. Receptors that activate Ca^{2+} proteases (e.g., insulin)	Peptide-regulated protein kinases and phosphatases
6. Receptor complexes that contain an endogenous phosphorylation system as part of their structure (e.g., EGF, nicotinic cholinergic)	Protein kinases that are regulated directly by hormone–receptor interaction and/or that phosphorylate receptor proteins

which phosphorylative activity is regulated directly by the neurohormone–receptor complex. Studies related to this possibility are summarized in Section 4. Earlier studies in this line of investigation have focused on the suggestion that the main role of protein phosphorylation in the series of events triggered by receptor stimulation is secondary to the activation of "second messenger" systems.

The various alternate routes by which receptor stimulation can lead to altered protein phosphorylation are summarized in Table I. The action of many hormones and neurotransmitters is mediated intracellularly via the activation of adenylate cyclase. As suggested in 1969 by Kuo and Greengard,[57] most of the functions of the newly generated cyclicAMP are exerted, in turn, by the regulation of a cyclic-AMP-dependent protein kinase activity. Other hormones are known to activate receptor-linked guanylate cyclase. The function of these receptors would be mediated by a cyclic-GMP-dependent protein kinase.[6]

Another ubiquitous second messenger for many receptors is calcium (or, more precisely, Ca^{2+} ions). The mechanism by which subtle alterations in intracellular Ca^{2+} concentrations induce changes in cellular function are believed to be mediated by the activity of Ca^{2+}/calmodulin-dependent protein kinase.[58] More recently, an enzyme designated C-kinase has been identified in neural and nonneural tissues.[59] This protein kinase is regulated by the combined interaction of Ca^{2+} and certain phospholipids and may serve as a "third messenger" for hormones whose direct effect is to induce changes in phospholipid metabolism. Finally, it has been reported that certain influences of insulin on cellular function are mediated by the generation of intermediary peptides that serve as intracellular second messengers[60,61] Regulation of protein phosphorylation by small peptides derived from brain membranes has been reported.[62]

However, whether such a mechanism is regulated by activation of neural receptors is yet to be investigated.

2.1. Model Systems in the Study of Protein Phosphorylation as "Third Messenger" in the Action of Neurohormones

Most of the information on the precise role of phosphoproteins in mediating cellular reponses to receptor activation by neurohormones is available not from studies of the nervous system but from research utilizing model systems such as cells grown in culture and from studies on the interaction of neurotransmitters with receptors localized in nonneural cells. For example, catecholamines, angiotensin, and vasopressin can act as neurotransmitters and/or neurohormones to regulate neural function in the CNS. The action of these hormones, however, has been studied in greater detail in relation to their effects on the activity of enzymes involved in glycogenolysis and gluconeogenesis in liver cells. Recent studies[63] of isolated hepatocytes provided data implying that angiotensin II, catecholamines, and vasopressin control hepatic carbohydrate metabolism through a Ca^{2+}-requiring, cyclic-AMP-independent pathway which involves the phosphorylation of key regulatory enzymes. On the other hand, in the same cells, the activity of the enzymes phosphorylase, glycogen synthetase, and pyruvate kinase are regulated by the hormone glucagon. Glucagon influences on these cells are mediated by the activity of a cyclic-AMP-dependent protein kinase.[63] This well-characterized regulatory pathway serves to best exemplify a specific role for protein phosphorylation in the regulation of intracellular metabolic responses following receptor-mediated activation of the well-known second messengers cyclic AMP and calcium ions. Similar experiments in neuronal cells have not been reported yet, although it is reasonable to assume that such regulatory mechanisms also operate in the nervous system.

The ability to dissect in detail a series of biochemical events activated by hormones and to determine accurately the specific enzymatic steps in which certain phosphoproteins exert regulatory influences, as exemplified in the studies cited above,[63] were provided in large part by the use of cultured, homogeneous cell populations. The physiological significance of these findings was underscored by the fact that in this research, the labeling of specific proteins had been achieved by incubating intact cells with [^{32}P] orthophosphate. Thus, the extent and selectivity of changes in phosphate incorporation detected in various specific phosphoproteins were determined "by the cell" and not by arbitrary assay conditions selected by the investigator.

The complexity of events that can be studied in such systems has become particularly evident in studies by Groppi and Browning,[64] who combined the techniques of cell culture, ^{32}Pi labeling, and high-resolution two-dimentional gel electrophoresis to demonstrate that homogeneous populations of C-6 glioma cells contain more than 200 different phosphoproteins. Exposure of intact cells to norepinephrine (10 μM NE) for 5 min rapidly increased [^{32}P]phosphate incorporation into seven of these proteins, characterized by estimated mol. wt., isoelectric point (IEP), and subcellular location. Only one of these proteins, however, was identified and shown to be an intermediate filament protein with mol. wt. of 58,000 and IEP of 6.7.[64] Three of these seven phosphoproteins were found to be cytosolic, one mitochondrial, one nuclear, and none in the plasma membrane. The effects of NE on their phosphorylation must have been mediated, therefore, by a second messenger, which in C-6 cells is known to be cyclicAMP.

With this investigative approach, future studies may be able to identify each of the 200 phosphoproteins in cells of glial origin and perhaps assign each a specific functional role. Investigation of primary cell cultures will add another dimension to this research, as exemplified by the identification of carbonic anhydrase (mol. wt. 30,000) in astroglia as a phosphoprotein regulated by exposure of intact cells to NE.[65] A recent study[66] of NE and kainic acid effects on protein phosphorylation in hippocampal slices that has identified ten specific phosphorylated proteins in such preparations demonstrated the feasibility of using the technique of ^{32}Pi labeling in this model system as well. This study also demonstrated the complexity of this model by pointing out the great extent of glial contribution to phosphorylation patterns observed in studies of neural tissue.

Investigation of phosphorylation events in all of the different cells of any part of the nervous system would be necessary in order to fully understand the role of protein phosphorlyation in mechanisms underlying the responses of a given neuronal pathway to receptor stimulation. Therefore, study of activity in brain is highly warranted. Nonetheless, investigation of protein phosphorylation in dissociated, isolated cells will undoubtedly prove to be essential for correct interpretation of the physiological significance of findings obtained with heterogeneous preparations of neural tissue.

Numerous biochemical mechanisms whose elucidation requires the study of a homogeneous cell population have been investigated in clonal lines of malignant cells derived from various tumors. Findings obtained in such studies are always subject to the question of whether a described process also occurs in the normal tissue or is tumor-specific. In spite of this reservation, the usefulness of such model systems is widely recognized. In studies of neuronal function, investigation of neuroblastoma cell lines have provided data pertinent to the understanding of several biochemical processes involved in neuronal differentiation,[67] including protein phosphorylation.[21,22] More recently, a study of the pheochromocytoma clonal cell line PC-12 enabled identification of protein phosphorylation systems responsive to receptor stimulation by nerve growth factor (NGF).[68] In the clonal cell line PC-G2, studies on the role of protein phosphorylation in the function of epidermal growth factor (EGF) as a neuroregulator have been initiated.[69] The detailed information available on the numerous phosphoproteins in a C-6 glioma cell line[64] now provides the basis for future studies of protein phosphorylation in mechanisms underlying the desensitization (tachyphylaxis) of β-adrenergic receptors.[70] Neuroblastoma × glioma hybrid cell lines,[71] which have provided a wealth of data on the function of prostaglandin, α_2-adrenergic, opiate, and muscarinic-cholinergic receptors,[71] may serve as an ideal model system in investigating the role of protein phosphorylation in the function of these neural receptors.[72]

An outstanding example of the successful use of a model system in studies of protein phosphorylation as a "third messenger" in the function of neurohormones has been the use of the turkey erythrocyte as a catecholamine-responsive cell in which isoproterenol (a β-adrenergic agonist) causes a tenfold activation of a "Na$^+$–K$^+$ cotransport" pump.[73] A series of reports from Greengard's laboratory[73–76] have demonstrated that in these cells, elevated cyclic AMP levels in response to catecholamines activate the phosphorylation of globin, a plasma membrane protein of mol. wt. 230,000. The kinetics of globin phosphorylation were shown to correlate well with the kinetics of this cotransport pump. The phosphorylation of distinct sites in globin by a cyclic-AMP- and Ca^{2+}/calmodulin-dependent protein kinase present in turkey erythrocytes indicated multiple modes by which receptor activation can lead to altered protein phosphorylation. A similar system has also been described in the frog erythrocyte.[77] The recent identification[27] of the α subunit of the voltage-sensitive sodium channel (purified from rat brain) as a substrate of cyclic-AMP-dependent protein kinase activity suggests that protein phosphorylation also acts as a mediator for hormone-regulated ion pumps in the nervous system. Continuation of these studies may provide the direct evidence necessary to support the hypothesis advanced by Greengard 10 years ago[10] that cyclic-AMP-dependent phosphorylation of specific proteins plays a role in receptor-mediated permeability changes of synaptic membranes.

2.2. "Protein I," a Neuronal-Specific Substrate for Cyclic-AMP- and Ca^{2+}/Calmodulin-Dependent Protein Kinases: Role as Intracellular Messenger for Neural Receptors

The most detailed and complete studies on a protein phosphorylation system that acts as a "third messenger" in the series of events triggered by neurohormone–receptor interaction in the CNS are those concerning "protein I," recently named Synapsin by P. Greengard. Several detailed reviews of these studies are available.[5–8,28] This specific, cyclic-AMP-dependent, endogenous phosphorylation system of synaptic membranes was identified by Johnson *et al.*[10] in 1972; its detailed phosphorylative properties were described by Ueda *et al.*[11] and confirmed by independent studies in several other laboratories.[12–14,19,78] Subsequent studies resulted in isolation and purification of this protein to apparent homogeneity.[79] Protein I is a collective name for two proteins with very similar properties but different mol. wts. (86,000 and 80,000, referred to as protein Ia and Ib, respectively). Parallel studies that have investigated protein phosphorylation in intact synaptosomes incubated *in vitro* with ^{32}Pi, have found that the phosphorylation of two similar proteins is stimulated by veratridine and high K$^+$ under conditions that induce Ca^{2+}-dependent depolarization.[80] These were later found[81] to be the same as proteins Ia and Ib. It is now clear that protein I is a substrate for both cyclic-AMP- and Ca^{2+}-dependent protein kinases, which phosphorylate this protein differentially at multiple sites.[82–84] Subcellular fractionation studies[85] detected protein I in presynaptic vesicles and, to a much lesser extent, in postsynaptic densities. Immunocytochemical studies[86–88] verified this localization and demonstrated the presence of protein I in nerve terminals throughout the nervous system, including the neuromuscular junction. Within the terminal, protein I is located primarily in synaptic vesicles.

The ability to determine accurately the state of phosphorylation of protein I by examining its derived specific peptides and to measure sensitively its amount by a detergent-based radioimmunoassay[89] has provided the means to investigate the function of this phosphoprotein under physiological conditions. With these techniques, it has been demonstrated that the state of phosphorylation of protein I in slices of the facial motor nucleus from rat brainstem is regulated by serotonin[90,91] and adenosine[92] and that both effects are blocked by the appropriate specific antagonists, indicating a receptor-mediated neurohormonal activation of phosphorylative activity. A study of rat cerebral cortex slices demonstrated an increase in the state of phosphorylation of protein I by stable cyclic-AMP analogues and by depolarizing agents.[93] Dopamine and depolarizing agents were shown to induce an increase in the state of phosphorylation of protein I in sections of bovine superior cervical sympathetic ganglion.[94]

Regulation of the state of phosphorylation of protein I by impulse conduction has recently been demonstrated in intact preparations of rabbit superior cervical ganglion maintained under physiological conditions.[95] The picture that has emerged from these studies is that certain neurotransmitters that stimulate adenylate cyclase and depolarizing agents that induce Ca^{2+} fluxes activate cyclic-AMP-dependent and Ca^{2+}/calmodulin-dependent protein kinases, respectively. These activations result in increased phosphorylation of protein I. The predominant localization of this phosphoprotein in vesicles of nerve endings,[85–88] taken together with the results of the studies cited above,[91–95] has led to the suggestion that alterations in the state of phosphorylation of protein I "might be involved in certain physiological mechanisms underlying potentiation or inhibition of neurotransmitter release by neuromodulatory agents."[91] With the availability of affinity-purified anti-protein-I antibody, which specifically inhibits the phosphorylation of this protein,[96] direct probing of the physiological function(s) of protein I should be feasible.

2.3. Protein Phosphorylation in Identifiable Neurons of Aplysia

The best direct evidence for the involvement of protein phosphorylation as intracellular messenger in processes in which receptor activation of adenylate cyclase plays a role in the regulation of specific neuronal functions comes from studies of invertebrate neurons. In these studies, the purified catalytic subunit of cyclic-AMP-dependent protein kinase has been microinjected into identified neurons of *Aplysia*. Such injection into the bag cell neurons resulted in enhanced calcium action potentials,[97] whereas microinjection of the catalytic subunit in sensory cells of the abdominal ganglion of *Aplysia* altered the duration of calcium action potentials and affected neurotransmitter release.[98] Whether these effects are mediated by phosphorylation of protein I or a similar phosphoprotein has not yet been determined.

A different experimental approach has been used by Lemos *et al.*[99] They microinjected [γ-^{32}P]ATP into living R15 cells of *Aplysia* and found that exposure of the cells to serotonin caused increased phosphorylation of proteins with mol. wts. of 230,000, 205,000, 135,000, and 26,000 and decreased phosphorylation of a protein with a mol. wt. of 43,000.

It may be expected that the combination of data obtained from studies of protein phosphorylation in mammalian brain, the availability of antibodies directed against cyclic-AMP- and Ca^{2+}-dependent phosphorylative activities specific for neuronal tissue, and the utilization of cultured cells from *Aplysia* and other sources, which enable accurate physiological determinations, would yield, in the near future, more complete information on the exact role of numerous specific phosphoproteins as intracellular messengers for various neurohormones.

2.4. Protein Phosphorylation as Intracellular Messenger for Receptors Coupled with Phospholipid Turnover

Activation of certain receptors by neurotransmitters, notably muscarinic-cholinergic and α-adrenergic receptors, results in a rapid increase of phosphatidylinositol turnover. The process involves cleavage of the phospholipid by phospholipase C to yield diacylglycerol (DG) and inositol phosphate. The DG is rapidly converted to phosphatidic acid, which then reacts with CTP to form CDP-DG, which in turn combines with inositol to reform phosphatidylinositol.[100,101] Studies by Nishizuka and his colleagues have provided evidence that the activity of this second messenger system is also mediated by a protein phosphorylation system.[102,103] The enzyme involved was named protein kinase C. It has been purified from brain[104] and shown to be activated in a Ca^{2+}-dependent manner by phospholipids.[105] This activation involves a calcium-dependent association of an inactive form of the enzyme with membrane phospholipids.[105,106] Although several membrane phospholipids can interact cooperatively in bringing about this activation, the only component effective by itself is phosphatidylserine.[107] A small amount of unsaturated diacylglycerol markedly increases the affinity of this enzyme for Ca^{2+} to the submicromolar range.[108,109]

The role of protein kinase C in cellular activation by receptor stimulation has been elucidated in studies of human platelets.[103,110,111] When platelets are stimulated by thrombin, DG is rapidly produced and this reaction is immediately followed by phosphorylation of a cytosolic 40-kilodalton protein. This phosphorylation reaction leads eventually to the release of serotonin from the platelets. The enzyme that catalyzes the phosphorylation of the 40-kilodalton protein was found to be protein kinase C. Its activation is believed to be mediated by thrombin-induced formation of DG. A similar chain of events was shown recently in human peripheral lymphocytes on stimulation with mitogens.[112,113] Since all of the above reactions can be inhibited by cyclicAMP and cyclicGMP[114,115] it is likely that cyclic nucleotides exert feedback control on this regulatory pathway. The fact that the brain contains the highest levels of C kinase of all tissues examined[105] suggests that receptor-mediated regulation of this enzyme may play a crucial role in neuronal function, but such studies have not been reported yet.

2.5. *Cyclic-GMP-Dependent Protein Phosphorylation in Nervous Tissue*

Although considerable evidence now suggests that cyclicGMP may play a role in neural function and that its activity is mediated by a cyclic-GMP-dependent protein kinase, very little is known concerning the biochemical processes that are regulated by this phosphorylation system. The neurotransmitter known best to interact with guanylate cyclase is acetylcholine, and, therefore, early studies of cyclic-GMP-dependent protein kinase have been carried out in smooth muscle preparations, where specific membrane-bound substrates for its activity have been identified (for a recent review, see Walter and Greengard[116]). In the brain, a cyclic-GMP-dependent protein kinase and a specific protein substrate with mol. wt. of 23,000 for its activity have been found in the cytosol from the cerebellum.[117] By use of the photoaffinity label 8-N_3-[^{33}P]cyclic IMP in conjunction with study of various mutants of mice, this endogenous phosphorylation system was shown to be enriched in Purkinje cells.[118] These findings were later verified by immunohistochemical techniques and radioimmunoassays using anti-cyclic-GMP-dependent protein kinase antiserum.[119] In a recent series of studies, the cerebellar substrate of this enzyme was purified,[120] the kinetics of its phosphorylation were characterized,[121] and the amino acid sequences around its two phosphorylation sites have been determined.[122] Preparation of synthetic peptides corresponding to these sites and raising antibodies against them could provide the tools for investigating the physiological role of this phosphorylation system.

3. *REGULATION OF NEUROTRANSMITTER METABOLISM AND RELEASE BY PROTEIN PHOSPHORYLATION*

The rate of the synthesis of neurotransmitter molecules, the regulation of their release into the synaptic cleft, and the mechanisms controlling their inactivation by metabolism or reuptake determine the extent and duration of the exposure of neuroreceptors to their naturally occurring agonist. Since these factors play a crucial role in the regulation of receptor sensitivity, this chapter would not be complete without the citation of studies that implicate protein phosphorylation in these processes.

3.1. *Protein Phosphorylation in the Regulation of Tyrosine Hydroxylase and Tryptophan Hydroxylase*

The rate-limiting enzyme in the biosynthesis of catecholamines is tyrosine hydroxylase (TH). The activity of this enzyme requires molecular oxygen and the reduced form of a cofactor, tetrahydrobiopterin. Early studies[123] demonstrated that cyclicAMP can activate TH *in vitro* by a process requiring ATP, Mg^{2+}, and cyclic-AMP-dependent protein kinase. It was subsequently shown that this mode of activation involves a reduction in the apparent K_m of TH to its pteridine cofactor and a significant increase in the K_i for the end product dopamine.[124,125] Studies of TH purified from several sources directly demonstrated that a subunit of the enzyme with mol. wt. of 62,000 is a substrate for cyclic-AMP-dependent protein kinase.[126–128] Localization of tetrahydrobiopterin in dopamine terminals of rat striatum and evaluation of its intraneuronal concentration[129] allowed the tentative conclusion to be drawn that the phosphorylation of TH may represent a mechanism for rapidly increasing the rate of transmitter synthesis in these neurons.

Tryptophan hydroxylase catalyzes the rate-limiting step in the biosynthesis of serotonin. Several laboratories have demonstrated that this enzyme is activated by protein phosphorylation in a process involving reduction in apparent K_m for a pteridine cofactor.[130–133] In contrast to the activation of TH, the phosphorylative activity involved in tryptophan hydroxylase regulation is independent of cyclic nucleotides, requires low (micromolar) concentrations of Ca^{2+} ions,[131] and may be dependent on calmodulin.[134]

3.2. Protein Phosphorylation and Neurotransmitter Release

As detailed in Section 2.2 above, identification of protein I as a phosphoprotein located primarily in the synaptic vesicle has led to the suggestion that cyclic-AMP- and Ca^{2+}-dependent protein phosphorylation play a role in the regulation of neurotransmitter release.[28] Earlier studies have indicated this possibility by providing evidence that both adenylate cyclase and membrane-bound cyclic-AMP-dependent protein kinase of rat brain are present in presynaptic nerve endings.[135] The suggestion that Ca^{2+}-dependent protein phosphorylation plays a mediatory role in neurotransmitter release was also advanced by DeLorenzo,[136] who has identified protein phosphorylation systems that are inhibited by the antiepileptic drug diphenylhydantoin (phenytoin, DPH; see reference 137 for a detailed review). The enzyme underlying this activity was found to be a Ca^{2+}/calmodulin-dependent protein kinase localized in synaptic vesicles.[138] The key study demonstrating the relevance of this system to neurotransmitter release utilized intact synaptosomes prelabeled with ^{32}Pi and demonstrated good correlation between stimulation of Ca^{2+}-dependent norepinephrine release and phosphorylation of specific substrates in these preparations.[139] Recent studies from the same laboratory suggest that the protein substrates involved may be the α and β subunits of vesicle-associated neurotubulin, phosphorylated by a calmodulin–kinase system.[140,141]

Addition of ATP (1 mM) to intact synaptosomes incubated in a Krebs–Ringerlike medium was shown to markedly reduce Ca^{2+}-dependent, K^+-induced release of preloaded [^3H]GABA.[142] Although the authors have suggested that this effect is mediated by protein phosphorylation, the intriguing possibility that such a phosphorylation system would be located on the outer surface of plasma membranes was not discussed in this report. A promising approach to the study of protein phosphorylation in the regulation of acetylcholine release by presynaptic muscarinic receptors has utilized pure cholinergic synaptosomes from *Torpedo ocellata*. In this preparation, a good correlation was obtained between the effects of oxotremorine (a cholinergic agonist) on ^{32}Pi incorporation into a specific protein with an approximate mol. wt. of 100,000 and the release of acetylcholine from these synaptosomes.[143]

Detailed studies on the possibility that reuptake of neurotransmitters is regulated by protein phosphorylation have not been carried out yet. A recent report[144] suggests, however, that such studies are warranted.

4. PHOSPHOPROTEINS IN RECEPTOR-MEDIATED SIGNAL TRANSDUCTION THROUGH PLASMA MEMBRANES

The phosphorylation and dephosphorylation of proteins provide the means to carry out reversible conformational alterations that may mediate functional changes. This regulatory system does not require *de novo* synthesis or degradation of preexisting molecules. Endogenous phosphorylation systems can operate within plasma membranes independently of cytoplasmic enzymes. Studies utilizing homogeneous preparations of plasma membranes have implicated phosphorylation and dephosphorylation of proteins in mechanisms underlying structural rearrangement of membrane components.[145] Depending on the nature of the regulation exercised by various modulating factors on protein phosphorylation in synaptic plasma membranes, changes in the phosphorylation state of specific proteins (which constitute part of the structure of these membranes) can be short-lived (seconds or less) or persist for minutes or even hours.[4,6,9,11,12,17,52–55]

All of the foregoing renders endogenous phosphorylation systems of plasma membranes as most suitable candidates for serving as a transducing mechanism in receptor function. Receptor activation is initiated by the binding of agonists to recognition sites on the outer surface of the plasma membrane. However, site occupancy alone cannot account for receptor function. It has been proposed that the binding of agonists induces a conformational change, which triggers the chain of events leading ultimately to the expression of cellular responses.[146] This section summarizes studies demonstrating that interaction of various hormones and neurotransmitters with their specific receptors can exert direct influences on the phosphorylation of membrane-bound phosphoproteins. The conformational changes brought about by these phosphorylative modifications can, in turn, cause alterations in activity of effector sites localized on the inner surface of plasma membranes.

4.1. Phosphorylation of Receptor Proteins

The best evidence available to date demonstrating that hormone–receptor interaction directly regulates membrane-bound protein phosphorylation activity comes from studies of nonneural systems. S. Cohen and his colleagues have shown that the binding of epidermal growth factor (EGF) to specific receptors in plasma membranes of the epidermoid cell line A-431 results in a marked stimulation of the *in vitro* phosphorylation of proteins in these membranes.[147,148] Maximal stimulation of protein phosphorylation by EGF was obtained at a concentration ($3-4 \times 10^{-8}$ M) that is very similar to that required to saturate receptor binding sites (1.6×10^{-8} M). Detergent solubilization of the membranes extracted both [^{125}I]EGF binding activity and EGF-stimulated endogenous phosphorylation activity. Affinity chromatography revealed that a specific protein component of molecular weight 150,000 serves both as a binding site for EGF and as a substrate of EGF-enhanced phosphorylative activity. These studies have not only demonstrated an inherent close relationship between receptor and protein phosphorylation but also suggest that hormone-binding and kinase activities may be present in the same molecule.[149] An important property of the EGF receptor–protein kinase complex is that it phosphorylates tyrosine residues (rather than serine or threonine) of the protein substrate.[150] Undoubtedly, the importance of this refined difference will have to be investigated and clarified in all future studies of receptor–protein phosphorylation interactions. In addition, it should be pointed out that the EGF-stimulated phosphorylation of its own receptor, as studied by *in vitro* assays, is supported by Mn^{2+} ions severalfold better than by Mg^{2+} ions. Whether this property of the endogenous phosphorylation system has any physiological relevance has not been determined. Nonetheless, in evaluating the data reported to date[147–150] it should be kept in mind that the standard reaction conditions utilized in these reports have included Mn^{2+} instead of Mg^{2+}.

Recent studies of cultured human lymphocytes and rat hepatoma cells prelabeled with inorganic ^{32}Pi have revealed that insulin in the nanomolar concentration range stimulates the phosphorylation of a 95-kilodalton protein that constitutes part of its own receptor.[151] The positive identification of this protein as a constituent of the receptor was achieved by using antibody directed against the receptor. After labeling, fractions enriched in receptor molecules were obtained from Triton-X-100-solubilized cells by chromatography on a wheat germ aglutinin–agarose column, and the receptor complex was quantitatively immunoprecipitated with the specific antiserum in the presence of protein A. With progress in studies attempting to purify various receptors for neurotransmitters and neurohormones and the production of antibodies against them, it should be possible to determine whether such regulatory mechanisms also operate in the central nervous system.

Direct evidence for phosphorylation of proteins of neurotransmitter receptors has been obtained to date only in the case of nicotinic cholinergic receptors. The high concentration of this receptor in electric organs has enabled two research groups to report this finding simultaneously using preparations from *Electrophorus electricus*[152] and from *Torpedo californica*.[153] Immunoelectrophoresis using respective antisera by both groups revealed that the major phosphorylated protein in receptor preparations from *Torpedo* has a mol. wt. of 65,000,[154] and that from preparations from *Electrophorus* 48,000.[155] It should be mentioned, however, that whereas the former study utilized Mg^{2+} in the *in vitro* reaction, the latter utilized Mn^{2+}. In preparations from *Torpedo marmorata*, the Changeux group also found a 66-kilodalton substrate.[156]

Studies by these groups[152–158] have confirmed that in addition to the protein substrates, these receptor preparations contain endogenous protein kinase and phosphoprotein phosphatase activities. Most interesting properties of this endogenous phosphorylation system are its stimulation by K^+ ions[153,156] and, as shown by Gordon and Diamond,[153,159] its inhibition by the cholinergic agonist carbachol. The functional significance of the phosphorylation of the acetylcholine receptor is yet to be determined. Changes in phosphorylation during receptor development[160] and effects of the phosphorylation state of receptor proteins on the susceptibility of the receptor to heat inactivation[161] have suggested a possible role in receptor stabilization. The observed effect of an agonist on protein phosphorylation led to the suggestion[158] that this activity may play a role in receptor desensitization. These possibilities, however, are not mutually exclusive and would be best resolved, in all likelihood, by studying the interaction of phosphorylative activity with the function of cholinergic receptors in intact, viable cells functioning under physiological conditions.

4.2. Protein Phosphorylation and Receptor Binding Activity

The desensitization of receptors by prolonged exposure to agonists, as mentioned above, involves in many cases not only decreased cellular responsivity but also a decrease in the number of available recognition sites detected by radioreceptor binding assays. Possible involvement of membrane-bound protein phosphorylation in this process has been suggested by Chuang and Costa.[162] They have demonstrated a temporal correlation among the downregulation of β-adrenergic receptors in frog erythrocytes, the internalization of receptors, and increased phosphorylation of membrane proteins.[162]

A more direct demonstration of a role for protein phosphorylation in receptor binding was attempted in studies of muscarinic-cholinergic receptors.[163-166] Specific ligand binding was measured using the high-affinity antagonist [^{3}H]quinuclidinyl benzilate (QNB). Preincubation of synaptic plasma membrane preparations with 1 mM ATP in the presence of Mg^{2+} resulted in 44% decrease of specific [^{3}H]QNB binding. Omitting Mg^{2+} prevented this effect, and addition of cyclicAMP stimulated it.[163] The inhibition of QNB binding by preincubation under phosphorylating conditions required the presence of Ca^{2+} and calmodulin; the concentrations of Ca^{2+} and calmodulin required for half-maximal effect on QNB binding were found to be almost identical to the concentrations required for half-maximal stimulation of endogenous protein phosphorylation activity in these membranes.[164] In support of the possible physiological significance of this finding, Burgoyne and Pearce[165] have demonstrated a temporal correlation between carachol-induced loss of [^{3}H]QNB binding sites in primary cultures of cerebellar cells from 8-day-old rats and carbachol-induced increase in the phosphorylation of three membrane-proteins with apparent mol. wts. of 75,000, 67,000, and 62,000.

The need for phosphotransferase action for obtaining this effect was recently supported by the demonstration that a nonhydrolyzable analogue of ATP, 5'-adenylyl imidodiphosphate (AppNHp) did not support receptor loss.[166] In the same report,[166] the author has presented data suggesting that there are differential effects of protein-phosphorylating conditions on the binding of the agonist carbachol to three separate subtypes of the muscarinic receptor in synaptic membranes from rat cortex and cerebellum. However, because of the great complexity of the system, it appears that studies with a more purified preparation of receptors could shed better light on the mechanisms involved.

Such studies have been carried out with the estradiol receptor. It was shown that specific binding of the hormone to estradiol receptors prepared from mouse uterus undergoes inactivation under dephosphorylating conditions and can be reactivated by a phosphorylating enzyme purified from calf uterus cytosol.[167] Future studies with solubilized receptors from brain membranes could possibly determine whether similar mechanisms also operate in the regulation of neurotransmitter binding. A possible candidate for such investigation may be the binding of agonists to dopamine receptors, which shows sensitivity to ATP in a Mg^{2+}-dependent fashion.[168-170] In this context, it may be mentioned that solubilized dopamine receptor preparations were shown by Y. Clement-Cormier[171] to contain endogenously phosphorylated proteins. However, the questions of whether the dopamine receptor itself is a phosphoprotein and whether phosphorylation/dephosphorylation cycles regulate binding activity in this receptor system are still open.

An intriguing mechanism through which protein phosphorylation may regulate receptor binding emerged from studies of GABA and benzodiazepine receptors. A thermostable polypeptide of mol. wt. 16,000 was isolated (from rat brain synaptic membrane preparations) that noncompetitively reduces the affinity of GABA for its Na^+-independent recognition sites.[172] This protein, termed GABA-modulin, was also found to reduce the apparent affinity of diazepam to its receptor. In addition, it was reported that benzodiazepines can modulate GABA affinity for its receptors by competing with GABA-modulin.[172] Since the same protein is also a potent inhibitor of brain protein kinase activity,[173] it has been suggested that protein phosphorylation plays a role in the regulation of the affinity of GABA and benzodiazepine receptors. Recently, GABA-modulin has been purified to homogeneity and shown to be itself a substrate for Ca^{2+}-dependent endogenous phosphorylative activity of cortical membranes.[175] Further research could determine whether the phosphorylation of this protein and/or its kinase-inhibitory activity plays a role in regulating the GABA–benzodiazepine receptor complex.

4.3. Direct Effects of Neurohormones on Membrane-Bound Protein Phosphorylation Systems

The studies summarized in Section 4.1 above, which have demonstrated direct effects of EGF and insulin on phosphorylation of their own membrane-bound receptors, were not the first to show effects of a peptide hormone on protein phosphorylation that do not appear to involve prior activation of an intracellular second messenger system. In 1976, Zwiers et al.[176] reported that incubation of synaptic plasma membranes from rat brain in the presence of an active fragment of adrenocorticotrophic hormone ($ACTH_{1-24}$) resulted in inhibition of the endogenous phosphorylation of specific proteins. The substrates whose phosphorylation was selectively inhibited by ACTH were a protein of mol. wt. 48,000, termed B-50, and a group of phosphoproteins in the mol. wt. range of 15–20,000. These appeared to be the proteins identified by Ehrlich and Routtenberg[12] and designated previously as bands F and H of mol. wts. 47,000 and 15–20,000, respectively. The phosphorylation of these proteins and its inhibition by ACTH were found to be Ca^{2+}-dependent and cyclic nucleotide independent.[177] Subsequent studies successfully isolated and purified B-50 and its ACTH-sensitive protein kinase[178,179] and provided data indicating that, in all likelihood, the inhibition is exerted by direct interaction of the peptide hormone with the inhibited enzyme. This membrane-bound protein kinase may in fact be the receptor that mediates some of the behavioral effects of ACTH.[180] B-50 was characterized as an acidic protein with an IEP of 4.5, whereas its ACTH-sensitive kinase has a mol. wt. of 70,000 and IEP of 5.5.[179] Subcellular fractionation[181] and immunohistochemical[182] studies provided evidence that B-50 is a synaptic phosphoprotein localized predominantly in presynaptic membranes and present exclusively in nervous tissue.[183] The mechanism through which ACTH inhibition of B-50 protein phosphorylation may exert its effects on neuronal function is discussed below (see Section 4.4.1.).

The demonstration of selective effects of *in vivo* morphine treatment on the phosphorylation of specific proteins with mol. wt. of 47,000 (designated F) and 15–20,000 (designated H) in neostriatal membranes[184,185] prompted us to investigate the possibility that opioid peptides exert direct effects on the phosphorylation of specific proteins in synaptic membranes.[9,62] These studies have revealed that methionine- and leucine-enkephalin specifically inhibit the endogenous phosphorylation of proteins F and H in Ca^{2+}-saturated neostriatal membranes incubated *in vitro* with $[\gamma\text{-}^{32}P]ATP$. The effects could also be produced by nanomolar concentrations of human β-endorphin under conditions in which the high-affinity μ and δ opiate receptors were blocked by pretreatment of the membranes with naloxone.[186] In addition, the potent opiate agonist etorphine as well as β-endorphin, in the nanomolar concentration range, were found to stimulate endogenous phosphorylation of all proteins that serve as endogenous substrates for Ca^{2+}/calmodulin-dependent protein kinase in synaptic membranes. This stimulation could be blocked by the opiate antagonist naloxone, indicating that the stimulatory effect is mediated by specific opiate receptors.[186] The major substrates for the opiate-stimulated activity were phosphoproteins of apparent mol. wt. 50–60,000. Stimulation of the phosphorylation of a 50-kilodalton protein by opiates that could be blocked by naloxone was shown also in hippocampal membranes.[55] Naloxone-insensitive inhibition of endogenous phosphorylation of a 48-kilodalton protein was recently shown also to be exerted by the opioid peptide dynorphin[187] and by another behaviorally active peptide, somatostatin.[188] Future studies should investigate the attractive possibility that protein phosphorylation may serve as a general mediator for certain actions of neuropeptides.

The regulation of protein phosphorylation by neurohormones is not limited to peptides but has also been shown for classical neurotransmitters such as the catecholamines. Adrenergic modification of membrane-bound protein phosphorylation was first demonstrated in human erythrocytes.[189] The use of ^{32}Pi-labeled cells and a series of adrenergic agonists and antagonists revealed an α-adrenergic receptor-mediated stimulation of protein phosphorylation that did not appear to involve modulation of the intracellular concentration of ATP, cyclic AMP, or cyclic GMP.[189] The principal protein substrate involved in this effect is spectrin. Stimulation of protein phosphorylation by catecholamines in preparation from the nervous system was shown for the neurotransmitter dopamine. Increased phosphorylation of a 50-kilodalton protein by 50–300 μM dopamine has been detected in a KC1 extract of calf striatal membranes containing solubilized dopamine receptors.[171] A similar effect was recently demonstrated in rat striatal membranes.[20] However, systematic investigation on the effects of neurotransmitters on endogenous protein phosphorylation in synaptic membranes has not yet been reported.

Numerous studies in the past few years have demonstrated that neural membranes contain specific receptors that bind various psychoactive drugs. Recently, receptors that specifically bind phenytoin in the micromolar range have been identified and shown to be identical to the micromolar benzodiazepine receptor of brain membranes.[190] Previous studies[191] have demonstrated that benzodiazepines in the micromolar range exert inhibitory effects on the Ca^{2+}/calmodulin-dependent kinase system of brain membranes. Future developments in this area of investigation may enable the design of novel psychoactive drugs based on their ability to exert selective effects on various endogenous phosphorylation systems of brain membranes.

4.4. Protein Phosphorylation and Receptor Coupling

Activation of receptors localized on the outer surface of plasma membranes involves coupling to a second-messenger-generating system operating intracellularly. The most extensively studied second messenger systems for neurotransmitters are adenylate cyclase, which generates cyclic AMP, and inositol-phospholipid metabolism, believed to play a role in the regulation of Ca^{2+} entry through the plasma membrane.[100,101] Recent data implicate membrane-bound protein phosphorylation activity in the regulation of both of these second messenger systems by neurohormones.

4.4.1. Regulation of Phospholipid Metabolism by Protein Phosphorylation

Studies on the role of protein phosphorylation in regulating inositol-phospholipid metabolism have stemmed from investigations of the ACTH-sensitive B-50 protein kinase detailed above.[177] In intact prelabeled synaptosomes, it was shown that ACTH stimulated [^{32}P]phosphate incorporation into phosphatidylinositol-4.5-diphosphate (TPI), which is formed by phosphorylation of phosphatidylinositol-4-phosphate (DPI).[192] In isolated membrane preparations, phosphorylation of DPI to TPI can be measured using [γ-^{32}P]ATP. In such preparations, simultaneous effects of ACTH$_{1-24}$ on lipid and protein phosphorylation could be measured.[193] Using a partially purified enzyme preparation, Jolles et al.[194] have demonstrated a sequence of events: inhibition of protein B-50 phosphorylation by ACTH preceded, and appeared to be the cause of, increased phosphorylation of DPI to TPI. It is possible, but yet to be proven, that protein B-50 is, or constitutes part of, the DPI lipid–kinase complex.

4.4.2. Regulation of Adenylate Cyclase by Protein Phosphorylation

The mechanisms underlying the regulation of adenylate cyclase activity by hormone–receptor interaction have been studied extensively in recent years[195] (see also chapters by K. A. Bonnet, in Volumes 2 and 4 of this *Handbook*). Briefly, it is known that the coupling of the receptor to adenylate cyclase involves activation of the enzyme by a GTP-regulatory system, binding of GTP, and GTPase activity. Nonetheless, not all of the details of adenylate cyclase regulation are fully understood. Constantapoulos and Najjar[196,197] were first to suggest that protein phosphorylation plays a role in adenylate cyclase regulation. Using membranes from rabbit polymorphonuclear granulocytes and from dog platelets, they demonstrated that the activation of adenylate cyclase in these preparations with fluoride (F^-) and prostaglandin E_1 (PGE_1), respectively, is inhibited by ATP. The findings that PGE_1 and F^- can stimulate phosphoprotein phosphatase activity in these preparations[198] and that the observed effect required hydrolyzable ATP and could not be exerted by AppNHp[199] prompted them to postulate that this regulatory mechanism involves protein phosphorylation.

In apparent contradiction to the above studies, Richards *et al.*[200] presented data indicating that protein phosphorylation activates adenylate cyclase rather than inhibiting it. It was shown that preincubation of purified rat liver plasma membranes with ATP, Mg^{2+}, and ATP-regenerating system for 48–96h at 4°C caused a four- to sevenfold increase in adenylate cyclase activity. Close examination of the data in this communication[200] revealed that although basal adenylate cyclase activity is indeed stimulated by this pretreatment, F^- stimulation of the enzyme is decreased after long preincubation with ATP and its regenerating system in agreement with the previous reports.[196–199] Again, AppNHp could not support the effects observed with ATP, and the authors have suggested a role for protein phosphorylation in the activation of basal adenylate cyclase of liver plasma membranes.[200]

Investigation of the possible regulation of basal adenylate cyclase activity by protein phosphorylation in brain membranes has been initiated recently in our laboratory.[20,201,202] In these studies, we have been preincubating preparations containing synaptic membranes from rat cerebral cortex under various phosphorylating conditions. The preincubation is followed by sedimentation and washes of the membranes under conditions that minimize dephosphorylation. The initial rate of adenylate cyclase is then measured in resuspended membranes at very short (10–60 s) reaction times. It was found that after preincubation under conditions that favor dephosphorylation of membrane-bound proteins, adenylate cyclase activity is inhibited.[201] Conversely, after preincubation for 5 min at 30°C with 1 mM ATP, the enzyme is stimulated[20] (also: S. R. Whittemore, S. G. Graber, R. H. Lenox, E. D. Hendley, and Y. H. Ehrlich, submitted). Longer preincubations resulted in decreased adenylate cyclase activity, suggesting that the activation of ATP is reversed by dephosphorylation. Indeed, irreversible thiophosphorylation of membrane proteins with the analogue ATP-γ-S resulted in sustained adenylate cyclase activation that was not reversed by prolonged preincubation time. Labeling of the membranes with ATP-γ-^{35}S implicated a phosphoprotein of apparent mol. wt. 54,000 in this activation.[202] The possible mechanism underlying this mode of regulation was indicated recently by studies that have identified a 54,000 mol. wt. protein in brain membranes as a principal substrate for an endogenous activity that prefers to utilize GTP over ATP as a phosphate donor in the phosphorylation of proteins.[44]

A recent report[203] from the laboratory of A. Gilman has demonstrated charge differences between the guanine nucleotide regulatory protein (the G/F factor) of adenylate cyclase from the uncoupled (UNC) mutant and wild-type S49 lymphoma cells. It has been pointed out that "phosphorylation of the substrates for cholera toxin in UNC is an attractive but (yet) unsubstantiated possibility" (Schleifer et al.,[203] p. 2643). Previous studies have demonstrated influences of the G/F factor on basal adenylate cyclase activity in a Mg^{2+}-dependent fashion.[204] Our results[44] on the effects of preincubation with GTP under phosphorylating conditions on adenylate cyclase activity are not yet conclusive, and perhaps mechanisms other than protein phosphorylation play a role in bringing about the observed activation. Nonetheless, we now have the data basis needed for continued investigation of the role of membrane-bound protein phosphorylation systems in the regulation of adenylate cyclase activity and its relationship to the function of the G/F factor. Such studies may be of particular relevance for elucidating mechanisms underlying the inhibition of adenylate cyclase, a process in which GTP cannot be replaced by its non-hydrolyzable analogues.[205]

Finite resolution of the question of whether protein phosphorylation plays a direct role in adenylate cyclase regulation will undoubtedly require solubilization and purification of the various components involved.[203,206] On the other hand, processes whereby protein phosphorylation may regulate hormone-sensitive adenylate cyclase in an indirect fashion may be best studied in intact membranes. For example, it has been shown that the binding of calmodulin to striatal membranes is regulated by protein phosphorylation and that this process plays a role in determining the sensitivity of adenylate cyclase to dopamine.[207–209] Another line of investigation has implicated lipid methylation in regulating receptor-coupled adenylate cyclase (for details, see Chapter 19 in this volume). A 40-kilodalton protein termed lipomodulin, which plays a regulatory role in this process, is itself a phosphoprotein whose activity is regulated by phosphorylation–dephosphorylation cycles.[210]

5. PROTEIN PHOSPHORYLATION IN RECEPTOR MODIFICATION AND NEURONAL ADAPTATION

Cells in the nervous system can undergo long-lasting alterations in response to various hormonal, pharmacological, and environmental stimulations. These persistent modifications are believed to involve subtle changes in the activity of receptors.[211] Thus, chronic exposure to agonists can result in subsensitivity, whereas chronic antagonist exposure may cause supersensitivity of receptors. Several lines of investigation provide data indicating that one of the sites of molecular adaptation underlying receptor modifications may be protein phosphorylative activity.

5.1. Effects of Psychoactive Drug Treatments on Protein Phosphorylation in Brain Membranes

Acute activation or inhibition of receptors that are coupled directly or indirectly to phosphorylative activity may be expected to cause transient changes in the phosphorylation state of specific proteins. Such changes can be detected by *post-hoc in vitro* assays provided that the sacrifice method would not induce further changes (see Section 1.2, above). Killing animals by immersion in liquid nitrogen enabled us to identify specific phosphoproteins responsive to ECS.[17] Using a similar experimental approach, Strombom *et al.*[212] demonstrated that anesthetic agents caused a decrease in the state of phosphorylation of protein I in mouse brain membranes and that convulsant agents caused an increase in the state of phosphorylation of this protein.

Following chronic drug treatment, changes in phosphorylative activity detected by *in vitro* assays may reflect alterations in the capacity of the enzymatic system and not necessarily only in the state of phosphorylation of the protein substrate. An initial report[213] has shown that chronic treatment (21 days) of rats with a neuroleptic drug (chlorpromazine) resulted in increased protein phosphorylation of striatal membranes. Recent studies[214] utilizing haloperidol have confirmed these findings and demonstrated that the proteins whose phosphorylation increased following chronic haloperidol administration are the substrates of Ca^{2+}/ calmodulin kinase. It has been suggested that these changes are involved in the drug-induced supersensitivity of dopamine receptors in the striatum.[214] However, whether these phosphorylative modifications represent a cause or a manifestation of the changes in receptor sensitivity cannot be stated.

Long-term morphine treatment, which induces a state of protracted narcotic dependence, was shown to produce a 50% decrease in the phosphorylation of proteins in rat neostriatal membranes.[215] This decrease was selective and appeared to involve reduced capacity of the membranes to phosphorylate endogenous proteins of mol. wts. 47,000 and 15–20,000, designated bands F and H.[184,216] These results were confirmed by independent studies in another laboratory.[217] Conversely, short-term morphine treatment resulted in an overall increase of Ca^{2+}-dependent protein phosphorylation in striatal membranes.[184] Thus, the development of tolerance on the one hand and of dependence on the other may be associated with two separate protein phosphorylation systems which differ in several parameters.[186] Protein I phosphorylation may be among the substrates responding to acute morphine,[218] but changes in its phosphorylation could not be detected after long-term exposure.[215]

It should be emphasized that because of the heterogeneity of cell populations in brain preparations, any suggestions regarding causal relationships between the effects of a drug on protein phosphorylation and on receptor function would be inconclusive. Such limitations could be overcome by studying model systems consisting of homogeneous cell populations. Conducting such studies with mast cells has enabled identification of a protein phosphorylation system that phosphorylates specifically a substrate of mol. wt. 78,000 involved in the action of antiasthmatic drugs.[219,220] Future studies of cultured neurons may provide such precise information regarding the mechanism of action of various psychoactive drugs.

5.2. Role for Protein Phosphorylation in the Desensitization of Receptor-Coupled Adenylate Cyclase

As mentioned above, persistent exposure of receptors to their agonists can cause a decrease in their responsivity, a process known as desensitization or tachyphylaxis. In the case of receptors coupled to adenylate cyclase, desensitization can be successfully studied *in vitro* in isolated membrane preparations. Several studies have demonstrated that preincubation of membranes with agonists desensitized the responsivity of adenylate cyclase.[221–226] In each of these studies, the presence of ATP or GTP in a hydrolyzable form and of Mg^{2+} or Mn^{2+} in the preincubation medium were essential requirements, suggesting involvement of phosphorylative activity. Only one of these studies,[221] however, directly tested the possibility that protein phosphorylation plays a role in desensitization. In this study, desensitization of luteinizing hormone (LH)-sensitive adenylate cyclase was achieved by preincubation of membranes from preovulatory porcine follicles with LH, ATP, and Mg^{2+}. The addition of phosphoprotein phosphatase partially purified from porcine follicular cytosol reversed the LH-induced desensitization in a concentration-dependent manner.[221] It is not yet known whether this mechanism of desensitization is general. Nonetheless, recent studies suggest that desensitization of neurotransmitter-sensitive adenylate cyclase in brain also involves protein phosphorylation. Induction of subsensitivity of adenylate cyclase to dopamine by incubating striatal slices *in vitro* with 10 μM of agonists was accompanied by a significant increase in protein phosphorylation.[227] The nature of the phosphate donor involved in this process and the identity of the phosphoprotein substrates are yet to be reported.

Ezra and Salomon,[223,224] who have studied the desensitization of LH-stimulated adenylate cyclase in ovarian membranes, demonstrated that in the *in vitro* desensitization reaction, GTP can substitute for ATP and that these membranes contain a GTP-preferring protein kinase activity.[228] These data are consistent with the possibility that the desensitization process may involve phosphorylation of proteins by GTP. The presence of a GTP-preferring kinase in striatal membranes and of specific protein substrates for this activity,[44] taken together with the procedure for inducing desensitization of dopamine-sensitive adenylate cyclase in slices from rat striatum,[227] now provide experimental means to study the role of specific protein phosphorylation systems in the tachyphylaxis of neurotransmitter-sensitive adenylate cyclase in brain tissue. The finding that the endogenous phosphorylation of specific proteins in brain membranes by GTP is inhibited by cyclic AMP[44] suggests the attractive possibility of feedback control in the regulation of receptor-coupled adenylate cyclase activity. It should be emphasized, however, that hormone effects on this phosphorylation system have not yet been demonstrated.

5.3. Involvement of Protein Phosphorylation in Neuronal Plasticity

Long-lasting alterations in neuronal function that underlie the ability of neurons to undergo plastic changes in response to various stimuli were shown to involve, in many cases, receptor-mediated transsynaptic induction of neural-specific enzymes such as tryosine hydroxylase (TH).[229] A series of studies at the laboratory of E. Costa has provided evidence that induction of TH synthesis by reserpine or cold stress involves an increase in RNA synthesis, in a process that is triggered by neurotransmitters, is mediated by cyclic AMP, and is specified by a cyclic-AMP-dependent protein kinase. According to these studies,[230–233] increased intracellular levels of cyclic AMP causes a dissociation of the cytoplasmic cyclic-AMP-dependent protein kinase to its regulatory and catalytic subunits. The free catalytic subunit translocates to the cell nucleus. There it catalyzes the phosphorylation of chromosomal proteins, which in turn promotes the synthesis of messenger RNA. By acting via this phosphorylation process, neurotransmitters that activate adenylate cyclase can cause a persistent alteration in metabolic processes within the affected cell.

These studies have used cells of peripheral tissue, the adrenal medulla, and neuroblastoma cells grown in culture as a target for the triggering stimulus. Induction of lactic dehydrogenase in C_6 glioma cells was shown to involve a similar sequence of events.[234] Moreover, in C_6 cells prelabeled with ^{32}Pi, it was demonstrated that the nuclear substrate whose phosphorylation is stimulated by exposing the cells to the β-adrenergic agonist isoproterenol is a protein subunit of RNA polymerase II.[235] Stimulation by isoproterenol of nuclear protein phosphorylation in cultured rat pineal glands was also reported[236] and suggested to play a role in the induction of serotonin-N-acetyltransferase. In this context, it could be mentioned that autophosphorylation of the regulator subunit of type II cyclic-AMP-dependent protein kinase was shown to be affected by steroid hormone administration.[237] Treatment with corticosterone has also been shown to increase the amount of protein I in rat hippocampus,[238] but whether this effect of steroid hormones involves nuclear translocation and increased transcriptional activity is not known.

Indication that similar processes may be involved in the differentiation and maturation of neurons came from studies of neuroblastoma cells grown in culture. It was found that in cytosol from differentiated neuroblastoma cells, the phophorylation of a protein with apparent molecular weight of 97,000 decreased to about 50% of that measured in multiplying, control cells.[21] The phosphorylation of a protein with apparent mol. wt. of 59,000, tentatively identified as the regulatory subunit of cyclic-AMP-dependent protein kinase, increased about twofold. The phosphorylation of the 59-kilodalton protein was found to be cyclic AMP dependent, and that of the 97-kilodalton protein independent. These bidirectional changes in cyctosolic protein phosphorylation systems were accompanied by a 50% increase in the phosphorylation of nonhistone nuclear proteins from the differentiated neuroblastoma cells.[22]

The ultimate expression of plasticity in the nervous system is manifested in the processes of learning and memory, which were shown to involve alterations on the cellular and molecular levels.[239] Phophorylation of specific proteins has been implicated in these processes as well. An early study[240] utilizing *post-hoc in vitro* assays has served to identify a specific protein (designated F) in striatal membranes whose phosphorylation was modified when rats or mice were subjected to training on a passive-avoidance conditioning task. Continued investigation by Morgan and Routtenberg[30,241] revealed that this protein is the α subunit of brain mitochondrial pyruvate dehydrogenase, whose phosphorylation as well as activity are influenced by training experience. Involvement of the same phosphoprotein in synaptic plasticity was also shown by Browning *et al.*[31,53,54] by demonstrating that changes in the phosphorylation of pyruvate dehydrogenase are associated with the long-term potentiation of synaptic strength in hippocampal slices subjected to high-frequency stimulation. A clue to the possible synaptic mechanism that may activate this adaptive process has been provided by the report[242] that this phosphorylation system is regulated by physiological concentrations of glutamate, which in the hippocampus is an excitatory neurotransmitter.

Major progress in understanding the molecular biology of learning and memory has been achieved by Kandel and his colleagues who carried out detailed analyses of physiological and biochemical mechanisms underlying the short-term sensitization of the gill and siphon reflex in identified neurons of *Aplysia californica.*[239] These studies have provided direct evidence that the modulation of synaptic action (via serotonin sensitive K^+ channels) by a cyclic-AMP-dependent protein kinase is a required step in this form of learning. Thus, it may be expected that continued investigation of protein phosphorylation systems could make a considerable contribution not only to our understanding of basic synaptic mechanisms but also to the elucidation of the most complex brain functions.

6. CONCLUSIONS AND FUTURE DIRECTIONS

Understanding of the ubiquitous role that protein phosphorylation systems play in the regulation of numerous cellular functions in general and of neuronal activity in particular has been advanced greatly by the identification of a multiplicity of protein substrates whose phosphorylation/dephosphorylation cycles are controled by the interaction of various hormones and neurotransmitters with their specific receptors. The most detailed information available to date on the phosphorylation systems that are involved in receptor-mediated neuronal functions concerns those protein substrates whose phosphorylation is regulated by second messengers such as cyclic AMP, cyclic GMP, and calcium ions. These intracellular protein-phosphorylation systems thus play the role of "third messengers" in receptor function. A major and concerted effort carried out in Paul Greengard's laboratory has been aiming at the isolation and purification of all of these proteins, their kinases, and phosphatases. Since many of these phosphorylation systems were found to be specific to the nervous

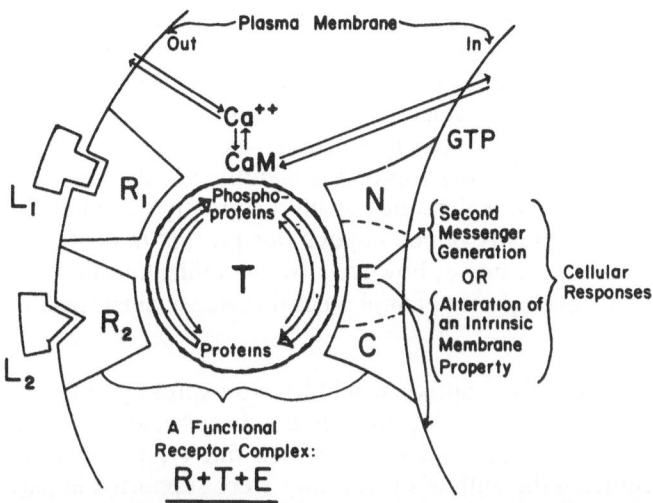

Fig. 1. A hypothetical role for protein phosphorylation in receptor-mediated signal transduction through the plasma membrane. The interaction of receptor ligands (L) with recognition sites (R) of the receptor results in activation or inhibition of an effector site (E), which initiates a chain of events leading to cellular responses. R confers the specificity to the binding of L. E may be, for example, adenylate cyclase, including catalytic (C) and regulatory (N) subunits, or an intrinsic membrane activity such as an ion channel. A transducing mechanism (T) is needed for the coupling of R with E. A phosphorylation/dephosphorylation system that is regulated by neurohormones could confer a reversible conformational change that carries a signal from R to E. Phosphorylation of a membrane-bound protein by a kinase that utilizes either ATP or GTP as a phosphate donor may play a regulatory role in the action of T. Since there is no requirement of specificity from T, a phosphorylation system may also play a modulatory role in balancing the responses to two hormones (L_1, L_2) interacting with separate recognition sites (R_1, R_2) present on the outer surface of the same cell. (Reproduced from Ehrlich,[216] with permission.)

system, their study will serve to elucidate most important steps in the chain of events triggered by the activation of neural receptors. A schematic illustration of this chain of biochemical events has been included in several reviews of these studies.[6–9] In addition to their function as third messengers in receptor function, protein phosphorylation systems that operate intrinsically within plasma membranes may play a direct role in the coupling of receptors to their effector sites. These hypothetical relationships are illustrated in Fig. 1 and detailed in its legend. Section 4 of this chapter summarizes recent studies that have demonstrated that certain membrane-bound protein phosphorylation systems are directly regulated by peptide hormones, neurotransmitters, trophic factors, and psychoactive drugs interacting with specific receptors localized on the outer surface of plasma membranes. In some cases it has been shown that the phosphoprotein substrates involved constitute part of the receptor complex. The working hypothesis presented in Fig. 1 is offered here as a possible guideline whose value should be judged primarily on the basis of the experimental tests that it will generate in future studies. Such a line of investigation may also shed light on mechanisms whereby phosphorylative modifications play a role in processes underlying alterations in receptor sensitivity.

In summary, two main directions for future research have emerged from current studies of protein phosphorylation systems: investigation of their role as intracellular messengers for neural receptors and the involvement of phosphoproteins in receptor-mediated signal transduction across plasma membranes. In either of these research directions, assignment of a defined function for each of the specific phosphoproteins identified in the nervous system is a major task. Such functional studies would be greatly complicated by the heterogeneity and complexity of cell populations present in neural tissues. These problems could be overcome, however, by the utilization of simplified model systems and the development of antibodies directed against specific brain phosphoproteins.

ACKNOWLEDGMENTS. The author would like to express gratitude to Dr. R. H. Lenox and Dr. E. D. Hendley for helpful discussions and to Mrs. Ginger McDowell and Ms. Vicki Sanderson for their fine typing support. Studies described herein from the author's laboratory were supported in part by USPHS grants DAO2747 and MH35735.

REFERENCES

1. Krebs, E. G., and Beavo, J. A., 1979, *Annu. Rev. Biochem.* **48**:923–959.
2. Cohen, P. (ed.), 1980, *Recently Discovered Systems of Enzyme Regulation by Reversible Phosphorylation. Molecular Aspects of Cellular Regulation*, Volume 1 Elsevier/North Holland, Amsterdam.
3. Rosen, O. M., and Krebs, E. G., (eds), 1981, *Protein Phosphorylation, Cold Spring Harbor Conferences on Cell Proliferation*, Volume 8, Cold Spring Harbor Laboratory, New York.
4. Weller, M., 1979, *Protein Phosphorylation*, Pion, London.
5. Rodnight, R., 1983, *Handbook of Neurochemistry*, 2nd ed. (A. Lajtha, ed.), Plenum Press, New York **4**:195–217.
6. Greengard, P., 1978, Cyclic Nucleotides, Phosphorylated Proteins and Neuronal Function, Raven Press, New York.
7. Greengard, P., 1979, *Fed. Proc.* **38**:2208–2217.
8. Greengard, P., 1981, *Harvey Lect.* Series 75, pp. 277–331.
9. Ehrlich, Y. H., 1979, *Adv. Exp. Med. Biol.* **116**:75–101.
10. Johnson, E. M., Ueda, T., Maeno, H., and Greengard, P., 1972, *J. Biol. Chem.* **247**:5650–5652.
11. Ueda, T., Maeno, H., and Greengard, P., 1973, *J. Biol. Chem.* **248**:8295–8305.
12. Ehrlich, Y. H., and Routtenberg, A., 1975, *FEBS Lett.* **45**:237–243.
13. Routtenberg, A., and Ehrlich, Y. H., 1975, *Brain Res.* **92**:415–430.
14. Rodnight, R., 1980, *Neurochem. Int.* **2**:113–122.
15. DeLorenzo, R. J., 1976, *Biophys. Res. Commun.* **71**:590–597.
16. Schulman, M., and Greengard, P., 1978, *Nature* **271**:478–479.
17. Ehrlich, Y. H., Reddy, M. N., Keen, P., and Davis, L. G., 1980, *J. Neurochem.* **34**:1327–1330.
18. Ehrlich, Y. H., Davis, L. G., Gilfoil, T., and Brunngraber, E. G., 1977, *Neurochem. Res.* **2**:533–548.
19. DeBlas, A. L. D., Wang, Y.-J., Sorensen, R., and Mahler, H. R., 1979, *J. Neurochem.* **33**:647–659.
20. Ehrlich, Y. H., Whittemore, S. R., Garfield, M. K., Graber, S. G., and Lenox, R. H., 1982, *Prog. Brain Res.* **56**:375–399.
21. Ehrlich, Y. H., Brunngraber, E. G., Sinah, P. K., and Prasad, K. N., 1977, *Nature* **265**:238–240.

22. Ehrlich, Y. H., Prasad, K. N., Davis, L. G., Sinah, P. K., and Brunngraber, E. G., 1978, *Neurochem. Res.* **3**:803–813.
23. Walter, U., Costa, M. R. C., Breakefield, X. O., and Greengard, P., 1979, *Proc. Natl. Acad. Sci. U.S.A.* **76**:3251–3255.
24. Lohmann, S. M., Ueda, T., and Greengard, P., 1978, *Proc. Natl. Acad. Sci. U.S.A.* **75**:4037–4041.
25. Holmes, H., and Rodnight, R., 1981, *Dev. Neurosci.* **4**:79–88.
26. Lohman, S. M., Walter, U., and Greengard, P., 1980, *J. Biol. Chem.* **253**:9985–9992.
27. Costa, M. R. C., Casnellie, J. E., and Catterall, W. A., 1982, *J. Biol. Chem.* **257**:7918–7921.
28. Greengard, P., 1978, *Advances in Pharmacology and Therapeutics*, Volume 3 (J. C. Stoclet, ed.), Pergamon Press, Oxford, pp. 231–250.
29. Walter, U., Kanof, P., Schulman, H., and Greengard, P., 1978, *J. Biol. Chem.* **253**:6275–6280.
30. Morgan, D. G., and Routtenberg, A., 1980, *Biochem. Biophys. Res. Commun.* **95**:569–576.
31. Browning, M., Bennett, W. F., Kelly, P., and Lynch, G., 1981, *Brain Res.* **218**:255–266.
32. Burke, B. E., and DeLorenzo, R. J., 1982, *Brain Res.* **236**:393–415.
33. Zwiller, J., Revel, M.-O., and Bassett, P., 1981, *Biochem. Biophys. Res. Commun.* **101**:1381–1387.
34. Miyamoto, E., and Kakiuchi, S., 1975, *Biochim. Biophys. Acta* **34**:458–465.
35. Miyamoto, E., Fukunaga, K., Matsui, K., and Iwasa, Y., 1981, *J. Neurochem.* **37**:1324–1330.
36. Rubin, C. S., and Rosen, O. M., 1975, *Annu. Rev. Biochem.* **44**:831–887.
37. Ehrlich, Y. H., Davis, L. G., and Brunngraber, E. G., 1977, *Soc. Neurosci. Abstr.* **3**:139.
38. Conway, R. G., and Routtenberg, A., 1978, *Brain Res.* **139**:366–373.
39. Ehrlich, Y. H., Davis, L. G., and Brunngraber, E. G., 1978, *Brain Res. Bull.* **3**:251–256.
40. Lenox, R. H., Kant, G. J., and Meyerhoff, J. L., *Handbook of Neurochemistry*, Volume 2, Second ed., (A. Lajtha, ed.), Plenum Press, New York **2**:77–102.
41. Duffy, T. E., Nelson, S. R., and Lowry, O. H., 1972, *J. Neurochem.* **19**:959–977.
42. Wiegant, V. M., Zwiers, H., Schotman, P., and Gispen, W. H., 1978, *Neurochem. Res.* **3**:443–453.
43. Ng, M., and Matus, A., 1979, *Neuroscience* **4**:1265–1274.
44. Ehrlich, Y. H., Whittemore, S. R., Lambert, R., Ellis, J., Graber, S. G., and Lenox, R. H., 1982, *Biochem. Biophys. Res. Commun.* **107**:699–706,
45. Manning, D. R., DiSalvo, J., and Stull, J. T., 1980, *Mol. Cell. Endocrinol.* **19**:1–19.
46. Berman, R. F., Hullihan, J. P., Kinnier, W. J., and Wilson, J. E., 1980, *J. Neurochem.* **34**:431–437.
47. Dunkley, P. R., and Robinson, P. J., 1981, *Biochem. Biophys. Res. Commun.* **102**:1196–1202.
48. Carstens, M., and Weller, M., 1981, *Mol. Cell. Biochem.* **40**:95–104.
49. Heald, P. J., 1957, *Biochem. J.* **66**:659–663.
50. Trevor, A. J., and Rodnight, R., 1965, *Biochem. J.* **95**:889–896.
51. Williams, M., Pavlik, A., and Rodnight, R., 1974, *J. Neurochem.* **22**:373–376.
52. Williams, M., and Rodnight, R., 1977, *Prog. Neurobiol.* **8**:183–250.
53. Browning, M., Dunwiddie, T., Bennett, W., Gispen, W., and Lynch, G., 1979, *Science* **203**:60–62.
54. Browning, M., Bennet, W., and Lynch, G., 1979, *Nature* **278**:273–275.
55. Bar, P. R., Schotman, P., and Gispen, W. H., 1980, *Eur. J. Pharmacol.* **65**:165–174.
56. Rodnight, R., 1980, *Synaptic Constituents in Health and Disease* (M. Brzin, D. Sket, and H. Bachelard, eds.), Pergamon Press, Oxford, pp. 81–96.
57. Kuo, J. E., and Greengard, P., 1969, *Proc. Natl. Acad. Sci. U.S.A.* **64**:1349–1355.
58. Schulman, H., and Greengard, P., 1978, *Proc. Natl. Acad. Sci. U.S.A.* **75**:5432–5436.
59. Nishizuka, Y., and Takai, Y., 1981, *Protein Phosphorylation, Cold Spring Harbor Conferences on Cell Proliferation*, Volume 8, Cold Spring Harbor Laboratory, New York, pp. 237–249.
60. Larner, J., Galasko, G., Cheng, K., Depaoli-Roach, A. A., Huang, L., Daggy, P., and Kellogg, J., 1979, *Science* **206**:1408–1410.
61. Jarett, L., and Seals, J. R., 1979, *Science* **206**:1407–1408.
62. Davis, L. G., and Ehrlich, Y. H., 1979, *Adv. Exp. Med. Biol.* **116**:233–244.

63. Garrison, J. C., Borland, M. K., Florio, V. A., and Twible, D. A., 1979, *J. Biol. Chem.* **254:**7147–7156.
64. Groppi, V. E., and Browning, E. T., 1980, *Mol. Pharmacol.* **18:**427–437.
65. Church, G. A., Kimelberg, H. K., and Sapirstein, V. A., 1980, *J. Neurochem.* **34:**873–879.
66. Hofstein, R., and Segal, M., 1982, *J. Neurochem.* **39:**478–485.
67. Prasad, K. N., 1980, *Regulation of Differentiation in Mammalian Nerve Cells*, Plenum Press, New York.
68. Halegoua, S., and Patrick, J., 1980, *Cell* **22:**571–581.
69. Goodman, R., and Ehrlich, Y. H., 1982, *Trans. Am. Soc. Neurochem.* **13:**240.
70. Salem, R., and DeVellis, J., 1980, *Eur. J. Biochem.* **107:**271–278.
71. Hamprecht, B., 1977, *Int. Rev. Cytol.* **49:**99–170.
72. Ehrlich, Y. H., Garfield, M., and Lenox, R. H., 1981, *Trans. Am. Soc. Neurochem.* **12:**229.
73. Palfrey, H. C., Alper, S. L., and Greengard, P., 1980, *J. Exp. Biol.* **89:**103–115.
74. Beam, K. G., Alper, S. L., Palade, G. E., and Greengard, P., 1979, *J. Cell Biol.* **83:**1–15.
75. Alper, S. L., Beam, K. G., and Greengard, P., 1980, *J. Biol. Chem.* **255:**4864–4871.
76. Alper, S. L., Palfrey, H. C., DeRiemer, S. A., Greengard, P., 1980, *J. Biol. Chem.* **255:**11029–11039.
77. Rudolph, S. A., and Greengard, P., 1980, *J. Biol. Chem.* **255:**8534–8540.
78. Mahler, H. R., 1977, *Neurochem. Res.* **2:**119–148.
79. Ueda, T, and Greengard, P., 1977, *J. Biol. Chem.* **252:**5155–5163.
80. Krueger, B. K., Forn, J., and Greengard, P., 1977, *J. Biol. Chem.* **252:**2764–2773.
81. Sieghart, W., Forn, J., and Greengard, P., 1979, *Proc. Natl. Acad. Sci. U.S.A.* **76:**2475–2479.
82. Huttner, W. B., and Greengard, P., 1979, *Proc. Natl. Acad. Sci. U.S.A.* **76:**5402–5406.
83. Schulman, H., Huttner, W. B., and Greengard, P., 1980, *Calcium and Cell Function*, Volume 1, Academic Press, New York. pp. 219–252.
84. Kennedy, M. B., and Greengard, P., 1980, *Proc. Natl. Acad. Sci. U.S.A.* **78:**1293–1297.
85. Ueda, T., Greengard, P., Berzins, K., Cohen, R. S., Blomberg, F., Grab, D. J., and Siekevitz, P., 1979, *J. Cell Biol.* **83:**308–319.
86. Bloom, F. E., Ueda, T., Battenberg, E., and Greengard, P., 1979, *Proc. Natl. Acad. Sci. U.S.A.* **76:**5977–5981.
87. DeCamilli, P., Ueda, T., Bloom, F. E., Battenberg, E., and Greengard, P., 1979, *Proc. Natl. Acad. Sci. U.S.A.* **76:**5977–5981.
88. DeCamilli, P., Cameron, R., and Greengard, P., 1980, *J. Cell Biol.* **87:**72a.
89. Goelz, S. E., Nestler, E. J., Chehrazi, B., and Greengard, P., 1981, *Proc. Natl. Acad. Sci. U.S.A.* **78:**2130–2134.
90. Dolphin, A. C., and Greengard, P., 1981, *Nature* **289:**76–69.
91. Dolphin, A. C., Goelz, S. E., and Greengard, P., 1981, *Pharmacol. Biochem. Behav.* **13:**169–174.
92. Dolphin, A. C., and Greengard, P., 1981, *J. Neurosci.* **1:**192–203.
93. Forn, J., and Greengard, P., 1978, *Proc. Natl. Acad. Sci. U.S.A.* **75:**5195–5199.
94. Nestler, E. J., and Greengard, P., 1980, *Proc. Natl. Acad. Sci. U.S.A.* **77:**7479–7483.
95 Nestler, E. J., and Greengard, P., 1982, *Nature* **296:**452–454.
96. Naito, S., and Ueda, T., 1981, *J. Biol. Chem.* **256:**10657–10663.
97. Kaxzmarek, L. K., Jennings, K. R., Strumwasser, R., Nairn, A. C., Walter, U., Wilson, F. E., and Greengard, P., 1980, *Proc. Natl. Acad. Sci. U.S.A.* **77:**7487–7491.
98. Castellucci, V. F., Kandel, E. R., Schwartz, J. H., Wilson, F., Nairn, A. C., and Greengard, P., 1980, *Proc. Natl. Acad. Sci. U.S.A.* **77:**7492–7496.
99. Lemos, J. R., Novak-Hofer, I., and Levitan, I. B., 1982, *Nature* **298:**64–65.
100. Michell, R. H., 1975, *Biochim. Biophys. Acta* **415:**81–147.
101. Michell, R. H., 1979, *Trends Biochem. Sci.* **4:**128–131.
102. Nishizuka, Y., 1980, *Molec. Biol. Biochem. Biophys.*, **32:**113–135.
103. Takai, Y., Minakuchi, R., Kikkawa, U., Sano, K., Kaibuchi, K., Yu, B., Matsubara, T., and Nishizuka, Y., 1982, *Prog. Brain Res.* **56:**287–301.
104. Yakai, T., Kishimoto, A., Inoue, A., and Nishizuka, Y., 1977, *J. Biol. Chem.* **252:**7603–7609.
105. Takai, Y., Kishimoto, A., Iwasa, Y., Kawahara, Y., Mori, T., and Nishizuka, Y., *J. Biol. Chem.* **254:**36⌐-3695.

106. Takai, T., Kishimoto, A., Iwasa, T., Kawahara, Y., Mori, T., Nishizuka, T., and Fujii, T., 1979, *J. Biochem.* (*Tokyo*) 86:575–578.
107. Kaibuchi, K., Takai, T., and Nishizuka, Y., 1981, *J. Biol. Chem.* 256:7146–7149.
108. Takai, T., Kishimoto, A., Kikkawa, U., Mori, T., and Nishizuka, Y., 1979, *Biochem. Biophys. Res. Commun.* 91:1218–1224.
109. Kishimoto, A., Takai, T., Mori, T., Kikkawa, U., and Nishizuka, T., 1980, *J. Biol. Chem.* 255:2273–2276.
110. Kawahara, Y., Takai, T., Minakuchi, R., Sano, K., and Nishizuka, Y., 1980, *J. Biochem.* (*Tokyo*) 88:913–916.
111. Kawahara, T., Takai, T., Minakuchi, R., Sano, K., and Nishizuka, Y., 1980, *Biochem. Biophys. Res. Commun.* 97:309–317.
112. Ogawa, Y., Takai, Y., Kawahara, ♔., Kimura, S., and Nishizuka, Y., 1981, *J. Immunol.* 127:1369–1374.
113. Ku, Y., Kishimoto, A., Takai, Y., Ogawa, Y., Kimura, S., and Nishizuka, Y., 1981, *J. Immunol.* 127:1375–1379.
114. Takai, Y., Kaibuchi, K., Matsubara, T., and Nishizuka, Y., 1981, *Biochem. Biophys. Res. Commun.* 101:61–67.
115. Takai, Y., Kaibuchi, K., Sano, K., and Nishizuka, Y., 1982, *J. Biochem.* (*Tokyo*) 91:403–406.
116. Walter, U., and Greengard, P., 1981, *Current Topics in Cellular Regulation, Volume 19*, (E. Stadtman and B. L. Horecker, eds.) Academic Press, New York, pp. 219–256.
117. Schlichter, D. J., Casnellie, J. E., and Greengard, P., 1978, *Nature* 273:61–62.
118. Schlichter, D. J., Detre, J. A., Aswad, D. W., Chehrazi, B., and Greengard, P., 1980, *Proc. Natl. Acad. Sci. U.S.A.* 77:5537–5541.
119. Lohmann, S. M., Walter, U., Miller, P. E., Greengard, P., and DeCamilli, P., 1981, Proc. Natl. Acad. Sci. U.S.A. 78:653–657.
120. Aswad, D. W., and Greengard, P., 1981, *J. Biol. Chem.* 256:3487–3493.
121. Aswad, D. W., and Greengard, P., 1981, *J. Biol. Chem.* 256:3494–3500.
122. Aitken, A., Bilham, T., and Cohen, P., 1981, *J. Biol. Chem.* 256:3501–3506.
123. Morgenroth, V. H., III, Hegstrand, L. R., Roth, R. H., and Greengard, P., 1975, *J. Biol. Chem.* 250:1946–1948.
124. Lovenberg, W., Bruckwick, E. A., and Hanbauer, I., 1975, *Proc. Natl. Acad. Sci. U.S.A.* 72:2955–2958.
125. Ames, M. M., Lerner, P., and Lovenberg, W., 1978, *J. Biol. Chem.* 253:27–31.
126. Joh, T. H., Park, D. H., and Reis, P. J., 1978, *Proc. Natl. Acad. Sci. U.S.A.* 75:4744–4748.
127. Raese, J. D., Edelman, A. M., Makk, G., Bruckwick, E. A., Lovenberg, W., and Barchas, J. D., 1979, *Commun. Phychopharmacol.* 3:295–301.
128. Vulliet, P. R., Langan, T. A., and Weiner, N., 1980, *Proc. Natl. Acad. Sci. U.S.A.* 1:92–96.
129. Levine, R. A., Miller, L. P., and Lovenberg, W., 1981, *Science* 214:919–921.
130. Hamon, M., Bourgoin, S., Artaud, F., and Hery, F., 1977, *J. Neurochem.* 28:811–818.
131. Kuhn, D. M., Vogel, R. L., and Lovenberg, W., 1978, *Biochem. Biophys. Res. Commun.* 82:759–766.
132. Hamon, M., Bourgoin, S., Hery, F., and Simonnet, G., 1978, *Mol. Pharmacol.* 14:99–110.
133. Lysz, T. W., and Sze, P. Y., 1978, *Trans. Am. Soc. Neurochem.* 9:128.
134. O'Callaghan, J. P., Dunn, L. A., and Lovenberg, W., 1980, *Proc. Natl. Acad. Sci. U.S.A.* 77:5812–5816.
135. Weller, M., 1977, *Biochim. Biophys. Acta* 469:350–354.
136. DeLorenzo, R. J., 1981, *Cell Calcium* 2:365–385.
137. DeLorenzo, R. J., 1980, *Antiepileptic Drugs: Mechanisms of Action* (G. H. Glaser, J. K., Penry, and D. M. Woodbury eds.), Raven Press, New York, pp. 399–414.
138. DeLorenzo, R. J., 1980, *Ann. N.Y. Acad. Sci.* 356:92–109.
139. DeLorenzo, R. J., Freedman, S. D., Yohe, W. B., and Maurer, S. C., 1979, *Proc. Natl. Acad. Sci. U.S.A.* 76:1838–1842.
140. Burke, B. E., and DeLorenzo, R. J., 1981, *Proc. Natl. Acad. Sci. U.S.A.* 78:991–995.
141. Burke, B. E., and DeLorenzo, R. J., 1982, *J. Neurochem.* 38:1205–1218.
142. Brennan, M. J. W., and Cantrill, R. C., 1980, *J. Neurochem.* 35:506–508.

143. Michaelson, D. M., Avissar, S., Kloog, Y., and Sokolovsky, M., 1979, *Proc. Natl. Acad. Sci. U.S.A.* **76:**6336–6340.
144. Ehrlich, Y. H., and Hendley, E. D., 1983, *Tran. Am. Soc. Neurochem.* **14:** p. 197
145. Gazitt, Y., Ohad, I., and Loyter, A., 1976, *Biochim. Biophys. Acta* **436:**1327–1330.
146. Goldstein, A., 1973, *Pharmacology and the Future of Man*, Kruger Basel, pp. 140–150.
147. Carpenter, G., King, L., and Cohen, S., 1978, *Nature* **276:**409–410.
148. Carpenter, G., King, L., Jr., and Cohen, S., 1979, *J. Biol. Chem.* **254:**4884–4891.
149. Cohen, S., Carpenter, G., and King, L., 1980, *J. Biol. Chem.* **255:**4834–4842.
150. Ushiro, H., and Cohen, S., 1981, *J. Biol. Chem.* **255:**8363–8365.
151. Kasuga, M., Karlsson, F. A., and Kahn, C. R., 1982, *Science* **215:**185–186.
152. Teichberg, V. I., and Changeux, J.-P., 1977, *FEBS Lett.* **74:**71–76.
153. Gordon, A. S., Davis, C. G., and Diamond, I., 1977, *Proc. Natl. Acad. Sci. U.S.A.* **74:**263–267.
154. Gordon, A. S., Davis, C. G., Milfay, D., and Diamond, I., 1977, *Nature* **267:**539–540.
155. Teichberg, V. I., Sobel, A., and Changeux, J.-P., 1977, *Nature* **267:**540–542.
156. Saitoh, T., and Changeux, J.-P., 1980, *Eur. J. Biochem.* **105:**51–62.
157. Gordon, A. S., Milfay, D., Davis, C. G., and Diamond, I., 1979, *Biochem. Biophys. Res. Commun.* **87:**876–883.
158. Gordon, A. S., Davis, C. G., Milfay, D., Kaur, J., and Diamond, I., 1980, *Biochim. Biophys. Acta* **600:**421–431.
159. Gordon, A. S., and Diamond, I., 1979, *Adv. Exp. Med. Biol.* **116:**175–198.
160. Saitoh, T., and Changeux, J.-P., 1981, *Proc. Natl. Acad. Sci. U.S.A.* **78:**4430–4434.
161. Saitoh, T., Wennogle, L. P., and Changeux, J.-P., 1979, *FEBS Lett.* **108:**489–494.
162. Chuang, D. M., and Costa, E., 1979, *Neurochem. Res.* **4:**777–793.
163. Burgoyne, R. D., 1980, *FEBS Lett.* **122:**288–292.
164. Burgoyne, R. D., 1981, *FEBS Lett.* **127:**144–147.
165. Burgoyne, R. D., and Pearce, B., 1982, *Dev. Brain Res.* **2:**55–63.
166. Burgoyne, R. D., 1983, *J. Neurochem.* **40:**324–331.
167. Auricchio, F., Migliaccio, A., Castoria, G., Lastoria, S., and Schiavone, E., 1981, *Biochem. Biophys. Res. Commun.* **101:**1171–1178.
168. Sano, K., and Maeno, H., 1976, *Biochem. Biophys. Res. Commun.* **73:**584–590.
169. Zahniser, N. R., Heidenreich, K. A., and Molinoff, P. E., 1981, *Mol. Pharmacol.* **19:**372–378.
170. Near, J. A., and Mahler, H. R., 1982, *J. Neurosci.* **2:**553–561.
171. Clement-Cormier, Y. C., 1980, *Receptors for Neurotransmitters and Peptide Hormones*, (G. Pepeu, M. J. Kuhar, and S. J. Enna, eds.), Raven Press, New York, pp. 159–167.
172. Guidotti, A., Toffano, G., and Costa, E., 1978, *Nature* **275:**553–555.
173. Szmigielski, A., Guidotti, A., and Costa, E., 1977, *J. Biol. Chem.* **252:**3848–3853.
174. Costa, E., and Guidotti, A., 1979, *Annu. Rev. Pharmacol. Toxicol.* **19:**531–545.
175. Wise, B. C., Guidotti, A., and Costa, E., 1982, *Soc. Neurosci. Abstr.* **8:**572.
176. Zwiers, H., Valdhuis, D., Schotman, P., and Gispen, 1976, *Neurochem. Res.* **1:**669–677.
177. Gispen, W. H., Zwiers, H., Wiegant, V. M., Schotman, P., and Wilson, J. E., 1979, *Adv. Exp. Med. Biol.* **116:**199–224.
178. Zwiers, H., Tonnaer, J., Wiegant, V. M., Schotman, P., and Gispen, W. H., 1979, *J. Neurochem.* **33:**247–256.
179. Zwiers, H., Schotman, P., and Gispen, W. H., 1980, *J. Neurochem.* **34:**1689–1700.
180. Zwiers, H., Wiegant, V. M., Schotman, P., and Gispen, W. H., 1978, *Neurochem. Res.* **3:**455–463.
181. Sorensen, R. G., Kleine, L. P., and Mahler, H. R., 1981, *Brain Res. Bull.* **7:**57–61.
182. Oestreicher, A. B., Zwiers, H., Schotman, P., and Gispen, W. H., 1981, *Brain Res. Bull.* **6:**145–153.
183. Kristjansson, G. I., Zwiers, H., Oestreicher, A. B., and Gispen, W. H., 1982, *J. Neurochem.* **39:**371–378.
184. Ehrlich, Y. H., Bonnet, K. A., Davis, L. G., and Brunngraber, E. G., 1978, *Life Sci.* **23:**137–146.
185. Ehrlich, Y. H., Bonnet, K. A., Davis, L. G., and Brunngraber, E. G., 1977, *Dev. Neurosci.* **2:**273–278.

186. Ehrlich, Y. H., Davis, L. G., Keen, P., and Brunngraber, E. G., 1980, *Life Sci.* **26**:1765–1772.
187. Zwiers, H., Aloyo, V. J., and Gispen, W. H., 1981, *Life Sci.* **28**:2545–2551.
188. Dokas, L. A., Zwiers, H., Coy, D. H., and Gispen, W. H., 1982, *Soc. Neurosci. Abstr.* **8**:983.
189. Nelson, M. J., Ferrell, J. E., Jr., and Huestis, W. H., 1979, *Biochim. Biophys. Acta* **558**:136–140.
190. Bowling, A. C., and DeLorenzo, R. J., 1982, *Soc. Neurosci. Abstr.* **8**:505.
191. DeLorenzo, R. J., Burndette, S., and Holderness, J., 1981, *Science* **213**:540–549.
192. Jolles, J., Wirtz, K. W. A., Schotman, P., and Gispen, W. H., 1979, *FEBS. Lett.* **105**:110–114.
193. Jolles, J., Zwiers, H., Dekker, A., Wirtz, K. W. A., and Gispen, W. H., 1981, *Biochem. J.* **194**:283–291.
194. Jolles, J., Zwiers, H., VanDongen, C., Schotman, P., Wirtz, K. W. A., and Gispen, W. H., 1980, *Nature* **286**:623–625.
195. Ross, E. M., and Gilman, A. G., 1980, *Annu. Rev. Biochem.* **49**:533–564.
196. Constantopoulos, A., and Najjar, V. A., 1973, *Biochem. Biophys. Res. Commun.* **53**:794–798.
197. Najjar, V. A., and Constantopoulos, A., 1973, *Mol. Cell. Biochem.* **1**:1–7.
198. Layne, P., Constantopoulos, A., Judge, J. F. X., Rauner, R., and Najjar, V. A., 1973, *Biochem. Biophys. Res. Commun.* **53**:800–804.
199. Schmidt, J. J., and Najjar, V. A., 1978, *Biochim. Biophys. Acta* **526**:276–288.
200. Richards, J. M., Tierney, J. M., and Swislocki, N. I., 1981, *J. Biol. Chem.* **256**:8889–8891.
201. Whittemore, S. R., Lenox, R. H., Hendley, E. D., and Ehrlich, Y. H., 1981, *Neurochem. Res.* **6**:777–787.
202. Whittemore, S. R., Lenox, R. H., Hendley, E. D., and Ehrlich, Y. H., 1981, *Soc. Neurosci. Abstr.* **7**:920.
203. Schleifer, L. S., Garrison, J. C., Sternweis, P. C., Northup, J. K., and Gilman, A. G., 1980, *J. Biol. Chem.* **255**:2641–2644.
204. Cech, S. Y., Broaddus, W. C., and Maquire, M. E., 1980, *Mol. Cell. Biochem.* **33**:67–92.
205. Jakobs, K. H., 1979, *Mol. Cell. Endocrinol.* **16**:147–156.
206. Neer, E. J., Exheverria, D., and Knox, S., 1980, *J. Biol. Chem.* **20**:9782–9789.
207. Gnegy, M. E., Uzunov, P., and Costa, E., 1976, *Proc. Natl. Acad. Sci. U.S.A.* **73**:3887–3890.
208. Gnegy, M., Uzunov, P., and Costa, E., 1977, *J. Pharmacol. Exp. Ther.* **202**:558–564.
209. Gnegy, M., and Treisman, G., 1981, *Mol. Pharmacol.* **19**:256–263.
210. Hirata, F. J., 1981, *J. Biol. Chem.* **256**:7730–7739.
211. Bonnet, K. A., 1979, *Adv. Exp. Med. Biol.* Vol. **116**:247–260.
212. Strombom, U., Forn, J., Dolphin, A. C., and Greengard, P., 1979, *Proc. Natl. Acad. Sci. U.S.A.* **76**:4687–4690.
213. Ehrlich, Y. H., and Brunngraber, E. G., 1976, *Trans. Am. Soc. Neurochem.* **7**:109.
214. Lau, Y.-S., and Gnegy, M. E., 1982, *Life Sci.* **30**:21–28.
215. Bonnet, K. A., Branchey, L. B., Friedhoff, A. J., and Ehrlich, Y. H., 1978, *Life Sci.* **22**:2003–2008.
216. Ehrlich, Y. H., 1982, *Adv. Neurol.* **34**:345–352.
217. O'Callaghan, J. P., Williams, N., and Clouet, D. H., 1979, *J. Pharmacol. Exp. Ther.* **200**:255–262.
218. Sieghart, W., Strombom, U., Walter, U., and Greengard, P., 1979, *Membrane Mechanisms of Drugs of Abuse*, Alan R. Liss, New York, pp. 123–134.
219. Theoharides, T. C., Sieghart, W., Greengard, P., and Douglas, W. W., 1980, *Science* **207**:80–82.
220. Sieghart, W., Theoharides, T. C., Douglas, W. W., and Greengard, P., 1981, *Biochem. Pharmacol.* **30**:2737–2738.
221. Hunzicker-Dunn, M., Derda, D., Swartz, T. L., and Birmbaumer, L., 1979, *Endocrinology* **104**:1785–1793.
222. Anderson, W. B., and Jaworski, C. J., 1979, *J. Biol. Chem.* **254**:4596–4601.
223. Ezra, E., and Salomon, Y., 1980, *J. Biol. Chem.* **255**:1252–1258.

224. Ezra, E., and Salomon, Y., 1981, *J. Biol. Chem.* **256:**5377–5382.
225. Iynegar, R., Minitz, P. W., Swartz, T. L., and Birnbaumer, L., 1980, *J. Biol. Chem.* **255:**11875–11882.
226. Salomon, Y., Ezra, E., and Amir-Zaltsman, Y., 1981, *Adv. Cyclic Nucleotide Res.* **14:**101–109.
227. Memo, M., Lovenberg, W., and Hanbauer, I., 1982, *Proc. Natl. Acad. Sci. U.S.A.* **79:**4456–4460.
228. Amir-Zaltsman, Y., Ezra, E., Walker, N., Lindner, H. R., and Salomon, Y., 1980, *FEBS Lett.* **122:**166–170.
229. Costa, E., Curosawa, A., and Guidotti, A., 1976, *Proc. Natl. Acad. Sci. U.S.A.* **73:**3887–3891.
230. Kurosawa, A., Guidotti, A., and Costa, E., 1979, *Mol. Pharmacol.* **15:**115–120.
231. Hollenbeck, R. A., Chuang, D. M., and Costa, E., 1979, *Brain Res.* **171:**481–487.
232. Schwartz, J. P., and Costa, E., 1980, *J. Biol. Chem.* **255:**2943–2948.
233. Chuang, D.-M., Hollenbeck, R. A., and Costa, E., 1977, *J. Biol. Chem.* **252:**8365–8373.
234. Derda, D. R., Miles, M. F., Schweppe, J. S., and Jungman, R. A., 1980, *J. Biol. Chem.* **255:**11112–11121.
235. Lee, S.-K., and Jungman, R. A., 1981, *Biochem. Biophys. Res. Commun.* **102:**538–544.
236. Winters, K. E., Morrissey, J. J., Loos, P. J., and Lovenberg, W., 1977, *Proc. Natl. Acad. Sci. U.S.A.* **74:**1928–1931.
237. Liu, A., T.-C., Walter, U., and Greengard, P., 1981, *Eur. J. Biochem.* **114:**539–548.
238. Nestler, E. J., Rainbow, T. C., McEwen, B. S., and Greengard, P., 1981, *Science* **212:**1162–1164.
239. Kandel, E. R., and Schwartz, J. H., 1982, *Science* **218:**433–442.
240. Ehrlich, Y. H., Rabjohns, R., and Routtenberg, A., 1977, *Pharmacol. Biochem. Behav.* **6:**169–174.
241. Morgan, D. G., and Routtenberg, A., 1981, *Science* **214:**470–471.
242. Sieghart, W., 1981, *J. Neurochem.* **37:**116–124.

Heterogeneity of Benzodiazepine Receptors

Kelvin W. Gee, Frederick J. Ehlert, William R. Roeske, and Henry I. Yamamura

1. INTRODUCTION

1.1. Neurotransmitter Receptor Heterogeneity

Evidence has accumulated from almost a decade of study of neurotransmitter and putative neurotransmitter receptors to support the conclusion that most neurotransmitters interact with a heterogeneous population of receptors. The existence of receptor subtypes for various neurotransmitters provides a means by which the functional capabilities of a neurotransmitter may be extended.

Receptor subtypes can be distinguished by their anatomic distribution and pharmacological and functional specificity. The cholinergic receptor was one of the initial receptors to be subclassified.[1] Receptors specifically sensitive to muscarine were termed muscarinic receptors, whereas those specifically responsive to nicotine were called nicotinic receptors. There appear to be at least two, possibly three, dopamine receptors.[2] The D_1 receptor is thought to be linked to adenylate cyclase, whereas the D_2 receptor is cyclase independent.

Noradrenergic receptors have been categorized as α- and β-adrenergic receptors.[3] The α-adrenergic receptor in the autonomic nervous system has been classified as α_1 and α_2. The α_2 receptors appear to be involved in the release of norepinephrine, suggesting a presynaptic localization for the α_2 receptor.[4,5] Opiate receptors have been classified into at least four different subtypes, designated δ, κ, μ, and σ.[6] Evidence for the δ, κ, and μ receptors was derived from studies of pharmacological effects produced by various morphine and benzomorphan drugs *in vivo*.[7,8] H-1 and H-2 histamine receptor subtypes have been identified.[9] H-1 receptors appear to mediate bronchoconstriction and smooth muscle contraction in the gut, whereas H-2 receptors mediate the release of gastric acid. The foregoing list of receptor subtypes was not meant

Kelvin W. Gee, Frederick J. Ehl· rt, William R. Roeske, and Henry I. Yamamura • Departments of Pharmacology, Internal Medicine, Biochemistry, Psychiatry, and the Arizona Research Laboratories, University of Arizona Health Sciences Center, Tucson, Arizona 85724.

to be exhaustive but was presented to simply illustrate the concept that receptor subtypes appear to be widespread and are of physiological, pharmacological, and functional significance.

Most of our understanding of neurotransmitter receptor subtypes has been derived from studies at all levels of organization ranging from whole-animal physiological experiments to the isolation and characterization of the receptor. Some of these methodologies have also been applied in the study of the brain benzodiazepine (BZD) receptor. In the present review, the evidence obtained from these studies will be presented to provide some insight into the molecular basis of benzodiazepine receptor heterogeneity and, more importantly, to evaluate the possible physiological and functional implications of BZD receptor heterogeneity.

1.2. Benzodiazepine Receptors in the Central Nervous System

The 1,4-benzodiazepines are a class of compounds with diverse pharmacological effects ranging from antianxiety and anticonvulsant effects to sedation. The compounds have a high therapeutic index and enjoy widespread clinical use, especially as anxiolytics. Consequently, the mechanism of action of the BZDs has been the topic of intense investigation in recent years. The diversity of the pharmacological effects of these compounds initially argued against a common neurochemical pathway mediating the various effects of the BZDs. This controversy was resolved when high-affinity, stereospecific and saturable binding sites for BZDs were discovered in the central nervous system (CNS) of vertebrates including humans.[10,11] This discovery provided a focal point on which the actions of the BZDs could be explained. Good correlations between the potencies of various BZDs as anxiolytics, anticonvulsants, and muscle relaxants and the potencies of these BZDs as inhibitors of [³H]BZD binding *in vitro* provided strong evidence in support of the hypothesis that these neuronally localized BZD receptors mediate the pharmacological actions of the BZDs. Additional evidence in support of this hypothesis was provided by the observation that the distribution of CNS BZD receptors correlated well with anatomic regions believed to be involved in the pharmacological effects of the BZDs.[12]

An extensive body of behavioral, pharmacological, and neurochemical evidence suggests that many of the pharmacological actions of the BZDs are primarily mediated through neuronal systems where γ-aminobutyric acid (GABA) acts as the neurotransmitter.[13-16] GABA will also influence the BZD receptor binding *in vitro*. This allosteric interaction between GABA and BZD receptors results in an increase in the affinity of BZD binding in the presence of GABA.[17-20] The allosteric modulation of BZD binding by GABA has been cited as evidence for the intimate association between GABA receptors and BZD binding sites.[21] The evidence suggests that all CNS BZD binding sites are coupled to a GABA receptor but that not all GABA receptors are associated with a BZD receptor.[22-24] Chloride ion also appears to be an allosteric modulator of BZD receptor affinity whose effect is additive to that produced by GABA.[25] Since the electrophysiological effects of GABA are thought to be exclusively mediated through changes in chloride ion conductance, the BZD binding site is believed to be a component of a GABA receptor–chloride ionophore complex.[26–28] Consequently, CNS BZD receptor affinity is subject to the regulatory influence of at least two factors, GABA and chloride ion.

2. PHARMACOLOGICAL BASIS FOR BENZODIAZEPINE HETEROGENEITY

2.1. The Triazolopyridazines

The initial characterization of BZD receptors in the mammalian brain suggested that a single homogeneous population of binding sites existed. This conclusion was based on equilibrium binding studies with [^{3}H]diazepam and [^{3}H]flunitrazepam ([^{3}H]FLU), which resulted in data consistent with simple Langmuir isotherms.[10,11,29] Hill slopes derived from BZD/[^{3}H]BZD competition curves did not differ significantly from 1, and K_i values for individual BZDs did not vary from one brain region to the next.[29]

The recent observation that a class of compounds known as the triazolopyridazines showed complex interactions with BZD receptors provided evidence for BZD receptor heterogeneity.[30] Competition curves derived from the inhibition of [^{3}H]FLU binding in rat cerebral cortex by the triazolopyridazine CL 218872 were characterized by an IC$_{50}$ value of 142 nM and shallow Hill slopes of approximately 0.7.[31,32] Hofstee plots of the CL 218872/[^{3}H]BZD competition data were curvilinear in some brain regions and suggested the presence of two sites with characteristic K_D and B_{max} values in these regions. CL 218872 possessed selective anxiolytic activity at doses comparable to diazepam in anticonflict paradigms and was effective against pentylenetetrazole-induced seizures.[31] Ten times the anxiolytic dose as determined in anticonflict tests did not produce ataxia or sedation. Thus, the triazolopyridazines are thought to be a class of compounds with selective anxiolytic and anticonvulsant properties, lacking the sedative–hypnotic properties of the BZDs.

The high- and low-affinity sites resolved by Hofstee analysis of CL 218872/[^{3}H]BZD competition data were designated as type I and type II receptors, respectively.[32] Recently, the ability of CL 218872 to discriminate BZD receptor subtypes was confirmed by equilibrium binding studies with [^{3}H]CL 218872.[33] Two binding sites with K_D values of 10–30 nM and 200–600 nM were resolved by Scatchard analysis. These K_Ds were similar to the K_i values derived from CL 218872/[^{3}H]diazepam competition experiments and may correspond to the type I and type II receptors. Regional distribution of CL 218872 binding has also been studied.[32] This triazolopyridazine was found to competitively inhibit specific [^{3}H]BZD binding with the greatest potency in the cerebellum and with the least potency in the dorsal hippocampus. The apparent differences in potency were the result of variations in the relative densities of type I and type II receptors in the various brain regions studied. The cerebellum was almost exclusively populated by type I receptors (90%), whereas only 40% of CL 218872 binding sites in the hippocampus were type I sites. Based on these observations, CL 218872 and other triazolopyridazines appear to have a high affinity for type I receptors and low affinity for type II receptors, whereas the BZDs have equal affinity for both receptor subtypes.

Consistent with differential distribution of type I and type II receptors, autoradiographic studies with CL 218872 showed that CL 218872 was more potent in displacing [³H]FLU binding in areas with a high density of type I receptors (i.e., cerebellum) and less potent in areas enriched with type II receptors.[34] In addition, CL 218872 inhibition of [³H]FLU in slide-mounted brain sections showed deviation from mass action behavior.

The theory has been proposed that the selective anxiolytic activity of the triazolopyridazine CL 218872 is a result of its high affinity for the type I receptor.[32] Implicit in this hypothesis is the contention that type I receptors mediate the anxiolytic effects of the BZDs whereas type II receptors mediate the anticonvulsant and sedative effects. In addition, anxiolytic activity is believed to be GABA independent since GABA enhancement of CL 218872 binding to type I was not observed.[32] However, type II receptors were regulated by GABA. Consequently, GABA is thought to be involved in the mediation of anticonvulsant and hypnotic effects of the BZDs.

In contrast, recent evidence supports the hypothesis that type II receptors are of primary importance in the mediation of anxiolytic activity as measured in a water-lick conflict paradigm.[35,36] A good correlation between receptor occupancy in the rat forebrain and hippocampus and the anticonflict effects of CL 218872, lorazepam, diazepam, and triazolam was observed. The proposal that type II receptors in the forebrain and hippocampus are important in the mediation of antianxiety effects of the BZD seems more plausible than the hypothesis based on type I receptors. A number of other inconsistencies make the hypothesis that type I receptors are primarily involved in anxiolytic activity difficult to substantiate. Perhaps the most prominent instance is the suggestion that type I receptors are independent of GABA regulation. There is abundant electrophysiological and biochemical evidence to support the hypothesis that BZDs act via GABAergic neuronal systems in the cerebellum, a brain region that appears to be populated almost exclusively by type I receptors. Indeed, there has been a direct demonstration of GABA enhancement of BZD and CL 218872 binding cerebellar homogenates.[19,37–39] The suggestion by Braestrup *et al.*[35] that type II receptors in hippocampus may be important in the mediation of anticonflict activity is especially appealing in view of the hypothesis that anxiety is a complex function mediated via the limbic system, the classical center of emotion.[40] Thus, it is difficult to rationalize why the cerebellum is rich in type I receptors when this brain region is more involved in the regulation of motor function than emotional states. Similarly, the hippocampus is predominantly populated by type II receptors, which are believed to mediate the sedation and ataxia produced by BZDs. Based on the foregoing reasons and the tenuous support for type I receptor mediation of anxiolytic activity, there is no compelling reason at this time to lend any credence to the postulate that type I and type II receptors mediate the anxiolytic and sedative–hypnotic effects of the BZDs, respectively.

2.2. The β-Carbolines

A series of compounds belonging to the B-carboline class were identified during the search for the endogenous ligand(s) for the BZD receptor.[41] These compounds were isolated from the brain and urine, and these potent alkyl esters were formed during the extraction procedure. However, this does not detract from the interesting pharmacological properties of these compounds. The methyl, ethyl, and propyl esters of β-carboline-3-carboxylate acid (MCC, ECC, and PCC, respectively) are potent inhibitors of specific [³H]FLU binding with IC_{50} values in the nanomolar range.[41-43] Competition curves for MCC, ECC, and PCC were shallow (Hill slopws < 1), reminiscent of the competition experiments using the triazolopyridazines as the inhibitors of [³H]BZD binding. Moreover these β-carbolines were found to be pharmacological antagonists of the BZDs *in vivo*.[44-47] The complex interactions of the alkyl β-carboline-3-carboxylates with BZD receptors have been attributed to the ability of these compounds to discriminate BZD receptor subtypes. Overall, the data have been consistent with the presence of two major classes of BZD receptors (i.e., type I and type II). Data derived from PCC/[³H]FLU competition curves yield two sites, a high-affinity site with a K_D of 0.5 nM, constituting 55% of BZD sites, and the remaining low-affinity sites with apparent K_D of 10 nM.[42] The ethyl and propyl esters of β-carboline-3-carboxylate are thought to have high affinity for the type I receptor and low affinity for the type II receptor.[48] Direct binding studies with [³H]PCC indicate the differences in stoichiometry of the total number of sites labeled when compared to [³H]FLU can be rationalized in terms of BZD receptor heterogeneity. In the cerebellum, a region populated mainly by type I receptors, the maximum number of sites labeled by [³H]PCC is equivalent to that observed with [³H]FLU. In contrast, [³H]PCC labels fewer sites in hippocampus than [³H]FLU, suggesting that [³H]PCC predominantly labels type I sites in the range of [³H]PCC concentrations used.[49]

Additional complexities have been described using PCC as an inhibitor of [³H]PCC binding in competition experiments.[42] When low concentrations of [³H]PCC (50 pM) are used to selectively label the high-affinity sites in rat cerebral cortex, the competition curves are too shallow to be described by a simple mass action curve. These data were consistent with the interpretation of the presence of a small population (3–6%) of super-high-affinity sites (K_D 30–50 pM).

Recently, the characteristics of [³H]PCC binding in mouse brain following photoaffinity labeling of BZD receptors with FLU have been presented as additional evidence for BZD receptor heterogeneity.[50,51] Under conditions in which the irreversible alkylation of BZD receptors by FLU results in an 80% reduction in specific [³H]diazepam binding, no significant reduction in [³H]PCC binding was observed. Based on these data, the proposal was made that a distinct "β-carboline site" may exist, which is a subset of BZD binding sites, since it was reasoned that if [³H]PCC and [³H]FLU interacted with the same binding site, then photoaffinity labeling of this site with FLU should lead to an equal reduction in binding of both ligands.

Table I
Effect of Photoaffinity Labeling of Cortical Membranes with Flunitrazepam on Binding of Various [³H]Ligands[a]

[³H]Ligand	Treatment	K_D (nM)	B_{max} (fmol/mg protein)[b]
[³H]Flunitrazepam	Control	1.70 ± 0.01	862.8 ± 2.3
	Photolabeled	1.67 ± 0.05	171.4 ± 9.5**
[³H]Propyl β-carbo-line-3-carboxylate	Control	0.32 ± 0.07	762.0 ± 40.5
	Photolabeled	0.23 ± 0.02	641 ± 24.0*
[³H]CGS 8216	Control	0.044 ± 0.005	1100.5 ± 30.8
	Photolabeled	0.043 ± 0.002	1013 ± 49.9
[³H]RO15 1788	Control	0.72 ± 0.07	873.6 ± 41.4
	Photolabeled	0.86 ± 0.04	858.4 ± 47.0

[a] Cortical membranes were prepared and photolabeled as described in the text. Six different concentrations of [³H]flunitrazepam (0.05–2.5 nM), [³H]CGS 8216, and eight different concentrations of [³H]propyl β-carboline-3-carboxylate (0.025–2.5 nM) were used. Specific binding was determined in duplicate; each value represents the mean ± SEM of 3–4 separate determinations. Data from Gee and Yamamura.[52]

[b] Significantly different from control at *P < 0.05 and **P < 0.001 by Student's *t*-test.

We recently performed a series of experiments to determine if (1) a distinct "β-carboline site" is present in rat cerebral cortex, (2) the pharmacological specificity of this "site," and (3) the relationship of this site to BZD receptor heterogeneity.[52] The procedure for the photoaffinity labeling of BZD binding sites followed the method described by Johnson and Yamamura.[53] Briefly, cortical homogenates were preincubated in the presence of 3 nM FLU or 3 nM FLU + 1 μM lorazepam at 0°C for 90 min in the dark. Subsequently, the tissue homogenate was irradiated by ultraviolet light for 4 min at 0–4°C. Tissue irradiated in the presence of FLU and lorazepam was defined as control tissue. Tissue irradiated under these conditions was no different from nonirradiated controls in the various binding studies using [³H]ligands except for an approximate 30% reduction in B_{max} values. Under conditions in which 80% of the BZD binding sites in cortical membranes were inactivated (measured by a decrease in specific [³H]FLU binding) by photolabeling with FLU, only a 16% reduction in the binding of [³H]PCC occurred (Table I). This observation was similar to that made by Hirsch *et al.*[51] in mouse brain. However, when the equilibrium binding characteristics of the [³H]BZD antagonists [³H]R015 1788 and [³H]CGS 8216 were examined, no significant differences in the apparent K_D or B_{max} values were observed (Table I). Moreover, competition curves using various BZD receptor agonists and antagonists as inhibitors of [³H]PCC binding demonstrated that photoaffinity labeling of BZD receptors caused a significant modification in the ability of BZD agonists to inhibit [³H]PCC binding as reflected by significant elevations in the IC_{50} values (Table II). In contrast, IC_{50} values for BZD receptor antagonists were similar in both control and photoaffinity-labeled membranes. These findings suggest that photolabeling of BZD receptors with FLU alters the ability of agonists to bind to the BZD receptor but does not affect the binding of antagonists. Thus, photoaffinity labeling of BZD receptors leads to an apparent "sparing" of BZD antagonist binding. Similar results have been observed by Mohler *et al.*[54]

Table II

Inhibition of [³H]Propyl β-Carboline-3-Carboxylate Binding by Various BZD Agonists and Antagonists in Control and Photolabeled Corticol Membranes[a]

Inhibitor	Control IC_{50} (nM)	Photolabeled IC_{50} (nM)[b]	IC_{50} (photolabeled)/ IC_{50} (control)
Ethyl β-carboline-3-carboxylate	0.37 ± 0.03	0.36 ± 0.01	0.98
Methyl β-carbo-line-3-carboxy-late	0.93 ± 0.07	0.99 ± 0.05	1.06
Propyl β-carbo-line-3-carboxy-late	0.61 ± 0.03	0.54 ± 0.03	0.88
RO15 1788	0.95 ± 0.07	0.97 ± 0.06	1.02
CGS 8216	0.047 ± 0.001	0.052 ± 0.002	1.10
CL 218872	34.2 ± 2.9	64.8 ± 3.8*	1.9
Clonazepam	0.51 ± 0.05	4.85 ± 0.07**	9.5
Diazepam	16.3 ± 2.7	411.0 ± 39.5***	25.2
Flunitrazepam	6.1 ± 0.8	77.6 ± 13.5***	12.7
Lorazepam	0.94 ± 0.05	13.7 ± 1.3***	14.6

[a] All incubations were performed at 0°C for 90 min in 50 mM Na/K phosphate buffer, pH 7.4. Nonspecific binding was defined as binding in the presence of 1 μM ethyl β-carboline-3-carboxylate. The IC_{50} value represents the concentration of inhibitor that caused half-maximal inhibition of specific [³H]propyl β-carboline-3-carboxylate binding. All values represent the mean ± S.E.M. of 3–6 separate determinations. Data from Gee and Yamamura.[52]

[b] Significantly different from control IC_{50} at *$P < 0.05$, **$P < 0.02$ and ***$P < 0.001$ by Student's t-test.

The results of the study examining the interaction of BZD agonist and antagonist in photolabeled membranes suggest that a distinct "β-carboline site" is not the basis for the differential effect of photoaffinity labeling on [³H]PCC binding.[50,51] If a β-carboline specific site separate from the BZD site was exposed following photolabeling, then a reduction in the binding of a BZD compound such as RO15 1788, which interacts competitively with BZD agonist at the receptor, would be predicted. Such an effect on [³H]RO15 1788 binding following the photolabeling of cortical membranes was not observed (Table I). In addition, it is difficult to reconcile the existence of distinct β-carboline sites when the potency of FLU measured by the inhibition of [³H]ligand binding in control membranes is considered. The characteristic IC_{50} values and the concentrations of FLU producing maximal inhibition of specific [³H]FLU or [³H]PCC binding are similar.[42] This apparent competitive interaction also suggests that the BZD site and the "β-carboline site" are one and the same site. In addition, PCC/[³H]RO15 1788 competition curves derived from control and photolabeled membranes showed similar IC_{50} values (2.0 nM) and Hill slopes (0.6). Thus, the heterogeneity of BZD receptors remains intact following photolabeling.

A reasonable explanation for the differential effect of photoaffinity labeling of BZD receptors on agonist and antagonist binding is based on the observations made by Thomas and Tallman.[55] These investigators found that in rat cortical membranes, four BZD sites were inactivated (loss of ability to bind [³H]FLU) for each site irreversibly alkylated by [³H]FLU. Thus, if 25% of the receptors are irreversibly alkylated, then the remaining 75% lose the ability to recognize FLU. If agonists and antagonists stabilize the BZD receptor in different conformations, then it is conceivable that the sites that are not irreversibly bound to FLU will retain the ability to bind antagonists but not agonists. Therefore, these "antagonist sites" may correspond to the "β-carboline site" proposed by Hirsch et al.[51] The small reduction (16%) in [³H]PCC binding observed following photoaffinity labeling may be a reflection of the number of sites irreversibly alkylated by FLU (Table I). Perhaps more convincing support for this hypothesis may be provided by examining agonist and antagonist binding to solubilized BZD binding sites that have been irreversibly alkylated with FLU. Under these conditions, an elimination of both agonist and antagonist binding would be predicted if the hypothesis is reasonable.

The differential effect of photoaffinity labeling on agonist and antagonist interactions with the BZD receptor may be used as an *in vitro* method to screen for BZD agonist and antagonist activity.[54] This method is feasible for the discrimination of agonist from antagonist but does not appear to predict convulsant activity in antagonists such as MCC (Table II). All β-carbolines tested have similar IC_{50} ratios, yet MCC produces frank convulsions, whereas PCC and ECC do not have any apparent direct effects at the doses studied.[56] However, this method may be used to identify selective anxiolytics such as CL 218872, which may have IC_{50} ratios greater than antagonists but less than agonists that possess the full spectrum of pharmacological activity (ranging from antianxiety effects to sedation).

3. BIOCHEMICAL BASIS FOR BENZODIAZEPINE RECEPTOR HETEROGENEITY

Thermal inactivation studies yielded some of the initial biochemical evidence for the existence of BZD receptor subtypes. Heat inactivation of [³H]FLU binding in 50 mM sodium phosphate buffer at 60°C revealed a biphasic disappearance of specific binding with half-lives of 10 min and 70 min for each component.[30] When the more heat labile of the two components was selectively inactivated by heating for 60 min at 60°C in sodium phosphate buffer, neither the Hill coefficient nor the K_i was altered for diazepam when compared to unheated membranes. In contrast, the K_i for CL 218872 was more than doubled, and the Hill coefficient was near unity in the thermally treated membranes. The suggestion was made that CL 218872 has a higher affinity for the heat-labile component and that this component corresponds to the type I receptor. The thermally protected sites were thought to be type II sites. Thermal inactivation in the presence of Tris-HCl buffer resulted in a monophasic decay curve with the eventual loss of all binding sites for [³H]FLU. Thus, phosphate ion appears to stabilize the type II receptor. Other factors have also been reported to afford differential protection of type I and type II receptors from heat inactivation. GABA has been shown to protect type II sites from thermal inactivation with only partial protection of type I sites.[57] Thus, the thermal inactivation data suggest that type I and type II sites may be differentially coupled to GABA receptors and ionophores. The ability of GABA and anions to confer thermal stability selectively to type II sites suggests that these sites may be closely coupled to a GABA site and an ionophore in a manner that allows an allosteric modification of the type II site conformation. Consequently, the conformational change induced may be resistant to thermal inactivation. In the case of type I receptors, coupling to GABA receptors or ionophores may not be present or complete.

Some insight into the biophysical basis of BZD receptor heterogeneity has been provided by studies in which the BZD receptor has been solubilized. At this time, the actual molecular basis for BZD receptor heterogeneity is unclear. It is not known whether BZD receptor subtypes are physically distinct and noninterconvertible recognition sites or different conformations of a single binding site whose equilibrium is regulated by various factors. It has been difficult to solubilize what may only be the BZD recognition site because of the intimate association of BZD receptors with the other components of the macromolecular complex (i.e., GABA receptor and chloride ionophore). Different components of the complex may be removed depending on the methods (i.e., different detergents) used to solubilize the receptor. Thus, accurate determinations of the molecular weight are difficult, which makes variations in molecular weight unreliable as evidence for structurally distinct BZD binding sites.

Both Triton X-100 and Lubrol-PX detergents have been utilized to solubilize the BZD binding site.[58-61] Lubrol-PX was used in the initial solubilization of the BZD receptor, which yielded a macromolecule of approximately 220,000 daltons.[58] Subsequent studies using various techniques including photoaffinity labeling of the binding site prior to solubilization and the use of different detergents yielded macromolecules of consistent molecular weight (200,000–300,000 daltons). These detergent-solubilized binding sites showed high-affinity binding of [³H]FLU and a similar rank order of potency for various BZDs as that observed with membrane-bound receptors.

Some evidence based on differences in the molecular weight of solubilized receptors have been presented to support the hypothesis that differences in the primary structures of BZD receptors form the basis for receptor heterogeneity.[62] Solubilized receptors photoaffinity labeled with [³H]FLU were analyzed by sodium dodecyl sulfate (SDS) polyacrylamide gel electrophoresis. Two binding proteins of 51,000 and 55,000 daltons were observed. The distributions of type I and type II receptors were well correlated with the distributions of the 51,000- and 55,000-molecular-weight species, respectively. Moreover, CL 218872 bound the 51,000-dalton site preferentially, which suggested that this site corresponds to the type I receptor. Several other laboratories using SDS polyacrylamide gel electrophoresis have found only a single binding site with a molecular weight of approximately 50,000.[63-65] These examples serve to emphasize the equivocal nature of the evidence for structurally distinct BZD receptor subtypes based on molecular weight.

Recently, the differential solubilization of two distinct BZD receptors has been accomplished.[66] CL 218872 and R022–7497 showed preferential affinities for the Triton X-100-insoluble fraction over the Triton X-100-soluble fraction as reflected by the inhibition of [³H]FLU binding. Scatchard analysis of equilibrium [³H]PCC binding exhibits a low-affinity, single-component curve in the Triton X-100-soluble fraction and high-affinity binding to a single class of receptors in the Triton X-100-insoluble fraction. Since CL 218872, PCC, and R022–7497 show greater affinity for type I receptors, the conclusion was reached that Triton X-100-insoluble and -soluble binding sites represent type I and type II receptors, respectively. The anatomic distribution of the insoluble and soluble fractions correlated with the distribution of type I and type II sites. The highest concentration of Triton X-100-soluble receptors was found in the hippocampus, whereas the cerebellum and striatum were enriched with Triton X-100-insoluble sites. Triton X-100-insoluble sites retained their pharmacological specificity following salt extraction (1 mM NaCl), and both detergent-soluble and -insoluble fractions have equal affinities for clonazepam, diazepam, flurazepam, and R07-1986/1. The authors of this study indicate that the Triton X-100-insoluble receptors (type I) may be intimately linked to cell cytoskeletal structures to account for their relative insolubility in mild detergents. This observation suggests that the membrane environment of type I receptors may be very different from that of type II sites (Triton X-100 soluble). Thus, it is plausible that different microenvironments of the receptor may impose different conformational constraints on the receptor, which may then be reflected as differences in apparent affinity for certain ligands.

4. REGULATORY INFLUENCES ON BENZODIAZEPINE RECEPTOR HETEROGENEITY

4.1. Temperature Effects

The uniform affinity of BZD receptor subtypes for various BZDs suggests the presence of a structurally invariant region on the BZD recognition site. If BZD receptor subtypes are structurally distinct entities, then the probability of such a region would be significantly lower. The assumption that BZD receptor subtypes represent different conformations of a single receptor implies that the equilibrium between these conformations may be regulated by various influences. Thus, apparent homogeneity or heterogeneity of binding sites may be influenced by manipulating these regulatory influences.

If the assumption is made that different conformations of the BZD receptor are responsible for BZD subtypes, then the ability of compounds such as CL 218872 and the β-carbolines to discriminate receptor subtypes may be affected by temperature (i.e., effect of temperatures on rates of isomerization, see discussion below). Evidence derived from binding studies in support of BZD receptor heterogeneity have been obtained under low-temperature conditions. Earlier studies indicate that the binding of BZDs to its receptor is temperature dependent.[67] The affinity of BZD binding decreases dramatically with increasing temperature. Therefore, binding studies are routinely carried out at 0–4°C in order to maintain conditions most favorable to high-affinity binding.

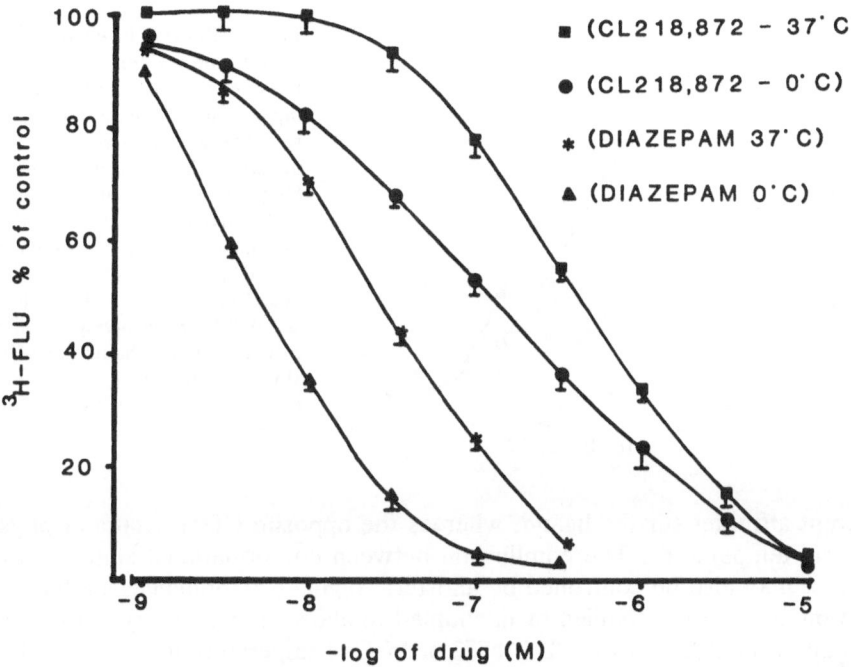

Fig. 1. Inhibition of 1 nM [³H]flunitrazepam binding by various concentrations of diazepam and CL 218872 in rat cerebral cortex at 0° and 37°C. Data from Gee *et al.*[68]

We have recently demonstrated that CL 218872 and PCC, ligands that discriminate BZD receptor subtypes under 0–4°C incubation conditions, lose this ability at physiological temperatures.[68,69] Figure 1 depicts diazepam/ [³H]FLU and CL 218872/[³H]FLU competition curves from rat cerebral cortex at 0° and 37°C. The apparent IC_{50} value for diazepam is reduced fourfold at 37°C and is reflected as a parallel shift of the competition curve. In contrast, the shallow competition curve for CL 218872 observed at 0°C becomes steep and shifted to the right at 37°C. The potency of CL 218872 is significantly less, and the Hill slope increases from 0.52 at 0°C to 0.91 at 37°C. Similarly, a heterogeneous population of receptors in hippocampus recognized by PCC at 0°C becomes homogeneous at 37°C (Fig. 2). The effect of temperature on receptor subtypes is reversible, since cortical or hippocampal membranes that have been preincubated for 30 min at 37°C yield shallow PCC/[³H]FLU and CL 218872/[³H]FLU competition curves under 0°C incubation conditions. A similar effect of temperature on the ability of two structurally distinct compounds to discriminate between receptor subtypes is most likely a consequence of temperature-induced changes (direct or indirect) in the conformational state of a BZD receptor subtype.

If various conformational states of the BZD receptor exist at equilibrium, then temperature may alter this equilibrium by influencing the rates of isomerization. At low temperatures, the isomerization rate may be slow enough that the different conformations may be detected as distinct receptors with

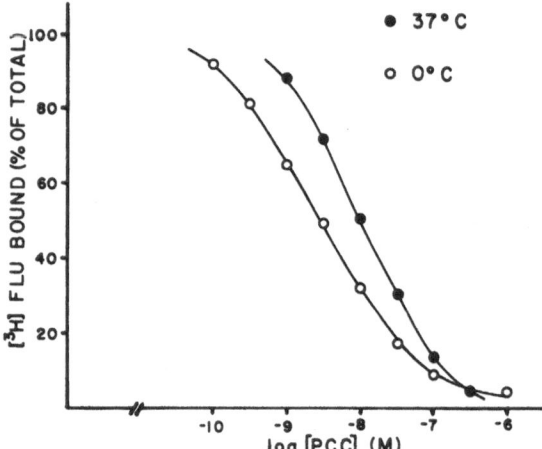

Fig. 2. Effect of incubation temperature on the competitive inhibition of [³H]FLU by PCC in rat hippocampus. Curves are representative of three separate determinations. The concentrations of [³H]FLU used was 0.05 nM. Incubation times were 90 min and 30 min at 0°C and 37°C, respectively. The IC₅₀ and Hill slope for PCC at 0°C were 5.0 ± 1.8 nM and 0.68 ± 0.001, respectively. Under 37°C conditions, the IC₅₀ and Hill slope were 13.7 ± 1.5 and 0.93 ± 0.005, respectively. Data from Gee *et al.*[39]

different affinities for the ligand, whereas the opposite effect occurs at physiological temperature. The equilibrium between conformational states of the receptor may also be controlled by allosteric regulatory influences so that the different states (i.e., coupled or uncoupled to allosteric regulatory units) may have different affinities for CL 218872 or PCC. Temperature may also modify this regulation and alter the equilibrium between conformations. We have also suggested that the change in the PCC/[³H]FLU competition curves by increasing temperature is associated with a large selective reduction in the affinity of the type I receptor.[70] Perhaps PCC induces a conformational change in the type I and not the type II receptor when it binds, so that the observed differences in affinity at 0°C may be related to the greater degree of conformational constraint imposed on type II receptors relative to type I. In summary, the temperature data suggest that differences in the primary structure of the BZD recognition site may not be the molecular basis of BZD receptor heterogeneity but may be related instead of the conformational state of a single receptor.

4.2. GABA Effects

The allosteric regulatory influence of GABA on the BZD receptor may be another determinant of BZD receptor heterogeneity. Benzodiazepine receptor subtypes may or may not be coupled to or regulated by GABA. We have recently demonstrated that the degree of influence of GABA on BZD binding is regionally dependent and appears to be mediated primarily through the type II receptor.[39,69] The regulation of BZD receptor subtypes of GABA was studied by assessing the effect of GABA on FLU binding to [³H]FLU- and [³H]PCC-labeled BZD receptors in the cerebral cortex, cerebellum, and hippocampus of the rat.

In the hippocampus, the type I BZD receptor (high affinity) was selectively labeled by a low concentration (0.04 nM) of [³H]PCC. Dissociation kinetics in the hippocampus support the hypothesis that very low concentrations of [³H]PCC (0.04 nM) will label a single high-affinity site whereas higher concentrations of [³H]PCC (0.5 nM) will interact with both high- and low-affinity sites.[39] The dissociation curve for 0.04 nM [³H]PCC binding in hippocampus is consistent with a simple bimolecular interaction. Dissociation was a linear first-order decay with a half-life of 3.3 min. The dissociation kinetics of 0.5 nM [³H]PCC was markedly different from that observed with 0.04 nM [³H]PCC. The half-life for decay of binding was approximately 1 min, and the decay curve appeared curvilinear. These data suggest that 0.5 nM [³H]PCC labels both high- and low-affinity sites, presumably type I and type II receptors.

Table III shows the results of the FLU/[³H]FLU and FLU/[³H]PCC competition experiments ($\pm 10^{-4}$ M GABA) in the hippocampus. Flunitrazepam inhibition of [³H]FLU was significantly enhanced (1.7-fold increase in affinity) by GABA. In contrast, when a low concentration of [³H]PCC (0.04 nM) was used to selectively label high-affinity BZD sites, the IC_{50} of FLU was unaltered by GABA (Table III). In order to establish that the absence of a GABA effect was not a peculiarity of the complex interaction between [³H]PCC and FLU, additional competition experiments were performed using a concentration of [³H]PCC (0.5 nM) that would label both high- and low-affinity sites. Under these conditions, GABA induced a significant 1.8-fold increase in affinity for FLU.

In the cerebral cortex, semilog plots of the dissociation of 0.04 nM and 0.5 nM [³H]PCC at 0°C yielded half-lives for the decay of binding of approximately 1 and 2 min, respectively.[39] Both decay curves showed some degree of curvature, suggesting that [³H]PCC at both concentrations labels type I and type II receptors. Extrapolation of the slowly dissociating components of the decay curves back to the ordinate suggested that approximately 32% of the [³H]PCC–receptor complex dissociates slowly when 0.5 nM [³H]PCC was used. At 0.04 nM [³H]PCC, slowly dissociating complexes accounted for 70% of the total receptors labeled. Collectively, these kinetic data suggest that low concentrations of [³H]PCC (0.04 nM) predominantly label one site whereas high concentrations of [³H]PCC (0.5 nM) label two types of BZD receptors in the rat cerebral cortex.

GABA enhanced (1.8-fold) FLU inhibition of [³H]FLU (0.05 nM) in cerebral cortex to the same degree as that observed in hippocampus (Table III). GABA also enhanced FLU inhibition of a low concentration of [³H]PCC, but the magnitude (1.47-fold) was significantly less ($P < 0.02$, Student's t-test) than the increase observed with [³H]FLU displacement. These findings were consistent with the dissociation kinetic data. Presumably, low concentrations of [³H]PCC labels predominantly type I but also some type II receptors. Since type II receptors were also labeled, some GABA effect would be expected, but somewhat less than that observed under conditions in which a greater proportion of type II receptors would be labeled (i.e., sites labeled by 0.5 nM [³H]PCC). The GABA effect observed at 0.5 nM [³H]PCC (Table III) supports this hypothesis: GABA enhancement was significantly increased (1.8-fold, $P < 0.05$ Student's t-test) over the effect observed with 0.04 nM [³H]PCC.

The findings from the cerebral cortex and hippocampus suggest that the type II or low-affinity BZD receptors in these brain regions are regulated by GABA whereas the type I receptors may not be subject to GABA regulation.

Table III

Effect of GABA on the Affinity of FLU for [³H]FLU- and [³H]PCC-Labeled Benzodiazepine Receptors in Various Regions of the Rat Brain[a]

Brain region	[³H]Ligand	[³H]Ligand concentration	Flunitrazepam		
			IC_{50} (control)	IC_{50} (+GABA)[b]	IC_{50} (control)/ IC_{50} (+GABA)[c]
Hippocampus	[³H]FLU	0.05	1.50 ± 0.09	0.91 ± 0.04**	1.67 ± 0.11**
	[³H]PCC	0.04	3.10 ± 0.20	2.75 ± 0.30	1.14 ± 0.08
	[³H]PCC	0.50	7.47 ± 0.70	4.23 ± 0.27*	1.80 ± 0.20*
Cortex	[³H]FLU	0.05	1.90 ± 0.20	1.04 ± 0.09*	1.84 ± 0.09***
	[³H]PCC	0.04	3.90 ± 0.60	2.65 ± 0.44	1.47 ± 0.07**
	[³H]PCC	0.05	8.40 ± 0.60	4.90 ± 0.40**	1.80 ± 0.12*
Cerebellum	[³H]FLU	0.05	2.24 ± 0.18	1.53 ± 0.06**	1.50 ± 0.15*
	[³H]PCC	0.04	4.10 ± 0.10	2.85 ± 0.20***	1.46 ± 0.14*

[a] [³H]Ligand concentrations and IC_{50} values are in nanomolar. All values represent the mean ± SEM of 4–7 determinations. The IC_{50} value represents the concentration of FLU that caused half-maximal inhibition of specific [³H]ligand binding. Tissue homogenates washed five times were used. Incubations were carried out in 50 mM Na/K phosphate buffer, pH 7.4, ± GABA (10^{-4} M) at 0°C and lasted 90 min. Data from Gee et al.[39]

[b] Significantly different from control at *$P < 0.01$, **$P < 0.005$ and ***$P < 0.001$ by Student's t-test.

[c] Significantly different from one at *$P < 0.05$, **$P < 0.01$, and ***$P < 0.005$ by Student's t-test.

There are several plausible explanations for the differential effect of GABA observed in the cerebral cortex and hippocampus. It is possible that type I receptors are not "coupled" to GABA receptors whereas the opposite applies for type II receptors. Perhaps different conformational constraints imposed by "coupling" may be manifest as a difference in the binding properties of [^{3}H]PCC at 0°C, but for thermodynamic considerations, this difference in affinity is not apparent at 37°C. Thus, at physiological temperatures, all BZD receptors may be functionally coupled to a GABA receptor. Another possibility may be that the type I receptors are already in a conformation with high affinity for FLU. If this conformation is similar to that induced by GABA, then the presence of GABA would have no further effect on the affinity of type I receptors for FLU. Alternatively, low concentrations of [^{3}H]PCC at 0°C may be labeling a BZD receptor subtype that is unresponsive to GABA or muscimol. This GABA receptor may have greater sensitivity for piperidine derivatives such as THIP and isoguvacine. The apparent heterogeneity of GABA receptors makes it plausible that BZD receptors may be associated with different subtypes of GABA receptors.[71,72]

The magnitudes of the effects of GABA on FLU/[^{3}H]FLU and FLU/[^{3}H]PCC competition experiments were indistinguishable in the cerebellum (Table III). GABA enhanced the affinity of FLU for both [^{3}H]FLU- (0.05 nM) and [^{3}H]PCC (0.04 nM)-labeled receptors approximately 1.5-fold. Both [^{3}H]FLU and [^{3}H]PCC label a homogeneous population of receptors (type I) in cerebellum.[32,48] Thus, the similar effects of GABA on [^{3}H]FLU- and [^{3}H]PCC-labeled receptors in the cerebellum are consistent with previous reports that BZD receptors in this brain region are almost exclusively type I receptors. An apparent GABA effect in the cerebellum, a brain area enriched with type I receptors,[32] suggests that type I receptors found in the cerebellum may be qualitatively different from those found in the hippocampus and cerebral cortex. Cerebellar type I receptors may be functionally associated with a GABA receptor at 0°C, unlike type I receptors in the hippocampus and cortex.

5. SUMMARY AND CONCLUSION

The most compelling evidence in support of BZD receptor heterogeneity is derived from study of the interaction of the triazolopyridazines and the β-carboline-3-carboxylates with BZD receptors. Studies on the binding of these novel non-BZD compounds suggest that there may be at least two BZD receptor subtypes (type I and type II) in the CNS. Heat denaturation studies in the presence and absence of GABA and anions indicate that differences in the allosteric regulation of type I and type II receptors may be a determinant of BZD heterogeneity. Receptor solubilization studies designed to establish structural heterogeneity of BZD receptor on the basis of differences in molecular weight have been equivocal. The difficulty may lie in the actual separation of the BZD recognition site devoid of other components of the GABA–BZD receptor–chloride ionophore complex. An apparent differential solubilization of type I and type II receptors has been achieved. The identification of these two sites was based on pharmacological specificities similar to the membrane-bound receptor subtypes and not on differences in molecular weight.

The influence of temperature on the binding of CL 218872 and PCC provides evidence that type I and type II receptors may actually represent different conformational states of the same receptor molecule. Different conformational constraints imposed by the membrane microenvironment may account for the difference in the binding properties of type I and type II receptors at 0°C. The proposal by Lo *et al.*[66] that type I and type II receptors are associated with different cytoskeletal environments is consistent with this hypothesis. Thus, different conformational states of the BZD receptor may provide a molecular basis of BZD receptor heterogeneity.

Another "receptor environment"-related determinant of BZD receptor heterogeneity may involve the nature of the functional coupling of BZD receptor subtypes with the GABA receptor. GABA does not appear to regulate the affinity of type I receptors in hippocampus and cerebral cortex at 0°C. In contrast, the affinity of type I receptors in cerebellum is enhanced by GABA at 0°C, suggesting regional differences in the type I receptor. The enhancement of BZD receptor binding by GABA in hippocampus and cerebral cortex appears to be mediated primarily through type II receptors. Thus, functional coupling of BZD receptors to GABA receptors may influence in conformational state of the BZD receptor in a way that makes them distinguishable from "uncoupled" BZD receptors. In addition, the allosteric regulatory role of GABA has provided a means to predict pharmacological efficacy of BZD receptor ligands and a basis for a model of BZD receptor function.[70]

Heterogeneity of binding and pharmacological efficacy of the ligand are well correlated in β-adrenergic and muscarinic receptor systems.[73,74] Such a correlation does not exist with drugs that interact with BZD receptors. Compounds such as the β-carbolines (MCC, ECC, and PCC), which antagonize the pharmacological effects of the BZDs, show heterogeneous binding properties. The triazolopyridazine CL 218872 also displays differential affinity for BZD receptor subtypes, yet it has pharmacological effects similar to the BZDs. In addition, the imidazodiazepine RO15 1788 antagonizes the effects of the BZDs by competitive inhibition at the receptor site, but it labels BZD receptor subtypes with equal affinity.[75] Based on these observations, a relationship between heterogeneous binding properties and the pharmacological effect of various ligands is not apparent. However, a correlation may exist between pharmacological activity of ligands and the characteristics of their interaction with photoaffinity-labeled BZD receptors. The differential interaction of agonists and antagonists with photolabeled BZD receptors may be related to the different conformational states of the receptor stabilized by agonists and antagonists. This phenomenon may be used as a screen for potential agonist or antagonist activity intrinsic to ligands that interact with BZD receptors.[54]

A functional significance has been attached to type I and type II receptors based on the unique spectrum of pharmacological effects displayed by CL 218872.[32] The hypothesis suggests that type I receptors mediate the anxiolytic effects of the BZDs whereas the anticonvulsant and sedative effects are mediated by type II receptors. CL 218872 is thought to be a selective anxiolytic because it has higher affinity for type I receptors. The most compelling argument against this theory is based on the observation that CL 218872 does not discriminate between type I and type II receptors at physiological temperatures (37°C). The selective anxiolytic effect of CL 218872 may be adequately explained by its "partial agonist" properties postulated by Ehlert *et al.*[70] At this time, we believe that there appears to be no overwhelming reason to suggest that BZD receptor subtypes mediate different pharmacological effects.

ACKNOWLEDGMENTS. We thank Alice Barrett for excellent secretarial assistance. Part of this work was supported by Public Health Service Grants MH-27257, MH-30626, HL-21486, and Program Project Grant HL-20984. K.W.G. is a recipient of a NRSA (NS-06923) from the National Institute of Neurological and Communicative Disorders and Stroke. W.R.R. is a recipient of a Research Career Development Award from the NHLBI (HL-00776). H.I.Y. is a recipient of a USPHS Research Scientist Development Award, Type II (MH-00095), from the National Institute of Mental Health.

REFERENCES

1. Dale, H. H., 1914, *J. Pharmacol. Exp. Ther.* **6**:147–190.
2. Seeman, P., 1980, *Pharmacol. Rev.* **32**:229–313.
3. Ahlquist, R. P., 1948, *Am. J. Physiol.* **153**:586–600.
4. Langer, S. Z., 1974, *Biochem. Pharmacol.* **23**:1793–1800.
5. Berthelsen, S., and Pettinger, W. A., 1977, *Life Sci.* **21**:595–606.
6. Chang, K. J., and Cuatrecasas, P., 1981, *Fed. Proc.* **40**:2729–2734.
7. Gilbert, P. E., and Martin, W. R., 1976, *J. Pharmacol. Exp. Ther.* **198**:66–82.
8. Martin, W. R., Eades, C. G., Thompson, J. A., Huppler, R. E., and Gilbert, P. E., 1976, *J. Pharmacol. Exp. Ther.* **197**:517–532.
9. Ash, A. S. V., and Schild, H. O., 1966, *Br. J. Pharmacol.* **27**:427–439.
10. Squires, R. F., and Braestrup, C., 1977, *Nature* **266**:732–734.
11. Mohler, H., and Okada, T., 1977, *Science* **198**:849–851.
12. Young, W. S., III, and Kuhar, M. J., 1980, *J. Pharmacol. Exp. Ther.* **212**:337–346.
13. Mao, C. C., Guidotti, A., and Costa, E., 1975, *Naunyn Schmiedebergs Arch. Pharmacol.* **289**:369–378.
14. Polc, P., Mohler, H., and Haefely, W., 1974, *Naunyn Schmiedebergs Arch. Pharmacol.* **284**:319–337.
15. Suria, A., and Costa, E., 1975, *Brain Res.* **87**:102–106.
16. Costa, E., Guidotti, A., Mao, C. C., and Suria, A., 1975, *Life Sci.* **17**:167–186.
17. Tallman, J. F., Thomas, J. W., and Gallager, D. W., 1978, *Nature* **274**:383–385.
18. Martin, I. L., and Candy, J. M., 1978, *Neuropharmacology* **17**:993–998.
19. Wastek, G., Speth, R., Reisine, T., and Yamamura, H. I., 1978, *Eur. J. Pharmacol.* **50**:445–447.
20. Karobath, M., and Sperk, G., 1979, *Proc. Natl. Acad. Sci. U.S.A.* **76**:1004–1008.
21. Unnerstall, J., Kuhar, M., Niehoff, D., and Palacios, J., 1981, *J. Pharmacol. Exp. Ther.* **218**:797–804.
22. Briley, M. S., and Langer, S. Z., 1978, *Eur. J. Pharmacol.* **52**:129–132.
23. Maggi, A., Satinover, J., Oberdorfer, M., Mann, E., and Enna, S. J., 1980, *Brain Res. Bull.* **5**(Suppl. 2):167–171.
24. Reisine, T. D., Overstreet, D., Gale, K., Rossor, M., Iversen, L. L., and Yamamura, H. I., 1980, *Brain Res.* **199**:79–88.

25. Costa, T., Rodbard, D., and Pert, C., 1979, *Nature* **277**:315–316.
26. Tallman, J. F., Paul, S. M., Skolnick, P., and Gallager, D. W., 1980, *Science* **207**:274–281.
27. Olsen, R. W., 1982, *Annu. Rev. Pharmacol. Toxicol.* **22**:245–277.
28. Speth, R. C., Guidotti, A., and Yamamura, H.I., 1981, *Nervous System and Behavioral Disorders* (G. Palmer ed.), Academic Press, New York, pp. 243–283.
29. Speth, R. C., Wastek, G. J., Johnson, P. C., and Yamamura, H. I., 1978, *Life Sci.* **22**:859–866.
30. Squires, R. F., Benson, P. I., Braestrup, C., Coupet, J., Klepner, C. A., Myers, V., and Beer, B., 1979, *Pharmacol. Biochem. Behav.* **10**:825–830.
31. Lippa, A. S., Coupet, J., Greenblatt, E. N., Klepner, C. A., and Beer, B., 1979, *Pharmacol. Biochem. Behav.* **11**:99–106.
32. Klepner, C. A., Lippa, A. S., Benson, D. I., Sano, M. C., and Beer, B., 1979, *Pharmacol. Biochem. Behav.* **11**:457–462.
33. Yamamura, H. I., Mimaki, T., Yamamura, S. H., Horst, W. D., Morelli, M., Bautz, G., and O'Brien, R. A., 1982, *Eur. J. Pharmacol.* **77**:351–354.
34. Young, W. S. III, Niehoff, D., Kuhar, M. J., Beer, B., and Lippa, A. S., 1981, *J. Pharmacol. Exp. Ther.* (E. Usdin, P. Skolnick, J. F. Tallman, D. Greenblatt, and S. M. Paul, eds.) **216**:425–430.
35. Braestrup, C., Schmiechen, R., Nielsen, M., and Petersen, E. N., 1982, *Pharmacology of Benzodiazepines* (S. M. Paul, eds.), Macmillian, London (in press).
36. Petersen, E. N., and Buus Lassen, J., 1981, *Psychopharmacology* **75**:236–239.
37. Regan, J., Roeske, W. R., Malick, J., Yamamura, S. H., and Yamamura, H. I., 1981, *Mol. Pharmacol.* **20**:477–483.
38. Niehoff, D. L., Marshall, R. D., Horst, W. D., O'Brien, R. A., Palacios, J. M., and Kuhar, M. J., 1982, *J. Pharmacol. Exp. Ther.* **221**:670–675.
39. Gee, K. W., Ehlert, F. J., and Yamamura, H. I., 1983, *J. Pharmacol. Exp. Ther.* **225**:132–137.
40. Papez, J. W., 1937, *Arch. Neurol. Psychiatry* **38**:725–743.
41. Braestrup, C., Nielsen, M., and Olsen, C., 1980, *Proc. Natl. Acad. Sci. U.S.A.* **77**:2288–2292.
42. Ehlert, F. J., Roeske, W. R., and Yamamura, H. I., 1981, *Life Sci.* **29**:235–248.
43. Nielsen, M., and Braestrup, C., 1981, *Nature* **286**:606–607.
44. Mitchell, R., and Martin, I. L., 1980, *Eur. J. Pharmacol.* **68**:513–514.
45. Oakley, N. R., and Jones, B. J., 1980, *Eur. J. Pharmacol.* **68**:381–382.
46. Tenen, S. S., and Hirsch, J. D., 1980, *Nature* **288**:609–610.
47. Cowen, P. J., Green, A. R., Nutt, D. J., and Martin, I. L., 1981, *Nature* **290**:54–55.
48. Nielsen, M., Schou, H., and Braestrup, C., 1981, *J. Neurochem.* **36**:276–285.
49. Braestrup, C., and Nielsen, M., 1980, *Trends Neurosci.* **3**:301–302.
50. Hirsch, J. D., 1982, *Pharmacol. Biochem. Behav.* **16**:245–248.
51. Hirsch, J. D., Kochman, R. L., and Sumner, P. R., 1982, *Mol. Pharmacol.* **21**:618–628.
52. Gee, K. W., and Yamamura, H. I., 1982, *Eur. J. Pharmacol.* **82**:239–241.
53. Johnson, R. W., and Yamamura, H. I., 1979, *Life Sci.* **25**:1613–1620.
54. Mohler, H., 1982, *Eur. J. Pharmacol.* **80**:435–436.
55. Thomas, J. W., and Tallman, J. F., 1981, *J. Biol. Chem.* **256**:9838–9842.
56. Braestrup, C., and Nielsen, M., 1981, *Nature* **294**:472–474.
57. Squires, R., Klepner, C., and Benson, E., 1980, *Receptors for Neurotransmitters and Peptide Hormones* (G. Pepeu, M. Kuhar, and S. Enna, eds.), Raven Press, New York, pp. 285–293.
58. Yousufi, M., Thomas, J., and Tallman, J., 1979, *Life Sci.* **25**:463–469.
59. Lang, B., Barnard, E., Change, L., and Dolly, D., 1979, *FEBS Lett.* **104**:149–153.
60. Gavish, M., Chang, R., and Snyder, S., 1979, *Life Sci.* **25**:783–789.
61. Asano, T., and Ogasawara, N., 1980, *Life Sci.* **26**:607–613.
62. Seighart, W., and Karobath, M., 1980, *Nature* **286**:285–287.
63. Mohler, H., Battersby, M. K., and Richards, J. G., 1980, *Proc. Natl. Acad. Sci. U.S.A.* **77**:1666–1670.
64. Tallman, J. F., Mallorga, P., Thomas, J. W., and Gallager, D. W., 1981, *GABA and Benzodiazepine Receptors* (E. Costa, G. DiChiara, and G. L. Gessa, eds.), Raven Press, New York, pp. 9–18.
65. Braestrup, C., Nielsen, M., Skovbjerg, H., and Gredal, O., 1981, *GABA and Benzodiazepine Receptors*, (E. Costa, G. DiChiara, and G. L. Gessa, eds.), Raven press, New York, pp. 147–156.

66. Lo, M. M. S., Strittmatter, S. M., and Snyder, S. H., 1982, *Proc. Natl. Acad. Sci. U.S.A.* **79:**680–684.
67. Speth, R. C., Wastek, G. J., and Yamamura, H. I., 1979, *Life Sci.* **24:**351–358.
68. Gee, K. W., Morelli, M., and Yamamura, H. I., 1982, *Biochem. Biophys. Res. Commun.* **105:**1532–1537.
69. Gee, K. W., Ehlert, F. J., and Yamamura, H. I., 1982, *Biochem. Biophys. Res. Commun.* **106:**1134–1140.
70. Ehlert, F. J., Roeske, W. R., Gee, K. W., and Yamamura, H. I., 1983, *Biochem. Pharmacol.* **32:**2375–2383.
71. Braestrup, C., Nielsen, M., Krosgaard-Larsen, P., and Falch, E., 1979, *Nature* **280:**331–333.
72. Karobath, M., and Sperck, G., 1979, *Proc. Natl. Acad. Sci. U.S.A.* **76:**1004–1008.
73. Birdsall, N. J. M., Burgen, A. S. V., and Hulme, E. C., 1977, *Cholinergic Mechanisms and Psychopharmacology* (D. J. Jenden, ed.), Plenum Press, New York, pp. 25–33.
74. Ehlert, F. J., Roeske, W. R., and Yamamura, H. I., 1983, *Handbook of Psychopharmacology,* (L. Iversen, S. Iversen, and S. H. Snyder, eds.), Plenum Press, New York **17:**241–284.
75. Mohler, H., Burkard, W. P., Keller, H. H., Richards, J. G., and Haefely, W., 1981, *J. Neurochem.* **37:**714–722.

Modulation of Catecholaminergic Receptors During Development and Aging

Benjamin Weiss, M. Blair Clark, and Louise H. Greenberg

1. INTRODUCTION

The density and nature of neural and hormonal receptors govern, to a large extent, the ability to which an organism can function in a changing environment. The observation that the very young and very old cannot respond as well to neural or hormonal stimuli as can those in the prime of life has been associated with changes in the density of receptors or in the coupling of these receptors to appropriate regulatory proteins and enzymes.

One means by which an organism can maintain essential functions in the face of changing circumstances is by adjusting the density of its receptors. In mature young animals, when neural input is decreased, there is a compensatory increase in the density of postjunctional receptors. This so-called denervation supersensitivity is essential for maintaining homeostasis. In immature and senescent animals, the adaptive mechanism by which animals adjust their receptors to changes in neural input appears to be impaired.

In the adrenergic nervous system, neural activity induces a complex series of short- and long-term biological events. Acutely, norepinephrine released from sympathetic nerve terminals causes an immediate activation of adrenergic receptors, leading to a postjunctional response. However, repeated stimulation of these receptors by the transmitter leads to a decreased density of functional receptors and a reduced response of the postjunctional cell on subsequent stimulation. Whereas the acute response probably does not involve changes in the synthesis of new receptor protein, the long-term effects of receptor activation may cause a decreased synthesis of new receptor protein or translocation of existing receptor molecules in the cell membrane.

Benjamin Weiss, M. Blair Clark, and Louise H. Greenberg • Department of Pharmacology, Medical College of Pennsylvania, Philadelphia, Pennsylvania 19129.

Receptors for catecholamines are influenced not only by neural activity but also by circulating hormones. Thus, hormones such as thyroxin and estrogen selectively alter the density of receptors for neurotransmitters, providing the biochemical means of controlling neural activity through the endocrine system. Recent evidence suggests that the mechanism by which some hormones influence catecholamine receptors may be lacking in immature animals and become defective in senescent animals.

In this chapter, we briefly review some of the factors that influence receptors for catecholamines, particularly those that may modify the development of those receptors. We also consider the experimental evidence suggesting that advanced age is associated with a loss of some of these receptors in specific brain areas and that there is a defect in the mechanism by which the organism can adjust its receptor density in response to changes in neural and hormonal input.

We have emphasized studies of the β-adrenergic receptors, using this as a model on which certain general principles may be derived. For a more extensive discussion of these and other types of receptors, the reader is referred to other recent reviews and monographs.[1-10]

2. MODULATION OF CATECHOLAMINERGIC RECEPTORS

Several factors have now been shown to alter the number and function of catecholaminergic receptors. These include neural activity, hormones, environmental factors, and pharmacological agents. Before discussing the influence of these factors on the developing and aged organism, it may be useful to consider their influence on catecholaminergic systems in young adult animals.

The rat pineal gland has been studied extensively with regard to β-receptor modulation, and many relevant points can be made with this system. Other tissues will be considered if different or unusual issues need be raised.

2.1. Modulation of β-Adrenergic Receptors by Neural Activity

It has been known since the late 1800s that denervation of the sympathetic input to a structure results in its increase responsiveness to adrenergic agonists.[11] The physiological and biochemical consequences of this phenomenon, which has been termed denervation supersensitivity, have been studied extensively by numerous investigators.[6,12-16] These studies have shown that the increased responsiveness of a tissue to adrenergic agonists following sympathetic denervation is caused by at least two distinct mechanisms, one presynaptic and one postsynaptic. Presynaptic supersensitivity may be explained by the loss of the catecholamine uptake mechanisms in the nerve terminals, which terminate the action of catecholamines, thereby allowing an increased interaction of the transmitter with postjunctional receptors. This type of supersensitivity, therefore, develops in coincidence with the loss of the nerve terminals.[17,18] Postsynaptic supersensitivity takes longer to develop and generally coincides with the development of additional postsynaptic receptors[19]; in the β-adrenergic nervous system, these receptors are linked to adenylate cyclase.[20] Receptor subsensitivity, on the other hand, is caused by a persistent interaction of adrenergic agonists with postsynaptic receptors. It may develop rapidly through a translocation of membrane-bound receptors or more slowly through a mechanism that causes a decreased density of receptors.

The initial studies showing that denervation of an adrenergically innervated structure causes a supersensitive response to catecholamines of the β-adrenergic receptor–adenylate cyclase complex was derived from studies on the rat pineal gland.[19,21] These studies showed that sympathetic denervation caused an increased responsiveness of adenylate cyclase to norepinephrine. Although denervation increased the maximum response to norepinephrine, it did not change the concentration of catecholamine required to half-maximally activate the enzyme. This suggested that the number of receptors was elevated but the affinity of the receptors for the agonist was unchanged.

More recent studies using radiolabeled ligands confirmed and extended these early observations and showed that decreased neural input to the pineal gland increased the density of β-adrenergic receptors without altering the affinity of the receptor for the ligand. Thus, denervation of the pineal gland by bilateral decentralization or bilateral removal of the superior cervical ganglia led to an increase in β-adrenergic receptors[22] and to a supersensitive response of β-adrenergic receptor-linked processes, such as the norepinephrine-induced increase in the accumulation of adenosine 3',5'-monophosphate (cyclic AMP),[23] and in the activation of adenylate cyclase[19,21,24,25] and serotonin N-acetyltransferase.[24]

A similar phenomenon following denervation is seen in other adrenergically innervated areas of the central nervous system. Administering 6-hydroxydopamine into rat brain, which destroys adrenergic nerve terminals, increased β-adrenergic receptors and increased the activity of catecholamine-sensitive adenylate cyclase.[23,26–28]

Subsensitivity of adrenergic responses, on the other hand, may be caused by increased adrenergic input. Lesions that increase the concentrations of catecholamines in a particular brain area decreased β-adrenergic receptors in that area. For example, specific lesions of the dorsal noradrenergic bundle increased the concentration of norepinephrine in the cerebellum and decreased the number of β-adrenergic receptors.[29,30] Lesions of these central noradrenergic efferents induce sprouting of surviving neurons,[31–33] leading to hyperinnervation of the cerebellum.[34] This may explain the increased concentration of norepinephrine, which, in turn, may cause the decrease of β-adrenergic receptors in the cerebellum.

Thus, the general principle that seems to apply for catecholaminergic systems, and perhaps for that of other biogenic amines, is that the degree to which a tissue will respond is inversely related to the prior degree to which it had been exposed to the neurotransmitter. One general mechanism by which the organism can effect these changes is through modulation of the receptors for these neurotransmitters.

2.2. Modulation of β-Adrenergic Receptors by Hormones

Although there is evidence that hormones alter neural responses, the mechanisms by which they act are less clear. One possibility is that hormones such as glucocorticoids, thyroid hormone, and estrogen alter responses to adrenergic stimuli by modifying the β-adrenergic receptor-linked adenylate cyclase system.

The effect of hormones on adrenergic receptors is complex and seems to be tissue specific.[35] For example, adrenalectomizing rats reduced the density of β-adrenergic receptors in lung[35] and increased their density in liver[36,37]; both effects were reversed by administering glucocorticoids.

Chronic treatment of rats with glucocorticoids increased β-adrenergic receptors in lung[35] and potentiated adrenergic responsiveness in several tissues including smooth muscle,[38,39] mast cells,[40] and adipose and cardiovascular tissue.[41] The demonstration that puromycin completely blocked the increase in receptor density induced by hydrocortisone in lung[42] suggested that hydrocortisone may increase the synthesis of new receptors. Furthermore, these effects of hydrocortisone appear to be direct, since they are also observed in culture preparations of lung.[42]

Although these studies suggest that glucocorticoids potentiate adrenergic responses by increasing β-adrenergic receptors through a mechanism involving protein synthesis, other factors may also be important, since hydrocortisone potentiated the epinephrine-induced accumulation of cyclic AMP in mast cells but had no effect on receptor density.[40]

Thyroid hormone also has been reported to alter adrenergic receptors. This hormone increased β-receptor density in rat heart[43-45] and pineal gland (L. H. Greenberg and B. Weiss, unpublished observations). Conversely, thyroid deficiency decreased the number of β-adrenergic receptors in rat heart.[43,46,47] In fact, hypothyroidism decreased β-adrenergic receptors in several species and tissues, including sheep[48] and rat[43,47] heart, rat skeletal muscle,[49] and turkey erythrocytes.[52] In adipose tissue, treatment with propylthiouracil decreased β-adrenergic receptors,[51] whereas thyroxine increased these receptors.[52] In contrast to its effect on β-adrenergic receptors, thyroid hormone decreased α-adrenergic receptors in rat heart.[43].

Adrenergic receptors also are influenced by estrogen and progesterone. The actions of these sex hormones are complex as evidenced by their differential effects on various tissues and brain regions and on different types of adrenergic receptors. Thus, treating ovariectomized rats with estrogen and progesterone decreased β-adrenergic receptors in rat pineal gland, cerebral cortex, and hippocampus.[53] However, the same treatment increased the density of α-adrenergic receptors of cerebral cortex and hypothalamus.[54] That these findings may have physiological significance is supported by the observation that similar cyclical variations are seen in adrenergic receptors during the estrous cycle of rats. During late proestrus and estrus, when estrogen levels are high,[55] β-adrenergic receptors of pineal gland and hippocampus are decreased, and α-adrenergic receptors of cerebral cortex and hypothalamus are increased. Conversely, during diestrus, when estrogen levels are low,[55] the β-adrenergic receptors are elevated, and the α-adrenergic receptors are depressed in these tissues.[53,54] These results are supported by those of Wagner *et al.*,[56] who reported that treating ovariectomized rats with estradiol decreased β-adrenergic receptors in several brain areas, including neocortex, olfactory bulb, and corpus striatum, and by those of Fuxe *et al.*[57] showing that estrogen increased α-receptors of cerebral cortex and hypothalamus. In contrast, other investigators[58,59] reported that treating ovariectomized rats with estradiol increased the density of β-adrenergic receptors in hypothalamus. These discrepancies may be explained by differential effects of estrogen on the β-receptors of the different brain areas[53,54] or by differences in the dose of estrogen employed.[60]

There is some evidence that estrogen can also increase dopamine receptors and receptor mechanisms. Estrogen increased dopamine-sensitive adenylate cyclase activity in the corpus striatum[61] and increased dopamine receptor-mediated stereotypic behavior.[62-65] These increased dopaminergic responses appear to result, in part, from an increase in dopamine receptors, since treating ovariectomized rats with estrogen and progesterone increased the binding of the dopamine agonist [³H]ADTN in the corpus striatum.[54] Additionally, [³H]ADTN binding was highest in the corpus striatum at late proestrus and early estrus, when steroid levels are high, and lowest at diestrus, when steroid levels are low.[54] Estrogen treatment also increased the binding of the dopamine antagonist [³H]spiroperidol in the corpus striatum of male rats[63] and in striatum, limbic forebrain, and frontal cortex but not anterior pituitary of female rats.[66] The relationship of these findings to sexual behavior has not been studied.

The effects of estrogen on adrenergic receptors may explain certain cyclical variations in tissue responsiveness to adrenergic stimuli. For example, in the pineal gland of ovariectomized female rats, there is an increased sensitivity of adenylate cyclase to norepinephrine, which can be decreased by treating these rats with estrogen.[67,68] Further, adrenergic responsiveness in pineal gland from female rats varies inversely with changes in the levels of estrogen and progesterone during the estrous cycle[67,68]; that is, norepinephrine-sensitive adenylate cyclase is greater during late proestrus, and early estrus when steroid levels are high, than during diestrus, when they are low.

2.3. Modulation of β-Adrenergic Receptors by Environmental Factors

The most extensively studied example of environmental control of a β-adrenergic response is the influence of light on melatonin production by the pineal gland.[69,70] Photic stimuli via a connection from the retina to the hypothalamus[71] decreases sympathetic input to the pineal gland[72] and decreases the release of norepinephrine from neurons innervating the gland.[73] It is believed that there is an endogenous clock or oscillator, located in the suprachiasmatic nucleus of the hypothalamus,[71,74] that controls the rhythm of neural activity of sympathetic nerves to the pineal gland. Neural activity is high during the night and low during daylight.[72,73]

Since light inhibits the activity of sympathetic nerves to the pineal gland, it might be expected that chronic exposure to light would, by decreasing sympathetic input to the gland, increase the density of β-adrenergic receptors and increase the responsiveness of the β-adrenergic receptor-linked adenylate cyclase system to norepinephrine. This is, in fact, the case. Chronic exposure of animals to light increases norepinephrine-stimulated adenylate cyclase activity of rat pineal gland[19] and increases the density of β-adrenergic receptors.[25,75–77] It has also been demonstrated that β-adrenergic receptors[76,78] and catecholamine-stimulated adenylate cyclase activity[19,79,80] increase during light and decrease during the dark period of a 12 h : 12 h light : dark cycle, a rhythm that is consistent with the circadian changes observed for adrenergically stimulated adenylate cyclase[21] and N-acetyltransferase activity.[81]

Temperature is another environmental factor that can vary considerably both diurnally and seasonally and which may be responsible for certain seasonal changes in an organism's responsiveness to adrenergic stimuli. Direct effects of temperature have been demonstrated on β receptors of heart.[82] Furthermore, suckling rats chronically exposed to either raised (34°C) or lowered (7°C) ambient temperatures have a decreased β-adrenergic response (i.e., N-acetyltransferase activity in pineal) compared to rats maintained at 22°C.[83] Perhaps in suckling rats, which are still insensitive to light, β-adrenergic responses are regulated by alterations in environmental temperature; the sensitivity to light may develop later in life.[84]

Since both light and temperature influence β-adrenergic receptor-mediated events, it is likely that seasonal changes, characterized by marked variations in temperature and length of day, alter β-adrenergic receptors. Such receptor alterations may explain the behavior of the organism during different seasons and at different times of the day.[85] In fact, the β-adrenergic receptor-linked production of melatonin varies with the season and with the length of daylight,[86–88] thereby altering sexual behavior.

2.4. Modulation of β-Adrenergic Receptors by Pharmacological Agents

A number of pharmacological agents can modify β-adrenergic receptors. These drugs fall into three general categories: (1) those that act directly at the β-adrenergic receptor, typified by β-adrenergic agonists and antagonists; (2) those that act indirectly on the β-adrenergic receptor by increasing or decreasing the availability of endogenous norepinephrine at the receptor, such as the tricyclic antidepressants[79,89]; and (3) those that modulate the activity of the receptor by acting at some step beyond the actual receptor protein. In this latter category are the phospholipase A_2 inhibitor mepacrine, which can prevent adrenergic desensitization,[90] and the antipsychotic agents, which bind to calmodulin and perhaps thereby prevent certain adrenergic responses.[91]

In rat pineal gland, chronic exposure to adrenergic agonists such as isoproterenol or norepinephrine decreases the number of β-adrenergic receptors[75,92] and decreases the responsiveness to catecholamines of the β-adrenergic receptor-linked adenylate cyclase–cyclic AMP system.[19,21,23,80,93–95]

Similar decreases in β-adrenergic receptors occur in other tissues after chronic exposure to adrenergic agonists.[96–99]

Chronic exposure to β-adrenergic antagonists such as *l*-propranolol, which prevent activation of β-adrenergic receptors, increases the number of these receptors and concomitantly increases the maximum responsiveness of adenylate cyclase to catecholamines in rat heart, erythrocytes,[100] and brain.[101]

Two different classes of antidepressant drugs have been shown to modify β-adrenergic receptors by increasing the availability of norepinephrine: the monoamine oxidase inhibitors, which prevent the metabolism of catecholamines,[102] and the tricyclic antidepressants, which act by preventing the reuptake of the catecholamines into adrenergic nerve terminals[103] or by increasing the release of norepinephrine by blocking presynaptic α-adrenergic receptors.[104]

Repeated administration of monoamine oxidase inhibitors, such as nialamide, reduced the density of β-adrenergic receptors and decreased the norepinephrine-stimulated accumulation of cyclic AMP in rat pineal gland.[105] Repeated, but not acute, treatment of rats with the tricyclic antidepressant compound desmethylimipramine decreased β-adrenergic receptors and the response to β-adrenergic agonists in pineal gland,[6,89,105–108] whole brain,[109] and cerebral cortex.[89,110–112] Other tricyclic antidepressants also have been shown to produce subsensitivity of β-adrenergic receptors in brain following their chronic administration. These include iprindole, doxepine, and trazadone.[113] Since trazadone reverses the central cardiovascular actions of clonidine,[114] it may block presynaptic α-receptors, an action that would increase release of norepinephrine from the presynaptic terminals and, therefore, lead to β-adrenergic receptor subsensitivity.

Reserpine and guanethidine are antihypertensive drugs that deplete catecholamines from sympathetic neurons. Repeated administration of reserpine results in long-term depletion of catecholamines from brain,[115,116] and guanethidine produces a permanent loss of catecholaminergic innervation of many central and peripheral tissues.[117] Administering reserpine for 3 days to adult rats increased β-adrenergic receptors in cerebral cortex and pineal gland[89,106] and, correspondingly, increased catecholamine-stimulated adenylate cyclase activity.[15,118–120] A single dose of reserpine, however, was not sufficient to alter β-adrenergic receptors[89] or β-adrenergic receptor-mediated mechanisms such as catecholamine-activated adenylate cyclase.[121] Similarly, repeated administration of guanethidine resulted in β-receptor supersensitivity in pineal gland.[25]

Other centrally active agents, such as amphetamine, also alter β-adrenergic receptors. Since amphetamine has been reported to (1) mimic catecholamines at adrenergic receptor sites,[122–124] (2) inhibit norepinephrine reuptake mechanisms,[125] and (3) enhance catecholamine release,[126] it would be expected that amphetamine would increase the availability of catecholamines at β-adrenergic receptors. Thus, after chronic drug administration, the receptors would become subsensitive. Indeed, repeated treatment of rats with *d*-amphetamine resulted in subsensitivity of β-adrenergic receptors in cerebral cortex[127] and in a decrease in norpinephrine-stimulated cyclic AMP accumulation in rat limbic forebrain.[128] These results are consistent with the prediction that repeated activation of β-receptors decreases their density.

However, there are reports that higher doses of amphetamine administered chronically increased β-receptors and that this supersensitivity is magnified on withdrawal of the drug.[111,129] These data may be explained by the amphetamine-induced depletion of catecholamines that occurs following long-term amphetamine treatment.[130]

These results demonstrate the complexity and plasticity of the β-adrenergic system. Acute treatment with amphetamine may decrease β-receptors by persistently activating them, whereas chronic treatment with high doses may produce a supersensitivity by depleting the catcholamines.

Ethanol is an example of a commonly abused drug that may produce a hyperadrenergic state in alcoholic patients during withdrawal. Recent evidence suggests that this hyperadrenergic state is associated with an increased density of β-adrenergic receptors during the withdrawal phase.[131] During and immediately after alcohol ingestion, β-adrenergic receptors are decreased in rat heart and brain. However, after withdrawal from chronic ethanol ingestion, the density of β-adrenergic receptors and the accumulation of cyclic AMP elicited by catecholamines are increased.[132–134]

Opioids also affect the adrenergic system and may alter adrenergic receptors when administered chronically. These compounds depress the electrical activity of adrenergic neurons in the locus coeruleus,[135] decrease the concentration and turnover of norepinephrine,[136] and inhibit the release of norepinephrine from slices of cerebral cortex.[137,138] These actions of opioid agents should decrease the availability of norepinephrine at the receptor site and result in supersensitivity after chronic exposure to opiates. As predicted, rats that had morphine pellets implanted subcutaneously for 10 days showed an increase in catecholamine-stimulated cyclic AMP production and an increase in β-adrenergic receptors in cerebral cortex.[139]

Similarly, chronic treatment of mice with phenobarbital increases β-adrenergic receptors in several areas of the brain[140]; the mechanism for these changes may also be related to a decreased availability of norepinephrine at certain receptor sites.[141,142]

Antipyschotic drugs such as trifluoperazine and chlorpromazine are generally thought to produce their antipsychotic effects by blockade of the dopaminergic system. However, neuroleptics also inhibit other catecholaminergic mechanisms, such as norepinephrine-induced stimulation of adenylate cyclase and accumulation of cyclic AMP.[143,144] In addition, the neuroleptics fluphenazine and flupenthixol reduce the ability of norepinephrine to inhibit the firing rate of cerebellar Purkinje cells.[145] Such actions at noradrenergic neurons could lead to a supersensitive increase in β-adrenergic receptors and norepinephrine-sensitive adenylate cyclase. In fact, chronic treatment of rats with trifluoperazine increased β-adrenergic receptors in the cerebral cortex and olfactory tubercle of rats.[106]

Trifluoperazine and other phenothiazine antipsychotic drugs have other biochemical actions, which might help explain their effects on β receptors. For example, they have been shown to inhibit calmodulin activity by selectively

Table I
Studies on the Ontogenetic Development of β-Adrenergic Receptors

Tissue	Species	References
Central nervous system		
Whole brain (without cerebellum)	Fetal rat	239
Brain cells (cultured)	Fetal rat	174
Cerebellum	Rat	157,172,180
Cerebrum	Rat	157,159
Pineal gland	Rat	6,22,157
Pineal gland (cultured)	Rat	164
Spinal cord	Chick	240
Blood cells	Tadpole	241
Blood cells	Rat	242,247
Fat cells	Rat	209
Heart	Rat	179,248–250
Heart	Rabbit	249
Heart	Mouse	251
Lung	Rat	179
Lung	Rabbit	252,253
Parotid gland	Rat	171
Placenta	Sheep	254
Skeletal muscle	Rabbit	255
Urinary bladder	Rabbit	256

binding to it.[146,147] Since calmodulin activates phospholipase A_2,[148] trifluoperazine, by inhibiting calmodulin, may block the activation of phospholipase A_2, thereby altering the metabolism of phospholipids. This, in turn, may lead to changes in the density of β-adrenergic receptors since phospholipids have been implicated in the modulation of these receptors.[90,149–153]

Clearly, many agents encompassing diverse pharmacological classes alter adrenergic responses, perhaps by inducing relatively long-term changes in adrenergic receptors. Although not all of these changes in receptors may prove to have pharmacological relevance, many of our long-held views on the action of certain drugs may need to be reevaluated in the light of the evidence that these drugs can induce alterations in the density of β-adrenergic receptors.

3. ONTOGENETIC DEVELOPMENT OF CATECHOLAMINERGIC RECEPTORS

3.1. Normal Development of Catecholaminergic Receptors

The ontogenetic development of β-adrenergic receptors has been studied in several tissues and species (Table I). Of these tissues, the rat pineal gland is a particularly good model for studying the neural and hormonal factors influencing the development of these receptors. When fully developed, it has a high density of β-receptors functionally linked to adenylate cyclase.[154–156] In addition, the gland is innervated after birth by adrenergic fibers arising from

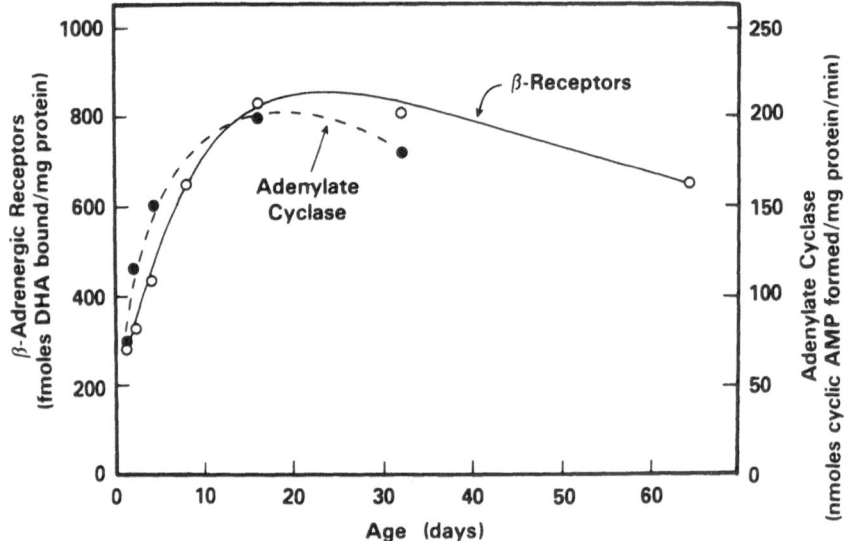

Fig. 1.   Ontogenetic development of β-adrenergic receptors and catecholamine-sensitive adenylate cyclase of rat pineal gland. The density of β-adrenergic receptors in the pineal gland from rats of various ages was determined from Scatchard analyses of the binding of [³H]dihydroalprenolol (DHA), Catecholamine-sensitive adenylate cyclase activity of pineal glands from rats of various ages was determined in the presence of 100 μM *l*-norepinephrine. Each point represents the mean of six experiments. Data from Cantor *et al.*[22] and Weiss.[155]

nerve cell bodies in the superior cervical ganglia; therefore, its sympathetic input can be readily altered or prevented by bilateral removal of these ganglia. The pineal gland is influenced by a number of neuroendocrine and environmental factors, and since it lies outside the blood–brain barrier, it is affected by circulating or injected catecholamines. It also can be readily studied *in vitro*, cultured as an entire gland or as dispersed cells derived from the gland. Characterization of receptor development in the pineal gland may provide the basis for principles of receptor development in many systems. Reference to β receptors in tissues other than pineal gland will be mentioned to emphasize concepts and to provide information not described for β-receptors in the pineal gland.

The hypothesis that the responsiveness of tissues to β-adrenergic agonists is dependent on the development of β-adrenergic receptors is supported by studies showing a temporal relationship between the ontogeny of the β-receptor and the responsiveness of adenylate cyclase to catecholamines.[154,155,157] Figure 1 illustrates the correlation between the development of β-adrenergic receptors and the developmental increase in the responsiveness of adenylate cyclase to norepinephrine stimulation. As may be seen, between 2 and 16 days of age there was nearly a three-fold increase in the density of β-adrenergic receptors in rat pineal gland and a three-fold increase in catecholamine-stimulated adenylate cyclase. Both norepinephrine-stimulated adenylate cyclase and β-receptor density declined after 16 days of age.

The development of catecholamine-sensitive adenylate cyclase activity also has been described for cerebellum, brainstem, and cerebral cortex.[155,157,158] As in the pineal gland, the development of β-adrenergic receptors in these other brain areas parallels the development of this enzyme's responsiveness to catecholamines.[157,159]

Studies on the prenatal development of rat pineal gland showed that the activity of N-acetyltransferase, the pineal enzyme that is activated by cyclic AMP,[160,161] is present in pineal glands from fetal rats 4 days before birth. The β-adrenergic blocking agent *l*-propranolol decreased N-acetyltransferase activity in the pineal glands of fetal rats,[162] suggesting the presence of a β-adrenergic receptor mechanism in the gland prior to birth. The presence of β-adrenergic receptors in pineal gland of fetal rat of gestational age 18 to 21 days has now been confirmed using radioligand binding techniques[163] (M. B. Clark and B. Weiss, unpublished observations). These results indicate that the several components of the β-adrenergic receptor–adenylate cyclase–cyclic AMP system responsible for activating N-acetyltransferase activity are present prior to birth but that the major development occurs after birth.

The development of β-adrenergic receptors has also been demonstrated *in vitro*. Pineal glands removed from 2-day-old rats showed a two- to threefold increase in β receptors after 48 h in organ culture. This increase was blocked if the β-adrenergic agonist *l*-isoproterenol or the protein synthesis inhibitor cycloheximide was included in the incubation medium. The pharmacologically inactive stereoisomer *d*-isoproterenol had no effect on the development of β receptors.[164] These results suggest that the *in vitro* developmental increase in β receptors requires the synthesis of new protein and can be modulated stereoselectively by catecholamines.

The question of whether the increase in adrenergic receptors was caused by an increased density of receptors per cell or by an increase in the total number of cells as a result of mitotic division was also addressed using the cultured pineal gland. It was found that the addition of the antimitotic agent cystosine-1-β-arabinofuranoside to cultured pineal glands failed to prevent the increase in receptors observed after 48 h in culture (M. B. Clark and B. Weiss, unpublished observations), suggesting that the number of receptors per cell increased.

3.2. Factors Influencing the Development of Catecholaminergic Receptors

A large number of factors, including neural activity, hormones, nutrition, drugs, and genetic influences, have been studied for their effects on the development of different types of catecholaminergic receptors in a variety of tissues. A summary of these studies is presented in Table II. We will consider in detail only those factors that concern the development of β-adrenergic receptors in the nervous system.

3.2.1. Innervation

An important question in neurobiology is to determine whether the innervation of a structure induces the development of specific receptors or

Table II

Factors Influencing the Development of Adrenergic Receptors

Receptor type	Factor[a]	Species	Tissue	Effect of receptor density[a]	References
α	Undernutrition	Rat	Whole brain	↓	182
α	PMSG	Rat	Cerebral cortex	↔	257
α	PMSG	Rat	Hypothalamus	↓ initially	257
β₂	Increased sympathetic innervation	Sheep	Placenta	↓	254
β	Surgical sympathectomy	Rat	Pineal gland	↔ initially	22
β	6-OH-DA-induced hypoinnervation	Rat	Cerebral cortex	↑	170
β	6-OH-DA-induced hyperinnervation	Rat	Cerebellum	↓	172
β	6-OH-DA	Rat	Parotid gland	↔ initially ↑ later	173
β	Guanethidine	Rat	Pineal gland	↑	25
β	Isoproterenol	Rat	Brain cell cultures	↓	174
β	DMI	Rat	Cerebral cortex	↓	181

				↔ initially	M. B. Clark and B. Weiss (unpublished data)
β	Estrogen	Rat	Pineal gland	→ later	53
β	PMSG	Rat	Cerebral cortex	←	257
β	PMSG	Rat	Hypothalamus	↕	257
β	Hyperthyroidism	Rat	Heart	←	177
β	T₄	Rat	Lung	←	179
β	T₃ analogue	Rabbit	Lung and heart	↕	176
β	PTU	Rat	Forebrain	→	178
β	PTU	Rat	Cerebellum	→	
β	PTU	Rat	Lung	→	179
β	Betamethasone	Rabbit	Heart	↕	176
β₁	Betamethasone	Rabbit	Lung	←	176
β	Dexamethasone	Rat	Heart	↕	250
β	Genetic hypertension	Rat (SHR)	Heart	→	258
β	Undernutrition	Rat	Whole brain	→	182
β	Increased cell surface area	Rat	Adipose tissue	←	209
Dopamine	6-OH-DA	Rat	Whole brain	→	259

[a] Abbreviations: DMI, desmethylimipramine; PMSG, pregnant mare serum gonadotrophin; PTU, propylthiouracil; 6-OH-DA, 6-hydroxydopamine; SHR, spontaneously hypertensive rat; T₃, triiodothyronine; T₄, tetraiodothyronine; ↓, decrease; ↑, increase; ↔, no change.

whether receptors can develop independently of the innervation. This question could be approached using the pineal gland since it is not innervated at birth; therefore, one can prevent its innervation by removing the superior cervical ganglia at birth or by administering to neonatal rats agents that destroy sympathetic nerve terminals.

The observation that in the pineal gland there is a temporal correlation between the maturation of its adrenergic innervation,[165–168] with the development of β-adrenergic receptors,[154] and the responsiveness of catecholamine-sensitive adenylate cyclase[155] suggested a relationship between the maturation of innervation and the developmental increase in β-adrenergic receptors. However, experiments designed to examine this relationship showed that adrenergic innervation is not required for receptor development. Thus, responsiveness of the β-adrenergic receptor–adenylate cyclase complex develops before adrenergic nerve terminals can be demonstrated in the pineal gland.[155,162] In addition, sympathetic denervation of the pineal gland by 6-hydroxydopamine treatment[169] or by bilateral removal of the superior cervical ganglia[22,155] failed to prevent the developmental increase of catecholamine-stimulated adenylate cyclase[155,169] or of β-adrenergic receptors.[22] Similarly, destruction of adrenergic input to the cerebral cortex[170] or parotid gland[171] by administering 6-hydroxydopamine to neonates failed to alter the normal rate of β-adrenergic receptor development. In fact, there is evidence for an inverse relationship between sympathetic input and the development of the β-adrenergic receptor-linked adenylate cyclase system. For example, there was an increased responsiveness of adenylate cyclase to norepinephrine in pineal glands of adult rats if these animals were administered 6-hydroxydopamine[169] or guanethidine[22] as neonates. Similar treatment of newborn rats increased the density of β-adrenergic receptors and increased the activation of adenylate cyclase by catecholamines in cerebral cortex of 16-day-old rats.[170] Conversely, treating neonatal rats with 6-hydroxydopamine resulted in hyperinnervation of the cerebellum and a concomitant reduction in β-adrenergic receptor density, norepinephrine-stimulated adenylate cyclase activity,[172] and cyclic AMP accumulation.[173]

Thus, although other, as yet undetermined, factors may control the ontogeny of β-adrenergic receptors, adrenergic innervation apparently is not essential for their normal development. However, the degree of innervation may regulate the final complement of β-adrenergic receptors and the sensitivity of adenylate cyclase to catecholamines.

3.2.2. Catecholamines

Recent data indicate that the ontogenetic increase of β-adrenergic receptors can be influenced by catecholamines. Thus, the addition of the catecholamine *l*-isoproterenol to cultured pineal glands from newborn rats[164] or to cultured cell aggregates of fetal rat brain[174] prevents the developmental increase in β-adrenergic receptors.

Although the results of these studies on tissues in culture suggest that catecholamines can regulate β-adrenergic receptor development, the question of whether catecholamines influence β-adrenergic receptor ontogeny *in vivo* is still not resolved. In neonates, the potential influence of circulating catecholamines on the ontogeny of β-adrenergic receptors of rat pineal gland has been studied by removing the adrenal medullae and superior cervical ganglia of rats at birth and measuring the ontogeny of β-adrenergic receptors. The results showed that the absence of circulating catecholamines arising from the adrenal medullae failed to alter the normal increase in β-adrenergic receptors.[22] It should be pointed out, however, that in these experiments there may still have been circulating catecholamines from extraadrenal chromaffin tissue. In other studies, it was found that administering the β-receptor-blocking drug *l*-propranolol to rats daily from birth increased β-adrenergic receptors in the pineal gland of 8-day-old rats (M. B. Clark and B. Weiss, unpublished observations), suggesting that circulating catecholamines may have an inhibitory influence on the ontogenetic development of the β-receptors *in vivo*.

3.2.3. Hormones

Several hormones including estrogen, glucocorticoids, and thyroid hormone have been found to influence the ontogeny of β-adrenergic receptors. The effects of estrogen apparently vary with the animal's stage of development. Animals treated with estradiol from birth to 20 days of age showed no change in receptor density in the pineal gland (M. B. Clark and B. Weiss, unpublished data), whereas estradiol decreased β-adrenergic receptors in pineal glands of adult rats.[53,175]

The effects of glucocorticoids on the development of β-adrenergic receptors vary with the tissue. Whereas glucocorticoids accelerated the developmental increase in β-adrenergic receptors of rabbit lung, they had no effect on β-receptors of rabbit heart.[176]

Thyroid hormone has been reported to increase the density of postsynaptic β-adrenergic receptors in adult animals. Similarly, in developing rats, hyperthyroidism increased β-adrenergic receptor development in rat heart,[177] and propylthiouracil-induced hypothyroidism inhibited β-adrenergic receptor development in forebrain, cerebellum[178] and lung.[179] Any differential effects that these hormones have on β-adrenergic receptors in various tissues may reflect a difference in the proportion of β-receptor subtypes found in each tissue.[180]

3.2.4. Pharmacological Factors

As discussed above, the β-adrenergic agonist isoproterenol inhibited β-adrenergic receptor development in culture preparations of neonatal rat pineal gland and cell aggregates of fetal rat brain. On the other hand, the β-adrenergic antagonist *l*-propranolol increased receptor development in rat pineal gland when administered chronically to neonatal rats. Agents acting indirectly on adrenergic receptors also influence receptor development. Thus, desmethylimipramine, which blocks the reuptake of norepinephrine into presynaptic adrenergic terminals,[103] and which has been shown to reduce the density of β-adrenergic receptors in brain of adult rats,[89,109,110] inhibited the developmental increase in β-adrenergic receptors in the brain of neonatal rats when given to their pregnant mothers.[181] Apparently, in some cases, administering drugs that alter adrenergic activity during the prenatal period causes biochemical changes in receptor development that extend into the postnatal period.

3.2.5. Environmental Factors

In addition to hormones and drugs, environmental factors such as nutrition, light, temperature, maternal behavior, and substances transferred by the mother may alter the development of adrenergic receptors or the systems linked to these receptors. For example, undernutrition between 14 days of gestation and 50 days postnatally resulted in a decreased density of adrenergic receptors when measured at 90 days of age.[182] Moreover, maternal activity or maternally transferred factors have been suggested as exogenous cues that regulate the development of rhythmic activity of N-acetyltransferase.[183-185] N-Acetyltransferase activity of pineal gland in suckling rats also is sensitive to changes in environmental temperature.[83]

4. EFFECT OF ADVANCED AGE ON THE DENSITY AND ADAPTABILITY OF CATECHOLAMINERGIC RECEPTORS

A variety of tissues from aged animals and humans show a reduced responsiveness to catecholamine agonists.[76,77,106,186-188] Recent studies summarized in Table III and discussed more fully below suggest that this age-associated reduction in the physiological and biochemical responsiveness of tissue to catecholamine stimulation may be caused, at least in part, by a decreased density of the receptor component of the receptor–adenylate cyclase–membrane complex.

Advanced age also is associated with a decreased ability of organisms to adapt their physiological responses to changes in their neural or hormonal environment.[77,89,157,189,190] This reduced adaptive capacity may be caused by a defect in the organism's ability to adjust its receptors in response to a changing environment. The following section reviews some of the experiments supporting these statements.

4.1. Loss of Catecholaminergic Receptors with Age

An age-related alteration in the density of α- and β-adrenergic receptors has been reported to occur in a number of tissues and species (Table III). In most brain areas, these receptors declined with advanced age. However, in studies of selective subtypes of β-adrenergic receptors in cerebellum, Pittman et al.[191] showed that whereas β_2-adrenergic receptors decreased with age, β_1-adrenergic receptors increased. Usually this change in density of adrenergic receptors is not accompanied by alterations in the affinity of the receptors for the ligand, suggesting that the quantity rather than the quality of the receptors changes with age.

Table III
Age-Related Alterations of Adrenergic Receptors of Brain

Receptor type	Radioligand[a]	Species	Strain[a]	Brain area	Effect on receptor	Age comparison (mo)	Reference
α	[³H]WB-4101	Rat	Fischer 344	Cerebral cortex	↓ density	5 vs. 25	198
α	[³H]Prazosin	Rat	S–D	Hypothalamus	↓ binding	2 to 5 vs. 9,18 to 23	175
α	[³H]Prazosin	Rat	S–D	Cerebral cortex	↓ binding	2 to 5 vs. 9,18 to 23	175
α	[³H]WB-4101	Rat	Fischer 344	Cerebellum	↓ binding	2 vs. 16,26, 30	L. H. Greenberg and B. Weiss (unpublished data)
α	[³H]WB-4101	Rat	Fischer 344	Hippocampus	↓ binding	16 vs. 26,30	L. H. Greenberg B. Weiss (unpublished data)
α	[³H]WB-4101	Rat	Fischer 344	Hypothalamus	↓ binding	4 to 26 vs. 30	L. H. Greenberg B. Weiss (unpublished data)
β	[³H]DHA	Rat	Fischer 344	Cerebellum	↓ density ↔ affinity	6 vs. 12,24	76,106
β	[³H]DHA	Rat	Fischer 344	Corpus striatum	↓ density ↔ affinity	6 vs. 12,24	76,106
β	[²H]DHA	Rat	Fischer 344	Pineal gland	↓ density ↔ affinity	1 vs. 3,12,24	76,106
β	[³H]DHA	Rat	Fischer 344	Cerebral cortex	↓ density ↔ affinity	3 vs. 24	77

(Continued)

Table III. (Continued)

Receptor type	Radioligand[a]	Species	Strain[a]	Brain area	Effect on receptor	Age comparison (mo)	Reference
β	[³H]DHA	Rat	Fischer 344	Cerebral cortex	↓ density ↔ affinity	5 vs. 25	198
β	[³H]DHA	Rat	Wistar	Cerebellum	↓ binding	3 vs. 24	260,261
β	[³H]DHA	Rat	Wistar	Brainstem	↓ binding	3 vs. 24	260,261
β	[³H]DHA	Rat	Wistar	Cerebral cortex	↔ binding	3 vs. 24	260,261
β	[³H]DHA	Rat	C–R	Corpus striatum	↓ binding	3 vs. 30	206
β	[³H]DHA	Rat	C–R	Pineal gland	↓ binding	3 vs. 30	206
β	[³H]DHA	Rat	C–R	Cerebral cortex	↓ binding	3 vs. 30	206
β	[125I]HYP	Rat	S–D	Cerebral cortex	↔ density	4.5 vs. 10,12,14	191
β₁	[125I]HYP	Rat	S–D	Cerebellum	↑ density	4.5 vs. 10,12,14	191
β₂	[125I]HYP	Rat	S–D	Cerebellum	↓ density	4.5 vs. 10,12,14	191
β	[³H]DHA	Human	—	Cerebral cortex	↔ binding	birth to 60 vs. 61 to 80 yr	260,261
β	[³H]DHA	Human	—	Cerebellum	↓ binding	birth to 60 vs. 61 to 80 yr	260, 261

[a] Abbreviations S–D, Sprague–Dawley; C–R, Charles River; [³H]DHA, [³H]dihydroalprenolol; [125I]HYP, [125I]hydroxybenzylpindolol; [³H]WB-4101, 2-(2,6-dimethoxy-phenoxyethyl)aminomethyl-1,4-benzodioxane [phenoxy-3-³H(n)].

In general, the binding of dopamine receptor ligands also has been reported to decline with advanced age (see Table IV), although in one study dopamine receptors increased with age.[192] There is still controversy as to whether the decreased binding of dopamine ligands results from a decreased density of receptors,[193-200] a decreased affinity of the binding sites for the ligands,[192,201,202] or a combination of both effects. Species and strain differences may account for some discrepancies, and methodologic differences related to subpopulations of dopamine binding sites may account for others. Thus, data varied depending on whether agonists or antagonists were used as the radioligands and depending on which unlabeled ligand was used to determine nonspecific binding.[203,204]

Regardless of the controversy about alterations in dopamine receptors with advanced age, it is well known that an age-related reduction in the sensitivity of adenylate cyclase to dopamine occurs in various brain areas of several species.[188,194-196,201,202,205,206] Further, the loss of binding sites for the dopamine agonist [³H]ADTN explains, at least in part, the age-related reduction in the sensitivity of adenylate cyclase to dopamine,[203] just as the loss of β-adrenergic binding sites correlates with decreased norepinephrine-stimulated cyclic AMP accumulation in aged rats.[187,188,207] An age-related decrease of dopaminergic receptors also may explain the reduced rotational behavior seen in aged rats following the administration of amphetamine[194] or dopamine to animals with unilateral lesions of the nigrostriatal tract.[203,204,208]

4.2. Mechanism for the Age-Related Loss of Catecholaminergic Receptors

Although the total density of β-adrenergic receptors in several areas of the brain declines with age when receptor density is expressed on a protein basis,[76,106] it is still not certain whether this decline is caused by a reduction in the number of receptors per cell or by a loss of specific cell types. However, in peripheral tissue such as human lymphocytes[186] and fat cells,[209] the number of β-adrenergic receptors per cell does decline with age.

In cerebellum, there is an age-related loss of Purkinje cells.[210] Since these cells receive the major noradrenergic input to the cerebellum,[211,212] the decrease in β-receptor density in cerebellum may reflect a loss of these target cells. However, a number of findings argue against this possibility and suggest that receptors are lost from individual cells. For example, in aged rats the decrease of β$_2$-adrenergic receptors in cerebellum was evident when expressed per milligram protein, per cerebellum, or per milligram DNA.[191] Moreover, in the nervous mouse, a mutant strain with an 85% loss of cerebellar Purkinje cells, there was no significant difference in the density of β-adrenergic receptors when compared to normal mice.[213]

The loss of β-adrenergic receptors with age may reflect a type of receptor subsensitivity since the reuptake mechanism for neuronally released norepinephrine is decreased in brain tissue from senescent rats.[214] Thus, chronic overexposure to norepinephrine may decrease receptors. Alternatively, aged

Table IV
Age-Related Alterations in Dopaminergic Receptors of Brain

Radioligand	Species	Strain	Brain area	Effect on receptor	Age comparison (mo)	Reference
[3H]Haloperidol	Rat	Wistar	Corpus striatum	↓ density ↔ affinity	6 vs. 25 to 29	193
[3H]Haloperidol	Rat	Wistar	Corpus striatum	↓ binding	10 vs. 30	201
[3H]Spiroperidol	Rat	Wistar	Corpus striatum	↔ density ↓ affinity	10 vs. 30	201
[3H]Spiroperidol	Rat	Fischer 344	Corpus striatum	↓ density ↔ affinity	5 vs. 25	198
[3H]Spiroperidol	Mouse	C57BL	Corpus striatum	↓ density ↔ affinity	3 or 8 vs. 28	199
[3H]Spiroperidol	Mouse	C57BL	Olfactory bulb	↓ density	3 or 8 vs. 28	199
[3H]Spiroperidol	Mouse	C57BL	Hypothalamus	↓ density	8 vs. 28	199
[3H]Spiroperidol	Mouse	C3HeB	Corpus striatum	↓ density ↔ affinity	8 vs. 18	199
[3H]Spiroperidol	Rat	S–D	Corpus striatum	↓ density ↔ affinity	7 vs. 25 to 29	199
[3H]ADTN[a]	Mouse	C57BL	Corpus striatum	↓ density ↔ affinity	8 or 10 vs. 18 to 26	199
[3H]ADTN	Mouse	C3HeB	Corpus striatum	↓ density ↔ affinity	8 or 10 vs. 18 to 26	199
[3H]Spiroperidol	Rat	S–D	Corpus striatum	↔ density ↓ affinity	3 to 4 vs. 24 to 30	202

Ligand	Species	Strain	Region	Measure	Change	Ages	Ref.
[³H]Spiroperidol	Rat	S-D	Olfactory tubercle	binding	↓	3 to 4 vs. 24 to 30	202
[³H]Spiroperidol	Rat	S-D	Pituitary gland	binding	←	3 to 4 vs. 24 to 30	202
[³H]Spiroperidol	Rat	Wistar	Corpus striatum	density / affinity	↓ / ↔	6 to 8 vs. 24	194
[³H]Haloperidol	Rat	C-R	Corpus striatum	density / affinity	↑ / ↔	6 vs. 26	192
[³H]Haloperidol	Mouse	C57BL	Corpus striatum	binding	↑	2 vs. 10 & 10 vs. 24	192
[³H]Haloperidol	Mouse	C57BL	Corpus striatum	binding	→	24 vs. 32	192
[³H]Spiroperidol	Rat	C-R	Corpus striatum	binding	↓	5 vs. 21 or 26	262
[³H]Spiroperidol	Rat	C-R	Corpus striatum	binding	↓	3 vs. 30	206
[³H]Spiroperidol	Rabbit	New Zealand	Corpus striatum	density / affinity	↓ / ↔	5 or 6 vs. 66	195–197
[³H]Spiroperidol	Rabbit	New Zealand	Frontal cortex	density / affinity	↓ / ↔	5 or 6 vs. 66	195–197
[³H]Spiroperidol	Rabbit	New Zealand	Limbic cortex	density / affinity	↓ / ↔	5 or 6 vs. 66	195–197
[³H]ADTN	Rabbit	New Zealand	Corpus striatum	density / affinity	↓ / ↔	5 vs. 66	195–197
[³H]Spiroperidol	Human	—	Caudate nucleus	density / affinity	↓ / ↔	Various ages	200
[³H]Spiroperidol	Human	—	Substantia nigra	binding	↓	Various ages	200
[³H]Spiroperidol	Human	—	Nucleus accumbens	binding	↓	Various ages	200

a Abbreviation: [³H]ADTN, amino-6,7-dihydroxy-1,2,3,4-tetrahydronaphthalene,2-[5,8-³H] tissues may have a decreased ability to synthesize new receptors during the normal process of receptor degradation and turnover.

Another proposal that may explain the age-related loss of catecholamine receptors in brain relates to the decrease in membrane fluidity seen in the aged brain.[215] Decreased membrane fluidity may result from a decrease in the activity of the methyltransferase enzymes[206] involved in the translocation of phospholipids in the membrane.[150,151,153] Support for this proposal is provided by data showing that decreased activity of methyltransferase and decreased membrane fluidity were accompanied by decreased β-adrenergic receptors in the corpus striatum and that chronic administration of S-adenosylmethionine, the methyl donor for the methyltransferase enzymes, increased membrane fluidity in senescent rats to the level of that found in brain tissue from 3-month-old rats and increased β-adrenergic receptors in the corpus striatum and pineal gland. This treatment with S-adenosylmethionine apparently was relatively selective, since it failed to reverse age-related reductions of spiroperidol binding sites and dopamine-sensitive adenylate cyclase activity in the corpus striatum.[206]

Regardless of the mechanism for the loss of catecholamine receptors with age, the evidence that these receptors decrease with advanced age correlates with data from earlier studies demonstrating that the sensitivity of adenylate cyclase to catecholamines in several areas of rat brain is decreased[187,188,207] and provides one explanation for the decline in certain neurally dependent functions in the elderly.

4.3. Adaptability of Catecholaminergic Receptors in Aging

4.3.1. Neural Influences

An age-associated decrease in axodendritic and axosomatic synapses in various areas of the brain of rats and humans[216,217] suggests that neural input to target cells may decrease with age. In young animals, the function of adrenergically innervated tissues may be maintained even when adrenergic input is decreased. This adaptive capacity is dependent upon an ability of these tissues to increase their density of adrenergic receptors (see Section 2.1). Since aged animals apparently cannot adapt their physiological responses to changing circumstances as readily as can young animals, the question arose as to whether aged animals have a decreased capacity to adapt their receptors to a changing neural environment.

As discussed in Section 2, exposure of young mature rats to light for a few hours decreases sympathetic input to the pineal gland and increases the density of β-adrenergic receptors and the responsiveness of adenylate cyclase to norepinephrine in pineal glands.[25,76,77,157] By contrast, exposure of aged rats to light failed to increase the density of β-adrenergic receptors in this gland even when the light exposure was continued for up to 30 days.[76,175]

Similarly, reserpine, which depletes norepinephrine from adrenergic nerve terminals, caused a compensatory increase in the sensitivity of brain adenylate cyclase to β-adrenergic agonists[15,77,118–120] and a dose-related increase in β-adrenergic receptors in pineal gland, cerebral cortex, and cerebellum of young rats.[77,89] In comparison to the effect of reserpine in young rats, the effect in aged rats was abolished in cerebellum and decreased in cerebral cortex and pineal gland, despite the finding of a severalfold higher concentration of reserpine in the pineal glands of the old animals.[89]

The ability of brain to increase the density of dopamine receptors in response to chronic blockade of dopamine receptor function also declines with advanced age. For example, chronic treatment with the dopamine antagonist haloperidol increased spiroperidol binding sites in the corpus striatum and increased apomorphine-induced stereotyped behavior in young (5-month) and middle-aged (12-month), but not in aged (24- to 26-month) mice.[218] By contrast, destruction of dopaminergic nerve terminals in the corpus striatum by injecting 6-hydroxydopamine into the substantia nigra produced a similar adaptive response in the striatum of aged and young rats.[194,195] In both age groups, denervation supersensitivity was evidenced by increased spiroperidol binding, increased sensitivity of adenylate cyclase to dopamine, and increased rotational behavior following administration of the dopamine agonist lergotril.

The discrepancy between these two studies may reflect species differences or differences in the methods used to induce supersensitivity. Lesions induced by 6-hydroxydopamine may produce a more severe and persistant disruption of dopaminergic function than that produced by chronic haloperidol treatment. In fact, in young rats, 6-hydroxydopamine has been found to induce a twofold greater increase in dopamine binding sites in the striatum than is produced by chronic haloperidol administration.[219,220] It is also possible that different populations of dopamine binding sites (e.g., pre- *vs.* postsynaptic sites, see refs. 203,204) are involved in the two methods for inducing supersensitivity and that these receptors show differential aging effects.

Another issue that should be addressed is whether aged tissue has an altered capacity to compensate for chronic increases in adrenergic stimulation. As discussed earlier (Section 2), chronic stimulation of β-adrenergic receptors by repeated administration of desmethylimipramine decreases the density of β-adrenergic receptors in certain brain areas, such as pineal gland and cerebral cortex, of mature young animals.[77,89,105,106,108-110,221] Studies in which the responses of young and old rats were compared showed that the β-adrenergic receptors declined equally in young and senescent rats following desmethylimipramine treatment, suggesting that there are no age-related changes in the ability of receptors to decrease in response to increased adrenergic input.[77,89] However, these data were confounded by the finding that the concentration of desmethylimipramine in pineal gland and cerebral cortex of aged rats was significantly higher than that found in those tissues of young rats.[89] Therefore, the question of the relative ability of brain tissues from young and old rats to develop a subsensitivity of adrenergic receptors is still open.

A related question is whether brain tissue from aged rates can restore its density of β-adrenergic receptors following desmethylimipramine-induced receptor subsensitivity. Recent studies showed that in young rats, the β-adrenergic receptors returned to control levels in cerebral cortex and pineal gland within 2 days after the last dose of desmethylimipramine. By contrast, in aged rats, it took 8 days for the β receptors to return to control values in the cortex and 16 days for them to return to normal in the pineal gland.[222]

These studies suggest that some areas of the brain in senescent rats have an impaired capacity to increase receptor density in response to reduced neural input or following induction of receptor subsensitivity. If new receptor synthesis is required for this increase in receptor density, perhaps aged brain tissue has a reduced capacity to synthesize receptors. Alternatively, an age-related loss of membrane fluidity may impair the insertion of performed receptors into the surface membrane of brain cells.

4.3.2. Hormonal Influences

The feedback regulation by sex steroids of gonadotropin secretion from the pituitary is impaired in aging female rodents as they go through various stages of reproductive decline.[223-226] Since catecholaminergic systems are involved in the positive and negative feedback mechanisms by which estrogen and progesterone regulate gonadotropin secretion (see refs. 227–230), a defect in the regulation of catecholamine receptors by estrogen and progesterone may explain the age-related loss of hypothalamic responsiveness to steroids.[226] Failure of catecholamine receptors to respond to estrogen and progesterone may, therefore, be an early manifestation of the progressive loss of neuroendocrine control of the reproductive system that occurs in the aging female.[231-234] In support of this hypothesis is the finding that the usual elevation of β-adrenergic receptors that occurs in the hippocampus, pineal gland, and cerebral cortex of young rats between early estrus and diestrus did not occur in any of these brain areas from aged rats.[175] This defect may have a functional correlate in that there also may be loss of the normal circadian variation in the synthesis and release of melatonin from the pineal gland. Indeed, melatonin rhythms in the aging female rat are markedly attenuated.[235] Since melatonin has antigonadotropic activity, persistent rather than phasic melatonin secretion from the pineal gland could contribute to estrous cycle abnormalities and reproductive decline.

Changes in β-adrenergic receptors in rats during middle and old age further support the hypothesis that receptor abnormalities may contribute to the acyclicity and reproductive decline of aging females.[175] α_1-Adrenergic receptors were significantly lower in the hypothalamus of middle-aged rats that exhibited estrous acyclicity than in young cycling rats in estrus. Furthermore, the usual increase in the binding of α_1-receptor ligands that occurs between diestrus and estrus in young cycling rats did not occur in the middle-aged rats that were in constant estrus or that were still cycling but had prolonged or irregular cycles. The proestrus surge of luteinizing hormone and follicle-stimulating hormone resulting from the positive feedback of estrogen and progesterone in the hypothalamus appears to be an α-adrenergic receptor-mediated event. The finding that the regulation of receptor density by estrogen and progesterone is impaired with age may partly explain why middle-aged rats exhibit a delayed and reduced gonadotropin surge at proestrus[236-238] and why ovariectomized middle-aged rats have a reduced surge of luteinizing hormone in response to estrogen and progesterone.[237]

Fig. 2. Reduced adaptability of β-adrenergic receptors in pineal gland of aged rats. Young (3- to 5-month-old) and aged (20- to 24-month-old) rats were maintained on a 12 h:12 h (males) or a 14:10h (females) light:dark cycle. The specific binding of [³H]dihydroalprenolol in pineal gland homogenates was determined to estimate the density of β-adrenergic receptors. Control: male rats or female rats in early estrus killed at the end of the dark cycle. Light: male or female rats killed at the end of the light cycle. Reserpine: male rats administered reserpine (0.25 µmol/kg, i.p.) once daily for 3 days and killed 24 h after the last dose (early in the light cycle). Diestrus: female rats in diestrus, as determined from vaginal smears, killed at the end of the dark cycle. Each value represents the mean of 4–8 experiments. Vertical brackets indicate the standard error. Data from Greenberg and Weiss.[76,89,175]

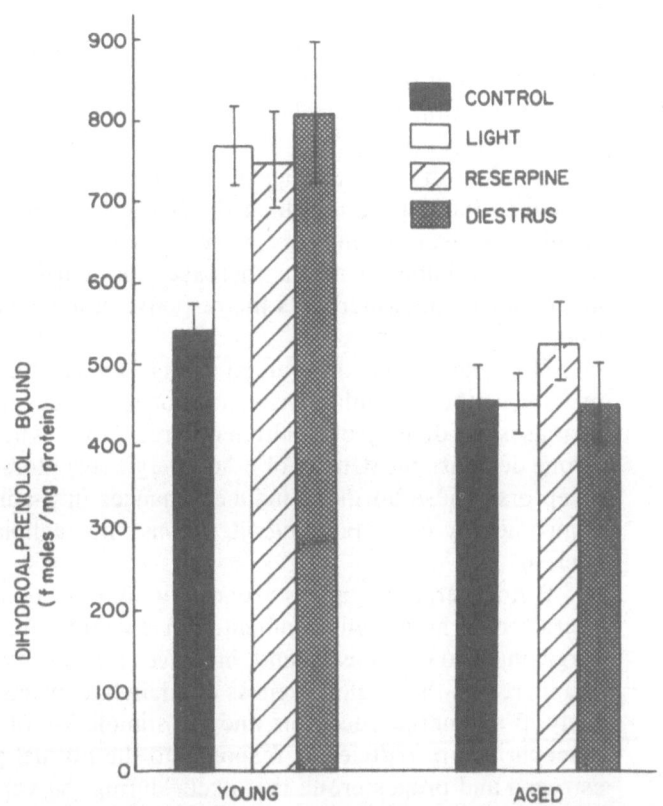

The studies reviewed here indicate that aged tissues have a reduced capacity to adapt their catecholaminergic receptors to changes in both neural and hormonal input. Figure 2 summarizes some of the data supporting these conclusions. As may be seen, light exposure, which reduces sympathetic input to the pineal gland, or reserpine treatment, which depletes catecholamine stores from adrenergic nerve terminals, increased the density of β-adrenergic receptors in pineal gland of young rats. Similarly, low concentrations of estrogen, as occur in diestrus, evoked an increase in β-adrenergic receptors in pineal glands of young rats. By contrast, neither reduced adrenergic input nor low concentrations of estrogen resulted in a significant increase in the density of β-adrenergic receptors in pineal glands of aged rats. This reduced capacity of aged tissues to modulate their adrenergic receptors may explain the relative inability of aged organisms to adapt their physiological responses to a changing environment.

5. SUMMARY

The density and function of catecholaminergic receptors are dependent upon the age of the animal and are influenced by a variety of neural, hormonal, and environmental factors. In most tissues, catecholaminergic receptors increase soon after birth and then decline with advanced age. The ontogenetic development of the β-adrenergic receptors and the loss of these receptors with age are functionally correlated with the responsiveness of adenylate cyclase to catecholamines.

In adult animals, the density of catecholaminergic receptors is inversely related to the degree to which the tissue has been stimulated by adrenergic agonists. Thus, reducing adrenergic input by any of a variety of surgical, chemical, or physiological means increases the density of β-adrenergic receptors and concomitantly increases the response of adenylate cyclase to norepinephrine.

Hormones have differential effects on catecholaminergic receptors depending on the particular tissue, hormone, and receptor type. Thyroid hormone increases the density of β-adrenergic receptors, whereas estrogen and progesterone decrease the density of β-adrenergic receptors but increase α-adrenergic receptors. These hormone-induced changes in β-adrenergic receptors are accompanied by corresponding alterations in catecholamine-sensitive adenylate cyclase.

β-Adrenergic receptors are also altered during normal fluctuations in the neural or hormonal environment. For example, light, which decreases adrenergic input to the pineal gland, increases the density of β-adrenergic receptors and increases the responsiveness of adenylate cyclase to norepinephrine. Similarly, β-adrenergic receptors and the stimulation of adenylate cyclase by norepinephrine are reduced in response to the normal physiological increases of estrogen and progesterone that occur during the various stages of the estrous cycle.

Very young animals as well as very old animals have a reduced capacity to adapt their receptors to changes in neural or hormonal input. For example, although the ontogeny of β-adrenergic receptors can be impeded by the presence of catecholamines, their rate of development is not markedly influenced by decreased sympathetic input or by increased circulatory concentrations of estrogen.

With advancing age, the number of catecholaminergic receptors declines in several areas of the brain. This reduction may explain the reduced ability of aged tissue to respond to adrenergic agonists. The mechanism for this decrease in catecholaminergic receptors with age may be related to a reduced ability of aged tissues to produce compensatory increases in their receptor density in the face of decreased adrenergic input. In middle-aged and senescent rats, the regulation of adrenergic receptors by sex steroids is also impaired. This impairment may initiate or contribute to the reproductive decline seen in aging female rats.

Thus, aged animals not only have a reduced density of adrenergic receptors but also have an impaired ability to adjust their adrenergic receptors to changes in their neural and hormonal environment. Loss of adrenergic receptors may explain certain age-related reductions in physiological responses, and the reduced adaptability of receptors may account for the relative inability of aged individuals to adjust their physiological responses to a changing environment.

ACKNOWLEDGMENTS. This work was supported by Grant MH30096 awarded by the National Institute of Mental Health, by Grant NS16242 awarded by the National Institute of Neurological and Communicative Diseases and Stroke, and by funds from the Department of Public Welfare, Commonwealth of Pennsylvania.

REFERENCES

1. Hollenberg, M. D., 1978, *Pharmacol. Rev.* **30**:393–410.
2. Bylund, D. B., 1979, *Adv. Exp. Med. Biol.* **116**:133–162.
3. Hoffman, B. B., and Lefkowitz, R. J., 1980, *Annu. Rev. Pharmacol. Toxicol.* **20**:581–608.
4. Miller, R. J., and Dawson, G., 1980, *Adv. Biochem. Psychopharmacol.* **21**:11–20.
5. Seeman, P., 1980, *Pharmacol. Rev.* **32**:229–313.
6. Weiss, B., Greenberg, L. H., and Clark, M. B., 1984, *Dynamics of Neurotransmitter Function* (I. Hanin, ed.), Raven Press, New York, pp. 319–329.
7. Creese, I., Sibley, D. R., Leff, S., and Hamblin, M., 1981, *Fed. Proc.* **40**:147–152.
8. Creese, I., and Sibley, D. R., 1981, *Annu. Rev. Pharmacol. Toxicol.* **21**:357–391.
9. Minneman, K. P., Pittman, R. N., and Molinoff, P. B., 1981, *Annu. Rev. Neurosci.* **4**:419–461.
10. Starke, K., 1981, *Ann. Rev. Pharmacol. Toxicol.* **21**:7–30.
11. Lewandowsky, M., 1898, *Arch. Physiol.* **73**:288–296.
12. Emmelin, N., 1961, *Pharmacol. Rev.* **13**:17–37.
13. Trendelenberg, U., 1963, *Pharmacol. Rev.* **15**:225–276.
14. Trendelenberg, U., 1966, *Pharmacol. Rev.* **18**:629–640.
15. Schwartz, J. C., Costentin, J., Martres, M. P., Protais, P., and Baudry, M., 1978, *Neuropharmacology* **17**:665–685.
16. Costa, E., 1980, *Adv. Biochem. Psychopharmacol.* **24**:363–377.
17. Van Orden, L. S., Bensch, K. G., and Giarman, N. J., 1967, *J. Pharmacol. Exp. Ther.* **155**:428–439.
18. Dempsey, P. J., and Cooper, T., 1968, *Am. J. Physiol.* **215**:1245–1249.
19. Weiss, B., 1969, *J. Pharmacol. Exp. Ther.* **168**:146–152.
20. Robison, G. A., Butcher, R. W., and Sutherland, E. W., 1967, *Ann. N.Y. Acad. Sci.* **139**:703–723.
21. Weiss, B., and Costa, E., 1967, *Science* **156**:1750–1752.
22. Cantor, E., Clark, M. B., and Weiss, B., 1981, *Brain Res. Bull.* **7**:243–247.
23. Strada, S. J., and Weiss, B., 1974, *Arch. Biochem. Biophys.* **160**:197–204.
24. Deguchi, T., and Axelrod, J., 1973, *Mol. Pharmacol.* **9**:612–618.
25. Cantor, E. H., Greenberg, L. H., and Weiss, B., 1981, *Mol. Pharmacol.* **19**:21–26.
26. Sporn, J. R., Harden, T. K., Wolfe, B. B., and Molinoff, P. B., 1976, *Science* **194**:624–625.
27. Sporn, J. R., Wolfe, B. B., Harden, T. K., and Molinoff, P. B., 1977, *Mol. Pharmacol.* **13**:1170–1180.
28. Skolnick, P., Stalvey, L. P., Daly, J. W., Hoyler, E., and Davis, J. N., 1978, *Eur. J. Pharmacol.* **47**:201–210.
29. U'Prichard, D. C., Reisine, T. D., Yamamura, S., Mason, S. T., Fibiger, H. C., Ehlert, F., and Yamamura, H. I., 1980, *Life Sci.* **26**:355–364.
30. U'Prichard, D. C., Reisine, T. D., Mason, S. T., Fibiger, H. C., and Yamamura, H. I., 1980, *Brain Res.* **187**:143–154.
31. Kostrzewa, R. M., and Garey, R. E., 1976, *J. Pharmacol. Exp. Ther.* **197**:105–118.
32. Kostrzewa, R. M., and Garey, R. E., 1977, *Brain Res.* **124**:385–391.
33. Clark, M. B., King, J. C., and Kostrzewa, R. M., 1979, *Neurosci. Lett.* **13**:331–336.
34. Pickel, V. M., Segal, M., and Bloom, F. E., 1974, *J. Comp. Neurol.* **155**:43–60.
35. Mano, K., Akbarzadeh, A., and Townley, R. G., 1979, *Life Sci.* **25**:1925–1930.
36. Wolfe, B. B., Harden, T. K., and Molinoff, P. B., 1976, *Proc. Natl. Acad. Sci. U.S.A.* **73**:1343–1347.

37. Guellaen, G., Yates-Aggerbeck, M., Vauquelin, G., Strosberg, D., and Hanoune, J., 1978, *J. Biol. Chem.* **253**:1114–1120.
38. Holgate, S. T., Baldwin, C. J., and Tattersfield, A. E., 1977, *Lancet* **2**:375–377.
39. Townley, R. G., Daley, D., and Selenke, W., 1970, *J. Allergy* **45**:71–86.
40. Tolone, G., Bonasera, L., and Sajeva, R., 1979, *Br. J. Exp. Pathol.* **60**:269–275.
41. Kalsner, S., 1970, *Can. J. Physiol. Pharmacol.* **48**:443–449.
42. Fraser, C. M., and Venter, J. C., 1980, *Biochem. Biophys. Res. Commun.* **94**:390–397.
43. Ciaraldi, T., and Marinetti, G. V., 1977, *Biochem. Biophys. Res. Commun.* **74**:984–991.
44. Kempson, S., Marinetti, G. V., and Shaw, A., 1978, *Biochim. Biophys. Acta* **540**:320–329.
45. Tse, J., Wrenn, R. W., and Kuo, J. F., 1980, *Endocrinology* **107**:6–16.
46. Williams, L. T., Lefkowitz, R. J., Watanabe, A. M., Hathaway, D. R., and Besch, H. R., Jr., 1977, *J. Biol. Chem.* **252**:2787–2789.
47. Banerjee, S. P., and Kung, L. S., 1977, *Eur. J. Pharmacol.* **43**:207–208.
48. Smith, R. M., Osborne-White, W. S., and King, R. A., 1978, *Biochem. Biophys. Res. Commun.* **80**:715–721.
49. Sharma, V. K., and Banerjee, S. P., 1978, *Biochim. Biophys. Acta* **539**:538–542.
50. Bilezikian, J. P., Loeb, J. N., and Gammon, D. E., 1979, *J. Clin. Invest.* **63**:184–192.
51. Malbon, C. C., Moreno, F. J., Cabelli, R. J., and Fain, J. N., 1978, *J. Biol. Chem.* **253**:671–678.
52. Ciaraldi, T. P., and Marinetti, G. V., 1978, *Biochim. Biophys. Acta* **541**:334–346.
53. Greenberg, L. H., and Weiss, B., 1981, *Fed. Proc.* **40**:260.
54. Greenberg, L. H., and Weiss, B., 1981, *Proceedings 8th International Congress of Pharmacology, Tokyo, Japan*, p. 338.
55. Smith, M. S., Freeman, M. E., and Neill, J. D., 1975, *Endocrinology* **96**:219–236.
56. Wagner, H. R., Crutcher, K. A., and Davis, J. N., 1979, *Brain Res.* **171**:147–151.
57. Fuxe, K., Andersson, F., Löfström, A., Hökfelt, T., Ferland, L., Agnati, L. F., Perez de la Mora, M., Schwarcz, R., Eneroth, P., Gustafsson, J. A., and Stett, P., 1979, *Central Regulation of the Endocrine System* (K. Fuxe, T. Hökfelt, and R. Luft, eds.), Plenum Press, New York, pp. 349–380.
58. Vacas, M. I., and Cardinali, D. P., 1980, *Neurosci. Lett.* **17**:73–77.
59. Wilkinson, M., Herdon, H., Pearce, M., and Wilson, C., 1979, *Brain Res.* **168**:652–655.
60. Cardinali, D. P., and Vacas, M. I., 1978, *J. Neural Transm.* **13**:175–201.
61. Kumakura, K., Hoffman, M., Cocchi, D., Trabucchi, M., Spano, P., and Muller, E. E., 1979, *Psychopharmacology* **61**:13–16.
62. Chiodo, L. A., Caggiula, A. R., and Saller, C. F., 1981, *Life Sci.* **28**:827–835.
63. Hruska, R. E., and Silbergeld, E. K., 1980, *Eur. J. Pharmacol.* **61**:397–400.
64. Lal, S., and Sourkes, T. L., 1972, *Arch. Int. Pharmacodyn.* **199**:289–301.
65. Nausieda, P. A., Koller, W. C., Weiner, W. J., and Klawans, H. L., 1979, *Life Sci.* **25**:521–526.
66. DiPaolo, T., Carmichael, R., Labrie F., and Raynaud, J. P., 1979, *Mol. Cell. Endocrinol.* **16**:99–112.
67. Weiss, B., and Crayton, J., 1970, *Endocrinology* **87**:527–533.
68. Weiss, B., and Crayton, J. W., 1970, *Adv. Biochem. Psychopharmacol.* **3**:217–239.
69. Wurtman, R. J., Axelrod, J., and Fischer, J. E., 1963, *Science* **143**:1328–1330.
70. Wurtman, R. J., Axelrod, J., and Phillips, L. S., 1963, *Science* **142**:1071–1073.
71. Moore, R. Y., and Klein, D. C., 1974, *Brain Res.* **71**:17–33.
72. Taylor, A. N., and Wilson, R. W., 1970, *Experientia* **26**:267–269.
73. Brownstein, M., and Axelrod, J., 1974, *Science* **184**:163–165.
74. Klein, D. C., and Moore, R. Y., 1979, *Brain Res.* **174**:245–262.
75. Kebabian, T. W., Zatz, M., Romero, J., and Axelrod, J., 1975, *Proc. Natl. Acad. Sci. U.S.A.* **72**:3735–3739.
76. Greenberg, L. H., and Weiss, B., 1978, *Science* **201**:61–63.
77. Weiss, B., Greenberg, L., and Cantor, E., 1979, *Fed. Proc.* **38**:1915–1921.
78. Romero, J. A., Zatz, M., Kebabian, J. W., and Axelrod, J., 1975, *Nature* **258**:435–436.
79. Romero, J. A., and Axelrod, J., 1974, *Science* **184**:1091–1092.
80. Romero, J. A., and Axelrod, J., 1975, *Proc. Natl. Acad. Sci. U.S.A.* **72**:1661–1665.

81. Deguchi, T., 1975, *J. Neurochem.* **25**:91-93.
82. Kunos, G., and Preiksaitis, H. G., 1978, *Recent Advances in the Pharmacology of Adrenoceptors* (E. Szabadi, C. M. Bradshaw, and P. Bevan, eds.), Elsevier/North Holland Biomedical Press, Amsterdam, pp. 209-216.
83. Ulrich, R., Yuwiler, A., Wetterberg, L., and Klein, D., 1973-74, *Neuroendocrinology* **13**:255-263.
84. Klein, D. C., and Weller, J. L., 1972, *Science* **177**:532-533.
85. Kafka, M. S., Wirz-Justice, A., and Naber, D., 1981, *Brain Res.* **207**:409-419.
86. Reiter, R. J., Rollag, M. D., Panke, E. S., and Banks, A. F., 1978, *J. Neural Transm.* **13**:209-223.
87. Reiter, R. J., Petterborg, L. J., and Philo, R. C., 1979, *Life Sci.* **25**:1571-1576.
88. Reiter, R. J., 1980, *Int. J. Biometeorol.* **24**:57-63.
89. Greenberg, L. H., and Weiss, B., 1979, *J. Pharmacol. Exp. Ther.* **211**:309-316.
90. Torda, T., Yamaguchi, I., Hirata, F., Kopin, I. J., and Axelrod, J., 1981, *Brain Res.* **205**:441-444.
91. Weiss, B., and Levin, R. M., 1978, *Adv. Cyclic Nucleotide Res.* **9**:285-303.
92. Zatz, M., Kebabian, J. W., Romero, J. A., and Axelrod, J., 1976, *J. Pharmacol. Exp. Ther.* **196**:714-722.
93. Klein, D., and Weller, J. L., 1973, *J. Pharmacol. Exp. Ther.* **186**:516-527.
94. Oleshansky, M. A., and Neff, N. H., 1975, *Life Sci.* **17**:1429-1432.
95. Deguchi, T., and Axelrod, J., 1973, *Proc. Natl. Acad. Sci. U.S.A.* **70**:2411-2414.
96. Makman, M. H., 1971, *Proc. Natl. Acad. Sci. U.S.A.* **68**:885-889.
97. Remold-O'Donnell, E., 1974, *J. Biol. Chem.* **249**:3615-3621.
98. Dismukes, R. K., and Daly, J. W., 1976, *J. Cyclic Nucleatide Res.* **2**:321-336.
99. Mickey, J. V., Tate, R., Mullikin, D., and Lefkowtiz, R. J., 1976, *Mol. Pharmacol.* **12**:409-419.
100. Glaubiger, S., and Lefkowitz, R. J., 1977, *Biochem. Biophys. Res. Commun.* **78**:720-725.
101. Palmer, G. C., and Greenberg, S., 1978, *Prog. Neuropsychopharmacol.* **2**:585-587.
102. Spector, S., 1963, *Ann. N.Y. Acad. Sci.* **107**:856-864.
103. Glowinski, J., and Axelrod, J., 1964, *Nature* **204**:1318-1319.
104. Svensson, T. H., and Usdin, T., 1978, *Science* **202**:1089-1091.
105. Moyer, J. A., Greenberg, L. H., Frazer, A., and Weiss, B., 1981, *Mol. Pharmacol.* **19**:187-193.
106. Greenberg, L. H., and Weiss, B., 1978, *Recent Advances in the Pharmacology of Adrenoceptors* (E. Szabadi, C. M. Bradshaw, and P. Bevan, eds.), Elsevier/North Holland Biomedical Press, Amsterdam, pp. 241-260.
107. Greenberg, L. H., and Weiss, B., 1978, *Fed. Proc.* **37**:878.
108. Moyer, J. A., Greenberg, L. H., Frazer, A., Brunswick, D. J., Mendels, J., and Weiss, B., 1979, *Life Sci.* **24**:2237-2244.
109. Banerjee, S. P., Kung, L. S., Riggi, S. J., and Chanda, S. K., 1977, *Nature* **268**:455-456.
110. Sarai, K., Frazer, A., Brunswick, D., and Mendels, J., 1978, *Biochem. Pharmacol.* **27**:2179-2181.
111. Banerjee, S. P., Sharma, V. K., Kung, L. S., and Chanda, S. K., 1978, *Nature* **271**:380-381.
112. Molinoff, P. B., Sporn, J. R., Wolfe, B. B., and Harden, T. K., 1978, *Adv. Cyclic Nucleatide Res.* **9**:465-483.
113. Clements-Jewery, S., 1978, *Neuropharmacology* **17**:779-781.
114. VanZwieten, P. A., 1977, *Pharmacology* **15**:331-336.
115. Holzbauer, M., and Vogt, M., 1956, *J. Neurochem.* **1**:8-11.
116. Brodie, B. B., Olin, J. S., Kuntzman, R. G., and Shore, P. A., 1957, *Science* **125**:1293-1294.
117. Johnson, E. M., Jr., Cantor, E., and Douglas, J. R., Jr., 1975, *J. Pharmacol. Exp. Ther.* **193**:503-512.
118. Williams, B. J., and Pirch, J. H., 1974, *Brain Res.* **68**:227-234.
119. Palmer, G. C., Sulser, F., and Robison, G. A., 1973, *Neuropharmacology* **12**:327-337.
120. Dismukes, R. K., and Daly, J. W., 1974, *Mol. Pharmacol.* **10**:933-940.
121. Baudry, M., Martres, M. P., and Schwartz, J. C., 1976, *Brain Res.* **116**:111-124.
122. Brodie, B. B., and Shore, P. A., 1957, *Ann. N.Y. Acad. Sci.* **66**:631.

123. Smith, C. B., 1963, *J. Pharmacol. Exp. Ther.* **142**:343–350.
124. Van Rossum, J. M., Van der Schoot, J. B., and Harkmans, J. T. M., 1962, *Experientia* **18**:229.
125. Snyder, S. H., Taylor, K. M., Horn, A. S., and Coyle, J. T., 1972, *Res. Publ. Assoc. Res. Nerv. Ment. Dis.* **50**:359–375.
126. Stein, L., 1964, *Fed. Proc.* **23**:836–850.
127. Sellinger-Barnette, M. M., Mendels, J., and Frazer, A., 1980, *Neuropharmacology* **19**:447–454.
128. Mobley, P. L., Sanders-Bush, E., Smith, H. E., and Sulser, F., 1979, *Nanuyn Schmiedebergs Arch. Pharmacol.* **306**:267–273.
129. Banerjee, S. P., Sharma, V. K., Kung-Cheung, L. S., Chanda, S. K., and Riggi, S. J., 1979, *Brain Res.* **175**:119–130.
130. Ellinwood, E. H., Jr., Sudilovsky, A., and Nelson, L. M., 1973, *Am. J. Psychiatry* **130**:1088–1093.
131. Banerjee, S. P., Sharma, V. K., and Khanna, J. M., 1978, *Nature* **276**:407–409.
132. Israel, M. A., Kimura, H., and Kuriyama, K., 1972, *Experientia* **28**:1322–1323.
133. French, S. W., and Palmer, D. S., 1973, *Res. Commun. Chem. Pathol. Pharmacol.* **6**:651–662.
134. French, S. W., Palmer, D. S., and Narod, M. E., 1975, *Can. J. Physiol. Pharmacol.* **53**:248–255.
135. Bird, S. J., and Kuhar, M. J., 1977, *Brain Res.* **122**:523–533.
136. Kuschinsky, K., 1979, *Klin. Wochenschr.* **57**:701–710.
137. Montel, H., Starke, K., and Taube, H. D., 1975, *Naunyn Schmiedebergs Arch. Pharmacol.* **288**:427–433.
138. Arbilla, S., and Langer, S. Z., 1978, *Nature* **271**:559–561.
139. Llorens, C., Martres, M. P., Baudry, M., and Schwartz, J. C., 1978, *Nature* **274**:603–605.
140. Waddingham, S., Riffee, W. R., Belknap, J. K., and Sheppard, J. R., 1978, *Res. Commun. Chem. Pathol. Pharmacol.* **20**:207–220.
141. Jaffe, J. H., and Sharpless, S. K., 1968, *Res. Publ. Assoc. Res. Nerv. Ment. Dis.* **46**:226–246.
142. Collier, H. O., 1965, *Nature* **205**:181–182.
143. Palmer, G. C., Robison, G. A., and Sulser, F., 1971, *Biochem. Pharmacol.* **20**:236–239.
144. Uzunov, P., and Weiss, B., 1972, *Adv. Cyclic Nucleotide Res.* **1**:435–453.
145. Freedman, R., and Hoffer, B. J., 1975, *J. Neurobiol.* **6**:277–288.
146. Levin, R. M., and Weiss, B., 1976, *Mol. Pharmacol.* **12**:581–589.
147. Levin, R. M., and Weiss, B., 1977, *Mol. Pharmacol.* **13**:690–697.
148. Wong, P. Y.-K., and Cheung, W. Y., 1979, *Biochem. Biophys. Res. Commun.* **90**:473–480.
149. Axelrod, J., and Hirata, F., 1980, *Pharmacol. Biochem. Behav.* **13**:167–168.
150. Hirata, F., and Axelrod, J., 1978, *Proc. Natl. Acad. Sci. U.S.A.* **75**:2348–2352.
151. Hirata, F., Viveros, H., DiliBerto, E. J., Jr., and Axelrod, J., 1978, *Proc. Natl. Acad. Sci. U.S.A.* **75**:1718–1721.
152. Hirata, F., Strittmatter, J., and Axelrod, J., 1979, *Proc. Natl. Acad. Sci. U.S.A.* **76**:368–372.
153. Strittmatter, W. J., Hirata, F., and Axelrod, J., 1979, *Science* **204**:1205–1207.
154. Cantor, E., and Weiss, B., 1978, *Fed. Proc.* **37**:524.
155. Weiss, B., 1971, *J. Neurochem.* **18**:469–477.
156. Weiss, B., and Costa, E., 1968, *J. Pharmacol. Exp. Ther.* **161**:310–319.
157. Weiss, B., Greenberg, L. H., and Cantor, E., 1980, *Adv. Biochem. Psychopharmacol.* **21**:461–472.
158. Perkins, J. P., and Moore, M. M., 1973, *Frontiers in Catecholamine Research* (E. Usdin and S. Snyder, eds.), Pergamon Press, New York, pp. 311–313.
159. Harden, T. K., Wolfe, B. B., Sporn, J. R., Perkins, J. P., and Molinoff, P. B., 1977, *Brain Res.* **125**:99–108.
160. Klein, D. C., Berg, G. R., and Weller, J., 1970, *Science* **168**:979–980.
161. Strada, S. J., Klein, D. C., Weller, J., and Weiss, B., 1972, *Endocrinology* **90**:1470–1475.
162. Yuwiler, A., Klein, D. C., Breda, M., and Weller, J. L., 1977, *Am. J. Physiol.* **233**:141–146.
163. Klein, D. C., Namboodiri, M. A. A., and Auerbach, D. A., 1980, *Life Sci.* **28**:1975–1986.
164. Clark, M. B., and Weiss, B., 1980, *Proc. Soc. Neurosci.* **6**:597.

165. Hakanson, R., Lombard des Gouttes, M. N., and Owman, C., 1967, *Life Sci.* **6**:2577–2585.
166. Machado, A. B., Machado, C. R., and Wragg, L. E., 1968, *Experientia* **24**:464–465.
167. Machado, C. R., Wragg, L. E., and Machado, A. B., 1968, *Brain Res.* **8**:310–318.
168. Eranko, L., 1972, *Histochem. J.* **4**:225–236.
169. Weiss, B., and Strada, S. J., 1973, *Fetal Pharmacology* (L. Boreus, ed.), Raven Press, New York, pp. 205–232.
170. Harden, T. K., Wolfe, B. B., Sporn, J. R., Poulos, B. K., and Molinoff, P. B., 1977, *J. Pharmacol. Exp. Ther.* **203**:132–143.
171. Ludford, J. M., and Talamo, B. R., 1980, *J. Biol. Chem.* **255**:4619–4627.
172. Harden, T. K., Mailman, R. B., Mueller, R. A., and Breese, G. R., 1979, *Brain Res.* **166**:194–198.
173. Jonnson, G., and Hallman, H., 1978, *Neurosci. Lett.* **9**:27–32.
174. Wolfe, B. B., Augustyn, D. H., Majocha, R. E., Dibner, M. D., Molinoff, P. B., Baldessarini, R. J., and Walton, K. G., 1981, *Brain Res.* **207**:174–177.
175. Greenberg, L. H., and Weiss, B., 1983, *Aging Brain and Ergot Alkaloids* (A. Agnoli, G. Crepaldi, P. F. Spano, and M. Trabucchi, eds.), Raven press, New York *Aging* **23**:37–52.
176. Cheng, J. B., Goldfien, A., Ballard, P. L., and Roberts, J. M., 1980, *Endocrinology* **107**:1646–1648.
177. Lau, C., and Slotkin, T. A., 1980, *J. Pharmacol. Exp. Ther.* **212**:126–130.
178. Smith, R. M., Patel, A. J., Kingsbury, A. E., Hunt, A., and Balázs, R., 1980, *Brain Res.* **198**:375–387.
179. Whitsett, J. A., Darovec-Beckerman, C., Adams, K., Pollinger, J., and Needelman, H., 1980, *Biochem. Biophys. Res. Commun.* **97**:913–917.
180. Pittman, R. N., Minneman, K. P., and Molinoff, P. B., 1980, *Brain Res.* **188**:357–368.
181. Jason, K. M., Cooper, T. B., and Friedman, E., 1981, *J. Pharmacol. Exp. Ther.* **217**:461–466.
182. Keller, E. A., Munaro, N. I., and Orsingher, O. A., 1982, *Science* **215**:1269–1270.
183. Deguchi, T., 1975, *Proc. Natl. Acad. Sci. U.S.A.* **76**:2814–2818.
184. Deguchi, T., 1977, *Am. J. Physiol.* **232**:E375–381.
185. Deguchi, T., 1979, *Biological Rhythms and their Central Mechanism* (M. Suda, O. Hayaishi, and H. Nakagawa, eds.), Elsevier/North Holland Biomedical Press, Amsterdam, pp. 159–168.
186. Schocken, D. D., and Roth, G. S., 1977, *Nature* **267**:856–858.
187. Schmidt, M. J., and Thornberry, J. F., 1978, *Brain Res.* **139**:169–177.
188. Walker, J. B., and Walker, J. P., 1973, *Brain Res.* **54**:391–396.
189. Roth, G. S., 1979, *Fed. Proc.* **38**:1910–1914.
190. Roth, G. S., 1979, *Mech. Ageing Dev.* **9**:497–514.
191. Pittman, R. N., Minneman, K. P., and Molinoff, P. B., 1980, *J. Neurochem.* **35**:273–275.
192. Marquis, J. K., Lippa, A. S., and Pelham, R. W., 1981, *Biochem. Pharmacol.* **30**:1876–1878.
193. Joseph, J. A., Berger, R. E., Engel, B. T., and Roth, G. S., 1978, *J. Gerontol.* **33**:643–649.
194. Joseph, J. A., Filburn, C. R., and Roth, G. S., 1981, *Life Sci.* **29**:575–584.
195. Makman, M. H., Ahn, H. S., Thal, L. H., Dvorkin, B., Horowitz, S. G., Sharpless, N. S., and Rosenfeld, M., 1978, *Parkinson's Disease—II: Aging: Neuroendocrine Relationships* (C. E. Finch, D. E. Potter, and A. D. Kenny, eds.), Plenum Press, New York, pp. 211–230.
196. Makman, M. H., Ahn, H. S., Thal, L. J., Sharpless, N. S., Dvorkin, B., Horowitz, S. G., and Rosenfeld, M., 1979, *Fed. Proc.* **38**:1922–1926.
197. Thal, L. J., Horowitz, S. G., Dvorkin, B., and Makman, M. H., 1980, *Brain Res.* **192**:185–194.
198. Misra, C. H., Shelat, H. S., and Smith, R. C., 1980, *Life Sci.* **27**:521–526.
199. Severson, J. A., and Finch, C. E., 1980, *Brain Res.* **192**:147–162.
200. Severson, J. A., and Finch, C. E., 1980, *Fed. Proc.* **39**:508.
201. Govoni, S., Spano, P. F., and Trabucchi, M., 1978, *J. Pharm. Pharmacol.* **30**:448–449.
202. Govoni, S., Memo, M., Spano, P. F., and Trabucchi, M., 1980, *Mech. Ageing Dev.* **12**:39–46.
203. Kebabian, J. W., and Calne, D. B., 1979, *Nature* **277**:93–96.
204. Sokoloff, P., Matres, M. P., and Schwartz, J. C., 1980, *Naunyn Schmiedebergs Arch. Pharmacol.* **315**:89–102.

205. Puri, S. K., and Volicer, L., 1977, *Mech. Ageing Dev.* **6**:53–58.
206. Cimino, M., Curatola, G., Pezzoli, C., Stramentinoli, G., Vantini, G., and Algeri, S., 1983, *Aging Brain and Ergot Alkaloids* (A. Agnoli, G. Crepaldi, P. F. Spano, and M. Trabucchi, eds.), Raven Press, New York *Aging* **23**:79–87.
207. Berg, A., and Zimmerman, I. D., 1975, *Mech. Ageing Dev.* **4**:377–383.
208. Cubells, J. F., and Joseph, J. A., 1981, *Life Sci.* **28**:1215–1218.
209. Guidicelli, Y., and Pecquery, R., 1978, *Eur. J. Biochem.* **90**:413–419.
210. Dayan, A. D., 1971, *Brain* **94**:31–42.
211. Bloom, F. E., Hoffer, B. J., and Siggins, G. R., 1971, *Brain Res.* **25**:501–521.
212. Hoffer, B. J., Siggins, G. R., Oliver, A. P., and Bloom, F. E., 1973, *J. Pharmacol. Exp. Ther.* **184**:553–569.
213. Schmidt, M. J., and Nadi, N. S., 1977, *J. Neurochem.* **29**:87–90.
214. Sun, A. Y., 1976, *Exp. Aging Res.* **2**:207–219.
215. Heron, D. S., Shinitzky, M., Hershkowitz, M., and Samuel, D., 1980, *Proc. Natl. Acad. Sci. U.S.A.* **25**:423–429.
216. Bondareff, W., 1980, *Neural Regulatory Mechanisms during Aging* (R. C. Adelman, J. Roberts, G. T. Baker, S. I. Baskin, and V. J. Cristofalo, eds.), Alan R. Liss, New York, pp. 143–158.
217. Scheibel, A. B., 1981, *Brain Neurotransmitters and Receptors in Aging and Age-Related Disorders* (S. J. Enna, T. Samorajski, and B. Beer, eds.), Raven press, New York, pp. 31–41.
218. Randall, P. K., Severson, J. A., and Finch, C. E., 1981, *J. Pharmacol. Exp. Ther.* **219**:690–700.
219. Burt, D. R., Creese, I., and Snyder, S. H., 1977, *Science* **196**:326–328.
220. Creese, I., Schneider, R., and Snyder, S. H., 1977, *Eur. J. Pharmacol.* **46**:377–381.
221. Greenberg, L. H., and Weiss, B., 1978, *Proc. 7th Int. Cong. Pharmacol.* **2**:863.
222. Greenberg, L. H., and Weiss, B., 1982, Proceedings of the Society for Neuroscience **8**:187.
223. Huang, H. H., Marshall, S., and Meites, J., 1976, *Biol. Reprod.* **14**:538–543.
224. Lu, K. H., Huang, H. H., Chen, H. T., Kurcz, M., Mioduszewski, R., and Meites, J., 1977, *Proc. Soc. Exp. Biol. Med.* **154**:82–85.
225. Shaar, C. J., Euker, J. S., Riegle, G. D., and Meites, J., 1975, *J. Endocrinol.* **66**:45–51.
226. Wise, P. H., and Ratner, A., 1980, *J. Gerontol.* **35**:506–511.
227. Sawyer, C. H., 1975, *Endocrinology* **17**:97–124.
228. Fuxe, K., Andersson, K., Agnati, L. F., Ferland, L., Hökfelt, T., Eneroth, P., Gustafsson, J. A., and Skett, P., 1979, *Catecholamines: Basic and Clinical Frontiers* (E. Usdin, I. J. Kopin, and J. Barchas, eds.), Pergamon Press, New York, pp. 1187–1203.
229. McCann, S. M., Krulich, L., Ojeda, S. R., Nigro-Vilar, A., and Vijayan, E., 1979, *Central Regulation of the Endocrine System* (K. Fuxe, T. Hökfelt, and R. Luft, eds.), Plenum Press, New York, pp. 329–347.
230. Barraclough, C. A., and Wise, P. M., 1982, *Endocrinol. Rev.* **3**:91–119.
231. Huang, H. H., Steger, R. W., Bruni, J. F., and Meites, J., 1978, *Endocrinology* **103**:1855–1859.
232. Finch, C. E., 1978, *The Aging Reproductive System* (E. L. Schneider, ed.), Raven Press, New York, pp. 193–212.
233. Meites, J., Huang, H. H., and Riegle, G. D., 1975, *Hypothalamic and Endocrine Functions* (F. Labrie, J. Meites, and G. Pelletier, eds.), Plenum Press, New York, pp. 3–20.
234. Meites, J., Huang, H. H., and Simpkins, J. W., 1978, *The Aging Reproductive System* (E. L. Schneider, ed.), Raven Press, New York, pp. 213–235.
235. Reiter, R. J., Craft, C. M., Johnson, J. E., Jr., King, T. S., Richardson, B. A., Vaughan, G. M., and Vaughan, M. K., 1981, *Endocrinology* **109**:1295–1297.
236. Cooper, R. L., Conn, P. M., and Walker, R. F., 1980, *Biol. Reprod.* **23**:611–615.
237. Gray, G. D., Tennent, B., Smith, E., and Davidson, M., 1980, *Endocrinology* **107**:187–194.
238. Van Der Schoot, P., 1975, *J. Endocrinol.* **69**:287–288.
239. Watanabe, Y., Itoh, T., and Yoshida, H., 1980, *Jpn. J. Pharmacol.* **30**:287–291.
240. Prozialeck, W. C., Pylypiw, A., and Ross, L., 1982, *Dev. Brain Res.* **3**:49–63.
241. Lefkowitz, R. J., Limbird, L. E., Mukherjee, C., and Caron, M. G., 1976, *Biochim. Biophys. Acta* **457**:1–39.

242. Bilezikian, J. P., Spiegel, A. M., Brown, E. M., and Aurbach, G. D., 1977, *Mol. Pharmacol.* **13**:775–785.
243. Bilezikian, J. P., Spiegel, A. M., Gammon, D. E., and Aurbach, G. D., 1977, *Mol. Pharmacol.* **13**:786–795.
244. Charness, M. E., Bylund, D. B., Beckman, B. S., Hollenberg, M. D., and Snyder, S. H., 1976, *Life Sci.* **19**:243–250.
245. Limbird, L. E., Gill, D. M., Stadel, J. M., Hickey, A. R., and Lefkowitz, R. J., 1980, *J. Biol. Chem.* **255**:1854–1861.
246. Shane, E., Gammon, D. E., and Bilezikian, J. P., 1981, *Arch. Biochem. Biophys.* **208**:418–425.
247. Spiegel, A. M., Bilezikian, J. P., and Aurbach, G. D., 1975, *Clin. Res.* **23**:390A.
248. Baker, S. P., and Potter, L. T., 1980, *Br. J. Pharmacol.* **68**:65–70.
249. Kunos, G., Brass, C., Kan, W. H., and Mucci, L., 1978, *Fed. Proc.* **37**:684.
250. Lau, C., and Slotkin, T. A., 1981, *J. Pharmacol. Exp. Ther.* **216**:6–11.
251. Chen, F. M., Yamamura, H. I., and Roeske, W. R., 1979, *Eur. J. Pharmacol.* **58**:255–264.
252. Giannapoulos, G., 1980, *Biochem. Biophys. Res. Commun.* **95**:388–394.
253. Whitsett, J. A., Manton, M. A., Darovec-Beckerman, C., Adams, K. G., and Moore, J. J., 1981, *Endocrinol. Metab.* **3**:E351–357.
254. Padbury, J. F., Hobel, C. j., Diakomanolis, E. S., Lam, R. W., and Fisher, D. A., 1981, *Am. J. Obstet. Gynecol.* **139**:459–464.
255. Smith, P. B., and Clark, G. F., 1980, *Biochim. Biophys. Acta* **633**:274–288.
256. Levin, R. M., Malkowicz, S. B., Jacobowitz, D., and Wein, A. J., 1981, *J. Pharmacol. Exp. Ther.* **219**:250–257.
257. Wilkinson, M., Herdon, H., Pearce, M., and Wilson, C., 1979, *Brain Res.* **167**:195–199.
258. Bhalla, R. C., Sharma, R. V., and Ramanathan, S., 1980, *Biochim. Biophys. Acta* **632**:497–506.
259. Deskin, R., Seidler, F. J., Whitmore, W. L., and Slotkin, T. A., 1981, *J. Neurochem.* **36**:1683–1690.
260. Maggi, A., Schmidt, M. J., Ghetti, B., and Enna, S. J., 1979, *Life Sci.* **24**:367–374.
261. Enna, S. J., and Strong, R., 1981, *Brain Neurotransmitters and Receptors in Aging and Age-Related Disorders* (S. J. Enna, T. Samorajski, and B. Beer, eds.), Raven Press, New York, pp. 133–142.
262. Algeri, S., Achilli, G., Cimino, M., Perego, C., Ponzio, F., and Vantini, G., 1984, *Physiological and Pathological Aspects of Aging Brain* (S. Hoyer, ed.), Springer-Verlag, Berlin, Heidelberg New York.

Receptor Regulation

S. J. Enna

1. INTRODUCTION

The ability to cope with a changing environment is a requirement for survival. Adaptive mechanisms can be observed in all areas of biology from the organism to the individual organ down to the cellular and molecular levels. For example, blood pressure, respiration, and body temperature are maintained within fairly narrow limits under a variety of conditions by the action of various hormones which orchestrate the appropriate biochemical and cellular responses. These in turn modify the organ system to meet the challenge.

With regard to the nervous system, research has revealed a great deal about the manner in which function is regulated at the cellular level. Early biochemical studies concentrated on defining presynaptic processes such as transmitter accumulation, synthesis, storage, and release. For example, in some cases the amount of transmitter present in a nerve terminal is regulated by the transmitter itself through a feedback inhibition of the enzyme necessary for synthesis.[1] For other transmitters, it appears that the active accumulation of precursor may be rate limiting.[2] Thus, by regulating the amount or activity of certain enzymes or transport molecules, neurons are capable of accommodating to changes in demand.

In recent years investigations have been directed towards understanding postsynaptic control mechanisms of neurotransmission. This was made possible by the development of biochemical techniques that facilitate an examination of neurotransmitter receptor structure and function at the molecular level.[3,4] As a result of these studies a great deal of information has been accumulated about the components of a receptor complex as well as the possible relationships between these components and receptor function. As more was learned about the structural and mechanical properties of receptors, new insights were gained about their role in the homeostatic control of neurotransmission.

S. J. Enna • Departments of Pharmacology, Neurobiology and Anatomy, University of Texas Medical School at Houston, Houston, Texas 77025.

The aim of the present chapter is to review some of the current concepts relating to mechanisms of receptor regulation. Although the emphasis will be on neurotransmitter receptors, results obtained with other systems will also be cited since developments in this field owe much to investigations using non-neural tissue. For the purposes of this chapter, a receptor is defined as that portion of the plasma membrane that recognizes and responds to the presence of neurotransmitter. This response may be in the form of either a direct alteration in membrane permeability, causing a change in the polarity of the cell, or an interaction with membrane-bound enzymes, which in turn catalyze specific reactions that alter cellular activity.[5]

2. MEMBRANE ARCHITECTURE

To conceptualize the manner in which receptors may be regulated, it is important to understand their milieu, the plasma membrane. Although most of the information obtained on this subject has been derived from studies on non-neural structures such as the erythrocyte, it appears that all plasma membranes have similar properties. Thus, membranes are composed of lipids, proteins, and carbohydrates in a ratio of approximately $45:45:10$. Although the lipids provide a flexible matrix, the membrane proteins are determinants of function. Because of their amphipatic nature, these structures are arranged in the form of a bilayer with the hydrophilic portion in contact with the aqueous surroundings. The protein molecules may either span the bilayer or be attached to other proteins that do.[6] Although proteins are capable of lateral movement within the membrane matrix,[7] they do not appear to move across the bilayer.[8] Because of this, the transport of substances and information across the membrane normally occurs by way of channels or pores or by a modification in the conformational structure of the protein. For example, in some systems stimulation of the receptor recognition site is thought to increase the probability of a collision between a membrane-bound enzyme and other membrane proteins, resulting in enzyme activation.[9] The active enzyme in turn opens an ion channel or initiates a response in the cytosol. Basically, then, a receptor response is dependent on the availability of a transmitter recognition site protein, its proximity to other membrane proteins such as enzymes, and their coupling to an ion channel or some other cellular component. Membrane fluidity, which is determined to a large extent by the lipids, must be maintained within a certain range for these systems to function properly.

Neurotransmitter receptors are thought to be intrinsic membrane proteins that are coupled to carbohydrates. These structures are synthesized in ribosomes and are transported within the neuron in small vesicles and then inserted into the membrane. Following endocytosis, older receptors are destroyed by lysosomal enzymes within the cell.[9] Although the rate of neurotransmitter receptor turnover has not yet been precisely defined, it is assumed that several hours or days are required to alter the receptor population by this mechanism, since the formation of new sites requires protein synthesis at a remote location.

Although it is possible that an ion channel associated with a recognition site is part of the same molecule, evidence suggests that in some instances they may be separate entitites.[10] With the latter arrangement, it may be possible that some channels are influenced by more than one recognition site, although conclusive evidence for this is lacking.[11] Also, separate molecules would mean that the rate of change in receptor number is a function of the turnover of two distinct proteins.

Data have also been presented to indicate that there may be membrane constituents, either phospholipids or proteins, that, although not part of the basic receptor molecule, can influence the affinity and number of recognition sites.[12–14] Such substances could provide the means for a rapid regulation of receptor availability and function.

3. RECOGNITION SITE REGULATION

3.1. Direct Influences

It has been known for over a century that neurotransmitter activity alters receptor sensitivity.[15] The early work demonstrated that chronic denervation of skeletal muscle leads to an enhancement in motor endplate sensitivity to exogenously applied acetylcholine. More recent studies have shown that, among other factors, this supersensitivity results from an increase in the number of nicotinic receptors and a migration of these sites along the muscle membrane.[16,17]

Central nervous system receptors are also regulated by the neurotransmitter itself. For example, several days following a lesion of the nigrostriatal dopamine pathway, there is an increase in the number and sensitivity of dopamine receptor recognition sites in the corpus striatum.[18] Also, chronic administration of receptor antagonists or of drugs that deplete biogenic amines also induces an increase in the number of postsynaptic receptor recognition sites for endogenous ligand.[19–21] These data with antagonists suggest that receptor occupancy alone is not sufficient for maintaining a constant number of recognition sites. Rather, this appears to be a function of agonist activity, suggesting that the signal for triggering an increase in the number of receptors is beyond the recognition site itself. For instance, an alteration in ion channel activity, or the amount of activated cyclase, may be the regulating factor. Accordingly, an increase in receptor number could theoretically occur in the presence of even a normal concentration of endogenous ligand if the signaling mechanism is defective.

Although a certain level of agonist activity is necessary for maintaining a constant number, supranormal amounts of agonist cause a decrease in recognition site number and response (desensitization). Thus, chronic administration of GABA receptor agonists decreases the number of GABA binding sites in the corpus striatum.[22–25] Also, administration of drugs such as antidepressants that increase the concentration of transmitter in the synaptic cleft decrease recognition site number and function.[26–28]

Because these changes are a consequence of neurotransmitter activity, any treatment that alters transmitter turnover or release has the potential to modify receptor sensitivity. Accordingly, chronic administration of receptor-active drugs or lesions of neuronal pathways may not only change receptor responsivity for the neurotransmitter system directly influenced by these treatments but may also, indirectly, modify receptor responses to other transmitters that are regulated by the affected system. For example, chronic administration of GABA agonists not only decreases the number of striatal GABA receptors but also increases the number and sensitivity of dopamine receptors in this brain region.[22] This change in dopamine receptors may result from a decrease in the firing rate of the nigrostriatal dopamine pathway. Thus, although the GABA receptor agonists themselves do not directly modify dopamine receptor function, they can influence these sites by altering dopamine activity.

Although the precise mechanism whereby the cell alters the number of recognition sites in response to a change in agonist activity is unknown, the fact that several days or weeks normally must elapse before they are apparent suggests that they are a function of the rate of protein synthesis or degradation. Thus, decreased agonist activity signals the cell to increase the production, or decrease the destruction, of receptor molecules. Moreover, at least with the nicotinic cholinergic system, denervation also alters the cellular response to potassium, changes the resting membrane potential, and modifies the storage and release of cellular calcium.[15] Thus, modifications in the amount of neurotransmitter present at the receptor causes an alteration in a number of cellular parameters, including recognition site binding. The sum total of these changes ultimately determines cellular and receptor sensitivity.

Another, perhaps more rapid, mechanism for altering receptor number is internalization.[29,30] This process has been found in association with both hormone and neurotransmitter receptors. With internalization, ligand attachment induces a lateral movement of the receptor protein to a region of the membrane (coated pits) where the receptor complex is taken into the cell by endocytosis. These receptors are subsequently degraded by lysosomes or, in some cases, may be returned to the cell membrane. Since this process takes only minutes to occur, it may explain the rapid densitization observed under certain conditions.[31-34] However, it is also possible that a pseudoirreversible attachment of agonist to the recognition site may partially explain the apparent decrease in receptor number observed in these experiments.[35] Several hours are required to replace the internalized receptors.

These data suggest that a change in presynaptic activity can induce an alteration in the number of receptor recognition sites. Furthermore, for long-term changes, receptor activity, rather than just occupancy, appears to be the major determinant with regard to regulating the number of recognition sites. On the other hand, receptor occupancy may be all that is necessary for short-term alterations such as those brought on by internalization. Moreover, a change in receptor number, in itself, is not sufficient evidence to prove that the alteration is primary, since receptor modifications can occur as the result of a dysfunction in a separate but functionally related system.

3.2. Indirect Influences

Receptor sites can also be influenced by events occuring at other sites on the membrane. For example, alterations in recognition site binding have been shown to result from activation of a different receptor located on the same membrane. This type of modification is termed heterospecific regulation.

Several examples of this phenomenon have been presented in recent years. Thus, carbachol, a muscarinic cholinergic receptor agonist, increases the affinity of cardiac α_1-adrenergic receptors.[36] Also, exposure of brain tissue to isoproterenol, a β-receptor agonist, increases the number of α_2-adrenergic binding sites.[34] The affinity of brain benzodiazepine receptors is increased as a consequence of GABA receptor activation.[37,38] Evidence that these receptor changes are mediated by activation of a separate neurotransmitter recognition site is provided by the fact that antagonists for the activating substance block the modification in receptor binding.

Data have also been presented to indicate that benzodiazepines increase GABA receptor binding.[39] Thus, under the proper conditions, the amount of [^{3}H]GABA bound is substantially higher in membranes coincubated with [^{3}H]GABA and diazepam than in those incubated in the absence of the benzodiazepine. This increase in GABA binding is thought to be caused by a benzodiazepine-induced alteration in the allosteric conformation of the GABA receptor.[14]

Hormones are also capable of influencing neurotransmitter recognition sites. For example, both adrenalectomy and ACTH administration cause a regionally selective increase in rat brain GABA receptor binding.[40] This hormone-induced modification appears to result from the appearance of a low-affinity, high-capacity, GABA receptor binding site.

Estrogen is capable of modifying the number of striatal dopamine receptors in male rat.[41-43] This effect is attenuated in hypophysectomized animals, suggesting that it may actually be mediated by pituitary hormones. Estrogenic substances also increase the number of hypothalamic cholinergic muscarinic receptors and decrease the number of cerebral cortical β-adrenergic receptors.[44-46] Estrogen also appears to play a permissive role with regard to drug-induced decreases in rat brain 5-HT$_2$ receptor binding.[47-49] Likewise, ACTH treatment shortens the amount of time necessary to observe a decrease in cerebral cortical β-receptor number following chronic treatment with some antidepressants.[49]

Guanine nucleotides are also known to influence recognition site binding.[50-52] Thus, GTP and related substances have been shown to decrease the affinity of the recognition site for agonists but not antagonists. This action is thought to be an important mechanism for controlling receptor activity, especially for those receptors coupled to a cyclase system.[5] Therefore, besides being an absolute requirement for activation of adenylate cyclase, GTP also functions to regulate recognition site affinity. By increasing the dissociation rate for the ligand, this nucleotide transiently diminishes receptor sensitivity.

Inorganic ions also serve as regulators of recognition site binding. For example, sodium decreases agonist affinity for the opiate receptor without altering the potency of antagonists.[53] Chaotropic agents such as thiocyanate and iodide increase the GABA receptor affinity for the antagonist bicuculline at concentrations that do not modify agonist binding.[54,55] These differential effects suggest that agonists and antagonists may attach to different portions of the same receptor molecule or that the recognition site fluctuates between two different states, one favoring agonist attachment, and the other favoring antagonists. This potential for shifting between agonist- and antagonist-favoring conformations endows the recognition site with another possible means for regulating activity.

Substances such as hormones, neurotransmitters, drugs, nucleotides, and ions can no doubt modify recognition site binding by a variety of mechanisms. Hormones may do so by enhancing or inhibiting neurotransmitter formation, metabolism, or release. Alternatively, these substances may directly influence the rate of receptor synthesis or degradation, or they may act by altering the viscosity of the plasma membrane.

It has been proposed that benzodiazepines activate GABA receptor binding by regulating an endogenous modulator of this recognition site.[14,39] Several studies have indicated that there is present, at or near the GABA receptor, some substance capable of either masking a group of higher-affinity sites or of allosterically modifying the conformation of exposed sites.[13,14] Alterations of this substance, either by drugs or disease states, change the affinity, number, and possibly the sensitivity of GABA receptors. Such substances would be capable of rapidly regulating this recognition site.

With regard to GTP, studies with the β-adrenergic receptor suggest that agonist attachment to the recognition site causes a conformational change in the molecule, exposing sites capable of interacting with a nucleotide-binding protein. This interaction between the recognition site and a GTP-binding protein causes a structural modification in the receptor, lowering the affinity for agonist by increasing the rate of dissociation.[56,57] Although nucleotides may not always be involved, it seems likely that regulation of recognition site affinity by agonist-induced conformational changes is probably a characteristic of all neurotransmitter receptors. In some cases, the change may be brought about by coupling with an ion.

These studies indicate that receptor recognition sites are subject to influence by a variety of factors. Given the complexity of plasma membranes and the dynamics of the system, these examples undoubtedly represent only a fraction of the ways in which receptors are regulated at the molecular level.

4. REGULATION OF OTHER RECEPTOR COMPONENTS

The ligand recognition site is only one element in the receptor complex. Activation of this site initiates a change in other receptor constituents, which ultimately leads to a modification in cellular activity. Ion channels and cyclic nucleotide systems appear to be two of the more common components linked to the recognition site. Thus, for example, the mobile receptor hypothesis states that in the absence of hormone or neurotransmitter, the recognition site and the catalytic unit of adenylate cyclase are freely, and independently, floating in the plane of the membrane. Activation of the recognition site induces a coupling between these components, resulting in a complex that spans the bilayer.[58] Since formation of this complex is crucial for proper receptor function, alterations in any one of these components can modify receptor activity. Thus, receptors may be subject to regulation at a point beyond the recognition site itself.

The concept of a second messenger was first developed by Sutherland and co-workers.[59,60] They proposed that the enzyme adenylate cyclase, a membrane constituent, catalyzes the formation of cyclic AMP from ATP. This enzyme is activated by the action of agonists at the receptor recognition site. The cyclic AMP subsequently formed is located on the inner surface of the membrane and acts as an intracellular messanger for transmitting information. This is accomplished by activation of protein kinases.

Kinases are found in virtually all cells, and many proteins are substrates for these enzymes. Kinase activation leads to a chain of events that amplify the response of the cell to the neurotransmitter or hormone. In order for the phosphorylase kinase reaction to occur, calcium must be bound to a component of the enzyme. This component is referred to as the δ subunit or calcium-binding protein (calmodulin).[61] The attachment of calcium causes a conformational change in the protein, which in turn influences the ability of the enzyme to interact with other molecules.

For some neurotransmitter receptors, agonist attachment to the recognition site induces the formation of a GTP-binding complex, which in turn activates adenylate cyclase. The cyclase is deactivated following the hydrolysis of GTP to GDP. The adenylate cyclase catalyzes the formation of cyclic AMP, which activates a protein kinase. Cyclic AMP is destroyed by phosphodiesterase. The activated kinase phosphorylates a membrane protein, which, for example, may be a part of the ion channel, leading to a change in membrane permeability and polarity.[5] This sequence also leads to the phosphorylation of intracellular proteins that regulate other processes such as cell division and microtubular activity.

Thus, there are various loci at which receptor activity may be regulated by events beyond the recognition site. As an example, studies have suggested that dopamine receptor sensitivity is related to the amount of membrane-bound calmodulin.[62] Since calmodulin increases the affinity of certain phosphodiesterases for cyclic AMP, a decrease in recognition site activation would lead to an increased retention of calmodulin on the plasma membrane, resulting in a less active form of the degradative enzyme. This allows for a greater accumulation of cyclic nucleotide per stimulus (supersensitivity). Conversely, persistent activation of the receptor may be associated with an increase in free (cytoplasmic) calmodulin and therefore more active phosphodiesterase and a more rapid destruction of cyclic AMP (desensitization).[9,62] Such changes in receptor sensitivity could occur without a significant alteration in recognition site binding.

Changes in the guanine nucleotide regulatory protein or in the rate at which GTP is converted to GDP could also modify receptor function. A decrease in the binding affinity for GTP would result in the activation of less adenylate cyclase and the formation of fewer molecules of cyclic AMP.

Changes in membrane fluidity, kinase activity, or the availability of calcium ion are other circumstances that could alter receptor action. The observation that under certain conditions norepinephrine is less capable of stimulating cyclic AMP formation in brain slices in the absence of any apparent change in recognition site binding may reflect a change in one of these parameters.[63] Tachyphylaxis is also thought to be the result of an alteration in one or more of these recognition site-coupled events.[5]

Less is known about the molecular characteristics of the ion channels associated with receptor recognition sites.[5] However, since the channel is coupled to the recognition site, if not actually a subunit of the same molecule, a perturbation in this component could affect the conformation of the recognition site. Furthermore, a change in the coupling mechanism between the channel and recognition site would also be manifest as a change in receptor sensitivity. However, it is difficult to speculate on the manner in which this part of the complex normally participates in receptor regulation until more has been learned about the molecular interactions between the two components.

5. SUMMARY AND CONCLUSIONS

The concept of a neurotransmitter receptor has changed appreciably over the past two decades. This is because a number of technical advances have made it possible to characterize more precisely the composition of these membrane constituents. As a result of these studies, insights have been gained into the manner in which receptor action is regulated. Thus, like other membrane proteins, receptor recognition sites are continuously being formed and degraded, and, therefore, factors that influence the rate of receptor turnover alter the cellular response to neurotransmitter. Also, recognition site number and affinity may be subject to local control by the actions of associated membrane proteins or phospholipids. In addition, since the receptor is a dynamic entity, a change in membrane fluidity could restrict the movement of the receptor in the lipid bilayer and thereby alter receptor function. Moreover, the properties of the recognition site are regulated, to some extent, by inorganic ions, guanyl nucleotides, and hormones.

Since recognition site activation is expressed through a coupling to other membrane constituents such as an ion channel or cyclic nucleotide system, modification of these components will also alter response. Thus, the affinity of phosphodiesterase for cyclic AMP appears to be regulated by a calcium-binding protein, the concentration of which may be determined by receptor occupancy. Likewise, the amount of active adenylate cyclase is controlled by a GTP-binding protein, and this component is regulated by the degree of recognition site activity. Accordingly, neurotransmitter receptor sensitivity can be modified in a number of ways. The intricacies of this mechanism make it an excellent target for drug action as well as increasing its vulnerability to disease. Clearly, a better definition of this system is crucial for understanding the homeostatic control of neurotransmission.

REFERENCES

1. Stjarne, L., Lishajko, F., and Roth, R. H., 1967, *Nature* **215**:770–772.
2. Simon, J. R., and Kuhar, M. J., 1975, *Nature* **255**:162–163.
3. Yamamura, H. I., Enna, S. J., and Kuhar, M. J., (eds.), 1978, *Neurotransmitter Receptor Binding,* Raven Press, New York.
4. Schulster, D., and Levitzki, A. (eds.), 1980, *Cellular Receptors for Hormones and Neurotransmitters,* John Wiley & Sons, New York.
5. Enna, S. J., and Strada, S. J., 1982, *Clinical Neurosciences* (R. Rosenberg, R. Grossman, S. Schochet, E. R. Heinz, and Willis, eds.), Churchill Livingston, Edinburgh (in press).
6. Weber, K., and Osborn, M., 1969, *J. Biol. Chem.* **244**:4406–4412.
7. Frye, L. D., and Edidin, M., 1970, *J. Cell Sci.* **7**:319–355.
8. Warren, G., and Houslay, M., 1980, *Cellular Receptors for Hormones and Neurotransmitters* (D. Schylster and A. Levitzki, eds.), John Wiley & Sons, New York, pp. 29–54.
9. Costa, E., 1980, *Modern Aging Research,* Volume 1 (R. C. Adelman, J. Roberts, G. T. Baker, S. I. Baskin, and V. J. Cristofalo, eds.), Alan R. Liss, New York, pp. 1–23.
10. Eldefrawi, M. E. and Eldefrawi, A. T., 1980, *Ann. N.Y. Acad. Sci.* **358**:239–252.
11. Coyle, J. T., 1980, *Neurotransmitter Receptors,* Part 1 (S. J. Enna and H. I. Yamamura, eds.), Chapman and Hall, London, pp. 5–40.
12. Enna, S. J., and Gallagher, J. P., 1983, *Int. Rev. Neurobiol.* **24**:181–212.
13. Johnston, G. A. R., and Kennedy, S. M. E., 1978, *Amino Acids as Chemical Transmitters* (F. Fonnum, ed.), Plenum Press, New York, pp. 507–516.
14. Massotti, M., Guidotti, A., and Costa, E., 1981, *J. Neurosci.* **1**:409–418.
15. Fleming, W. W., McPhillips, J. J., and Westfall, D. P., 1973, *Rev. Physiol. Biochem. Exp. Ther.* **68**:55–119.
16. Fleming, W. W., 1976, *Rev. Neurosci.* **2**:43–90.
17. Fleming, W. W., 1981, *Trends Pharmacol. Sci.* **2**:152–154.
18. Creese, I., Burt, D. R., and Snyder, S. H., 1977, *Science* **197**:596–598.
19. Burt, D. R., Creese, I., and Snyder, S. H., 1977, *Science* **197**:326–328.
20. Moore, K. E., and Thornburg, J. E., 1975, *Adv. Neurol.* **9**:93–104.
21. Ungerstedt, U., Ljungberg, T., Hoffer, B. and Siggins, G., 1975, *Adv. Neurol.* **9**:57–66.
22. Ferkany, J. W., Strong, R., and Enna, S. J., 1980, *J. Neurochem.* **34**:247–249.
23. Enna, S. J., Ferkany, J. W., and Strong, R., 1980, *Receptors for Neurotransmitters and Peptide Hormones* (G. Pepeu, M. J. Kuhar, and S. J. Enna, eds.), Raven Press, New York, pp. 253–263.
24. Ferkany, J. W., and Enna, S. J., 1980, *Life Sci.* **27**:143–149.
25. Enna, S. J., and Ferkany, J. W., 1980, *Psychopharmacology and Biochemistry of Neurotransmitter Receptors* (H. I. Yamamura, R. Olsen, and E. Usdin, eds.), Elsevier/North Holland, New York, pp. 525-535.
26. Sulser, F., and Vetulani, J., 1977, *Animal Models in Psychiatry and Neurology* (E. Usdin and I. Hanin, eds.), Pergamon Press, New York, pp. 189–199.
27. Peroutka, S. J., and Snyder, S. H., 1980, *Science* **210**:88–90.
28. Enna, S. J., and Kendall, D. A., 1981, *J. Clin. Psychopharmacol.* **1**:125–165.
29. Goldstein, J. L., Anderson, R. G., and Brown, M. S., 1979, *Nature* **279**:679–681.
30. Chuang, D. M., and Costa, E., 1979, *Proc. Natl. Acad. Sci. U.S.A.* **76**:3024–3028.
31. Mickey, J., Tate, R., and Lefkowitz, R., 1975, *J. Biol. Chem.* **250**:5727–5729.
32. Mukherjee, C., and Lefkowitz, R. J., 1977, *Mol. Pharmacol.* **13**:291–303.
33. Dibner, M. D., and Molinoff, P. B., 1979, *J. Pharmacol. Exp. Ther.* **210**:433–439.
34. Maggi, A., U'Prichard, D. C., and Enna, S. J., 1980, *Science* **207**:645–647.
35. Williams, L. T., and Lefkowitz, R. J., 1977, *J. Biol. Chem.* **252**:7207–7213.
36. Yamada, S., Yamamura, H. I., and Roeske, W. R., 1980, *Eur. J. Pharmacol.* **63**:239–241.

37. Gallager, D. W., Thomas, J. W., and Tallman, J. F., 1978, *Biochem. Pharmacol.* **27:**2745–2749.
38. Maggi, A., Satinover, J., Oberdorfer, M., Mann, E., and Enna, S. J., 1980, *Brain Res. Bull.* **5:**167–171.
39. Guidotti, A., Toffano, G., and Costa, E., 1978, *Nature* **275:**553–555.
40. Kendall, D. A., McEwen, B. S., and Enna, S. J., 1982, *Brain Res.* **236:**365–374.
41. Chioda, L. A., Caggiula, A. R., and Saller, C. F., 1981, *Life Sci.* **28:**827–835.
42. Hruska, R. E., and Silbergeld, E. K., 1980, *Science* **208:**1466–1568.
43. Savageau, M. M., and Beatty, W. W., 1981, *Pharmacol. Biochem. Behav.* **14:**17–21.
44. Rainbow, T. C., Degroff, V., Luine, V. N., and McEwen, B. S., 1980, *Brain Res.* **198:**239–243.
45. Wagner, H. R., Crutcher, K. A., and Davis, J. N., 1979, *Brain Res.* **171:**147–151.
46. Wagner, H. R., and Davies, J. N., 1980, *Brain Res.* **201:**235–239.
47. Kendall, D. A., Stancel, G. M., and Enna, S. J., 1981, *Science* **211:**1183–1185.
48. Kendall, D. A., Stancel, G. M., and Enna, S. J., 1982, *J. Neurosci.* **2:**354–360.
49. Kendall, D. A., Slopis, J., Duman, R., Stancel, G. M., and Enna, S. J., 1982, *Proteins of the Nervous System—Structure and Function* (B. Haber, J. R. Perez-Polo, and J. Coulter, eds.), Alan R. Liss, New York, pp. 193–207.
50. Blume, A. J., 1978, *Proc. Natl. Acad. Sci. U.S.A.* **75:**1713–1717.
51. U'Prichard, D. C., and Snyder, S. H., 1978, *J. Biol. Chem.* **253:**3444–3452.
52. Williams, L. T., and Lefkowitz, R. J., 1977, *J. Biol. Chem.* **252:**7207–7213.
53. Pert, C. B., Pasternak, G. W., and Snyder, S. H., 1973, *Science* **182:**1359–1361.
54. Enna, S. J., and Snyder, S. H., 1977, *Mol. Pharmacol.* **13:**442–453.
55. Browner, M., Ferkany, J. W., and Enna, S. J., *J. Neurosci.* **1:**514–518.
56. Vauquelin, G., Bottari, S., Andre, C., Jacobsson, B., and Strosberg, A. D., 1980, *Proc. Natl. Acad. Sci. U.S.A.* **77:**3801–3805.
57. Minneman, K. P., 1981, *Neurotransmitter Receptors Part 2* (H. I. Yamamura and S. J. Enna, eds.), Chapman and Hall, London, pp. 185–268.
58. Cuatrecasas, P., Hollenberg, M. D., Chang, K. J., and Bennett, V., 1975, *Rec. Prog. Hormone Res.* **31:**37–94.
59. Sutherland, E. W., 1972, *Science* **177:**401–408.
60. Robison, G. A., Butcher, R. W., and Sutherland, E. W., 1967, *Ann. N.Y. Acad. Sci.* **139:**703–723.
61. Cheung, W. Y. (ed.), 1980, *Calcium and Cell Function.* Volume 1, *Calmodulin,* Academic Press, New York.
62. Gnegy, M. E., Uzunov, P., and Costa, E., 1977, *J. Pharmacol. Exp. Ther.* **202:**558–564.
63. Mishra, R., Janowsky, A., and Sulser, F., 1980, *Neuropharmacology* **19:**983–987.

Receptor Adaptation to Psychotropic Drugs

Jack W. Schweitzer, Kenneth A. Bonnet, and Arnold J. Friedhoff

1. INTRODUCTION

The membrane receptor is a site at which specific chemicals can regulate cellular function; in turn, cells can regulate their sensitivity to many chemicals to compensate for persistent changes in agonist supply. The terms up- and down-regulation are used to describe shifts in receptor responsivity to drugs or to endogenous transmitters. Chronic antagonist administration, for example, can lead to a proliferation of receptor sites, a change viewed as an attempt by the cell to compensate for the loss of agonist resulting from the blockade by antagonist. In similar fashion, down-regulation can take place with chronic agonist oversupply. Thus, the receptor can act as both a regulated and regulatory unit. In this chapter an attempt is made to identify specific neuronal receptors that are acted on and ultimately altered by neuroleptics, antidepressants, and anxiolytic benzodiazepines and to discuss the application of receptor sensitivity modification (RSM) as a new approach to treatment of disorders that are responsive to alterations of CNS monoaminergic activity.

2. DOPAMINE RECEPTORS

Dopamine (DA) receptors are of great interest because they are believed to be the primary target for neuroleptic drugs. Neuroleptics presumably exert a behavioral normalizing effect through their interaction with DA receptors because of the important role of these receptors in the regulation of behavior. Dopamine receptors regulate postural and other reflexes controlled by the extrapyramidal system but also appear to play some role in the maintenance of

Jack W. Schweitzer, Kenneth A. Bonnet, and Arnold J. Friedhoff • Department of Psychiatry, Millhauser Laboratories, New York University School of Medicine, New York, New York 10016.

normal thinking and emotional processes. The decrease in dopaminergic activity resulting from acute neuroleptic blockade appears to be the effect that produces the therapeutic action of neuroleptics, inasmuch as all effective neuroleptics share this property. Neuroleptics antagonize DA receptors, and, with persistent exposure, most, if not all, produce an increase in the number of receptors in both caudate and mesolimbic areas. The effect of such increases can be masked by the continued presence of the drug; however, after neuroleptic washout, the supersensitive condition becomes apparent. It would seem likely that the compensatory increase in receptors would offset the blocking action of neuroleptics, yet neuroleptics do not lose their potency at a time when a compensatory up-regulation would be expected. The reason why tolerance to neuroleptics does not occur is not clear. It may be that compensatory up-regulation is more prominent in the striatal motor areas, whereas the therapeutic action of these drugs is mediated by the limbic cortex, which is more resistant to compensatory effects. Alternatively, other adjustments may occur that are not presently known that prevent the offsetting effects of the compensatory changes.

Maintenance of neuroleptic therapy in patients brings with it the potential for producing uncontrollable movements of limbs and mouth, collectively known as tardive dyskinesia. These symptoms are believed to be related to the compensatory increase in DA receptors in striatal motor areas; however, these changes, in rats at least, occur soon after treatment is initiated and in all rats treated. Tardive dyskinesia, on the other hand, occurs in only some patients and, as the name suggests, is a late-appearing phenomenon.

Thus, defining the complex properties of DA receptors is important in understanding the regulation of normal motor behavior and the modulation of thinking and feeling. The DA system is also involved in mediating the therapeutic effects and some of the more invidious side effects of neuroleptic drugs. The role of this system in mood-altering drugs is less clear.

2.1. Postsynaptic Effects of Antagonists on Dopamine Receptors

As summarized by Seeman,[1] striatal DA receptors of three distinct types have been recognized. These include a presynaptic autoreceptor (D_3), a postsynaptic receptor linked to adenylate cyclase (D_1), and a postsynaptic receptor found on or closely related to cholinergic interneurons (D_2). In 1977, Burt *et al.*[2] demonstrated that administration of a neuroleptic for at least 1 week produced a compensatory increase in DA receptors in striatum. Following the initial report by this group, this effect has been repeatedly demonstrated with a variety of neuroleptics. In addition to a number of studies wtih haloperidol, Seeman[1] has documented reports of the up-regulation of the D_2 receptor with chronically administered metoclopromide, clozapine, trifluoperizine, and thioridazine, to which can be added *cis*-flupenthixol,[3] fluphenazine,[4] and teflutixol (unpublished data). It is of interest that clozapine, which has a low affinity for the D_2 receptor[5] and a shorter half-life than haloperidol (HAL),[6] is nontheless capable of up-regulating D_2 sites after 3 weeks of treatment.[7] This finding may reflect the fact that some receptors are regulated indirectly as well as by their own ligands.

The number of specific receptors can be determined by several techniques. Frequently, this is estimated by measuring the number of recognition sites— the sites on the membrane to which the specific agonist or antagonist will bind. This can be accomplished by incubating a radiolabeled agonist or antagonist with a membrane fraction in the presence of a specific unlabeled agonist or antagonist in relatively high concentration. It is assumed that the unlabeled compound will displace the ligand only from specific binding sites. Thus, the number of sites can be determined by subtracting the amount bound in the presence of displacer from the amount bound in the absence of displacer.

A receptor, of course, is a functional unit, not just a binding site. Thus, one cannot always, with complete confidence, describe the properties of receptors from those of the binding or recognition sites. Intervening coupling effects and response mechanisms must also be considered.

Although up-regulation can readily be detected by radioreceptor assay when [^{3}H]antagonists are used in the assay, the use of [^{3}H]agonists such as apomorphine (APO), ADTN, N-propylnorapomorphine (NPA), and DA as a means for measuring DA sites have produced somewhat conflicting results. In some cases, divergent findings can be traced to assay conditions. These include the membrane preparation, where insufficient washing may permit endogenous DA to influence binding.[8] In addition, the presence or absence of physiological concentrations of salts and the choice of drug needed to estimate nonspecific binding can have profound effects.[9] Since most agonists are catechols, there has been a tendency to include an antioxidant, such as ascorbate, in the binding assay. This agent, presenting a catechol-like structure in one of its resonance forms and which undergoes mono-0-methylation *in vivo*,[10] has been shown to interfere with [^{3}H]agonist binding.[11] Moreover, the strong influence of GTP on [^{3}H]agonist binding[12] may reflect the fact that, in part, such ligands bind to the D_1 receptor as well as the D_2.

In parallel with the terms V_{max} and K_m used in describing enzymes, maximum receptor density (B_{max}) and dissociability (K_d) of a given ligand have become useful parameters for characterizing receptors. Values for these constants are often obtained by means of Scatchard[13] analysis of saturation data. The Scatchard equation:

$$B/F = -B(1/K_d) + B_{max}/K_d$$

is a reconceptualization and linear rearrangement of the Michaelis–Menten equation. B is obtained by subtracting nonspecific binding from total binding, and F is obtained by subtracting total binding from the concentration of introduced ligand. Theoretical and practical aspects of this type of analysis have been discussed in several books and reviews.[13–16]

Among other things, saturation studies can provide information on the competitive or noncompetitive nature of drugs. One recent study[17] deserves attention. It was shown that, as expected, the addition of fluphenazine to bovine anterior pituitary membranes subjected to [^{3}H]spiroperidol binding led to an apparent reduction in affinity but no change in B_{max}. Scatchard analysis of the effect of N-propylnorapomorphine on [^{3}H]spiroperidol binding revealed no change in affinity but a decrease in B_{max}, as though NPA had bound irreversibly to a portion of the D_2 receptors. However, since [^{3}H]NPA binding has been shown to be reversible, Sibley and Creese[17] have suggested that two receptors are present in this tissue, each with equal affinity for [^{3}H]spiroperidol but with radically different affinities for NPA. A situation similar to this, they warn, might also be applicable when a rapid loss of β-adrenergic receptors is apparently demonstrated shortly after tissue has been exposed to a high concentration of agonist and then subjected to (possibly inadequate) washing prior to binding.

Another potentially confounding variable in studies of receptor sensitivity is the possible effect of seasonal, daily, and even hourly changes in receptor responsivity.[18] Cools[19] has observed that the anterodorsal part of the head of the caudate nucleus of the cat

> . . . is marked by a rhythmically changing susceptibility to catecholaminergic agents with crests for NE agonists and DA antagonists, but troughs for NE antagonists and DA agonists during November, December, January and July and troughs for NE agonists and DA antagonists but crests for NE antagonists and DA agonists during March, April, May, August, September and October. . . .

In a series of reports, Wirz-Justice and colleagues[20-24] have indicated that dopaminergic, noradrenergic, and cholinergic receptors undergo circadian and seasonal rhythms, although these findings have not yet been replicated. [^{3}H]Opioid agonist and antagonist binding as well have been shown to undergo seasonal changes.[25]

In regard to daily rhythms in [^{3}H]spiroperidol binding to striatal DA receptors, Naber *et al.*[20] have reported a 70% increase in binding at 2 p.m. when compared with a 10 a.m. nadir for animals maintained for 3 weeks in a strict 12-h lights on–lights off environment. This change is two to three times the maximum effect inducible with chronic (2- to 4-week) treatment with neuroleptic. In a pilot study (unpublished) designed specifically to test their observation, we were not able to detect a significant difference between the two mentioned time periods; however, it is possible that we chose the wrong time of year for this examination, since the ultradian cycle shifts during the year,[21] and chronic imipramine reportedly attenuates the amplitude of the daily fluctuations.[20] These studies clearly require additional investigation.

The minimum time required for producing detectable up-regulation of striatal DA receptors in the rat is reported to be 1 week.[2] In the mouse, however, behavioral supersensitivity (APO-induced climbing) was significantly increased after one dose of HAL measured 3 days after administration,[26] although [^{3}H]pimozide binding had not changed. After 7 days of treatment, both behavior and binding were elevated. Apomorphine-induced enhancement of stereotypy has also been observed days after the administration of a single dose of chlorpromazine (CPZ),[27] suggesting that a single dose can trigger a change in neuronal sensitivity although receptor binding, a less sensitive measure, has undergone too small an increase to be detected. It is also of interest that a single

ECS treatment to rats desensitized nigral responses to iontophoretically applied DA 7 days later.[28]

In our own studies we have found that the degree of neuroleptic (HAL)-induced increase in [^{3}H]spiroperidol ([^{3}H]SPIRO) binding is dependent mainly on the duration of treatment. In 11 out of 12 comparisons at different washout periods, rats treated daily for 30 days exhibited greater elevations in binding than those given HAL for only 14 days. Dose, although less important, is also a contributing factor. In 13 out 14 studies, rats receiving 5 mg/kg of HAL exhibited greater elevations in receptor number, measured at various periods during withdrawal, than those receiving 0.5 mg/kg. In a drug holiday study (unpublished data), rats receiving 0.5 mg/kg on alternate days over a 27-day period did not exhibit significantly lower [^{3}H]SPIRO binding elevations than rats given daily doses over the same period. It has been reported that three drug holidays, each of 1 month's duration, interspersed during the chronic administration of either trifluoperizine or *cis*-flupenthixol over a 9-month period failed to prevent increases in striatal [^{3}H]SPIRO binding or of enhanced APO-induced stereotypy.[3]

We have also examined the rate of decline of receptor binding after neuroleptic treatment was terminated. Rats returned to base-line number of DA ([^{3}H]spiroperidol) sites in 0.7 of the number of days of treatment. This is in close agreement with an earlier estimate of Muller and Seeman[29] in which they found that days to return to base line/days on drug was 0.6. On this basis, dopaminergic receptors in psychiatric patients receiving neuroleptic for 2 years might require as many as 450 days to return to their predrug state, which might explain the persistence of tardive dyskinesia (TD) symptoms after neuroleptic withdrawal.[30]

With more severe chronic treatment, 6 months or longer, some of the symptoms associated with TD in man can be elicited in rats. Vacuous chewing movements "not directed towards physical material, distinct from APO-induced gnawing" have been noted after 6 months of depot fluphenazine decanoate treatment.[31] As might be expected, since medication was not stopped,[32] APO-inducible stereotypy was largely absent. Dyskinetic behaviors have also been observed in monkeys treated with HAL. After the tenth weekly dose, oral movements, peculiar postures, writhing, and stretching could be seen.[33] On increasing the dose to 0.5 mg/kg, these abnormal behaviors appeared regularly after each weekly dose for almost 2 years. In an additional report,[34] it was noted that monkeys treated for several months with HAL at 5 injections/week and then withdrawn for 508 days exhibited severe dyskinetic behaviors after a single dose of HAL. This is puzzling inasmuch as, in man, TD symptoms tend to disappear on neuroleptic challenge. These and other studies of effects produced by long-term neuroloptic treatment are listed in Table I.

Some of the results, especially behaviors induced by APO, appear to be contradictory. However, the dose of APO used by one group[3,35-38] was 0.5 mg/kg, whereas others used 0.15[31,39,40] or 1 mg/kg.[40] In a number of these studies neuroleptic was not withdrawn; therefore, high doses of APO might be required to override neuroleptic blockade, as is illustrated by the second of the two *cis*-flupenthixol studies described in Table I. A further confounding factor

Table I
Effects of Long-Term Neuroleptic Administration

Neuroleptic	Duration of treatment	Washout	Major effects[a]	Reference
Trifluoperizine	9 mo.	None	D_2 ↑	3
cis-Flupenthixol	9 mo.	None	D_2 ↑	3
Haloperidol	10 wk.–2 yr.	—[b]	Dyskinesias	33
Haloperidol[b]	6 mo.	508 days	Dyskinesias	34
Fluphenazine	6 mo.	None	VC	31
Fluphenazine	6 mo.	6 mo.	St. ↓	32
Trifluoperizine	6 mo.	None	St. ↓, SB ↑, OM ↑	44
Trifluoperizine	4–6 mo.	None	[^{3}H]DA ↑, St. ↑	42
Haloperidol	3.5 mo.	None	St. ↓, SB ↓	39
Trifluoperizine	1 yr.	None	D_2 ↑, AC ↑	35,37
Thioridazine	1 yr.	None	D_2 ↑, AC ↑	35,37
Trifluoperizine	6 mo.	None	D_2 ↑, AC ↑, St. ↑	36
Trifluoperizine	6 mo.	None	D_2 ↑	38
Trifluoperizine or thioridazine	1 yr.	None	Prol. ↑, St. ↑, D_2 ↑	45
cis-Flupenthixol	6 mo.	None	St: LD APO ↓, HD APO ↑	40
Haloperidol	9 mo.	None	SL ↓, St. ↓	46
Haloperidol	9 mo.	7–10 days	SL ↑, St. ↑	46

[a] Abbreviations: D_2, postsynaptic DA receptor not linked to adenylate cyclase; ↑ or ↓, increased or decreased activity (or binding), respectively; St., apomorphine (APO)-induced stereotypy; SB, spontaneous behavior; AC, dopamine-sensitive adenylate cyclase activity; SL, spontaneous locomotion; VC, vacuous chewing; OM, oral movements; [^{3}H]DA, high-affinity [^{3}H]DA binding; Prol., prolactin level; LD, low dose; HD, high dose.
[b] See text for details of this study.

is the potential for unique neuroleptic–APO interactions. It has been shown[41] that a single dose of HAL, cis-flupenthixol, metoclopramide, or reserpine can block the accumulation of injected APO into both dopaminergic and nondopaminergic areas in the CNS, and ADTN levels following an injection of dibenzoyl-ADTN were decreased with HAL but not with cis-flupenthixol. Thus, the effects of individual neuroleptics on APO testing procedures may have to be evaluated before it is possible to assign behaviors to specific DA receptors.

From recent evidence, it appears that D_1 as well as D_2 receptors may be involved in APO-induced stereotypic responses. In a study with rats treated for 4–6 months with trifluoperizine,[42] APO-induced stereotypy was found to be significantly enhanced, although the animals were still on drug. Since trifluoperizine binds more strongly to the D_2 receptor than to the D_1 receptor,[5] it is possible that the induced stereotypy was mediated by D_1 receptors. [^{3}H]Dopamine binding was also examined in these rats. When domperidone or ADTN was used to estimate nonspecific binding, neuroleptic-treated rats exhibited increases in binding. However, the addition of GTP eradicated differences between control and neuroleptic-treated rats. Since GTP is required for the coupling of DA to the D_1 receptor[43] but also accelerates its departure,[12,43] net binding is suppressed by the cofactor. Thus, the increase in [^{3}H]DA binding resulted largely from D_1 receptor supersensitivity. In addition, whereas administered amphetamine and nomifensine, drugs that increase synaptic DA, were also effective in revealing enhanced stereotypic responses, ergots, which act primarily at D_2 receptors, failed to enhance the behaviors in the neuroleptic-treated group.

Although effects of very-long-term treatment appear to model TD hyper-responsive DA receptors cannot be the sole reason for the appearance of TD in humans, inasmuch as symptoms are rarely seen before 6 months to 1 year of treatment, whereas DA receptor supersensitivity should have already developed, and, whereas receptor density increases in all rats so treated, only a fraction of patients receiving neuroleptic medication develop dyskinetic symptoms.[30]

It has been shown that D_1 and D_2 receptors in the striatum are largely independent of each other. Kainic acid lesions, for example, destroy considerably more of the DA-sensitive adenylate cyclase[48] and also more [^{3}H]*cis*-flupenthixol binding[47] (predominantly to the D_1 receptor) than of [^{3}H]spiroperidol binding.[47,48] Moreover, whereas D_2 receptors appear to be associated with cholinergic function in the striatum, DA-sensitive adenylate cyclase appears not to be related to such activity[49,50]; however, the loss in [^{3}H]*cis*-flupenthixol binding[47] correlated significantly with [^{3}H]spiroperidol binding decreases after kainic acid with a slope that suggests that some [^{3}H]spiroperidol binding sites might be associated with the D_1 receptor. Recently, a putative relationship between these receptors was demonstrated by Kebabian and co-workers.[51,52] These investigators took advantage of two novel drugs, SKF 38393 and LY 141865, agonists specific for the D_1 and D_2 receptors, respectively. Both DA and SKF 38393 stimulated the release (and therefore, presumably, the formation) of cyclic AMP from blocks of rat striatum. Whereas $(-)$-sulpiride, a D_2 antagonist with essentially no activity at D_1 receptors, enhanced agonist induction of cyclic AMP efflux, [$(-)$-sulpiride alone had no effect on basal cyclic AMP efflux], LY 141865 suppressed agonist-induced cyclic AMP efflux. This suppression was reversed by the addition of $(-)$-sulpiride. Thus, stimulation of D_2 receptors reduced D_1 activity, whereas antagonism of D_2 receptors enhanced D_1 activity. These studies suggest that both receptors may be located on the same neuron, but, obviously, further examinations of these exciting preliminary observations will be necessary.

2.2. Presynaptic Involvement in D_2 Receptor Regulation

It has been shown recently[53] that the concurrent administration of a lithium salt and HAL can prevent the up-regulation of D_2 receptors seen with chronic HAL alone. Lithium blockade of supersensitivity inducible by HAL has also been demonstrated by electrophysiological monitoring of the activity of nigral DA cell bodies.[54] In the HAL-treated group, DA or APO rapidly stopped spontaneous firing, presumably because of HAL-induced supersensitivity of inhibitory autoreceptors, whereas Li^+ appeared to prevent this action of HAL; however, this mode of action of Li^+ could not be confirmed by Meller and Friedman.[55] These investigators, examining acute HAL-induced DOPAC elevations and behavioral supersensitivity in rats subsequent to chronic treatment with HAL plus Li^+, found that this drug combination reduced APO-induced stereotypy but failed to prevent tolerance to the DOPAC elevations seen with chronic HAL alone.

From these studies, it appears that the modifications of presynaptic mechanisms for control of DA turnover induced by neuroleptics may not be related to the neuroleptic-induced increase in D_2-related behavioral supersensitivity, and, furthermore, an increase in D_2 receptor number may not be the only factor in neuroleptic-induced supersensitivity. In this regard, Staunton et al.[56] found that the combination of Li^+ plus HAL was effective in attenuating APO-induced stereotypy but were unable to reproduce the reported ability to Li^+ to resist HAL-induced D_2 receptor up-regulation. It must be recognized, however, that conclusions drawn from the above observations are complicated by the fact that the increase in D_2 receptor density after a few weeks of neuroleptic treatment is approximately 15–30%, whereas APO-induced stereotypy is increased three- to fourfold. Thus, the manipulation of APO-inducible stereotypy by drugs may be accompanied by alterations in D_2 receptor density that are less discernibly different. In our hands, changes of $\pm$ 7–10% in D_2 receptor binding with groups of six to eight animals frequently fail to reach statistical significance. By extrapolation, such small changes in binding could be accompanied by halved or doubled APO-induced behaviors.[56]

The relative inefficacy of DA in preventing neuroleptic-induced D_2 receptor density up-regulation can be seen from a study (H. Rosengarten and A. J. Friedhoff, unpublished data) in which 320 times as much l-DOPA as HAL was required to prevent coadministered HAL from increasing D_2 receptor number. Thus, HAL, which has a much greater affinity for DA sites than DA itself, easily resists displacement by DA. In contrast, when l-DOPA therapy is started after neuroleptic withdrawal,[4,57] smaller amounts are effective in accelerating the return to the control density, from which it appears that DA can effect down-regulation only when it can reach the receptor site.

Changes in the responsivity of DA receptors have also been demonstrated with nonnueroleptics. Morphine, for example, can induce prolactin excretion in humans, and this effect can be blocked by APO[58] and methadone.[59] The rise in prolactin may, however, result from the release of ACTH as a stress response to morphine.[60] The indirect action of morphine on DA receptors is also revealed by the failure of opiates to reverse DA inhibition of prolactin release in isolated anterior pituitary incubations[61]; however, Met-enkephalins, implicated in the mediation of euphoria and drive reduction reward,[62] are elevated in rat striatum after chronic HAL treatment.[63] The relationship of dopaminergic and opiate pathways may lie in their ability to control cholinergic activity, since opiates can also inhibit the release of ACh.[64] Thus, increasing cholinergic activity in humans by means of physostigmine administration led to elevated plasma β-endorphin levels, which rose in parallel with signs of increasing depressive symptomatology.[65] Effects of some other nonneuroleptics are outlined in Table II.

Of potential interest is a recent observation that l-prolyl-l-leucylglycinamide, a neuropeptide derived from the carboxyl terminal of oxytocin, when coad-

Table II

Chronic Treatments Other than Neuroleptics that Lead to Altered DA Receptor Sensitivity[a]

Treatment	Change seen	Method	Reference
Ethanol (mice)	Down-reg.	AC	66,67
Morphine	Down-reg.	[³H]SPIRO	68
Morphine (mice)	Up-reg.	[³H]DOM	69
Estradiol	Up-reg.	[³H]SPIRO	70
Exptl. diabetes	Up-reg.	[³H]SPIRO	71
3-week strict light–dark	Ultradian rhythm	[³H]SPIRO	20,21
Cocaine	Up-reg.	[³H]SPIRO	72
Bromocriptine	Down-reg.	[³H]SPIRO	73
Amphetamine	Down-reg.	[³H]SPIRO	74
Amphetamine	Up-reg.	[³H]SPIRO	75

[a] Studies were carried out with rats except where indicated. Reg., regulation; AC, DA-stimulated adenylate cylase; SPIRO, spiroperidol; DOM, domperidone.

ministered with HAL or CPZ, antagonized their ability to elevate striatal D_2 receptors.[76] The investigators have suggested that there may be a specific receptor for the tripeptide. Effects of this and other drugs on up-regulated DA receptor sensitivity are described in Table III.

There are, apparently, regional differences in the regulation of transsynaptic dopaminergic activity in different regions of the brain. Thus, DA neurons projecting to the frontal cortex are not believed to be regulated by autoreceptors. Despite these differences, there is a basic regulatory scheme capable of producing short-term responses and acute and chronic compensatory responses. In general, blockade of postsynaptic receptors produces a reduction in DA inhibitory effects and an increase in electrical activity of postpsynaptic

Table III

Treatments that Prevent or Reverse DA Receptor Supersensitivity Induced by Neuroleptics or Other Drugs[a]

Chronic treatment	Prophylactic treatment	Assessment measure	Reference
Morphine	Pretreat with cyclo(leu-gly)	APO-ind. locomotor act.	77
Haloperidol	Concurrent Li⁺	APO-ind. stereo; [³H]SPIRO	53
Haloperidol or CPZ	Concurrent L-leucyl-glycinamide	[³H]SPIRO	76
Haloperidol	Concurrent Li⁺	Electrophys.	54
Haloperidol or flu-phenazine	Posttreat with *l*-DOPA	[³H]DA; AC	4
Haloperidol	Posttreat with *l*-DOPA	[³H]SPIRO	78
Haloperidol	Posttreat with *l*-DOPA	APO-ind. stereo.	57

[a] APO, apomorphine; ind, induced; act., activity; AC, DA-stimulated adenylate cyclase activity; stim., stimulated; stereo., stereotypy; [³H]SPIRO, [³H]spiroperidol binding in caudate; [³H]DA, [³H]dopamine binding in caudate.

neurons. This is accompanied by an increase in DA release from presynaptic terminals, presumably a compensatory effort to overcome the blockade. After one or more weeks, this compensatory increase subsides, perhaps because of the development of presynaptic autoreceptor supersensitivity, allowing synaptic DA to more efficiently shut down release. Closely following the normalization of DA release, a second compensatory effect occurs, the increase in number of postsynaptic DA sites, which increases the efficiency of the DA attempting to act on postsynaptic DA receptors in the face of neuroleptic antagonism. Other adjustments occur in biosynthetic enzymes and probably in coupling mechanisms and downstream response elements.

Despite this potential for compensatory adjustment, effective neuroleptic treatment appears to be dependent on the maintenance of effective blockade. One important side effect, parkinsonianlike extrapyramidal effects, is directly contingent on the blockade but disappears as a result of the compensatory adjustments. Tardive dyskinesia may be a manifestation of the compensatory effects or perhaps a result of a breakdown of these mechanisms. Thus, it is not entirely clear whether neuroleptic drugs can be designed with enhanced therapeutic effects but without undesirable motoric side effects.

3. ANTIDEPRESSANTS

A common characteristic of all drugs presently used as antidepressants (ADs) is their ability to down-regulate norepinephrine (NE)-stimulated, β-receptor-mediated adenylate cyclase activity[79–81]; Sulser and associates have attributed the therapeutic action of these drugs to their down-regulating potential. In the case of most known antidepressants, this down-regulating effect is believed to be secondary to an acute blockade of reuptake of NE. This results in an increase in NE in the synapse, producing over time a secondary decrease in β-receptor number. Similar blocking effects have been observed for serotonin (5-HT). Various ADs have effects on both transmitters, but some have more effect on NE reuptake, others on 5-HT. Although the reuptake blockade and resultant β-receptor down-regulation appear to be causally related to most ADs, an equally clear association has not been found for 5-HT and 5-HT receptors. Other types of drugs with some antidepressant properties such as amphetamine[82] and some neuroleptics[83] also have the down-regulating ability. Most ADs block reuptake of DA, 5-HT, or both as well as NE; however, mianserine[84] and iprindole[85] may be exceptions. Moreover, α_2 receptors, believed to be autoreceptors regulating NE reuptake, have been variously reported to remain unchanged,[86] to undergo an increase in density,[87] and to become subsensitive[88] depending on the AD used and the brain area under study (see refs. 89,90 for review).

Cortical serotonergic receptors have also been studied in relation to chronic AD treatment. Of the two types of such receptors, 5-HT$_1$, which binds [^{3}H]5-HT, is less down-regulated than the 5-HT$_2$ receptor, whose density can be measured with [^{3}H]SPIRO. All of the small number of ADs so far tested chronically appear to down-regulate the 5-HT$_2$ receptor.[91–93] In most cases, the magnitude of the receptor number decrease was greater than β-receptor down-regulation. It is of interest that significant decreases in CSF 5-HIAA levels have been reported for patients treated chronically with imipramine, amitriptyline, or clomipramine,[94-96] suggesting that ADs, or at least these drugs, induce a net reduction in serotonergic activity. However, the failure of repeated ECS treatment to lower CSF 5-HIAA[97] although similar treatment elevates 5-HT$_2$ receptors in rats[98] suggests that the putative relationship between noradrenergic and serotonergic systems must be mediated by an as yet unknown interactive system. In electrophysiological studies (see review 95), it has been found that chronic AD treatment increased the sensitivity of 5-HT postsynaptic receptors, suggesting that 5-HT$_2$ receptors are located presynaptically and that their reduction following chronic AD treatment is paralleled by increasing postsynaptic serotonergic responsivity.

The involvement of the serotonergic system in depression is further confounded by recent results with [^{3}H]ADs used as ligands in binding assays. Thus, [^{3}H]imipramine binding, possibly associated with a serotonin uptake site,[99] is reduced in cortex following antidepressant treatment in rats,[100] in platelets of nontreated depressed patients,[101,102] and in frontal cortex of suicides.[103] Binding of other [^{3}H]ADs has also been studied, including [^{3}H]DMI and [^{3}H]mianserine.[104] Each has its own unique binding characteristics and tissue specificity. Curiously, [^{3}H]imipramine has been found to bind to platelets of Wistar but not of Fawn-hooded rats.[90] Although [^{3}H]mianserine failed to bind in a specific manner to platelets, its profile of inhibition by various uptake inhibitors and 5-HT$_2$ antagonists in cortical membranes resembled that of [^{3}H]SPIRO binding in cortex. The relationship between these AD binding sites and sites for serotonin and NE is presently unclear.

Potentiation of AD effect has been attempted by simultaneous treatment with α-antagonists and AD. In general, whereas daily administration of a single dose of AD requires 2–3 weeks for β-receptor down-regulation to take place, combined treatments, as shown in Table IV, can achieve this effect in just a few days or less. Desmethylimipramine, for example, when coadministered with phenoxybenzamine, an α$_1$ blocker, induces significant β-receptor down-regulation after a single injection,[105] but coadministration of DMI with yohimbine up-regulates α$_2$ receptors, often after a single injection, without significant alteration in β-receptor number.[106] It is possible that the rapidity of receptor alterations obtained with combined treatments may be caused in part by the consistancy of receptor occupancy by NE. Thus, the administration of DMI twice rather than once daily produces significant β-receptor down-regulation after only three to five doses.[107,108]

Perhaps the most interesting finding in regard to AD mediation of receptor alterations is the reported ability of ACTH to augment 5-HT$_2$ receptor down-regulation by iprindole and mianserine.[109] ACTH release is one consequence of NE stimulation or of α$_2$ inhibition, the latter controlling NE release. Regulation of 5-HT$_2$ receptors by drugs that affect the noradrenergic system should provide an interesting new avenue for studying interactions between biogenic amine pathways.

Reinhardt and Roth[110] investigated the effects of the α$_2$ agonist clonidine on 5-HT and NE metabolism. The administration of as little as 30 μg/kg de-

Table IV
Effects of Combined Treatments on Cortical Receptors

Agent	Antidepressant	Days of combined treatment	Effect[a]	Reference
Phenoxybenzamine	DMI	4	β ↓	109
	DMI	1	β ↓	105
	Trazodone	4	5-HT$_2$ ↓	112
	Tranylcypromine	1	β ↓	105
		2	β ↓	105
Yohimbine	DMI	1	α$_2$ ↑, β ↔	109
	DMI	3	β ↓	109
	DMI	4	β ↓, α$_2$ ↑	106
	DMI	4	β ↓, α$_2$ ↑	115
	IMIP	3	β ↓	109
	AMI	3	5-HT$_2$ ↓, β ↔	109
	AMI	4	β ↓	106
	Iprindole	3	5-HT$_2$ ↓, β ↔	109
	Iprindole	4	β ↓	106
	Mianserine	3	5-HT$_2$ ↓, β ↔	109
ACTH	DMI	3	β ↓	109
	IMIP	3	β ↓	109
	Iprindole	3	5-HT$_2$ ↓	109
	Mianserine	3	5-HT$_2$ ↓	109
Amphetamine	Iprindole	3	β ↓, α$_2$ ↑	87

[a] Direction of arrow indicates up- or down-regulation, and ↔ indicates no significant change in receptor number.

creased MHPG levels, whereas 5-HIAA levels did not decrease until the dose of clonidine reached 300 μg/kg. Furthermore, l-amphetamine, at doses that do not alter the level of either metabolite, suppressed the inhibitory action of clonidine, suggesting that clonidine decreases 5-HT metabolism by reducing NE release.

Possibly, then, the augmentation of down-regulation by ACTH may be mediated by its effects on corticosteroid release. Indeed, corticosterone administration can reverse the increase in NE-stimulated adenylate cyclase activity in cortical slices produced by adrenalectomy.[111] A similar effect on the noradrenergic system has been demonstrated in the hippocampus. Adrenalectomy performed on chemically lesioned rats raised [^{3}H]DHA binding significantly beyond that achieved with 6-OH-DA alone. Corticosterone reduced the adrenalectomy-induced increase back to that seen in the lesioned animals.[112] Thus, steroid hormonal control of receptors sensitive to NE may be an important characteristic of the noradrenergic system.

Duman et al.[109] have suggested, on the basis of their studies with ACTH (see above), that the speed of onset of AD therapy may be a function of the hormonal state of the patient. Stone[113] has offered a correlative hypothesis in regard to the various therapies for depression. Noting that footshock and immobilization stresses as well as AD treatment decrease NE-stimulated adenylate cyclase activity in cortical slices, he has concluded that the development of subsensitivity may be one of the biochemical factors underlying adaptation to stress. Since ADs also decrease β-receptor density and NE-stimulated cyclase, the biological response to AD therapy may actually be an adaptation to stress produced by a drug without the necessity of experiencing the stress. Thus, treatments that increase resistance to stress in patients, regardless of the mechanism, facilitate recovery from depression. Globally, the views of Duman *et al.*[109] and Stone[113] suggest that, to a degree, we all possess an endogenous system for the control of depression. In this regard, abnormalities in cortical function described in depressed patients are of considerable interest.

Adaptive responses to AD treatments are less predictable than those induced by neuroleptics. However, progress in the matching of selected ADs to specific depressive syndromes may help to elucidate brain mechanisms associated with mania, depression, and other categories of abnormal behaviors.

Although much biological psychiatric research hinges on an initial assumption that a particular neuronal system may be the primary site of action of drugs used in the treatment of specific psychiatric illnesses, our steadily increasing knowledge of neuron and receptor interactions stands in defiance of single neuron, single receptor hypotheses.

4. BENZODIAZEPINE RECEPTOR ADAPTATION

The primary actions of the benzodiazepines are anxiolytic, anticonvulsant, and muscle relaxant. It appears that these are regulated through different receptor types. Brain-specific benzodiazepine sites selectively recognize pharmacologically active benzodiazepines, have high affinity, are saturable, and are heat labile and sensitive to proteolytic enzymes.[116–122] Sites for benzodiazepines appear to be localized predominantly on neurons but are also found on C_6 astrocytoma cells.[123,124] The receptor can be solubilized with good recovery and does not require sulfhydryl groups for binding.[125–127] Thermal degradation studies carried out at 60°C in Tris-HCl buffer are consistent with a homogeneous receptor population; however, thermal degradation is biphasic in sodium phosphate buffer.[128]

There is increasing evidence for multiple benzodiazepine sites. The dissociation constants for various benzodiazepines are the same and are monophasic for most brain regions. However, the dissociation rate for the cortex and hippocampus is biphasic. The triazolopyridazine CL 218,872 displaces [³H]diazepam from specific binding sites in brain and is as potent as diazepam in anxiolytic and anticonvulsant properties but exhibits no ataxic or muscle-relaxing properties.[129] Triazolopyridazines exhibit higher affinity for the heat-labile receptor sites than diazepam. Regional differences in the relative proportion of receptor types was found, with higher triazolopyridazine displacement of [³H]flunitrazepam binding in cerebellar vermis, frontal cortex, and dorsal hippocampus. Approximately 40% of the [³H]flunitrazepam binding sites are GABA-independent and correspond to the triazolopyridazine binding sites which appear to be homogeneous in the cerebellum. GABA, isoguanine, or

THIP (a rigid analogue of GABA) inhibits [^{3}H]flunitrazepam binding to the GABA-linked receptors.

Benzodiazepine receptors appear to be located primarily on neurons in the cerebellum. Mice with genetic cerebellar degeneration resulting in selective loss of Purkinje cells show loss of benzodiazepine receptors as well, whereas the heterozygotes, or mice with other types of atrophy, do not evidence loss of cerebellar benzodiazepine receptors.[130] Similarly, lesions of the olfactory tract in rats resulted in a decline of benzodiazepine binding receptor number that correlated with the loss of mitral cells and retrograde degeneration.[131]

The demonstration of GABA-associated benzodiazepine receptors has relevance to nervous system function including modulation of seizure susceptibility. Lesions of the nigrostriatal fibers containing GABA axons result in a 50% loss of GABA content and a 40% loss of benzodiazepine binding sites in the substantia nigra; however, in this "GABA-deafferented" nigra, the ability of GABA to potentiate the *in vitro* binding of [^{3}H]diazepam was increased threefold over the control (intact) nigral receptor system. The negative feedback control of dopamine release from the nigrostriatal fibers by GABA at the dopamine cell bodies in the nigra is capable of adaptation and may involve GABA-linked benzodiazepine receptors in the tonic regulation of dopamine release in the substantia nigra but not the ventral tegmentum, where such long feedback loops appear not to occur.

The injection of kainic acid into the substantia nigra causes a long-term decrease in the total number of binding sites for benzodiazepines.[132] Curiously, the decrease was restored to prelesion levels by the injection of kainic acid into the caudate nucleus homolateral to the original lesion. These GABA sites associated with the benzodiazepine receptors appear to be depleted in the substantia nigra in humans with Parkinson's disease[133,134] and in the neostriatum of terminal cases of Huntington's disease.[135,136] The same types of sites are increased in the substantia nigra in Huntington's disease.[137,138]

Several reports have appeared describing endogenous ligands for the benzodiazepine receptors, although none has been identified and sequenced as yet.[139] In addition to these peptidyl or proteinaceous ligands, it has been reported by several groups that the purine nucleotides act as low-affinity but effective agonist ligands at benzodiazepine receptors.[140] Most notable among these are inosine and hypoxanthine.[141] It is possible that the persistent, high-level presence of these compounds in Lesch–Nyhan disease may account for some of the muscular and behavioral anomalies accompanying the disorder by down-regulation of the benzodiazepine receptors.[142]

Benzodiazepine receptors appear to be sensitive to regulation by the rate of ligand encounter, just as described in the dopaminergic and adrenergic systems. Chiu and Rosenberg[143] reported that the chronic administration of benzodiazepines resulted in a decrease in the number, but not affinity, of benzodiazepine binding sites in brain. Similarly, the denervation studies above appear to indicate that the increased number of receptors results from the chronic decrease in ligand encounter by the benzodiazepine receptors. However, there has been almost no potential for studying the functional up-regulation of benzodiazepine receptors until recent reports of two compounds that selectively inhibit benzodiazepine ligand from interacting with the receptor. Ro5-3663 appears to be an antagonist to the GABA-linked receptors and is reported to be an active convulsant.[144] Similarly, Ro15-1788 is reported to be a selective antagonist of benzodiazepine interaction at the receptor in the central nervous system but not in the periphery, where benzodiazepines act at the kidney and schistosomes. The chronic administration of these antagonist compounds would be very informative in regard to the functional up-regulation of the benzodiazepine receptors. Such studies would have implications for restoring receptor sensitivity in clinical states, since the antagonist Ro15-1788 has no behavioral or physiological effects akin to the benzodiazepines themselves.[145] No such studies have been conducted to date.

A number of the actions of the benzodiazepines in the GABA-linked sites occur through inhibition of the action of a GABA-inhibiting protein termed GABA-modulin.[146] Tofizopam, a 3,4-benzodiazepine, potentiates benzodiazepine receptor binding but has no action at the receptor itself. It appears that tofizopam exerts its action through truncating the action of GABA-modulin.[147]

Recent reports indicate that persistent stress or viral infection can substantially alter the number of forebrain benzodiazepine receptors in animals and, presumably, in man.[148,149] It is not uncommon to encounter clinically the onset of seizure disorders following viral infection. Therefore, one might conceive of the viral-induced loss of benzodiazepine receptors as being a condition in which the purposeful manipulation of GABA-linked benzodiazepine receptors might be an efficacious strategy through which to restore normal seizure threshold and attenuate the persistent or recurrent disorder.

Exposure to diazepam at early gestational ages *in utero* results in behavioral anomalies that are indicative of changes in benzodiazepine receptor function. Kellogg and co-workers have reported the loss of normal locomotor development and of normal acoustic startle reflex in animals exposed to diazepam *in utero* from the 13th gestational day.[150] These studies have important ramifications for the *in utero* exposure to diazepam and for the exposure to repeated stress by the mother. Such studies imply that the benzodiazepine receptor system is regulated by the exposure to ligand prenatally, much as has been described by the studies of Rosengarten and Friedhoff[151] for the prenatal exposure to dopamine receptor active agents.

5. USE OF RECEPTOR ADAPTATION AS A TREATMENT

Neuroleptics were used in schizophrenia because they worked. Only subsequently was it discovered that their most consistent action was the blocking of DA receptors and presumably a reduction in dopaminergic activity. From these observations it was concluded that production of hypodopaminergia is beneficial to many schizophrenic patients. Neuroleptics also provoke a compensatory increase in receptor number, which, as has been mentioned, is believed to be a factor in tardive dyskinesia. We were, therefore, interested in developing an approach to treatment that would reduce dopaminergic activity by down-regulating the number of DA receptors rather than by DA receptor blockade, with the hope that this approach would not subject the patient to the risk of tardive dyskinesia. We have, therefore, developed an approach to treatment that attempts, directly, to produce the desired compensatory effect using the principle of receptor sensitivity modification.[152-154] This approach involves a novel use of drugs in that the treatment, initially, worsens the condition to some degree, therapeutic benefit taking place after the drug treatment is terminated.

As a basis for the investigation of the possibility that RSM might be useful as a treatment, the assumption was made that a change in DA postsynaptic receptor density would have an effect similar to that resulting from a change in agonist supply. Since what is wanted in the treatment of schizophrenia is an unresponsive postsynaptic DA receptor, treatment with an agonist might induce that state. Several studies of the treatment of patients with schizophrenia have been carried out using the RSM approach. Alpert et al.[155] administered CPZ or *l*-DOPA to a small number of patients. *l*-DOPA initially exacerbated the symptoms, as had been reported earlier[156]; however, in contrast to the earlier finding, much of the worsening appeared to occur from an overlying toxic psychosis rather than from an increase in the primary symptoms of schizophrenia. Soon, however, tolerance developed toward some of the toxic effects. Patients were only observed for 4 days following discontinuation of drug, and significant overall improvement would not have been expected and was not observed at that time.

Beramendi et al.[157] treated 12 patients with *l*-DOPA–carbidopa (Sinemet®), a combination that prevents peripheral decarboxylation of *l*-DOPA, nine with haloperidol, and ten with placebo. Clinically significant improvement was seen in approximately half of the *l*-DOPA group, although overall, the haloperidol group fared better. Gutierrez et al.[158] studied muscle rigidity in these patients. They found, as expected, that some patients became more rigid during the haloperidol treatment, presumably from the extrapyramidal effects of this drug. Surprisingly, some of the *l*-DOPA group also became rigid. Patients who became most rigid with HAL showed the least improvement from HAL, whereas those who became most rigid with *l*-DOPA had the most improvement from *l*-DOPA.

Alpert et al.,[159] using resting finger tremor as a measure of extrapyramidal function, found a negative correlation between a tremor index of the effectiveness of neuroleptic blockade and a reduction in schizophrenic symptomatology after 3 weeks of treatment. Again, individuals who showed evidence of the most effective blockade tended to respond least well to neuroleptic medication. Since the neuroleptics used in these studies are potent D_2 receptor blockers but only weakly active against the D_1 receptor, and since neuroleptics cause an increase in the release of DA, it is tempting to speculate that such release and the availability of a receptor site (D_1) only partially blocked by neuroleptic are somehow involved in neuroleptic therapy. The results of Stoof and Kebabian,[51] discussed earlier, are entirely relevant to this possibility.

As indicated earlier, TD may stem from an exaggerated up-regulation of DA receptors resulting from chronic exposure to neuroleptics. In the rat, receptor number increases induced by HAL can be reversed more rapidly by *l*-

DOPA than by the normal restitutive process that occurs during washout.[4,78] Inasmuch as *l*-DOPA can be tolerated in large doses in man, it was decided to try this potential therapeutic procedure on patients showing pronounced symptoms of TD.[154,159] Patients were observed for 6 weeks prior to entry into the study, and those whose symptoms regressed during this period were dropped. The remaining patients were treated with Sinemet® in gradually increasing doses to a maximum of the equivalent of 6 g of *l*-DOPA alone over a period of 4 weeks and then maintained at this dose for 4 weeks. Sinemet® treatments produced a significant decrease in TD symptoms when compared to patients maintained on their prior treatment or no treatment at all.

We have also begun investigating the use of *l*-DOPA in the treatment of Gilles de la Tourette syndrome, an illness characterized by abnormal motoric behavior and uncontrollable bursts of cursing. To date, six patients have been treated with gradually increasing doses of Sinemet®. Some worsening was seen during medication, followed by general remission of symptoms after termination of treatment; however, in all of the cases, a recurrence of symptoms was seen within 4 months to 2 years. Some patients have been retreated with good effect.

The use of deliberate receptor sensitivity modification in the treatment of disorders that appear to involve spontaneous or drug-induced receptor supersensitivity seems promising, especially for TD and Tourette syndrome. With a better understanding of the molecular mechanisms involved in the regulation of receptor sensitivity, it may be possible to achieve effects such as those described above without the transient worsening of symptoms.

6. CONCLUSIONS

In this chapter, we have discussed aspects of several receptor systems and have emphasized those aspects of the system that have been found to be involved in the immediate or long-term effects of neuroleptics, antidepressants, and benzodiazepines. Previously, the neuronal receptor was viewed as a target site for a specific transmitter or exogenous substance that was compatible with its structure. From that viewpoint, when the specific ligand interacted with the receptor molecule, a specific response was initiated. Based on a large number of recent findings, our current view of transsynaptic activity has changed dramatically. Both the presynaptic and postsynaptic elements have been found to be highly regulated. The presence of peptides in presynaptic DA terminals, for instance, appears to reflect a modulatory role for these substances on the effects of DA. Similarly, the postsynaptic receptors are regulated at many loci including the coupling and response mechanisms. Undoubtedly, some of these aspects of the synapse also play important roles in the mediation of psychotropic drug effects; however, these are only presently being investigated.

In addition to the complex interactions occurring in the process of neuronal transmission, long-term interventions with pharmacological agents provoke adaptive responses of various kinds. These responses produce relatively durable changes in the sensitivity of the system that persist after the pharmacological treatment is terminated. Thus, down-regulation mimics the decrease in transsynaptic function produced by pharmacological blockade, and up-regulation the enhancement of activity produced by increasing transmitter turnover. Although these long-term adjustments have been shown to have some of the behavioral effects produced by acute drug intervention, it is not known whether their effects may differ in other regards. It is clear that manipulation of the adaptive capacity of the central nervous system opens up new and important possibilities for understanding the regulation of normal and abnormal behavior. Equally, this provides opportunities for the development of new therapies that are mediated by deliberate manipulation of synaptic adaptive mechanisms.

REFERENCES

1. Seeman, P., 1980, *Pharmacol. Rev.* **32**:229–313.
2. Burt, D. R., Creese, I., and Snyder, S. H., 1977, *Science* **196**:326–328.
3. Clow, A., Jenner, P., Marsden, C. D., Murugaiah, K., and Theodorou, A., 1980, *Proc. Br. Psychiatr. Soc.* 242P.
4. Friedhoff, A. J., Bonnet, K., and Rosengarten, H., 1977, *Res. Commun. Chem. Pathol. Pharmacol.* **16**:411–423.
5. Hyttel, J., 1978, *Life Sci.* **23**:551–556.
6. Faustman, W., Fowler, S., and Walker, C., 1981, *Eur. J. Pharmacol.* **70**:65–70.
7. Kobayashi, R. M., Fields, J. Z., Hruska, R. E., Beaumont, K., and Yamamura, H. I., 1977, *Animal Models in Psychiatry and Neurology* (I. Hanin and E. Usdin, eds.), Pergamon Press, New York, pp. 405–409.
8. Bacopoulos, N. G., 1981, *Biochem. Pharmacol.* **30**:2037–2040.
9. Leysen, J. E., and Gommeren, W., 1981, *J. Neurochem.* **36**:201–219.
10. Blaschke, E., and Hertting, G., 1971, *Biochem. Pharmacol.* **20**:1363–1370.
11. Kayaalp, S. O., Rubenstein, J. S., and Neff, N. H., 1981, *Neuropharmacology* **20**:409–410.
12. Zahniser, N. R., and Molinoff, P. B., 1978, *Nature* **275**:453–455.
13. Scatchard, G., 1949, *Ann. N.Y. Acad. Sci.* **53**:660–672.
14. Feldman, H. A., 1972, *Anal. Biochem.* **48**:317–338.
15. Williams, L. T., and Lefkowitz, R. J., 1978, *Receptor Binding Studies in Adrenergic Pharmacology*, Raven Press, New York.
16. Yamamura, H. I., Enna, S. J., and Kuhar, M. J., 1978, *Neurotransmitter Receptor Binding*, Raven Press, New York.
17. Sibley, D. R., and Creese, I., 1980, *Eur. J. Pharmacol.* **65**:131–133.
18. Wehr, T. A., Wirz-Justice, A., Goodwin, F. K., Duncan, W., and Gillin, J. C., 1979, *Science* **206**:710–713.
19. Cools, A. R., 1978, *Life Sci.* **23**:2475–2484.
20. Naber, D., Wirz-Justice, A., Kafka, M. S., and Wehr, T. A., 1980, *Psychopharmacology* **68**:1–5.
21. Naber, D., Wirz-Justice, A., Kafka, M. S., Tobler, I., and Borbely, A. A., 1981, *Biol. Psychiatry* **16**:831–835.
22. Wirz-Justice, A., Kafka, M. S., Naber, D., and Wehr, T. A., 1980, *Life Sci.* **27**:341–347.
23. Kafka, M. S., Wirz-Justice, A., Naber, D., and Wehr, T. A., 1981, *Neuropharmacology* **20**:421–425.
24. Kafka, M. S., Wirz-Justice, A., and Naber, D., 1981, *Brain Res.* **207**:409–419.
25. Codd, E. E., and Byrne, W. L., 1981, *Life Sci.* **28**:2577–2583.
26. Schwartz, J.-C., Baudry, M., Martres, M.-P., Costentin, J., and Protais, P., 1978, *Life Sci.* **23**:1785–1790.
27. Christensen, A. V., Fjalland, B., and Moller-Nielsen, I., 1976, *Psychopharmacology* **48**:1–6.

28. Chiodo, L. A., and Antelman, S. M., 1980, *Science* **210**:799–801.
29. Muller, P., and Seeman, P., 1978, *Psychopharmacology* **60**:1–11.
30. Baldessarini, R. J., and Tarsy, D., 1978, *Psychopharmacology: A Generation of Progress* (M. A. Lipton, A. DiMascio, and K. F. Killam, eds.), Raven Press, New York, pp. 993–1004.
31. Waddington, J. L., and Gamble, S. J., 1980, *Eur. J. Pharmacol.* **68**:387–388.
32. Waddington, J. L., and Gamble, S. J., 1981, *Lancet* **1**:1375.
33. Weiss, B., and Santelli, S., 1978, *Science* **200**:799–801.
34. Weiss, B., Santelli, S., and Lusink, G., 1977, *Psychopharmacologia* **53**:289.
35. Clow, A., Theodorou, A., Jenner, P., and Marsden, C. D., 1980, *Eur. J. Pharmacol.* **63**:135–144.
36. Clow, A., Jenner, P., Theodorou, A., and Marsden, C. D., 1979, *Nature* **278**:59–61.
37. Clow, A., Theodorou, A., Jenner, P., and Marsden, C. D., 1980, *Eur. J. Pharmacol.* **63**:145–157.
38. Theodorou, A., Gommeren, W., Clow, A., Leysen, J., Jenner, P., and Marsden, C. D., 1981, *Life Sci.* **28**:1621–1627.
39. Waddington, J. L., and Gamble, S. J., 1980, *Psychopharmacology* **71**:75–77.
40. Waddington, J. L., Gamble, S. J., and Bourne, R. C., 1981, *Eur. J. Pharmacol.* **69**:511–513.
41. Westerlink, B. H. C., and Horn, A. S., 1979, *Eur. J. Pharmacol.* **58**:39–48.
42. Dawbarn, D., Long, S. K., and Pycock, C. J., 1980, *Proc. Br. Psychiatr. Soc.* 240P.
43. Kebabian, J. W., Chen, T. C., and Cote, T. E., 1979, *Comm. Psychopharmacol.* **3**:421–428.
44. Gamble, S. J., and Waddington, J. L., 1980, *Proc. Br. Psychiatr. Soc.* 240P.
45. Dyer, R. G., Murugaiah, K., Theodorou, A., Clow, A., Jenner, P., and Marsden, C. D., 1981, *Life Sci.* **28**:167–174.
46. Owen, F., Cross, A. J., Waddington, J. L., Poulter, M., Gamble, S. J. and Crow, T. J., 1980, *Life Sci.* **26**:55–59.
47. Cross, A. J., and Waddington, J. L., 1981, *Eur. J. Pharmacol.* **71**:327–332.
48. Schwarcz, R., Creese, I., Coyle, J. T., and Snyder, S. H., 1978, *Nature* **271**:766–768.
49. Euvrard, C., Premont, J., Oberlander, C., Boissier, J. R., and Bockaert, J., 1979, *Naunyn Schmiedebergs Arch. Pharmacol.* **309**:241–245.
50. Marchais, D., and Bockaert, J., 1980, *Biochem. Pharmacol.* **29**:1331–1336.
51. Stoof, J. C. and Kebabian, J. W., 1981, *Nature* **294**:366–368.
52. Tsuruta, K., Frey, E. A., Grewe, C. W., Cote, T. E., Eskay, R. L., and Kebabian, J. W., 1981, *Nature* **292**:463–465.
53. Pert, A., Rosenblatt, J. E., Sivit, C., Pert, C. B., Bunney, W. E. Jr., 1978, *Science* **201**:171–173.
54. Gallager, D. W., Pert, A., and Bunney, W. E. Jr., 1978, *Nature* **273**:309–312.
55. Meller, E., and Friedman, E., 1981, *Eur. J. Pharmacol.* **76**:25–29.
56. Staunton, D. A., Magistretti, P. J., Deyo, S. N., Shoemaker, W. J., and Bloom, F. E., 1981, *Soc. Neurosci. Abstr.* **7**:163.
57. Ezrin-Waters, C., and Seeman, P., 1978, *Life Sci.* **22**:1027–1032.
58. Tolis, C., Hickey, J., and Guyda, H., 1975, *J. Clin. Endocrinol. Metab.* **41**:797.
59. Clemens, J. A., and Sawyer, B. D., 1974, *Endocrine Res. Commun.* **1**:373.
60. Rossier, J., French, E., Rivier, C., Shibasaki, T., Guillemin, R., and Blood, F. E., 1980, *Proc. Natl. Acad. Sci. U.S.A.* **77**:666–669.
61. Login, I. S., and MacLeod, R. M., 1979, *Eur. J. Pharmacol.* **60**:253–255.
62. Belluzzi, J., and Stein, L., 1977, *Nature* **266**:556–558.
63. Hong, J. S., Yang, H.-Y. T., Fratta, W., and Costa, E., 1978, *J. Pharmacol. Exp. Ther.* **205**:141–147.
64. Moroni, F., Cheney, D. L., and Costa, E., 1977, *Nature* **267**:267.
65. Risch, S. C., Cohen, R. M., Janowsky, D. S., Kalin, N. H., and Murphy, D. L, 1980, *Science* **209**:1545–1546.
66. Hoffman, P. L., and Tabakoff, B., 1977, *Nature* **268**:551–553.
67. Creese, I., and Sibley, D. R., 1981, *Annu. Rev. Pharmacol. Toxicol.* **21**:357–391.
68. Puri, S. K., Spaulding, T. C., and Montione, C. R., 1978, *Life Sci.* **23**:637–642.
69. Baume, S., Patey, G., Marcois, H., Protais, P., Costentin, J., and Schwartz, J.-C., 1979, *Life Sci.* **24**:2333–2342.

70. Hruska, R. E., and Silbergeld, E. K., 1980, *Eur. J. Pharmacol.* **61**:397–400.
71. Lozovsky, D., Saller, C. F., and Kopin, I. J., 1981, *Science* **214**:1031–1033.
72. Taylor, D. L., Ho, B. T., and Fagan, J. D., 1979, *Commun. Psychopharmacol.* **3**:137–142.
73. Quik, M., and Iversen, L. L., 1979, *Naunyn Schmiedebergs Arch. Pharmacol.* **304**:141–145.
74. Howlett, D. R., and Nahorski, S. R., 1979, *Brain Res.* **161**:173–178.
75. Robertson, H. A., 1979, *Soc. Neurosci. Abstr.* **5**:570.
76. Chiu, S., Paulose, C. S., and Mishra, R. K., 1981, *Science* **214**:1261–1262.
77. Ritzmann, R. F., Walter, R., Bhargava, H. N., and Flexner, L. B., 1979, *Proc. Natl. Acad. Sci. U.S.A.* **76**:5997–5998.
78. List, S., and Seeman, P., 1979, *Life Sci.* **24**:1447–1452.
79. Vetulani, J., and Sulser, F., 1975, *Nature* **257**:495–497.
80. Wolfe, B., Harden, T., Sporn, J., and Molinoff, P. B., 1978, *J. Pharmacol. Exp. Ther.* **207**:446–457.
81. Mishra, R., Janowsky, A., and Sulser, F., 1979, *Eur. J. Pharmacol.* **60**:379–382.
82. Baudry, M., Martres, M. P., and Schwartz, J.-C., 1976, *Brain Res.* **116**:111–124.
83. Schmidt, M. J., and Thornberry, J. F., 1977, *Arch. Int. Pharmacodyn. Ther.* **229**:42–51.
84. Leonard, B. E., 1978, *Br. J. Pharmacol.* **5**:115–125.
85. Zis, A. P., and Goodwin, A. P., 1979, *Arch. Gen. Psychiatry* **36**:1097.
86. Svensson, T., and Usdin, T., 1979, *Catecholamine: Basic and Clinical Frontiers* (E. Usdin, I. Kopir, and J. Barchas, eds.), Pergamon Press, New York, pp. 672–674.
87. Reisine, T. D., U'Prichard, D. C., Wiech, N., Ursillo, R., and Yamamura, H. I., 1980, *Brain Res.* **188**:587–592.
88. Crews, F. T., and Smith, C. B., 1978, *Science* **202**:322–324.
89. Reisine, T., 1981, *Neuroscience* **6**:1471–1502.
90. Charney, D. S., Menkes, D. B., and Heninger, G. R., 1981, *Arch. Gen. Psychiatry* **38**:1160–1180.
91. Mann, E., and Enna, S. J., 1980, *Soc. Neurosci. Abstr.* **6**:800.
92. Tang, S. W., Seeman, P., and Kwan, S., 1980, *Soc. Neurosci. Abstr.* **6**:861.
93. Blackshear, M. A., Steranke, R., and Sanders-Bush, E., 1980, *Soc. Neurosci. Abstr.* **6**:860.
94. Bertilsson, L., Asberg, M., and Thoren, P., 1976, *Eur. J. Clin. Pharmacol.* **21**:194–200.
95. Post, R. M., and Goodwin, F. K., 1974, *Arch. Gen. Psychiatry* **30**:234–239.
96. Bowers, M. B., 1974, *Clin. Pharmacol. Ther.* **15**:167–170.
97. Abrams, R., Essman, W. B., and Taylor, N. A., 1970, *Biol. Psychiatry* **11**:85–90.
98. Kellar, K. J., Cascio, C. S., Butler, J. A., and Kurtzke, R. N., 1981, *Eur. J. Pharmacol.* **69**:515–518.
99. Langer, S. Z., Moret, C., Raisman, R., Dubocovich, M. L., and Briley, M., 1980, *Science* **210**:1133–1135.
100. Raisman, R., Briley, M., and Langer, S. Z., 1979, *Eur. J. Pharmacol.* **61**:373–380.
101. Briley, M., Langer, S. Z., Raisman, R., Sechter, D., and Zarijian, E., 1980, *Science* **209**:303–304.
102. Asarch, K. B., Shih, J. C., and Kulcsar, A., 1981, *Soc. Neurosci. Abstr.* **7**:712.
103. Stanley, M., Virgilio, J., and Gershon, S., 1982, *Science* **216**:1337–1339.
104. Dunbrille-Ross, A., and Tang, S. W., 1981, *Soc. Neurosci. Abstr.* **7**:8.
105. Crews, F. T., Paul, S. M., and Goodwin, F. K., 1981, *Nature* **290**:787–789.
106. Ursillo, R., Wiech, N., Reisine, T., and Yamamura, H. I., 1980, *Psychopharmacology and Biochemistry of Neurotransmitter Receptors* (H. I. Yamamura, R. W. Olson, and E. Usdin, eds.), Elsevier, Amsterdam, p. 189.
107. Schweitzer, J. W., Schwartz, R., and Friedhoff, A. J., 1979, *J. Neurochem.* **33**:377–379.
108. Sarai, K., Frazer, A., Brunswick, D., and Mendels, J., 1978, *Biochem. Pharmacol.* **27**:2179–2181.
109. Duman, R., Slopis, J., Kendall, D., and Enna, S. J., 1981, *Soc. Neurosci. Abstr.* **7**:104.
110. Reinhard, J. F., and Roth, R. H., 1981, *Soc. Neurosci. Abstr.* **7**:151.
111. Mobley, P. L., and Sulser, F., 1980, *Nature* **286**:608–609.
112. Roberts, D. C. S., and Bloom, F. E., 1981, *Eur. J. Pharmacol.* **74**:37–41.
113. Stone, E. A., 1979, *Res. Commun. Psychol. Psychiatry Behav.* **4**:241–255.
114. Taylor, D. P., Allen, L. E., Ashworth, E. M., Becker, J. A., Hyslop, D. K., and Riblet, L. A., 1981, *Neuropharmacology* **20**:513–516.

115. Johnson, R. W., Reisine, T., Spotnitz, S., Wiech, N., Ursillo, R., and Yamamura, H. I., 1980, *Eur. J. Pharmacol.* **67**:123–127.
116. Basmann, H. B., 1978, *FEBS. Lett.* **87**:199–202.
117. Mackerer, C., 1978, *J. Pharmacol. Exp. Ther.* **206**:405–411.
118. Mohler, H., and Okada, T., 1978, *Life Science* **20**:2101–2110.
119. Mohler, H., and Okada, T., 1977, *Science* **198**:849–851.
120. Nielson, M., Braestrup, C., and Squires, R. F., 1978, *Brain Res.* **141**:342–346.
121. Speth, R. C., and Wester, H., 1978, *Life Sci.* **22**:859.
122. Speth, R. C., Wastek, C. J., and Yamamura, H. I., 1979, *Life Sci.* **24**:351–358.
123. Lippa, A. S., Sano, M. C., Coupet, J., Klepner, C. A., and Beer, B., 1978, *Life Sci.* **23**:2213–2218.
124. Mallorga, P., Hamburg, M., Tallman, J. F., and Gallager, D. W., 1980, *Neuropharmacology* **19**:405–408.
125. Thomas, J., Yousufi, M., and Tallman, J., 1979, *Soc. Neurosci. Abstr.* **5**:2269.
126. Gavish, M., and Snyder, S. H., 1980, *Life Sci.* **26**:579–582.
127. Sieghart, W., and Karobath, M., 1980, *Nature* **286**:285–287.
128. Squires, R., Benson, D. I., Braestrup, C., Coupet, J., and Klepner, C. A., 1979, *Pharm. Biochem. Behav.* **10**:825–839.
129. Lippa, A. S., Critchett, D. J., Sano, M. C., Klepner, C. A., Greenblatt, E. N., Coupet, J., and Beer, B., 1979, *Pharmacol. Biochem. Behav.* **10**:831–843.
130. Skolnick, P., Marangos, P., Syapin, P., Goodwin, F., and Paul, S., 1979, *Pharmacol. Biochem. Behav.* **10**:815–823.
131. Meyerson, L. R., Sano, M. C., Critchett, D. J., Beer, B., and Lippa, A. S., 1981, *Eur. J. Pharmacol.* **71**:147–150.
132. Biggio, G., Corda, M. G., and Gessa, G. L., 1981, *Brain Res.* **220**:344–349.
133. Lloyd, K. G., Shemen, L., and Hornykiewicz, O., 1977, *Brain Res.* **127**:269–275.
134. Rinne, U. K., Koskinen, V., Laaksonen, H., Lonnberg, P., and Sonninen, V., 1978, *Life Sci.* **22**:2225–2228.
135. Lloyd, K. G., Dreksler, S., and Bird, E. D., 1977, *Life Sci.* **21**:747–754.
136. Olsen, R. W., Van Ness, P. C., Napias, C., Bergman, M., and Tourtellette, W. W., 1980, *Adv. Biochem. Psychopharmacol.* **21**:451–460.
137. Enna, S. J., Bennett, J. P., Bylund, D. B., Snyder, S. H., Bird, E. D., and Iversen, L. L., 1976, *Brain Res.* **116**:531–537.
138. Waddington, J. L., Cross, A. J., 1980, *Brain Res. Bull.* **5**(Suppl. 2):825–828.
139. Davis, L. G., and Cohen, R. K., 1980, *Biochem. Biophys. Res. Commun.* **92**:141–148.
140. Marangos, P. J., Paul, S. M., Goodwin, F. S., Syapin, P., and Skolnick, P., 1979, *Life Sci.* **24**:851–858.
141. Marangos, P. J., Martino, A. M., Paul, S., and Skolnick, P., 1981, *Psychopharmacology* **72**:269–273.
142. Anderson, L. T., David, R., Bonnet, K. A., and Dancis, J., 1979, *Life Sci.* **24**:905–910.
143. Chiu, T. H., and Rosenberg, H. C., 1979, *Eur. J. Pharmacol.* **56**:337–345.
144. Schlosger, W., and Franco, S., 1979, *J. Pharmacol. Exp. Ther.* **21**:290–295.
145. Hunkeler, W., 1981, *Nature* **290**:514–516.
146. Guidotti, A., Massotti, M., and Costa, E., 1980, *Psychopharmacology and Biochemistry of Neurotransmitter Receptors,* (H. I. Yamamura, R. W. Olsen, and E. Usdin, eds.), Elsevier, Amsterdam, New York, pp. 655–660.
147. Saano, V., Urtti, A., and Airaksinen, M. M., 1981, *Pharmacol. Res. Commun.* **13**:75–85.
148. Simantov, R., Oster-Granite, M. L., Herndon, R. M., and Snyder, S. H., 1976, *Brain Res.* **105**:365–371.
149. Shibuya, H., Gale, K., and Pert, C. B., 1980, *Eur. J. Pharmacol.* **62**:243–244.
150. Kellogg, C., Tervo, D., Ison, J., and Parisi, T., 1980, *Science* **207**:205–207.
151. Rosengarten, H., and Friedhoff, A. J., 1979, *Science* **203**:1133–1135.
152. Friedhoff, A. J., 1977, *Comp. Psychiatry* **18**:309–317.
153. Friedhoff, A. J., and Alpert, M., 1978, *Psychopharmacology: A Generation of Progress* (M. Lipton, A. DiMascio and K. F. Killam, eds.), Raven Press, New York, pp. 797–801.

154. Alpert, M., and Friedhoff, A. J., 1980, *Tardive Dyskinesia: Research and Treatment* (W. E. Fann, J. M. Davis, R. C. Smith, and E. F. Domino, eds.), Spectrum, New York, pp. 471–474.
155. Alpert, M., Friedhoff, A. J., Marcos, L. R., and Diamond, F., 1978, *Am. J. Psychiatry* **135**:1329–1332.
156. Yaryura-Tobias, J. A., Wolpert, A., Dana, L., and Merlis, S., 1970, *Dis. Nerv. Syst.* **31**:60–63.
157. Beramendi, V., Alpert, M., Guimon, J., Friedhoff, A. J., and Gutierrez, M., 1980, *Arch. Neurobiol.* **43**:107–124.
158. Gutierrez, M., Alpert, M., Guimon, M., Friedhoff, A. J., and Bermendi, V., 1979, *Informaciones Psiquiatricas 2°-Trimestre*, No. 75, Barcelona, pp. 99–107.
159. Alpert, M., Friedhoff, A. J., and Diamond, F., 1983, in *Advances in Neurology, Vol. 37: Experimental Therapeutics of Movement Disorders* (S. Fahn, D. B. Calne, and I. Shoulson, eds.), Raven Press, New York, pp. 253–258.

Index